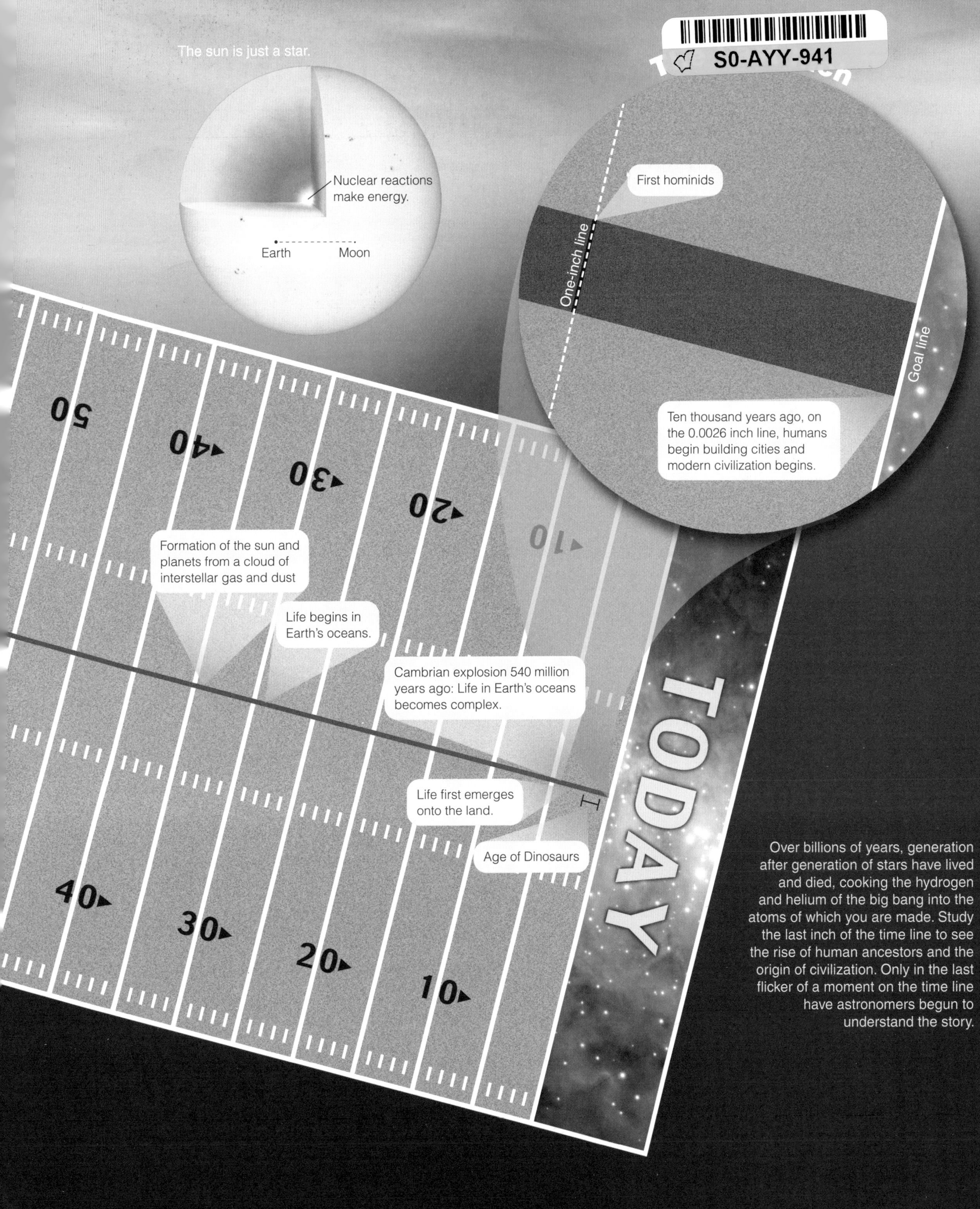
The sun is just a star.
Nuclear reactions make energy.
Earth
Moon
First hominids
One-inch line
Goal line
Ten thousand years ago, on the 0.0026 inch line, humans begin building cities and modern civilization begins.
Formation of the sun and planets from a cloud of interstellar gas and dust
Life begins in Earth's oceans.
Cambrian explosion 540 million years ago: Life in Earth's oceans becomes complex.
Life first emerges onto the land.
Age of Dinosaurs
TODAY
Over billions of years, generation after generation of stars have lived and died, cooking the hydrogen and helium of the big bang into the atoms of which you are made. Study the last inch of the time line to see the rise of human ancestors and the origin of civilization. Only in the last flicker of a moment on the time line have astronomers begun to understand the story.

# Build your perfect course solution with our flexible package options

Because every course is as unique as its instructor—and its students—Mike Seeds and Dana Backman's textbooks give you the flexibility to build the course solution that best suits your needs. Increase the value of your course package by including one of the resources below with your text at no additional charge.

## Planetarium Software Options

Don't be forced to use whatever CD comes with your text—select the software that works for you. Choose either **Starry Night**™ or **TheSky Student Edition**™ based on your own personal preference.

Starry Night 6.0 contains data on 1 million stars and 110 deep-space objects including galaxies, star clusters, and nebulae. Choose an exact location on Earth or from elsewhere within 20,000 light-years of our solar system and view the sky from there. This user-friendly and fun software even allows students to create personal QuickTime® movies of the experience.

TheSkyX Student Edition for Mac or Windows turns students' computers into powerful personal planetariums. Loaded with data on 1 million stars and 13,000 deep-sky objects with images, it lets students view the universe from any location on Earth at any point in time from 6,000 years ago to 10,000 years in the future. Students will see the sky in motion, view constellations, see moon phases, search the database for objects, display both lunar and solar eclipses, print star charts, locate and simulate Iridium flares and more.

Enhance students' understanding of the scientific method with **Virtual Astronomy Laboratories**. Focusing on 20 of the most important concepts in astronomy, these labs offer students hands-on exercises that complement text topics. Instructors can set up classes online and view student results or students can print their reports for submission, making the labs ideal for homework assignments, lab exercises, and extra-credit work.

## Night Sky Planisphere

This planisphere, available in 3 specific latitude settings, is an ideal portable reference for observational exercises.

Instant feedback and ease of use are just two reasons that **WebAssign**® is the most widely used homework system in higher education. The WebAssign Homework Delivery System lets you deliver, collect, grade, and record assignments via the web. And now, this proven system has been enhanced to include select material from ***Foundations of Astronomy*, Eleventh Edition**. With **Enhanced WebAssign** you can assign select end-of-chapter problems and collect student scores in an online gradebook. Special problems ask students to analyze photos and graphs, manipulate *Active Figures*, or assess their conceptual knowledge.

*ALSO AVAILABLE!*

### *Foundations of Astronomy* eBook: *Powered by* Enhanced WebAssign

Provide your students with a digital version of the text, completely integrated with the **Enhanced WebAssign** homework management system! Whether you want to assign homework in an online environment, or just want your students to have access to an affordable and interactive online text (or both), this is the ideal solution to meet your course needs.

---

Visit
**www.cengage.com/community/seeds**
to learn more about Seeds and Backman's textbooks and technology offerings.

## Student Textbook Ordering Options

Student edition of text: ISBN 13: 978-1-4390-5036-1 ISBN 10: 1-4390-5036-8

Student edition of text with **Virtual Astronomy Labs**: ISBN 13: 978-1-1111-1644-6 ISBN 10: 1-1111-1644-X

Student edition of text with **Starry Night** Planetarium software: ISBN 13: 978-1-1111-1401-5 ISBN 10: 1-1111-1401-3

Student edition of text with **SkyX College Edition** Planetarium software: ISBN 13: 978-1-1111-1403-9 ISBN 10: 1-1111-1403-X

Student edition of the text with **The Night Sky Planisphere Latitude 20-30**: ISBN 13: 978-1-1111-1633-0 ISBN 10: 1-1111-1633-4

Student edition of the text with **The Night Sky Planisphere Latitude 30-40**: ISBN 13: 978-1-1111-1634-7 ISBN 10: 1-1111-1634-2

Student edition of the text with **The Night Sky Planisphere Latitude <40**: ISBN 13: 978-1-1111-1632-3 ISBN 10: 1-1111-1632-6

Student edition of text with **Enhanced WebAssign**: ISBN 13: 978-1-1111-1424-4 ISBN 10: 1-1111-1424-2

# Flash Reference: H–R Diagram

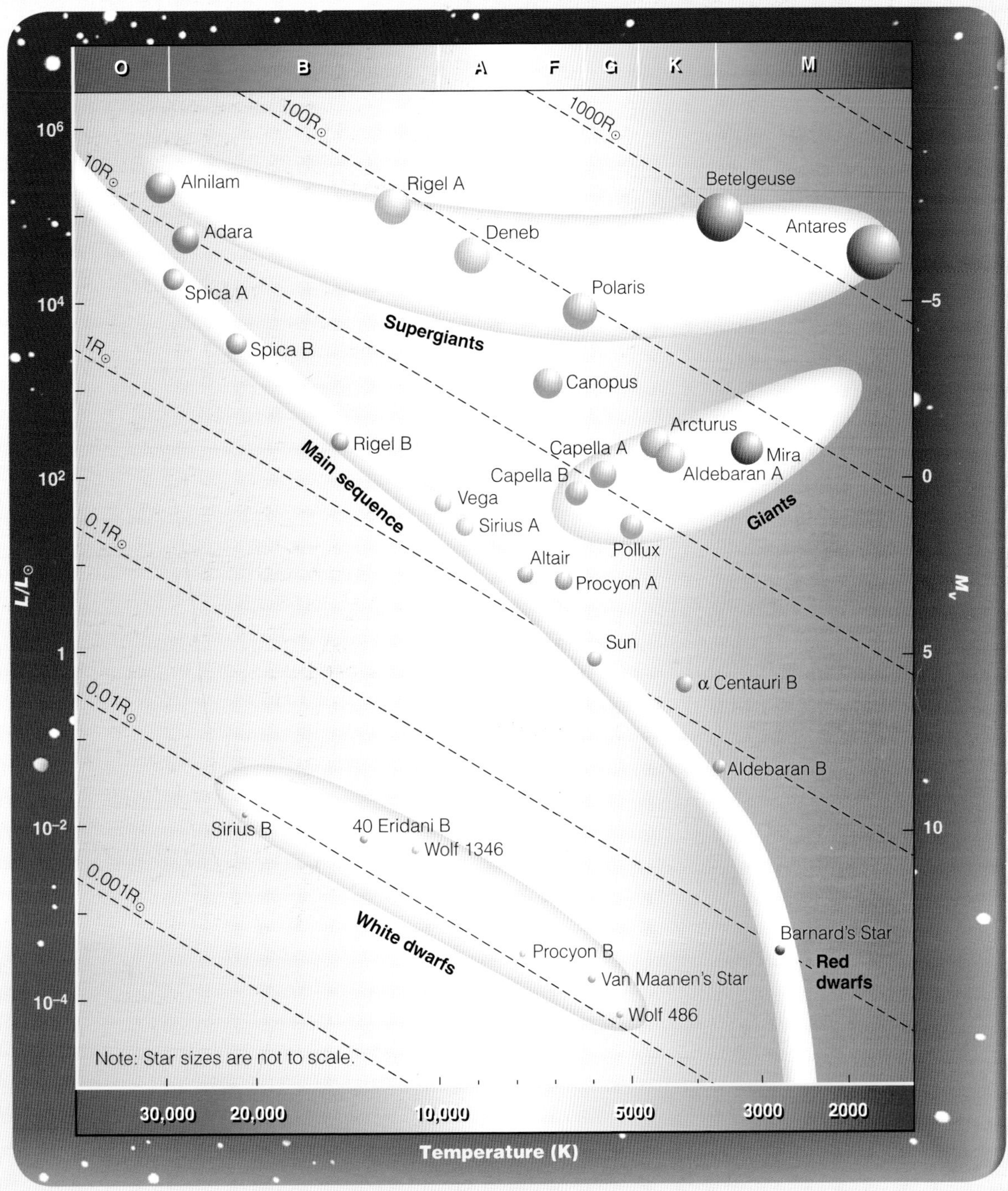


**The H–R diagram is the key to understanding stars, their birth, their long lives, and their eventual deaths. Luminosity ($L/L_\odot$) refers to the total amount of energy that a star emits in terms of the sun's luminosity, and the temperature refers to the temperature of its surface. Together, the temperature and luminosity of a star locate it on the H–R diagram and tell astronomers its radius, its family relationships with other stars, and a great deal about its history and fate.**

# Flash Reference: Comparative Planetology

The terrestrial or Earthlike planets lie very close to the sun, and their orbits are hardly visible in a diagram that includes the outer planets.

Mercury, Venus, Earth and its moon, and Mars are small worlds made of rock and metal with little or no atmospheric gases.

The outer worlds of our solar system orbit far from the sun. Jupiter, Saturn, Uranus, and Neptune are Jovian or Jupiter-like planets much bigger than Earth. They contain large amounts of low-density gases.

Pluto is one of a number of small, icy worlds orbiting beyond Neptune. Astronomers have concluded that Pluto is not really a planet and now refer to it as a dwarf planet.

The Terrestrial Worlds

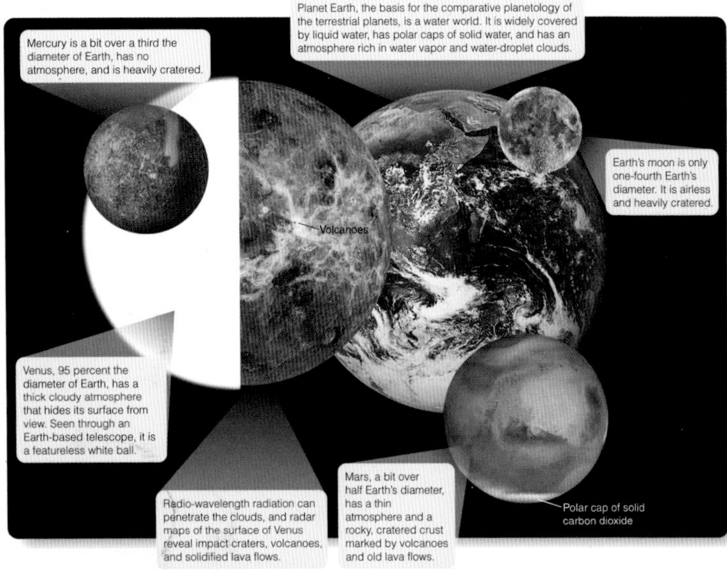
Planet Earth, the basis for the comparative planetology of the terrestrial planets, is a water world. It is widely covered by liquid water, has polar caps of solid water, and has an atmosphere rich in water vapor and water-droplet clouds.
Mercury is a bit over a third the diameter of Earth, has no atmosphere, and is heavily cratered.
Earth's moon is only one-fourth Earth's diameter. It is airless and heavily cratered.
Volcanoes
Venus, 95 percent the diameter of Earth, has a thick cloudy atmosphere that hides its surface from view. Seen through an Earth-based telescope, it is a featureless white ball.
Radio-wavelength radiation can penetrate the clouds, and radar maps of the surface of Venus reveal impact craters, volcanoes, and solidified lava flows.
Mars, a bit over half Earth's diameter, has a thin atmosphere and a rocky, cratered crust marked by volcanoes and old lava flows.
Polar cap of solid carbon dioxide

Planetary Orbits

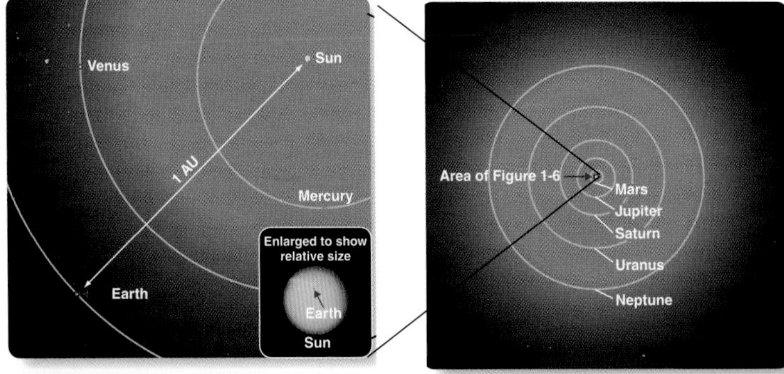
Venus
Sun
1 AU
Mercury
Earth
Enlarged to show relative size
Earth
Sun
Area of Figure 1-6
Mars
Jupiter
Saturn
Uranus
Neptune

The Outer Worlds

Jupiter, more than 11 times Earth's diameter, is the largest planet in our solar system.
The cloud belts and zones on Saturn are less distinct than those on Jupiter.
Uranus and Neptune are both both about four times Earth's diameter.
Shadow of one of Jupiter's many moons
Earth is the largest of the Terrestrial worlds, but it is small compared with the Jovian planets.
Uranus and Neptune are green- and blue-colored because of small amounts of methane in their hydrogen-rich atmospheres.

• See page 425 for the terrestrial planets. See page 4 for the two orbital diagrams. See page 494 for the outer worlds.

# Flash Reference: Arrows

This book is designed to use arrows to alert you to important concepts in diagrams and graphs. Some arrows point things out, but others represent motion, force, or even the flow of light. Look at arrows in the book carefully and use this Flash Reference card to catch all of the arrow clues.

**Point at things:**

You are here

**Force:**

**Process flow:**

**Measurement:**

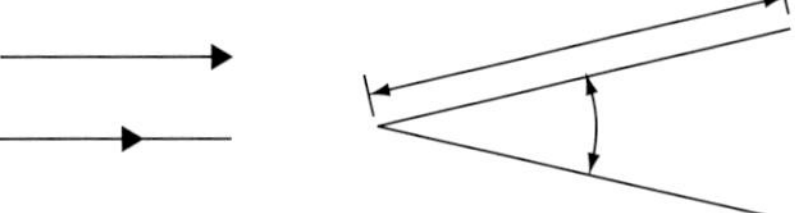

**Direction:**

**Radio waves, infrared, photons:**

**Motion:**

*Rotation 2-D*

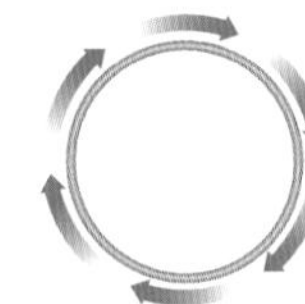

*Rotation 3-D*

*Linear*

**Light flow:**
Updated arrow style

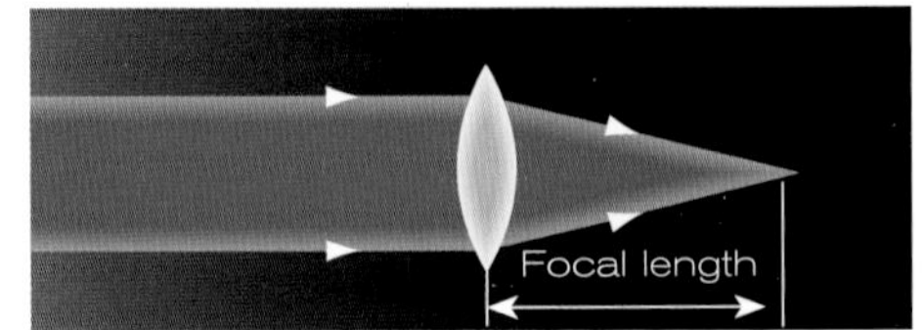
Focal length

# The Solar System

# About the Authors

**Mike Seeds** has been a Professor of Physics and Astronomy at Franklin and Marshall College in Lancaster, Pennsylvania, since 1970. In 1989 he received F&M College's Lindback Award for Distinguished Teaching. Mike's love for the history of astronomy led him to create upper-level courses on "Archaeoastronomy" and "Changing Concepts of the Universe." His research interests focus on variable stars and the automation of astronomical telescopes. Mike is coauthor with Dana Backman of *Horizons: Exploring the Universe*, Eleventh Edition (2009), *Astronomy: The Solar System and Beyond*, Sixth Edition (2009), and *Perspectives on Astronomy* (2008), all published by Cengage. He was Senior Consultant for creation of the 20-episode telecourse accompanying the book *Horizons, Exploring the Universe.*

**Dana Backman** taught in the physics and astronomy department at Franklin and Marshall College in Lancaster, Pennsylvania, from 1991 until 2003. He invented and taught a course titled "Life in the Universe" in F&M's interdisciplinary Foundations program. Dana now teaches introductory astronomy, astrobiology, and cosmology courses in Stanford University's Continuing Studies Program. His research interests focus on infrared observations of planet formation, models of debris disks around nearby stars, and evolution of the solar system's Kuiper belt. Dana is employed by the SETI Institute of Mountain View, California as Director of Outreach for the SOFIA (Stratospheric Observatory for Infrared Astronomy) mission at NASA Ames Research Center. He is coauthor with Mike Seeds of *Horizons: Exploring the Universe*, Eleventh Edition (2009), *Astronomy: The Solar System and Beyond*, Sixth Edition (2009), and *Perspectives on Astronomy* (2008), all published by Cengage.

7

SEVENTH EDITION

# The Solar System

**Michael A. Seeds**

Joseph R. Grundy Observatory
Franklin and Marshall College

**Dana E. Backman**

SOFIA (Stratospheric Observatory
for Infrared Astronomy)
SETI Institute & NASA Ames
Research Center

Australia • Brazil • Canada • Mexico • Singapore • Spain
United Kingdom • United States

**The Solar System, Seventh Edition**
**Michael A. Seeds, Dana E. Backman**

Editor in Chief: Michelle Julet
Acquisitions Editor: Kilean Kennedy
Development Editor: Teri Hyde
Editorial Assistant: Laura Bowen
Senior Media Editor: Rebecca Berardy Schwartz
Marketing Manager: Nicole Mollica
Marketing Coordinator: Kevin Carroll
Marketing Communications Manager: Belinda Krohmer
Associate Content Project Manager: Jill Clark
Senior Art Director: Cate Rickard Barr
Senior Manufacturing Buyer: Diane Gibbons
Senior Rights Account Manager: Bob Kauser
Production Service: Integra Onshore
Text Designer: Liz Harasymczuk Design
Photo Manager: Amanda Groszko
Cover Designer: Irene Morris
Cover Image: NASA
Compositor: Integra Offshore



For product information and technology assistance, contact us at
**Cengage Learning Customer & Sales Support, 1-800-354-9706.**
For permission to use material from this text or product,
Submit all requests online at **www.cengage.com/permissions.**
Further permissions questions can be emailed to
**permissionrequest@cengage.com.**

Library of Congress Control Number: 2009937789

ISBN-13: 978-1-4390-5036-1
ISBN-10: 1-4390-5036-8

**Brooks/Cole**
20 Channel Center Street
Boston, MA 02210
USA

Cengage Learning is a leading provider of customized learning solutions with office locations around the globe, including Singapore, the United Kingdom, Australia, Mexico, Brazil, and Japan. Locate your local office at **international.cengage.com/region**

Cengage Learning products are represented in Canada by Nelson Education, Ltd.

For your course and learning solutions, visit **www.cengage.com.**

Purchase any of our products at your local college store or at our preferred online store **www.ichapters.com**.

Printed in China
3 4 5 6 7 13 12 11

# Dedication

*For Emery & Helen Seeds and Edward & Antonette Backman*

# Brief Contents

## Part 1 Exploring the Sky



## Part 2 The Stars



## Part 4 The Solar System



## Part 5 Life

# Contents

## Part 1: Exploring the Sky

### How Do We Know?



### Concept Art Portfolios

# Part 2: The Stars



## How Do We Know?



## Concept Art Portfolios



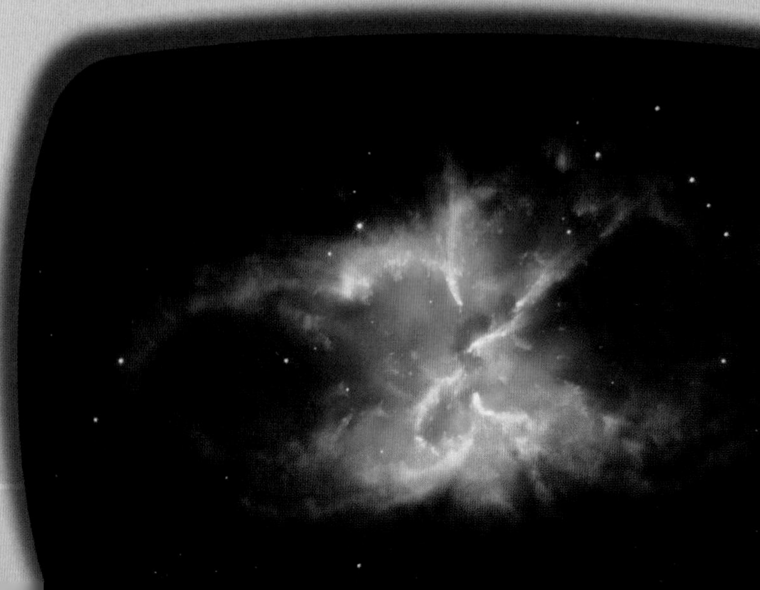

## Part 4: The Solar System



### How Do We Know?



### Concept Art Portfolios

*The longest journey begins with a single step.*

— LAO TSE

## 1-1 Where Are You?

TO FIND YOUR PLACE among the stars, you can take a cosmic zoom, a ride out through the universe to preview the kinds of objects you are about to study.

You can begin with something familiar. ■ Figure 1-1 shows a region about 50 feet across occupied by a human being, a sidewalk, and a few trees—all objects whose size you can understand. Each successive picture in this cosmic zoom will show you a region of the universe that is 100 times wider than the preceding picture. That is, each step will widen your **field of view**, the region you can see in the image, by a factor of 100.

Widening your field of view by a factor of 100 allows you to see an area 1 mile in diameter (■ Figure 1-2). People, trees, and sidewalks have become too small to see, but now you see a college campus and surrounding streets and houses. The dimensions of houses and streets are familiar. This is still the world you know.

Before leaving this familiar territory, you should make a change in the units you use to measure sizes. All scientists, including astronomers, use the metric system of units because it is well understood worldwide and, more important, because it simplifies calculations. If you are not already familiar with the metric system, or if you need a review, study Appendix A before reading on.

■ Figure 1-1

Michael A. Seeds

■ Figure 1-2

This box ■ represents the relative size of the previous frame. (USGS)

The photo in Figure 1-2 is 1 mile across, which equals about 1.6 kilometers. You can see that a kilometer (abbreviated km) is a bit under two-thirds of a mile—a short walk across a neighborhood. But when you expand your field of view by a factor of 100, the neighborhood you saw in the previous photo vanishes (■ Figure 1-3). Now your field of view is 160 km wide, and you see cities and towns as patches of gray. Wilmington, Delaware, is visible at the lower right. At this scale, you can see some of the natural features of Earth's surface. The Allegheny Mountains of southern Pennsylvania cross the image in the upper left, and the Susquehanna River flows southeast into Chesapeake Bay. What look like white bumps are a few puffs of clouds.

Figure 1-3 is an infrared photograph in which healthy green leaves and crops show up as red. Human eyes are sensitive to only a narrow range of colors. As you explore the universe, you will learn to use a wide range of other "colors," from X-rays to radio waves, to reveal sights invisible to unaided human eyes. You will learn much more about infrared, X-rays, and radio energy in later chapters.

At the next step in your journey, you can see your entire planet, which is nearly 13,000 km in diameter (■ Figure 1-4). At any particular moment, half of Earth's surface is exposed to sunlight, and half is in darkness. As Earth rotates on its axis, it carries you through sunlight and then through darkness, producing the cycle of day and night. The blurriness you see at the extreme right of the photo is the boundary between day and night—the sunset line. This is a good example of how a photo can give you

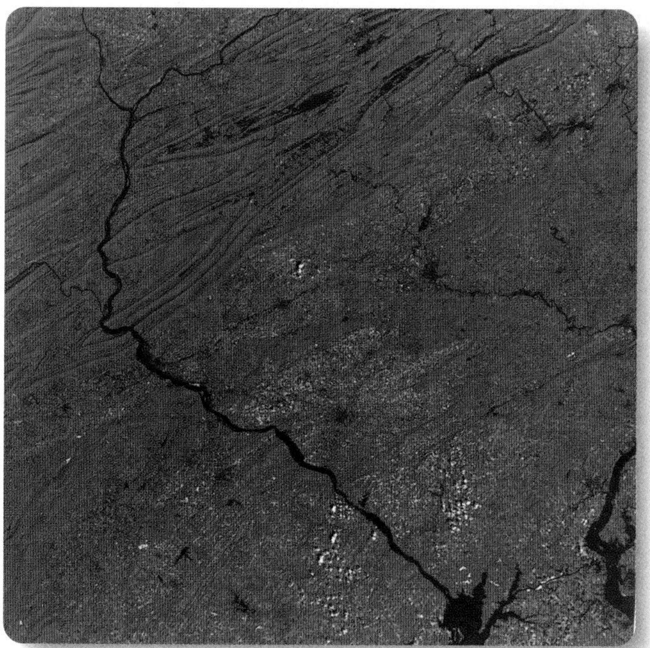

■ Figure 1-3

NASA

visual clues to understanding a concept. Special questions called "Learning to Look" at the end of each chapter give you a chance to use your own imagination to connect images with theories about astronomical objects.

Enlarge your field of view by a factor of 100, and you see a region 1,600,000 km wide (■ Figure 1-5). Earth is the small blue dot in the center, and the moon, whose diameter is only one-fourth that of Earth, is an even smaller dot along its orbit 380,000 km away.

These numbers are so large that it is inconvenient to write them out. Astronomy is sometimes known as the science of big numbers, and soon you will be using numbers much larger than these to discuss the universe. Rather than writing out these numbers as in the previous paragraph, it is more convenient to write them in **scientific notation.** This is nothing more than a simple way to write very big or very small numbers without using lots of zeros. In scientific notation, 380,000 becomes $3.8 \times 10^5$. If you are not familiar with scientific notation, read the section on powers of 10 notation in Appendix A. The universe is too big to discuss without using scientific notation.

When you once again enlarge your field of view by a factor of 100, Earth, the moon, and the moon's orbit all lie in the small red box at lower left of ■ Figure 1-6. Now you can see the sun and two other planets that are part of our solar system. Our **solar system** consists of the sun, its family of planets, and some smaller bodies, such as moons and comets.

Earth, Venus, and Mercury are **planets,** small, spherical, nonluminous bodies that orbit a star and shine by reflected light. Venus is about the size of Earth, and Mercury is just over a third of Earth's diameter. On the diagram, they are both too small to be seen as anything but tiny dots. The sun is a **star,** a self-luminous

■ Figure 1-4

NASA

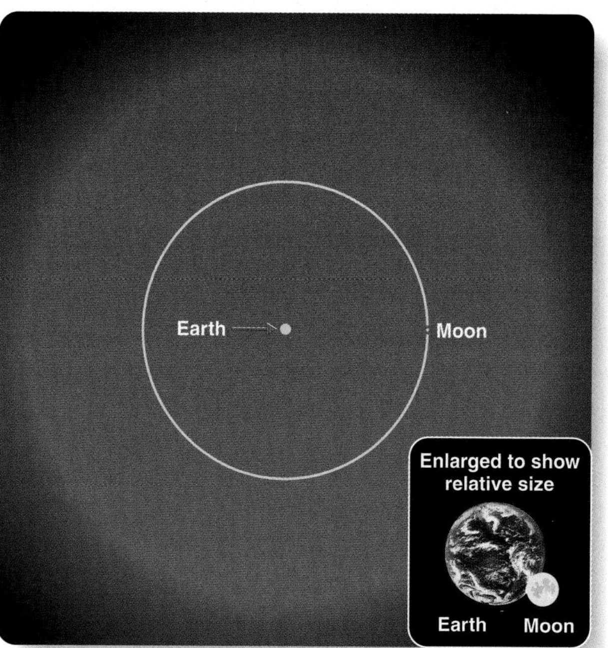


■ Figure 1-5

NASA

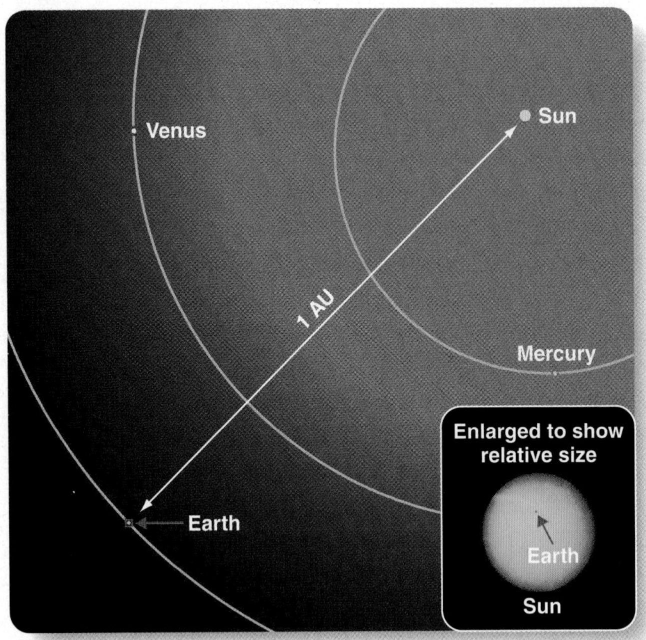


**■ Figure 1-6**

NOAO

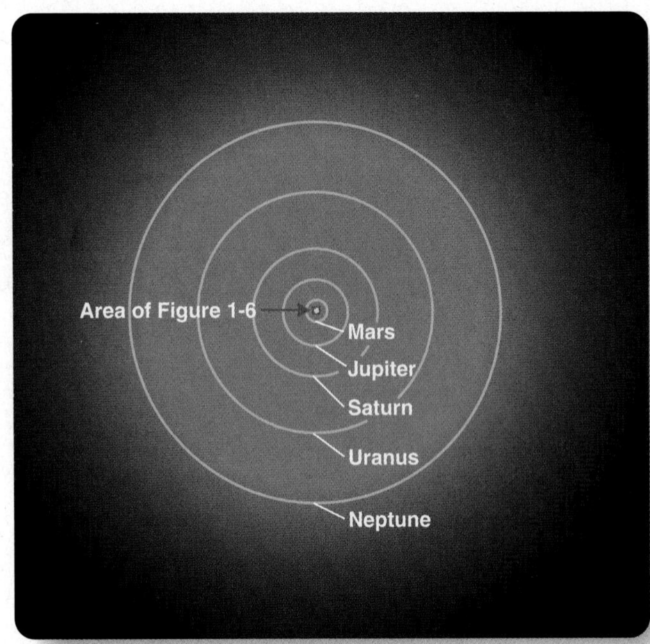


**■ Figure 1-7**

ball of hot gas that generates its own energy. Even though the sun is 109 times larger in diameter than Earth (inset), it too is nothing more than a dot in this diagram.

This diagram represents an area with a diameter of $1.6 \times 10^8$ km. One way astronomers simplify calculations using large numbers is to define larger units of measurement. For example, the average distance from Earth to the sun is a unit of distance called the **astronomical unit (AU)**, which is equal to $1.5 \times 10^8$ km. Using that, you can express the average distance from Venus to the sun as about 0.72 AU and the average distance from Mercury to the sun as about 0.39 AU.

These distances are averages because the orbits of the planets are not perfect circles. This is particularly apparent in the case of Mercury. Its orbit carries it as close to the sun as 0.307 AU and as far away as 0.467 AU. You can see the variation in the distance from Mercury to the sun in Figure 1-6. Earth's orbit is more circular, and its distance from the sun varies by only a few percent.

Enlarge your field of view again, and you can see the entire solar system (■ Figure 1-7). The sun, Mercury, Venus, and Earth lie so close together that you cannot see them separately at this scale, and they are lost in the red square at the center of this diagram. You can see only the brighter, more widely separated objects such as Mars, the next planet outward. Mars is only 1.5 AU from the sun, but Jupiter, Saturn, Uranus, and Neptune are farther from the sun and so are easier to place in this diagram. They are cold worlds far from the sun's warmth. Light from the sun reaches Earth in only 8 minutes, but it takes over 4 hours to reach Neptune.

You can remember the order of the plants from the sun outward by remembering a simple sentence: *My Very Educated Mother Just Served Us Noodles.* The first letter of each word reminds you of a planet: Mercury, Venus, Earth, Mars, Jupiter, Saturn, Uranus, Neptune.

When you again enlarge your field of view by a factor of 100, the solar system vanishes (■ Figure 1-8). The sun is only a point of light, and all the planets and their orbits are now crowded into the small red square at the center. The planets are too small and too faint to be visible so near the brilliance of the sun.

Nor are any stars visible except for the sun. The sun is a fairly typical star, and it seems to be located in a fairly average neighborhood in the universe. Although there are many billions of stars like the sun, none are close enough to be visible in this diagram, which shows a region only 11,000 AU in diameter. Stars in the sun's neighborhood are typically separated by distances about 30 times larger than that.

In ■ Figure 1-9, your field of view has expanded to a diameter of a bit over 1 million AU. The sun is at the center, and at this scale you can see a few of the nearest stars. These stars are so distant that it is not convenient to give their distances in astronomical units. To express distances so large, astronomers define a new unit of distance, the light-year. One **light-year (ly)** is the distance that light travels in one year, roughly $10^{13}$ km or 63,000 AU. It is a **Common Misconception** that a light-

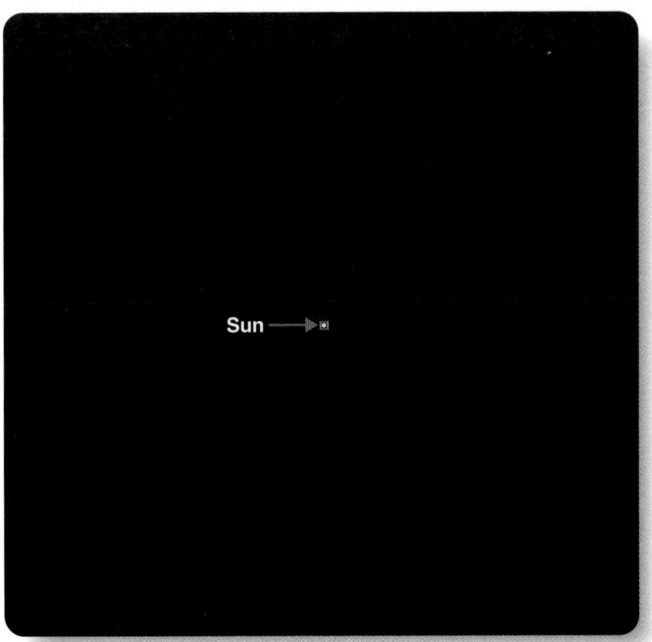


■ Figure 1-8

year is a unit of time, and you can sometimes hear the term misused in science fiction movies and TV shows. The next time you hear someone say, "It will take me light-years to finish my history paper," you could tell the person that a light-year is a distance, not a time. The diameter of your field of view in Figure 1-9 is 17 ly.

Another **Common Misconception** is that stars look like disks when seen through a telescope. Although stars are roughly the same size as the sun, they are so far away that astronomers cannot see them as anything but points of light. Even the closest star to the sun—Alpha Centauri, only 4.2 ly from Earth—looks like a point of light through even the biggest telescopes on Earth. Furthermore, planets that circle other stars are much too small, too faint, and too close to the glare of their star to be easily visible. Astronomers have used indirect methods to detect over 300 planets orbiting other stars, but very few have been photographed directly.

Figure 1-9 follows the astronomical custom of making the sizes of the dots represent not the sizes of the stars but their brightness. This is how star images are recorded on photographs. Bright stars make larger spots on a photograph than faint stars, so the size of a star image in a photograph tells you not how big the star is but only how bright it looks.

In ■ Figure 1-10, you expand your field of view by another factor of 100, and the sun and its neighboring stars vanish into the background of thousands of other stars. The field of view is now 1700 ly in diameter. Of course, no one has ever journeyed thousands of light-years from Earth to look back and photograph the solar neighborhood, so this is a representative photograph of the sky. The sun is a relatively faint star that would not be easily located in a photo at this scale.

If you again expand your field of view by a factor of 100, you see our galaxy, a disk of stars about 80,000 ly in diameter (■ Figure 1-11). A **galaxy** is a great cloud of stars, gas, and dust

■ Figure 1-9

■ Figure 1-10

NOAO

**Figure 1-11**

© Mark Garlick/space-art.com

**Figure 1-12**

held together by the combined gravity of all of its matter. Galaxies range from 1500 to over 300,000 ly in diameter, and some contain over 100 billion stars. In the night sky, you can see our galaxy as a great, cloudy wheel of stars ringing the sky. This band of stars is known as the **Milky Way**, and our galaxy is called the **Milky Way Galaxy**.

How does anyone know what our galaxy looks like if no one can leave it and look back? Astronomers use evidence to guide their explanations as they imagine what the Milky Way looks like. Artists can then use those scientific descriptions to create a painting. Many images in this book are artists' renderings of objects and events that are too big or too dim to see clearly, emit energy your eyes cannot detect, or happen too slowly or too rapidly for humans to sense. These images are not just guesses; they are scientifically based illustrations guided by the best information astronomers can gather. As you explore, notice how astronomers use the methods of science to imagine, understand, and depict cosmic events.

The artist's conception of the Milky Way reproduced in Figure 1-11 shows that our galaxy, like many others, has graceful **spiral arms** winding outward through its disk. In a later chapter, you will learn that the spiral arms are places where stars are formed from clouds of gas and dust. Our own sun was born in one of these spiral arms; and, if you could see the sun in this picture, it would be in the disk of the galaxy about two-thirds of the way out from the center.

Ours is a fairly large galaxy. Only a century ago astronomers thought it was the entire universe—an island cloud of stars in an otherwise empty vastness. Now they know that our galaxy is not unique; it is only one of many billions of galaxies scattered throughout the universe.

You can see a few of these other galaxies when you expand your field of view by another factor of 100 (■ Figure 1-12). Our galaxy appears as a tiny luminous speck surrounded by other specks in a region 17 million light-years in diameter. Each speck represents a galaxy. Notice that our galaxy is part of a cluster of a few dozen galaxies. Galaxies are commonly grouped together in such clusters. Some galaxies have beautiful spiral patterns like our own galaxy, but others do not. Some are strangely distorted. In a later chapter, you will learn what produces these differences among the galaxies.

Now is a chance for you to correct another **Common Misconception**. People often say "galaxy" when they mean "solar system," and they sometimes confuse both terms with "universe." Your cosmic zoom has shown you the difference. The solar system is the sun and its planets. Our galaxy contains our solar system plus billions of other stars and whatever planets orbit around them, in other words, billions of planetary systems. The universe includes everything: all of the galaxies, stars, and planets, including our own galaxy and a very small part of it, our solar system.

If you expand your field of view one more time, you can see that clusters of galaxies are connected in a vast network (■ Figure 1-13). Clusters are grouped into superclusters—clusters of clusters—and the superclusters are linked to form long filaments and walls outlining nearly empty voids. These filaments and walls appear to be the largest structures in the universe. Were you to expand your field of view another time, you would probably see a uniform fog of filaments and walls. When you puzzle over the origin of these structures, you are at the frontier of human knowledge.

**■ Figure 1-13**

(Based on data from M. Seldner, B. L. Siebers, E. J. Groth, and P. J. E. Peebles, *Astronomical Journal 82* [1977].)

## 1-2 When Is Now?

NOW THAT YOU HAVE AN IDEA where you are in space, you need to know where you are in time. The stars shined for billions of years before the first human looked up and wondered what they were. To get a sense of your place in time, all you need is a long red ribbon.

Imagine stretching a ribbon from goal line to goal line down the center of a football field, as shown on the inside front cover of this book. Imagine that one end of the ribbon is *today* and that the other end represents the beginning of the universe—the moment of beginning that astronomers call the *big bang*. In a later chapter, "Modern Cosmology," you will learn all about the big bang and see evidence that the universe is about 14 billion years old. Your long red ribbon represents 14 billion years, the entire history of the universe.

Imagine beginning at the goal line labeled *Big Bang* and replaying the entire history of the universe as you walk along your ribbon toward the goal line labeled *Today*. Observations tell astronomers that the big bang filled the entire universe with hot, glowing gas, but as the gas cooled and dimmed the universe went dark. All that happened along the first half inch of the ribbon. There was no light for the first 400 million years, until gravity was able to pull some of the gas together to form the first stars. That seems like a lot of years, but if you stick a little flag beside the ribbon to mark the birth of the first stars, it would be not quite 3 yards from the goal line where the universe began.

You have to walk only about 5 yards along the ribbon before galaxies formed in large numbers. Our home galaxy would be one of those taking shape. By the time you cross the 50-yard line, the universe is full of galaxies, but the sun and Earth have not formed yet. You need to walk past the 50-yard line down to the 35-yard line before you can finally stick a flag beside the ribbon to mark the formation of the sun and planets—our solar system.

You can carry your flags a few yards farther to the 29-yard line to mark the appearance of the first life on Earth–microscopic creatures in the oceans—and you have to walk all the way to the 3-yard line before you can mark the emergence of life on land. Your dinosaur flag goes just inside the 2-yard line. Dinosaurs go extinct as you pass the one-half-yard line.

What about people? You can put a little flag for the first humanlike creatures only about an inch—four million years–from the goal line labeled *Today*. Civilization, the building of cities, began about 10,000 years ago, so you have to try to fit that flag in only 0.0026 inch from the goal line. That's half the thickness of a sheet of paper. Compare the history of human civilization with the history of the universe. Every war you have ever heard of, every person whose name is recorded, every structure ever built from Stonehenge to the building you are in right now fits into that 0.0026 inch.

Humanity is very new to the universe. Our civilization on Earth has existed for only a flicker of an eyeblink in the history of the universe. As you will discover in the chapters that follow, only in the last hundred years or so have astronomers begun to understand where we are in space and in time.

## 1-3 Why Study Astronomy?

YOUR EXPLORATION OF THE UNIVERSE will help you answer two fundamental questions:

What are we?
How do we know?

The question, "What are we?" is the first organizing theme of this book. Astronomy is important to you because it will tell you what you are. Notice that the question is not "*Who* are we?" If you want to know who we are, you may want to talk to a sociologist, theologian, paleontologist, artist, or poet. "*What* are we?" is a fundamentally different question.

As you study astronomy, you will learn how you fit into the history of the universe. You will learn that the atoms in your body had their first birthday in the big bang when the universe began. Those atoms have been cooked and remade inside generations of stars, and now, after billions of years, they are inside you. Where will they be in another billion years? This is a story

# How Do We Know? 1-1

## The So-Called Scientific Method

**How do scientists learn about nature?** You have probably heard of the scientific method as the process by which scientists form hypotheses and test them against evidence gathered by experiment or observation. Scientists use the scientific method all the time, and it is critically important, but they rarely think of it at all, and they certainly don't think of it as a numbered list of steps. It is such an ingrained way of thinking and understanding nature that it is almost invisible to the people who use it most.

Scientists try to form hypotheses that explain how nature works. If a hypothesis is contradicted by evidence from experiments or observations, it must be revised or discarded. If a hypothesis is confirmed, it must be tested further. In that very general way, the scientific method is a way of testing and refining ideas to better describe how nature works.

For example, Gregor Mendel (1822–1884) was an Austrian abbot who liked plants. He formed a hypothesis that offspring usually inherit traits from their parents not as a smooth blend, as most scientists of the time believed, but in discrete units according to strict mathematical rules. Mendel cultivated and tested over 28,000 pea plants, noting which produced smooth peas and which produced wrinkled peas and how that trait was inherited by successive generations. His study of pea plants confirmed his hypothesis and allowed the development of a series of laws of inheritance. Although the importance of his work was not recognized in his lifetime, Mendel is now called the "father of modern genetics."

The scientific method is not a simple, mechanical way of grinding facts into understanding. It is, in fact, a combination of many ways of analyzing information, finding relationships, and creating new ideas. A scientist needs insight and ingenuity to form and test a good hypothesis. Scientists use the scientific method almost automatically, forming, testing, revising, and discarding hypotheses almost minute by minute as they discuss a new idea. Sometimes, however, a scientist will spend years studying a single promising hypothesis. The so-called scientific method is a way of thinking and a way of knowing about nature. The "How Do We Know?" essays in the chapters that follow will introduce you to some of those methods.

*Whether peas are wrinkled or smooth is an inherited trait* (Inspirestock/jupiterimages).

everyone should know, and astronomy is the only course on campus that can tell you that story.

Every chapter in this book ends with a short segment titled "What Are We?" This summary shows how the astronomy in the chapter relates to your role in the story of the universe.

The question, "How do we know?" is the second organizing theme of this book. It is a question you should ask yourself whenever you encounter statements made by so-called experts in any field. Should you swallow a diet supplement recommended by a TV star? Should you vote for a candidate who warns of a climate crisis? To understand the world around you and to make wise decisions for yourself, for your family, and for your nation, you need to understand how science works.

You can use astronomy as a case study in science. In every chapter of this book, you will find short essays titled "How Do We Know?" They are designed to help you think not about *what* is known but about *how* it is known. To do that, they will explain different aspects of scientific reasoning and in that way help you understand how scientists know about the natural world.

Over the last four centuries, scientists have developed a way to understand nature that has been called the **scientific method** (**How Do We Know? 1-1**). You will see this process applied over and over as you read about exploding stars, colliding galaxies, and alien planets. The universe is very big, but it is described by a small set of rules, and we humans have found a way to figure out the rules—a method called *science*.

## What Are We? Part of the Story

Astronomy will give you perspective on what it means to be here on Earth. This chapter has helped you locate yourself in space and time. Once you realize how vast our universe is, Earth seems quite small. People on the other side of the world seem like neighbors. And, in the entire history of the universe, the human story is only the blink of an eye. This may seem humbling at first, but you can be proud of how much we humans have understood in such a short time.

Not only does astronomy locate you in space and time, it places you in the physical processes that govern the universe. Gravity and atoms work together to make stars, light the universe, generate energy, and create the chemical elements in your body. The chapters that follow will show how you fit into that cosmic process.

Although you are very small and your kind have existed in the universe for only a short time, you are an important part of something very large and very beautiful.

# Study and Review

## Summary

- You surveyed the universe by taking a cosmic zoom in which each **field of view (p. 2)** was 100 times wider than the previous field of view.
- Astronomers use the metric system because it simplifies calculations and use **scientific notation (p. 3)** for very large or very small numbers.
- You live on a **planet (p. 3),** Earth, which orbits our **star (p. 3),** the sun, once a year. As Earth rotates once a day, you see the sun rise and set.
- The moon is only one-fourth the diameter of Earth, but the sun is 109 times larger in diameter than Earth—a typical size for a star.
- The **solar system (p. 3)** includes the sun at the center, all of the planets that orbit around it—Mercury, Venus, Earth, Mars, Jupiter, Saturn, Uranus, and Neptune—plus the moons of the planets, plus other objects bound to the sun by its gravity.
- The **astronomical unit (AU) (p. 4)** is the average distance from Earth to the sun. Mars, for example, orbits 1.5 AU from the sun. The **light-year (ly) (p. 4)** is the distance light can travel in one year. The nearest star is 4.2 ly from the sun.
- Many stars seem to have planets, but such small, distant worlds are difficult to detect. Only a few hundred have been found so far, but planets seem to be common, so you can probably trust that there are lots of planets in the universe, including some like Earth.
- The **Milky Way (p. 6),** the hazy band of light that encircles the sky, is the **Milky Way Galaxy (p. 6)** seen from inside. The sun is just one out of the billions of stars that fill the Milky Way Galaxy.
- **Galaxies (p. 5)** contain many billions of stars. Our galaxy is about 80,000 ly in diameter and contains over 100 billion stars.
- Some galaxies, including our own, have graceful **spiral arms (p. 6)** bright with stars, but some galaxies are plain clouds of stars.
- Our galaxy is just one of billions of galaxies that fill the universe in great clusters, clouds, filaments, and walls—the largest structures in the universe.
- The universe began about 14 billion years ago in an event called the big bang, which filled the universe with hot gas.
- The hot gas cooled, the first galaxies began to form, and stars began to shine only about 400 million years after the big bang.
- The sun and planets of our solar system formed about 4.6 billion years ago.
- Life began in Earth's oceans soon after Earth formed but did not emerge onto land until only 400 million years ago. Dinosaurs evolved not long ago and went extinct only 65 million years ago.
- Humanlike creatures developed on Earth only about 4 million years ago, and human civilizations developed only about 10,000 years ago.
- Although astronomy seems to be about stars and planets, it describes the universe in which you live, so it is really about you. Astronomy helps you answer the question, "What are we?"
- As you study astronomy, you should ask "How do we know?" and that will help you understand how science gives us a way to understand nature.
- In its simplest outline, science follows the **scientific method (p. 8),** in which scientists expect statements to be supported by evidence compared with hypotheses. In fact, science is a complex and powerful way to think about nature.

## Review Questions

1. What is the largest dimension of which you have personal knowledge? Have you run a mile? Hiked 10 miles? Run a marathon?
2. What is the difference between our solar system, our galaxy, and the universe?
3. Why are light-years more convenient than miles, kilometers, or astronomical units for measuring certain distances?
4. Why is it difficult to detect planets orbiting other stars?
5. What does the size of the star image in a photograph tell you?
6. What is the difference between the Milky Way and the Milky Way Galaxy?
7. What are the largest known structures in the universe?
8. How does astronomy help answer the question, "What are we?"
9. **How Do We Know?** How does the scientific method give scientists a way to know about nature?

*The Southern Cross I saw every night abeam. The sun every morning came up astern; every evening it went down ahead. I wished for no other compass to guide me, for these were true.*

— CAPTAIN JOSHUA SLOCUM
*SAILING ALONE AROUND THE WORLD*

THE NIGHT SKY is the rest of the universe as seen from our planet. When you look up at the stars, you are looking out through a layer of air only a little more than a hundred kilometers deep. Beyond that, space is nearly empty, and the stars are scattered light-years apart.

As you read this chapter, keep in mind that you live on a planet in the midst of these scattered stars. Because Earth turns on its axis once a day, the sky appears to revolve around you in a daily cycle. Not only does the sun rise in the eastern part of the sky and set in the western part, but so do the stars.

## 2-1 The Stars

ON A DARK NIGHT far from city lights, you can see a few thousand stars. The ancients organized what they saw by naming stars and groups of stars. Some of those names survive today.

### Constellations

All around the world, ancient cultures celebrated heroes, gods, and mythical beasts by giving their names to groups of stars—**constellations** (■ Figure 2-1). You should not be surprised that the star patterns do not look like the creatures they represent any more than Columbus, Ohio, looks like Christopher Columbus. The constellations simply celebrate the most important mythical figures in each culture. The oldest constellations named by Western cultures originated in Assyria over 3000 years ago, and others were added by Babylonian and Greek astronomers during the classical age. Of these ancient constellations, 48 are still in use.

Different cultures grouped stars and named constellations differently. The constellation you know as Orion was known as Al Jabar, the giant, to the ancient Syrians, as the White Tiger to the Chinese, and as Prajapati in the Form of a Stag in India. The Pawnee Indians knew the constellation Scorpius as two groupings. The long tail of the scorpion was the Snake, and the two bright stars at the tip of the scorpion's tail were the Two Swimming Ducks.

Many ancient cultures, including the Greeks, northern Asians, and Native Americans, associated the stars of the Big Dipper with a bear. The concept of the celestial bear may have crossed the land bridge into North America with the first Americans roughly 12,000 years ago. The names of some of the constellations you see in the sky may be among the oldest surviving traces of human culture.

To the ancients, a constellation was a loose grouping of stars. Many of the fainter stars were not included in any constellation, and the stars of the southern sky not visible to the ancient astronomers of northern latitudes were not grouped into constellations. Constellation boundaries, when they were defined at all, were only approximate (■ Figure 2-2a), so a star like Alpheratz could be thought of as part of Pegasus or part of Andromeda. To correct these gaps and ambiguities, astronomers have added 40 modern constellations, and in 1928 the International Astronomical Union established 88 official constellations with clearly defined boundaries (Figure 2-2b). Consequently, a constellation now represents not a group of stars but an area of the sky, and any star within the region belongs to one and only one constellation. Alpheratz belongs to Andromeda.

■ **Figure 2-1**

The constellations are an ancient heritage handed down for thousands of years as celebrations of great heroes and mythical creatures. Here Sagittarius and Scorpius hang above the southern horizon.

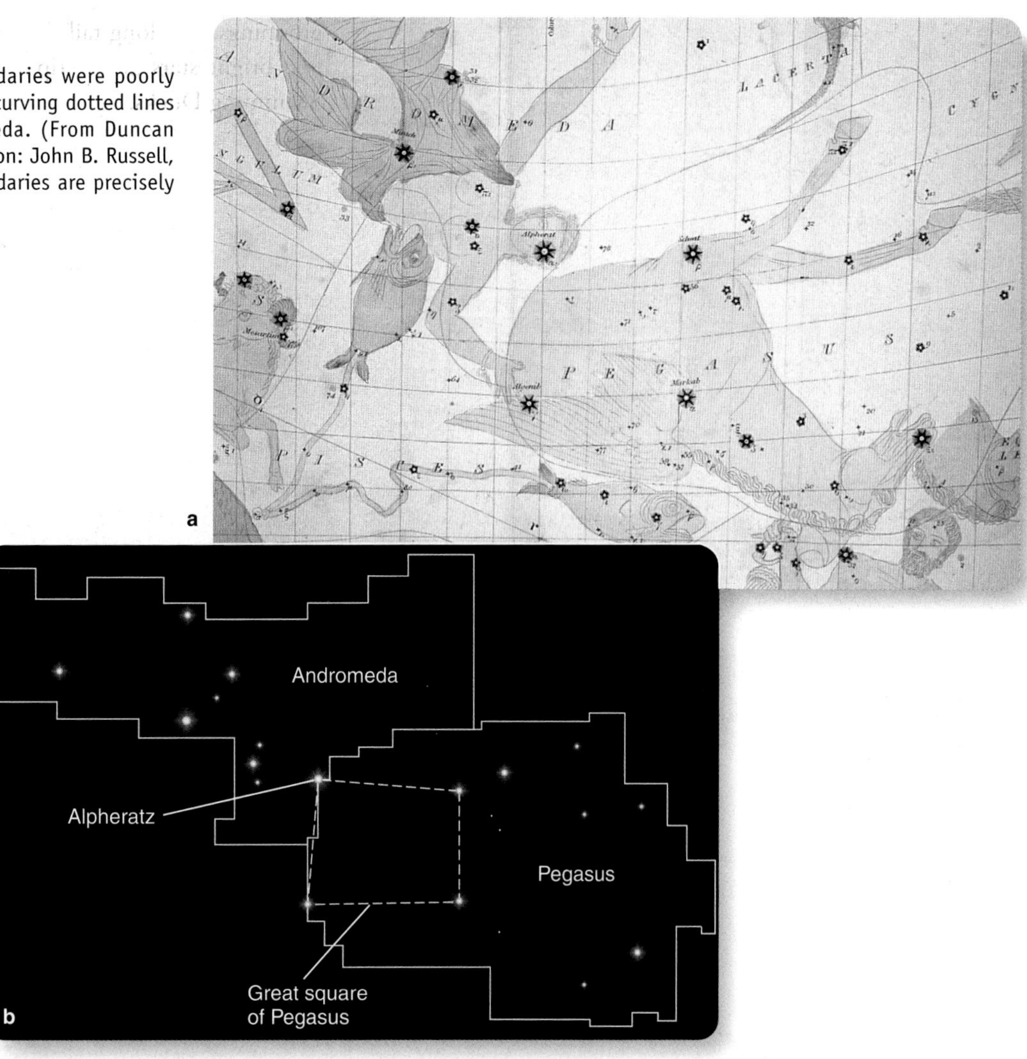


**■ Figure 2-2**

(a) In antiquity, constellation boundaries were poorly defined, as shown on this map by the curving dotted lines that separate Pegasus from Andromeda. (From Duncan Bradford, *Wonders of the Heavens,* Boston: John B. Russell, 1837.) (b) Modern constellation boundaries are precisely defined by international agreement.

In addition to the 88 official constellations, the sky contains a number of less formally defined groupings called **asterisms.** The Big Dipper, for example, is a well-known asterism that is part of the constellation Ursa Major (the Great Bear). Another asterism is the Great Square of Pegasus (Figure 2-2b), which includes three stars from Pegasus plus Alpheratz from Andromeda. The star charts at the end of this book will introduce you to the brighter constellations and asterisms.

Although constellations and asterisms are groups of stars that appear close together in the sky, it is important to remember that most are made up of stars that are not physically associated with one another. Some stars may be many times farther away than others and moving through space in different directions. The only thing they have in common is that they happen to lie in approximately the same direction from Earth (■ Figure 2-3).

## The Names of the Stars

In addition to naming groups of stars, ancient astronomers gave individual names to the brightest individual stars. Modern astronomers still use many of those ancient names. Although the constellation names came from Greek translated into Latin—the language of science until the 19th century—most star names come from ancient Arabic, though much altered by the passing centuries. The name of Betelgeuse, the bright orange star in Orion, for example, comes from the Arabic *yad al jawza,* meaning "shoulder of Jawza [Orion]." Names such as Sirius (the Scorched One) and Aldebaran (the Follower of the Pleiades) are beautiful additions to the mythology of the sky.

Naming individual stars is not very helpful because you can see thousands of them. How many names could you remember? Also, a simple name gives you no clues to the location of the star in the sky or to its brightness. A more useful way to identify stars is to assign letters to the bright stars in a constellation in approximate order of brightness. Astronomers use the Greek alphabet for this purpose. Thus, the brightest star in a constellation is usually designated alpha, the second brightest beta, and so on. Often the name of the Greek letter is spelled out, as in "alpha," but sometimes the actual Greek letter is used, especially in charts. You will find the Greek alphabet in Appendix A. For many constellations, the letters follow the order of brightness, but some constellations, by tradition, mistake, or the personal preferences of early chart makers, are exceptions (■ Figure 2-4).

To identify a star by its Greek-letter designation, you would give the Greek letter followed by the possessive (genitive) form of the constellation name; for example, the brightest star in the constellation Canis Major is alpha Canis Majoris, which can also be written $\alpha$ Canis Majoris. This both identifies the star and the constellation and gives a clue to the relative brightness of the star. Compare this with the ancient name for this star, Sirius, which tells you nothing about location or brightness.

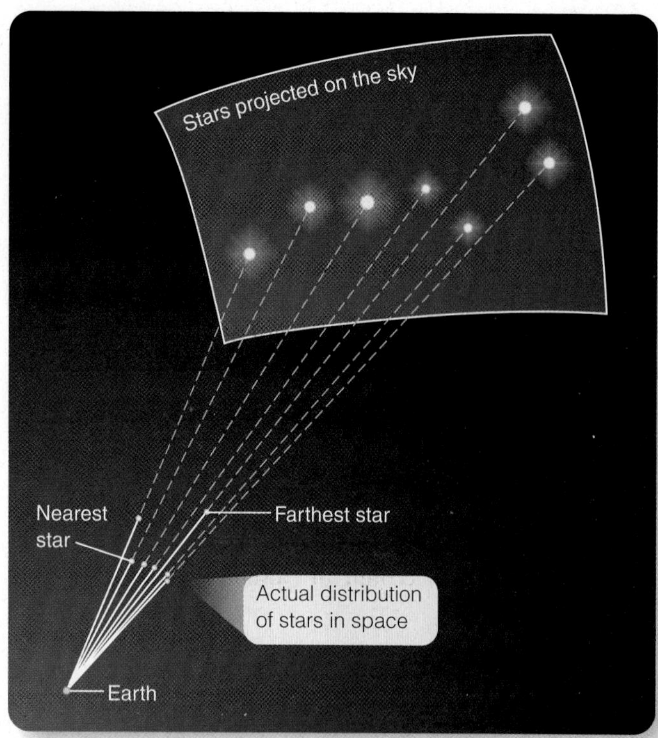


**Figure 2-3**

You see the Big Dipper in the sky because you are looking through a group of stars scattered through space at different distances from Earth. You see them as if they were projected on a screen, and they form the shape of the Dipper.

## Favorite Stars

It is fun to know the names of the brighter stars, but they are more than points of light in the sky. They are glowing spheres of gas much like the sun, each with its unique characteristics. ■ Figure 2-5 identifies eight bright stars that you can adopt as Favorite Stars. As you study astronomy you will discover their peculiar personalities and enjoy finding them in the evening sky. You will learn, for example, that Betelgeuse is not just an orange point of light but is an aging, cool star over 800 times larger than the sun. As you learn more in later chapters, you may want to add more Favorite Stars to your list.

You can use the star charts at the end of this book to help you locate these Favorite Stars. You can see Polaris year-round, but Sirius, Betelgeuse, Rigel, and Aldebaran are in the winter sky. Spica is a summer star, and Vega is visible evenings in later summer and fall. Alpha Centauri, only 4 ly away, is the nearest star to the sun, and you will have to travel as far south as southern Florida to glimpse it above the southern horizon.

## The Brightness of Stars

Besides naming individual stars, astronomers need a way to describe their brightness. Astronomers measure the brightness of stars using the **magnitude scale,** a system that first appeared in the writings of the ancient astronomer Claudius Ptolemy about AD 140. The system probably originated even earlier, and most astronomers attribute it to the Greek astronomer Hipparchus (about 190–120 BC). Hipparchus compiled the first known star catalog, and he may have used the magnitude system in that catalog. Almost 300 years later, Ptolemy used the magnitude system in his own catalog, and successive generations of astronomers have continued to use the system.

The ancient astronomers divided the stars into six classes. The brightest were called first-magnitude stars and those that

The brighter stars in a constellation are usually given Greek letters in order of decreasing brightness.
Orion
α Orionis is also known as Betelgeuse.
β Orionis is also known as Rigel.
In Orion β is brighter than α, and κ is brighter than η. Fainter stars do not have Greek letters or names, but if they are located inside the constellation boundaries, they are part of the constellation.
α
λ
γ
δ
ε
ζ
η
ι
τ
κ
β

**Figure 2-4**

Stars in a constellation can be identified by Greek letters and by names derived from Arabic. The spikes on the star images in the photograph were produced by the optics in the camera. (William Hartmann)

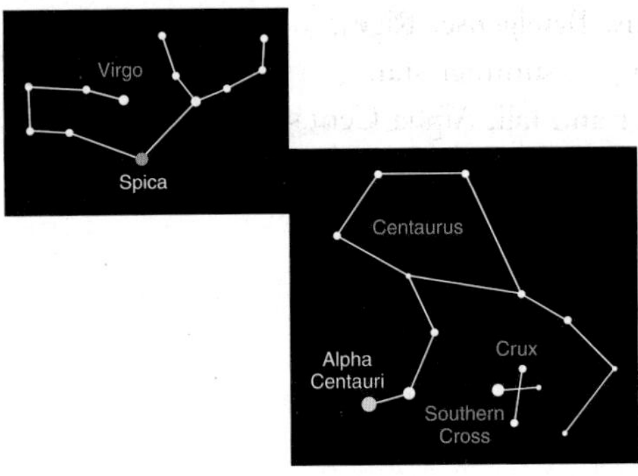


| | | |
|---|---|---|
| **Sirius** | **Brightest star in the sky** | **Winter** |
| **Betelgeuse** | **Bright red star in Orion** | **Winter** |
| **Rigel** | **Bright blue star in Orion** | **Winter** |
| **Aldebaran** | **Red eye of Taurus the Bull** | **Winter** |
| **Polaris** | **The North Star** | **Year round** |
| **Vega** | **Bright star overhead** | **Summer** |
| **Spica** | **Bright southern star** | **Summer** |
| **Alpha Centauri** | **Nearest star to the sun** | **Spring, far south** |

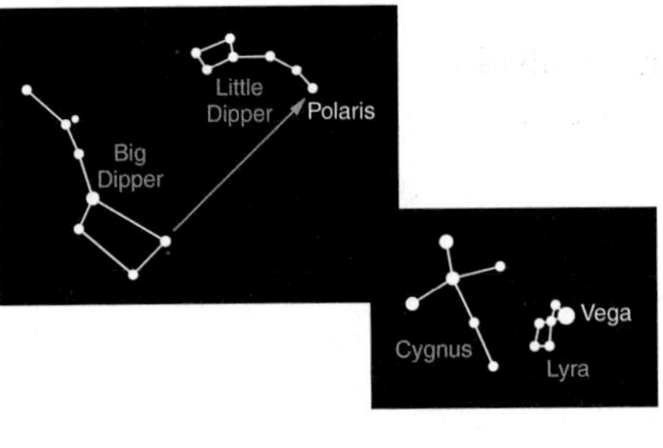


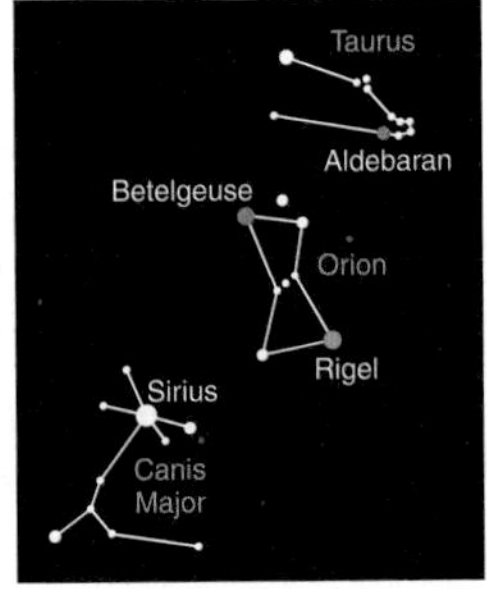


**■ Figure 2-5**

Favorite Stars: Locate these bright stars in the sky and learn why they are interesting.

were fainter, second-magnitude. The scale continued downward to sixth-magnitude stars, the faintest visible to the human eye. Thus, the larger the magnitude number, the fainter the star. This makes sense if you think of the bright stars as first-class stars and the faintest visible stars as sixth-class stars.

Ancient astronomers could only estimate magnitudes, but modern astronomers can measure the brightness of stars to high precision, so they have made adjustments to the scale of magnitudes. Instead of saying that the star known by the charming name Chort (Theta Leonis) is third magnitude, they can say its magnitude is 3.34. Accurate measurements show that some stars are actually brighter than magnitude 1.0. For example, Favorite Star Vega (alpha Lyrae) is so bright that its magnitude, 0.04, is almost zero. A few are so bright the modern magnitude scale must extend into negative numbers (■ Figure 2-6). On this scale, our Favorite Star Sirius, the brightest star in the sky, has a magnitude of −1.47. Modern astronomers have had to extend the faint end of the magnitude scale as well. The faintest stars you can see with your unaided eyes are about sixth magnitude, but if you use a telescope, you will see stars much fainter. Astronomers must use magnitude numbers larger than 6 to describe these faint stars.

These numbers are known as **apparent visual magnitudes** **($m_V$)**, and they describe how the stars look to human eyes observing from Earth. Although some stars emit large amounts of infrared or ultraviolet light, human eyes can't see those types of radiation, and they are not included in the apparent visual magnitude. The subscript "V" stands for "visual" and reminds you that only visible light is included. Apparent visual magnitude also does not take into account the distance to the stars. Very distant stars look fainter, and nearby stars look brighter. Apparent visual magnitude ignores the effect of distance and tells you only how bright the star looks as seen from Earth.

## Magnitude and Intensity

Your interpretation of brightness is quite subjective, depending on both the physiology of human eyes and the psychology of perception. To be accurate you should refer to **flux**—a measure of the light energy from a star that hits one square meter in one second. Such measurements precisely define the intensity of starlight, and a simple relationship connects apparent visual magnitudes and intensity.

Astronomers use a simple formula to convert between magnitudes and intensities. If two stars have intensities $I_A$ and $I_B$, then the ratio of their intensities is $I_A/I_B$. To make today's measurements agree with ancient catalogs, astronomers have defined the modern magnitude scale so that two stars that differ by five magnitudes have an intensity ratio of exactly 100. Then two stars that differ by one magnitude must have an intensity ratio that equals the fifth root of 100, $\sqrt[5]{100}$, which equals about 2.512—that is, the light of one star must be 2.512 times more intense.

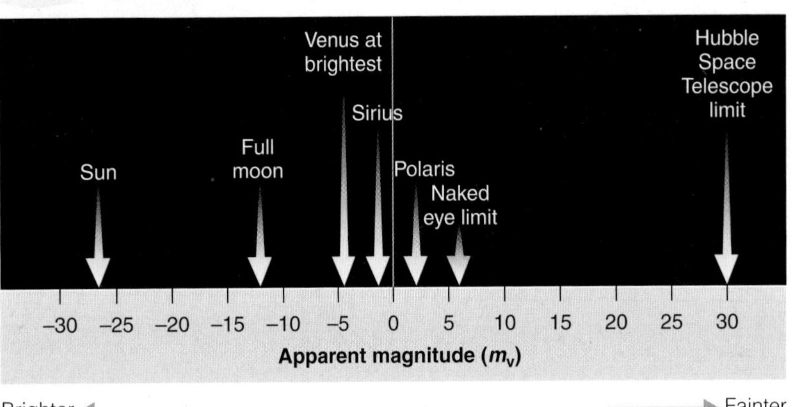


**■ Figure 2-6**

The scale of apparent visual magnitudes extends into negative numbers to represent the brightest objects and to positive numbers larger than 6 to represent objects fainter than the human eye can see.

This allows you to compare the brightness of two stars. If they differ by two magnitudes. they will have an intensity ratio of 2.512 × 2.512, or about 6.3. If they differ by three magnitudes, they will have an intensity ratio of 2.512 × 2.512 × 2.512, which is $2.512^3$ or about 6.3, and so on (■Table 2-1).

For example, suppose one star is third magnitude, and another star is ninth magnitude. What is the intensity ratio? In this case, the magnitude difference is six magnitudes, and the table shows the intensity ratio is 250. Therefore light from one star is 250 times more intense than light from the other star.

A table is convenient, but for more precision you can express the relationship as a simple formula. The intensity ratio $I_A/I_B$ is equal to 2.512 raised to the power of the magnitude difference $m_B - m_A$:

$$\frac{I_A}{I_B} = (2.512)^{(m_B - m_A)}$$

If, for example, the magnitude difference is 6.32 magnitudes, then the intensity ratio must be $2.512^{6.32}$. A pocket calculator tells you the answer: 337. One of the stars is 337 times brighter than the other.

On the other hand, when you know the intensity ratio of two stars and want to find their magnitude difference, it is convenient to rearrange the formula above and write it as

$$m_B - m_A = 2.5 \log\left(\frac{I_A}{I_B}\right)$$

For example, the light from Sirius is 24.2 times more intense than light from Polaris. The magnitude difference is 2.5 log(24.2). Your pocket calculator tells you the logarithm of 24.2 is 1.38, so the magnitude difference is 2.5 × 1.38, which equals 3.4 magnitudes. Sirius is 3.4 magnitudes brighter than Polaris.

The modern magnitude system has some big advantages. It compresses a tremendous range of intensity into a small range of magnitudes, as you can see in Table 2-1. More important, it allows modern astronomers to measure and report the brightness of stars to high precision while still making comparisons to observations of apparent visual magnitude that go back to the time of Hipparchus.

**■ Table 2-1 | Magnitude and Intensity**

| Magnitude Difference | Approximate Intensity Ratio |
|---|---|
| 0 | 1 |
| 1 | 2.5 |
| 2 | 6.3 |
| 3 | 16 |
| 4 | 40 |
| 5 | 100 |
| 6 | 250 |
| 7 | 630 |
| 8 | 1600 |
| 9 | 4000 |
| 10 | 10,000 |
| ⋮ | ⋮ |
| 15 | 1,000,000 |
| 20 | 100,000,000 |
| 25 | 10,000,000,000 |
| ⋮ | ⋮ |

## 2-2 The Sky and Its Motion

THE SKY ABOVE seems to be a great blue dome in the daytime and a sparkling ceiling at night. It was this ceiling that the first astronomers observed long ago as they tried to understand the night sky.

### The Celestial Sphere

Ancient astronomers believed the sky was a great sphere surrounding Earth with the stars stuck on the inside like thumbtacks in a ceiling. Modern astronomers know that the stars are scattered through space at different distances, but it is still convenient to think of the sky as a great starry sphere enclosing Earth.

The Concept Art Portfolio **The Sky Around You** on pages 18–19 takes you on an illustrated tour of the sky. Throughout this book, these two-page art spreads introduce new concepts and new terms through photos and diagrams. These concepts and new terms are not discussed elsewhere, so examine the art spreads carefully. Notice that The Sky Around You introduces you to three important principles and 16 new terms that will help you understand the sky:

1. The sky appears to rotate westward around Earth each day, but that is a consequence of the eastward rotation of Earth. That rotation produces day and night. Notice how reference points on the *celestial sphere* such as the *zenith, nadir, horizon, celestial equator,* and the *north celestial pole* and *south celestial pole* define the four directions, *north point, south point, east point,* and *west point.*

2. Astronomers measure *angular distance* across the sky as angles and express them as degrees, *arc minutes,* and *arc seconds.* The same units are used to measure the *angular diameter* of an object.

3. What you can see of the sky depends on where you are on Earth. If you lived in Australia, you would see many constellations and asterisms invisible from North America, but you would never see the Big Dipper. How many *circumpolar constellations* you see depends on where you are. Remember your Favorite Star Alpha Centauri? It is in the southern sky and isn't visible from most of the United States. You could just glimpse it above the southern horizon if you were in Miami, Florida, but you could see it easily from Australia.

Pay special attention to the new terms on pages 18–19. You need to know these terms to describe the sky and its motions, but don't fall into the trap of just memorizing new terms. The goal of

# How Do We Know? 2-1

## Scientific Models

**How can a scientific model be useful if it isn't entirely true?** A scientific model is a carefully devised conception of how something works, a framework that helps scientists think about some aspect of nature, just as the celestial sphere helps astronomers think about the motions of the sky.

Chemists, for example, use colored balls to represent atoms and sticks to represent the bonds between them, kind of like Tinkertoys. Using these molecular models, chemists can see the three-dimensional shape of molecules and understand how the atoms interconnect. The molecular model of DNA proposed by Watson and Crick in 1953 led to our modern understanding of the mechanisms of genetics. You have probably seen elaborate ball-and-stick models of DNA, but does the molecule really look like Tinkertoys? No, but the model is both simple enough and accurate enough to help scientists think about their theories.

A scientific model is not a statement of truth; it does not have to be precisely true to be useful. In an idealized model, some complex aspects of nature can be simplified or omitted. The ball-and-stick model of a molecule doesn't show the relative strength of the chemical bonds, for instance. A model gives scientists a way to think about some aspect of nature but need not be true in every detail.

When you use a scientific model, it is important to remember the limitations of that model. If you begin to think of a model as true, it can be misleading instead of helpful. The celestial sphere, for instance, can help you think about the sky, but you must remember that it is only a model. The universe is much larger and much more interesting than this ancient scientific model of the heavens.

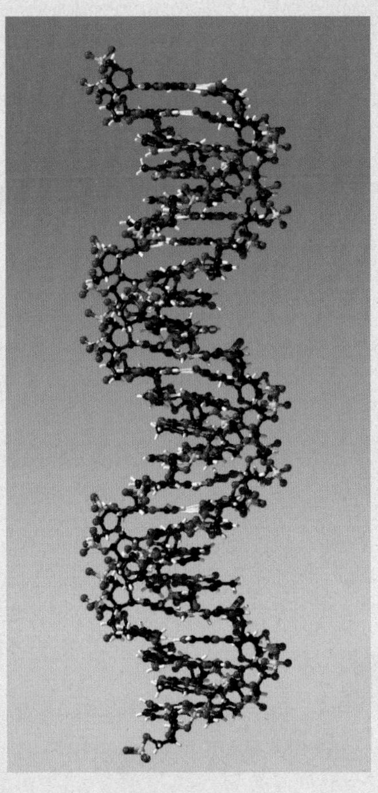

*Balls represent atoms and rods represent chemical bonds in this model of a DNA molecule.* (Digital Vision/Getty Images)

science is to understand nature, not to memorize definitions. Study the diagrams and see how the geometry of the celestial sphere and its motions produce the sky you see above you.

The celestial sphere is an example of a **scientific model,** a common feature of scientific thought (**How Do We Know? 2-1**). Notice that a scientific model does not have to be true to be useful. You will encounter many scientific models in the chapters that follow, and you will discover that some of the most useful models are highly simplified descriptions of the true facts.

This is a good time to eliminate a couple of **Common Misconceptions.** Lots of people, without thinking about it much, assume that the stars are not in the sky during the daytime. The stars are actually there day and night; they are just invisible during the day because the sky is lit up by sunlight. Also, many people insist that Favorite Star Polaris is the brightest star in the sky. It is actually the 51st visually brightest star. You now know that Polaris is important because of its position, not because of its brightness.

In addition to causing the obvious daily motion of the sky, Earth's rotation conceals a very slow celestial motion that can be detected only over centuries.

## Precession

Over 2000 years ago, Hipparchus compared a few of his star positions with those recorded nearly two centuries earlier and realized that the celestial poles and equator were slowly moving across the sky. Later astronomers understood that this motion is caused by the top-like motion of Earth.

If you have ever played with a gyroscope or top, you have seen how the spinning mass resists any sudden change in the direction of its axis of rotation. The more massive the top and the more rapidly it spins, the more it resists your efforts to twist it out of position. But you probably recall that even the most rapidly spinning top slowly swings its axis around in a circle... The weight of the top tends to make it tip over, and this combines with its rapid rotation to make its axis sweep out the shape of a cone in a motion called **precession** (■ Figure 2-7a). In later chapters, you will learn that many celestial bodies precess.

Earth spins like a giant top, but it does not spin upright in its orbit; it is tipped 23.5° from vertical. Earth's large mass and rapid rotation keep its axis of rotation pointed toward a spot near the star Polaris, and the axis would not wander if it were not for precession.

# The Sky Around You

**1** The eastward rotation of Earth causes the sun, moon, and stars to move westward in the sky as if the **celestial sphere** were rotating westward around Earth. From any location on Earth you see only half of the celestial sphere, the half above the **horizon.** The **zenith** marks the top of the sky above your head, and the **nadir** marks the bottom of the sky directly under your feet. The drawing at right shows the view for an observer in North America. An observer in South America would have a dramatically different horizon, zenith, and nadir.

The apparent pivot points are the **north celestial pole** and the **south celestial pole** located directly above Earth's north and south poles. Halfway between the celestial poles lies the **celestial equator.** Earth's rotation defines the directions you use every day. The **north point** and **south point** are the points on the horizon closest to the celestial poles. The **east point** and the **west point** lie halfway between the north and south points. The celestial equator always touches the horizon at the east and west points.

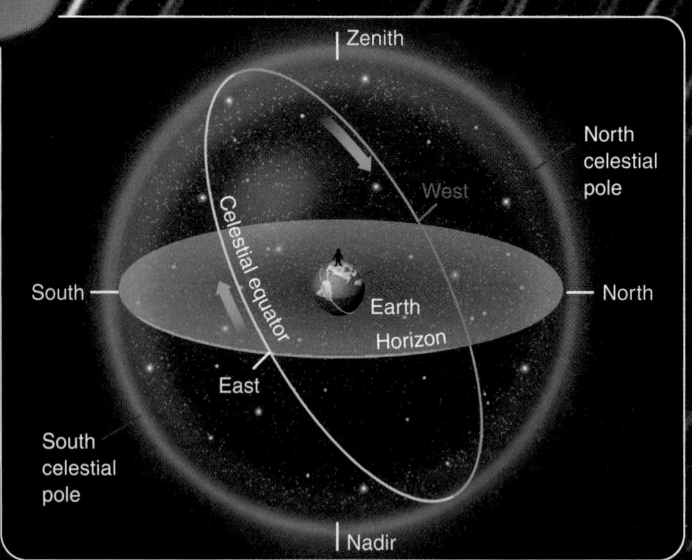


AURA/NOAO/NSF

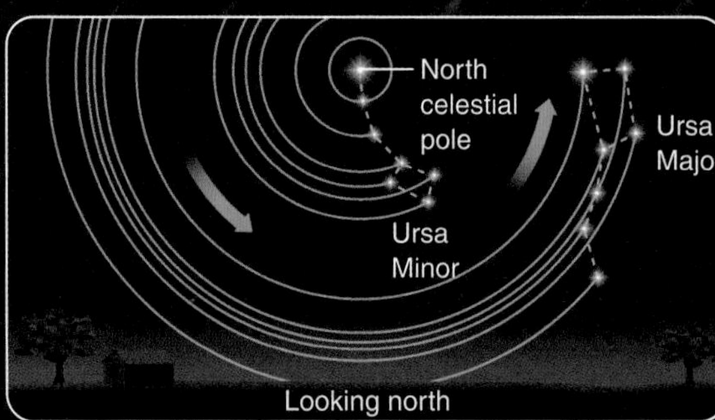


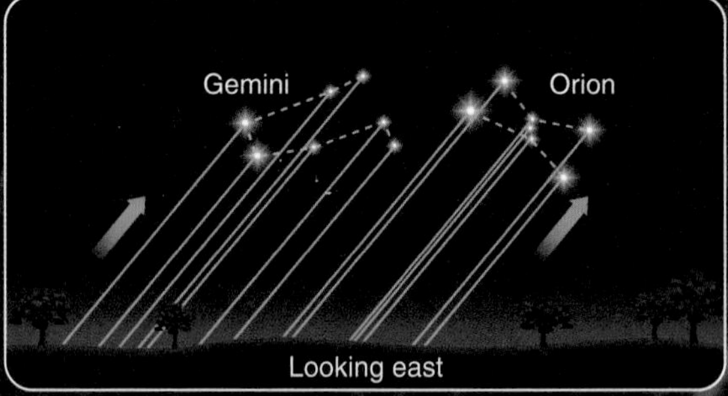


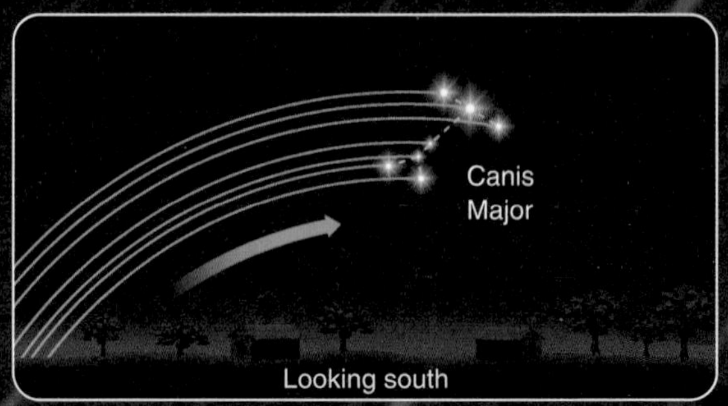


**1a** This time exposure of about 30 minutes shows stars as streaks, called star trails, rising behind an observatory dome. The camera was facing northeast to take this photo. The motion you see in the sky depends on which direction you look, as shown at right. Looking north, you

**Astronomers measure distance across the sky as angles.**

Angular distance

**2** Astronomers might say, "The star was only 2 degrees from the moon." Of course, the stars are much farther away than the moon, but when you think of the celestial sphere, you can measure distances *on the sky* as **angular distances** in degrees, minutes of arc, and seconds of arc. An **arc minute** is 1/60th of a degree, and an **arc second** is 1/60th of a minute of arc. Then the **angular diameter** of an object is the angular distance from one edge to the other. The sun and moon are each about half a degree in diameter, and the bowl of the Big Dipper is about 10° wide. In order to distinguish these units from the similarly named units of time, astronomers refer to small angles in arc minutes or arc seconds. The word *arc* indicates that the numbers refer to angles and not time.

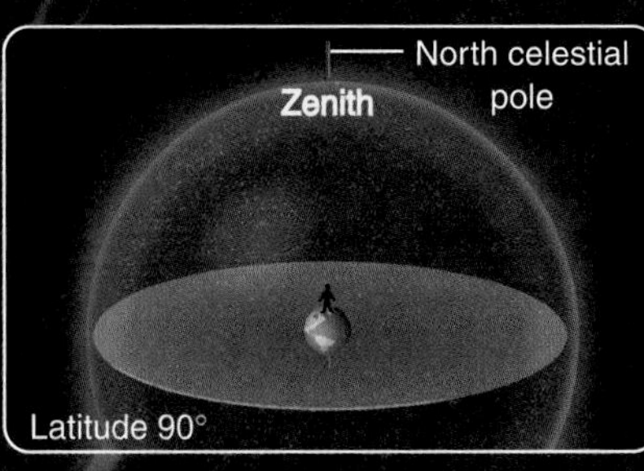


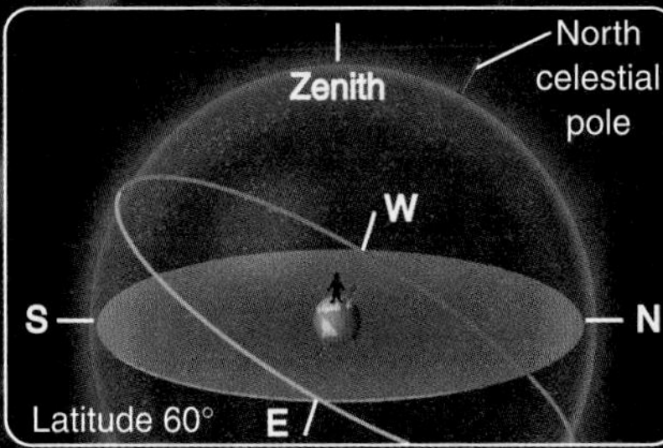


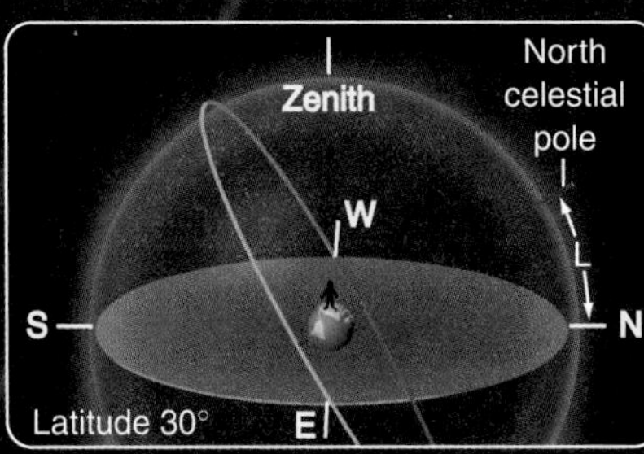


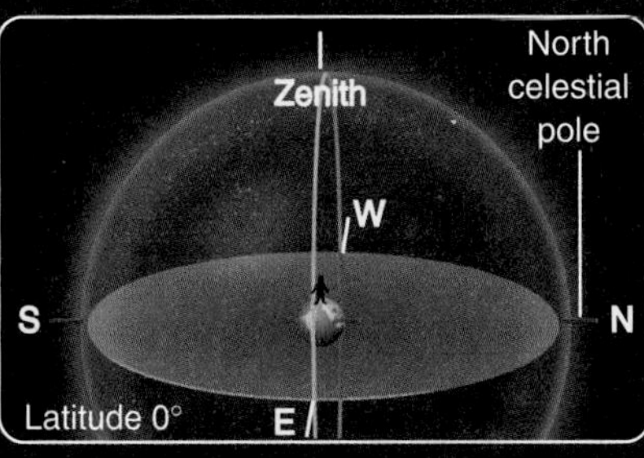


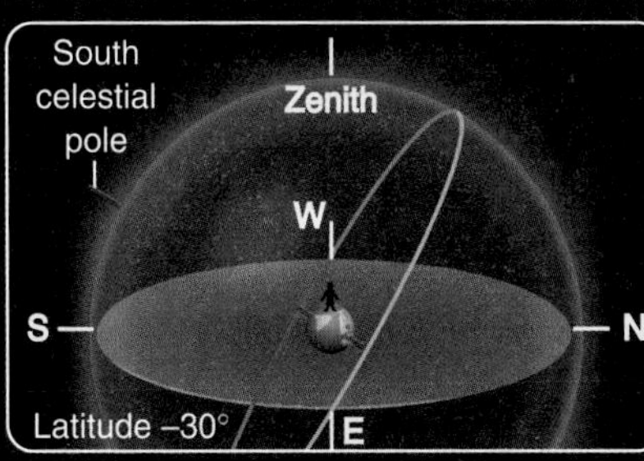


**3** What you see in the sky depends on your latitude as shown at right. Imagine that you begin a journey in the ice and snow at Earth's North Pole with the north celestial pole directly overhead. As you walk southward, the celestial pole moves toward the horizon, and you can see farther into the southern sky. The angular distance from the horizon to the north celestial pole always equals your latitude (L)—the basis for celestial navigation. As you cross Earth's equator, the celestial equator would pass through your zenith, and the north celestial pole would sink below your northern horizon.

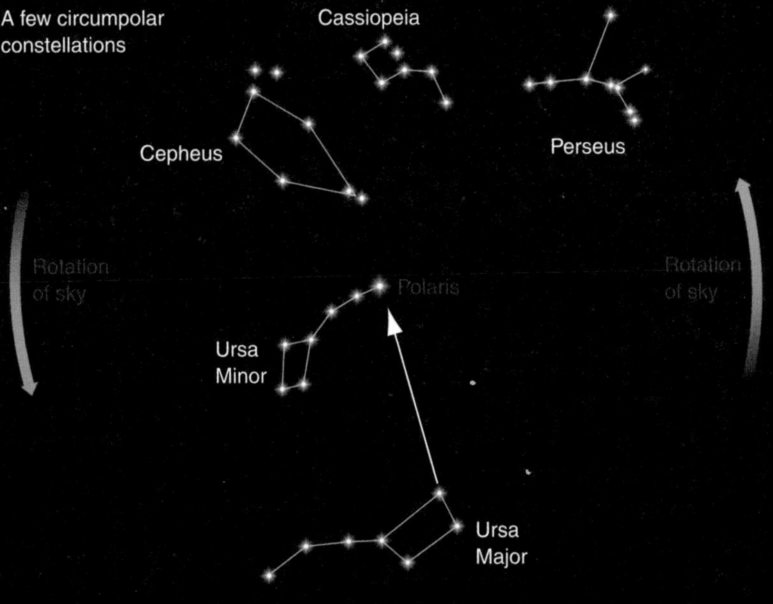


**3a** **Circumpolar constellations** are those that never rise or set. From mid-northern latitudes, as shown at left, you see a number of familiar constellations circling Polaris and never dipping below the horizon. As the sky rotates, the pointer stars at the front of the Big Dipper always point toward Polaris. Circumpolar constellations near the south celestial pole never rise as seen from mid-northern latitudes. From a high latitude such as Norway, you would have more circumpolar constellations, and from Quito, Ecuador, located on Earth's equator, you would have no circumpolar constellations at all.

Because of its rotation, Earth has a slight bulge around its middle, and the gravity of the sun and moon pull on this bulge, tending to twist Earth's axis "upright" relative to its orbit. If Earth were a perfect sphere, it would not get twisted. Notice that gravity tends to make the top fall over, but it tends to twist Earth upright in its orbit. In both cases, the twisting of the axis of rotation combines with the rotation of the object and causes precession. Earth's axis precesses, taking about 26,000 years for one cycle (Figure 2-7b).

Because the locations of the celestial poles and equator are defined by Earth's rotational axis, precession slowly moves these reference marks. You would notice no change at all from night to night or year to year, but precise measurements can reveal the slow precession of the celestial poles and the resulting change in orientation of the celestial equator.

Over centuries, precession has dramatic effects. Egyptian records show that 4800 years ago the north celestial pole was near the star Thuban (alpha Draconis). The pole is now approaching Polaris and will be closest to it in about the year 2100. In about 12,000 years, the pole will have moved to within 5° of Vega (alpha Lyrae). Next time you glance at Favorite Star Vega, remind yourself that it will someday be a very impressive north star. Figure 2-7c shows the path followed by the north celestial pole.

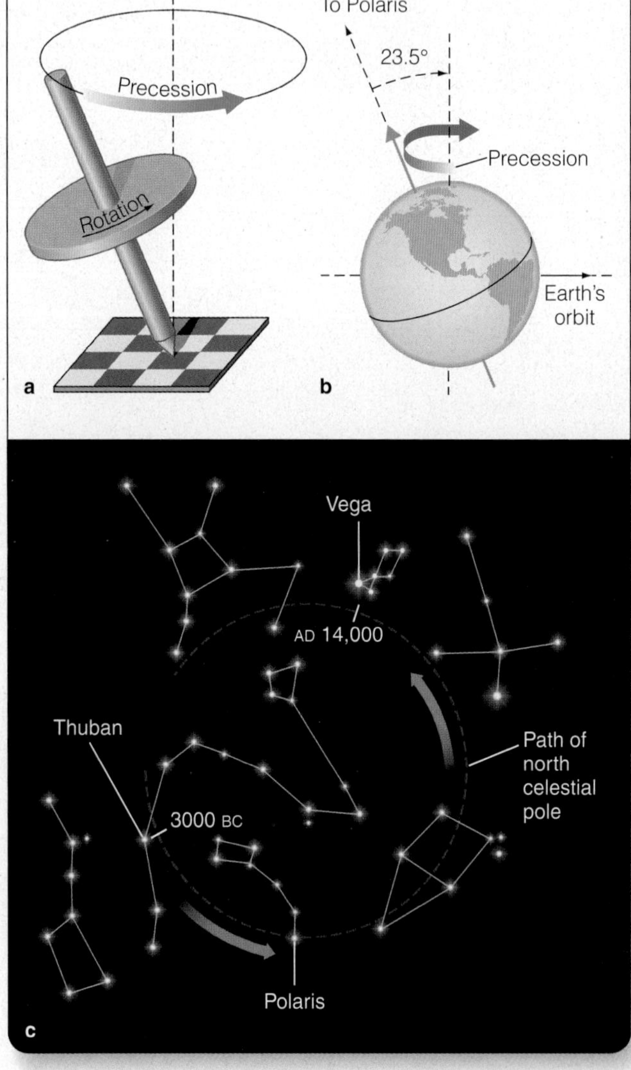


**■ Figure 2-7**

Precession. (a) A spinning top precesses in a conical motion around the perpendicular to the floor because its weight tends to make it fall over. (b) Earth precesses around the perpendicular to its orbit because the gravity of the sun and moon acting on Earth's equatorial bulge tend to twist it upright. (c) Precession causes the north celestial pole to move slowly among the stars, completing a circle in 26,000 years.

## 2-3 The Cycles of the Sun

EARTH'S ROTATION ON ITS AXIS causes the cycle of day and night, but it is its motion around the sun in its orbit that defines the year. Notice an important distinction. **Rotation** is the turning of a body on its axis, but **revolution** means the motion of a body around a point outside the body. Consequently, astronomers are careful to say Earth rotates once a day on its axis and revolves once a year around the sun.

Because day and night are caused by the rotation of Earth, your time of day depends on your location on Earth. You can see this if you watch live international news. It may be lunchtime where you are, but for a newscaster in the Middle East, it can already be dark. In ■ Figure 2-8, you can see that four people in different places on Earth have different times of day.

### The Annual Motion of the Sun

Even in the daytime, the sky is filled with stars, but the glare of sunlight fills Earth's atmosphere with scattered light, and you can see only the brilliant sun. If the sun were fainter, you would be able to see it rise in the morning in front of the stars. During the day, you would see the sun and the stars moving westward, and the sun would eventually set in front of the same stars. If you watched carefully as the day passed, you would notice that the sun was creeping slowly eastward against the background of stars.

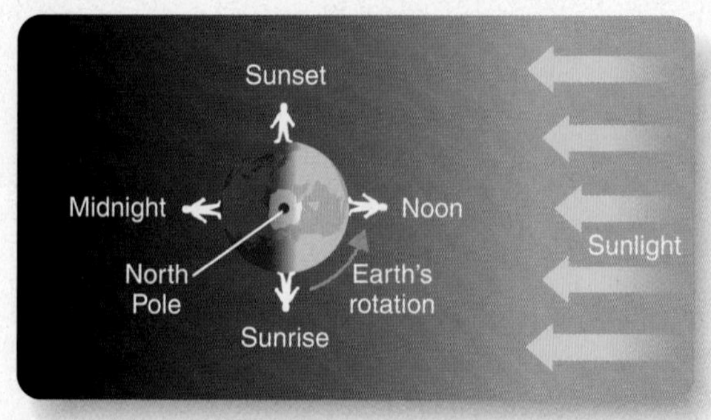


**■ Figure 2-8**

Looking down on Earth from above the North Pole shows how the time of day or night depends on your location on Earth.

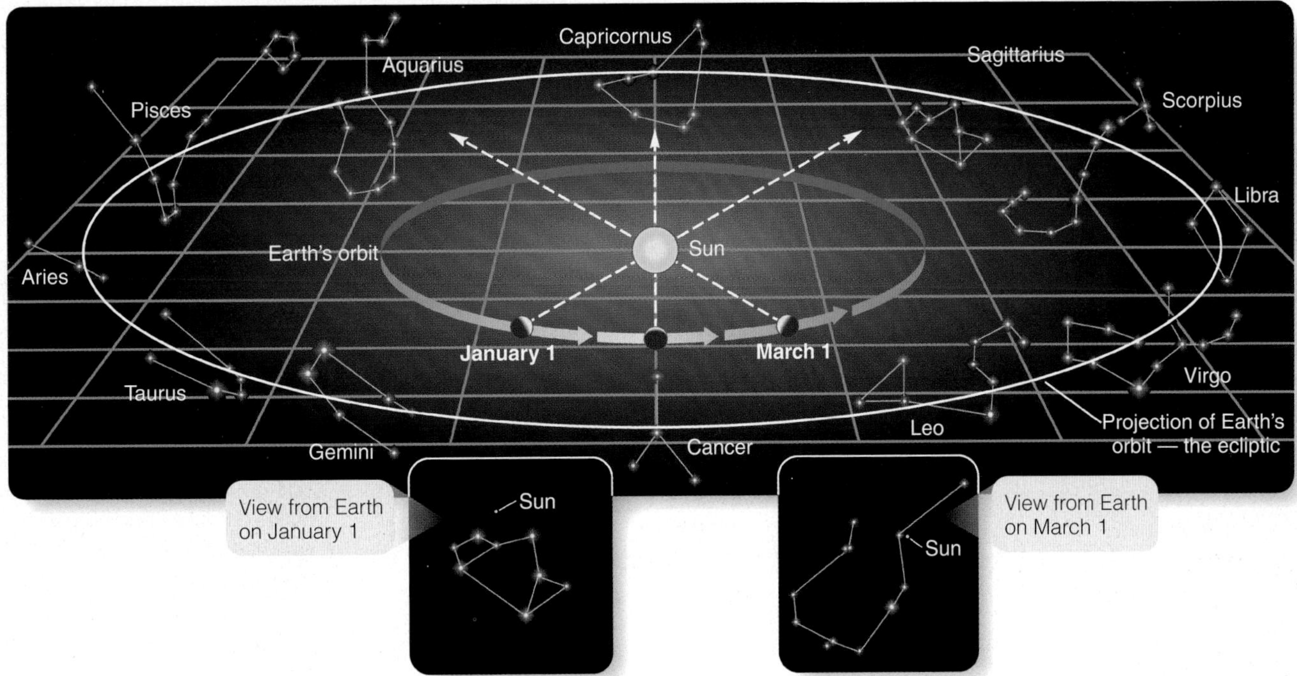


■ **Figure 2-9**

Earth's orbit is a nearly perfect circle, but it is shown in an inclined view in this diagram and consequently looks oval. Earth's motion around the sun makes the sun appear to move against the background of the stars. Earth's circular orbit is thus projected on the sky as the circular path of the sun, the ecliptic. If you could see the stars in the daytime, you would notice the sun crossing in front of the distant constellations as Earth moves along its orbit.

It would move a distance roughly equal to its own diameter between sunrise and sunset. This motion is caused by the motion of Earth in its nearly circular orbit around the sun.

For example, in January, you would see the sun in front of the constellation Sagittarius (■ Figure 2-9). As Earth moves along its circular orbit, the sun appears to move eastward among the stars. By March, you would see it in front of Aquarius.

Although people often say the sun is "in Sagittarius" or "in Aquarius," it isn't really correct to say the sun is "in" a constellation. The sun is only 1 AU away, and the stars visible in the sky are at least a million times more distant. Nevertheless, in March of each year, the sun crosses in front of the stars that make up Aquarius, and people conventionally use the expression, "The sun is in Aquarius."

The apparent path of the sun against the background of stars is called the **ecliptic.** If the sky were a great screen, the ecliptic would be the shadow cast by Earth's orbit. That is why the ecliptic is often called the projection of Earth's orbit on the sky.

Earth circles the sun in 365.25 days and, consequently, the sun appears to circle the sky in the same period. That means the sun, traveling 360° around the ecliptic in 365.25 days, travels about 1° eastward in 24 hours, about twice its angular diameter. You don't notice this apparent motion of the sun because you can't see the stars in the daytime, but it does have an important consequence that you do notice—the seasons.

## The Seasons

Earth would not experience seasons if it rotated upright in its orbit, but because its axis of rotation is tipped 23.4° from the perpendicular to its orbit, it has seasons. Study **The Cycle of the Seasons** on pages 22–23 and notice two important principles and six new terms:

1. Because Earth's axis of rotation is inclined 23.4°, the sun moves into the northern sky in the spring and into the southern sky in the fall. That causes the cycle of the seasons. Notice how the *vernal equinox,* the *summer solstice,* the *autumnal equinox,* and the *winter solstice* mark the beginning of the seasons. Further, notice the very minor effects of Earth's slightly elliptical orbit as it travels from *perihelion* to *aphelion.*

2. Earth goes through a cycle of seasons because of changes in the amount of solar energy that Earth's northern and southern hemispheres receive at different times of the year. Because of circulation patterns in Earth's atmosphere, the northern and southern hemispheres are mostly isolated from each other and exchange little heat. When one hemisphere receives more solar energy than the other, it grows rapidly warmer.

Notice that the seasons in Earth's southern hemisphere are reversed with respect to those in the northern hemisphere. Australia and other lands in the southern hemisphere experience

# The Cycle of the Seasons

**1** You can use the celestial sphere to help you think about the seasons. The celestial equator is the projection of Earth's equator on the sky, and the ecliptic is the projection of Earth's orbit on the sky. Because Earth is tipped in its orbit, the ecliptic and equator are inclined to each other by 23.5° as shown at right. As the sun moves eastward around the sky, it spends half the year in the southern half of the sky and half the year in the northern half. That causes the seasons.

The sun crosses the celestial equator going northward at the point called the **vernal equinox.** The sun is at its farthest north at the point called the **summer solstice.** It crosses the celestial equator going southward at the **autumnal equinox** and reaches its most southern point at the **winter solstice.**

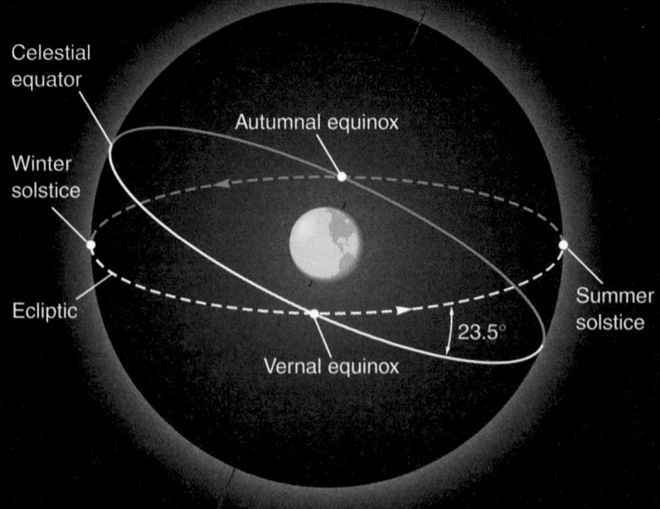


**1a** The seasons are defined by the dates when the sun crosses these four points, as shown in the table at the right. *Equinox* comes from the word for "equal"; the day of an equinox has equal amounts of daylight and darkness. *Solstice* comes from the words meaning "sun" and "stationary." *Vernal* comes from the word for "green." The "green" equinox marks the beginning of spring.

| Event | Date* | Season |
|---|---|---|
| Vernal equinox | March 20 | Spring begins |
| Summer solstice | June 22 | Summer begins |
| Autumnal equinox | September 22 | Autumn begins |
| Winter solstice | December 22 | Winter begins |

* Give or take a day due to leap year and other factors.

**1b** On the day of the summer solstice in late June, Earth's northern hemisphere is inclined toward the sun, and sunlight shines almost straight down at northern latitudes. At southern latitudes, sunlight strikes the ground at an angle and spreads out. North America has warm weather, and South America has cool weather.

Earth's axis of rotation points toward Polaris, and, like a top, the spinning Earth holds its axis fixed as it orbits the sun. On one side of the sun, Earth's northern hemisphere leans toward the sun; on the other side of its orbit, it leans away. However, the direction of the axis of rotation does not change.



**Earth at summer solstice**

**1c** Light striking the ground at a steep angle spreads out less than light striking the ground at a shallow angle. Light from the summer solstice sun strikes northern latitudes from nearly overhead and is concentrated.

Light from the winter solstice sun strikes northern latitudes at a much shallower angle and spreads out. The same amount of energy is spread over a larger area, so the ground receives less energy from the winter sun.

**2** The two causes of the seasons are shown at right for someone in the northern hemisphere. First, the noon summer sun is higher in the sky and the winter sun is lower, as shown by the longer winter shadows. Thus winter sunlight is more spread out. Second, the summer sun rises in the northeast and sets in the northwest, spending more than 12 hours in the sky. The winter sun rises in the southeast and sets in the southwest, spending less than 12 hours in the sky. Both of these effects mean that northern latitudes receive more energy from the summer sun, and summer days are warmer than winter days.

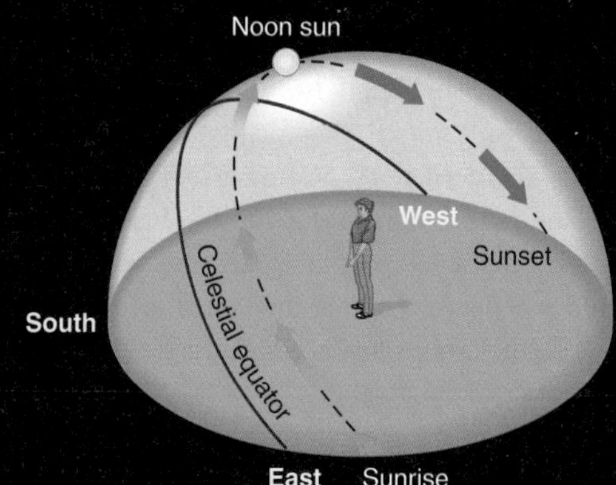


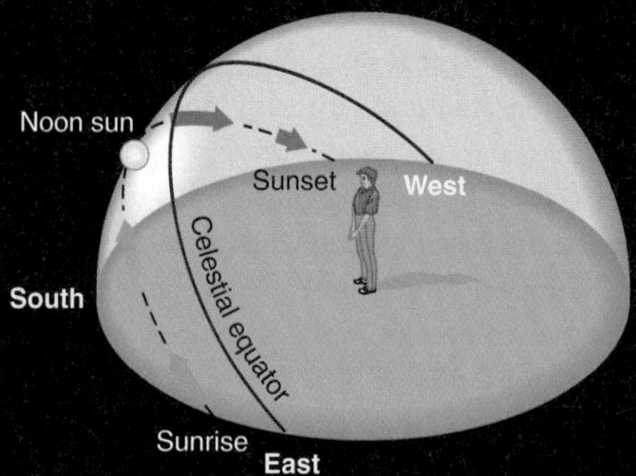


23.5°
To Polaris
Sunlight spread out on northern latitudes
40° N latitude
Equator
To sun
Sunlight nearly direct on southern latitudes
40° S latitude

**1d** On the day of the winte solstice in late December Earth's northern hemisphere is inclined away from the sun, anc sunlight strikes the ground a an angle and spreads out. A southern latitudes, sunligh shines almost straigh down and does not spreac out. North America has cool weather and South America has warm weather

Earth's orbit is only very slightly elliptical. Abou January 3, Earth is a **perihelion**, its closes point to the sun, when i is only 1.7 percent closer than average. About July 5 Earth is at **aphelion,** its mos distant point from the sun when it is only 1.7 percen farther than average. This smal variation does not significantly affec the seasons

winter from June 22 to September 22, and summer from December 22 to March 20.

Now you can set your friends straight if they mention two of the most **Common Misconceptions** about the seasons. First, the seasons don't occur because Earth moves closer to or farther from the sun. If that were the cause, both of Earth's hemispheres would experience winter at the same time, and that's not what happens. Earth's orbit is nearly circular. Earth is only 1.7 percent closer to the sun in January and 1.7 percent farther away in July, and that small variation isn't enough to cause the seasons. Rather, the seasons arise because Earth's axis is not perpendicular to its orbit.

Here's a second **Common Misconception:** It is not easier to stand a raw egg on end on the day of the vernal equinox. Have you heard that one? Radio and TV personalities love to talk about it, but it just isn't true. It is one of the silliest misconceptions in science. You can stand a raw egg on end any day of the year if you have steady hands. (*Hint:* It helps to shake the egg really hard to break the yoke inside so it can settle to the bottom.)

In ancient times, the cycle of the seasons, especially the solstices and equinoxes, were celebrated with rituals and festivals. Shakespeare's play *A Midsummer Night's Dream* describes the enchantment of the summer solstice night. (In Shakespeare's time, the equinoxes and solstices were taken to mark the midpoints of the seasons.) Many North American Indians marked the summer solstice with ceremonies and dances. Early church officials placed Christmas day in late December to coincide with an earlier pagan celebration of the winter solstice.

## The Motion of the Planets

The planets of our solar system produce no visible light of their own; they are visible only by reflected sunlight. Mercury, Venus, Mars, Jupiter, and Saturn are all easily visible to the unaided eye and look like stars, but Uranus is usually too faint to be seen, and Neptune is never bright enough.

All of the planets of the solar system, including Earth, move in nearly circular orbits around the sun. If you were looking down on the solar system from the north celestial pole, you would see the planets moving in the same counterclockwise direction around their orbits, with the planets farthest from the sun moving the slowest. Seen from Earth, the outer planets move slowly eastward* along the ecliptic. In fact, the word *planet* comes from the Greek word meaning "wanderer." Mars moves completely around the ecliptic in slightly less than 2 years, but Saturn, being farther from the sun, takes nearly 30 years.

Mercury and Venus also stay near the ecliptic, but they move differently from the other planets. They have orbits inside Earth's orbit, and that means they are never seen far from the sun in the sky. Observed from Earth, they move eastward away from the sun and then back toward the sun, crossing the near part of their orbit. They continue moving westward away from the sun and then move back crossing the far part of their orbit before they move out east of the sun again. To find one of these planets, you need to look above the western horizon just after sunset or above the eastern horizon just before sunrise. Venus is easier to locate because it is brighter and because its larger orbit carries it higher above the horizon than does Mercury's (■ Figure 2-10). Mercury's orbit is so small that it can never get farther than 28° from the sun. Consequently, it is hard to see against the sun's glare and is often hidden in the clouds and haze near the horizon.

By tradition, any planet visible in the evening sky is called an **evening star,** even though planets are not stars. Similarly, any planet visible in the sky shortly before sunrise is called a **morning star.** Perhaps the most beautiful is Venus, which can become as bright as magnitude –4.7. As Venus moves around its orbit, it can dominate the western sky each evening for many weeks, but eventually its orbit carries it back toward the sun, and it is lost in the haze near the horizon. In a few weeks, it reappears in the dawn sky, a brilliant morning star.

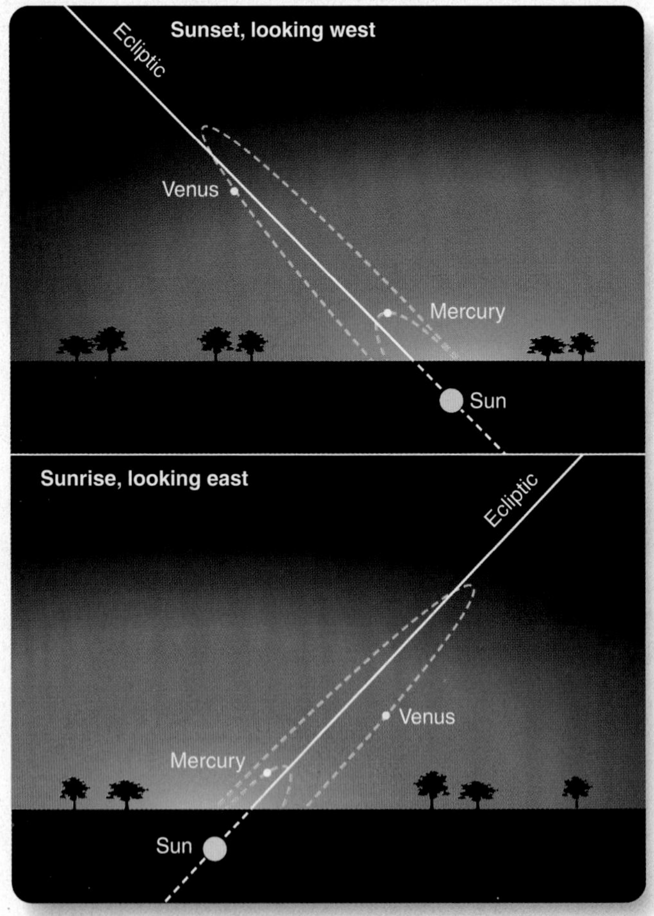


**■ Figure 2-10**

Mercury and Venus follow orbits that keep them near the sun, and they are visible only soon after sunset or before sunrise when the brilliance of the sun is hidden below the horizon. Venus takes 584 days to move from the morning sky to the evening sky and back again, but Mercury zips around in only 116 days.

*You will discover occasional exceptions to this eastward motion in Chapter 4.

## Pseudoscience

**What is the difference between a science and a pseudoscience?** Astronomers have a low opinion of beliefs such as astrology, mostly because they are groundless but also because they *pretend* to be a science. They are pseudosciences, from the Greek *pseudo,* meaning false.

A **pseudoscience** is a set of beliefs that appear to be based on scientific ideas but that fail to obey the most basic rules of science. For example, in the 1970s a claim was made that pyramidal shapes focus cosmic forces on anything underneath and might even have healing properties. For example, it was claimed that a pyramid made of paper, plastic, or other materials would preserve fruit, sharpen razor blades, and do other miraculous things. Many books promoted the idea of the special power of pyramids, and this idea led to a popular fad.

A key characteristic of science is that its claims can be tested and verified. In this case, simple experiments showed that any shape, not just a pyramid, protects a piece of fruit from airborne spores and allows it to dry without rotting. Likewise, any shape allows oxidation to improve the cutting edge of a razor blade. Because experimental evidence contradicted the claim and because supporters of the theory declined to abandon or revise their claims, you can recognize pyramid power as a pseudoscience. Disregard of contradictory evidence and alternate theories is a sure sign of a pseudoscience.

Pseudoscientific claims can be self-fulfilling. For example, some believers in pyramid power slept under pyramidal tents to improve their rest. There is no logical mechanism by which such a tent could affect a sleeper, but because people wanted and expected the claim to be true they reported that they slept more soundly. Vague claims based on personal testimony that cannot be tested are another sign of a pseudoscience.

Astrology is another pseudoscience. It has been tested over and over for centuries, and it doesn't work. Nevertheless, many people believe in astrology despite contradictory evidence.

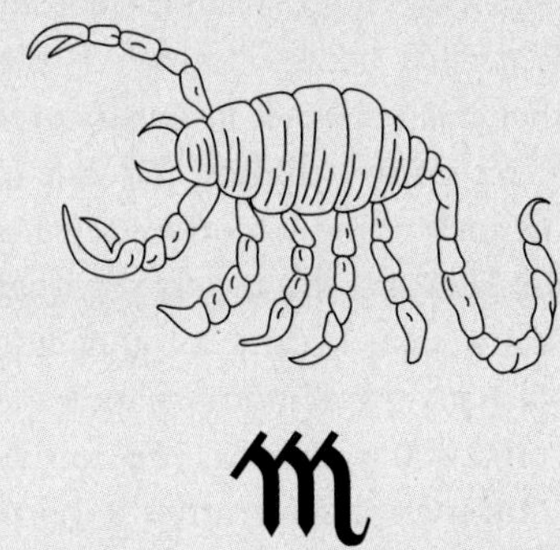

*Astrology may be the oldest pseudoscience.*

Many pseudosciences appeal to our need to understand and control the world around us. Some such claims involve medical cures, ranging from using magnetic bracelets and crystals to focus mystical power to astonishingly expensive, illegal, and dangerous treatments for cancer. Logic is a stranger to pseudoscience, but human fears and needs are not.

The cycles of the sky are so impressive that it is not surprising that people have strong feelings about them. Ancient peoples saw the motion of the sun around the ecliptic as a powerful influence on their daily lives, and the motion of the planets along the ecliptic seemed similarly meaningful. The ancient superstition of astrology is based on the cycles of the sun and planets around the sky. You have probably heard of the **zodiac,** a band around the sky extending 9 degrees above and below the ecliptic. The signs of the zodiac take their names from the 12 principal constellations along the ecliptic. A **horoscope** is just a diagram showing the location of the sun, moon, and planets around the ecliptic and their position above or below the horizon for a given date and time. Centuries ago, astrology was an important part of astronomy, but the two are now almost exact opposites—astronomy is a science that depends on evidence, and astrology is a superstition that survives in spite of evidence (**How Do We Know? 2-2**). The signs of the zodiac are no longer important in astronomy.

## 2-4 Astronomical Influences on Earth's Climate

The seasons are produced by the annual motion of Earth around the sun, but subtle changes in that motion can have dramatic affects on climate. You don't notice these changes during your lifetime; but, over thousands of years, they can bury continents under glaciers.

Earth has gone through ice ages when the worldwide climate was cooler and dryer and thick layers of ice covered polar regions, extending partway to the equator. Between ice ages, Earth is warmer and there are no ice sheets even at the poles. Scientists have found evidence of at least four ice ages in Earth's past. One occurred 2.5 billion years ago, but the other three that have been identified have all occurred in the last billion years. There were probably others, but evidence of early ice ages is usually erased by more recent ice sheets. The lengths of ice ages range from a few tens of millions of years to a few hundred million years. The most recent ice age began only about 3 million years ago and is still going on. You are living during one of the periodic episodes during an ice age when Earth grows slightly warmer and the glaciers melt back and do not extend as far from the poles. The current warm period began about 12,000 years ago.

Ice ages seem to occur with a period of roughly 250 million years, and glaciers advance and melt back during ice ages in a complicated pattern that involves cycles of about 40,000 years and 100,000 years. (These cycles have no connection with global warming, which can produce changes in Earth's climate over just a few decades. Global warming is discussed in Chapter 20.) Evidence shows that these slow cycles of the ice ages have an astronomical origin.

## The Hypothesis

Sometimes a theory or hypothesis is proposed long before scientists can find the critical evidence to test it. That happened in 1920 when the Yugoslavian meteorologist Milutin Milankovitch proposed what became known as the **Milankovitch hypothesis**—that small changes in Earth's orbit, precession, and inclination affect Earth's climate and can cause ice ages. You should examine each of these motions separately.

First, Earth's orbit is only very slightly elliptical, but astronomers know that, because of gravitational interactions with the other planets, the elliptical shape varies slightly over a period of about 100,000 years. As you learned earlier in this chapter, Earth's orbit at present carries it 1.7 percent closer than average to the sun during northern hemisphere winters and 1.7 percent farther away in northern hemisphere summers. This makes northern winters very slightly warmer, and that is critical—most of the landmass where ice can accumulate is in the northern hemisphere. When Earth's orbit becomes more elliptical, northern summers might be too cool to melt all of the snow and ice from the previous winter. That would make glaciers grow larger.

A second factor is also at work. Precession causes Earth's axis to sweep around a cone with a period of about 26,000 years, and that gradually changes the points in Earth's orbit where a given hemisphere experiences the seasons. Northern hemisphere summers now occur when Earth is 1.7 percent farther from the sun, but in 13,000 years northern summers will occur on the other side of Earth's orbit where Earth is 1.7 percent closer to the sun. Northern summers will be warmer, which could melt all of the previous winter's snow and ice and prevent the growth of glaciers.

The third factor is the inclination of Earth's equator to its orbit, currently at 23.4°. Because of gravitational tugs of the moon, sun, and even the other planets, this angle varies from 22° to 24°, with a period of roughly 41,000 years. When the inclination is greater, seasons are more severe.

In 1920, Milankovitch proposed that these three factors cycle against each other to produce complex periodic variations in Earth's climate and the advance and retreat of glaciers (■ Figure 2-11a). No evidence was available to test the theory in 1920, however, and scientists treated it with skepticism. Many thought it was laughable.

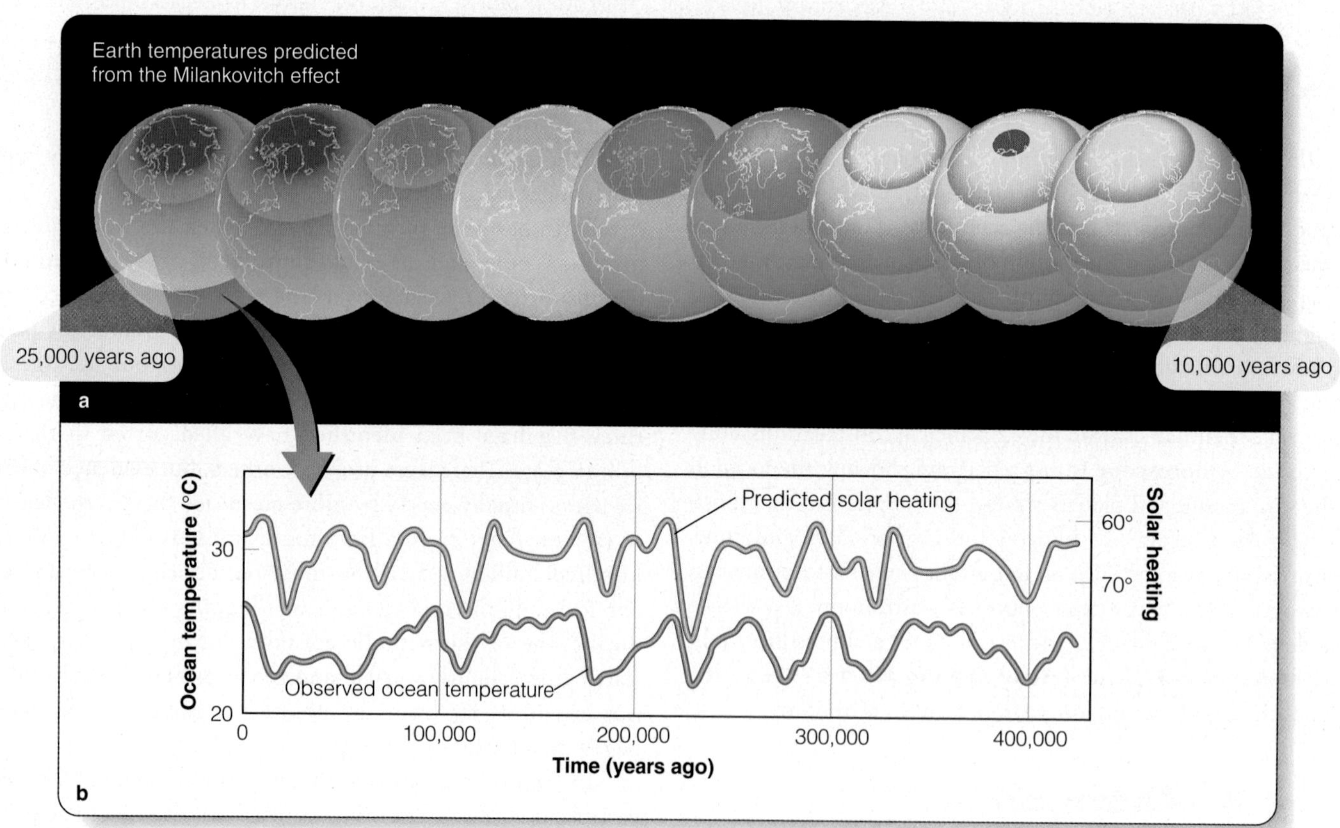


■ **Figure 2-11**

(a) Mathematical models of the Milankovitch effect can be used to predict temperatures on Earth over time. In these Earth globes, cool temperatures are represented by violet and blue and warm temperatures by yellow and red. These globes show the warming that occurred beginning 25,000 years ago, which ended the last ice age about 12,000 years ago. (Courtesy Arizona State University, Computer Science and Geography Departments) (b) Over the last 400,000 years, changes in ocean temperatures measured from fossils found in sediment layers from the seabed approximately match calculated changes in solar heating. (Adapted from Cesare Emiliani)

## How Do We Know? 2-3

### Evidence as the Foundation of Science

**Why is evidence critical in science?** From colliding galaxies to the inner workings of atoms, scientists love to speculate and devise theories, but all scientific knowledge is ultimately based on evidence from observations and experiments. Evidence is reality, and scientists constantly check their ideas against reality.

When you think of evidence, you probably think of criminal investigations in which detectives collect fingerprints and eyewitness accounts. In court, that evidence is used to try to understand the crime, but there is a key difference in how lawyers and scientists use evidence. A defense attorney can call a witness and intentionally fail to ask a question that would reveal evidence harmful to the defendant. In contrast, the scientist must be objective and not ignore any known evidence. The attorney is presenting only one side of the case, but the scientist is searching for the truth. In a sense, the scientist must deal with the evidence as both the prosecution and the defense.

It is a characteristic of scientific knowledge that it is supported by evidence. A scientific statement is more than an opinion or a speculation because it has been tested objectively against reality.

As you read about any science, look for the evidence in the form of observations and experiments. Every theory or conclusion should have supporting evidence. If you can find and understand the evidence, the science will make sense. All scientists, from astronomers to zoologists, demand evidence. You should, too.

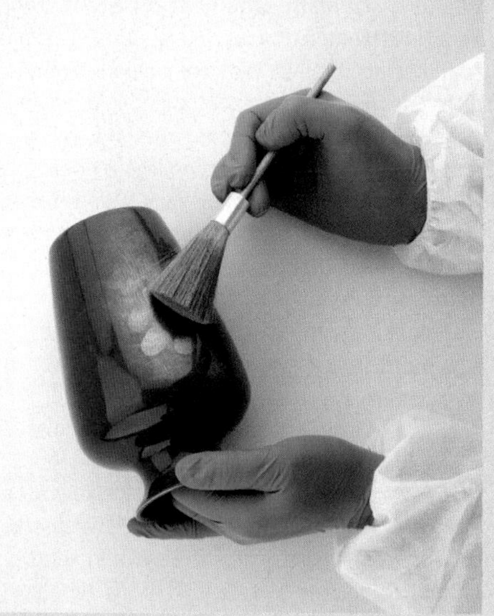

*Fingerprints are evidence to past events.*
(Dorling Kindersly/Getty Images)

### The Evidence

By the middle 1970s, Earth scientists were able to collect the data that Milankovitch had lacked. Oceanographers drilled deep into the seafloor to collect long cores of sediment. In the laboratory, geologists could take samples from different depths in the cores and determine the age of the samples and the temperature of the oceans when they were deposited on the seafloor. From this, scientists constructed a history of ocean temperatures that convincingly matched the predictions of the Milankovitch hypothesis (Figure 2-11b).

The evidence seemed very strong, and by the 1980s the Milankovitch hypothesis was widely considered the leading hypothesis. But science follows a mostly unstated set of rules that holds that a hypothesis must be tested over and over against all available evidence (**How Do We Know? 2-3**). In 1988, scientists discovered contradictory evidence.

For 500,000 years rainwater has collected in a deep crack in Nevada called Devil's Hole. That water has deposited the mineral calcite in layer on layer on the walls of the crack. It isn't easy to get to, and scientists had to dive with scuba gear to drill out samples of the calcite, but it was worth the effort. Back in the laboratory, they could determine the age of each layer in their core samples and the temperature of the rainwater that had formed the calcite in each layer. That gave them a history of temperatures at Devil's Hole that spanned many thousands of years, and the results were a surprise. The evidence seemed to show that Earth had begun warming up thousands of years too early for the last ice age to have been caused by the Milankovitch cycles.

These contradictory findings are irritating because we humans naturally prefer certainty, but such circumstances are common in science. The disagreement between the ocean floor samples and the Devil's Hole samples triggered a scramble to understand the problem. Were the age determinations of one or the other set of samples wrong? Were the calculations of ancient temperatures wrong? Or were scientists misunderstanding the significance of the evidence?

In 1997, a new study of the ages of the samples confirmed that those from the ocean floor are correctly dated. But the same study found that the ages of the Devil's Hole samples are also correct. Evidently the temperatures at Devil's Hole record local climate changes in the region that became the southwestern United States. The ocean floor samples record global climate changes, and they fit well with the Milankovitch hypothesis. This has given scientists renewed confidence in the Milankovitch hypothesis, and although it is widely accepted today, it is still being tested whenever scientists can find more evidence.

As you review this section, notice that it is a **scientific argument,** a careful presentation of theory and evidence in a logical discussion. **How Do We Know? 2-4** expands on the ways

# How Do We Know? | 2-4

## Scientific Arguments

**How is a scientific argument different from an advertisement?** Advertisements sometimes sound scientific, but they are fundamentally different from scientific arguments. An advertisement is designed to convince you to buy a product. "Our shampoo promises 85 percent shinier hair." The statement may sound like science, but it isn't a complete, honest discussion. "Shinier than what?" you might ask. An advertiser's only goal is a sale.

Scientists construct arguments because they want to test their own ideas and give an accurate explanation of some aspect of nature. For example, in the 1960s, biologist E. O. Wilson presented a scientific argument to show that ants communicate by smells. The argument included a description of his careful observations and the ingenious experiments he had conducted to test his theory. He also considered other evidence and other theories for ant communication. Scientists can include any evidence or theory that supports their claim, but they must observe one fundamental rule of science: They must be totally honest—they must include all of the evidence and all of the theories.

Scientists publish their work in scientific arguments, but they also think in scientific arguments. If, in thinking through his argument, Wilson had found a contradiction, he would have known he was on the wrong track. That is why scientific arguments must be complete and honest. Scientists who ignore inconvenient evidence or brush aside other theories are only fooling themselves.

A good scientific argument gives you all the information you need to decide for yourself whether the argument is correct. Wilson's study of ant communication is now widely understood and is being applied to other fields such as pest control and telecommunications networks.

*Scientists have discovered that ants communicate with a large vocabulary of smells.*
(Eye of Science/Photo Researchers, Inc.)

scientists organize their ideas in logical arguments. Throughout this book, many chapter sections end with short reviews entitled "Scientific Argument." These feature a review question, which is then analyzed in a scientific argument. A second question gives you a chance to build your own scientific argument on a related topic. You can use these "Scientific Argument" features to review chapter material, but also to practice thinking like a scientist.

### SCIENTIFIC ARGUMENT

*Why was it critical in testing the Milankovitch hypothesis to determine the ages of ocean sediment?*

Ocean floors accumulate sediment year after year in thin layers. Scientists can drill into the ocean floor and collect cores of those sediment layers, and from chemical tests they can find the temperature of the seawater when each layer was deposited. Those observations of temperature could be used as reality checks for the climate temperatures predicted by the Milankovitch theory, but that meant the ages of the sediment layers had to be determined correctly. When a conflict arose with evidence from Devil's Hole in Nevada, the age determinations of the samples were carefully reexamined and found to be correct.

After reviewing all of the evidence, scientists concluded that the ocean core samples do indeed support the Milankovitch hypothesis. Now construct your own argument. **What might the temperatures recorded at Devil's Hole represent?**

### What Are We?
### Along for the Ride

Human civilization is spread over the surface of planet Earth like a thin coat of paint. Great cities of skyscrapers and tangles of superhighways may seem impressive, but if you use your astronomical perspective, you can see that we humans are confined to the surface of our world.

The rotation of Earth creates a cycle of day and night that controls everything from TV schedules to the chemical workings of our brains. We wake and sleep within that 24-hour cycle of light and dark. Furthermore, Earth's orbital motion around the sun, combined with the inclination of its axis, creates a yearly cycle of seasons, and we humans, along with every other living thing on Earth, have evolved to survive within those extremes of temperature. We protect ourselves from the largest extremes and have spread over most of Earth, hunting, gathering, and growing crops within the cycle of the seasons.

In recent times, we have begun to understand that conditions on Earth's surface are not entirely stable. Slow changes in its motions produce a cycle of ice ages with irregular pulses of glaciation. All of recorded history, including all of the cities and superhighways that paint our globe, has occurred since the last glacial retreat only 12,000 years ago, so we humans have no recorded experience of Earth's harshest climate. We have never experienced our planet's icy personality. We are along for the ride and enjoying Earth's good times.

# Study and Review

## Summary

- Astronomers divide the sky into 88 **constellations (p. 12).** Although the constellations originated in Greek and Middle Eastern mythology, the names are Latin. Even the modern constellations, added to fill in the spaces between the ancient figures, have Latin names. Named groups of stars that are not constellations are called **asterisms (p. 13).**
- The names of stars usually come from ancient Arabic, though modern astronomers often refer to a star by its constellation and a Greek letter assigned according to its brightness within the constellation.
- Astronomers refer to the brightness of stars using the **magnitude scale (p. 14).** First-magnitude stars are brighter than second-magnitude stars, which are brighter than third-magnitude stars, and so on. The magnitude you see when you look at a star in the sky is its **apparent visual magnitude, $m_v$ (p. 15)**, which includes only types of light visible to the human eye and does not take into account the star's distance from Earth.
- **Flux (p. 15)** is a measure of light energy related to intensity. The magnitude of a star is related directly to the flux of light received on Earth and so to its intensity.
- The **celestial sphere (p. 18)** is a **scientific model (p. 17)** of the sky, to which the stars appear to be attached. Because Earth rotates eastward, the celestial sphere appears to rotate westward on its axis.
- The **north** and **south celestial poles (p. 18)** are the pivots on which the sky appears to rotate, and they define the four directions around the **horizon (p. 18)**: the **north, south, east,** and **west points (p. 18).** The point directly over head is the **zenith (p. 18)**, and the point on the sky directly underfoot is the **nadir (p. 18).**
- The **celestial equator (p. 17)**, an imaginary line around the sky above Earth's equator, divides the sky into northern and southern halves.
- Astronomers often refer to distances "on" the sky as if the stars, sun, moon, and planets were equivalent to spots painted on a plaster ceiling. These **angular distances (p. 19)**, measured in degrees, **arc minutes (p. 19)**, and **arc seconds (p. 19)**, are unrelated to the true distance between the objects in light-years. The angular distance across an object is its **angular diameter (p. 19).**
- What you see of the celestial sphere depends on your latitude. Much of the southern hemisphere of the sky is not visible from northern latitudes. To see that part of the sky, you would have to travel southward over Earth's surface. **Circumpolar constellations (p. 19)** are those close enough to a celestial pole that they do not rise or set.
- The angular distance from the horizon to the north celestial pole always equals your latitude. This is the basis for celestial navigation.
- **Precession (p. 17)** is caused by the gravitational forces of the moon and sun acting on the equatorial bulge of the spinning Earth and causing its axis to sweep around in a conical motion like the motion of a top's axis. Earth's axis of rotation precesses with a period of 26,000 years, and consequently the celestial poles and celestial equator move slowly against the background of the stars.
- The **rotation (p. 20)** of Earth on its axis produces the cycle of day and night, and the **revolution (p. 20)** of Earth around the sun produces the cycle of the year.
- Because Earth orbits the sun, the sun appears to move eastward along the **ecliptic (p. 21)** through the constellations completing a circuit of the sky in a year. Because the ecliptic is tipped 23.4° to the celestial equator, the sun spends half the year in the northern celestial hemisphere and half in the southern celestial hemisphere.
- In each hemisphere's summer, the sun is above the horizon longer and shines more directly down on the ground. Both effects cause warmer weather in that hemisphere and leave Earth's other hemisphere cooler. In each hemisphere's winter, the sun is above the sky fewer hours than in summer and shines less directly, so the winter hemisphere has colder weather, and the opposite hemisphere has summer. Consequently the seasons are reversed in Earth's southern hemisphere relative to the northern hemisphere.
- The beginning of spring, summer, winter, and fall are marked by the **vernal equinox (p. 22)**, the **summer solstice (p. 22)**, the **autumnal equinox (p. 22)**, and the **winter solstice (p. 22).**
- Earth is slightly closer to the sun at **perihelion (p. 23)** in January and slightly farther away from the sun at **aphelion (p. 23)** in July. This has almost no effect on the seasons.
- The planets move generally eastward along the ecliptic, and all but Uranus and Neptune are visible to the unaided eye looking like stars. Mercury and Venus never wander far from the sun and are sometimes visible in the evening sky after sunset or in the dawn sky before sunrise.
- Planets visible in the sky at sunset are traditionally called **evening stars (p. 24)**, and planets visible in the dawn sky are called **morning stars (p. 24)** even though they are not actually stars.
- The locations of the sun and planets along the **zodiac (p. 25)** are diagramed in a **horoscope (p. 25)**, which is the bases for the ancient **pseudoscience (p. 25)** known as astrology.
- According to the **Milankovitch hypothesis (p. 26)**, changes in the shape of Earth's orbit, in its precession, and in its axial tilt can alter the planet's heat balance and cause the cycle of ice ages. Evidence found in seafloor samples support the hypothesis, and it is widely accepted today.
- Scientists routinely test their own ideas by organizing theory and evidence into a **scientific argument (p. 27).**

## Review Questions

1. Why are most modern constellations composed of faint stars or located in the southern sky?
2. How does the Greek-letter designation of a star give you clues to its location and brightness?
3. From your knowledge of stars names and constellations, which of the following stars in each pair is the brighter? Explain your answers.
   a. Alpha Ursae Majoris; Epsilon Ursae Majoris
   b. Epsilon Scorpii; Alpha Pegasus
   c. Alpha Telescopium; Alpha Orionis
4. How did the magnitude system originate in a classification of stars by brightness?
5. What does the word *apparent* mean in *apparent visual magnitude?*
6. What does the word *visual* mean in *apparent visual magnitude*?
7. In what ways is the celestial sphere a scientific model?
8. If Earth did not rotate, could you define the celestial poles and celestial equator?
9. Where would you go on Earth if you wanted to be able to see both the north celestial pole and the south celestial pole at the same time?
10. Why does the number of circumpolar constellations depend on the latitude of the observer?
11. Explain two reasons why winter days are colder than summer days.
12. How do the seasons in Earth's southern hemisphere differ from those in the northern hemisphere?
13. Why should the eccentricity of Earth's orbit make winter in Earth's northern hemisphere different from winter in Earth's southern hemisphere?

*Even a man who is pure in heart and says his prayers by night*

*May become a wolf when the wolfbane blooms and the moon shines full and bright.*

— PROVERB FROM OLD WOLFMAN MOVIES

DID ANYONE EVER WARN YOU, "Don't stare at the moon—you'll go crazy"? For centuries, the superstitious have associated the moon with insanity. The word *lunatic* comes from a time when even doctors thought that the insane were "moonstruck." Of course, the moon does not cause madness, but it is so bright and moves so rapidly along the ecliptic that people *expect* it to affect them. It is, in fact, one of the most beautiful and dramatic objects in the sky.

## 3-1 The Changeable Moon

STARTING THIS EVENING, begin looking for the moon in the sky. As you watch for the moon on successive evenings, you will see it following its orbit around Earth and cycling through its phases as it has done for billions of years.

### The Motion of the Moon

Just as the planets revolve counterclockwise around the sun as seen from the direction of the celestial north pole, the moon revolves counterclockwise around Earth. Because the moon's orbit is tipped a little over 5 degrees from the plane of Earth's orbit, the moon's path takes it slightly north and then slightly south, but it is always somewhere near the ecliptic.

The moon moves rapidly against the background of the constellations. If you watch the moon for just an hour, you can see it move eastward by slightly more than its angular diameter. In the previous chapter, you learned that the moon is about 0.5° in angular diameter, and it moves eastward a bit more than 0.5° per hour. In 24 hours, it moves 13°. Thus, each night you see the moon about 13° eastward of its location the night before.

As the moon orbits around Earth, its shape changes from night to night in a month-long cycle.

### The Cycle of Phases

The changing shape of the moon as it revolves around Earth is one of the most easily observed phenomena in astronomy. Study **The Phases of the Moon** on pages 34–35 and notice three important points and two new terms:

1. The moon always keeps the same side facing Earth. "The man in the moon" is produced by the familiar features on the moon's near side, but you never see the far side of the moon.

2. The changing shape of the moon as it passes through its cycle of phases is produced by sunlight illuminating different parts of the side of the moon you can see.

3. Notice the difference between the orbital period of the moon around Earth (*sidereal period*) and the length of the lunar phase cycle (*synodic period*). That difference is a good illustration of how your view from Earth is produced by the combined motions of Earth and other heavenly bodies, such as the sun and moon.

You can figure out where the moon will be in the sky by making a moon-phase dial from the middle diagram on page 34. Cover the lower half of the moon's orbit with a sheet of paper and align the edge of the paper to pass through the word "Full" at the left and the word "New" at the right. Push a pin through the edge of the paper at Earth's North Pole to make a pivot and, under the word "Full," write on the paper "Eastern Horizon." Under the word "New," write "Western Horizon." The paper now represents the horizon you see when you stand facing south. You can set your moon-phase dial for a given time by rotating the diagram behind the horizon-paper. For example, set the dial to sunset by turning the diagram until the human figure labeled "sunset" is standing at the top of the Earth globe; the dial shows, for example, that the full moon at sunset would be at the eastern horizon.

The phases of the moon are dramatic, and they have attracted a number of peculiar ideas. You have probably heard a number of **Common Misconceptions** about the moon. Sometimes people are surprised to see the moon in the daytime sky, and they think something has gone wrong! No, the gibbous moon is often visible in the daytime, although quarter moons and especially crescent moons can also be in the daytime sky but are harder to see when the sun is above the horizon. You may hear people mention "the dark side of the moon," but you will be able to assure them that this is a misconception; there is no permanently dark side. Any location on the moon is sunlit for two weeks and is in darkness for two weeks as the moon rotates in sunlight. Finally, you have probably heard one of the strangest misconceptions about the moon: that people tend to act up at full moon. Actual statistical studies of records from schools, prisons, hospitals, and so on, show that it isn't true. There are always a few people who misbehave; the moon has nothing to do with it.

For billions of years, the man in the moon has looked down on Earth. Ancient civilizations saw the same cycle of phases that you see (■ Figure 3-1), and even the dinosaurs may have noticed the changing phases of the moon. Occasionally, however, the moon displays more complicated moods when it turns copper-red in a lunar eclipse.

■ **Figure 3-1**

In this sequence of lunar phases, the moon cycles through its phases from crescent to full to crescent. You see the same face of the moon, the same mountains, craters, and plains, but the changing direction of sunlight produces the lunar phases.

## 3-2 Lunar Eclipses

In cultures all around the world, the sky is a symbol of order and power, and the moon is the regular counter of the passing days. So it is not surprising that people are startled and sometimes worried when they see the moon grow dark and angry-red. Such events are neither mysterious nor frightening once you understand how they arise. To begin, you can think about Earth's shadow.

### Earth's Shadow

The orbit of the moon is tipped only a few degrees from the plane of Earth's orbit around the sun, and Earth's shadow points directly away from the sun also in the plane of Earth's orbit. A **lunar eclipse** can occur at full moon if the moon's path carries it through the shadow of Earth. That is unusual because most full moons pass north or south of Earth's shadow and there is no eclipse. If the moon enters Earth's shadow, sunlight is cut off, and the moon grows dim.

The shadow consists of two parts (■ Figure 3-2). The **umbra** is the region of total shadow. If you were drifting in your spacesuit in the umbra of Earth's shadow, the sun would be completely hidden behind Earth, and you would not be able to see any part of the sun's bright disk. If you drifted into the **penumbra,** however, you would see part of the sun peeking around the edge of Earth, so you would be in partial shadow. In the penumbra, sunlight is dimmed but not extinguished.

The umbra of Earth's shadow is over three times longer than the distance to the moon and points directly away from the sun. A giant screen placed in the shadow at the average distance of the moon would reveal a dark umbral shadow about 2.5 times the diameter of the moon. The faint outer edges of the penumbra would mark a circle about 4.6 times the diameter of the moon. Consequently, when the moon's orbit carries it through the umbra, it has plenty of room to become completely immersed in shadow.

### Total Lunar Eclipses

Once or twice a year, the orbit of the moon carries it through the umbra of Earth's shadow; and, if no part of the moon remains outside the umbra in the partial sunlight of the penumbra, you

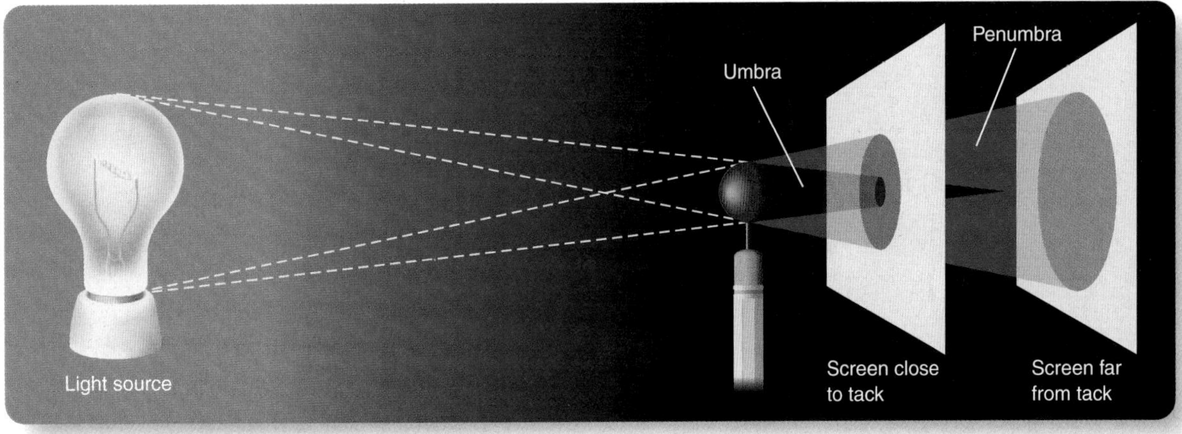


■ **Figure 3-2**

The shadows cast by a map tack resemble the shadows of Earth and the moon. The umbra is the region of total shadow; the penumbra is the region of partial shadow.

# The Phases of the Moon

**1** As the moon orbits Earth, it rotates to keep the same side facing Earth as shown at right. Consequently you always see the same features on the moon, and you never see the far side of the moon. A mountain on the moon that points at Earth will always point at Earth as the moon revolves and rotates.

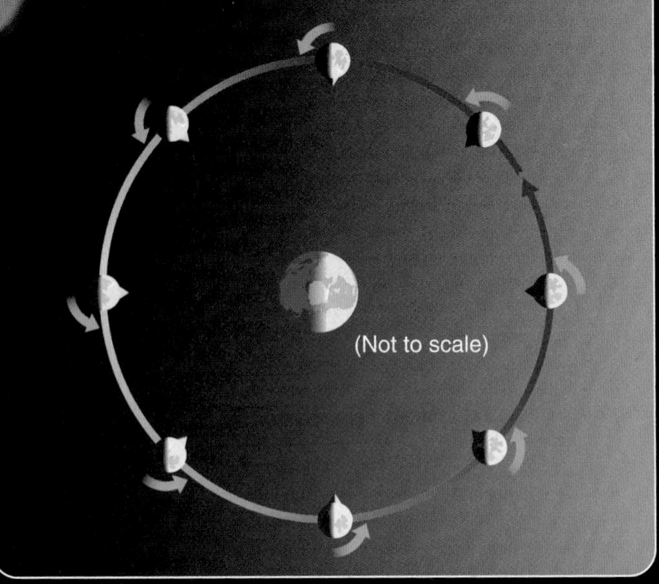


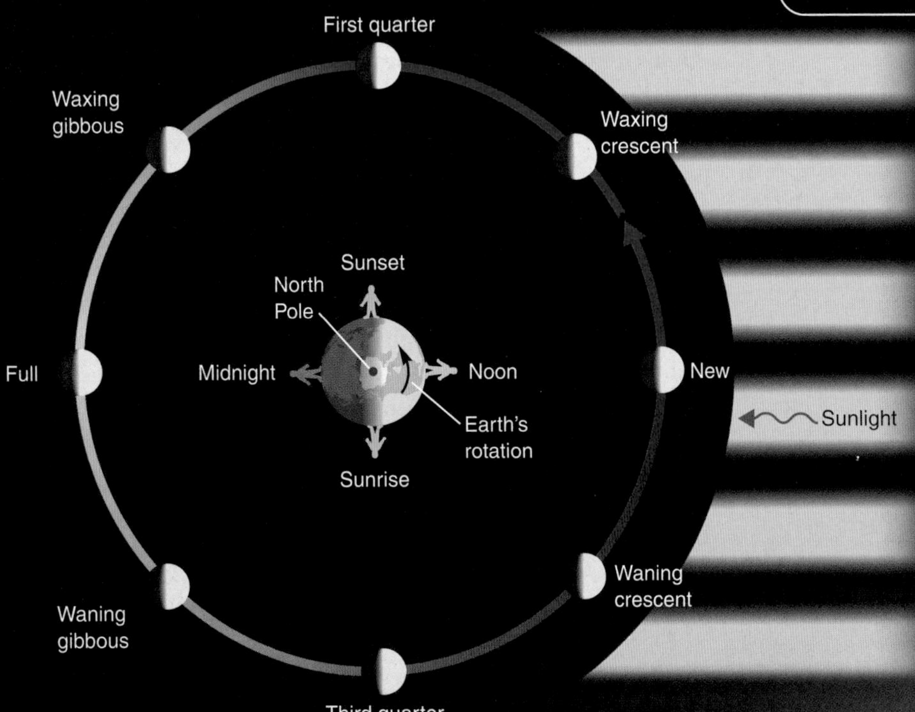


**2** As seen at left, sunlight always illuminates half of the moon. Because you see different amounts of this sunlit side, you see the moon cycle through phases. At the phase called "new moon," sunlight illuminates the far side of the moon, and the side you see is in darkness. At new moon you see no moon at all. At full moon, the side you see is fully lit, and the far side is in darkness. How much you see depends on where the moon is in its orbit.

Notice that there is no such thing as the "dark side of the moon." All parts of the moon experience day and night in a month-long cycle.

In the diagram at the left, you see that the new moon is close to the sun in the sky, and the full moon is opposite the sun. The time of day depends on the observer's location on Earth.

**2a** The first 2 weeks of the cycle of the moon are shown below by its position at sunset on 14 successive evenings. As the moon grows fatter from new to full, it is said to wax.

The first quarter moon is one week through its 4-week cycle.

Gibbous comes from the Latin word for humpbacked.

Waxing gibbous

Waxing crescent

The full moon is two weeks through its 4-week cycle.

New moon is invisible near the sun

14 13 12 11 10 9 8 7 6 5 4 3 2 1

THE SKY AT SUNSET

Days since new moon

Full moon rises at sunset

East South West

3 The moon orbits eastward around Earth in 27.32 days, its **sidereal period.** This is how long the moon takes to circle the sky once and return to the same position among the stars.

A complete cycle of lunar phases takes 29.53 days, the moon's **synodic period.** (Synodic comes from the Greek words for "together" and "path.")

To see why the synodic period is longer than the sidereal period, study the star charts at the right.

Although you think of the lunar cycle as being about 4 weeks long, it is actually 1.53 days longer than 4 weeks. The calendar divides the year into 30-day periods called months (literally "moonths") in recognition of the 29.53 day synodic cycle of the moon.

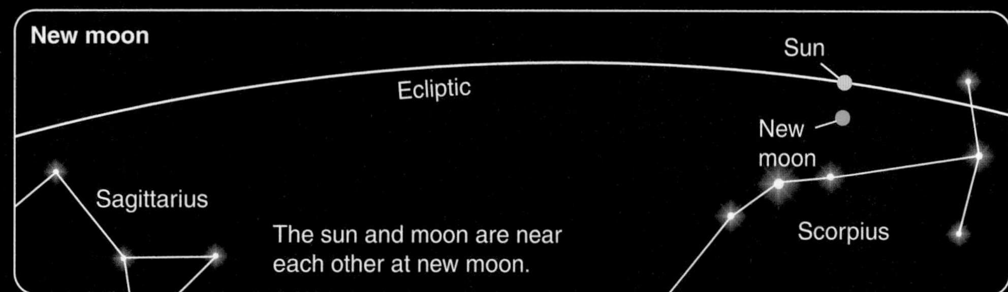


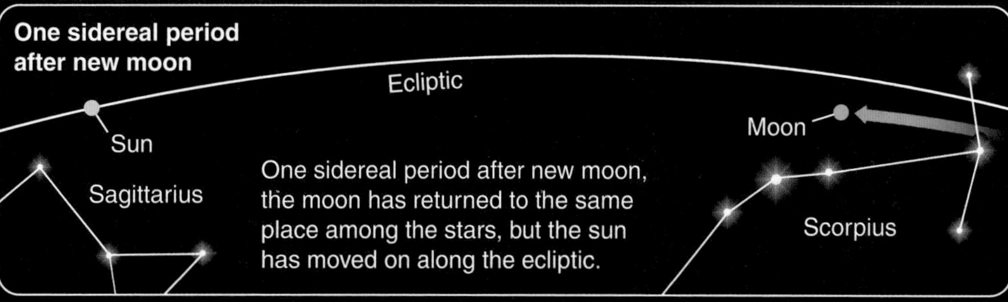


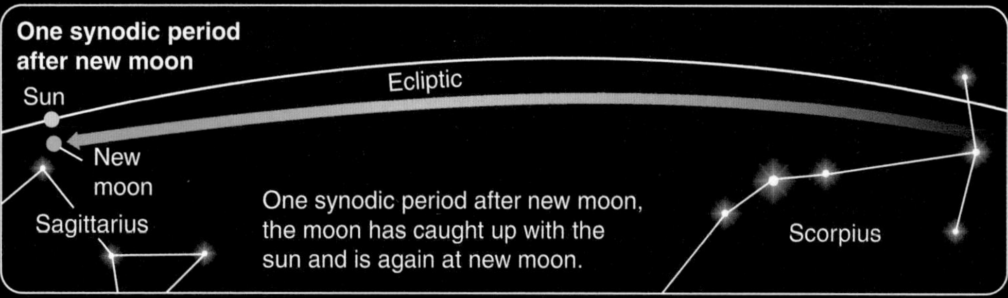


You can use the diagram on the opposite page to determine when the moon rises and sets at different phases.

**TIMES OF MOONRISE AND MOONSET**

| Phase | Moonrise | Moonset |
|---|---|---|
| New | Dawn | Sunset |
| First quarter | Noon | Midnight |
| Full | Sunset | Dawn |
| Third quarter | Midnight | Noon |

**2b** The last two weeks of the cycle of the moon are shown below by its position at sunrise on 14 successive mornings. As the moon shrinks from full to new, it is said to wane.

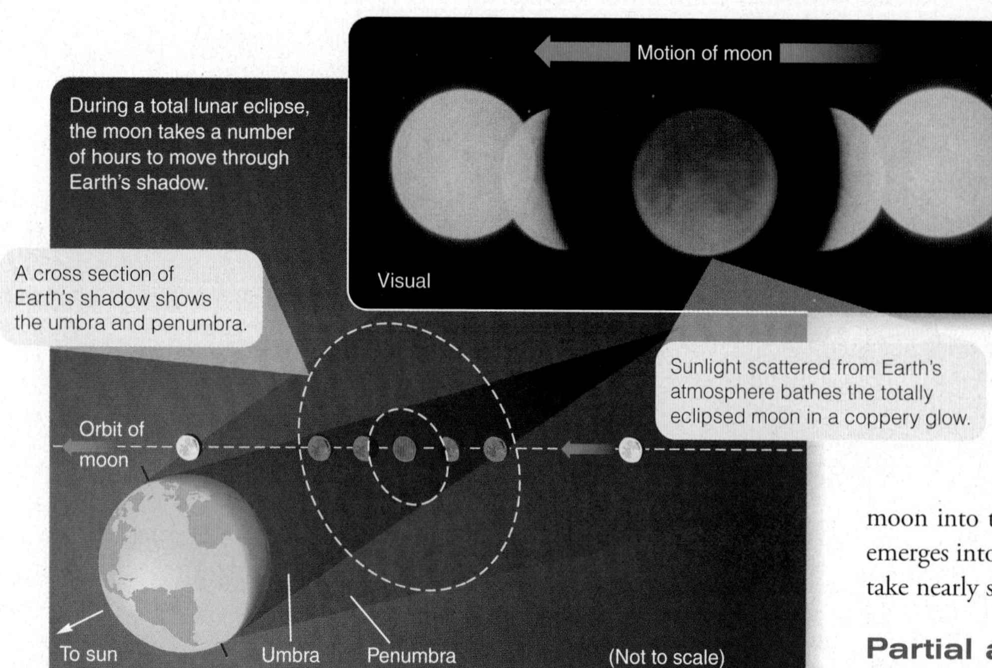
Motion of moon
During a total lunar eclipse, the moon takes a number of hours to move through Earth's shadow.
A cross section of Earth's shadow shows the umbra and penumbra.
Visual
Sunlight scattered from Earth's atmosphere bathes the totally eclipsed moon in a coppery glow.
Orbit of moon
To sun
Umbra
Penumbra
(Not to scale)

■ **Figure 3-3**

In this diagram of a total lunar eclipse, the moon passes from right to left through Earth's shadow. A multiple-exposure photograph shows the moon passing through the umbra of Earth's shadow. A longer exposure was used to record the moon while it was totally eclipsed. The moon's path appears curved in the photo because of photographic effects. (© 1982 Dr. Jack B. Marling)

see a **total lunar eclipse** (■ Figure 3-3). As you watch the eclipse begin, the moon first moves into the penumbra and dims slightly; the deeper it moves into the penumbra, the more it dims. Eventually, the moon reaches the umbra, and you see the umbral shadow first darken part, then all, of the moon.

When the moon is totally eclipsed, it does not disappear completely. Although it receives no direct sunlight, the moon in the umbra does receive some sunlight that is refracted (bent) through Earth's atmosphere. If you were on the moon during **totality,** you would not see any part of the sun because it would be entirely hidden behind Earth. However, you would see Earth's atmosphere illuminated from behind by the sun. The red glow from this ring of "sunsets" and "sunrises" illuminates the moon during totality and makes it glow coppery red, as shown in ■ Figure 3-4.

How dim the totally eclipsed moon becomes depends on a number of things. If Earth's atmosphere is especially cloudy in the regions that would bend light into the umbra, the moon will be darker than usual. An unusual amount of dust in Earth's atmosphere (from volcanic eruptions, for instance) also causes a dark eclipse. Also, total lunar eclipses tend to be darkest when the moon's orbit carries it through the center of the umbra.

The exact timing of a lunar eclipse depends on where the moon crosses Earth's shadow. If it crosses through the center of the umbra, the eclipse will have maximum length. For such an eclipse, the moon spends about an hour crossing the penumbra and then another hour entering the darker umbra. Totality can last as long as 1 hour 45 minutes, followed by the emergence of the moon into the penumbra, plus another hour as it emerges into full sunlight. A total lunar eclipse can take nearly six hours from start to finish.

### Partial and Penumbral Lunar Eclipses

Because the moon's orbit is inclined by a bit over 5° to the plane of Earth's orbit, the moon does not always pass through the center of the umbra. If the moon passes a bit too far north or south, it may only partially enter the umbra, and you see a **partial lunar eclipse.** The part of the moon that

Visual

■ **Figure 3-4**

During a total lunar eclipse, the moon turns coppery red. In this photo, the moon is darkest toward the lower right, the direction toward the center of the umbra. The edge of the moon at upper left is brighter because it is near the edge of the umbra. (Celestron International)

**■ Table 3-1 | Total and Partial Eclipses of the Moon, 2010 to 2017***

| Year | Time** of Mid-eclipse (GMT) | Length of Totality (Hr: Min) | Length of Eclipse† (Hr: Min) |
|---|---|---|---|
| 2010 June 26 | 11:40 | Partial | 2:42 |
| 2010 Dec. 21 | 8:18 | 1:12 | 3:28 |
| 2011 June 15 | 20:13 | 1:40 | 3:38 |
| 2011 Dec. 10 | 14:33 | 0:50 | 3:32 |
| 2012 June 4 | 11:03 | Partial | 2:08 |
| 2013 April 25 | 20:10 | Partial | 0:28 |
| 2014 April 15 | 7:48 | 1:18 | 3:34 |
| 2014 Oct. 8 | 10:55 | 0:58 | 3:18 |
| 2015 April 4 | 12:02 | Partial | 3:28 |
| 2015 Sept. 28 | 2:48 | 72 | 3:30 |
| 2017 Aug. 7 | 18:22 | Partial | 1:54 |

*There are no total or partial lunar eclipses during 2016.

**Times are Greenwich Mean Time. Subtract 5 hours for Eastern Standard Time, 6 hours for Central Standard Time, 7 hours for Mountain Standard Time, and 8 hours for Pacific Standard Time. From your time zone, lunar eclipses that occur between sunset and sunrise will be visible, and those at midnight will be best placed.

†Does not include penumbral phase.

remains in the penumbra receives some direct sunlight, and the glare is usually great enough to prevent your seeing the faint coppery glow of the part of the moon in the umbra. Partial eclipses are interesting, but they are not as beautiful as a total lunar eclipse.

If the orbit of the moon carries it far enough north or south of the umbra, the moon may pass through only the penumbra and never reach the umbra. Such **penumbral lunar eclipses** are not dramatic at all. In the partial shadow of the penumbra, the moon is only partially dimmed. Most people glancing at a penumbral eclipse would not notice any difference from a full moon.

Total, partial, or penumbral, lunar eclipses are interesting events in the night sky and are not difficult to observe. When the full moon passes through Earth's shadow, the eclipse is visible from anywhere on Earth's dark side. Consult ■ Table 3-1 to find the next lunar eclipse visible in your part of the world.

## SCIENTIFIC ARGUMENT

***What would a total lunar eclipse look like if Earth had no atmosphere?***

As a way to test and improve their understanding, scientists often experiment with their ideas by changing one thing and trying to imagine what would happen. In this case, the absence of an atmosphere around Earth would mean that no sunlight would be bent toward the eclipsed moon, and it would not glow red. It would be very dark in the sky during totality.

Now change a different parameter. **What would a lunar eclipse look like if the moon and Earth were the same diameter?**

## 3-3 Solar Eclipses

For millennia, cultures worldwide have understood that the sun is the source of life, so you can imagine the panic people felt at the fearsome sight of the sun gradually disappearing in the middle of the day. Many imagined that the sun was being devoured by a monster (■ Figure 3-5). Modern scientists must

**■ Figure 3-5**

(a) A 12th-century Mayan symbol believed to represent a solar eclipse. The black-and-white sun symbol hangs from a rectangular sky symbol, and a voracious serpent approaches from below. (b) The Chinese representation of a solar eclipse shows a monster usually described as a dragon flying in front of the sun. (From the collection of Yerkes Observatory) (c) This wall carving from the ruins of a temple in Vijayanagaara in southern India symbolizes a solar eclipse as two snakes approach the disk of the sun. (T. Scott Smith)

## How Do We Know? 3-1

### Scientific Imagination

**How do scientists produce theories to test?** Good scientists are invariably creative people with strong imaginations who can look at raw data about some invisible aspect of nature such as an atom and construct mental pictures as diverse as a plum pudding or a solar system. These scientists share the same human impulse to understand nature that drove ancient cultures to imagine eclipses as serpents devouring the sun.

As the 20th century began, physicists were busy trying to imagine what an atom was like. No one can see an atom, but English physicist J. J. Thomson used what he knew from his experiments and his powerful imagination to create an image of what an atom might be like. He suggested that an atom was a ball of positively charged material with negatively charged electrons distributed throughout like plums in a plum pudding.

The key difference between using a plum pudding to represent the atom and a hungry serpent to represent an eclipse is that the plum pudding model was based on experimental data and could be tested against new evidence. As it turned out, Thomson's student, Ernest Rutherford, performed ingenious new experiments and showed that atoms can't be made like plum puddings. Rather, he imagined an atom as a tiny positively charged nucleus surrounded by negatively charged electrons much like a tiny solar system with planets circling the sun. Later experiments confirmed that Rutherford's description of atoms is closer to reality, and it has become a universally recognized symbol for atomic energy.

Ancient cultures pictured the sun being devoured by a serpent. Thomson, Rutherford, and scientists like them used their scientific imaginations to visualize natural processes and then test and refine their ideas with new experiments and observations. The critical difference is that scientific imagination is continually tested against reality and is revised when necessary.

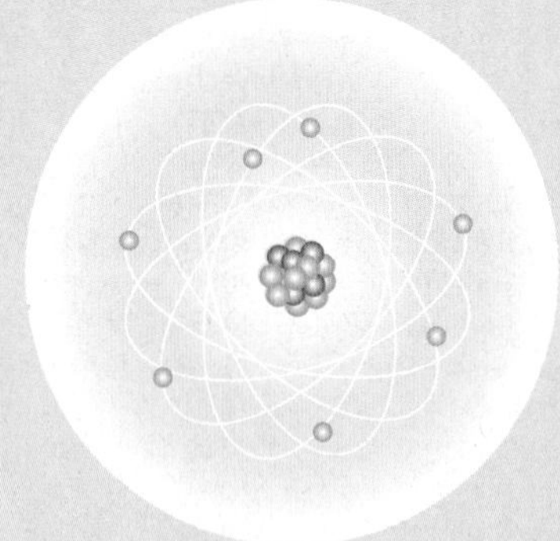

*A model image of the atom as electrons orbiting a small nucleus has become the symbol for atomic energy.*

use their imaginations to visualize how nature works, but with a key difference; they test their ideas against reality (**How Do We Know? 3-1**). You can take comfort that today's astronomers explain solar eclipses without imagining celestial monsters.

A **solar eclipse** occurs when the moon moves between Earth and the sun. If the moon covers the disk of the sun completely, you see a spectacular **total solar eclipse** (■Figure 3-6). If, from your location, the moon covers only part of the sun, you see a less dramatic **partial solar eclipse.** During a solar eclipse, people in one place on Earth may see a total eclipse while people only a few hundred kilometers away see a partial eclipse.

The geometry of a solar eclipse is quite different from that of a lunar eclipse. You can begin by considering how big the sun and moon look in the sky.

## The Angular Diameter of the Sun and Moon

Solar eclipses are spectacular because Earth's moon happens to have nearly the same angular diameter as the sun, so it can cover the sun's disk almost exactly. You learned about angular diameter in Chapter 2; now you need to think carefully about how the size and distance of an object like the moon determine its angular diameter.

**■ Figure 3-6**

Solar eclipses are dramatic. In June 2001, an automatic camera in southern Africa snapped pictures every 5 minutes as the afternoon sun sank lower in the sky. From upper right to lower left, you can see the moon crossing the disk of the sun. A longer exposure was needed to record the total phase of the eclipse. (©2001 F. Espenak)

Linear diameter is simply the distance between an object's opposite sides. You use linear diameter when you order a 16-inch pizza—the pizza is 16 inches across. The linear diameter of the moon is 3476 km.

The angular diameter of an object is the angle formed by lines extending toward you from opposite edges of the object and meeting at your eye (■ Figure 3-7). Clearly, the farther away an object is, the smaller its angular diameter.

To find the angular diameter of the moon, you need to use the **small-angle formula.** It gives you a way to figure out the angular diameter of any object, whether it is a pizza or the moon. In the small-angle formula, you should always express angular diameter in arc seconds and always use the same units for distance and linear diameter:

$$\frac{\text{angular diameter}}{206{,}265} = \frac{\text{linear diameter}}{\text{distance}}$$

You can use this formula* to find any one of these three quantities if you know the other two; in this case, you are interested in finding the angular diameter of the moon.

The moon has a linear diameter of 3476 km and a distance from Earth of about 384,000 km. Because the moon's linear diameter and distance are both given in the same units, kilometers, you can put them directly into the small-angle formula:

$$\frac{\text{angular diameter}}{206{,}265} = \frac{3476 \text{ km}}{384{,}000 \text{ km}}$$

To solve for angular diameter, multiply both sides by 206,265. You will find that the angular diameter of the moon is 1870 arc seconds. If you divide by 60, you get 31 arc minutes, or, dividing by 60 again, about 0.5°. The moon's orbit is slightly elliptical, so the moon can sometimes look a bit larger or smaller, but its angular diameter is always close to 0.5°. It is a **Common Misconception** that the moon is larger when it is on the horizon. Certainly the rising full moon looks big when you see it on the horizon, but that is an optical illusion. In reality, the moon is the same size on the horizon as when it is high overhead.

Now repeat this small-angle calculation to find the angular diameter of the sun. The sun is $1.39 \times 10^6$ km in linear diameter and $1.50 \times 10^8$ km from Earth. If you put these numbers into the small-angle formula, you will discover that the sun has an angular diameter of 1900 arc seconds, which is 32 arc minutes, or about 0.5°. Earth's orbit is slightly elliptical, and consequently the sun can sometimes look slightly larger or smaller in the sky, but it, like the moon, is always close to 0.5° in angular diameter.

By fantastic good luck, you live on a planet with a moon that is almost exactly the same angular diameter as your sun. Thanks to that coincidence, when the moon passes in front of the sun, it is almost exactly the right size to cover the sun's brilliant surface but leave the sun's atmosphere visible.

*The number 206,265 is the number of arc seconds in a radian. When you divide by 206,265, you convert the angle from arc seconds to radians.

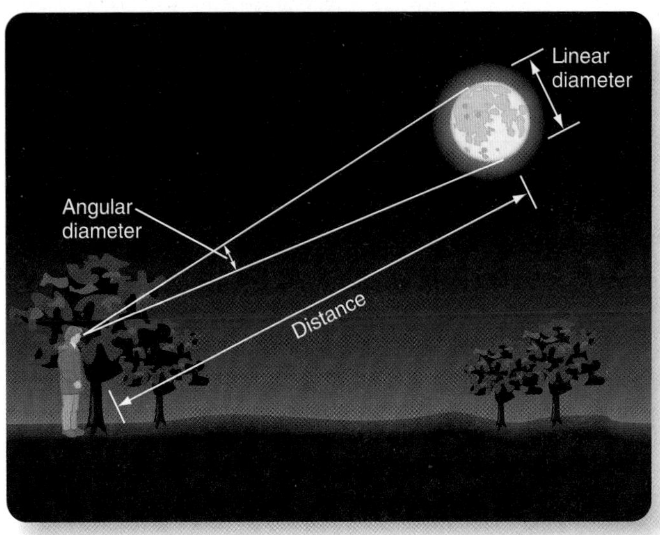


■ **Figure 3-7**

The angular diameter of an object is related to both its linear diameter and its distance.

## The Moon's Shadow

To see a solar eclipse, you have to be in the moon's shadow. Like Earth's shadow, the moon's shadow consists of a central umbra of total shadow and a penumbra of partial shadow. The moon's umbral shadow produces a spot of darkness about 270 km ( 170 mi) in diameter on Earth's surface. (The exact size of the umbral shadow depends on the location of the moon in its elliptical orbit and the angle at which the shadow strikes Earth.) Because of the orbital motion of the moon and the rotation of Earth, the moon's shadow rushes across Earth at speeds of at least 1700 km/h (1060 mph) sweeping out a **path of totality** (■ Figure 3-8). People lucky enough to be in the path of totality will see a total eclipse of the sun when the umbral spot sweeps over them. Observers just outside the path of totality will see a partial solar eclipse as the penumbral shadow sweeps over their location. Those living even farther from the path of totality will see no eclipse.

Sometimes the moon's umbral shadow is not long enough to reach Earth, and, seen from Earth's surface, the disk of the moon is not big enough to cover the sun. The orbit of the moon is slightly elliptical, and its distance from Earth varies. When it is at **apogee,** its farthest point from Earth, its angular diameter is about 6 percent smaller than average, and when it is at **perigee,** its closest point to Earth, its angular diameter is 6 percent larger than average. Another factor is Earth's slightly elliptical orbit around the sun. When Earth is closest to the sun in January, the sun looks 1.7 percent larger in angular diameter; and, when Earth is at its farthest point in July, the sun looks 1.7 percent smaller. If the moon crosses in front of the sun when the moon's disk is smaller in angular diameter than the sun's, it produces an **annular eclipse,** a solar eclipse in which a ring (or annulus) of light is visible around the disk of the moon (■ Figure 3-9). An annular eclipse swept across the United States on May 10, 1994.

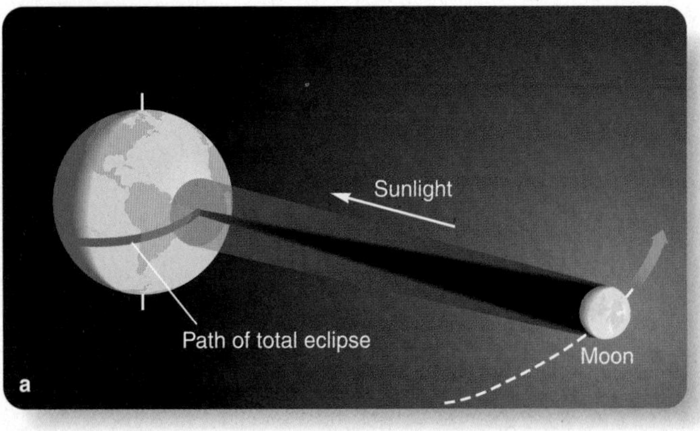


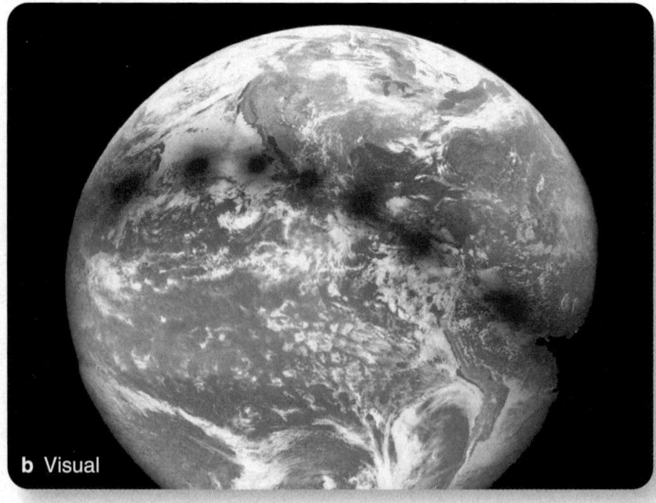


**Figure 3-8**

(a) The umbra of the moon's shadow sweeps from west to east across Earth, and observers in the path of totality see a total solar eclipse. Those outside the umbra but inside the penumbra see a partial eclipse. (b) Eight photos made by a weather satellite have been combined to show the moon's shadow moving across Mexico, Central America, and Brazil. (NASA GOES images courtesy of MrEclipse.com)

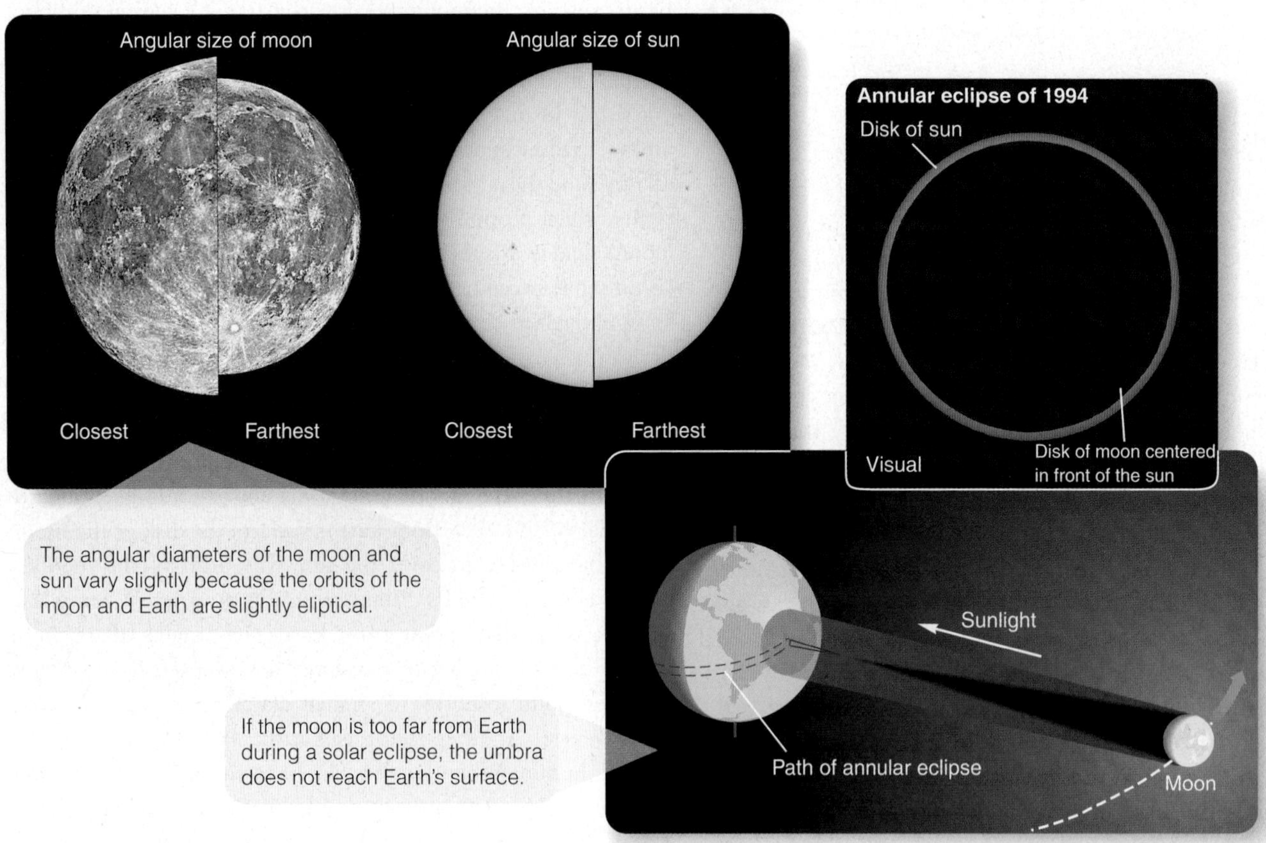


**Figure 3-9**

Because the angular diameter of the moon and the sun vary slightly, the disk of the moon is sometimes too small to cover the disk of the sun. This means the umbra of the moon does not reach Earth, and the eclipse is annular. From Earth, you see an annular eclipse because the moon's angular diameter is smaller than the angular diameter of the sun. In the photograph of the annular eclipse of 1994, the dark disk of the moon is almost exactly centered on the bright disk of the sun. (Daniel Good)

**Table 3-2 | Total and Annular Eclipses of the Sun, 2010 to 2019***

| Date | Total/ Annular (T/A) | Time of Mid-eclipse** (GMT) | Maximum Length of Total or Annular Phase (Min:Sec) | Area of Visibility |
|---|---|---|---|---|
| 2010 Jan. 15 | A | $7^h$ | 11:10 | Africa, Indian Ocean |
| 2010 July 11 | T | $20^h$ | 5:20 | Pacific, South America |
| 2012 May 20 | A | $23^h$ | 5:46 | Japan, North Pacific, western USA |
| 2012 Nov. 13 | T | $22^h$ | 4:02 | Australia, South Pacific |
| 2013 May 10 | A | $0^h$ | 6:04 | Australia, Pacific |
| 2013 Nov. 3 | AT | $13^h$ | 1:40 | Atlantic, Africa |
| 2015 March 20 | T | $10^h$ | 2:47 | North Atlantic, Arctic |
| 2016 March 9 | T | $2^h$ | 4:10 | Borneo, Pacific |
| 2016 Sept. 1 | A | $9^h$ | 3:06 | Atlantic, Africa, Indian Ocean |
| 2017 Feb. 26 | A | $15^h$ | 1:22 | S. Pacific, S. America, Africa |
| 2017 Aug. 21 | T | $18^h$ | 2:40 | Pacific, USA, Atlantic |
| 2019 July 2 | T | $19^h$ | 4:32 | Pacific, South America |
| 2019 Dec. 26 | A | $5^h$ | 3:40 | Southeast Asia, Pacific |

The next major total solar eclipse visible from the United States will occur on August 21, 2017, when the path of totality will cross the United States from Oregon to South Carolina.

*There are no total or partial solar eclipses in 2011, 2014, or 2018.

**Times are Greenwich Mean Time. Subtract 5 hours for Eastern Standard time, 6 hours for Central Standard Time, 7 hours for Mountain Standard Time, and 8 hours for Pacific Standard Time.

$^h$ hours.

Total solar eclipses are rare as seen from any one place. If you stay in one location, you will see a total solar eclipse about once in 360 years. Some people are eclipse chasers. They plan years in advance and travel halfway around the world to place themselves in the path of totality. ■ Table 3-2 shows the date and location of solar eclipses over the next few years.

## Features of Solar Eclipses

A solar eclipse begins when you first see the edge of the moon encroaching on the sun. This is the moment when the edge of the penumbra sweeps over your location.

During the partial phases of a solar eclipse, the moon gradually covers the bright disk of the sun (■ Figure 3-10). Totality begins as the last sliver of the sun's bright surface disappears behind the moon. This is the moment when the edge of the umbra sweeps over your location. So long as any of the sun is visible, the countryside is bright; but, as the last of the sun disappears, dark falls in a few seconds. Automatic streetlights come on, car drivers switch on their headlights, and birds go to roost. The darkness of totality depends on a number of factors, including the weather at the observing site, but it is usually dark enough to make it difficult to read the settings on cameras.

The totally eclipsed sun is a spectacular sight. With the moon covering the bright surface of the sun, called the **photosphere,*** you can see the sun's faint outer atmosphere, the **corona,** glowing with a pale, white light so faint you can safely look at it directly. The corona is made of hot, low-density gas that is given a wispy appearance by the solar magnetic field, as shown in the last frame of Figure 3-10. Also visible just above the photosphere is a thin layer of bright gas called the **chromosphere.** The chromosphere is often marked by eruptions on the solar surface, called **prominences** (■ Figure 3-11a), that glow with a clear, pink color due to the high temperature of the gases involved. The small-angle formula reveals that a large prominence is about 3.5 times the diameter of Earth.

Totality during a solar eclipse cannot last longer than 7.5 minutes under any circumstances, and the average is only 2 to 3 minutes. Totality ends when the sun's bright surface reappears at the trailing edge of the moon. Daylight returns quickly, and the corona and chromosphere vanish. This corresponds to the moment when the trailing edge of the moon's umbra sweeps over the observer.

Just as totality begins or ends, a small part of the photosphere can peek through a valley at the edge of the lunar disk. Although it is intensely bright, such a small part of the photosphere does not completely drown out the fainter corona, which forms a silvery ring of light with the brilliant spot of photosphere gleaming like a diamond (■ Figure 3-11b). This **diamond ring effect** is one of the most spectacular of astronomical sights, but it is not visible during every solar eclipse. Its occurrence depends on the exact orientation and motion of the moon.

## Observing an Eclipse

Not too many years ago, astronomers traveled great distances to forbidding places to get their instruments into the path of totality so they could study the faint outer corona that is visible only during the few minutes of a total solar eclipse. Now, many of those observations can be made every day by solar telescopes in

*The photosphere, corona, chromosphere, and prominences will be discussed in detail in Chapter 8. Here the terms are used as the names of features you see during a total solar eclipse.

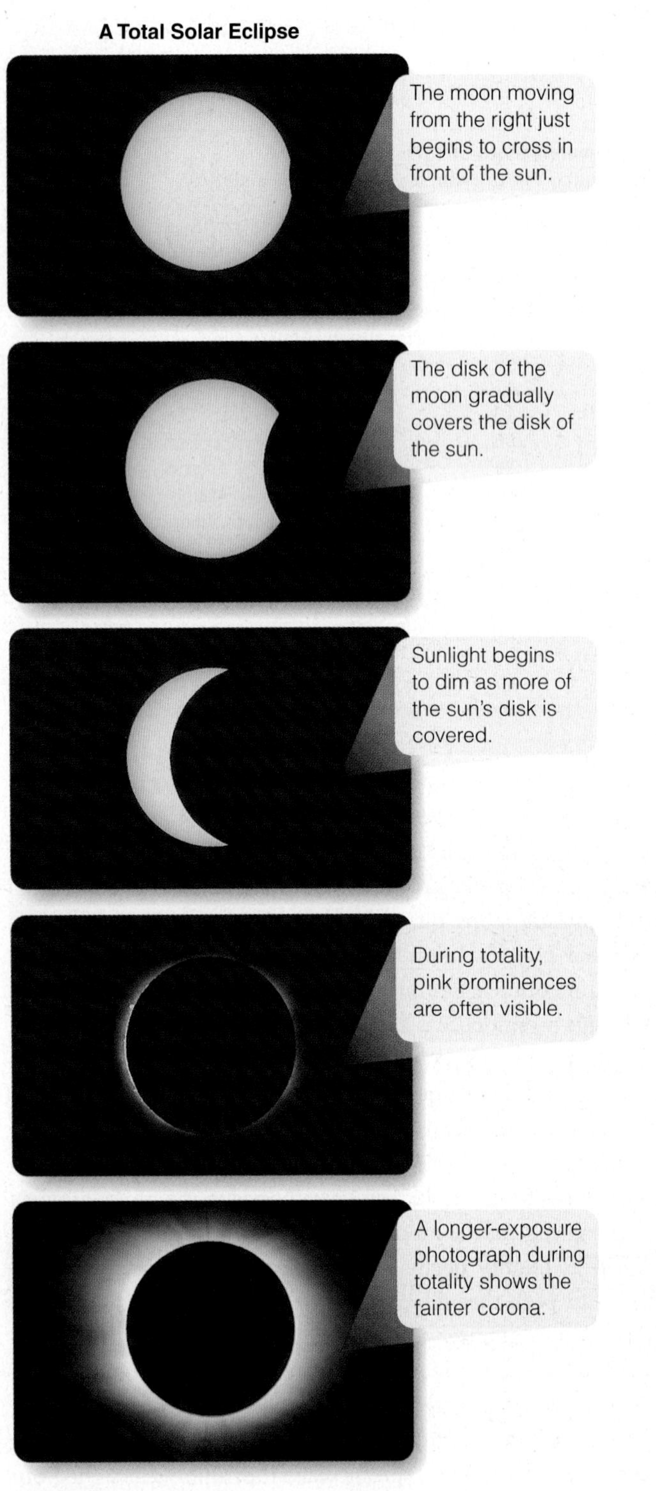
A Total Solar Eclipse
The moon moving from the right just begins to cross in front of the sun.
The disk of the moon gradually covers the disk of the sun.
Sunlight begins to dim as more of the sun's disk is covered.
During totality, pink prominences are often visible.
A longer-exposure photograph during totality shows the fainter corona.

**■ Figure 3-10**

This sequence of photos shows the first half of a total solar eclipse. (Daniel Good)

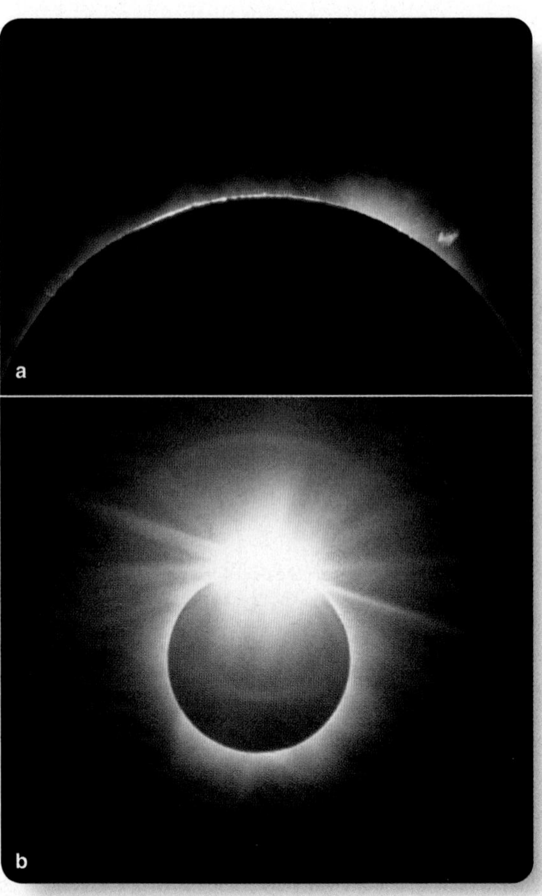
a
b

**■ Figure 3-11**

(a) During a total solar eclipse, the moon covers the photosphere, and the ruby-red chromosphere and prominences are visible. Only the lower corona is visible in this image. (©2005 by Fred Espenak) (b) The diamond ring effect can sometimes occur momentarily at the beginning or end of totality if a small segment of the photosphere peeks out through a valley at the edge of the lunar disk. (National Optical Astronomy Observatory)

space, but eclipse enthusiasts still journey to exotic places for the thrill of seeing a total solar eclipse.

No matter how thrilling a solar eclipse is, you must be cautious when viewing it. During the partial phase, part of the sun remains visible, so it is hazardous to look at the eclipse without protection. Dense filters and exposed film do not necessarily provide protection because some filters do not block the invisible heat radiation (infrared) that can burn the retina of your eyes. Dangers like these have led officials to warn the public not to look at solar eclipses at all and have even frightened some people into locking themselves and their children into windowless rooms. It is a **Common Misconception** that sunlight is somehow more dangerous during an eclipse. In fact, it is always dangerous to look at the sun. The danger posed by an eclipse is that people are tempted to ignore common sense and look at the sun directly, which can burn their eyes.

The safest and simplest way to observe the partial phases of a solar eclipse is to use pinhole projection. Poke a small pinhole in a sheet of cardboard. Hold the sheet with the hole in sunlight and allow light to pass through the hole and onto a second sheet of cardboard (■ Figure 3-12). On a day when there is no eclipse, the result is a small, round spot of light that is an image of the sun. During the partial phases of a solar eclipse, the image will show the dark silhouette of the moon obscuring part of the sun. Pinhole

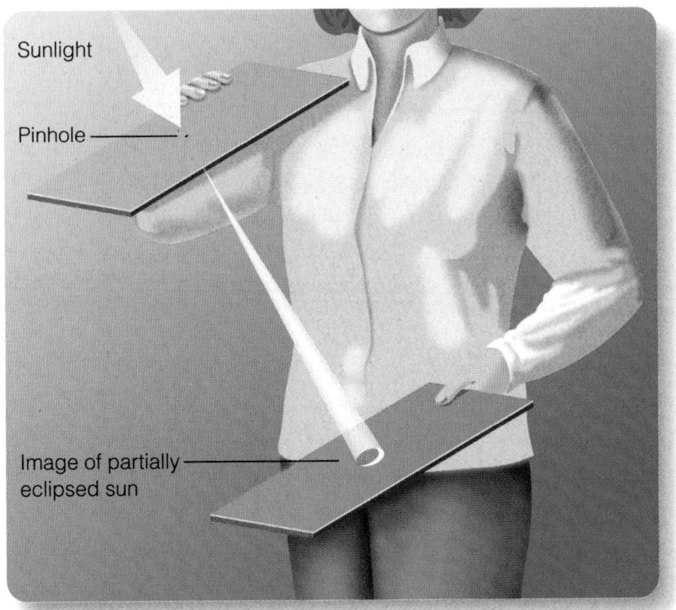


**Figure 3-12**

A safe way to view the partial phases of a solar eclipse. Use a pinhole in a card to project an image of the sun on a second card. The greater the distance between the cards, the larger (and fainter) the image will be.

images of the partially eclipsed sun can also be seen in the shadows of trees as sunlight peeks through the tiny openings between the leaves and branches. This can produce an eerie effect just before totality as the remaining sliver of sun produces thin crescents of light on the ground under trees. Once totality begins, it is safe to look directly. The totally eclipsed sun is fainter than the full moon.

### SCIENTIFIC ARGUMENT

***If you were on Earth watching a total solar eclipse, what would astronauts on the moon see when they looked at Earth?***
Building this argument requires that you change your point of view and imagine seeing the eclipse from a new location. Scientists commonly imagine seeing events from multiple points of view as a way to develop and test their understanding. Astronauts on the moon could see Earth only if they were on the side that faces Earth. Because solar eclipses always happen at new moon, the near side of the moon would be in darkness, and the far side of the moon would be in full sunlight. The astronauts on the near side of the moon would be standing in darkness, and they would be looking at the fully illuminated side of Earth. They would see Earth at full phase. The moon's shadow would be crossing Earth; and, if the astronauts looked closely, they might be able to see the spot of darkness where the moon's umbral shadow touched Earth. It would take hours for the shadow to cross Earth.

Standing on the moon and watching the moon's umbral shadow sweep across Earth would be a cold, tedious assignment. Perhaps you can imagine a more interesting assignment for the astronauts. **What would astronauts on the moon see while people on Earth were seeing a total lunar eclipse?**

## 3-4 Predicting Eclipses

A Chinese story tells of two astronomers, Hsi and Ho, who were too drunk to predict the solar eclipse of October 22, 2137 BC. Or perhaps they failed to conduct the proper ceremonies to scare away the dragon that, according to Chinese tradition, was snacking on the sun's disk. When the emperor recovered from the terror of the eclipse, he had the two astronomers beheaded.

Making exact eclipse predictions requires a computer and proper software, but some ancient astronomers could make educated guesses as to which full moons and which new moons might result in eclipses. There are three good reasons to reproduce their methods. First, it is an important chapter in the history of science. Second, it will illustrate how apparently complex phenomena can be analyzed in terms of cycles. Third, eclipse prediction will exercise your scientific imagination and help you visualize Earth, the moon, and the sun as objects moving through space.

### Conditions for an Eclipse

You can predict eclipses by thinking about the motion of the sun and moon in the sky. Imagine that you can look up into the sky from your home on Earth and see the sun moving along the ecliptic and the moon moving along its orbit. Because the orbit of the moon is tipped slightly over 5 degrees to the plane of Earth's orbit, you see the moon follow a path tipped by the same angle to the ecliptic. Each month, the moon crosses the ecliptic at two points called **nodes.** It crosses at one node going southward, and two weeks later it crosses at the other node going northward.

Eclipses can only occur when the sun is near one of the nodes of the moon's orbit. Only then can the new moon cross in front of the sun and produce a solar eclipse, as shown in ■ Figure 3-13a. Most new moons pass too far north or too far south of the sun to cause an eclipse. When the sun is near one node, Earth's shadow points near the other node, and a full moon can enter the shadow and be eclipsed. A lunar eclipse doesn't happen at every full moon because most full moons pass too far north or too far south of the ecliptic and miss Earth's shadow. You can see a partial lunar eclipse illustrated in ■ Figure 3-13b.

So, there are two conditions for an eclipse: The sun must be near a node, and the moon must be crossing either the same node (solar eclipse) or the other node (lunar eclipse). That means, of course, that solar eclipses can occur only when the moon is new, and lunar eclipses can occur only when the moon is full.

Now you know the ancient secret of predicting eclipses. An eclipse can occur only in a period called an **eclipse season,** during which the sun is close to a node in the moon's orbit. For solar eclipses, an eclipse season is about 32 days long. Any new moon during this period will produce a solar eclipse. For lunar eclipses, the eclipse season is a bit shorter, about 22 days. Any full moon in this period will encounter Earth's shadow and be eclipsed.

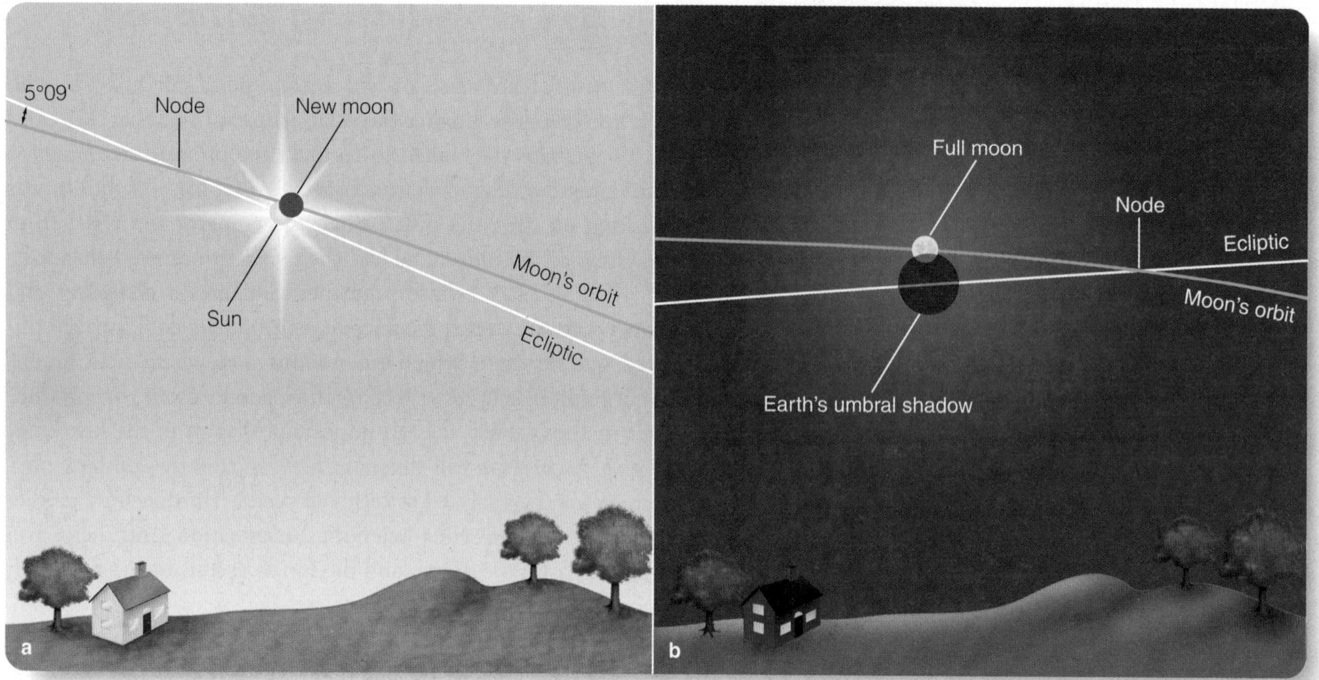


**■ Figure 3-13**

Eclipses can occur only when the sun is near one of the nodes of the moon's orbit. (a) A solar eclipse occurs when the moon meets the sun near a node. (b) A lunar eclipse occurs when the sun and moon are near opposite nodes. Partial eclipses are shown here for clarity.

This makes eclipse prediction easy. All you have to do is keep track of where the moon crosses the ecliptic (where the nodes of its orbit are). This system works fairly well, and ancient astronomers such as the Maya may have used such a system. You could have been a very successful ancient Mayan astronomer with what you know about eclipse seasons, but you can do even better if you change your point of view.

## The View from Space

Change your point of view and imagine that you are looking at the orbits of Earth and the moon from a point far away in space. You can imagine the moon's orbit tipped at an angle to Earth's orbit. As Earth orbits the sun, the moon's orbit remains fixed in direction. The nodes of the moon's orbit are the points where it passes through the plane of Earth's orbit; an eclipse season occurs each time the line connecting these nodes, the **line of nodes,** points toward the sun. Look at ■ Figure 3-14a and notice that the line of nodes does not point at the sun in the example at lower left, and no eclipses are possible; the shadows miss. At lower right, the line of nodes points toward the sun, and the shadows produce eclipses.

The shadows of Earth and moon are long and thin, as shown in ■ Figure 3-14b. That is why it is so easy for them to miss their mark at new moon or full moon and fail to produce an eclipse. Only during an eclipse season do the long, skinny shadows produce eclipses.

If you watched for years from your point of view in space, you would see the orbit of the moon precess like a hubcap spinning on the ground. This precession is caused mostly by the gravitational influence of the sun, and it makes the line of nodes rotate once every 18.6 years. People back on Earth see the nodes slipping westward along the ecliptic 19.4° per year. Consequently, the sun does not need a full year to go from a node all the way around the ecliptic and back to that same node. Because the node is moving westward to meet the sun, it will cross the node after only 346.6 days (an **eclipse year**). This means that, according to our calendar, the eclipse seasons begin about 19 days earlier every year (■ Figure 3-15). If you see an eclipse in late December one year, you will see eclipses in early December the next year, and so on.

Once you see a few eclipses, you know when the eclipse seasons are occurring, and you can predict next year's eclipse seasons by subtracting 19 days. New moons and full moons near those dates are candidates for causing eclipses.

The cyclic pattern of eclipses shown in Figure 3-15 gives you another way to predict eclipses. Eclipses follow a pattern, and if you were an ancient astronomer and understood the pattern, you could predict eclipses without ever knowing what the moon was or how an orbit works.

## The Saros Cycle

Ancient astronomers could predict eclipses in an approximate way using eclipse seasons, but they could have been much more accurate if they had recognized that eclipses occur following certain patterns. The most important of these is the **saros cycle**

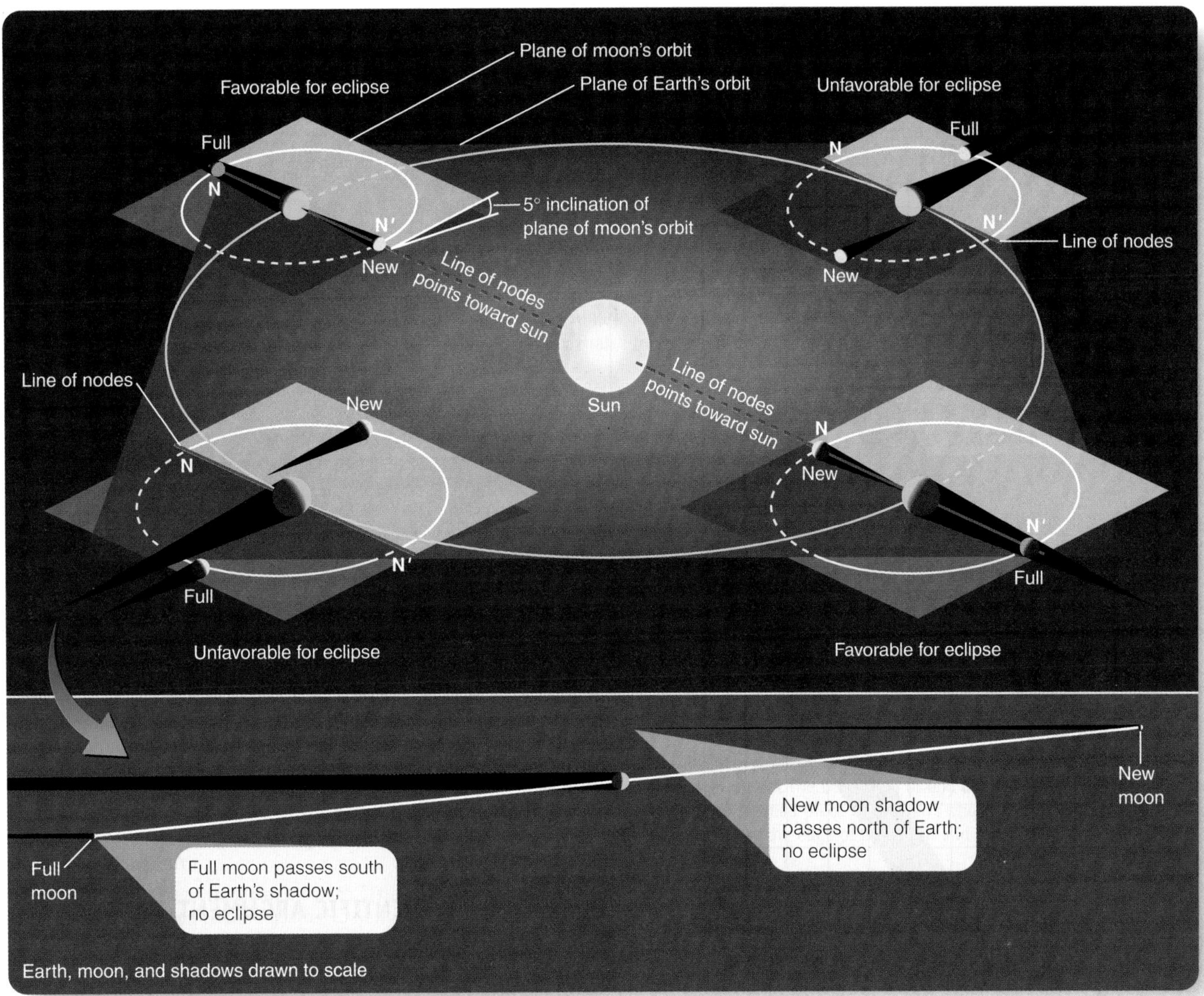


**■ Figure 3-14**

The moon's orbit is tipped a bit over 5° to Earth's orbit. The nodes N and N′ are the points where the moon passes through the plane of Earth's orbit. If the line of nodes does not point at the sun, the long narrow shadows miss, and there are no eclipses at new moon and full moon. At those parts of Earth's orbit where the line of nodes points toward the sun, eclipses are possible at new moon and full moon.

(sometimes referred to simply as the saros). After one saros cycle of 18 years $11\frac{1}{3}$ days, the pattern of eclipses repeats. In fact, *saros* comes from a Greek word that means "repetition."

One saros contains 6585.321 days, which is equal to 223 lunar months. Therefore, after one saros cycle, the moon is back to the same phase it had when the cycle began. But one saros is also equal to 19 eclipse years. After one saros cycle, the sun has returned to the same place it occupied with respect to the nodes of the moon's orbit when the cycle began. If an eclipse occurs on a given day, then 18 years $11\frac{1}{3}$ days later the sun, the moon, and the nodes of the moon's orbit return to nearly the same relationship, and the eclipse occurs all over again.

Although the eclipse repeats almost exactly, it is not visible from the same place on Earth. The saros cycle is one-third of a day longer than 18 years 11 days. When the eclipse happens again, Earth will have rotated one-third of a turn farther east, and the eclipse will occur one-third of the way westward around Earth (■ Figure 3-16). That means that after three saros cycles—a period of 54 years 1 month—the same eclipse occurs in the same part of Earth.

One of the most famous predictors of eclipses was the Greek philosopher Thales of Miletus (about 640–546 BC), who supposedly learned of the saros cycle from the Chaldeans. No one knows which eclipse Thales predicted, but some scholars suspect the eclipse of May 28, 585 BC. In any case, the eclipse occurred at the height of a battle between the Lydians and the Medes, and the mysterious darkness in midafternoon so startled the two armies that they concluded a truce.

## Review Questions

1. Which lunar phases would be visible in the sky at dawn? At midnight?
2. If you looked back at Earth from the moon, what phase would Earth have when the moon was full? New? A first-quarter moon? A waxing crescent?
3. If a planet has a moon, must that moon go through the same phases that Earth's moon displays?
4. Could a solar-powered spacecraft generate any electricity while passing through Earth's umbral shadow? Through Earth's penumbral shadow?
5. If a lunar eclipse occurred at midnight, where in the sky would you look to see it?
6. Why do solar eclipses happen only at new moon? Why not every new moon?
7. Why isn't the corona visible during partial or annular solar eclipses?
8. Why can't the moon be eclipsed when it is halfway between the nodes of its orbit?
9. Why aren't solar eclipses separated by one saros cycle visible from the same location on Earth?
10. How could Thales of Miletus have predicted the date of a solar eclipse without observing the location of the moon in the sky?
11. **How Do We Know?** Some people think science is like a grinder that cranks data into theories. What would you tell them about the need for scientists to be creative and imaginative?

## Discussion Questions

1. How would eclipses be different if the moon's orbit were not tipped with respect to the plane of Earth's orbit?
2. Are there other planets in our solar system from whose surface you could see a "lunar" (moon) eclipse? A total solar eclipse? Which ones and why?

## Problems

1. Identify the phases of the moon if on March 20 the moon is located at the point on the ecliptic called (a) the vernal equinox, (b) the autumnal equinox, (c) the summer solstice, (d) the winter solstice.
2. Identify the phases of the moon if at sunset the moon is (a) near the eastern horizon, (b) high in the southern sky, (c) in the southeastern sky, (d) in the southwestern sky.
3. About how many days must elapse between first-quarter moon and third-quarter moon?
4. If on March 1 the full moon is near the Favorite Star Spica, when will the moon next be full? When will it next be near Spica?
5. How many times larger than the moon is the diameter of Earth's umbral shadow at the moon's distance? (*Hint:* See the photo in Figure 3-3.)
6. Use the small-angle formula to calculate the angular diameter of Earth as seen from the moon.
7. During solar eclipses, large solar prominences are often seen extending 5 arc minutes from the edge of the sun's disk. How far is this in kilometers? In Earth diameters?
8. If a solar eclipse occurs on October 3: (a) Why can't there be a lunar eclipse on October 13? (b) Why can't there be a solar eclipse on December 28?
9. A total eclipse of the sun was visible from Canada on July 10, 1972. When did eclipse occur next with the same sun-moon-Earth geometry? From what part of Earth was it total?
10. When will the eclipse described in Problem 9 next be total as seen from Canada?
11. When will the eclipse seasons occur during the current year? What eclipse(s) will occur?

## Learning to Look

1. Use the photos in Figure 3-1 as evidence to show that the moon always keeps the same side facing Earth.
2. Draw the umbral and penumbral shadows onto the diagram in the middle of page 34. Use the diagram to explain why lunar eclipses can occur only at full moon and solar eclipses can occur only at new moon.
3. Can you detect the saros cycle in Figure 3-15?
4. The stamp at right shows a crescent moon. Explain why the moon could never look this way.

5. The photo at right shows the annular eclipse of May 30, 1984. How is it different from the annular eclipse shown in Figure 3-9?

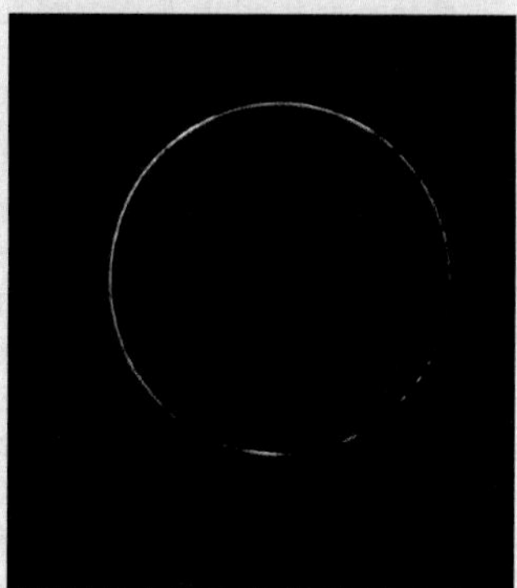

# 4 The Origin of Modern Astronomy

Galileo's telescope revealed such things as craters on the moon, and he explained how that evidence could be used to test the prevailing Earth-centered model of the universe. He was condemned by the Inquisition in 1633. (*Science* magazine [www.sciencemag.org, 16 January 2009, page 326 [top].)

## Guidepost

The preceding chapters gave you a modern view of Earth. You can now imagine how Earth, the moon, and the sun move through space and how that produces the sights you see in the sky. But how did humanity first realize that we live on a planet moving through space? That required the revolutionary overthrow of an ancient and honored theory of Earth's place.

By the 16th century, many astronomers were uncomfortable with the ancient theory that Earth sat at the center of a spherical universe. In this chapter, you will discover how a Polish astronomer named Nicolaus Copernicus changed the old theory, how a German astronomer named Johannes Kepler discovered the laws of planetary motion, and how the Italian Galileo Galilei changed the way we know about nature. Here you will find answers to four essential questions:

- **How did classical philosophers describe Earth's place in the universe?**
- **How did Copernicus revise that ancient theory?**
- **How did astronomers discover the laws of planetary motion?**
- **Why was Galileo condemned by the Inquisition?**

This chapter is not just about the history of astronomy. As they struggled to understand Earth and the heavens, the astronomers of the Renaissance invented a new way of understanding nature—a way of thinking that is now called science.

*How you would burst out laughing, my dear Kepler, if you would hear what the greatest philosopher of the Gymnasium told the Grand Duke about me...*

— FROM A LETTER BY GALILEO GALILEI

ABOUT FOUR CENTURIES AGO, Galileo was condemned by the Inquisition for his part in a huge controversy over the nature of the universe, a controversy that focused on two problems. The place of the Earth was the most acrimonious issue; was it the center of the universe or was the sun at the center? A related issue was the nature of planetary motion. Ancient astronomers could see the sun, moon, and planets moving along the ecliptic, but they couldn't describe those motions precisely. To solve the first problem of the place of the Earth, astronomers also had to solve the second problem of planetary motion.

## 4-1 The Roots of Astronomy

ASTRONOMY HAS ITS ORIGIN in a noble human trait, curiosity. Just as modern children ask their parents what the stars are and why the moon changes, so did ancient humans ask themselves those same questions. Their answers, often couched in mythical or religious terms, reveal great reverence for the order of the heavens.

### Archaeoastronomy

Most of the history of astronomy is lost forever. You can't go to a library or search the Internet to find out what the first astronomers thought about their world because they left no written record. The study of the astronomy of ancient peoples has been called **archaeoastronomy,** and it clearly shows that trying to understand the heavens is part of human nature.

Perhaps the best-known example of archaeoastronomy is also a huge tourist attraction. Stonehenge, standing on Salisbury Plain in southern England, was built in stages from about 3000 BC to about 1800 BC, a period extending from the late Stone Age into the Bronze Age. Though the public is most familiar with the monument's massive stones, they were added late in its history. In its first stages, Stonehenge consisted of a circular ditch slightly larger in diameter than the length of a football field, with a concentric bank just inside the ditch and a long avenue leading away toward the northeast. A massive stone, the Heelstone, stood then, as it does now, outside the ditch in the opening of the avenue.

As early as AD 1740, the English scholar W. Stukely suggested that the avenue pointed toward the rising sun at the summer solstice, but few historians accepted that it was intentional. Nevertheless, seen from the center of the monument, the summer solstice sun does rise behind the Heelstone. More recently, astronomers have recognized other significant astronomical alignments at Stonehenge. For example, sight lines point toward the most northerly and most southerly risings of the moon (■ Figure 4-1).

The significance of these alignments has been debated. Some have claimed that the Stone Age people who built Stonehenge were using it as a device to predict lunar eclipses. After studying eclipse prediction in the previous chapter, you understand that predicting eclipses is easier than most people assume, so perhaps it was used in that way, but the truth may never be known. The builders of Stonehenge had no written language and left no records of their intentions. Nevertheless, the presence of solar and lunar alignments at Stonehenge and at many other Stone Age monuments dotting England and continental Europe shows that so-called primitive peoples were paying close attention to the sky. The roots of astronomy lie not in sophisticated science and logic but in human curiosity and wonder.

Astronomical alignments in sacred structures are common all around the world. For example, many tombs are oriented toward the rising sun, and Newgrange, a 5000-year-old passage-grave in Ireland (■ Figure 4-2), faces southeast so that, at dawn on the day of the winter solstice, light from the rising sun shines into its long passageway and illuminates the central chamber. No one today knows what the alignment meant to the builders of Newgrange, and many experts doubt that it was actually intended to be a tomb. Whatever its original purpose, Newgrange is clearly a sacred site linked by its alignment to the order and power of the sky.

Building astronomical alignments into structures gives them meaning by connecting them with the heavens. Navajo Indians of the American southwest, for example, have for centuries built their hogans with the door facing east so they can greet the rising sun each morning with prayers and offerings.

Some alignments may have served calendrical purposes. The 2000-year-old Temple of Isis in Dendera, Egypt, was build to align with the rising point of the bright star Sirius. The first appearance of this star in the dawn twilight marked the flooding of the Nile, so it was an important calendrical indicator. The link between Sirius and the Nile was described in Egyptian mythology; the goddess Isis was associated with the star Sirius, and her husband, Osiris, was linked to the constellation you know as Orion and also to the Nile, the source of Egypt's agricultural fertility.

An intriguing American site in New Mexico known as the Sun Dagger unfortunately has no surviving mythology to tell its

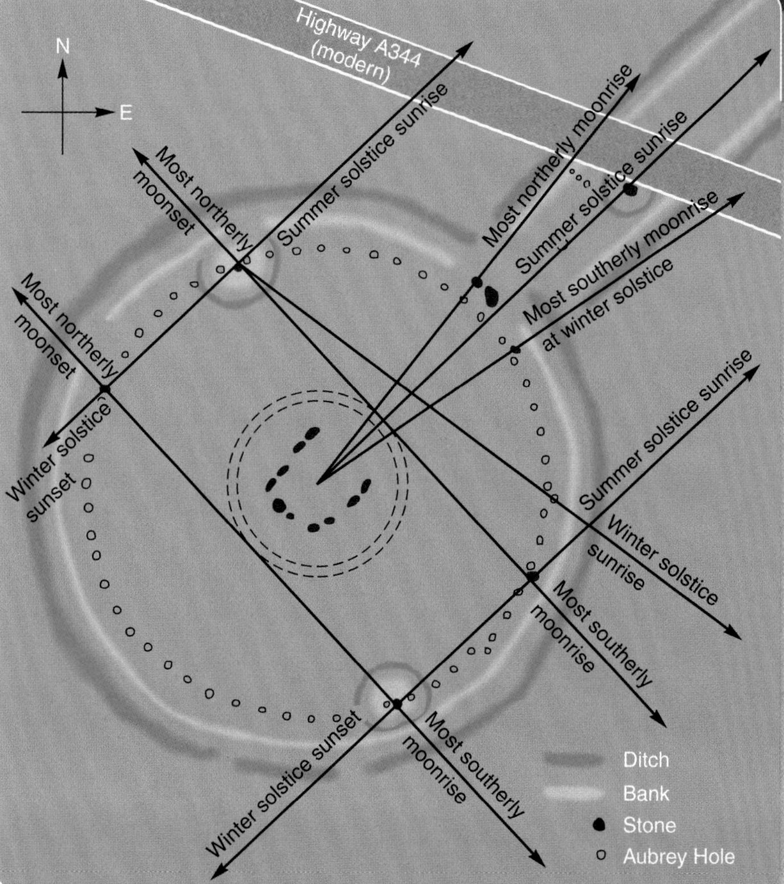


## ■ Figure 4-1

The central horseshoe of upright stones is only the most obvious part of Stonehenge. The best-known astronomical alignment at Stonehenge is the summer solstice sun rising over the Heelstone. Although a number of astronomical alignments have been found at Stonehenge, experts debate their significance. (Photo: Jamie Backman)

## ■ Figure 4-2

Newgrange was built on a small hill in Ireland about 3200 BC. A long passageway extends from the entryway back to the center of the mound, and sunlight shines down the passageway into the central chamber at dawn on the day of the winter solstice. Other passage graves have similar alignments, but their purpose is unknown. (Newgrange: Benelux Press/Index Stock Imagery)

High on Fajada Butte, the Sun Dagger is off limits to visitors.
The spiral pattern is the size of a dinner plate.

■ **Figure 4-3**

In the ancient Native American settlement known as Chaco Canyon, New Mexico, sunlight shines between two slabs of stone high on the side of 440-foot-high Fajada Butte to form a dagger of light on the cliff face. About noon on the day of the summer solstice, the dagger of light slices through the center of a spiral pecked into the sandstone.
(NPS Chaco Culture National Historic Park)

story. At noon on the day of the summer solstice, a narrow dagger of sunlight shines across the center of a spiral carved on a cliff face high above the desert floor (■ Figure 4-3). The purpose of the Sun Dagger is open to debate, but similar examples have been found throughout the American Southwest. It may have been more a symbolic and ceremonial marker than a precise calendrical indicator. In any case, it is just one of the many astronomical alignments that ancient people built into their structures to link themselves with the sky.

Some scholars are examining not ancient structures but small artifacts made thousands of years ago. Scratches on certain bone and stone implements follow a pattern that may record the phases of the moon (■ Figure 4-4). Some scientists contend that humanity's first attempts at writing were stimulated by a desire to record and predict lunar phases.

Archaeoastronomy is uncovering the earliest roots of astronomy and simultaneously revealing some of the first human efforts at systematic inquiry. The most important lesson of archaeoastronomy is that humans don't have to be technologically sophisticated to admire and study the universe. Efforts to understand the sky have been part of human cultures for a long time.

Although the methods of archaeoastronomy can show how ancient people observed the sky, their thoughts about their universe are in many cases lost. Many cultures had no written language. In other cases, the written record has been lost. Dozens, perhaps hundreds, of beautiful Mayan manuscripts, for instance, were burned by Spanish missionaries who believed that the books were the work of Satan. Only four of these books have survived, and all four contain astronomical references. One contains sophisticated tables that allowed the Maya to predict the motion of Venus and eclipses of the moon. No one will ever know what was burnt.

The fate of the Mayan books illustrates one reason why histories of astronomy usually begin with the Greeks. Some of their writing has survived, and from it you can discover what they thought about the shape and motion of the heavens.

## The Astronomy of Greece

Greek astronomy was derived from Babylon and Egypt, but the Greek philosophers took a new approach. Rather than relying on religion and astrology, the Greeks proposed a rational universe whose secrets could be understood through logic and reason. Before you begin your study of the history of

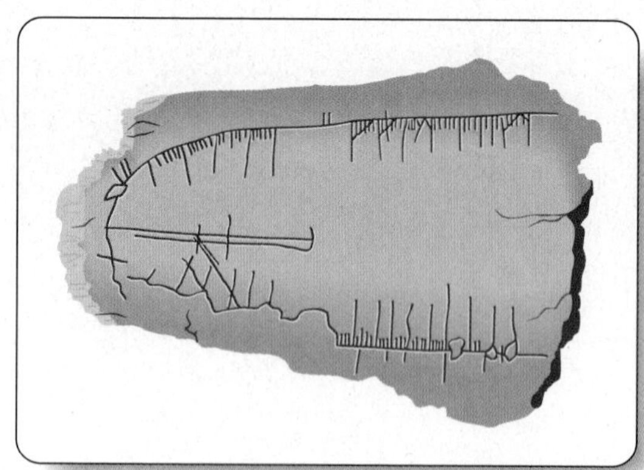

■ **Figure 4-4**

A fragment of a 27,000-year-old mammoth tusk found at Gontzi in Ukraine contains scribe marks on its edge, simplified in this drawing. These markings have been interpreted as a record of four cycles of lunar phases. Although controversial, such finds suggest that some of the first human attempts at recording events in written form were stimulated by astronomical phenomena.

astronomical thought from the classical Greek era through the Renaissance, you should remember the point made in Chapter 1 that the terms *solar system, galaxy,* and *universe* have very different meanings. You know now that our solar system, consisting of Earth, the moon, the sun, Earth's sibling planets, their moons, and so on, is your very local neighborhood, much smaller than the Milky Way Galaxy, which in turn is tiny compared with the observable universe. However, from ancient times up through Copernicus's day, it was thought that the whole universe, everything that exists, did not extend much beyond the farthest planet of our solar system. Asking back then whether Earth or the sun is the center of the solar system was the same question as asking whether Earth or the sun is the center of the universe. As you read this chapter, but only this chapter, you can pretend to have the old-fashioned view in which "solar system" and "universe" meant much the same thing.

This new Greek attitude toward studying the heavens with logic and reason was a first step toward modern science, and it was made possible by two early Greek philosophers. Thales of Miletus (c. 624–547 BC) lived and worked in what is now Turkey. He taught that the universe is rational and that the human mind can understand why the universe works the way it does. This view contrasts sharply with that of earlier cultures, which believed that the ultimate causes of things are mysteries beyond human understanding. To Thales and his followers, the mysteries of the universe were mysteries only because they were unknown, not because they were unknowable.

The other philosopher who made the new scientific attitude possible was Pythagoras (c. 570–500 BC). He and his students noticed that many things in nature seem to be governed by geometrical or mathematical relations. Musical pitch, for example, is related in a regular way to the lengths of plucked strings. This led Pythagoras to propose that all nature was underlain by musical principles, by which he meant mathematics. One result of this philosophy was the later belief that the harmony of the celestial movements produced actual music, the music of the spheres. But, at a deeper level, the teachings of Pythagoras made Greek astronomers look at the universe in a new way. Thales said that the universe could be understood, and Pythagoras said that the underlying rules were mathematical.

In trying to understand the universe, Greek astronomers did something that Babylonian astronomers had never done—they tried to describe the universe using geometrical forms. Anaximander (c. 611–546 BC) described a universe made up of wheels filled with fire: The sun and moon were holes in the wheels through which the flames could be seen. Philolaus (fifth century BC) argued that Earth moved in a circular path around a central fire (not the sun), which was always hidden behind a counter-Earth located between the fire and Earth. This, by the way, was the first theory to suppose that Earth is in motion.

The great philosopher Plato (428–347 BC) was not an astronomer, but his teachings influenced astronomy for 2000 years. Plato argued that the reality humans see is only a distorted shadow of a perfect, ideal form. If human observations are distorted, then observation can be misleading, and the best path to truth, said Plato, is through pure thought on the ideal forms that underlie nature.

Plato argued that the most perfect geometrical form was the sphere, and therefore, he said, the perfect heavens must be made up of spheres rotating at constant rates and carrying objects around in circles. Consequently, later astronomers tried to describe the motions of the heavens by imagining multiple rotating spheres. This became known as the principle of **uniform circular motion.**

Eudoxus of Cnidus (409–356 BC), a student of Plato, applied this principle when he devised a system of 27 nested spheres that rotated at different rates about different axes to produce a mathematical description of the motions of the universe (■ Figure 4-5).

At the time of the Greek philosophers, it was common to refer to systems such as that of Eudoxus as descriptions of the world, where the word *world* included not only Earth but all of the heavenly spheres. The reality of these spheres was open to debate. Some thought of the spheres as nothing more than mathematical ideas that described motion in the world model, while others began to think of the spheres as real objects made of perfect celestial material. Aristotle, for example, seems to have thought of the spheres as real.

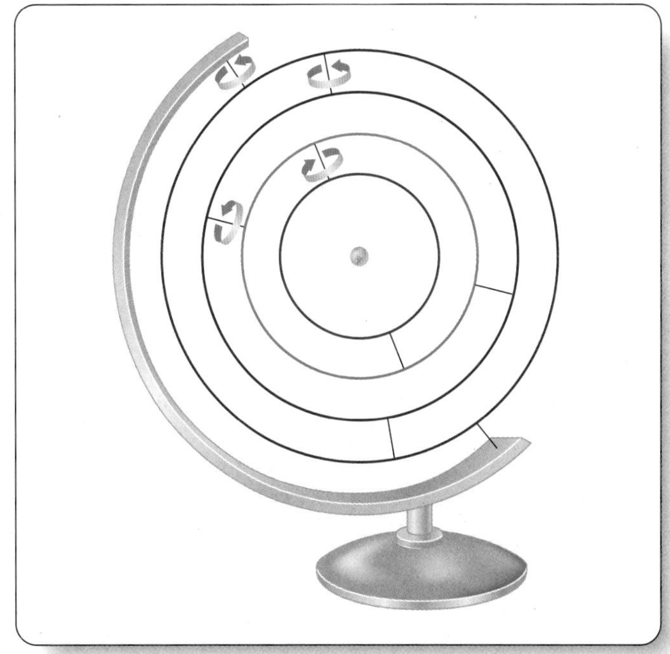

**■ Figure 4-5**

The spheres of Eudoxus explain the motions in the heavens by means of nested spheres rotating about various axes at different rates. Earth is located at the center. In this illustration, only four of the 27 spheres are shown.

## Aristotle and the Nature of Earth

Aristotle (384–322 BC), another of Plato's students, made his own unique contributions to philosophy, history, politics, ethics, poetry, drama, and other subjects (■ Figure 4-6). Because of his sensitivity and insight, he became the greatest authority of antiquity, and his astronomical model was accepted with minor variations for almost 2000 years.

Much of what Aristotle wrote about scientific subjects was wrong, but that is not surprising. The scientific method, with its insistence on evidence and hypothesis, had not yet been invented. Aristotle, like other philosophers of his time, attempted to understand his world by reasoning logically and carefully from first principles. A first principle is something that is held to be obviously true. The perfection of the heavens, meaning that the heavens had to be composed of rotating perfect spheres, was, for Aristotle, a first principle. Once a principle is recognized as true, whatever can be logically derived from it must also be true.

Aristotle believed that the universe was divided into two parts: Earth, imperfect and changeable; and the heavens, perfect and unchanging. Like most of his predecessors, he believed that Earth was the center of the universe, so his model is called a **geocentric universe.** The heavens surrounded Earth, and he added to the model proposed by Eudoxus to bring the total to 55 crystalline spheres turning at different rates and at different angles to carry the sun, moon, and planets across the sky. The lowest sphere, that of the moon, marked the boundary between the changeable imperfect region of Earth and the unchanging perfection of the celestial realm above the moon.

Because he believed Earth to be immobile, Aristotle had to make this entire nest of spheres whirl westward around Earth every 24 hours to produce day and night. The spheres also had to move more slowly with respect to one another to produce the motions of the sun, moon, and planets against the background of the stars. Because his model was geocentric, he taught that Earth could be the only center of motion, meaning that all of the whirling spheres had to be centered on Earth.

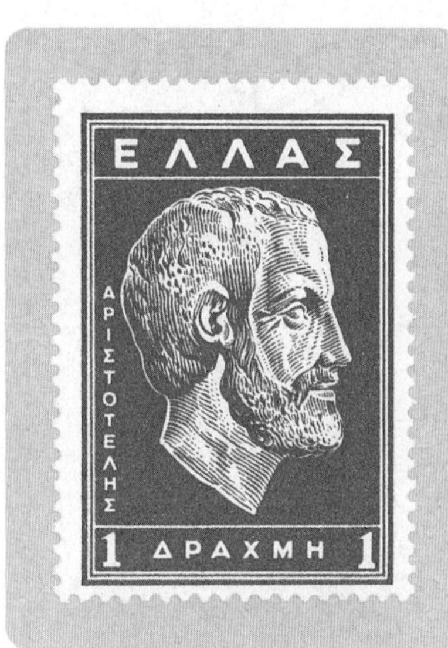


■ **Figure 4-6**

Aristotle, honored on this Greek stamp, wrote on such a wide variety of subjects and with such deep insight that he became the great authority on all matters of learning. His opinions on the nature of Earth and the sky were widely accepted for almost two millennia.

About a century after Aristotle, the Alexandrian philosopher Aristarchus proposed that Earth rotated on its axis and revolved around the sun. This theory is, of course, generally correct, but most of the writings of Aristarchus were lost, and his theory was not well known. Later astronomers rejected any suggestion that Earth could move, because it conflicted with the teachings of the great philosopher Aristotle.

Aristotle had taught that Earth had to be a sphere because it always casts a round shadow during lunar eclipses, but he could only estimate its size. About 200 BC, Eratosthenes, working in the great library in the Egyptian city of Alexandria, found a way to measure Earth's radius. He learned from travelers that the city of Syene (Aswan) in southern Egypt contained a well into which sunlight shone vertically on the day of the summer solstice. This told him that the sun was at the zenith at Syene; but, on that same day in Alexandria, he noted that the sun was 1/50 of the circumference of the sky (about 7°) south of the zenith.

Because sunlight comes from such a great distance, its rays arrive at Earth traveling almost parallel. That allowed Eratosthenes to use simple geometry to conclude that the distance from Alexandria to Syene was 1/50 of Earth's circumference (■ Figure 4-7).

To find Earth's circumference, Eratosthenes had to learn the distance from Alexandria to Syene. Travelers told him it took 50 days to cover the distance, and he knew that a camel can travel about 100 stadia per day. That meant the total distance was about 5000 stadia. If 5000 stadia is 1/50 of Earth's circumference, then Earth must be 250,000 stadia in circumference, and, dividing by $2\pi$, Eratosthenes found Earth's radius to be 40,000 stadia.

How accurate was Eratosthenes? The stadium (singular of *stadia*) had different lengths in ancient times. If you assume 6 stadia to the kilometer, then Eratosthenes's result was too big by only 4 percent. If he used the Olympic stadium, his result was 14 percent too big. In any case, this was a much better measurement of Earth's radius than Aristotle's estimate, which was much too small, about 40 percent of the true radius.

You might think this is just a disagreement between two ancient philosophers, but it is related to a **Common Misconception**. Christopher Columbus did not have to convince Queen Isabella that the world was round. At the time of Columbus, all educated people knew that the world was round and not flat, but they weren't sure how big it was. Columbus, like many others, adopted Aristotle's diameter for Earth, so he thought Earth was small enough that he could sail west and reach Japan and the Spice Islands of the East Indies. If he had accepted Eratosthenes' diameter, Columbus would never have risked the voyage. He and his crew were lucky that America was in the way; if there had been open ocean all the way to Japan, they would have starved to death long before they reached land.

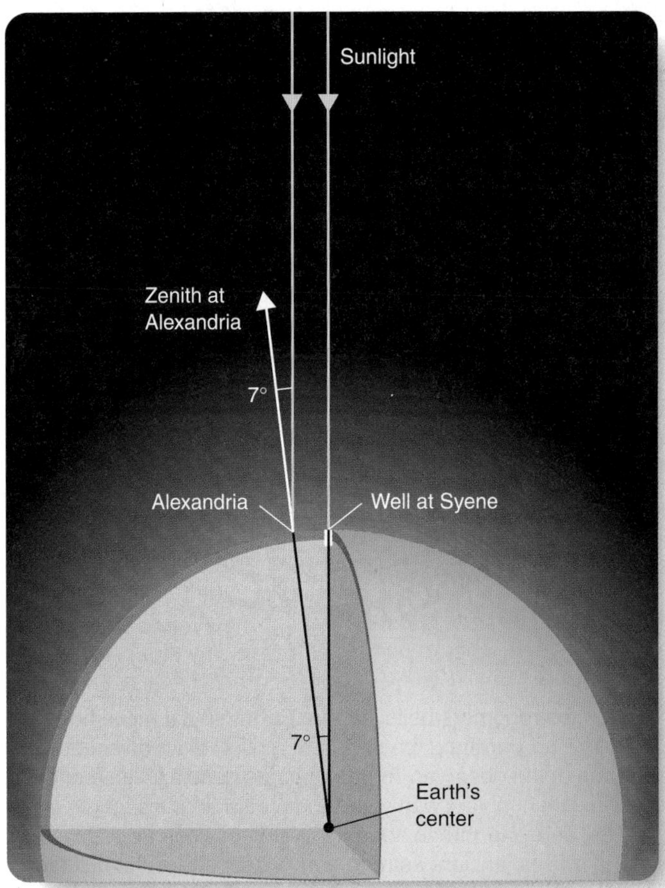


**Figure 4-7**

On the day of the summer solstice, sunlight fell to the bottom of a well at Syene, but the sun was about $\frac{1}{50}$ of a circle (7°) south of the zenith at Alexandria. From this, Eratosthenes was able to calculate Earth's radius.

Aristotle, Aristarchus, and Eratosthenes were philosophers, but the next person you will meet was a real astronomer who observed the sky in detail. Little is known about Hipparchus, who lived during the second century BC, about two centuries after Aristotle. He is usually credited with the invention of trigonometry, the creation of the first star catalog, and the discovery of precession (Chapter 2). Hipparchus also described the motion of the sun, moon, and planets as following circular paths with Earth near, but not at, their centers. These off-center circles are now known as **eccentrics.** Hipparchus recognized that he could reproduce this motion in a model where the celestial bodies traveled around a small circle that followed a larger circle around Earth. The compounded circular motion that he devised became the key element in the masterpiece of the last great astronomer of classical times, Claudius Ptolemy.

## The Ptolemaic Universe

Claudius Ptolemaeus was one of the great astronomer-mathematicians of antiquity. His nationality and birth date are unknown, but around AD 140 he lived and worked in the Greek settlement at Alexandria in what is now Egypt. He ensured the continued acceptance of Aristotle's universe by transforming it into a sophisticated mathematical model.

When you read **The Ancient Universe** on pages 56–57, notice three important ideas and five new terms that show how first principles influenced early descriptions of the universe and it motions:

1. Ancient philosophers and astronomers accepted as first principles that the heavens were geocentric with Earth located at the center and sun, moon, and planets moving in uniform circular motion. It seemed clear to them that Earth was not moving because they saw no *parallax* in the positions of the stars.

2. The observed motion of the planets, the evidence, did not fit the theory very well. The *retrograde motion* of the planets was very difficult to explain using geocentrism and uniform circular motion.

3. Ptolemy attempted to explain the motion of the planets by devising a small circle, an *epicycle,* that rotated along the edge of a larger circle, and the *deferent,* which enclosed a slightly off-center Earth. An *equant* was a point from which the center of an epicycle appeared to move at a constant rate. That meant the speed of the planets would vary slightly as viewed from Earth.

Ptolemy lived roughly five centuries after Aristotle, and although Ptolemy based his work on the Aristotelian universe, he was interested in a different problem—the motion of the planets. He was a brilliant mathematician, and he was mainly interested in creating a mathematical description of the motions he saw in the heavens. For him, first principles took second place to mathematical precision.

Aristotle's universe, as embodied in the mathematics of Ptolemy, dominated ancient astronomy, but it was wrong. The planets don't follow circles at uniform speeds. At first, the Ptolemaic system predicted the positions of the planets well; but, as centuries passed, errors accumulated. If your watch gains only one second a year, it will keep time well for many years, but the error will gradually become noticeable. So, too, did the errors in the Ptolemaic system gradually accumulate as the centuries passed, but, because of the deep respect people had for the writings of Aristotle, the Ptolemaic system was not abandoned. Islamic and later European astronomers tried to update the system, computing new constants and adjusting epicycles. In the middle of the 13th century, a team of astronomers supported by King Alfonso X of Castile studied the *Almagest* for 10 years. Although they did not revise the theory very much, they simplified the calculation of the positions of the planets using the Ptolemaic system and published the result as *The Alfonsine Tables,* the last great attempt to make the Ptolemaic system of practical use.

## SCIENTIFIC ARGUMENT

***How did the astronomy of Hipparchus and Ptolemy violate the principles of the early Greek philosophers Plato and Aristotle?***
Today, scientific arguments depend on evidence and theory; but, in classical times, they started from first principles. Hipparchus and Ptolemy lived very late in the history of classical astronomy, and they concentrated more on the mathematical problems and less on philosophical principles. They replaced the perfect spheres of Plato with nested circles in the form of epicycles and deferents. Earth was moved slightly away from the center of the deferent, so their models of the universe were not exactly geocentric, and the epicycles moved uniformly only as seen from the equant. The celestial motions were no longer precisely uniform, and the principles of geocentrism and uniform circular motion were weakened.

The work of Hipparchus and Ptolemy led eventually to a new understanding of the heavens, but first astronomers had to abandon uniform circular motion. Construct an argument in the classical style based on first principles to answer the following: **Why did Plato argue for uniform circular motion?**

# 4-2 The Copernican Revolution

YOU WOULD NOT HAVE EXPECTED NICOLAUS COPERNICUS to trigger a revolution in astronomy and science. He was born in 1473 to a merchant family in Poland. Orphaned at the age of 10, he was raised by his uncle, an important bishop, who sent him to the University of Cracow and then to the best universities in Italy. There he studied law and medicine before pursuing a lifelong career as an important administrator in the Church. Nevertheless, he had a passion for astronomy (■ Figure 4-8).

■ **Figure 4-8**

Nicolaus Copernicus (1473–1543) pursued a lifetime career in the Church, but he was also a talented mathematician and astronomer. His work triggered a revolution in human thought. These stamps were issued in 1973 to mark the 500th anniversary of his birth.

## The Copernican Model

If you had sat beside Copernicus in his astronomy classes, you would have studied the Ptolemaic universe. The central location of Earth was widely accepted, and everyone knew that the heavens moved in uniform circular motion. For most scholars, questioning these principles was not an option because, over the course of centuries, Aristotle's proposed geometry had become linked with Christian teachings. According to the Aristotelian universe, the most perfect region was in the heavens and the most imperfect at Earth's center. This classical geocentric universe matched the commonly held Christian geometry of heaven and hell, so anyone who criticized the Ptolemaic model was not only questioning Aristotle's geometry but also indirectly challenging belief in heaven and hell.

For this reason, Copernicus probably found it difficult at first to consider alternatives to the Ptolemaic universe. Throughout his life, he was associated with the Catholic Church, which had adopted many of Aristotle's ideas. His uncle was an important bishop in Poland, and, through his uncle's influence, Copernicus was appointed a canon at the cathedral in Frauenberg at the unusually young age of 24. (A canon was not a priest but a Church administrator.) This gave Copernicus an income, although he continued his studies at the universities in Italy. When he left the universities, he joined his uncle and served as his secretary and personal physician until his uncle died in 1512. At that point, Copernicus moved into quarters adjoining the cathedral in Frauenburg, where he served as canon for the rest of his life.

His close connection with the Church notwithstanding, Copernicus began to consider an alternative to the Ptolemaic universe, probably while he was still at university. Sometime before 1514, he wrote an essay proposing a model of a **heliocentric universe** in which the sun, not Earth, was the center of that universe. To explain the daily and annual cycles of the sky, he proposed that Earth rotates on its axis and revolves around the sun. He distributed this commentary in handwritten form, without a title, and in some cases anonymously, to friends and astronomical correspondents. He may have been cautious out of modesty, or out of respect for the Church, or out of fear that his revolutionary ideas would be attacked unfairly. After all, the place of Earth was a controversial theological subject. Although this early essay discusses every major aspect of his later work, it did not include observations and calculations. His ideas needed supporting evidence, so he began gathering observations and making detailed calculations to be published as a book that would demonstrate the truth of his revolutionary idea.

### *De Revolutionibus*

Copernicus worked on his book *De Revolutionibus Orbium Coelestium (The Revolutions of the Celestial Spheres)* over a period of many years and was essentially finished by about 1529; yet he hesitated to publish it even though other astronomers already knew of his theories. Even Church officials, concerned about the reform of the calendar, sought his advice and looked forward to the publication of his book.

One reason he hesitated was that he knew that the idea of a heliocentric universe would be highly controversial. This was a time of rebellion in the Church—Martin Luther (1483–1546) was speaking harshly about fundamental Church teachings, and others, both scholars and scoundrels, were questioning the Church's authority. Even matters as abstract as astronomy could stir controversy. Remember, too, that Earth's place in astronomical theory was linked to the geometry of heaven and hell, so moving Earth from its central place was a controversial and perhaps heretical idea.

Another reason Copernicus may have hesitated to publish was that his work was incomplete. His model could not accurately predict planetary positions, so he continued to refine it. Finally in 1540 he allowed the visiting astronomer Joachim Rheticus (1514–1576) to publish an account of the Copernican universe in Rheticus's book *Prima Narratio (First Narrative).* In 1542, Copernicus sent the manuscript for *De Revolutionibus* off to be printed. He died in the spring of 1543 before the printing was completed.

The most important idea in the book was the location of the sun at the center of the universe. That single innovation had an astonishing consequence—the retrograde motion of the planets was immediately explained in a straightforward way without the large epicycles that Ptolemy had used.

In the Copernican system, Earth moves faster along its orbit than the planets that lie farther from the sun. Consequently, Earth periodically overtakes and passes these planets. To visualize this, imagine that you are in a race car, driving rapidly along the inside lane of a circular racetrack. As you pass slower cars driving in the outer lanes, they fall behind, and if you did not realize you were moving, it would look as if the cars in the outer lanes occasionally slowed to a stop and then backed up for a short interval. ■ Figure 4-9 shows how the same thing happens as Earth passes a planet such as Mars. Although Mars moves steadily along its orbit, as seen from Earth it appears to slow to a stop and move westward (retrograde) as Earth passes it. This happens to any planet whose orbit lies outside Earth's orbit, so the ancient astronomers saw Mars, Jupiter, and Saturn occasionally move retrograde along the ecliptic. Because the planetary orbits do not lie in precisely the same plane, a planet does not resume its eastward motion in precisely the same path it followed earlier. Consequently, it describes a loop whose shape depends on the angle between the orbital planes.

Copernicus could explain retrograde motion without epicycles, and that was impressive. The Copernican system was elegant and simple compared with the whirling epicycles and off-center equants of the Ptolemaic system. You can see Copernicus's own diagram for his heliocentric system in ■ Figure 4-10a. However, *De Revolutionibus* failed in one critical way—the Copernican model could not predict the positions of the planets any more accurately than the Ptolemaic system could. To understand why it failed, you must understand Copernicus and his world.

Copernicus proposed a revolutionary idea when he made the planetary system heliocentric, but he was a classical astronomer with tremendous respect for the old concept of uniform circular motion. In fact, Copernicus objected strongly to Ptolemy's use of the equant. It seemed arbitrary to Copernicus, an obvious violation of the elegance of Aristotle's philosophy of the heavens. Copernicus called equants "monstrous" because they undermined both geocentrism and uniform circular motion. In devising his model, Copernicus demonstrated a strong belief in uniform circular motion.

Although he did not need epicycles to explain retrograde motion, Copernicus quickly discovered that the sun, moon, and planets suffered other small variations in their motions that he could not explain using uniform circular motion centered on the sun. Today astronomers recognize those variations as the result of planets following elliptical orbits, but because Copernicus held firmly to uniform circular motion, he had to introduce small epicycles to try to reproduce these minor variations in the motions of the sun, moon, and planets.

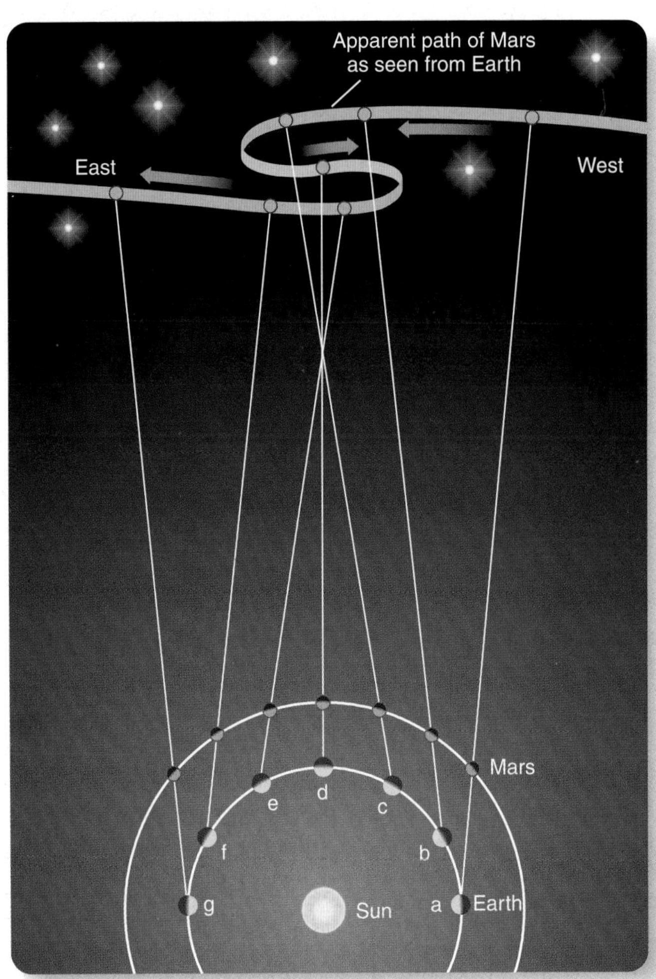


■ **Figure 4-9**

The Copernican explanation of retrograde motion. As Earth overtakes Mars (a–c), Mars appears to slow its eastward motion. As Earth passes Mars (d), Mars appears to move westward. As Earth draws ahead of Mars (e–g), Mars resumes its eastward motion against the background stars. The positions of Earth and Mars are shown at equal intervals of one month.

NICOLAI COPERNICI

net, in quo terram cum orbe lunari tanquam epicyclo contineri diximus. Quinto loco Venus nono menſe reducitur. Sextum denique locum Mercurius tenet, octuaginta dierum ſpacio circū currens. In medio uero omnium reſidet Sol. Quis enim in hoc

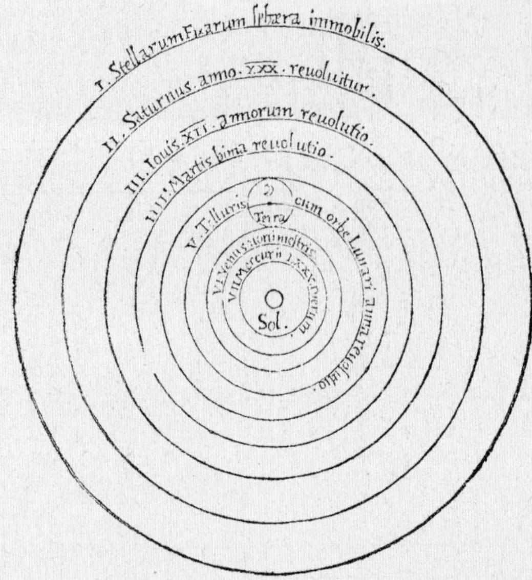


pulcherrimo templo lampadem hanc in alio uel meliori loco poneret, quàm unde totum ſimul poſsit illuminare? Siquidem non inepte quidam lucernam mundi, alij mentem, alij rectorem uocant. Trimegiſtus uiſibilem Deum, Sophoclis Electra intuentē omnia. Ita profecto tanquam in ſolio regali Sol reſidens circum agentem gubernat Aſtrorum familiam. Tellus quoque minime fraudatur lunari miniſterio, ſed ut Ariſtoteles de animalibus ait, maximam Luna cum terra cognationē habet. Cōcipit interea à Sole terra, & impregnatur annuo partu. Inuenimus igitur ſub hac

a

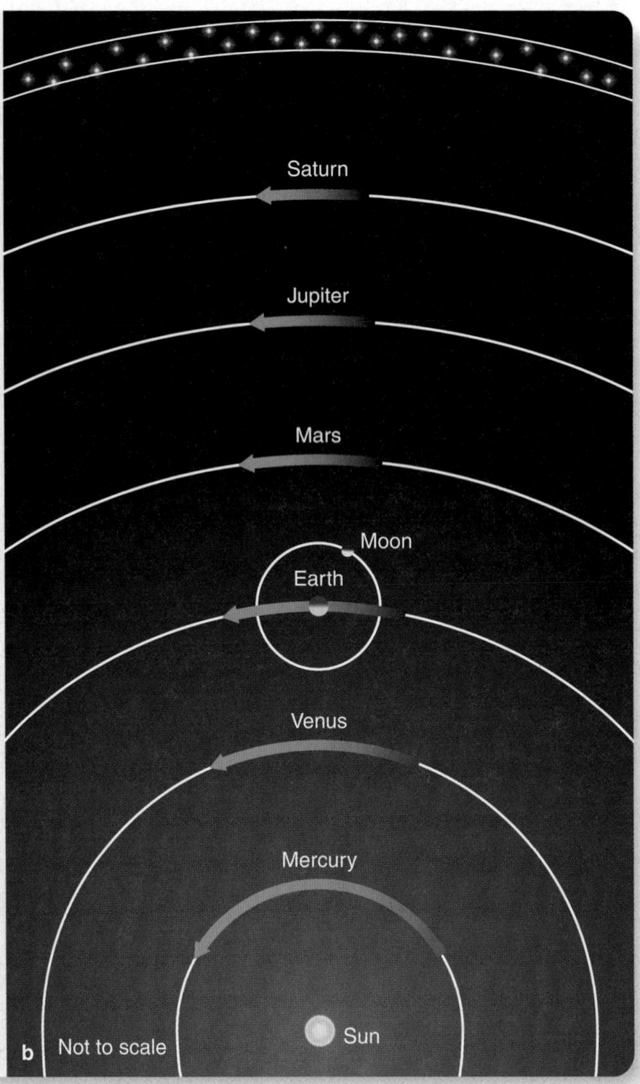


**Figure 4-10**

(a) The Copernican universe as reproduced in *De Revolutionibus*. Earth and all the known planets revolve in separate circular orbits about the sun (Sol) at the center. The outermost sphere carries the immobile stars of the celestial sphere. Notice the orbit of the moon around Earth (Terra). (Yerkes Observatory) (b) The model was elegant not only in its arrangement of the planets but also in their motions. Orbital speed (blue arrows) decreased from Mercury, the fastest, to Saturn, the slowest. Compare the elegance of this model with the complexity of the Ptolemaic model on page 57.

Because Copernicus imposed uniform circular motion on his model, it could not accurately predict the motions of the planets. *The Prutenic Tables* (1551) were based on the Copernican model, and they were not significantly more accurate than the 13th-century *Alfonsine Tables* that were based on Ptolemy's model. Both could be in error by as much as 2°, which is four times the angular diameter of the full moon.

The Copernican *model* is inaccurate. It includes uniform circular motion and consequently does not precisely describe the motions of the planets. But the Copernican *hypothesis* that the universe is heliocentric is correct, considering how little astronomers of the time knew of other stars and galaxies. The planets circle the sun, not Earth, so the universe that Copernicus knew was heliocentric.

Although astronomers throughout Europe read and admired *De Revolutionibus,* they did not immediately accept the Copernican hypothesis. The mathematics was elegant, and the astronomical observations and calculations were of tremendous value; but few astronomers believed, at first, that the sun actually was the center of the planetary system and that Earth moved. How the Copernican hypothesis was gradually recognized as correct has been called the Copernican Revolution because it was

## How Do We Know? 4-1

### Scientific Revolutions

**How do scientific revolutions occur?** You might think from what you know of the scientific method that science grinds forward steadily as new theories are tested against evidence and accepted or rejected. In fact, science sometimes leaps forward in scientific revolutions. The Copernican Revolution is often cited as the perfect example; in a few decades, astronomers rejected the 2000-year-old geocentric model and adopted the heliocentric model. Why does that happen? It's all because scientists are human.

The American philosopher of science Thomas Kuhn has referred to a commonly accepted set of scientific ideas and assumptions as a scientific **paradigm**. The pre-Copernican astronomers shared a geocentric paradigm that included uniform circular motion, geocentrism, and the perfection of the heavens. Although they were intelligent, they were prisoners of that paradigm. A scientific paradigm is powerful because it shapes your perceptions. It determines what you judge to be important questions and what you judge to be significant evidence. Consequently, the ancient astronomers could not recognize how their geocentric paradigms limited what they understood.

You will see here how the work of Copernicus, Galileo, and Kepler overthrew the geocentric paradigm. Scientific revolutions occur when the deficiencies of the old paradigm build up until finally a scientist has the insight to think "outside the box." Pointing out the failings of the old ideas and proposing a new paradigm with supporting evidence is like poking a hole in a dam; suddenly the pressure is released, and the old paradigm is swept away.

Scientific revolutions are exciting because they give you a dramatic new understanding of nature, but they are also times of conflict as new insights sweep away old ideas.

*The ancients believed the stars were attached to a starry sphere.* (NOAO and Nigel Sharp)

not just the adoption of a new idea but a total change in the way astronomers, and, in fact, all of humanity, thought about the place of the Earth (**How Do We Know? 4-1**).

There are probably a number of reasons why the Copernican hypothesis gradually won support, including the revolutionary temper of the times, but the most important factor may have been the elegance of the idea. Placing the sun at the center of the universe produced a symmetry among the motions of the planets that is pleasing to the eye as well as to the intellect (Figure 4-10b). In the Ptolemaic model, Mercury and Venus were treated differently from the rest of the planets; their epicycles had to remain centered on the Earth–sun line. In the Copernican model, all of the planets were treated the same. They all followed orbits that circled the sun at the center. Furthermore, their speed depended in an orderly way on their distance from the sun, with those closest moving fastest.

The most astonishing consequence of the Copernican hypothesis was not what it said about the sun but what it said about Earth. By placing the sun at the center, Copernicus made Earth into a planet, moving along an orbit like the other planets. By making Earth a planet, Copernicus revolutionized humanity's view of its place in the universe and triggered a controversy that would eventually bring the astronomer Galileo Galilei before the Inquisition. This controversy over the apparent conflict between scientific knowledge and philosophical and theological ideals continues even today.

### SCIENTIFIC ARGUMENT

***Why would you say the Copernican hypothesis was correct but the model was inaccurate?***

To build this argument, you must distinguish carefully between a hypothesis and a model. The Copernican hypothesis was that the sun and not Earth was the center of the universe. Given the limited knowledge of the Renaissance astronomers about distant stars and galaxies, that hypothesis was correct.

The Copernican model, however, included not only the heliocentric hypothesis but also uniform circular motion. The model is inaccurate because the planets don't really follow circular orbits, and the small epicycles that Copernicus added to his model never quite reproduced the motions of the planets.

Now build a new argument. The Copernican hypothesis won converts because it is elegant and can explain retrograde motion. **How does its explanation of retrograde motion work, and how is it more elegant than the Ptolemaic explanation?**

## 4-3 Planetary Motion

The Copernican hypothesis solved the problem of the place of Earth, but it didn't explain planetary motion. If planets don't move in uniform circular motion, how do they move? The puzzle of planetary motion was solved during the century following the death of Copernicus through the work of two men. One compiled the observations, and the other did the analysis.

## Tycho Brahe

Tycho Brahe (1546–1601) was not a churchman like Copernicus but rather a nobleman from an important family, educated at the finest universities. He was well known for his vanity and his lordly manners, and by all accounts he was a proud and haughty nobleman. Tycho's disposition was not improved by a dueling injury from his university days. His nose was badly disfigured, and for the rest of his life he wore false noses made of gold and silver, stuck on with wax (■ Figure 4-11).

Although Tycho officially studied law at the university, his real passions were mathematics and astronomy, and early in his university days he began measuring the positions of the planets in the sky. In 1563, Jupiter and Saturn passed very near each other in the sky, nearly merging into a single point on the night of August 24. Tycho found that the *Alfonsine Tables* were a full month in error and that the *Prutenic Tables* were in error by a number of days.

In 1572, a "new star" (now called Tycho's supernova) appeared in the sky, shining more brightly than Venus, and Tycho carefully measured its position. According to classical astronomy, the new star represented a change in the heavens and therefore had to lie below the sphere of the moon. To Tycho, who still believed in a geocentric universe, that meant that the new star should show parallax, meaning that it would appear slightly too far east as it rose and slightly too far west as it set (■ Figure 4-12). But Tycho saw no parallax in the position of the new star, so he concluded that it must lie above the sphere of the moon and was probably on the starry sphere itself. This contradicted Aristotle's conception of the starry sphere as perfect and unchanging.

No one before Tycho could have made this discovery because no one had ever measured the positions of celestial objects so accurately. Tycho had great confidence in the precision of his measurements, and he had studied astronomy thoroughly, so when he failed to detect parallax for the new star, he knew it was important evidence against the Ptolemaic theory. He announced his discovery in a small book, *De Stella Nova (The New Star),* published in 1573.

The book attracted the attention of astronomers throughout Europe, and soon Tycho's family introduced him to the court of the Danish King Frederik II, where he was offered funds to build an observatory on the island of Hveen just off the Danish coast. To support his observatory, Tycho was given a steady income as lord of a coastal district from which he collected rents. (He was not a popular landlord.) On Hveen, Tycho constructed

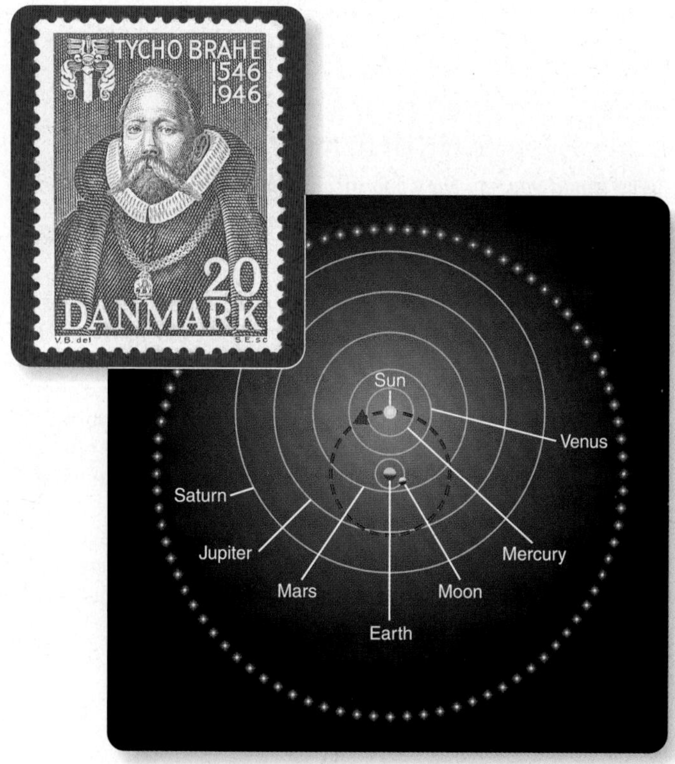


**■ Figure 4-11**

Tycho Brahe (1546–1601) was, during his lifetime, the most famous astronomer in the world. Proud of his noble rank, he wears the elephant medal awarded him by the king of Denmark. His artificial nose is suggested in this engraving. Tycho Brahe's model of the universe retained the first principles of classical astronomy; it was geocentric with the sun and moon revolving around Earth, but the planets revolved around the sun. All motion was along circular paths.

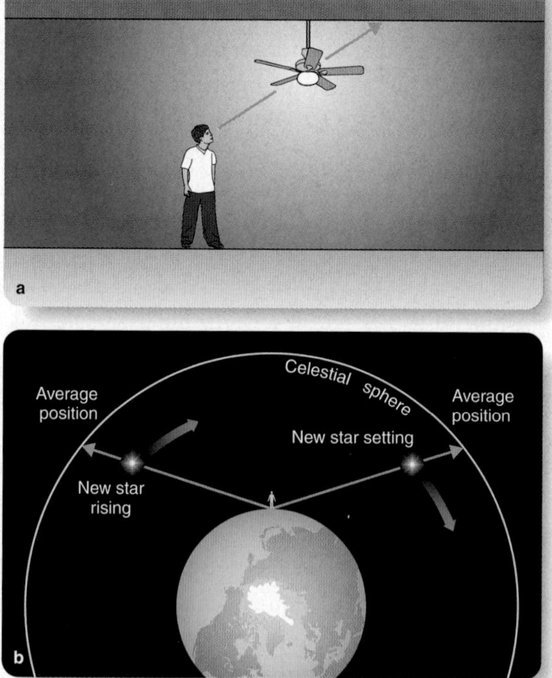


**■ Figure 4-12**

(a) A fan hanging below a ceiling displays parallax when seen from the side. (b) According to Aristotle, the new star of 1572 should have been located below the sphere of the moon; consequently, reasoned Tycho, it should display parallax and be seen east of its average position as it was rising and west of its average position when it was setting. Because he did not detect this daily parallax, he concluded that the new star of 1572 had to lie on the celestial sphere.

a luxurious home with six towers especially equipped for astronomy and populated it with servants, assistants, and a dwarf to act as jester. Soon Hveen was an international center of astronomical study.

## Tycho Brahe's Legacy

Tycho made no direct contribution to astronomical theory. Because he could measure no parallax for the stars, he concluded that Earth had to be stationary, thus rejecting the Copernican hypothesis. However, he also rejected the Ptolemaic model because of its inaccuracy. Instead he devised a complex model in which Earth was the immobile center of the universe around which the sun and moon moved. The other planets circled the sun (Figure 4-11). The model thus incorporated part of the Copernican model, but in it Earth—not the sun—was stationary. In this way, Tycho preserved the central immobile Earth. Although Tycho's model was very popular at first, the Copernican model replaced it within a century. The true value of Tycho's work was observational. Because he was able to devise new and better instruments, he was able to make highly accurate observations of the position of the stars, sun, moon, and planets. Tycho had no telescopes—they were not invented until the next century—so his observations were made by the unaided eye peering along sight-lines. He and his assistants made precise observations for 20 years at Hveen.

Unhappily for Tycho, King Fredrik II died in 1588, and his young son took the throne. Suddenly, Tycho's temper, vanity, and noble presumptions threw him out of favor. In 1596, taking most of his instruments and books of observations, he went to Prague, the capital of Bohemia, and became imperial mathematician to the Holy Roman Emperor Rudolph II. His goal was to revise the *Alfonsine Tables* and publish the result as a monument to his new patron. It would be called the *Rudolphine Tables.*

Tycho did not intend to base the *Rudolphine Tables* on the Ptolemaic system but rather on his own Tyconic system, proving once and for all the validity of his hypothesis. To assist him, he hired a few mathematicians and astronomers, including one Johannes Kepler. Then, in November 1601, Tycho collapsed at a nobleman's home. Before he died, 11 days later, he asked Rudolph II to make Kepler imperial mathematician. The newcomer became Tycho's replacement (though at one-sixth Tycho's salary).

## Kepler: An Astronomer of Humble Origins

No one could have been more different from Tycho Brahe than Johannes Kepler (■ Figure 4-13). Kepler was born in 1571 to a poor family in a region that is now part of southwest Germany. His father was unreliable and shiftless, principally employed as a mercenary soldier fighting for whoever paid enough. He was often absent for long periods and finally failed to return from a military

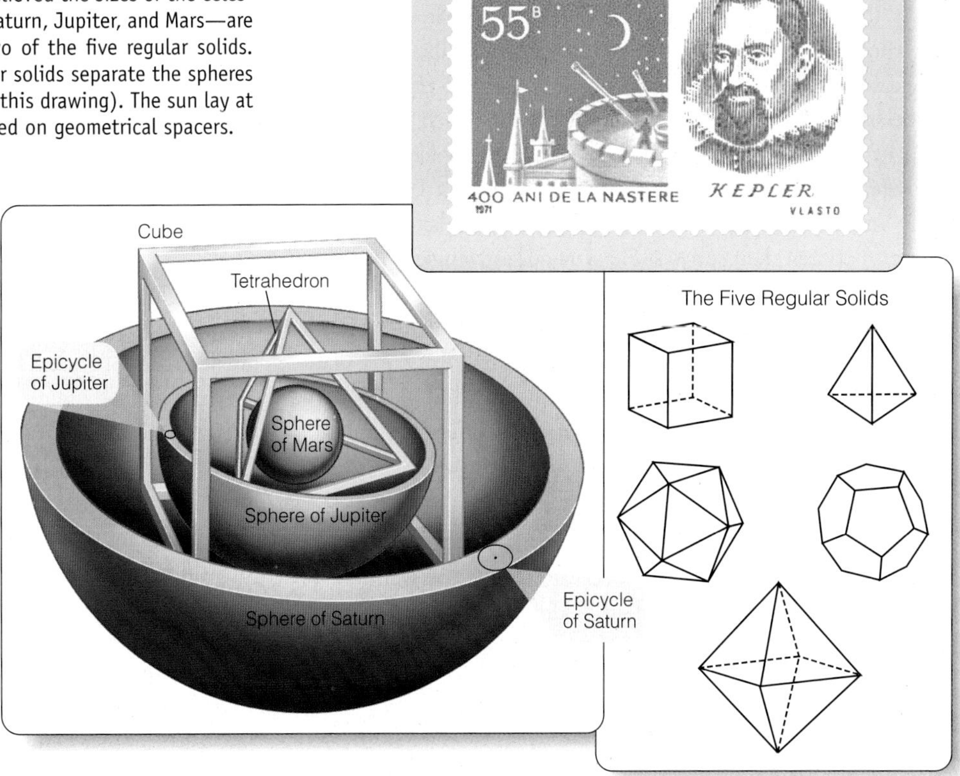


**■ Figure 4-13**

Johannes Kepler (1571–1630) was Tycho Brahe's successor. This diagram, based on one drawn by Kepler, shows how he believed the sizes of the celestial spheres carrying the outer three planets—Saturn, Jupiter, and Mars—are determined by spacers (blue) consisting of two of the five regular solids. Inside the sphere of Mars, the remaining regular solids separate the spheres of the Earth, Venus, and Mercury (not shown in this drawing). The sun lay at the very center of this Copernican universe based on geometrical spacers.

expedition. Kepler's mother was apparently an unpleasant and unpopular woman. She was accused of witchcraft in later years, and Kepler had to defend her in a trial that dragged on for three years. She was finally acquitted, but died the following year.

In spite of family disadvantages and chronic poor health, Kepler did well in school, winning promotion to a Latin school and eventually a scholarship to the university at Tübingen, where he studied to become a Lutheran pastor. During his last year of study, Kepler accepted a job in Graz teaching mathematics and astronomy, a job he resented because he knew little about the subjects. Evidently he was not a good teacher—he had few students his first year and none at all his second. His superiors put him to work teaching a few introductory courses and preparing an annual almanac that contained astronomical, astrological, and weather predictions. Through good luck, in 1595 some of his weather predictions were fulfilled, and he gained a reputation as an astrologer and seer. Even in later life he earned money from his almanacs.

While still a college student, Kepler had become a believer in the Copernican hypothesis, and at Graz he used his extensive spare time to study astronomy. By 1596, the same year Tycho arrived in Prague, Kepler was sure he had solved the mystery of the universe. That year he published a book called *The Forerunner of Dissertations on the Universe, Containing the Mystery of the Universe.* The book, like nearly all scientific works of that age, was written in Latin and is now known as *Mysterium Cosmographicum.*

By modern standards, the book contains almost nothing of value. It begins with a long appreciation of Copernicanism and then goes on to speculate on the reasons for the spacing of the planetary orbits. Kepler assumed that the heavens could be described by only the most perfect of shapes. Therefore he felt that he had found the underlying architecture of the universe in the sphere plus the five regular solids.* In Kepler's model, the five regular solids became spacers for the orbits of the six planets which were represented by nested spheres (Figure 4-13). In fact, Kepler concluded that there could be only six planets (Mercury, Venus, Earth, Mars, Jupiter, and Saturn) because there were only five regular solids to act as spacers between their spheres. He provided astrological, numerological, and even musical arguments for his theory.

The second half of the book is no better than the first, but it has one virtue—as Kepler tried to fit the five solids to the planetary orbits, he demonstrated that he was a talented mathematician and that he was well versed in astronomy. He sent copies of his book to Tycho on Hveen and to Galileo in Rome.

## Joining Tycho

Life was unsettled for Kepler because of the persecution of Protestants in the region, so when Tycho Brahe invited him to Prague in 1600, Kepler went readily, eager to work with the famous Danish astronomer. Tycho's sudden death in 1601 left Kepler, the new imperial mathematician, in a position to use the observations from Hveen to analyze the motions of the planets and complete *The Rudolphine Tables.* Tycho's family, recognizing that Kepler was a Copernican and guessing that he would not follow the Tychonic system in completing *The Rudolphine Tables,* sued to recover the instruments and books of observations. The legal wrangle went on for years. Tycho's family did get back the instruments Tycho had brought to Prague, but Kepler had the books, and he kept them.

Whether Kepler had any legal right to Tycho's records is debatable, but he put them to good use. He began by studying the motion of Mars, trying to deduce from the observations how the planet moved. By 1606, he had solved the mystery, this time correctly. The orbit of Mars is an ellipse and not a circle, he realized, and with that he abandoned the 2000-year-old belief in the circular motion of the planets. But even this insight was not enough to explain the observations. The planets do not move at uniform speeds along their elliptical orbits. Kepler's analysis showed that they move faster when close to the sun and slower when farther away. With those two brilliant discoveries, Kepler abandoned uniform circular motion and finally solved the puzzle of planetary motion. He published his results in 1609 in a book called *Astronomia Nova (New Astronomy).*

In spite of the abdication of Rudolph II in 1611, Kepler continued his astronomical studies. He wrote about a supernova that appeared in 1604 (now known as Kepler's supernova) and about comets, and he wrote a textbook about Copernican astronomy. In 1619, he published *Harmonice Mundi (The Harmony of the World),* in which he returned to the cosmic mysteries of *Mysterium Cosmographicum.* The only thing of note in *Harmonice Mundi* is his discovery that the radii of the planetary orbits are related to the planets' orbital periods. That and his two previous discoveries are so important that they have become known as the three most fundamental rules of orbital motion.

## Kepler's Three Laws of Planetary Motion

Although Kepler dabbled in the philosophical arguments of his day, he was at heart a mathematician, and his triumph was his explanation of the motion of the planets. The key to his solution was the ellipse.

An **ellipse** is a figure that can be drawn around two points, called the *foci,* in such a way that the distance from one focus to any point on the ellipse and back to the other focus equals a constant. This makes it easy to draw ellipses using two thumbtacks and a loop of string. Press the thumbtacks into a board, loop the string about the tacks, and place a pencil in the loop. If you keep the string taut as you move the pencil, it traces out an ellipse (■ Figure 4-14).

The geometry of an ellipse is described by two simple numbers. The **semimajor axis, *a*,** is half of the longest diameter, as you can see in Figure 4-14. The **eccentricity, *e*,** of an ellipse is half the distance between the foci divided by the semimajor axis.

*The five regular solids, also known as the Platonic solids, are the tetrahedron, cube, octahedron, dodecahedron, and icosahedron. They were considered perfect because the faces and the angles between the faces are the same at every corner.

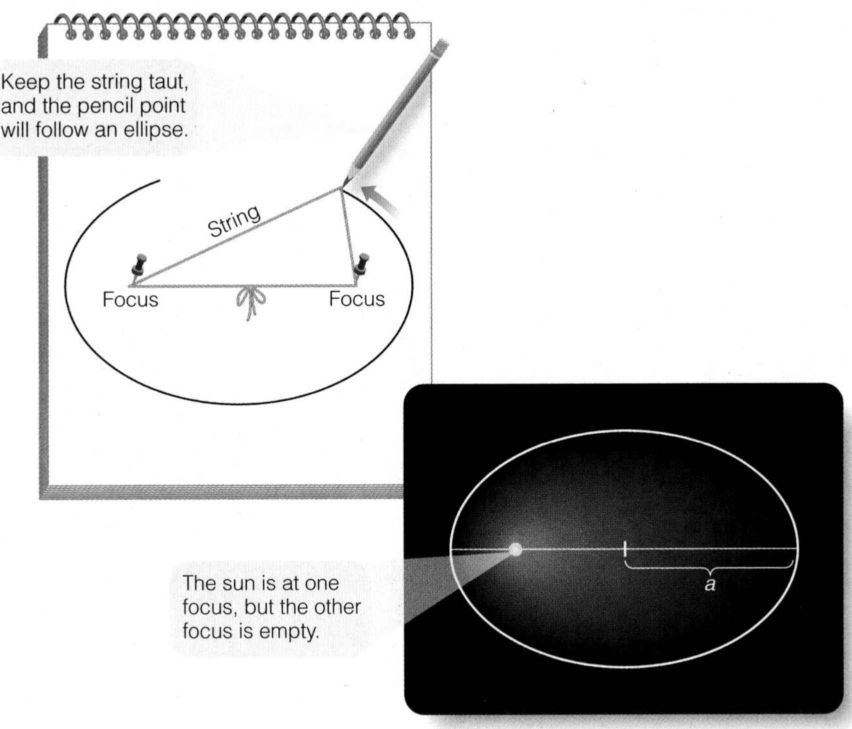

Keep the string taut, and the pencil point will follow an ellipse.
String
Focus
Focus
The sun is at one focus, but the other focus is empty.
a

**■ Figure 4-14**

The geometry of elliptical orbits: Drawing an ellipse with two tacks and a loop of string is easy. The semimajor axis, *a*, is half of the longest diameter. The sun lies at one of the foci of the elliptical orbit of a planet.

The eccentricity of an ellipse tells you its shape; if *e* is nearly equal to one, the ellipse is very elongated. If *e* is closer to zero, the ellipse is more circular. To draw a circle with the string and tacks shown Figure 4-14, you would have to move the two thumbtacks together because a circle is really just an ellipse with eccentricity equal to zero. Try fiddling with real thumbtacks and string, and you'll be surprised how easy it is to draw graceful, smooth ellipses with various eccentricities.

Ellipses are a prominent part of Kepler's three fundamental rules of planetary motion. Those rules have been tested and confirmed so many times that astronomers now refer to them as natural laws (**How Do We Know? 4-2**). They are commonly called Kepler's laws of planetary motion (■ Table 4-1).

**■ Table 4-1 | Kepler's Laws of Planetary Motion**

I. The orbits of the planets are ellipses with the sun at one focus.
II. A line from a planet to the sun sweeps over equal areas in equal intervals of time.
III. A planet's orbital period squared is proportional to its average distance from the sun cubed:

$$P^2_{yr} = a^3_{AU}$$

Kepler's first law says that the orbits of the planets around the sun are ellipses with the sun at one focus. Thanks to the precision of Tycho's observations and the sophistication of Kepler's mathematics, Kepler was able to recognize the elliptical shape of the orbits even though they are nearly circular. Mercury has the most elliptical orbit, but even it deviates only slightly from a circle (■ Figure 4-15).

Kepler's second law says that an imaginary line drawn from the planet to the sun always sweeps over equal areas in equal intervals of time. This means that when the planet is closer to the sun and the line connecting it to the sun is shorter, the planet moves more rapidly, and the line sweeps over the same area that is swept over when the planet is farther from the sun. You can see how the planet in Figure 4-15 would move from point *A* to point *B* in one month, sweeping over the area shown. But when the planet is farther from the sun, one month's motion would be shorter, from *A'* to *B'*, and the area swept out would be the same.

Kepler's third law relates a planet's orbital period to its average distance from the sun. The orbital period, *P*, is the time a planet takes to travel around the sun once. Its average distance from the sun turns out to equal the semimajor axis of its orbit, *a*. Kepler's third law says that a planet's orbital period squared is proportional to the semimajor axis of its orbit cubed. Measuring *P* in years and *a* in astronomical units, you can summarize the third law as

$$P^2_{y} = a^3_{AU}$$

For example, Jupiter's average distance from the sun is roughly 5.2 AU. The semimajor axis cubed is about 140, so the period must be the square root of 140, which equals just under 12 years.

Notice that Kepler's three laws are empirical. That is, they describe a phenomenon without explaining why it occurs. Kepler derived the laws from Tycho's extensive observations, not from any first principle, fundamental assumption, or theory. In fact, Kepler never knew what held the planets in their orbits or why they continued to move around the sun.

### The Rudolphine Tables

Kepler continued his mathematical work on *The Rudolphine Tables,* and at last, in 1627, it was ready. He financed the printing himself, dedicating the book to the memory of Tycho Brahe. In fact, Tycho's name appears in larger type on the title page than Kepler's own. This is surprising because the tables were not based on the Tyconic system but on the heliocentric model of Copernicus and the elliptical orbits of Kepler. The reason for

## How Do We Know? 4-2

### Hypothesis, Theory, and Law

**Why is a theory much more than just a guess?** Scientists study nature by devising and testing new hypotheses and then developing the successful ideas into theories and laws that describe how nature works. A good example is the connection between sour milk and the spread of disease.

A scientist's first step in solving a natural mystery is to propose a reasonable explanation based on what is known so far. This proposal, called a **hypothesis,** is a single assertion or statement that must be tested through observation and experimentation. From the time of Aristotle philosophers believed that food spoils as a result of the spontaneous generation of life—for example, mold growing out of drying bread. French chemist Louis Pasteur (1822–1895) hypothesized that microorganisms were not spontaneously generated but were carried through the air. To test his hypothesis he sealed an uncontaminated nutrient broth in glass completely protecting it from the microorganisms on dust particles in the air. No mold grew, effectively disproving spontaneous generation. Although others had argued against spontaneous generation before Pasteur, it was Pasteur's meticulous testing of his hypothesis through experimentation that finally convinced the scientific community.

A **theory** generalizes the specific results of well-confirmed hypotheses to give a broader description of nature, which can be applied to a wide variety of circumstances. For instance, Pasteur's specific hypothesis about mold growing in broth contributed to a broader theory that disease is caused by microorganisms transmitted from sick people to well people. This theory, called the germ theory of disease, is a cornerstone of modern medicine. It is a **Common Misconception** that the word "theory" means a tentative idea, a guess. As you have just learned, scientists actually use the word "theory" to mean an idea that is widely applicable and confirmed by abundant evidence.

Sometimes, when a theory has been refined, tested, and confirmed so often that scientists have great confidence in it, it is called a **natural law.** Natural laws are the most fundamental principles of scientific knowledge. Kepler's laws of planetary motion are good examples.

Confidence is the key. In general, scientists have more confidence in a theory than in a hypothesis and the most confidence in a natural law. However, there is no precise distinction among a hypothesis, a theory, and a law, and use of these terms is sometimes a matter of tradition. For instance, some textbooks refer to the Copernican "theory" of heliocentrism, but it had not been well tested when Copernicus proposed it, and it is more rightly called the Copernican hypothesis. At the other extreme, Darwin's "theory" of evolution, containing many hypotheses that have been tested and confirmed over and over for nearly 150 years, might more correctly be called a natural law.

*A fossil of a 500-million-year-old trilobite: Darwin's theory of evolution has been tested many times and is universally accepted in the life sciences, but by custom it is called Darwin's theory and not Darwin's law.*
(From the collection of John Coolidge III)

Kepler's evident deference was Tycho's family, still powerful and still intent on protecting Tycho's reputation. They even demanded a share of the profits and the right to censor the book before publication, though they changed nothing but a few words on the title page and added an elaborate dedication to the emperor.

*The Rudolphine Tables* was Kepler's masterpiece. It could predict the positions of the planets 10 to 100 times more accurately than previous tables. Kepler's tables were the precise model of planetary motion that Copernicus had sought but failed to find. The accuracy of *The Rudolphine Tables* was strong evidence that both Kepler's laws of planetary motion and the Copernican hypothesis for the place of Earth were correct. Copernicus would have been pleased.

Kepler died in 1630. He had solved the problem of planetary motion, and his *Rudolphine Tables* demonstrated his solution. Although he did not understand why the planets moved or why they followed ellipses, insights that had to wait half a century for Isaac Newton, Kepler's three laws worked. In science the only test of a theory is, "Does it describe reality?" Kepler's laws have been used for almost four centuries as a true description of orbital motion.

### SCIENTIFIC ARGUMENT

***How was Kepler's model with regular solids based on first principles? How were his three laws based on evidence?***

When he was younger, Kepler accepted Plato's argument for the perfection of the heavens. Furthermore, Kepler argued that the five regular solids were perfect geometrical figures and should be part of the perfect heavens along with spheres. He then arranged the five regular solids to produce the approximate spacing among the spheres that carried the planets in the Copernican model. Kepler's model was thus based on a first principle—the perfection of the heavens.

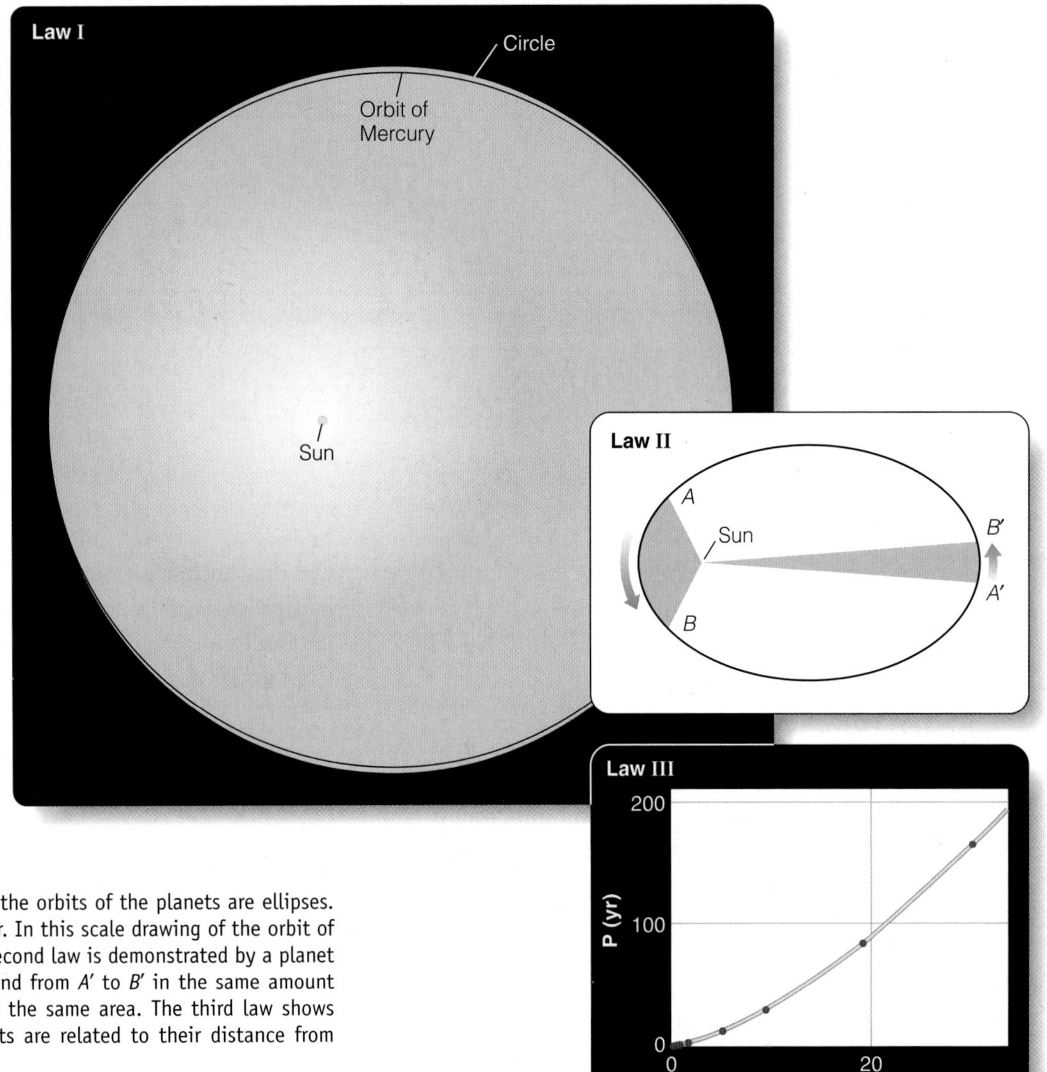


■ **Figure 4-15**

Kepler's three laws: The first law says the orbits of the planets are ellipses. The orbits, however, are nearly circular. In this scale drawing of the orbit of Mercury, it looks nearly circular. The second law is demonstrated by a planet that moves from *A* to *B* in 1 month and from *A′* to *B′* in the same amount of time. The two blue segments have the same area. The third law shows that the orbital periods of the planets are related to their distance from the sun.

Much later, Kepler derived his three laws of motion from the observations made by Tycho Brahe during 20 years on Hveen. The observations were the evidence, and they gave Kepler a reality check each time he tried a new calculation. He chose ellipses because they fit the data and not because he thought ellipses had any special significance.

The Copernican model was a poor predictor of planetary motion, but *The Rudolphine Tables* was much more accurate. **What first principle did Copernicus follow that was abandoned when Kepler looked at the evidence?**

## 4-4 Galileo Galilei

MOST PEOPLE THINK they know two facts about Galileo, but both facts are wrong; they are **Common Misconceptions,** so you have probably heard them. Galileo did not invent the telescope, and he was not condemned by the Inquisition for believing that Earth moved around the sun. Then why is Galileo so famous? Why did the Vatican reopen his case in 1979, almost 400 years after his trial? As you learn about Galileo, you will discover that his trial concerned not just the place of Earth and the motion of the planets but also a new and powerful method of understanding nature, a method called science.

### Telescopic Observations

Galileo Galilei (■ Figure 4-16) was born in 1564 in Pisa, a city in what is now Italy, and he studied medicine at the university there. His true love, however, was mathematics, and, although he had to leave school early for financial reasons, he returned only four years later as a professor of mathematics. Three years after that he became professor of mathematics at the university at Padua, where he remained for 18 years.

**■ Figure 4-16**

Galileo, remembered as the great defender of Copernicanism, also made important discoveries in the physics of motion. He was honored here on an Italian 2000-lira note.

During this time, Galileo seems to have adopted the Copernican model, although he admitted in a 1597 letter to Kepler that he did not support Copernicanism publicly. At that time, the Copernican hypothesis was not officially considered heretical, but it was hotly debated among astronomers, and Galileo, living in a region controlled by the Church, cautiously avoided trouble. It was the telescope that finally drove Galileo to publicly defend the heliocentric model.

The telescope was apparently invented around 1608 by lens makers in Holland. Galileo, hearing descriptions in the fall of 1609, was able to build telescopes in his workshop. In fact, Galileo was not the first person to look at the sky through a telescope, but he was the first person to apply telescopic observations to the theoretical problem of the day—the place of Earth.

What Galileo saw through his telescopes was so amazing that he rushed a small book into print. *Sidereus Nuncius (The Sidereal Messenger)* reported three major discoveries. First, the moon was not perfect. It had mountains and valleys on its surface, and Galileo even used some of the mountains' shadows to calculate their height. Aristotle's philosophy held that the moon was perfect, but Galileo showed that it was not only imperfect but was a world with features like Earth's.

The second discovery reported in the book was that the Milky Way was made up of myriad stars too faint to see with the unaided eye. While intriguing, this could not match Galileo's third discovery. Galileo's telescope revealed four new "planets" circling Jupiter, objects known today as the **Galilean moons** of Jupiter (■ Figure 4-17).

The moons of Jupiter were strong evidence for the Copernican model. Critics of Copernicus had said Earth could not move because the moon would be left behind, but Galileo's discovery showed that Jupiter, which everyone agreed was moving, was able to keep its satellites. That suggested that Earth, too, could move and keep its moon. Aristotle's philosophy also included the belief that all heavenly motion was centered on Earth. Galileo's observations showed that Jupiter's moons revolve around Jupiter, suggesting that there could be other centers of motion besides Earth.

Some time after *Sidereus Nuncius* was published, Galileo noticed something else that made Jupiter's moons even stronger evidence for the Copernican model. When he measured the orbital periods of the four moons, he found that the innermost moon had the shortest period and that the moons farther from Jupiter had proportionally longer periods. Jupiter's moons made up a harmonious system ruled by Jupiter, just as the planets in the Copernican universe were a harmonious system ruled by the sun.

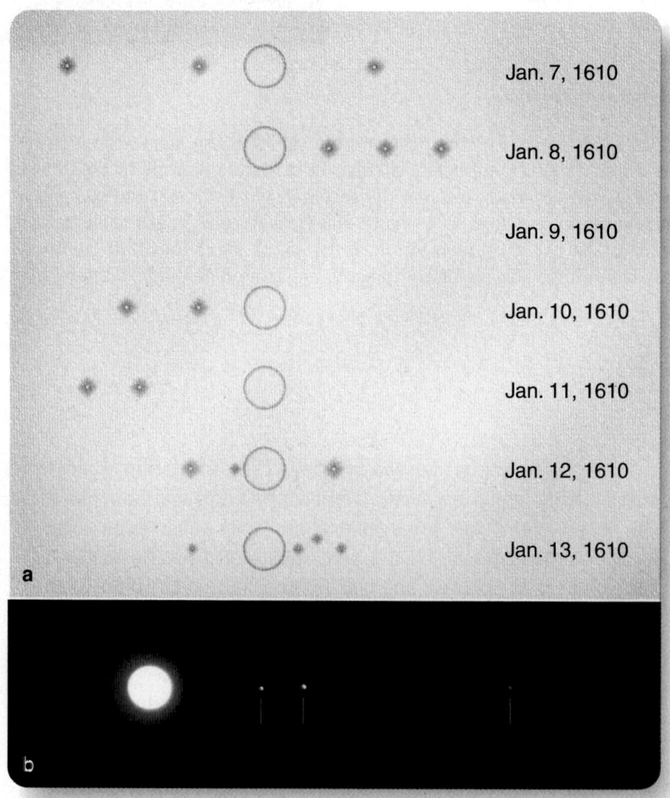


**■ Figure 4-17**

(a) On the night of January 7, 1610, Galileo saw three small "stars" near the bright disk of Jupiter and sketched them in his notebook. On subsequent nights (excepting January 9, which was cloudy), he saw that the stars were actually four moons orbiting Jupiter. (b) This photo taken through a modern telescope shows the overexposed disk of Jupiter and three of the four Galilean moons. (Grundy Observatory)

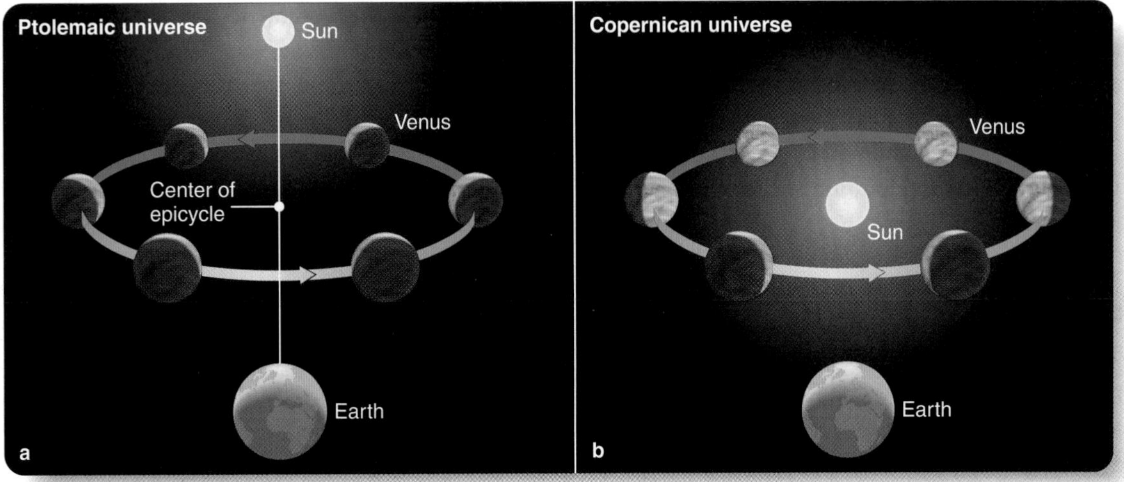


**■ Figure 4-18**

(a) If Venus moved in an epicycle centered on the Earth–sun line (see page 57), it would always appear as a crescent. (b) Galileo's telescope showed that Venus goes through a full set of phases, proving that it must orbit the sun.

(See Figure 4-10b.) The similarity isn't proof, but Galileo saw it as an argument that the solar system could be sun centered rather than Earth centered.

In the years following publication of *Sidereus Nuncius,* Galileo made two additional discoveries. When he observed the sun, he discovered sunspots, raising the suspicion that the sun was less than perfect. Further, by noting the movement of the spots, he concluded that the sun was a sphere and that it rotated on its axis.

His most dramatic discovery came when he observed Venus. Galileo saw that it was going through phases like those of the moon. In the Ptolemaic model, Venus moves around an epicycle centered on a line between Earth and the sun. That means it would always be seen as a crescent (■ Figure 4-18a). But Galileo saw Venus go through a complete set of phases, which proved that it did indeed revolve around the sun (Figure 4-18b). There is no way the Ptolemaic model could produce those phases. This was the strongest evidence that came from Galileo's telescope; but, when controversy erupted, it focused more on the perfection of the sun and moon and the motion of the satellites of Jupiter.

*Sidereus Nuncius* was very popular and made Galileo famous. He became chief mathematician and philosopher to the Grand Duke of Tuscany in Florence. In 1611, Galileo visited Rome and was treated with great respect. He had long, friendly discussions with the powerful Cardinal Barberini, but he also made enemies. Personally, Galileo was outspoken, forceful, and sometimes tactless. He enjoyed debate, but most of all he enjoyed being right. In lectures, debates, and letters he offended important people who questioned his telescopic discoveries.

By 1616, Galileo was the center of a storm of controversy. Some critics said he was wrong, and others said he was lying. Some refused to look through a telescope lest it mislead them, and others looked and claimed to see nothing (hardly surprising, given the awkwardness of those first telescopes) (■ Figure 4-19). Pope Paul V decided to end the disruption, so when Galileo visited Rome in 1616 Cardinal Bellarmine interviewed him privately and ordered him to cease debate. There is some controversy today about the nature of Galileo's instructions, but he did not pursue astronomy for some years after the interview. Books

**■ Figure 4-19**

Galileo's telescope made him famous, and he demonstrated his telescope and discussed his observations with powerful people. Some thought the telescope was the work of the devil and would deceive anyone who looked. In any case, Galileo's discoveries produced intense and, in some cases, angry debate. (Yerkes Observatory)

relevant to Copernicanism were banned in all Catholic lands, although *De Revolutionibus,* recognized as an important and useful book in astronomy, was only suspended pending revision. Everyone who owned a copy of the book was required to cross out certain statements and add handwritten corrections stating that Earth's motion and the central location of the sun were only theories and not facts.

## *Dialogo* and Trial

In 1621 Pope Paul V died, and his successor, Pope Gregory XV, died in 1623. The next pope was Galileo's friend Cardinal Barberini, who took the name Urban VIII. Galileo rushed to Rome hoping to have the prohibition of 1616 lifted; and, although the new pope did not revoke the orders, he did apparently encourage Galileo. Soon after returning home, Galileo began to write his great defense of Copernicanism, finally completing it at the end of 1629. After some delay, the book was approved by both the local censor in Florence and the head censor of the Vatican in Rome. It was printed in February 1632.

Called *Dialogo Sopra i Due Massimi del Mondo (Dialogue Concerning the Two Chief World Systems),* it confronts the ancient astronomy of Aristotle and Ptolemy with the Copernican model and with telescopic observations as evidence. Galileo wrote the book in the form of a debate among three friends. Salviati, a swift-tongued defender of Copernicus, dominates the book; Sagredo is intelligent but largely uninformed. Simplicio, the dismal defender of Ptolemy, makes all the old arguments and sometimes doesn't seem very bright.

The publication of *Dialogo* created a storm of controversy, and it was sold out by August 1632, when the Inquisition ordered sales stopped. The book was a clear defense of Copernicus, and, probably unintentionally, Galileo exposed the pope's authority to ridicule. Urban VIII was fond of arguing that, as God was omnipotent, He could construct the universe in any form while making it appear to humans to have a different form, and thus its true nature could not be deduced by mere observation. Galileo placed the pope's argument in the mouth of Simplicio, and Galileo's enemies showed the passage to the pope as an example of Galileo's disrespect. The pope thereupon ordered Galileo to face the Inquisition.

Galileo was interrogated by the Inquisition four times and was threatened with torture. He must have thought often of Giordano Bruno, a philospher, poet, and Dominican monk, who was tried, condemned, and burned at the stake in Rome in 1600. One of Bruno's offenses had been Copernicanism. However, Galileo's trial did not center on his belief in Copernicanism. *Dialogo* had been approved by two censors. Rather, the trial centered on the instructions given Galileo in 1616. From his file in the Vatican, his accusers produced a record of the meeting between Galileo and Cardinal Bellarmine that included the statement that Galileo was "not to hold, teach, or defend in any way" the principles of Copernicus. Some historians believe that this document, which was signed neither by Galileo nor by Bellarmine nor by a legal secretary, was a forgery. Others suspect it may be a draft that was never used. It is quite possible that Galileo's actual instructions were much less restrictive; but, in any case, Bellarmine was dead and could not testify at Galileo's trial.

The Inquisition condemned Galileo not for heresy but for disobeying the orders given him in 1616. On June 22, 1633, at the age of 70, kneeling before the Inquisition, Galileo read a recantation admitting his errors. Tradition has it that as he rose he whispered *"E pur si muove"* ("Still it moves"), referring to Earth.

Although he was sentenced to life imprisonment, he was, perhaps through the intervention of the pope, confined at his villa for the next ten years. He died there on January 8, 1642, 99 years after the death of Copernicus.

Galileo was not condemned for heresy, nor was the Inquisition interested when he tried to defend Copernicanism. He was tried and condemned on a charge you might call a technicality. Nevertheless, in his recantation he was forced to abandon all belief in heliocentrism. His trial has been held up as an example of the suppression of free speech and free inquiry and as a famous attempt to deny reality. Some of the world's greatest authors, including Bertolt Brecht, have written about Galileo's trial. That is why Pope John Paul II created a commission in 1979 to reexamine the case against Galileo.

To understand the trial, you must recognize that it was the result of a conflict between two ways of understanding the universe. Since the Middle Ages, biblical scholars had taught that the only path to true understanding was through religious faith. St. Augustine (AD 354–430) wrote *"Credo ut intelligame,"* which can be translated as "Believe in order to understand." Galileo and other scientists of the Renaissance, however, used their own observations as evidence to try to understand nature. When their observations contradicted Scripture, they assumed that it was their observations that truly represented reality. Galileo paraphrased Cardinal Baronius in saying, "The Bible tells us how to go to heaven, not how the heavens go." The trial of Galileo was not really about the place of Earth in the universe. It was not about Copernicanism. It wasn't even about the instructions Galileo received in 1616. It was, in a larger sense, about the birth of modern science as a rational way to understand the universe (■ Figure 4-20).

The commission appointed by John Paul II in 1979, reporting its conclusions in October 1992, said of Galileo's inquisitors, "This subjective error of judgment, so clear to us today, led them to a disciplinary measure from which Galileo 'had much to suffer.'" Galileo was not found innocent in 1992 so much as the Inquisition was forgiven for having charged him in the first place.

■ Figure 4-20

Although he did not invent it, Galileo will always be associated with the telescope because it was the source of the observational evidence he used to try to understand the universe. By depending on observation of reality instead of first principles of philosophy and theology, Galileo led the way to the invention of modern science as a way to know about the natural world.

### SCIENTIFIC ARGUMENT

***How were Galileo's observations of the moons of Jupiter evidence against the Ptolemaic model?***

Scientific arguments are based on evidence, and reasoning from evidence was Galileo's fundamental way of knowing about the heavens. Galileo presented his arguments in the form of evidence and conclusions, and the moons of Jupiter were key evidence. Ptolemaic astronomers argued that Earth could not move or it would lose its moon, but even in the Ptolemaic universe Jupiter moved, and the telescope showed that it had moons and kept them. Evidently, Earth could move and not leave its moon behind. Furthermore, moons circling Jupiter did not fit the classical belief that all motion was centered on Earth. Obviously there could be other centers of motion. Finally, the orbital periods of the moons were related to their distance from Jupiter, just as the orbital periods of the planets were, in the Copernican system, related to their distance from the sun. This similarity suggested that the sun rules its harmonious family of planets just as Jupiter rules its harmonious family of moons.

Of all of Galileo's telescopic observations, the moons of Jupiter caused the most debate, but the craters on the moon and the phases of Venus were also critical evidence. Build an argument to discuss that evidence. **How did craters on the moon and the phases of Venus argue against the Ptolemaic model?**

## 4-5 Modern Astronomy

THE SCIENCE KNOWN AS MODERN ASTRONOMY began during the 99 years between the deaths of Copernicus and Galileo (1543 to 1642). It was an age of transition that marked the change from the Ptolemaic model of the universe to the Copernican model with the attendant controversy over the place of the Earth. But that same period also marked a transition in the nature of astronomy in particular and of science in general, a transition illustrated in the solution of the puzzle of planetary motion. The puzzle was not solved by philosophical arguments about the perfection of the heavens or by debate over the meaning of scripture. It was solved by precise observation and careful computation, techniques that are the foundation of modern science.

The discoveries made by Kepler and Galileo found acceptance in the 1600s because the world was in transition. Astronomy was not the only thing changing during this period. The Renaissance is commonly taken to be the period between 1300 and 1600, and these 99 years of astronomical history lie at the culmination of the reawakening of learning in all fields (■ Figure 4-21). Ships were sailing to new lands and encountering new cultures. The world was open to new ideas and new observations. Martin Luther reformed the Christian religion, and other philosophers and scholars rethought their areas of human knowledge. Had Copernicus not published his hypothesis, someone else would have suggested that the universe is heliocentric. History was ready to shed the Ptolemaic system.

This period marks the beginning of the modern scientific method. Beginning with Copernicus, scientists such as Tycho, Kepler, and Galileo depended more and more on evidence, observation, and measurement rather than on first principles. This, too, is coupled to the Renaissance and its advances in metalworking and lens making. At the time of Copernicus, no astronomer had looked through a telescope, because one could not be made. By 1642, not only telescopes but also other sensitive measuring instruments had transformed science into something new and precise. As you can imagine, scientists were excited by these discoveries, and they founded scientific societies that increased the exchange of observations and hypotheses and stimulated more and better work. The most important advance, however, was the application of mathematics to scientific questions. Kepler's work demonstrated the power of mathematical analysis; and, as the quality of these numerical techniques improved, the progress of science accelerated. This story of the birth of modern astronomy is actually the story of the birth of modern science as well.

# Orbiting Earth

**1** You can understand orbital motion by thinking of a cannonball falling around Earth in a circular path. Imagine a cannon on a high mountain aimed horizontally as shown at right. A little gunpowder gives the cannonball a low velocity, and it doesn't travel very far before falling to Earth. More gunpowder gives the cannonball a higher velocity, and it travels farther. With enough gunpowder, the cannonball travels so fast it never strikes the ground. Earth's gravity pulls it toward Earth's center, but Earth's surface curves away from it at the same rate it falls. It is in orbit. The velocity needed to stay in a circular orbit is called the **circular velocity.** Just above Earth's atmosphere, circular velocity is 7780 m/s or about 17,400 miles per hour, and the orbital period is about 90 minutes.

A satellite above Earth's atmosphere feels no friction and will fall around Earth indefinitely.

North Pole

Earth satellites eventually fall back to Earth if they orbit too low and experience friction with the upper atmosphere.

**1a** A **geosynchronous satellite** orbits eastward with the rotation of Earth and remains above a fixed spot — ideal for communications and weather satellites.

**A Geosynchronous Satellite**

At a distance of 42,250 km (26,260 miles) from Earth's center, a satellite orbits with a period of 24 hours.

The satellite orbits eastward, and Earth rotates eastward under the moving satellite.

The satellite remains fixed above a spot on Earth's equator.

**1b** According to Newton's first law of motion, the moon should follow a straight line and leave Earth forever. Because it follows a curve, Newton knew that some force must continuously accelerate it toward Earth — gravity. Each second the moon moves 1020 m (3350 ft) eastward and falls about 1.6 mm (about 1/16 inch) toward Earth. The combination of these motions produces the moon's curved orbit. The moon is falling.

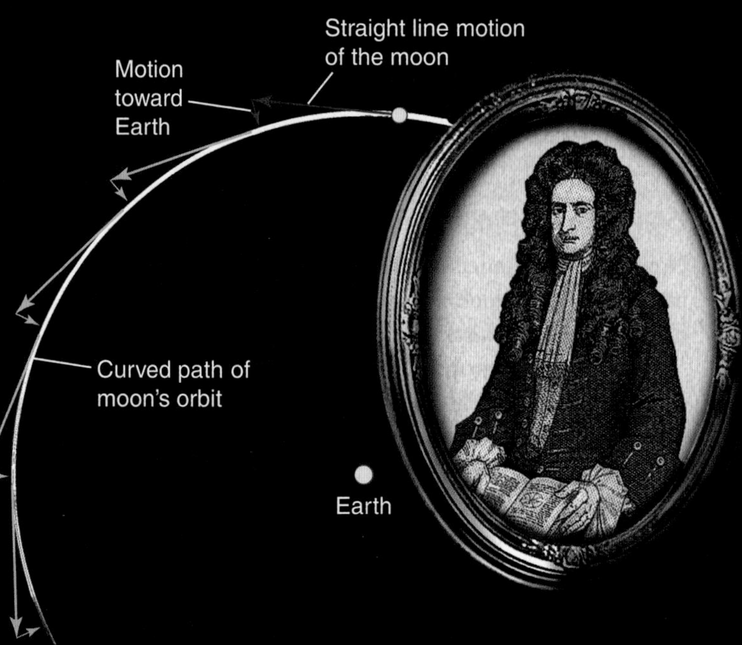

**1c** Astronauts in orbit around Earth feel weightless, but they are not "beyond Earth's gravity," to use a term from old science fiction movies. Like the moon, the astronauts are accelerated toward Earth by Earth's gravity, but they travel fast enough along their orbits that they continually "miss the Earth." They are literally falling around Earth. Inside or outside a spacecraft, astronauts feel weightless because they and their spacecraft are falling at the same rate. Rather than saying they are weightless, you should more accurately say they are in free fall.

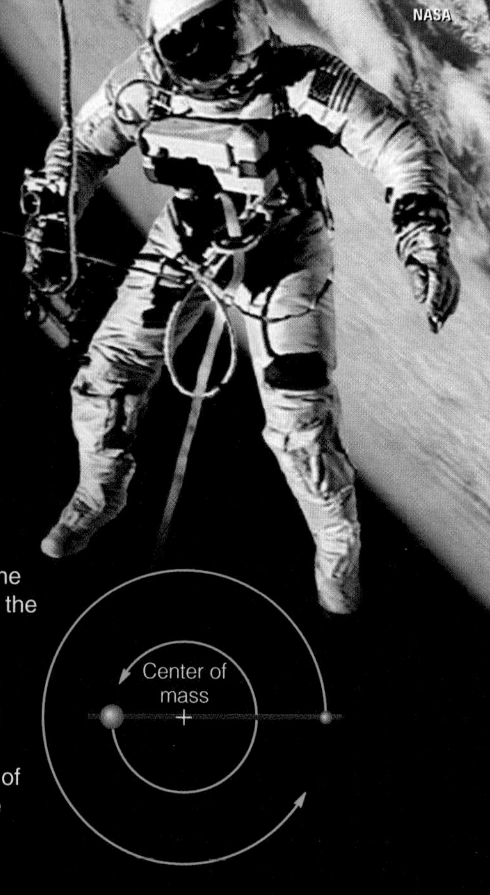

**2** To be precise you should not say that an object orbits Earth. Rather the two objects orbit each other. Gravitation is mutual, and if Earth pulls on the moon, the moon pulls on Earth. The two bodies revolve around their common **center of mass,** the balance point of the system.

**2a** Two bodies of different mass balance at the center of mass, which is located closer to the more massive object. As the two objects orbit each other, they revolve around their common center of mass as shown at right. The center of mass of the Earth–moon system lies only 4708 km (2926 miles) from the center of Earth — inside the Earth. As the moon orbits the center of mass on one side, the Earth swings around the center of mass on the opposite side.

Center of mass

**3** **Closed orbits** return the orbiting object to its starting point. The moon and artificial satellites orbit Earth in closed orbits. Below, the cannonball could follow an elliptical or a circular closed orbit. If the cannonball travels as fast as **escape velocity**, the velocity needed to leave a body, it will enter an open orbit. An **open orbit** does not return the cannonball to Earth. It will escape.

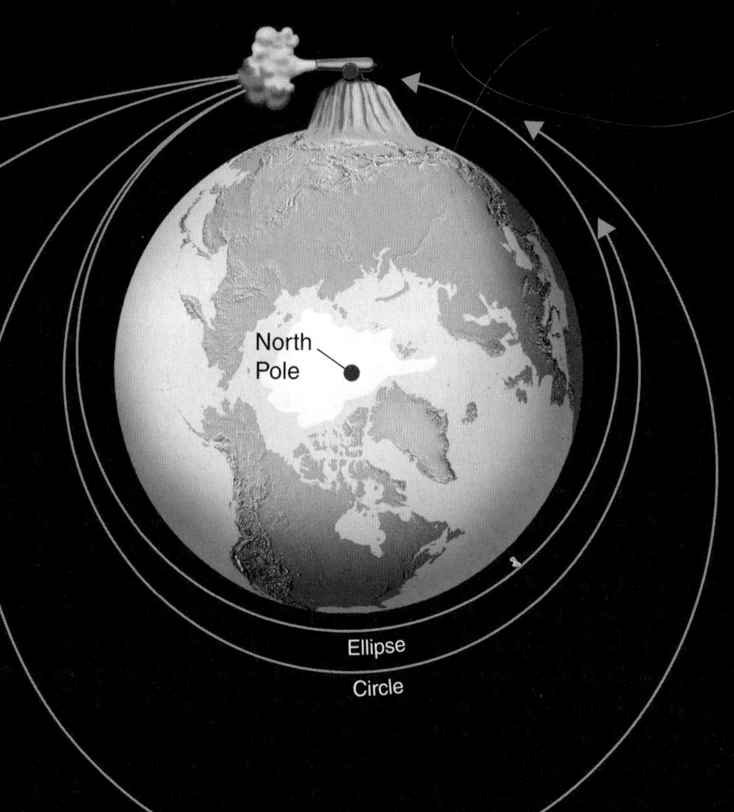


A cannonball with a velocity greater than escape velocity will follow a hyperbola and escape from Earth.

A cannonball with escape velocity will follow a parabola and escape.

**3a** As described by Kepler's second law, an object in an elliptical orbit has its lowest velocity when it is farthest from Earth (apogee), and its highest velocity when it is closest to Earth (perigee). Perigee must be above Earth's atmosphere, or friction will rob the satellite of energy and it will eventually fall back to Earth.

# Focus on Fundamentals 2

## Energy

Physicists define **energy** as the ability to do work, but you might paraphrase that definition as the ability to produce a change. A moving body has energy called **kinetic energy.** A planet moving along its orbit, a cement truck rolling down the highway, and a golf ball sailing down the fairway all have the ability to produce a change. Imagine colliding with any of these objects!

Energy need not be represented by motion. Sunlight falling on a green plant, on photographic film, or on unprotected skin can produce chemical changes, and thus light is a form of energy. Batteries and gasoline are examples of chemical energy, and uranium fuel rods contain nuclear energy. A tank of hot water contains thermal energy.

**Potential energy** is the energy an object has because of its position in a gravitational field. A bowling ball on a shelf above your desk has potential energy. It is only potential, however, and doesn't produce any changes until the bowling ball descends onto your desk. The higher the shelf, the more potential energy the ball has.

Energy constantly flows through nature and produces changes. Sunlight (energy) is absorbed by ocean plants and stored as sugars and starches (energy). When the plant dies, it and other ocean life are buried and become oil (energy), which gets pumped to the surface and burned in automobile engines to produce motion (energy).

Aristotle believed that all change originated in the motion of the starry sphere and flowed down to Earth. Modern science has found a more sophisticated description of the continual change you see around you. In a way, science is the study of the way energy flows through the world and produces change. Energy is the pulse of the natural world.

Using the metric system (Appendix A), energy is expressed in **joules** (abbreviated **J**). One joule is about as much energy as that given up when an apple falls from a table to the floor.

*Energy is the ability to cause change.*

MASS | **ENERGY** | TEMPERATURE AND HEAT | DENSITY | PRESSURE

energy (**Focus on Fundamentals 2**). A planet orbiting the sun has a specific amount of energy that depends only on its average distance from the sun. That energy can be divided between energy of motion and energy stored in the gravitational attraction between the planet and the sun. The energy of motion depends on how fast the planet moves, and the stored energy depends on the size of its orbit. The relation between these two kinds of energy is determined by Newton's laws. That means there has to be a fixed relationship between the rate at which a planet moves around its orbit and the size of the orbit—between its orbital period *P* and the orbit's semimajor axis *a*. You can even derive Kepler's third law from Newton's laws of motion as shown in the next section.

**Figure 5-7**

Skaters demonstrate conservation of angular momentum when they spin faster by drawing their arms and legs closer to their axis of rotation.

### Newton's Version of Kepler's Third Law

The equation for circular velocity is actually a version of Kepler's third law, as you can prove with three lines of simple algebra. The result is one of the most useful formulas in astronomy.

The equation for circular velocity, as you have seen, is:

$$V_c = \sqrt{\frac{GM}{r}}$$

The orbital velocity of a planet is simply the circumference of its orbit divided by the orbital period:

$$V = \frac{2\pi r}{P}$$

If you substitute this for *V* in the first equation and solve for $P^2$, you will get:

$$P^2 = \frac{(4\pi^2)}{(GM)} r^3$$

Here $M$ is just the total mass of the system in kilograms. For a planet orbiting the sun, you can use the mass of the sun for $M$, because the mass of the planet is negligible compared to the mass of the sun. (In a later chapter, you will apply this formula to two stars orbiting each other, and then the mass $M$ will be the sum of the two masses.) For a circular orbit, $r$ equals the semimajor axis $a$, so this formula is a general version of Kepler's third law, $P^2 = a^3$. In Kepler's version, you used astronomical units (AU) and years, but in Newton's version of the formula, you should use units of meters, seconds, and kilograms. $G$, of course, is the gravitational constant.

This is a powerful formula. Astronomers use it to calculate the masses of bodies by observing orbital motion. If, for example, you observed a moon orbiting a planet and you could measure the size of the moon's orbit, $r$, and its orbital period, $P$, you could use this formula to solve for $M$, the total mass of the planet plus the moon. There is no other way to find masses of objects in the universe, and, in later chapters, you will see this formula used over and over to find the masses of stars, galaxies, and planets.

This discussion is a good illustration of the power of Newton's work. By carefully defining motion and gravity and by giving them mathematical expression, Newton was able to derive new truths, among them Newton's version of Kepler's third law. His work transformed the mysterious wanderings of the planets into understandable motions that follow simple rules. In fact, his discovery of gravity explained something else that had mystified philosophers for millennia—the ebb and flow of the oceans.

## Tides and Tidal Forces

Newton understood that gravity is mutual—Earth attracts the moon, and the moon attracts Earth—and that means the moon's gravity can explain the ocean tides.

Tides are caused by small differences in gravitational forces. For example, Earth's gravity attracts your body downward with a force equal to your weight. The moon is less massive and more distant, so it attracts your body with a force that is a tiny percent of your weight. You don't notice that little force, but Earth's oceans respond dramatically.

The side of Earth that faces the moon is about 4000 miles closer to the moon than is the center of Earth. Consequently, the moon's gravity, tiny though it is at the distance of Earth, is just a bit stronger when it acts on the near side of Earth than on the center. It pulls on the oceans on the near side of Earth a bit more strongly than on Earth's center, and the oceans respond by flowing into a bulge of water on the side of Earth facing the moon. There is also a bulge on the side of Earth that faces away from the moon because the moon pulls more strongly on Earth's center than on the far side. Thus the moon pulls Earth away from the oceans, which flow into a bulge away from the moon as shown at the top of ■ Figure 5-8.

You might wonder: If Earth and moon accelerate toward each other, why don't they smash together? The answer is that they would collide in about two weeks except that they are orbiting around their common center of mass. The ocean tides are caused by the accelerations Earth and its oceans feel as they orbit around that center of mass.

A **Common Misconception** holds that the moon's effect on tides means that the moon has an affinity for water—including the water in your body—and, according to some people, that's how the moon affects you. That's not true. If the moon's gravity affected only water, then there would be only one tidal bulge, the one facing the moon. As you know, the moon's gravity acts on the rock of Earth as well as on water, and that produces the tidal bulge in the oceans on the far side of Earth. In fact, small tidal bulges occur in the rocky bulk of Earth as it is deformed by the moon's gravity, and although you don't notice it, as Earth rotates, its mountains and plains rise and fall by a few centimeters. The moon has no special affinity for water, and, because your body is so much smaller than Earth, any tides the moon raises in your body are immeasurably small. Ocean tides are large because oceans are large.

You can see dramatic evidence of tides if you watch the ocean shore for a few hours. As Earth rotates on its axis, the tidal bulges remain fixed with respect to the moon. As the turning Earth carries you and your beach into a tidal bulge, the ocean water deepens, and the tide crawls up the sand. The tide does not so much "come in" as you are carried into the tidal bulge. Later, when Earth's rotation carries you out of the bulge, the ocean becomes shallower, and the tide falls. Because there are two bulges on opposite sides of Earth, the tides rise and fall twice a day on an ideal coast.

In reality, the tidal cycle at any given location can be quite complex because it is affected by the latitude of the site, shape of the shore, winds, and so on. Tides in the Bay of Fundy (New Brunswick, Canada), for example, occur twice a day and can exceed 40 feet. In contrast, the northern coast of the Gulf of Mexico has only one tidal cycle a day of roughly 1 foot.

Gravity is universal, so the sun also produces tides on Earth. The sun is roughly 27 million times more massive than the moon, but it lies almost 400 times farther from Earth. Tides on Earth caused by the sun are less than half as high as those caused by the moon. Twice a month, at new moon and at full moon, the moon and sun produce tidal bulges that add together and produce extreme tidal changes; high tide is exceptionally high, and low tide is exceptionally low. Such tides are called **spring tides.** Here the word *spring* does not refer to the season of the year but to the rapid welling up of water. At first- and third-quarter moons, the sun and moon pull at right angles to each other, and the sun's tides cancel out some of the moon's tides. These less-extreme tides are called **neap tides,** and they do not rise very high or fall very low. The word *neap* comes from an obscure Old English word, *nep,* that seems to have meant "lacking power to advance." Spring tides and neap tides are illustrated in Figure 5-8.

Galileo tried to understand tides, but it was not until Newton described gravity that astronomers could analyze tidal

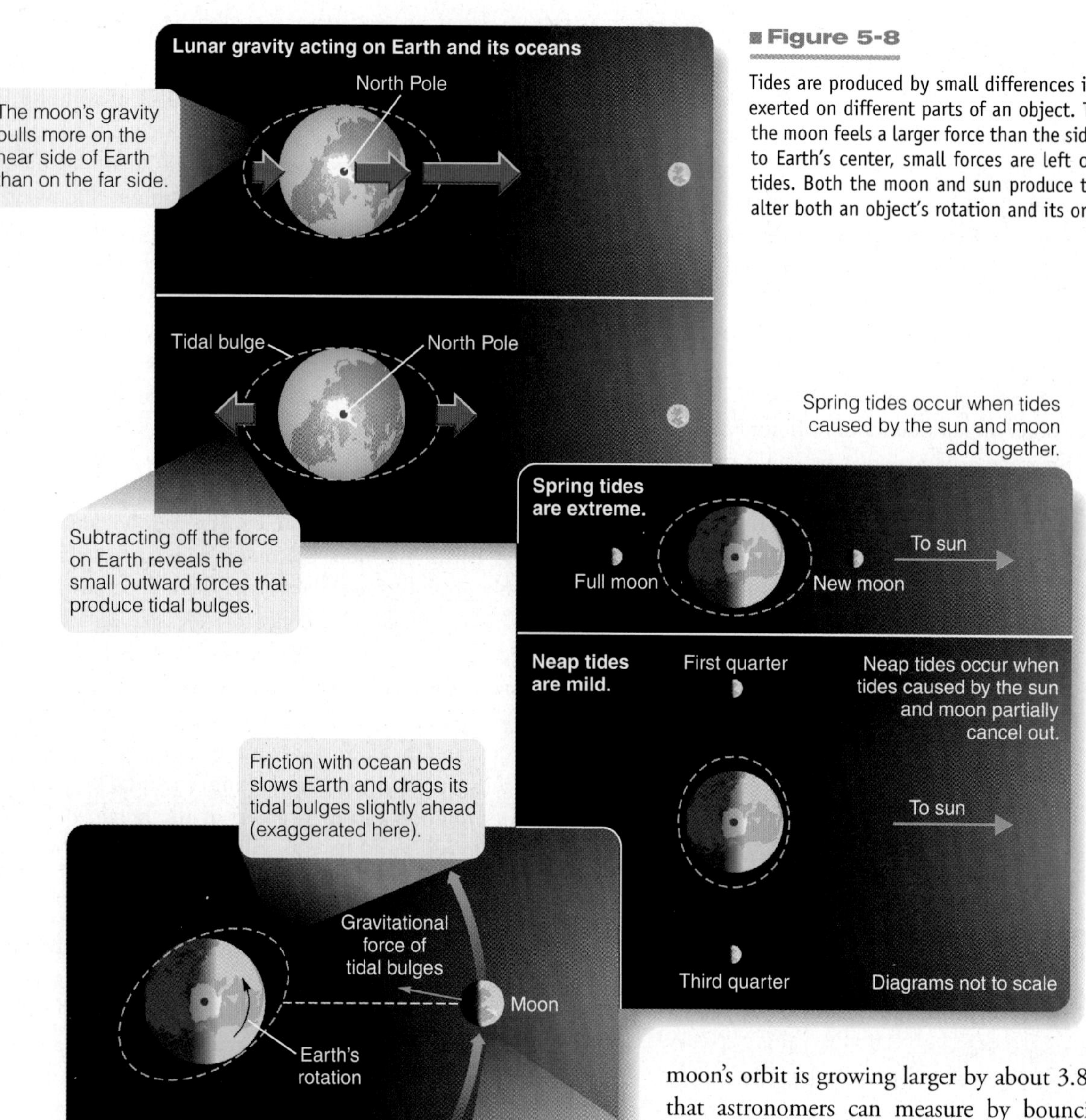
Lunar gravity acting on Earth and its oceans
North Pole
The moon's gravity pulls more on the near side of Earth than on the far side.
Tidal bulge
North Pole
Subtracting off the force on Earth reveals the small outward forces that produce tidal bulges.
Spring tides occur when tides caused by the sun and moon add together.
Spring tides are extreme.
Full moon
New moon
To sun
Neap tides are mild.
First quarter
Neap tides occur when tides caused by the sun and moon partially cancel out.
To sun
Third quarter
Diagrams not to scale
Friction with ocean beds slows Earth and drags its tidal bulges slightly ahead (exaggerated here).
Gravitational force of tidal bulges
Moon
Earth's rotation
Gravity of tidal bulges pulls the moon forward and alters its orbit.

**■ Figure 5-8**

Tides are produced by small differences in the gravitational force exerted on different parts of an object. The side of Earth nearest the moon feels a larger force than the side farthest away. Relative to Earth's center, small forces are left over, and they cause the tides. Both the moon and sun produce tides on Earth. Tides can alter both an object's rotation and its orbital motion.

forces and recognize their surprising effects. For example, the moving water in tidal bulges experiences friction with the ocean beds and resistance as it runs into continents, and that slows Earth's rotation and makes the length of a day grow by 0.0023 second per century. Thin layers of silt laid down millions of years ago where rivers empty into the oceans record tidal cycles as well as daily, monthly, and annual cycles. Those confirm that only 620 million years ago Earth's day was less than 22 hours long. Tidal forces can also affect orbital motion. Earth rotates eastward, and friction with the ocean beds drags the tidal bulges slightly eastward out of a direct Earth–moon line. These tidal bulges are massive, and their gravitational field pulls the moon forward in its orbit, as shown at the bottom of Figure 5-8. As a result, the moon's orbit is growing larger by about 3.8 cm a year, an effect that astronomers can measure by bouncing laser beams off reflectors left on the lunar surface by the Apollo astronauts.

Earth's gravitation exerts tidal forces on the moon, and although there are no bodies of water on the moon, friction within the flexing rock has slowed the moon's rotation to the point that it now keeps the same face toward Earth.

Newton's gravitation is much more than just the force that makes apples fall. In later chapters, you will see how tides can pull gas away from stars, rip galaxies apart, and melt the interiors of small moons orbiting near massive planets. Tidal forces produce some of the most surprising and dramatic processes in the universe.

## Astronomy After Newton

Newton published his work in 1687 in a book called *Philosophiae Naturalis Principia Mathematica (Mathematical Principles of Natural Philosophy),* now known simply as *Principia* (pronounced *Prin-KIP-ee-uh*) (■ Figure 5-9). It is one of the most important books ever written. *Principia* changed astronomy, changed science, and changed the way people think about nature.

# How Do We Know? 5-2

## Testing a Hypothesis by Prediction

**How are the predictions of a hypothesis useful in science? (Note the careful use of language here, to distinguish a tentative hypothesis from a theory that is well confirmed.)** Scientific hypotheses face in two directions. They look back into the past and explain phenomena previously observed. For example, Newton's laws of motion and gravity explained observations of the movements of the planets made over many centuries. But hypotheses also look forward in that they make predictions about what you should find as you explore further. For example, Newton's laws allowed astronomers to calculate the orbits of comets, predict their return, and eventually understand their origin.

Scientific predictions are important in two ways. First, if a prediction of a hypothesis is confirmed, scientists gain confidence that the hypothesis is a true description of nature. But predictions are important for a second reason. They can point the way to unexplored avenues of knowledge.

Particle physics is a field in which predictions have played a key role in directing research. In the early 1970s, physicists proposed a hypothesis about the fundamental forces and particles in atoms called the Standard Model. This hypothesis explained what scientists had already observed in experiments, but it also predicted the existence of particles that hadn't yet been observed. To test the hypothesis, scientists focused their efforts on building more and more powerful particle accelerators in the hopes of detecting the predicted particles.

A number of these particles have since been discovered, and they do match the characteristics predicted by the Standard Model, further confirming the hypothesis. One predicted particle, the Higgs boson, has not yet been found, as of this writing, but an even larger accelerator may allow its detection. Will the Higgs boson be found, or will someone come up with a better hypothesis? This is only one of many cliff-hangers in modern science.

You learned in an earlier chapter that a hypothesis that has passed many tests and made successful predictions can "graduate" to being considered a theory. As you read about any scientific hypothesis or theory, think about both what it can explain and what it can predict.

*Physicists build huge accelerators to search for sub-atomic particles predicted by their theories.* (Brookhaven National Laboratory)

*Principia* changed astronomy by ushering in a new age. No longer did people have to appeal to the whim of the gods to explain things in the heavens. No longer did they speculate on why the planets wander across the sky. After *Principia* astronomers understood that the motions of the heavenly bodies are governed by simple, universal rules that describe the motions of everything from orbiting planets to falling apples. Suddenly the universe was understandable in simple terms, and astronomers could accurately predict future planetary motions (**How Do We Know? 5-2**).

*Principia* also changed science in general. The works of Copernicus and Kepler had been mathematical, but no book before had so clearly demonstrated the power of mathematics as a language of precision. Newton's arguments in *Principia* were so powerful an illustration of the quantitative study of nature that scientists around the world adopted mathematics as their most powerful tool.

Also, *Principia* changed the way people thought about nature. Newton showed that the rules that govern the universe are simple. Particles move according to three rules of motion and attract each other with a force called gravity. These motions are predictable, and that make's the universe a vast machine based on a few simple rules. It is complex only in that it contains a vast number

**Figure 5-9**

Newton, working from the discoveries of Galileo and Kepler, derived three laws of motion and the principle of mutual gravitation. He and some of his discoveries were honored on this old English pound note. Notice the diagram of orbital motion in the background and the open copy of *Principia* in Newton's hands.

of particles. In Newton's view, if he knew the location and motion of every particle in the universe, he could, in principle, derive the past and future of the universe in every detail. This mechanical determinism has been undermined by modern quantum mechanics, but it dominated science for more than two centuries during which scientists thought of nature as a beautiful clockwork that would be perfectly predictable if they knew how all the gears meshed.

Most of all, Newton's work broke the last bonds between science and formal philosophy. Newton did not speculate on the good or evil of gravity. He did not debate its meaning. Not more than a hundred years before, scientists would have argued over the "reality" of gravity. Newton didn't care for these debates. He wrote, "It is enough that gravity exists and suffices to explain the phenomena of the heavens."

Newton's laws dominated astronomy for two centuries. Then, early in the 20th century, Albert Einstein proposed a new way to describe gravity. The new theory did not replace Newton's laws but rather showed that they were only approximately correct and could be seriously in error under special circumstances. Einstein's theories further extend the scientific understanding of the nature of gravity. Just as Newton had stood on the shoulders of Galileo, Einstein stood on the shoulders of Newton.

### SCIENTIFIC ARGUMENT

***How do Newton's laws of motion explain the orbital motion of the moon?***

The key here is to build your argument step by step. If Earth and the moon did not attract each other, the moon would move in a straight line in accord with Newton's first law of motion and vanish into deep space. Instead, gravity pulls the moon toward Earth's center, and the moon accelerates toward Earth. This acceleration is just enough to pull the moon away from its straight-line motion and cause it to follow a curve around Earth. In fact, it is correct to say that the moon is falling, but because of its lateral motion it continuously misses Earth.

Every orbiting object is falling toward the center of its orbit but is also moving laterally fast enough to compensate for the inward motion, and it follows a curved orbit. That is an elegant argument, but it raises a question: **How can astronauts float inside spacecraft in a "weightless" state? Why might "free fall" be a more accurate term?**

## 5-3 Einstein and Relativity

IN THE EARLY YEARS OF THE LAST CENTURY, Albert Einstein (1879–1955) (■ Figure 5-10) began thinking about how motion and gravity interact. He soon gained international fame by showing that Newton's laws of motion and gravity were only partially correct. The revised theory became known as the theory of relativity. As you will see, there are really two theories of relativity.

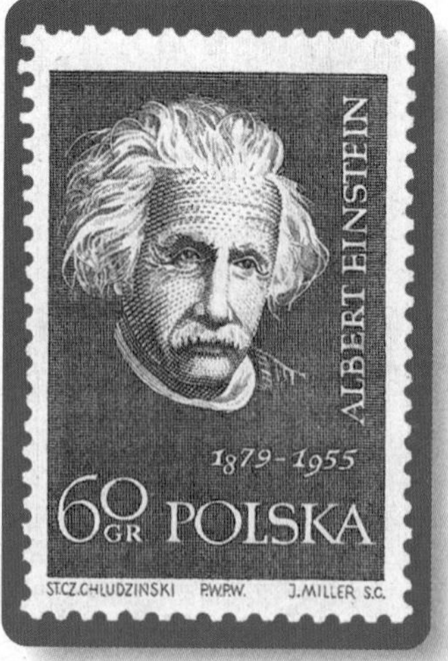


**■ Figure 5-10**

Einstein has become a symbol of the brilliant scientist. His fame began when he was a young man and thought deeply about the nature of motion. That led him to revolutionary insights into the meaning of space and time and a new understanding of gravity.

### Special Relativity

Einstein began by thinking about how moving observers see events around them. His analysis led him to the first postulate of relativity, also known as the principle of relativity:

> **First postulate** (the principle of relativity): Observers can never detect their *uniform* motion except relative to other objects.

You may have experienced the first postulate while sitting on a train in a station. You suddenly notice that the train on the next track has begun to creep out of the station. However, after several moments you realize that it is your own train that is moving and that the other train is still motionless on its track. You can't tell which train is moving until you look at external objects such as the station platform.

Consider another example. Suppose you are floating in a spaceship in interstellar space, and another spaceship comes coasting by (■ Figure 5-11a). You might conclude that it is moving and you are not, but someone in the other ship might be equally sure that you are moving and it is not. Of course, you could just look out a window and compare the motion of your spaceship with a nearby star, but that just expands the problem. Which is moving, your spaceship or the star? The principle of relativity says that there is no experiment you can perform to decide which ship is moving and which is not. This means that there is no such thing as absolute rest—all motion is relative.

Because neither you nor the people in the other spaceship could perform any experiment to detect your absolute motion through space, the laws of physics must have the same form in both spaceships. Otherwise, experiments would produce different results in the two ships, and you could decide who was moving. So, a more general way of stating the first postulate refers to these laws of physics:

> **First postulate** (alternate version): The laws of physics are the same for all observers, no matter what their motion, so long as they are not *accelerated.*

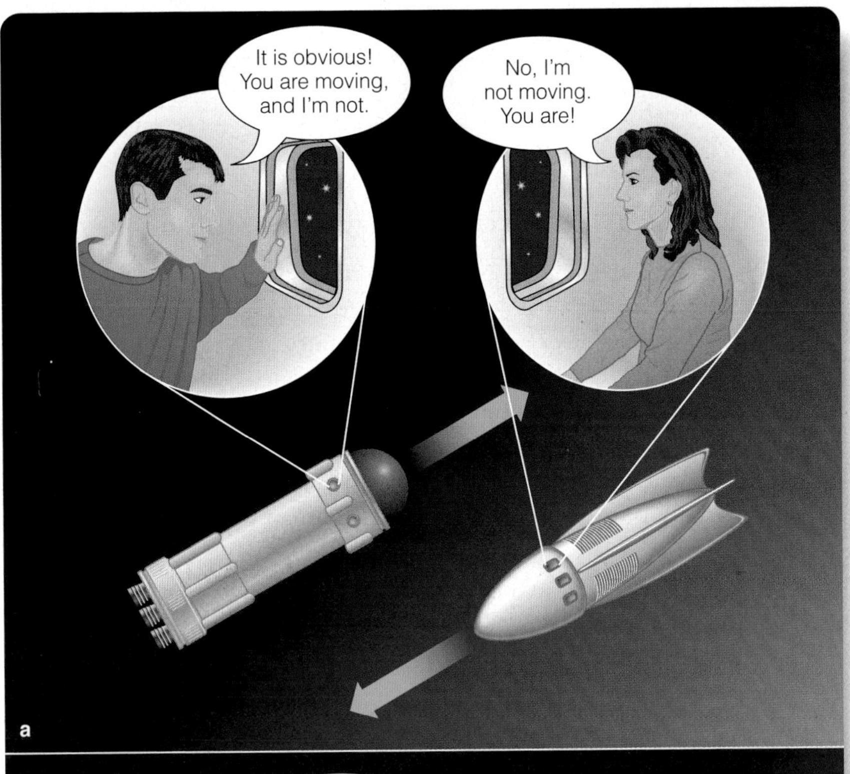


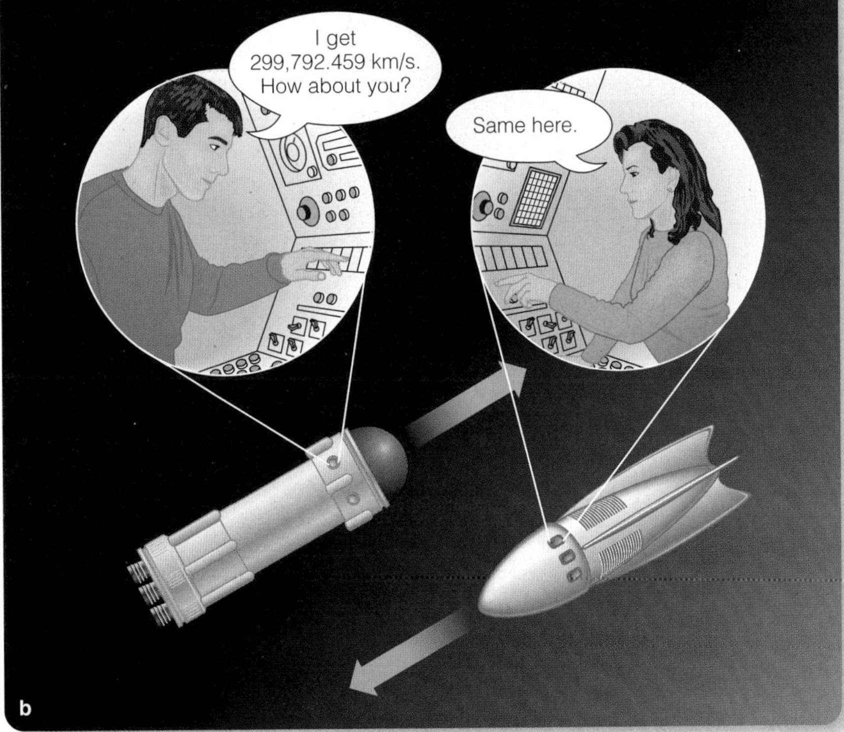


**Figure 5-11**

(a) The principle of relativity says that observers can never detect their uniform motion, except relative to other objects. Thus, neither of these travelers can decide who is moving and who is not. (b) If the velocity of light depended on the motion of the observer through space, then these travelers could perform measurements inside their spaceships to discover who was moving. If the principle of relativity is correct, then the velocity of light must be a constant when measured by any observer.

The word *accelerated* is important. If either spaceship were to fire its rockets, then its velocity would change. The crew of that ship would know it because they would feel the acceleration pressing them into their couches. Accelerated motion, therefore, is different—the pilots of the spaceships can always tell which ship is accelerating and which is not. The postulates of relativity discussed here apply only to the special case of observers in *uniform* motion, which means *unaccelerated* motion. That is why the theory is called the **special theory of relativity.**

The first postulate led Einstein to the conclusion that the speed of light must be constant for all observers. No matter how you are moving, your measurement of the speed of light has to give the same result (Figure 5-11b). This became the second postulate of special relativity:

**Second postulate:** The speed of light is constant and will be the same for all observers independent of their motion relative to the light source.

Remember, this is required by the first postulate; if the speed of light were not constant, then the pilots of the spaceships could measure the speed of light inside their spaceships and decide who was moving.

Once Einstein had accepted the basic postulates of relativity, he was led to some startling discoveries. Newton's laws of motion and gravity worked well as long as distances were small and velocities were low. But when Einstein began to think about very large distances or very high velocities, he realized that Newton's laws were no longer adequate to describe what happens. Instead, the postulates led Einstein to derive a more accurate description of nature now known as the special theory of relativity. It predicts some peculiar effects. For example, special relativity shows that the observed mass of a moving particle depends on its velocity. The higher the velocity, the greater the mass of the particle. This is not significant at low velocities, but it becomes very important as the velocity approaches the speed of light. Such increases in mass are actually observed whenever physicists accelerate atomic particles to high velocities (■ Figure 5-12).

This discovery led to yet another insight. The relativistic equations that describe the energy of a moving particle predict

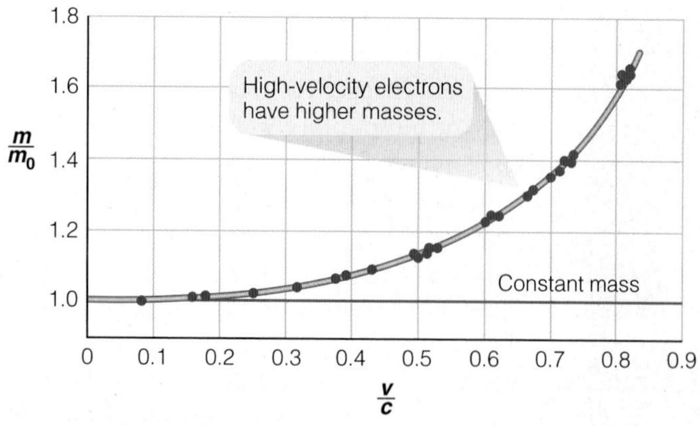


**■ Figure 5-12**

The observed mass of moving electrons depends on their velocity. As the ratio of their velocity to the velocity of light, $v/c$, gets larger, the mass of the electrons in terms of their mass at rest, $m/m_0$, increases. Such relativistic effects are quite evident in particle accelerators, which accelerate atomic particles to very high velocities.

that the energy of a motionless particle is not zero. Rather, its energy at rest is $m_0c^2$. This is of course the famous equation:

$$E = m_0c^2$$

The constant $c$ is the speed of light, and $m_0$ is the mass of the particle when it is at rest. This simple formula shows that mass and energy are related, and you will see in later chapters how nature can convert one into the other inside stars.

For example, suppose that you convert 1 kg of matter into energy. The speed of light is $3 \times 10^8$ m/s, so your result is $9 \times 10^{16}$ joules (J) (approximately equal to a 20-megaton nuclear bomb). Recall that a joule is a unit of energy roughly equivalent to the energy given up when an apple falls from a table to the floor. This simple calculation shows that the energy equivalent of even a small mass is very large.

Other relativistic effects include the slowing of moving clocks and the shrinkage of lengths measured in the direction of motion. A detailed discussion of the major consequences of the special theory of relativity is beyond the scope of this book, but you can have confidence that these strange effects have been confirmed many times in experiments. Einstein's work is called the special *theory* of relativity because it meets the scientific definition of a *theory*: it is very well understood, has been checked many times in many ways, and is widely applicable.

## The General Theory of Relativity

In 1916, Einstein published a more general version of the theory of relativity that dealt with accelerated as well as uniform motion. This **general theory of relativity** contained a new description of gravity.

Einstein began by thinking about observers in accelerated motion. Imagine an observer sitting in a windowless spaceship. Such an observer cannot distinguish between the force of gravity and the inertial forces produced by the acceleration of the spaceship (■ Figure 5-13). This led Einstein to conclude that gravity

**■ Figure 5-13**

(a) An observer in a closed spaceship on the surface of a planet feels gravity. (b) In space, with the rockets smoothly firing and accelerating the spaceship, the observer feels inertial forces that are equivalent to gravitational forces.

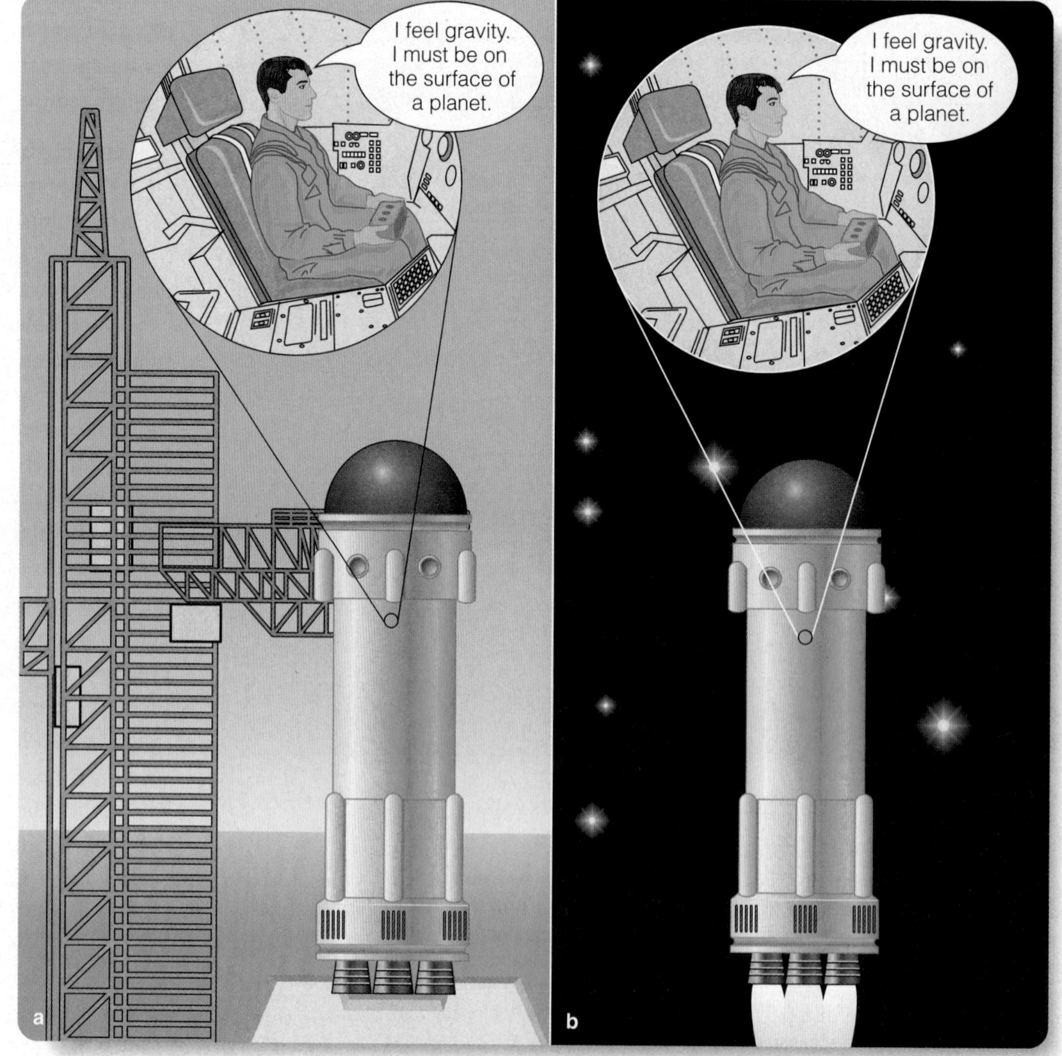

and acceleration are related, a conclusion now known as the equivalence principle:

> **Equivalence principle:** Observers cannot distinguish locally between inertial forces due to acceleration and uniform gravitational forces due to the presence of a massive body.

This should not surprise you. Earlier in this chapter you read that Newton concluded that the mass that resists acceleration is the same as the mass that exerts gravitational forces. He even performed an elegant experiment with pendulums to test the equivalence of the mass related to motion and the mass related to gravity.

The importance of the general theory of relativity lies in its description of gravity. Einstein concluded that gravity, inertia, and acceleration are all associated with the way space is related to time in what is now referred to as space-time. This relation is often referred to as curvature, and a one-line description of general relativity explains a gravitational field as a curved region of space-time:

> **Gravity according to general relativity:** Mass tells space-time how to curve, and the curvature of space-time (gravity) tells mass how to accelerate.

So, you feel gravity because Earth's mass causes a curvature of space-time. The mass of your body responds to that curvature by accelerating toward Earth's center, and that presses you downward in your chair. According to general relativity, all masses cause curvature, and the larger the mass, the more severe the curvature. That's gravity.

## Confirmation of the Curvature of Space-Time

Einstein's general theory of relativity has been confirmed by a number of experiments, but two are worth mentioning here because they were among the first tests of the theory. One involves Mercury's orbit, and the other involves eclipses of the sun.

Johannes Kepler understood that the orbit of Mercury is elliptical, but it wasn't until 1859 that astronomers discovered that the long axis of its orbit sweeps around the sun in a motion that is an example of precession (■ Figure 5-14). The total observed precession is a little over 1.5° per century. This precession is produced by the gravitation of Venus, Earth, and the other planets and by the precession of Earth's axis. However, when astronomers take all known effects into account and use Newton's description of gravity to account for the gravitational influence of all of the planets, they are left with a small excess. Mercury's orbit is advancing 43 seconds of arc per century faster than Newton's laws predict.

This is a tiny effect. Each time Mercury returns to perihelion, its closest point to the sun, it is about 29 km (18 mi) past the position predicted by Newton's laws. This is such a small distance compared with the planet's diameter of 4850 km that it could never have been detected had it not been cumulative. Each orbit, Mercury gains 29 km, and in a century it gains over 12,000 km—more than twice its own diameter. This tiny effect, called the advance of perihelion of Mercury's orbit, accumulated into a serious discrepancy in the Newtonian description of the universe.

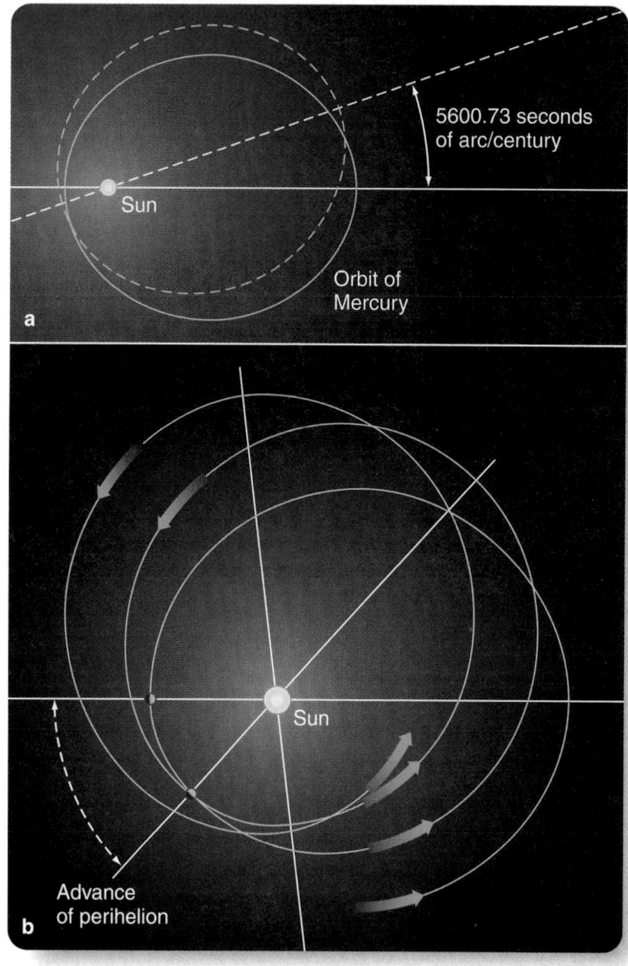


**■ Figure 5-14**

(a) Mercury's orbit precesses 43.11 arc seconds per century faster than predicted by Newton's laws. (b) Even when you ignore the influences of the other planets, Mercury's orbit is not a perfect ellipse. Curved space-time near the sun distorts the orbit from an ellipse into a rosette. The advance of Mercury's perihelion is exaggerated about a million times in this figure.

The advance of perihelion of Mercury's orbit was one of the first problems to which Einstein applied the principles of general relativity. First he calculated how much the sun's mass curves space-time in the region of Mercury's orbit, and then he calculated how Mercury moves through the space-time. The theory predicted that the curved space-time should cause Mercury's orbit to advance by 43.03 seconds of arc per century, well within the observational accuracy of the excess (Figure 5-14b).

When his theory matched observations, Einstein was so excited he could not return to work for three days. He would be

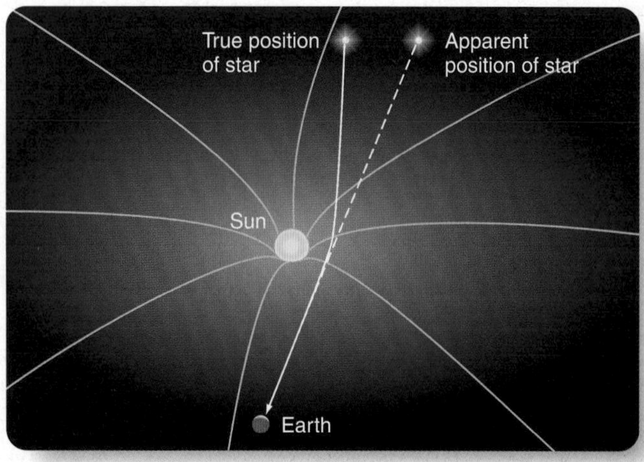


■ **Figure 5-15**

Like a depression in a putting green, the curved space-time near the sun deflects light from distant stars and makes them appear to lie slightly farther from the sun than their true positions.

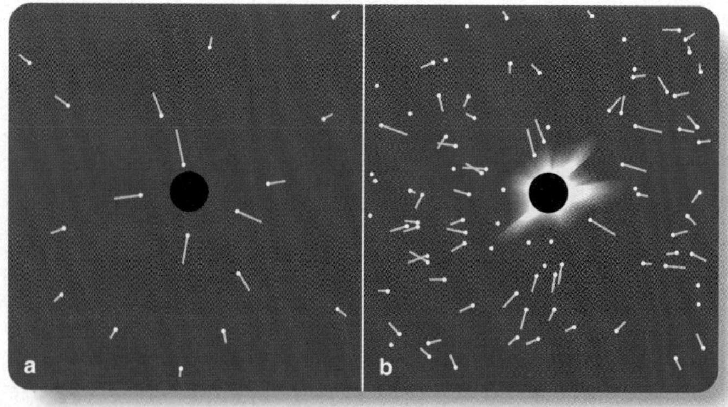


■ **Figure 5-16**

(a) Schematic drawing of the deflection of starlight by the sun's gravity. Dots show the true positions of the stars as photographed months before the eclipse. Lines point toward the positions of the stars during the eclipse. (b) Actual data from the eclipse of 1922. Random uncertainties of observation cause some scatter in the data, but in general the stars appear to move away from the sun by 1.77 arc seconds at the edge of the sun's disk. The deflection of stars is magnified by a factor of 2300 in both (a) and (b).

even happier with modern studies that have shown that Mercury, Venus, Earth, and even Icarus, an asteroid that comes close to the sun, have orbits observed to be slipping forward due to the curvature of space-time near the sun. This same effect has been detected in pairs of stars that orbit each other.

A second test of general relativity was directly related to the motion of light through the curved space-time near the sun. Because light has a limited speed, Newton's laws predict that the gravity of an object should slightly bend the paths of light beams passing nearby. The equations of general relativity indicated that light would have an extra deflection caused by curved space-time, just as a rolling golf ball is deflected by undulations in a putting green. Einstein predicted that starlight grazing the sun's surface would be deflected by 1.75 arc seconds, twice the deflection that Newton's law of gravity would predict (■ Figure 5-15). Starlight passing near the sun is normally lost in the sun's glare, but during a total solar eclipse stars beyond the sun can be seen. As soon as Einstein published his theory, astronomers rushed to observe such stars and test the curvature of space-time.

The first solar eclipse following Einstein's announcement in 1916 occurred on June 8, 1918. It was cloudy at some observing sites, and results from other sites were inconclusive. The next occurred on May 29, 1919, only months after the end of World War I, and was visible from Africa and South America. British teams went to both Brazil and Príncipe, an island off the coast of Africa. Months before the eclipse, they photographed the part of the sky where the sun would be located during the eclipse and measured the positions of the stars on the photographic plates. Then, during the eclipse, they photographed the same star field with the eclipsed sun located in the middle. After measuring the plates, they found slight changes in the positions of the stars. During the eclipse, the positions of the stars on the plates were shifted outward, away from the sun (■ Figure 5-16). If a star had been located at the edge of the solar disk, it would have been shifted outward by about 1.8 arc seconds. This represents good agreement with the theory's prediction.

Because the angles are so small, this is a very delicate observation, and it has been repeated at many total solar eclipses since 1919, with similar results. The most accurate results were obtained in 1973 when a Texas–Princeton team measured a deflection of 1.66 ± 0.18 arc seconds—good agreement with Einstein's theory.

The general theory of relativity is critically important in modern astronomy. You will meet the theory again in the discussions of black holes, distant galaxies, and the big bang universe. Einstein revolutionized modern physics by providing an explanation of gravity based on the geometry of curved space-time. Galileo's inertia and Newton's mutual gravitation are shown to be not just descriptive rules but fundamental properties of space and time.

## SCIENTIFIC ARGUMENT

***What does the equivalence principle tell you?***

The equivalence principle says that there is no observation you can make inside a closed spaceship to distinguish between uniform acceleration and gravitation. Of course, you could open a window and look outside, but then you would no longer be in a closed spaceship. As long as you make no outside observations, you can't tell whether your spaceship is firing its rockets and accelerating through space or resting on the surface of a planet where gravity gives you weight.

Einstein took the equivalence principle to mean that gravity and acceleration through space-time are somehow related. The general theory of relativity gives that relationship mathematical form and shows that gravity is really a distortion in space-time that physicists refer to as curvature. Consequently, you can say "mass tells space-time how to curve, and space-time tells mass how to accelerate." The equivalence principle led Einstein to an explanation for gravity.

Einstein began his work by thinking carefully about common things such as what you feel when you are moving uniformly or accelerating. This led him to deep insights now called postulates. Special relativity sprang from two postulates. **Why does the second postulate have to be true if the first postulate is true?**

### What Are We? Falling

Everything in the universe is falling. The moon is falling around Earth. Earth is falling along its orbit around the sun, and the sun and every other star in our galaxy are falling along their orbits around the center of our galaxy. Stars in other galaxies are falling around the center of those galaxies, and every galaxy in the universe is falling as it feels the gravitational tugs of every bit of matter that exists.

Newton's explanation of gravity as a force between two unconnected masses was action at a distance, and it offended many of the scientists of his time. They thought Newton's gravity seemed like magic. Einstein explained that gravity is a curvature of space-time and that every mass accelerates according to the curvature it feels around it. That's not action at a distance, and it can give you a new insight into how the universe works.

The mass of every atom in the universe contributes to the curvature, creating a universe filled with three-dimensional hills and valleys of curved space-time. You and your world, your sun, your galaxy, and every other object in the universe are falling through space guided by the curvature of space-time.

# Study and Review

## Summary

- Aristotle argued that the universe was composed of four elements: earth at the center, with water, air, and fire in layers above. **Natural motion (p. 77)** occurred when a displaced object returned to its natural place. **Violent motion (p. 77)** was motion other than natural motion and had to be sustained by a force.
- Galileo found that a falling object is accelerated; that is, it falls faster and faster with each passing second. The rate at which it accelerates, termed the **acceleration of gravity (p. 77)**, is 9.8 m/s$^2$ (32 ft/s$^2$) at Earth's surface and does not depend on the weight of the object, contrary to what Aristotle said.
- According to tradition, Galileo demonstrated this by dropping balls of iron and wood from the Leaning Tower of Pisa to show that they would fall together. Air resistance would have ruined the experiment, but a feather and a hammer dropped on the airless moon by an astronaut did fall together.
- Galileo stated the law of inertia: In the absence of friction, a moving body on a horizontal plane will continue moving forever.
- The first of Newton's three laws of motion was based on Galileo's law of inertia: A body continues at rest or in uniform motion in a straight line unless it is acted on by some force.
- **Momentum (p. 79)** is the tendency of a moving body to continue moving.
- **Mass (p. 79)** is the amount of matter in a body.
- Newton's second law says that an **acceleration (p. 79)** (a change in velocity) must be caused by a force. A **velocity (p. 79)** is a directed speed, so a change in speed or direction is an acceleration.
- Newton's third law says that forces occur in pairs acting in opposite directions.
- Newton realized that the curved path of the moon meant that it was being accelerated toward Earth and away from a straight-line path. That required the presence of a force—gravity.

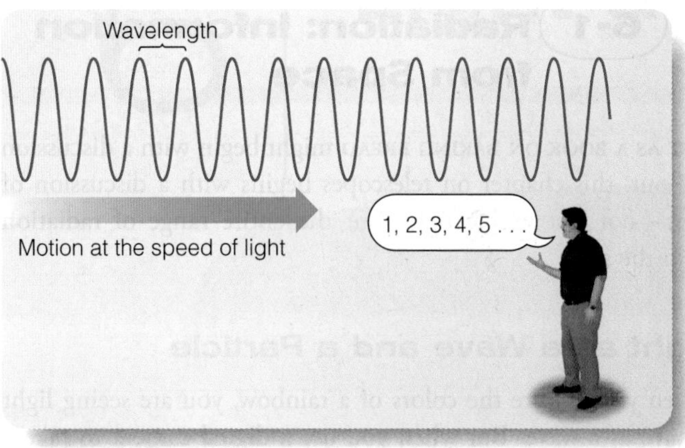


■ **Figure 6-2**

All electromagnetic waves travel at the speed of light. The wavelength is the distance between successive peaks. The frequency of the wave is the number of peaks that pass you in one second.

light does not require a medium, and so it can travel through a perfect vacuum. There is no sound on the moon, but there is plenty of sunlight. There is a **Common Misconception** that radio waves are related to sound. Actually, radio waves are a type of light, electromagnetic radiation that your radio converts into sound so you can listen. Radio communication works just fine between astronauts standing on the moon.

Although electromagnetic radiation can behave as a wave, it can also behave as a flood of particles. A particle of electromagnetic radiation is called a **photon,** and you can think of a photon as a bundle of waves.

The amount of energy a photon carries depends inversely on its wavelength. That is, shorter-wavelength photons carry more energy, and longer-wavelength photons carry less. A simple formula expresses this relationship:

$$E = \frac{hc}{\lambda}$$

Here $h$ is Planck's constant ( $6.63 \times 10^{-34}$ joule s), $c$ is the speed of light ($3.00 \times 10^{8}$ m/s), and $\lambda$ is the wavelength in meters. This book will not use this formula for calculations; the important point is the inverse relationship between the energy $E$ and the wavelength $\lambda$. As $\lambda$ gets smaller, $E$ gets larger. A photon of long wavelength carries a very small amount of energy, but a photon with a very short wavelength can carry much more energy.

## The Electromagnetic Spectrum

A **spectrum** is an array of electromagnetic radiation displayed in order of wavelength. You are most familiar with the spectrum of visible light, which you see in rainbows. The colors of the visible spectrum differ in wavelength, with red having the longest wavelength and violet the shortest. The visible spectrum is shown at the top of ■ Figure 6-3.

The average wavelength of visible light is about 0.0005 mm. You could put roughly 50 light waves end to end across the thickness of a sheet of household plastic wrap. It is awkward to describe such short distances in millimeters, so scientists give the wavelength of light using **nanometer (nm)** units, equal to one-billionth of a meter ($10^{-9}$ m). Another unit that astronomers commonly use is called the **angstrom** (**Å**) (named after the Swedish astronomer Anders Jonas Ångström). One angstrom is $10^{-10}$ m, one-tenth of a nanometer. The wavelength of visible light ranges from about 400 to 700 nm (4000 to 7000 Å). Just as you sense the wavelength of sound as pitch, you sense the wavelength of light as color. Light near the short-wavelength end of the visible spectrum (400 nm) looks violet to your eyes, and light near the long-wavelength end (700 nm) looks red.

Figure 6-3 shows that the visible spectrum makes up only a small part of the entire electromagnetic spectrum. Beyond the red end of the visible spectrum lies **infrared radiation,** where wavelengths range from 700 nm to about 1 mm. Your eyes are not sensitive to this radiation, but your skin senses it as heat. A "heat lamp" warms you by giving off infrared radiation.

Beyond the infrared part of the electromagnetic spectrum lie radio waves. The radio radiation used for AM radio transmissions has wavelengths of a few kilometers down to a few hundred meters, while FM, television, military, government, cell phone, and ham radio transmissions have wavelengths that range down to a few centimeters. Microwave transmission, used for radar and some long-distance telephone communications as well as for cooking food in a microwave oven, has wavelengths from a few centimeters down to about 1 mm.

The boundaries between the sections of the spectrum are not sharply defined. Long-wavelength infrared radiation blends smoothly into the shortest microwave radio waves. Similarly, there is no natural division between the short-wavelength infrared and the long-wavelength part of the visible spectrum.

Now look at the other end of the electromagnetic spectrum in Figure 6-3 and notice that electromagnetic waves shorter than violet are called **ultraviolet.** Electromagnetic waves that are even shorter are called X-rays, and the shortest are gamma rays. Again, the boundaries between these wavelength ranges are defined only by conventional usage, not by natural divisions.

Recall the formula for the energy of a photon. Extremely short-wavelength photons such as X-rays and gamma rays have high energies and can be dangerous. Even ultraviolet photons have enough energy to do you harm. Small doses of ultraviolet produce a suntan, and larger doses cause sunburn and skin cancers. Contrast this to the lower-energy infrared photons. Individually they have too little energy to affect skin pigment, a fact that explains why you can't get a tan from a heat lamp. Only by concentrating many low-energy photons in a small area, as in a microwave oven, can you transfer significant amounts of energy.

Astronomers are interested in electromagnetic radiation from space because it carries clues to the nature of stars, planets, and other celestial objects. Earth's atmosphere is opaque to most electromagnetic radiation, as shown in the graph at the bottom

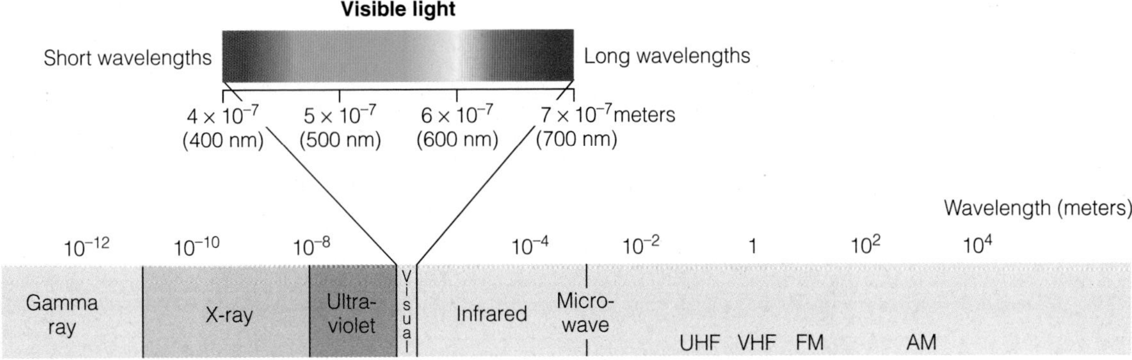


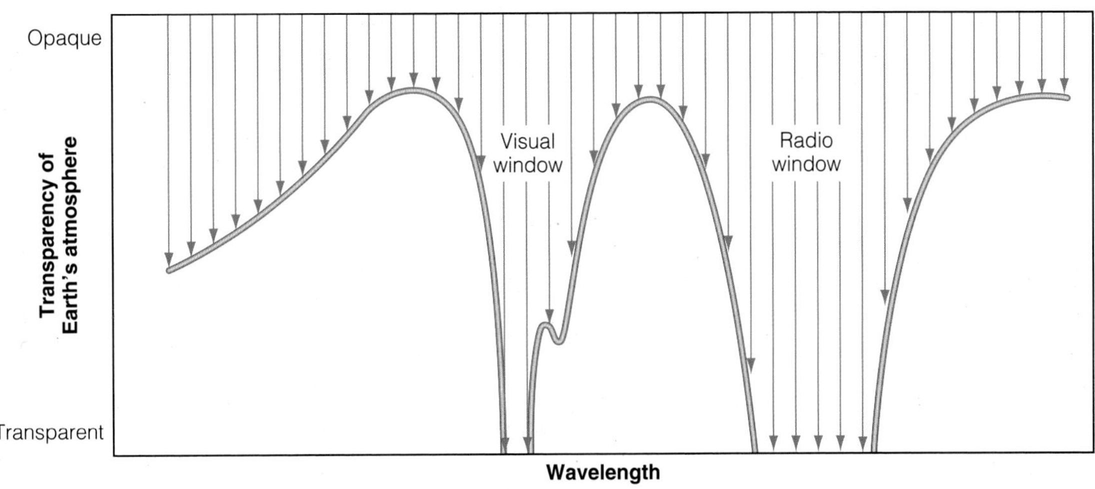


**■ Figure 6-3**

The spectrum of visible light, extending from red to violet, is only part of the electromagnetic spectrum. Most radiation is absorbed in Earth's atmosphere, and only radiation in the visual window and the radio window can reach Earth's surface.

of Figure 6-3. Gamma rays, X-rays, and some radio waves are absorbed high in Earth's atmosphere, and a layer of ozone ($O_3$) at altitudes of about 15–30 km absorbs ultraviolet radiation. Water vapor in the lower atmosphere absorbs the longer-wavelength infrared radiation. Only visible light, some shorter-wavelength infrared, and some radio waves reach Earth's surface through two wavelength regions called **atmospheric windows.** Obviously, if you wish to study the sky from Earth's surface, you must look out through one of these windows.

### SCIENTIFIC ARGUMENT

***What would you see if your eyes were sensitive only to X-rays?*** Sometimes the best way to understand a concept is to build a scientific argument and change one factor so you must imagine a totally new situation. In this case, you might at first expect to be able to see through walls, but remember that your eyes detect only light that already exists. There are almost no X-rays bouncing around at Earth's surface, so if you had X-ray eyes, you would be in the dark and would be unable to see anything. Even when you looked up at the sky, you would see nothing, because Earth's atmosphere is not transparent to X-rays. If Superman can see through walls, it is not because his eyes can detect X-rays.

But now imagine a slightly different situation and modify your argument. **Would you be in the dark if your eyes were sensitive only to radio wavelengths?**

## 6-2 Optical Telescopes

EARTH HAS TWO ATMOSPHERIC WINDOWS, so there are two main types of ground-based telescopes—optical telescopes and radio telescopes. You can start with optical telescopes, which gather light and focus it into sharp images. This requires sophisticated optical and mechanical designs, and it leads astronomers to build gigantic telescopes on the tops of high mountains.

### Two Kinds of Optical Telescopes

Optical telescopes can focus light into an image by using either a lens or a mirror, as shown in ■ Figure 6-4. In a **refracting telescope,** the **primary** (or **objective**) **lens** bends (refracts) the light as it passes through the glass and brings it to a focus to form a small inverted image. In a **reflecting telescope,** the **primary**

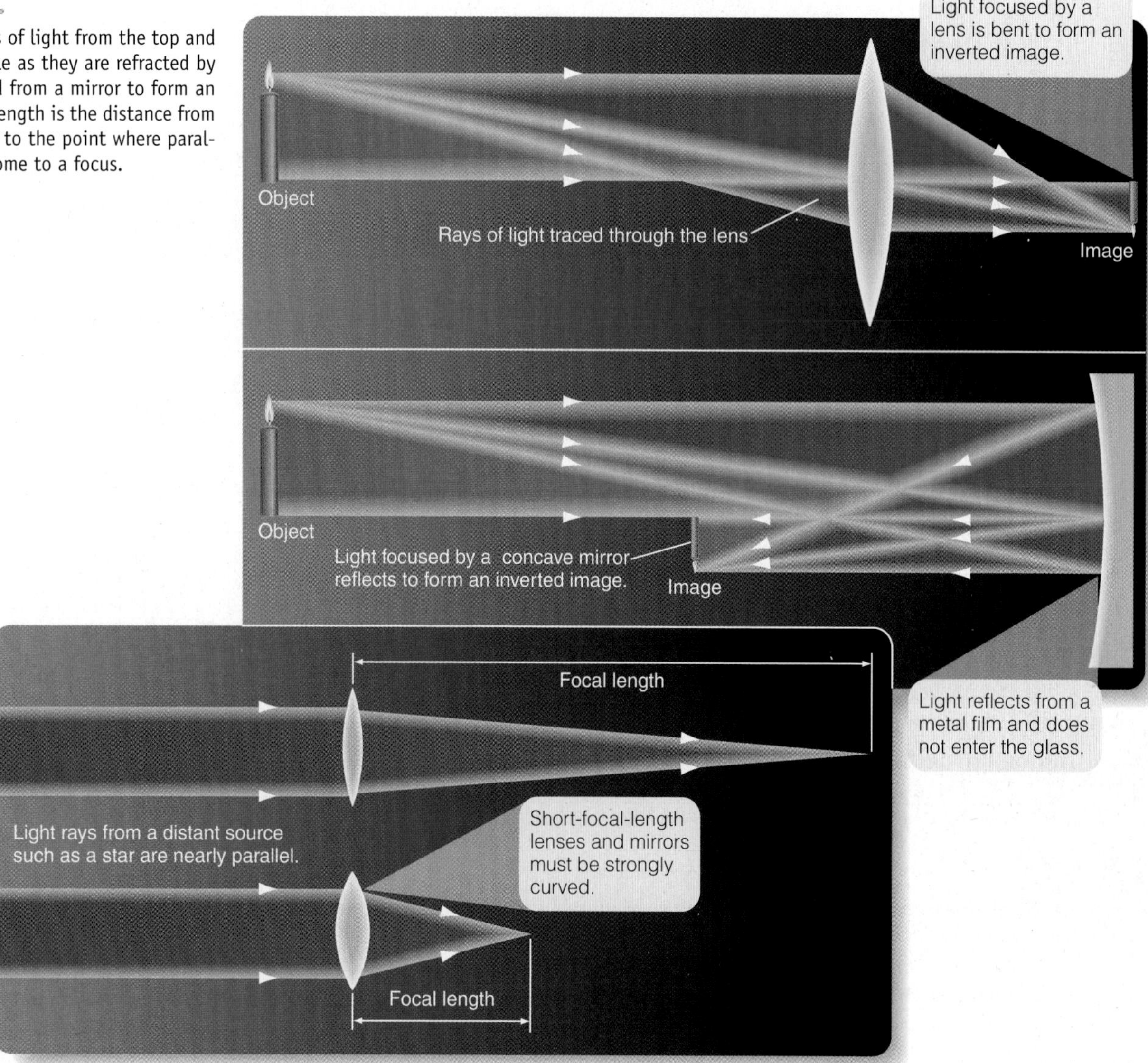


■ **Figure 6-4**

You can trace rays of light from the top and bottom of a candle as they are refracted by a lens or reflected from a mirror to form an image. The focal length is the distance from the lens or mirror to the point where parallel rays of light come to a focus.

(or **objective**) **mirror**—a concave piece of glass with a reflective surface—forms an image by reflecting the light. In either case, the **focal length** is the distance from the lens or mirror to the image of a distant light source such as a star. Short-focal-length lenses and mirrors must be strongly curved, and long-focal-length lenses and mirrors are less strongly curved. Grinding the proper shape on a lens or mirror is a delicate, time-consuming, and expensive process.

The image formed by the primary lens or primary mirror of a telescope is small, inverted, and difficult to view directly. Astronomers use a small lens called the **eyepiece** to magnify the image and make it convenient to view (■ Figure 6-5).

Refracting telescopes suffer from a serious optical distortion that limits their usefulness. When light is refracted through glass, shorter wavelengths bend more than longer wavelengths, so blue light, for example, having shorter wavelengths, comes to a focus closer to the lens than does red light (■ Figure 6-6a). That means if you focus the eyepiece on the blue image, the other colors are out of focus, and you see a colored blur around the image. If you focus on the red image, all the other colors blur. This color separation is called **chromatic aberration.** Telescope designers can grind a telescope lens of two components made of different kinds of glass and so bring two different wavelengths to the same focus (Figure 6-6b). This does improve the image, but these **achromatic lenses** are not totally free of chromatic aberration. Even though two colors have been brought together, the other wavelengths still blur. Telescopes made with achromatic lenses were popular until the end of the 19th century.

The primary lens of a refracting telescope is more expensive than a mirror of the same size. The lens must be achromatic, so it must be made of two different kinds of glass with four precisely ground surfaces. Also, the glass must be pure and flawless because the light passes through it. The largest refracting telescope in the world was completed in 1897 at Yerkes Observatory in Wisconsin.

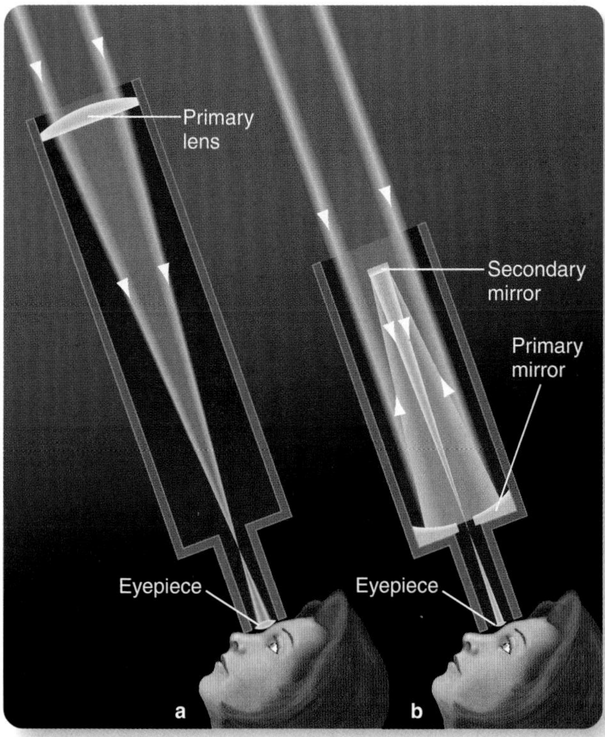


■ **Figure 6-5**

(a) A refracting telescope uses a primary lens to focus starlight into an image that is magnified by a lens called an eyepiece. The primary lens has a long focal length, and the eyepiece has a short focal length. (b) A reflecting telescope uses a primary mirror to focus the light by reflection. A small secondary mirror reflects the starlight back down through a hole in the middle of the primary mirror to the eyepiece.

Its lens is 1 m (40 in.) in diameter and weighs half a ton. Larger refracting telescopes are prohibitively expensive.

The primary mirrors of reflecting telescopes are much less expensive because the light reflects off the front surface of the mirror. This means that only the front surface needs to be ground to precise shape. This front surface is coated with a highly reflective surface of an aluminum alloy, and the light reflects from this front surface without entering the glass. Consequently, the glass of the mirror need not be perfectly transparent, and the mirror can be supported over its back surface to reduce sagging. Most important, reflecting telescopes do not suffer from chromatic aberration because the light is reflected before it enters the glass. For these reasons, every large astronomical telescope built since the beginning of the 20th century has been a reflecting telescope.

## The Powers of a Telescope

Astronomers build large telescopes because a telescope can aid your eyes in three ways—the three powers of a telescope—and the two most important of these powers depends on the diameter of the telescope.

Nearly all of the interesting objects in the sky are faint sources of light, so astronomers need telescopes that can gather large amounts of light to produce bright images. **Light-gathering power** refers to the ability of a telescope to collect light. Catching light in a telescope is like catching rain in a bucket—the bigger the bucket, the more rain it catches (■ Figure 6-7). Light-gathering power is proportional to the area of the telescope

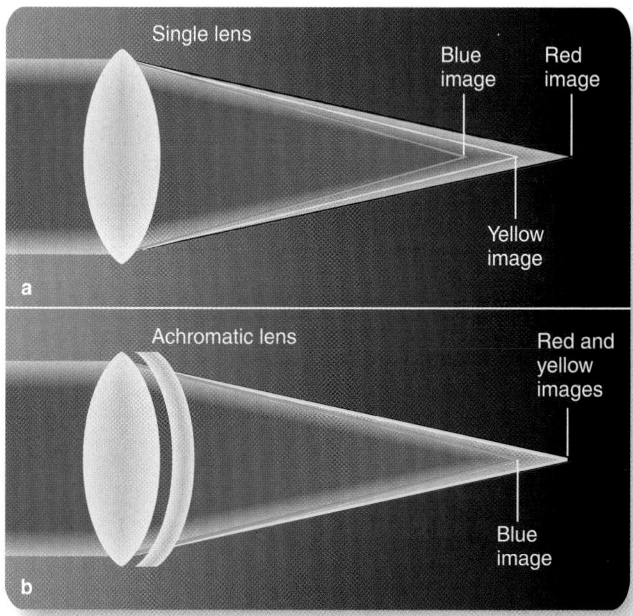


■ **Figure 6-6**

(a) A normal lens suffers from chromatic aberration because short wavelengths bend more than long wavelengths. (b) An achromatic lens, made in two pieces of two different kinds of glass, can bring any two colors to the same focus, but other colors remain slightly out of focus.

■ **Figure 6-7**

Gathering light is like catching rain in a bucket. A large-diameter telescope gathers more light and has a brighter image than a smaller telescope of the same focal length.

objective. A lens or mirror with a large area gathers a large amount of light. The area of a circular lens or mirror is $\pi r^2$, or, written in terms of the diameter, the area is $\pi(D/2)^2$. To compare the relative light-gathering powers *(LGP)* of two telescopes *A* and *B,* you can calculate the ratio of the areas of their objectives, which reduces to the ratio of their diameters *(D)* squared.

$$\frac{LGP_A}{LGP_B} = \left(\frac{D_A}{D_B}\right)^2$$

For example, suppose you compared a telescope 24 cm in diameter with a telescope 4 cm in diameter. The ratio of the diameters is 24/4, or 6, but the larger telescope does not gather six times as much light. Light-gathering power increases as the ratio of diameters squared, so it gathers 36 times more light than the smaller telescope. This example shows the importance of diameter in astronomical telescopes. Even a small increase in diameter produces a large increase in light-gathering power and allows astronomers to study much fainter objects.

The second power, **resolving power,** refers to the ability of the telescope to reveal fine detail. Because light acts as a wave, it produces a small **diffraction fringe** around every point of light in the image, and you cannot see any detail smaller than the fringe (■ Figure 6-8). Astronomers can't eliminate diffraction fringes, but the larger a telescope is in diameter, the smaller the diffraction fringes are. That means the larger the telescope, the better its resolving power.

If you consider only optical telescopes, you can estimate the resolving power by calculating the angular distance between two stars that are just barely visible through the telescope as two separate images. Astronomers say the two images are "resolved," meaning they are separated from each other. The resolving power, $\alpha$, in arc seconds, equals 11.6 divided by the diameter of the telescope in centimeters:

$$\alpha = \frac{11.6}{D}$$

For example, the resolving power of a 25.0 cm telescope is 11.6 divided by 25.0, or 0.46 arc seconds. No matter how perfect the telescope optics, this is the smallest detail you can see through that telescope.

This calculation gives you the best possible resolving power of a telescope of diameter *D,* but the actual resolution can be limited by two other factors—lens quality and atmospheric conditions. A telescope must contain high-quality optics to achieve its full potential resolving power. Even a large telescope reveals little detail if its optics are marred with imperfections. Also, when you look through a telescope, you are looking up through miles of turbulent air in Earth's atmosphere, which makes the image dance and blur, a condition called **seeing.** A related phenomenon is the twinkling of stars. The twinkles are caused by turbulence in Earth's atmosphere, and a star near the horizon, where you look through more air, will twinkle and blur more than a star overhead.

On a night when the atmosphere is unsteady, the images are blurred, and the seeing is bad (■ Figure 6-9). Even under good seeing conditions, the detail visible through a large telescope is limited, not by its diffraction fringes but by the air through which the telescope must look. A telescope performs better on a high mountaintop where the air is thin and steady, but even there Earth's atmosphere limits the detail the best telescopes can reveal to about 0.5 arc seconds. You will learn later in this chapter about telescopes that orbit above Earth's atmosphere and are not limited by seeing.

Seeing and diffraction limit the amount of information in an image, and that limits the accuracy of any measurement made based on that image. Have you ever tried to magnify a newspaper photo to distinguish some detail? Newspaper photos are made up of tiny dots of ink, and no detail smaller than a single dot will be visible no matter how much you magnify the photo. In an astronomical image, the resolution is often limited by seeing. You can't see a detail in the image that is smaller than the resolution.

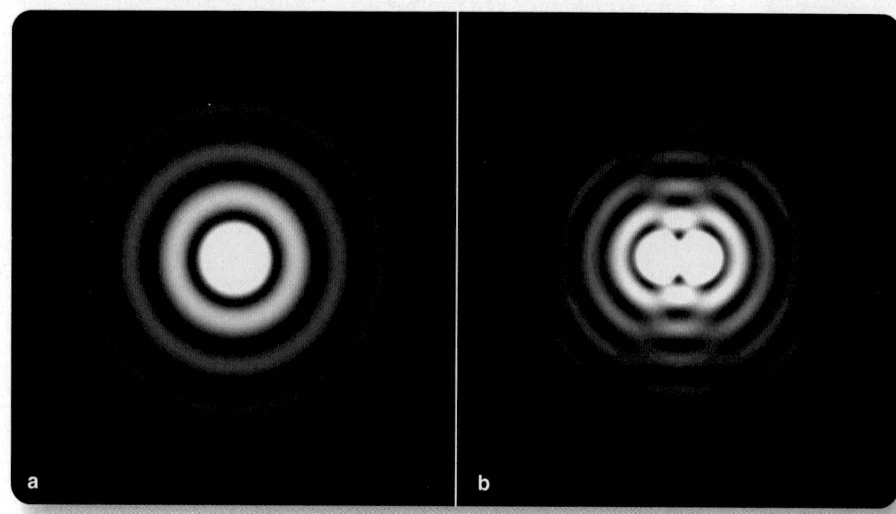


**■ Figure 6-8**

(a) Stars are so far away that their images are points, but the wave nature of light surrounds each star image with diffraction fringes (much magnified in this computer model). (b) Two stars close to each other have overlapping diffraction fringes and become impossible to detect separately. (Computer model by M. A. Seeds)

## How Do We Know? 6-1

### Resolution and Precision

**What limits the detail you can see in an image?** All images have limited resolution. You see this on your computer screen because images there are made up of picture elements, pixels. If your screen has large pixels, the resolution is low, and you can't see much detail. In an astronomical image, the size of a picture element is set by seeing and by diffraction in the telescope. You can't see detail smaller than that resolution limit.

This limitation on the detail in an image is related to the limited precision of any measurement. Imagine a zoologist trying to measure the length of a live snake by holding it along a meter stick. The wriggling snake is hard to hold, so it is hard to measure accurately. Also, meter sticks are usually not marked finer than millimeters. Both factors limit the precision of the measurement. If the zoologist said the snake was 43.28932 cm long, you might be suspicious. The resolution of the measurement technique does not justify the accuracy implied by all those digits.

Whenever you make a measurement you should ask yourself how accurate that measurement can be. The accuracy of the measurement is limited by the resolution of the measurement technique, just as the amount of detail in a photograph is limited by its resolution. If you photographed a star, you would not be able to see details on its surface for the same reason the zoologist can't measure the snake to high precision.

*A high-resolution image of Mars reveals details such as mountains, craters, and the southern polar cap.* (NASA)

That's why stars look like fuzzy points of light no matter how big your telescope. All measurements have some built-in uncertainty (**How Do We Know? 6-1**), and scientists must learn to work within those limitations.

**Figure 6-9**

The left half of this photograph of a galaxy is from an image recorded on a night of poor seeing. Small details are blurred. The right half of the photo is from an image recorded on a night when Earth's atmosphere above the telescope was steady and the seeing was better. Much more detail is visible under good seeing conditions. (Courtesy William Keel)

It is a **Common Misconception** that the purpose of an astronomical telescope is to magnify the image. In fact, the **magnifying power** of a telescope, its ability to make the image bigger, is the least important of the three powers. Because the amount of detail you can see is limited by the seeing conditions and the resolving power, very high magnification does not necessarily show more detail. You can change the magnification by changing the eyepiece, but you cannot alter the telescope's light-gathering power or resolving power without changing the diameter of the objective lens or mirror, and that would be so expensive that you might as well build a whole new telescope.

You can calculate the magnification of a telescope by dividing the focal length of the objective by the focal length of the eyepiece:

$$M = \frac{F_o}{F_e}$$

For example, if a telescope has an objective with a focal length of 80 cm and you use an eyepiece whose focal length is 0.5 cm, the magnification is 80/0.5, or 160 times.

Notice that the two most important powers of the telescope, light-gathering power and resolving power, depend on the diameter of the telescope. This explains why astronomers refer to telescopes by diameter and not by magnification. Astronomers will refer to a telescope as an 8-meter telescope or a 10-meter telescope, but they would never identify a research telescope as being a 200-power telescope.

The quest for light-gathering power and high resolution explains why nearly all major observatories are located far from big cities and usually on high mountains. Astronomers avoid cities because **light pollution,** the brightening of the night sky by light scattered from artificial outdoor lighting, can make it impossible to see faint objects (■ Figure 6-10a). In fact, many residents of cities are unfamiliar with the beauty of the night sky because they can see only the brightest stars. Even far from cities, nature's own light pollution, the moon, is sometimes so bright it drowns out fainter objects, and astronomers are often unable to observe on the nights near full moon. On such nights, faint objects cannot be detected even with the largest telescopes on high mountains.

Astronomers prefer to place their telescopes on carefully selected high mountains for a number of reasons. The air there is thin, very dry, and more transparent, especially in the infrared. For the best seeing, astronomers select mountains where the air flows smoothly and is not turbulent. Building an observatory on top of a high mountain far from civilization is difficult and expensive, as you can imagine from the photo in Figure 6-10b, but the dark sky and steady seeing make it worth the effort.

## Buying a Telescope

Thinking about how to shop for a new telescope will not only help you if you decide to buy one but will also illustrate some important points about astronomical telescopes.

Assuming you have a fixed budget, you should buy the highest-quality optics and the largest-diameter telescope you can afford. You can't make the atmosphere less turbulent, but you can choose good optics. If you buy a telescope from a toy store and it has plastic lenses, you shouldn't expect to see very much. Also, you want to maximize the light-gathering power of your telescope, so you want to purchase the largest-diameter telescope you can afford. Given a fixed budget, that means you should buy a reflecting telescope rather than a refracting telescope. Not only will you get more diameter per dollar, but your telescope will not suffer from chromatic aberration.

You can safely ignore magnification. Department stores and camera shops may advertise telescopes by quoting their magnification, but it is not an important number. What you can see is determined by light-gathering power, optical quality, and Earth's atmosphere. You can change the magnification simply by changing eyepieces.

Other things being equal, you should choose a telescope with a solid mounting that will hold the telescope steady and allow you to point it at objects easily. Computer-controlled

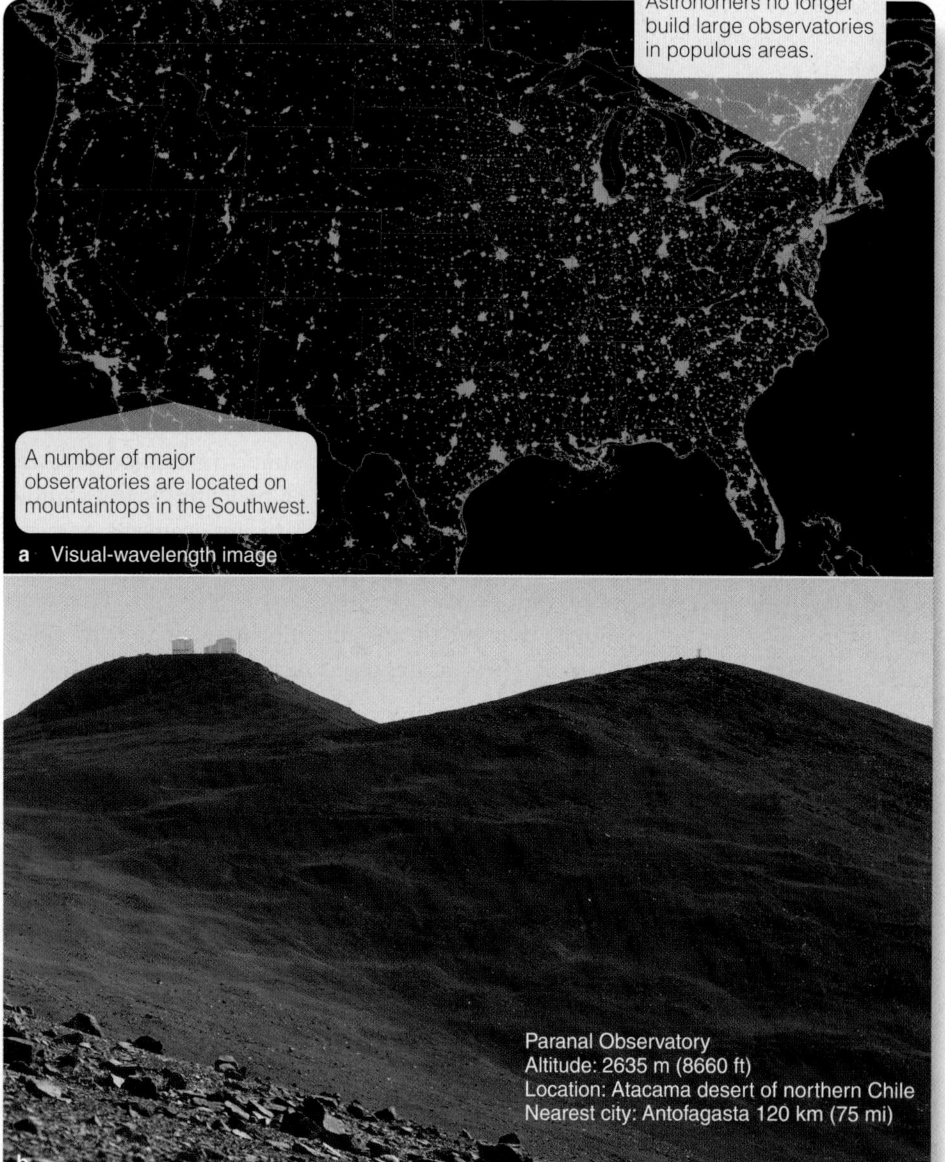


■ **Figure 6-10**

(a) This satellite view of the continental United States at night shows the light pollution and energy waste produced by outdoor lighting. Observatories cannot be located near large cities. (NASA) (b) The domes of four giant telescopes are visible at upper left at Paranal Observatory, built by the European Southern Observatory. The Atacama Desert is believed to be the driest place on Earth. (ESO)

pointing systems are available for a price on many small telescopes. A good telescope on a poor mounting is almost useless.

You might be buying a telescope to put in your backyard, but you must think about the same issues astronomers consider when they design giant telescopes to go on mountaintops.

## Observing Beyond the Ends of the Visible Spectrum

Telescopes in mountain-top observatories usually observe at visual wavelengths, but important observations also can be made from Earth at some infrared and ultraviolet wavelengths.

Beyond the red end of the visible spectrum, some infrared radiation leaks through the atmosphere in narrow, partially open atmospheric windows ranging from wavelengths of 1200 nm to about 20,000 nm. Infrared astronomers usually measure wavelength in micrometers ($10^{-6}$ meters), so they refer to this wavelength range as 1.2 to 30 micrometers (or microns for short). Even in this range, much of the radiation from celestial sources is absorbed by water vapor, carbon dioxide, and ozone molecules. Nevertheless, some infrared observations can be made from mountaintops where the air is thin and dry. For example, a number of important infrared telescopes observe from the 4200-m (13,800-ft) summit of Mauna Kea in Hawaii. At this altitude, the telescopes are above much of the water vapor in Earth's atmosphere (■ Figure 6-11).

Infrared telescopes have flown to high altitudes under balloons and in airplanes to get above absorption by water vapor. NASA is now testing the Stratospheric Observatory for Infrared Astronomy (SOFIA), a Boeing 747SP that will carry a 2.5-m telescope, control systems, and a team of astronomers, technicians, and educators into the dry fringes of the atmosphere. Once at that altitude, they can open a door above the telescope and make infrared observations for hours as the plane flies a precisely calculated path. You can see the door in the photo in Figure 6-11.

To reduce internal noise, the light-sensitive detectors in astronomical telescopes are cooled to very low temperatures, usually with liquid nitrogen, as shown in Figure 6-11. This is especially necessary for a telescope observing at infrared wavelengths. Infrared radiation is emitted by heated objects, and if the telescope is warm it will emit many times more infrared radiation than that coming from a distant object. Imagine trying to look for something at night through binoculars that are themselves glowing.

Beyond the other end of the visible spectrum, astronomers can observe in the near-ultraviolet at wavelengths of about 290 to 400 nm. Your eyes don't detect this radiation, but it can be recorded by specialized detectors. Wavelengths shorter than about 290 nm, the far-ultraviolet, are completely absorbed by the ozone layer extending from about 15 km to 30 km above Earth's surface. No mountaintop is that high, and no airplane can fly to such an altitude. To observe in the far-ultraviolet or beyond at X-ray or gamma-ray wavelengths, telescopes must be in space above the atmosphere.

Infrared astronomers can often observe with the dome lights on. Their instruments are not usually sensitive to visible light.
SOFIA will fly at roughly 12 km (over 40,000 ft) to get above most of Earth's atmosphere.
NASA
DLR
Adding liquid nitrogen to the camera on a telescope is a familiar task for astronomers.

**■ Figure 6-11**

Comet Hale–Bopp hangs in the sky over the 3-meter NASA Infrared Telescope Facility (IRAF) atop Mauna Kea. The air at high altitudes is so dry that it is transparent to shorter infrared photons. SOFIA will fly at altitudes up to 14 km ( 45,000 feet) where it will be able to observe infrared wavelengths that cannot be observed from mountaintops. Most astronomical CCD cameras must be cooled to low temperatures, and this is especially true for infrared cameras. (IRTF: William Keel; SOFIA: NASA; Camera: Kris Koenig/Coast Learning Systems)

## New-Generation Telescopes

For most of the 20th century, astronomers faced a serious limitation on the size of astronomical telescopes. Traditional telescope mirrors were made thick to avoid sagging that would distort the reflecting surface, but those thick mirrors were heavy. The 5-m (200-in.) mirror on Mount Palomar weighs 14.5 tons. These traditional telescopes were big, heavy, and expensive.

Today's astronomers have solved these problems in a number of ways. Read **Modern Astronomical Telescopes** on pages 110–111 and notice three important points about telescope design and ten new terms that describe astronomical telescopes and their operation:

1. Traditional telescopes use large, solid, heavy mirrors to focus starlight to a *prime focus*, or, by using a *secondary mirror*, to a *Cassegrain focus*. Some small telescopes have a *Newtonian focus* or a *Schmidt-Cassegrain focus*.

2. Telescopes must have a *sidereal drive* to follow the stars, and an *equatorial mounting* with easy motion around a *polar axis* is the traditional way to provide that motion. Today, astronomers can build simpler, lighter-weight telescopes on *alt-azimuth mountings* that depend on computers to move the telescope so that it follows the westward motion of the stars as Earth rotates.

3. *Active optics*, computer control of the shape of telescope mirrors, allows the use of thin, lightweight mirrors—either "floppy" mirrors or segmented mirrors. Reducing the weight of the mirror reduces the weight of the rest of the telescope and makes it stronger and less expensive. Also, thin mirrors cool faster at nightfall and produce better images.

High-speed computers have allowed astronomers to build new, giant telescopes with unique designs. A few are shown in ■ Figure 6-12. The European Southern Observatory has built the Very Large Telescope (VLT) high in the remote Andes Mountains of northern Chile. The VLT actually consists of four telescopes, each with a computer-controlled mirror 8.2 m in diameter and only 17.5 cm (6.9 in.) thick. The four telescopes can work singly or can combine their light to work as one large telescope. Italian and American astronomers have built the Large Binocular Telescope, which carries a pair of 8.4-m mirrors on a single mounting. The Gran Telescopio Canarias, located atop a volcanic peak in the Canary Islands, carries a segmented mirror 10.4 meters in diameter and holds, for the moment, the record as the largest single telescope in the world

Other giant telescopes are being planned with segmented mirrors or with multiple mirrors (■ Figure 6-13). The Giant Magellan Telescope will carry seven thin mirrors, each 8.4 meters in diameter, on a single mounting. It will be located in the Chilean Andes and is planned to have the light-gathering power of a 22-m telescope. The Thirty Meter Telescope, now under development by American astronomers, will have a mirror 30 meters in diameter, composed of 492 hexagonal segments. The European Extremely Large Telescope is being planned by an international team. It will carry 906 segments, making up a mirror 42 meters in diameter. Other very large telescopes are being proposed with completion dates of 2016 or later.

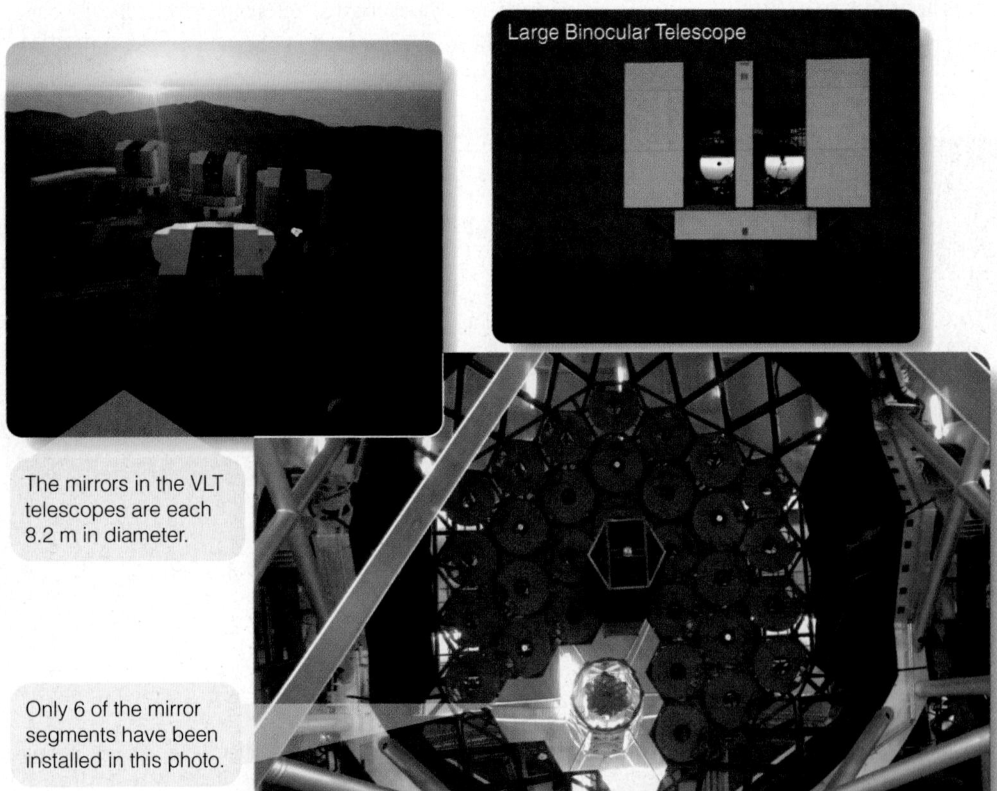

Large Binocular Telescope
The mirrors in the VLT telescopes are each 8.2 m in diameter.
Only 6 of the mirror segments have been installed in this photo.


**■ Figure 6-12**

The four telescopes of the VLT are housed in separate domes at Paranal Observatory in Chile (Figure 6-10). The Large Binocular Telescope (LBT) carries two 8.4-m mirrors that combine their light. The entire building rotates as the telescope moves. The Gran Telescopio Canarias contains 36 hexagonal mirror segments in its 10.4-m primary mirror. (VLT: ESO; LBT: Large Binocular Telescope Project and European Industrial Engineer; GMT: ESO; GTC: Gara Mora, IAC)

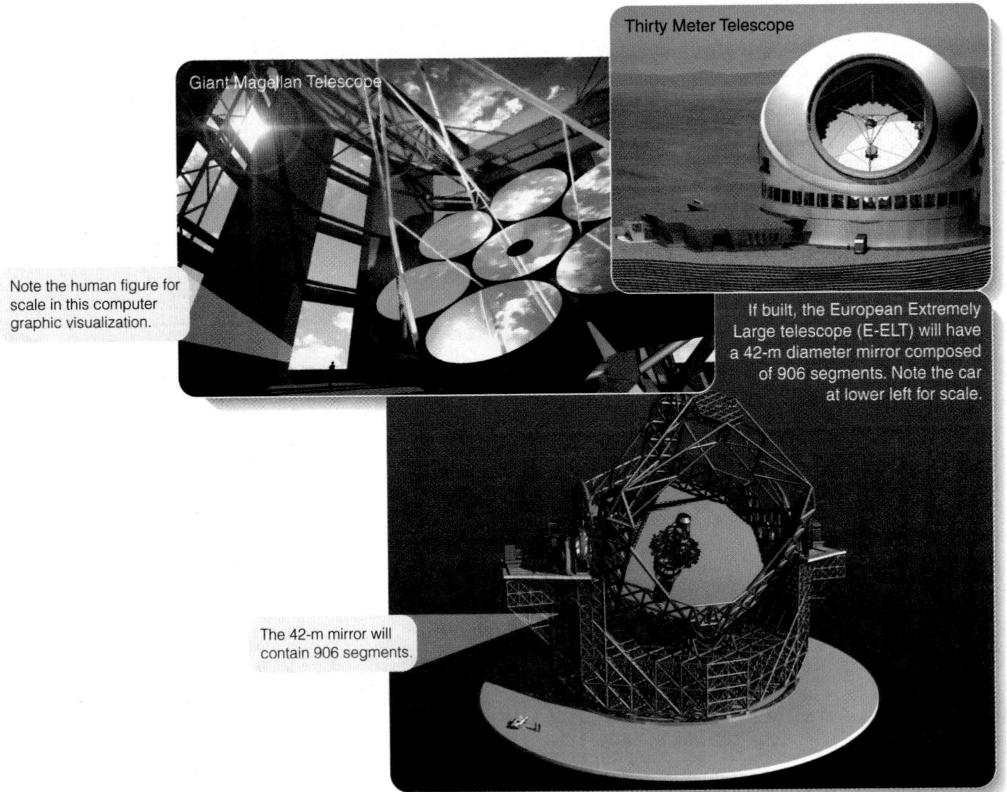

Thirty Meter Telescope
Giant Magellan Telescope
Note the human figure for scale in this computer graphic visualization.
If built, the European Extremely Large telescope (E-ELT) will have a 42-m diameter mirror composed of 906 segments. Note the car at lower left for scale.
The 42-m mirror will contain 906 segments.


**■ Figure 6-13**

The proposed Giant Magellan Telescope will have the resolving power of a telescope 24.5 meters in diameter when it is finished in about 2016. The Thirty Meter Telescope (TMT) is planned to occupy a specially designed dome. Like nearly all of the newest large telescopes, the European Extremely Large Telescope will be on an alt-azimuth mounting. (GMT: ESO; TMT: TMT: Thirty-Meter Telescope; E-ELT: ESO)

Modern computers have revolutionized telescope design and operation. Nearly all large telescopes are operated by astronomers and technicians working at computers in a control room, and some telescopes can be operated by astronomers thousands of miles from the observatory. Some telescopes are fully automated and observe without direct human supervision. This has made possible huge surveys of the sky in which millions of objects are observed. The Sloan Digital Sky Survey, for example, mapped the sky, measuring the position and brightness of 100 million stars and galaxies at a number of wavelengths. The Two-Micron All Sky Survey (2MASS) has mapped the entire sky at three wavelengths in the infrared. Other surveys are being made at other wavelengths. Astronomers will study those data banks for decades to come.

## Adaptive Optics

Not too many years ago, astronomers thought it was pointless to build more large telescopes on Earth's surface because of seeing distortion caused by the atmosphere. In the 1990s, computers became fast enough to allow astronomers to correct for some of that distortion, and that has made the new generation of giant telescopes possible.

**Adaptive optics** uses high-speed computers to monitor the distortion produced by turbulence in Earth's atmosphere and then correct the telescope image to sharpen a fuzzy blob into a crisp picture. The resolution of the image is still limited by diffraction in the telescope, but removing much of the seeing distortion produces a dramatic improvement in the detail visible (■ Figure 6-14). Don't confuse adaptive optics with the slower-speed active optics that controls the overall shape of a telescope mirror.

To monitor the distortion in an image, adaptive optics systems must look at a fairly bright star in the field of view, and there isn't always such a star properly located near a target object such as a faint galaxy. In that case, astronomers can point a laser at a spot in the sky very close to their target object, and where the laser excites gas in Earth's upper atmosphere, it produces a glowing

**■ Figure 6-14**

In these images of the center of our galaxy, the adaptive optics system was turned off for the left image and on for the right image. Not only are the images of stars sharper, but because the light is focused into smaller images, fainter stars are visible. The laser beam shown leaving one of the Keck Telescopes produces an artificial star in the field of view, and the adaptive optics system uses the laser-produced point of light to reduce seeing distortion in the entire image. (left, CFHT; right, Paul Hirst)

artificial star in the field of view. The adaptive optics system can use the artificial star to correct the image of the fainter target.

Today astronomers are planning huge optical telescopes composed of segmented mirrors tens of meters in diameter. Those telescopes would be almost useless without adaptive optics.

### Interferometry

One of the reasons astronomers build big telescopes is to increase resolving power, and astronomers have been able to achieve very high resolution by connecting multiple telescopes together to work as if they were a single telescope. This method of synthesizing a larger telescope is known as **interferometry** (■ Figure 6-15). One expert said, "We combine the light from separate telescopes and fool the waves into thinking they were collected by one big 'scope." The images from such a virtual telescope are not limited by the diffraction fringes of the individual small telescopes but rather by the diffraction fringes of the much larger virtual telescope.

In an interferometer, the light from the separate telescopes must be combined as if it had been collected by a single large mirror. That means that a network of small, high-precision mirrors must bring the light beams together, and the path that each light beam travels must be controlled so that it does not vary more than a small fraction of the wavelength. Turbulence in Earth's atmosphere constantly distorts the light, and high-speed computers must continuously adjust the light paths. Recall that the wavelength of light is very short, roughly 0.0005 mm, so building optical interferometers is one of the most difficult technical problems that astronomers face. Infrared- and radio-wavelength interferometers are slightly easier to build because the wavelengths are longer. In fact, as you will discover later in this chapter, the first astronomical interferometers were built by radio astronomers.

The VLT shown in Figure 6-12 consists of four 8.2-m telescopes that can operate separately but can also be linked together through underground tunnels with three 1.8-m telescopes on the same mountaintop. The resulting optical interferometer provides the resolution (but, of course, not the light gathering power) of a telescope 200 meters in diameter. Other telescopes, such as the two Keck 10-m telescopes, can work as interferometers. The CHARA array on Mt. Wilson combines six 1-meter telescopes to create the resolving power equivalent of a telescope one-fifth of a mile in diameter. The Large Binocular Telescope shown in Figure 6-12 can be used as an interferometer.

Although turbulence in Earth's atmosphere can be partially averaged out in an interferometer, plans are being made to put interferometers in space to avoid atmospheric turbulence altogether. The Space Interferometry Mission, for example, will work at visual wavelengths and study everything from the cores of erupting galaxies to planets orbiting nearby stars.

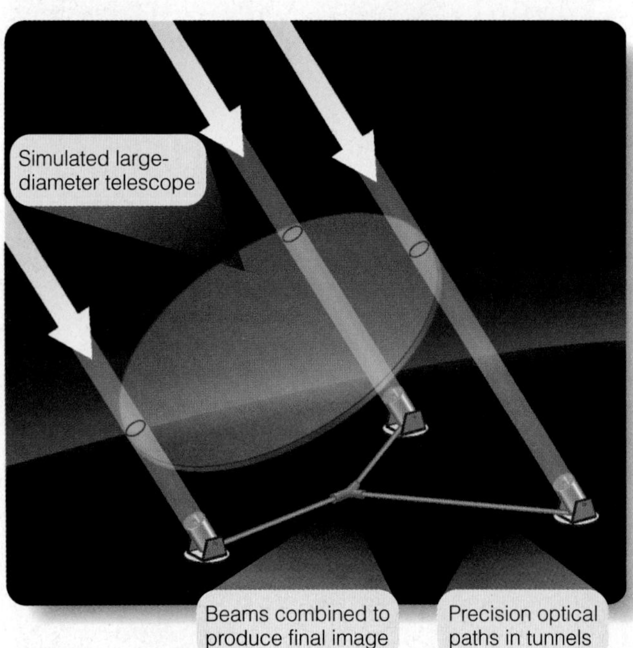


**■ Figure 6-15**

In an astronomical interferometer, smaller telescopes can combine their light through specially designed optical tunnels to simulate a larger telescope with a resolution set by the separation of the smaller telescopes.

#### SCIENTIFIC ARGUMENT

***Why do astronomers build observatories at the tops of mountains?***
Measurement accuracy is so important that scientists often take extreme steps to gather accurate information. It certainly isn't easy to build a large, delicate telescope at the top of a high mountain, but it is worth the effort. A telescope on top of a high mountain is above the densest part of Earth's atmosphere. There is less air to dim the light, and there is less water vapor to absorb infrared radiation. Even more important, the thin air on a mountaintop causes less disturbance to the image, and consequently the seeing is better. A large telescope on Earth's surface has a resolving power much better than the distortion caused by Earth's atmosphere. So, it is limited by seeing, not by its own diffraction. It really is worth the trouble to build telescopes atop high mountains.

Astronomers not only build telescopes on mountaintops, they also build gigantic telescopes many meters in diameter. Revise your argument to focus on telescope design. **What are the problems and advantages in building such giant telescopes?**

## 6-3 Special Instruments

Just looking through a telescope doesn't tell you much. A star looks like a point of light. A planet looks like a little disk. A galaxy looks like a hazy patch. To use an astronomical telescope to learn about the universe, you must be able to analyze the light the telescope gathers. Special instruments attached to the telescope make that possible.

## Imaging Systems

The original imaging device in astronomy was the photographic plate. It could record images of faint objects in long time exposures and could be stored for later analysis. But photographic plates have been almost entirely replaced by electronic imaging systems.

Most modern astronomers use **charge-coupled devices (CCDs)** to record images. A CCD is a specialized computer chip containing millions of microscopic light detectors arranged in an array about the size of a postage stamp. Although CCDs for astronomy are extremely sensitive and therefore expensive, less sophisticated CCDs are used in video and digital cameras. Not only can CCD chips replace photographic plates, but they have some dramatic advantages. They can detect both bright and faint objects in a single exposure, are much more sensitive than photographic plates, and can be read directly into computer memory for later analysis.

You can easily sharpen and enhance images from your digital camera because the image from a CCD is stored as numbers in computer memory. Astronomers can also manipulate images to bring out otherwise invisible details. Astronomers can produce **false-color images** in which different colors represent different levels of intensity and are not related to the true colors of the object. Or they can use false colors to represent different wavelengths not otherwise visible to the human eye, as in ■ Figure 6-16. False-color images are so useful they are commonly used in many other fields, such as medicine and meteorology.

In the past, measurements of intensity and color were made using specialized light meters attached to a telescope or on photographic plates. Today, nearly all such measurements are made more easily and more accurately with CCD images.

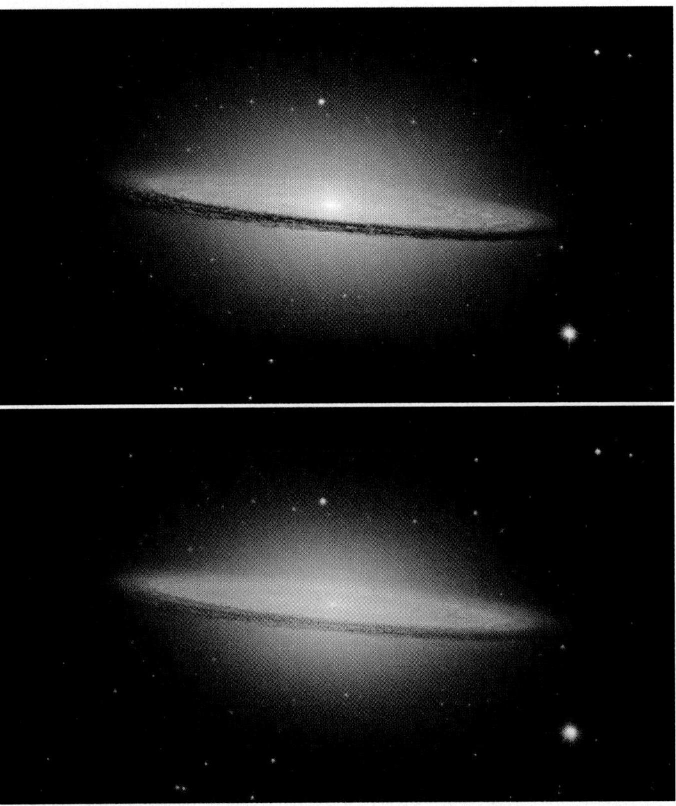

**■ Figure 6-16**

(a) The Sombrero Galaxy is 29 million light-years from Earth and contains many billions of stars. This visual wavelength image shows the clouds of dust in its disk. (NASA, Hubble Heritage Team; STScI/AURA) (b) This infrared image is reproduced in false color to show different wavelengths of infrared emission. Red represents the longest wavelengths and blue the shortest. (NASA/JPL-Caltech, Hubble Heritage Team; STScI/AURA)

## The Spectrograph

To analyze light in detail, astronomers need to spread the light out according to wavelength to form a spectrum, a task performed by a **spectrograph.** You can understand how this instrument works if you imagine reproducing an experiment performed by Isaac Newton in 1666. Newton bored a small hole in the window shutter of his bedroom to admit a thin beam of sunlight. When he placed a prism in the beam, it spread the light into a beautiful spectrum that splashed across his bedroom wall. From this and related experiments Newton concluded that white light is made of a mixture of all the colors.

Light passing through a prism is bent at an angle that depends on its wavelength. Violet (short wavelength) bends most, red (long wavelength) least, so the white light is spread into a spectrum (■ Figure 6-17). You could build a simple spectrograph using a prism to spread the light and a lens to guide the light into a camera.

Nearly all modern spectrographs use a grating in place of a prism. A **grating** is a piece of glass with thousands of microscopic parallel grooves scribed onto its surface. Different wavelengths of light reflect from the grating at slightly different angles, so white light is spread into a spectrum. You have probably noticed this effect when you look at the closely spaced lines etched onto a compact disk; as you tip the disk, different colors flash across its surface. You could build a modern spectrograph by using a high-quality grating to spread light into a spectrum and a CCD camera to record the spectrum.

The spectrum of an astronomical object can contain hundreds of **spectral lines**—dark or bright lines that cross the spectrum at specific wavelengths. The sun's spectrum, for instance, contains hundreds of dark spectral lines produced by the atoms in the sun's hot gases. To measure the precise wavelengths of individual lines and identify the atoms that produced them, astronomers use a **comparison spectrum** as a calibration. Special bulbs built into the spectrograph produce bright lines given off by such atoms as thorium, argon, or neon. The wavelengths of these spectral lines have been measured to high precision in the laboratory, so astronomers can use spectra of these light sources as guides to measure wavelengths and identify spectral lines in the spectrum of a star, galaxy, or planet.

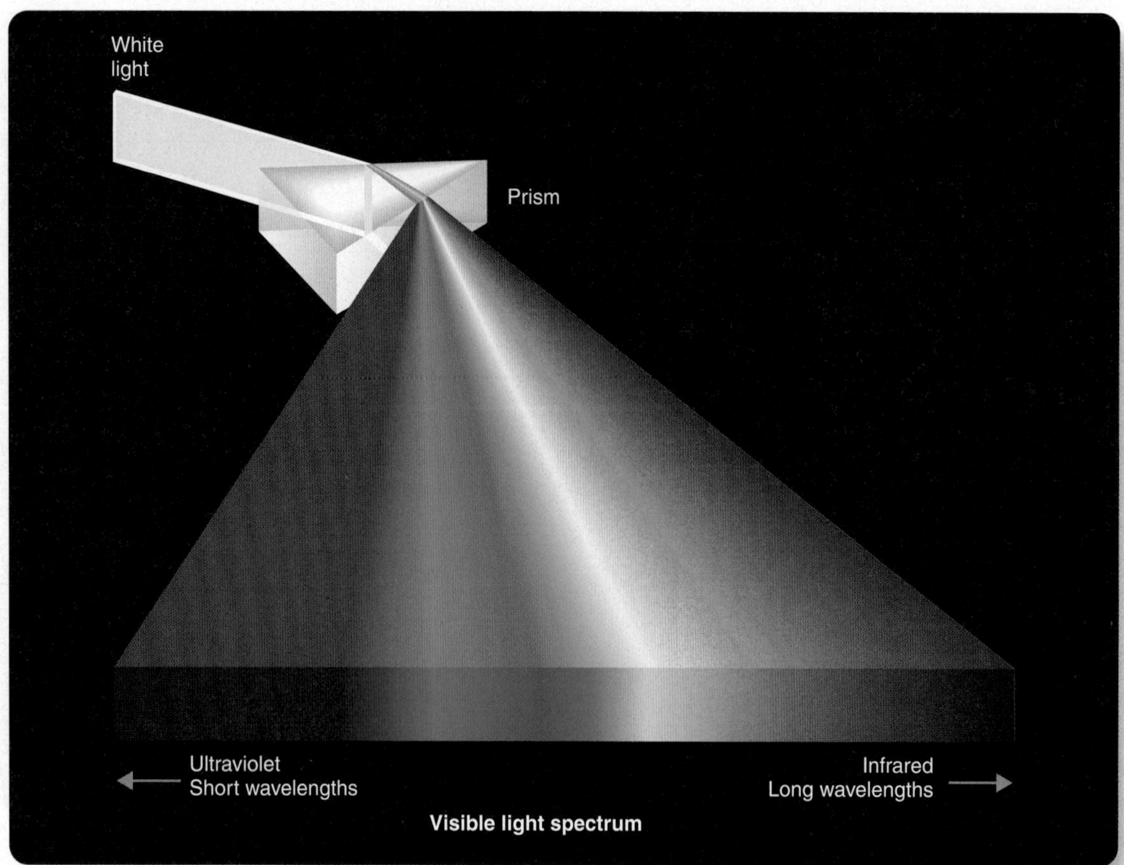


**Figure 6-17**

A prism bends light by an angle that depends on the wavelength of the light. Short wavelengths bend most and long wavelengths least. Thus, white light passing through a prism is spread into a spectrum.

Because astronomers understand how light interacts with matter, a spectrum carries a tremendous amount of information, and that makes a spectrograph the astronomer's most powerful instrument. In the next chapter, you will learn more about the information astronomers can extract from a spectrum. An astronomer once remarked, "We don't know anything about an object till we get a spectrum," and that is only a slight exaggeration.

## 6-4 Radio Telescopes

Celestial objects such as clouds of gas and erupting stars emit radio energy, and astronomers on Earth can study such objects by observing at wavelengths in the radio window where Earth's atmosphere is transparent to radio waves (see Figure 6-3). You might think an erupting star would produce a strong radio signal, but the signals arriving on Earth are astonishingly weak—a million to a billion times weaker than the signal from an FM radio station. Detecting such weak signals calls for highly sensitive equipment.

### The Operation of a Radio Telescope

A radio telescope usually consists of four parts: a dish reflector, an antenna, an amplifier, and a recorder (■ Figure 6-18). These components, working together, make it possible for astronomers to detect radio radiation from celestial objects.

The dish reflector of a radio telescope, like the mirror of a reflecting telescope, collects and focuses radiation. Because radio waves are much longer than light waves, the dish need not be as smooth as a mirror; wire mesh will reflect all but the shortest wavelength radio waves.

Though a radio telescope's dish may be many meters in diameter, the antenna may be as small as your hand. Like the antenna on a TV set, its only function is to absorb the radio energy collected by the dish. Because the radio energy from celestial objects is so weak, it must be strongly amplified before it can be measured and its strength is recorded in computer memory.

Radio telescopes do not produce images. A single observation with a radio telescope measures the amount of radio energy coming from a specific point on the sky. So the radio telescope

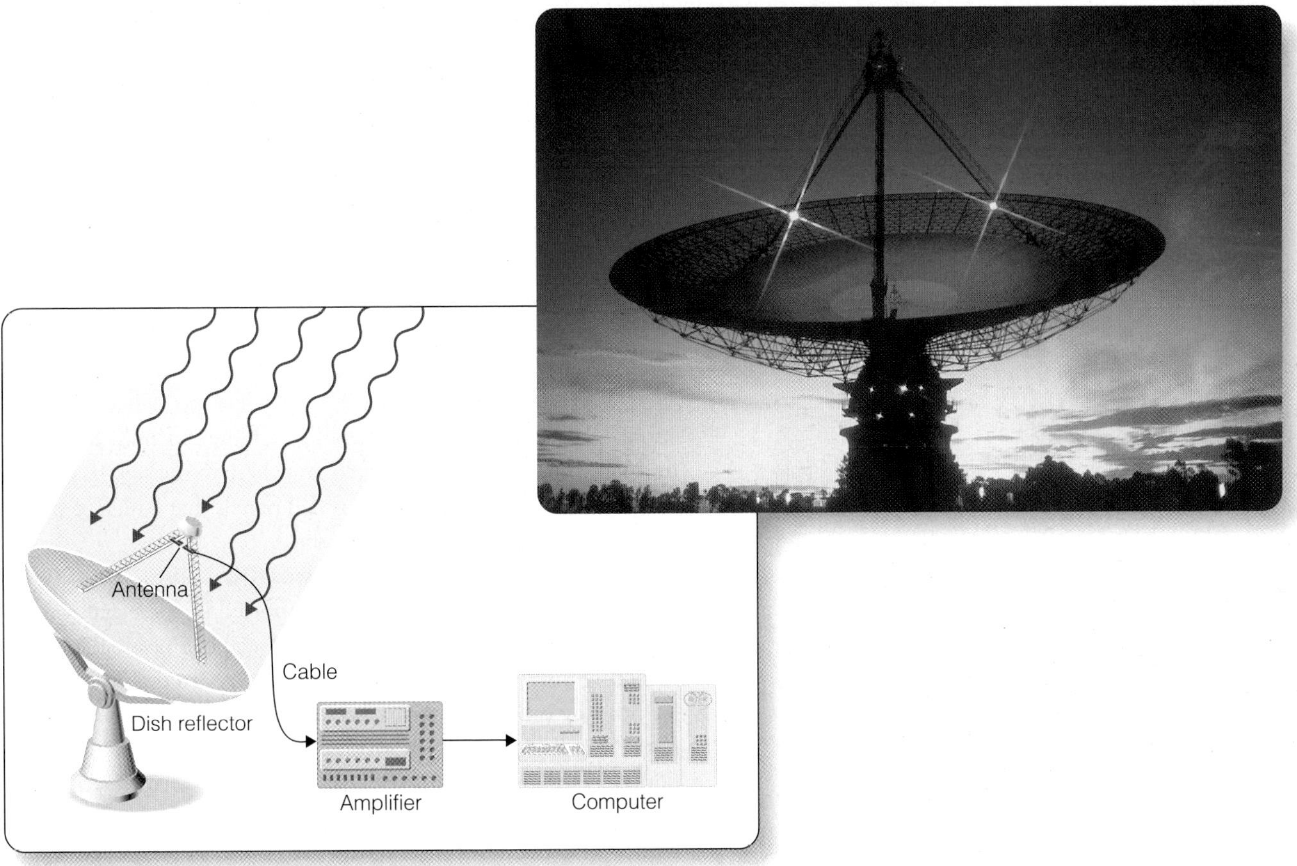


**Figure 6-18**

In most radio telescopes, a dish reflector concentrates the radio signal on the antenna. The signal is then amplified and recorded. For all but the shortest radio waves, wire mesh is an adequate reflector.

must be scanned over an object to produce a map of the radio intensity at different points. Such radio maps are usually represented using contours to mark areas of similar radio intensity, much like a weather map where contours filled with color indicate areas of precipitation (■ Figure 6-19).

## Limitations of a Radio Telescope

A radio astronomer works under three handicaps: poor resolution, low intensity, and interference. Recall that the resolving power of an optical telescope depends on the diameter of the objective lens or mirror. It also depends on the wavelength of the radiation because, at very long wavelengths like those of radio waves, the diffraction fringes are very large. That means a radio map can't show fine detail. As with an optical telescope, there is no way to improve the resolving power without building a bigger telescope. Consequently, radio telescopes generally have large diameters to minimize the diffraction fringes.

Even so, the resolving power of a radio telescope is not good. A dish 30 m in diameter receiving radiation with a wavelength of 21 cm has a resolving power of about 0.5°. Such a radio telescope would be unable to detect any details in the sky smaller than the moon. Fortunately, radio astronomers can combine two or more radio telescopes to form a **radio interferometer** capable of much higher resolution. Just as in the case of optical interferometers, the radio astronomer combines signals from two or more widely separated dishes and "fools the waves" into behaving as if they were collected by a much bigger radio telescope.

Radio interferometers can be quite large. The Very Large Array (VLA) consists of 27 dish antennas spread in a Y-shape across the New Mexico desert (■ Figure 6-20). In combination, they have the resolving power of a radio telescope 36 km (22 mi) in diameter. The VLA can resolve details smaller than 1 arc second. Eight new dish antennas being added across New Mexico will give the VLA ten times better resolving power. Another large radio interferometer, the Very Long Baseline Array (VLBA), consists of matched radio dishes spread from Hawaii to the Virgin Islands and has an effective diameter almost as large as Earth. The Allen Telescope Array being built in California will eventually include 350 separate radio dishes. Radio astronomers are now planning the Square Kilometer Array, which will contain a huge number of radio dishes totaling a square kilometer of collecting area and spread to a diameter of at least 6000 kilometers. These

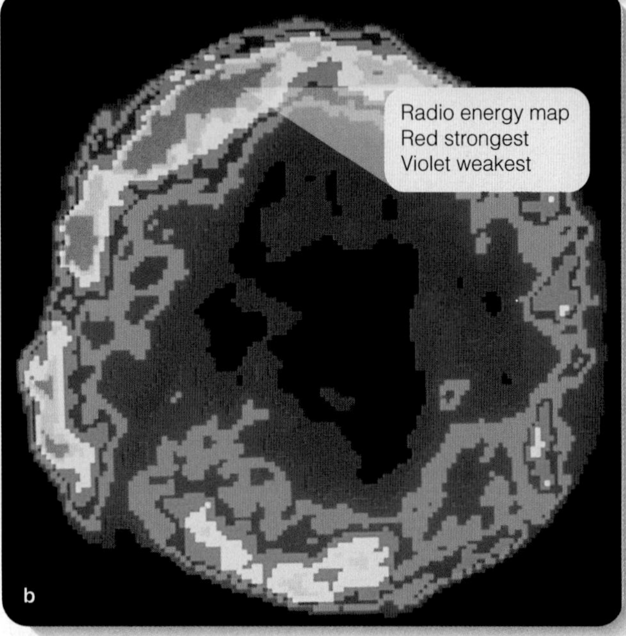


■ **Figure 6-19**

(a) A typical weather map uses contours with added color to show which areas are likely to receive precipitation. (b) A false-color-image radio map of Tycho's supernova remnant, the expanding shell of gas produced by the explosion of a star in 1572. The radio contour map has been color-coded to show intensity. (Courtesy NRAO)

giant radio interferometers depend on state-of-the-art, high-speed computers to combine signals and create radio maps.

The second handicap radio astronomers face is the low intensity of the radio signals. You learned earlier that the energy of a photon depends on its wavelength. Photons of radio energy have such long wavelengths that their individual energies are quite low. To get detectable signals focused on the antenna, the radio astronomer must build large collecting areas either as single large dishes or arrays of smaller dishes.

The largest fully steerable radio telescope in the world is at the National Radio Astronomy Observatory in Green Bank, West Virginia (■ Figure 6-21a). The telescope has a reflecting surface 100 meters in diameter, big enough to hold an entire football field, and can be pointed anywhere in the sky. Its surface consists of 2004 computer-controlled panels that adjust to maintain the shape of the reflecting surface.

The largest radio dish in the world is 300 m (1000 ft) in diameter. So large a dish can't be supported in the usual way, so it is built into a mountain valley in Arecibo, Puerto Rico. The reflecting dish is a thin metallic surface supported above the valley floor by cables attached near the rim, and the antenna hangs above the dish on cables from three towers built on three mountain peaks that surround the valley (Figure 6-21b). By moving the antenna above the dish, radio astronomers can point the telescope at any object that passes within 20 degrees of the zenith as Earth rotates. Since completion in 1963, the telescope has been an international center of radio astronomy research.

The third handicap the radio astronomer faces is interference. A radio telescope is an extremely sensitive radio receiver listening to faint radio signals. Such weak signals are easily drowned out by interference that includes everything from poorly designed transmitters in Earth satellites to automobiles with faulty ignition systems. A few narrow radio bands in the electromagnetic spectrum are reserved for radio astronomy, but even those are often contaminated by radio noise. To avoid interference, radio astronomers locate their telescopes as far from civilization as possible. Hidden deep in mountain valleys, they are able to listen to the sky protected from human-made radio noise.

■ **Figure 6-20**

(a) The Very Large Array uses 27 radio dishes, which can be moved to different positions along a Y-shaped set of tracks across the New Mexico desert. They are shown here in the most compact arrangement. Signals from the dishes are combined to create very-high-resolution radio maps of celestial objects. (NRAO) (b) The proposed Square Kilometer Array will have a concentration of detectors and radio dishes near its center with more dishes scattered up to 3000 km away. (Xilostudios/SKA Program Development Office)

**■ Figure 6-21**

(a) The largest steerable radio telescope in the world is the GBT located in Green Bank, West Virginia. With a diameter of 100 m, it stands higher than the Statue of Liberty. (Mike Bailey: NRAO/AUII) (b) The 300-m (1000-ft) radio telescope in Arecibo, Puerto Rico, hangs from cables over a mountain valley. The Arecibo Observatory is part of the National Astronomy and Ionosphere Foundation operated by Cornell University and the National Science Foundation. (David Parker/Science Photo Library)

### Advantages of a Radio Telescope

Building large radio telescopes in isolated locations is expensive, but three factors make it all worthwhile. First, and most important, a radio telescope can reveal where clouds of cool hydrogen and other atoms and molecules are located. These clouds are important because, for one thing, they are the places where stars are born. Although cool clouds of gas are completely invisible to normal telescopes because they produce no visible light of their own and reflect too little to be detected in photographs, some gas atoms and molecules do emit radio photons. Cool hydrogen, for example, emits radio energy at the specific wavelength of 21 cm. (You will see how the hydrogen produces this radiation when you learn about the gas clouds in space in Chapter 10.) Other gas molecules emit radio energy with their own characteristic wavelengths. The only way astronomers can detect these clouds is with a radio telescope.

Another reason radio telescopes are important is related to dust in space. Astronomers observing at visual wavelengths can't see through the dusty clouds in space. Light waves are so short that they are scattered by the tiny dust grains and never get through the dust to reach optical telescopes on Earth. However, radio signals have wavelengths much longer than the diameters of dust grains, so radio waves from far across the galaxy pass unhindered, giving radio astronomers an unobscured view. (The same is also true, to a large extent, of infrared radiation.)

Finally, radio telescopes are important because they can detect objects that are more luminous at radio wavelengths than at visible wavelengths. This includes, for example, intensely hot gas orbiting black holes. Some of the most violent events in the universe are detectable at radio wavelengths.

## 6-5 Astronomy from Space

You have learned about the observations that ground-based telescopes can make through the two atmospheric windows in the visible and radio parts of the electromagnetic spectrum. Most of the rest of the electromagnetic radiation—infrared, ultraviolet, X-ray, and gamma ray—never reaches Earth's surface; it is absorbed high in Earth's atmosphere. To observe at these wavelengths, telescopes must go above the atmosphere.

### The Hubble Space Telescope

Named after Edwin Hubble, the astronomer who discovered the expansion of the universe, the Hubble Space Telescope is the most successful telescope ever to orbit Earth (■ Figure 6-22). It was launched in 1990 and contains a 2.4-m (95-in.) mirror with which it can observe visible, near-ultraviolet, and near-infrared light. It is controlled from a research center on Earth and observes continuously. Nevertheless, the telescope has time to complete only a fraction of the projects proposed by astronomers from around the world.

Most of the observations Hubble makes are at visual wavelengths, so its greatest advantage in being above Earth's

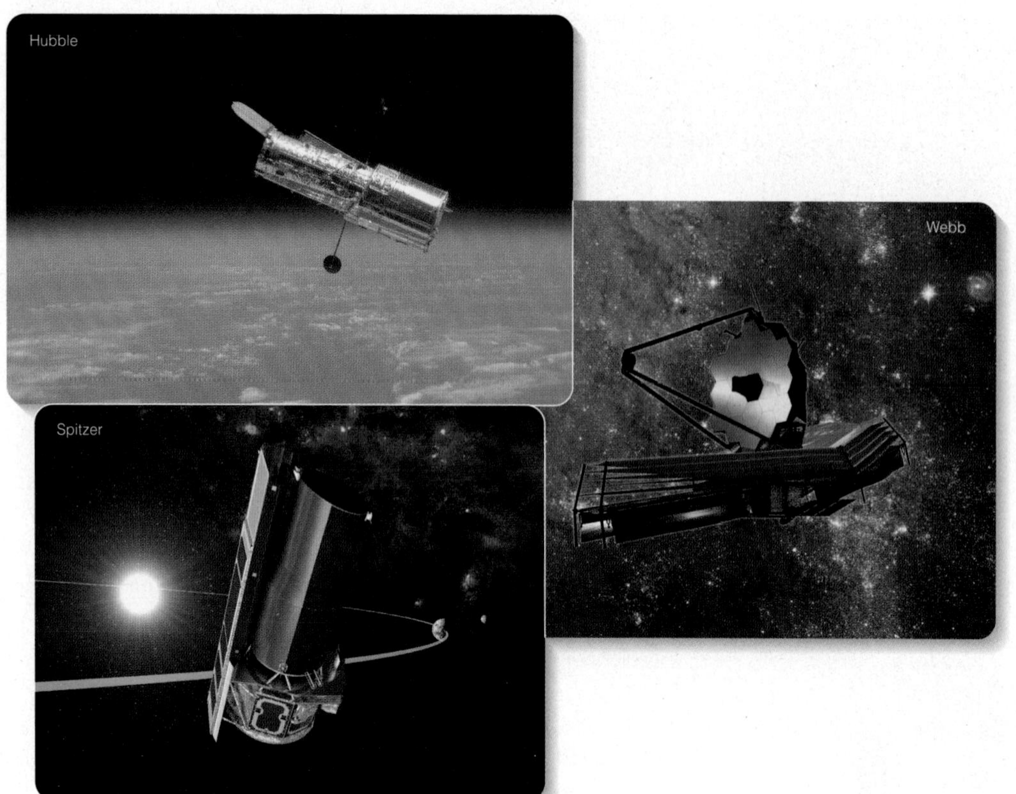


**■ Figure 6-22**

The Hubble Space Telescope orbits Earth only 569 km (353 mi) above the surface. Here it is looking to the upper left. The James Webb Space Telescope, planned to replace Hubble, will be over six times larger in collecting area. It will not have a tube but will observe from behind a sun screen. The infrared Spitzer Space Telescope orbits the sun slightly more slowly than Earth and gradually falls behind as it uses up its liquid helium coolant. (NASA/JPL-Caltech)

atmosphere is the lack of seeing distortion. It can detect fine detail, and, because it concentrates light into sharp images, it can see faint objects.

The telescope is as big as a large bus and has been visited a number of times by the space shuttle so that astronauts can maintain its equipment and install new cameras and spectrographs. Astronomers hope that it will last until it is replaced by the James Webb Space Telescope expected to launch no sooner than 2013. The Webb telescope will carry a cluster of beryllium segments that will open to form a 6.5-m (256-in.) mirror once in space.

## Infrared Astronomy from Orbit

Telescopes that observe in the far-infrared must be protected from heat and must get above Earth's absorbing atmosphere. They have limited lifetimes because they must carry coolant to chill their optics. The Infrared Astronomical Satellite (IRAS) was a joint project of the United Kingdom, the United States, and the Netherlands. IRAS was launched in January of 1983 and carried liquid helium coolant to keep its telescope cold. It made 250,000 observations and, for example, discovered disks of dust around stars where planets are now thought to have formed. Its coolant ran out after 300 days of observation.

The most sophisticated of the infrared telescopes put in orbit, the Spitzer Space Telescope, like IRAS, is cooled to –269°C (–452°F). Launched in 2003, it observes from behind a sunscreen. In fact, it could not observe from Earth orbit because Earth is such a strong source of infrared radiation, so the telescope was sent into an orbit around the sun that will carry it slowly away from Earth as its coolant is used up. Named after theoretical physicist Lyman Spitzer Jr. who originally suggested that space telescopes would be useful, it has made important discoveries concerning star formation, planets orbiting other stars, distant galaxies, and more.

## High-Energy Astrophysics

High-energy astrophysics refers to the use of X-ray and gamma-ray observations of the sky. Making such observations is difficult but can reveal the secrets of processes such as the explosive deaths of massive stars and eruptions of supermassive black holes.

The first astronomical satellite, Ariel 1, was launched by British astronomers in 1962 and made solar observations in the ultraviolet and X-ray part of the spectrum. Since then many space telescopes have made high-energy observations from orbit. Some of these satellites have been general-purpose telescopes that can observe many different kinds of objects. ROSAT, for example, was an X-ray observatory developed by an international consortium of European astronomers. Some space telescopes are designed to study a single problem or a single object. The Japanese satellite Hinode, for example, studies the sun continuously at visual, ultraviolet, and X-ray wavelengths.

**Figure 6-23**

From space, the Chandra X-Ray Observatory recorded an X-ray image of galaxy M101. It is superimposed here on a visual wavelength image from the Hubble Space Telescope and an infrared image from the Spitzer Space Telescope. Blue shows hot gas around newborn stars, while red shows cool, dusty clouds of gas. (X-ray: NASA/CSC/JHU/K. Kuntz et al.; Optical: NASA/ESA/STScI/JHU/K. Kuntz et al.; IR: NASA/JPL-Caltech/STScI/K. Gordon)

The largest X-ray telescope to date, the Chandra X-ray Observatory, was launched in 1999. Chandra orbits a third of the way to the moon and is named for the late Indian-American Nobel laureate Subrahmanyan Chandrasekhar, who was a pioneer in many branches of theoretical astronomy. Focusing X-rays is difficult because they penetrate into most mirrors, so astronomers devised cylindrical mirrors in which the X-rays reflect from the polished inside of the cylinders and form images on special detectors. The telescope has made important discoveries about everything from star formation to monster black holes in distant galaxies (■ Figure 6-23) that will be described in later chapters.

One of the first gamma-ray observatories was the Compton Gamma Ray Observatory, launched in 1991. It mapped the entire sky at gamma-ray wavelengths. The European INTEGRAL satellite was launched in 2002 and has been very productive in the study of violent eruptions of stars and black holes. The GLAST (Gamma-Ray Large Area Space Telescope), launched in 2008, is capable of mapping large areas of the sky to high sensitivity.

Modern astronomy has come to depend on observations that cover the entire electromagnetic spectrum. More orbiting space telescopes are planned that will be more versatile and more sensitive.

## Cosmic Rays

All of the radiation you have read about in this chapter has been electromagnetic radiation, but there is another form of energy raining down from space, and scientists aren't sure where it comes from. **Cosmic rays** are subatomic particles traveling through space at tremendous velocities. Almost no cosmic rays reach the ground, but they do smash gas atoms in the upper atmosphere, and fragments of those atoms shower down on you day and night over your entire life. These secondary cosmic rays are passing through you as you read this sentence.

Some cosmic-ray research can be done from high mountains or high-flying aircraft; but, to study cosmic rays in detail, detectors must go into space. A number of cosmic-ray detectors have been carried into orbit, but this area of astronomical research is just beginning to bear fruit.

Astronomers can't be sure what produces cosmic rays. Because they are atomic particles with electric charges, they are deflected by the magnetic fields spread through our galaxy, and that means astronomers can't tell where their original sources are located. The space between the stars is a glowing fog of cosmic rays. Some lower-energy cosmic rays come from the sun, and observations show that at least some high-energy cosmic rays are produced by the violent explosions of dying stars and by supermassive black holes at the centers of galaxies.

At present, cosmic rays largely remain an exciting mystery. You will meet them again in future chapters.

### What Are We? Curious

Telescopes are creations of curiosity. You look through a telescope to see more and to understand more. The unaided eye is a limited detector, and the history of astronomy is the history of bigger and better telescopes gathering more and more light to search for fainter and more distant objects.

The old saying, "Curiosity killed the cat," is an insult to the cat and to curiosity. We humans are curious, and curiosity is a noble trait—the mark of an active, inquiring mind. At the limits of human curiosity lies the fundamental question, "What are we?" Telescopes extend and amplify our senses, but they also extend and amplify our curiosity about the universe around us.

When people find out how something works, they say their curiosity is *satisfied.* Curiosity is an appetite like hunger or thirst, but it is an appetite for understanding. As astronomy expands our horizons and we learn how distant stars and galaxies work, we feel satisfaction because we are learning about ourselves. We are beginning to understand what we are.

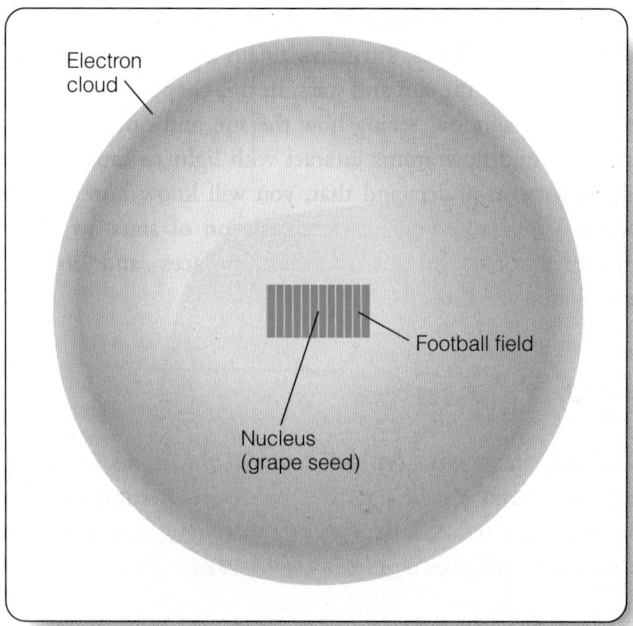


**Figure 7-2**

Magnifying a hydrogen atom by $10^{12}$ makes the nucleus the size of a grape seed and the diameter of the electron cloud about 4.5 times larger than a football field.

of a sphere 4.5 football fields in diameter, you can see that an atom is mostly empty space.

Now you can consider a **Common Misconception.** Most people, without thinking about it much, imagine that matter is solid, but you have seen that atoms are mostly empty space. The chair you sit on, the floor you walk on, are mostly not there. When you study the deaths of stars in a later chapter, you will see what happens to a star when most of the empty space gets squeezed out of its atoms.

## Different Kinds of Atoms

There are over a hundred chemical elements. The number of protons in the nucleus of an atom identifies which element it is. For example, a carbon atom has six protons in its nucleus. An atom with one more proton than this is nitrogen, and an atom with one fewer proton is boron.

Although an atom of a given element always has the same number of protons in its nucleus, the number of neutrons is less restricted. For instance, if you added a neutron to a carbon nucleus, it would still be carbon, but it would be slightly heavier. Atoms that have the same number of protons but a different number of neutrons are **isotopes**. Carbon has two stable isotopes. One contains six protons and six neutrons for a total of 12 particles and is thus called carbon-12. Carbon-13 has six protons and seven neutrons in its nucleus.

The number of electrons in an atom of a given element can vary. Protons and neutrons are bound tightly into the nucleus, but the electrons are held loosely in the electron cloud. Running a comb through your hair creates a static charge by removing a few electrons from their atoms. An atom that has lost or gained one or more electrons is said to be **ionized** and is called an **ion.** A neutral carbon atom has six electrons that balance the positive charge of the six protons in its nucleus. If you ionize the atom by removing one or more electrons, the atom is left with a net positive charge. Under some circumstances, an atom may capture one or more extra electrons, giving it more negative charges than positive. Such a negatively charged atom is also considered an ion.

Atoms that collide may form bonds with each other by exchanging or sharing electrons. Two or more atoms bonded together form a **molecule.** Atoms do collide in stars, but the high temperatures cause violent collisions that are unfavorable for chemical bonding. Only in the coolest stars are the collisions gentle enough to permit the formation of chemical bonds. You will see later that the presence of molecules such as titanium oxide (TiO) in a star is a clue that the star is very cool. In later chapters, you will see that molecules also can form in cool gas clouds in space and in the atmospheres of planets.

## Electron Shells

So far you have been thinking of the cloud of the whirling electrons in a general way, but now it is time to be more specific about how the electrons behave within the cloud.

Electrons are bound to the atom by the attraction between their negative charge and the positive charge on the nucleus. This attraction is known as the **Coulomb force,** after the French physicist Charles-Augustin de Coulomb (1736–1806). To ionize an atom, you need a certain amount of energy to pull an electron completely away from the nucleus. This energy is the electron's **binding energy,** the energy that holds it to the atom.

The size of an electron's orbit is related to the energy that binds it to the atom. If an electron orbits close to the nucleus, it is tightly bound, and a large amount of energy is needed to pull it away. Consequently, its binding energy is large. An electron orbiting farther from the nucleus is held more loosely, and less energy is needed to pull it away. That means it has less binding energy.

Nature permits atoms only certain amounts (quanta) of binding energy, and the laws that describe how atoms behave are called the laws of **quantum mechanics** (**How Do We Know? 7-1**). Much of this discussion of atoms is based on the laws of quantum mechanics.

Because atoms can have only certain amounts of binding energy, your model atom can have orbits only of certain sizes, called **permitted orbits.** These are like steps in a staircase: You can stand on the number-one step or the number-two step, but not on the number-one-and-one-quarter step–there isn't one. The electron can occupy any permitted orbit, but there are no orbits in between.

■ **Figure 6-23**

From space, the Chandra X-Ray Observatory recorded an X-ray image of galaxy M101. It is superimposed here on a visual wavelength image from the Hubble Space Telescope and an infrared image from the Spitzer Space Telescope. Blue shows hot gas around newborn stars, while red shows cool, dusty clouds of gas. (X-ray: NASA/CSC/JHU/K. Kuntz et al.; Optical: NASA/ESA/STScI/JHU/K. Kuntz et al.; IR: NASA/JPL-Caltech/STScI/K. Gordon)

The largest X-ray telescope to date, the Chandra X-ray Observatory, was launched in 1999. Chandra orbits a third of the way to the moon and is named for the late Indian-American Nobel laureate Subrahmanyan Chandrasekhar, who was a pioneer in many branches of theoretical astronomy. Focusing X-rays is difficult because they penetrate into most mirrors, so astronomers devised cylindrical mirrors in which the X-rays reflect from the polished inside of the cylinders and form images on special detectors. The telescope has made important discoveries about everything from star formation to monster black holes in distant galaxies (■ Figure 6-23) that will be described in later chapters.

One of the first gamma-ray observatories was the Compton Gamma Ray Observatory, launched in 1991. It mapped the entire sky at gamma-ray wavelengths. The European INTEGRAL satellite was launched in 2002 and has been very productive in the study of violent eruptions of stars and black holes. The GLAST (Gamma-Ray Large Area Space Telescope), launched in 2008, is capable of mapping large areas of the sky to high sensitivity.

Modern astronomy has come to depend on observations that cover the entire electromagnetic spectrum. More orbiting space telescopes are planned that will be more versatile and more sensitive.

## Cosmic Rays

All of the radiation you have read about in this chapter has been electromagnetic radiation, but there is another form of energy raining down from space, and scientists aren't sure where it comes from. **Cosmic rays** are subatomic particles traveling through space at tremendous velocities. Almost no cosmic rays reach the ground, but they do smash gas atoms in the upper atmosphere, and fragments of those atoms shower down on you day and night over your entire life. These secondary cosmic rays are passing through you as you read this sentence.

Some cosmic-ray research can be done from high mountains or high-flying aircraft; but, to study cosmic rays in detail, detectors must go into space. A number of cosmic-ray detectors have been carried into orbit, but this area of astronomical research is just beginning to bear fruit.

Astronomers can't be sure what produces cosmic rays. Because they are atomic particles with electric charges, they are deflected by the magnetic fields spread through our galaxy, and that means astronomers can't tell where their original sources are located. The space between the stars is a glowing fog of cosmic rays. Some lower-energy cosmic rays come from the sun, and observations show that at least some high-energy cosmic rays are produced by the violent explosions of dying stars and by supermassive black holes at the centers of galaxies.

At present, cosmic rays largely remain an exciting mystery. You will meet them again in future chapters.

### What Are We? Curious

Telescopes are creations of curiosity. You look through a telescope to see more and to understand more. The unaided eye is a limited detector, and the history of astronomy is the history of bigger and better telescopes gathering more and more light to search for fainter and more distant objects.

The old saying, "Curiosity killed the cat," is an insult to the cat and to curiosity. We humans are curious, and curiosity is a noble trait—the mark of an active, inquiring mind. At the limits of human curiosity lies the fundamental question, "What are we?" Telescopes extend and amplify our senses, but they also extend and amplify our curiosity about the universe around us.

When people find out how something works, they say their curiosity is *satisfied.* Curiosity is an appetite like hunger or thirst, but it is an appetite for understanding. As astronomy expands our horizons and we learn how distant stars and galaxies work, we feel satisfaction because we are learning about ourselves. We are beginning to understand what we are.

# Study and Review

## Summary

- Light is the visible form of **electromagnetic radiation (p. 99)**, an electric and magnetic disturbance that transports energy at the speed of light. The electromagnetic **spectrum (p. 100)** includes gamma rays, X-rays, ultraviolet radiation, visible light, infrared radiation, and radio waves.
- You can think of a particle of light, a **photon (p. 100)**, as a bundle of waves that sometimes acts as a particle and sometimes acts as a wave.
- The energy a photon carries depends on its **wavelength (p. 99)**. The wavelength of visible light, usually measured in **nanometers (p. 100)** ($10^{-9}$ m) or **angstroms (p. 100)** ($10^{-10}$ m), ranges from 400 nm to 700 nm (4000 to 7000 Å). Radio and **infrared radiation (p. 100)** have longer wavelengths and carry less energy. X-ray, gamma-ray, and **ultraviolet** radiation **(p. 100)** have shorter wavelengths and more energy.
- **Frequency (p. 99)** is the number of waves that pass a stationary point in 1 second. Wavelength equals the speed of light divided by the frequency.
- Earth's atmosphere is fully transparent in only two **atmospheric windows (p. 101)**—visible light and radio.
- Astronomical telescopes use a **primary lens (p. 101)** or **mirror (pp. 101–102)** (also called an **objective lens** or **mirror [pp. 101–102])** to gather light and focus it into a small image, which can be magnified by an **eyepiece (p. 102)**. Short-**focal-length (p. 102)** lenses and mirrors must be more strongly curved and are more expensive to grind to shape.
- A **refracting telescope (p. 101)** uses a lens to bend the light and focus it into an image. Because of **chromatic aberration (p. 102)**, refracting telescopes cannot bring all colors to the same focus, resulting in color fringes around the images. An **achromatic lens (p. 102)** partially corrects for this, but such lenses are expensive and cannot be made much larger than about 1 m in diameter.
- **Reflecting telescopes (p. 101)** use a mirror to focus the light and are less expensive than refracting telescopes of the same diameter. Also, reflecting telescopes do not suffer from chromatic aberration. Most large telescopes are reflectors.
- **Light-gathering power (p. 103)** refers to the ability of a telescope to produce bright images. **Resolving power (p. 104)** refers to the ability of a telescope to resolve fine detail. **Diffraction fringes (p. 104)** in the image limit the detail visible. **Magnifying power (p. 105)**, the ability to make an object look bigger, is less important because it can be changed by changing the eyepiece.
- Astronomers build observatories on remote, high mountains for two reasons. Turbulence in Earth's atmosphere blurs the image of an astronomical telescope, a phenomenon that astronomers refer to as **seeing (p. 104)**. Atop a mountain, the air is steady, and the seeing is better. Observatories are located far from cities to avoid **light pollution (p. 106)**.
- In a reflecting telescope, light first comes to a focus at the **prime focus (p. 110)**, but **secondary mirrors (p. 110)** can direct light to other focus locations such as a **Cassegrain focus (p. 110)** or a **Newtonian focus (p. 110)**. The **Schmidt-Cassegrain focus (p. 110)** is popular for small telescopes.
- Because Earth rotates, telescopes must have a **sidereal drive (p. 111)** to follow the stars. An **equatorial mounting (p. 111)** with a **polar axis (p. 111)** makes this possible, but **alt-azimuth mountings (p. 111)** are becoming more popular.
- Very large telescopes can be built with **active optics (p. 111)**, maintaining the shape of floppy mirrors that are thin or in segments. Such thin mirrors weigh less, are easier to support, and cool faster at nightfall.
- High-speed **adaptive optics (p. 109)** can monitor distortions caused by turbulence in Earth's atmosphere and partially cancel out the blurring caused by seeing.
- **Interferometry (p. 112)** refers to connecting two or more separate telescopes together to act as a single large telescope that has a resolution equivalent to that of a telescope as large in diameter as the separation between the telescopes.
- For many decades astronomers used photographic plates to record images at the telescope, but modern electronic systems such as **charge-coupled devices (CCDs) (p. 113)** have replaced photographic plates in most applications.
- Astronomical images in digital form can be computer enhanced and reproduced as **false-color images (p. 113)** to bring out subtle details.
- **Spectrographs (p. 113)** using prisms or a **grating (p. 113)** spread starlight out according to wavelength to form a spectrum revealing hundreds of **spectral lines (p. 113)** produced by atoms in the object being studied. A **comparison spectrum (p. 113)** containing lines of known wavelength allows astronomers to measure wavelengths in spectra of astronomical objects.
- Astronomers use radio telescopes for three reasons: They can detect cool hydrogen and other atoms and molecules in space; they can see through dust clouds that block visible light; and they can detect certain objects invisible at other wavelengths.
- Most radio telescopes contain a dish reflector, an antenna, an amplifier, and a data recorder. Such a telescope can record the intensity of the radio energy coming from a spot on the sky. Scans of small regions are used to produce radio maps.
- Because of the long wavelength, radio telescopes have very poor resolution, and astronomers often link separate radio telescopes together to form a **radio interferometer (p. 115)** capable of resolving much finer detail.
- Earth's atmosphere absorbs gamma rays, X-rays, ultraviolet, and far-infrared. To observe at these wavelengths, telescopes must be located in space.
- Earth's atmosphere distorts and blurs images. Telescopes in orbit are above this seeing distortion and are limited only by diffraction in their optics.
- **Cosmic rays (p. 119)** are not electromagnetic radiation; they are subatomic particles such as electrons and protons traveling at nearly the speed of light. They can best be studied from above Earth's atmosphere.

## Review Questions

1. Why would you not plot sound waves in the electromagnetic spectrum?
2. If you had limited funds to build a large telescope, which type would you choose, a refractor or a reflector? Why?
3. Why do nocturnal animals usually have large pupils in their eyes? How is that related to astronomical telescopes?
4. Why do optical astronomers often put their telescopes at the tops of mountains, while radio astronomers sometimes put their telescopes in deep valleys?
5. Optical and radio astronomers both try to build large telescopes but for different reasons. How do these goals differ?
6. What are the advantages of making a telescope mirror thin? What problems does this cause?

7. Small telescopes are often advertised as "200 power" or "magnifies 200 times." As someone knowledgeable about astronomical telescopes, how would you improve such advertisements?
8. Not too many years ago an astronomer said, "Some people think I should give up photographic plates." Why might she change to something else?
9. What purpose do the colors in a false-color image or false-color radio map serve?
10. How is chromatic aberration related to a prism spectrograph?
11. Why would radio astronomers build identical radio telescopes in many different places around the world?
12. Why do radio telescopes have poor resolving power?
13. Why must telescopes observing in the far-infrared be cooled to low temperatures?
14. What might you detect with an X-ray telescope that you could not detect with an infrared telescope?
15. The moon has no atmosphere at all. What advantages would you have if you built an observatory on the lunar surface?
16. **How Do We Know?** How is the resolution of an astronomical image related to the precision of a measurement?

## Discussion Questions

1. Why does the wavelength response of the human eye match so well the visual window of Earth's atmosphere?
2. Most people like beautiful sunsets with brightly glowing clouds, bright moonlit nights, and twinkling stars. Astronomers don't. Why?

## Problems

1. The thickness of the plastic in plastic bags is about 0.001 mm. How many wavelengths of red light is this?
2. What is the wavelength of radio waves transmitted by a radio station with a frequency of 100 million cycles per second?
3. Compare the light-gathering powers of one of the 10-m Keck telescopes and a 0.5-m telescope.
4. How does the light-gathering power of one of the Keck telescopes compare with that of the human eye? (*Hint:* Assume that the pupil of your eye can open to about 0.8 cm.)
5. What is the resolving power of a 25-cm telescope? What do two stars 1.5 arc seconds apart look like through this telescope?
6. Most of Galileo's telescopes were only about 2 cm in diameter. Should he have been able to resolve the two stars mentioned in Problem 5?
7. How does the resolving power of a 5-m telescope compare with that of the Hubble Space Telescope? Why does the HST outperform a 5-m telescope?
8. If you build a telescope with a focal length of 1.3 m, what focal length should the eyepiece have to give a magnification of 100 times?
9. Astronauts observing from a space station need a telescope with a light-gathering power 15,000 times that of the human eye, capable of resolving detail as small as 0.1 arc seconds and having a magnifying power of 250. Design a telescope to meet their needs. Could you test your design by observing stars from Earth?
10. A spy satellite orbiting 400 km above Earth is supposedly capable of counting individual people in a crowd. Roughly what minimum-diameter telescope must the satellite carry? (*Hint:* Use the small-angle formula.)

## Learning to Look

1. The two images at the right show a star before and after an adaptive optics system was switched on. What causes the distortion in the first image, and how does adaptive optics correct the image?

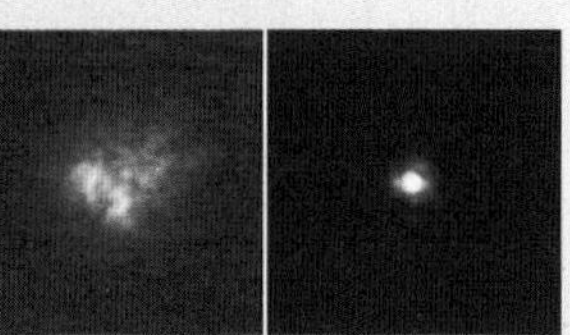
ESO

2. The star images in the photo at the right are tiny disks, but the diameter of these disks is not related to the diameter of the stars. Explain why the telescope can't resolve the diameter of the stars.

NASA, ESA and G. Meylan

3. The X-ray image at right shows the remains of an exploded star. Explain why images recorded by telescopes in space are often displayed in false color rather than in the "colors" received by the telescope.

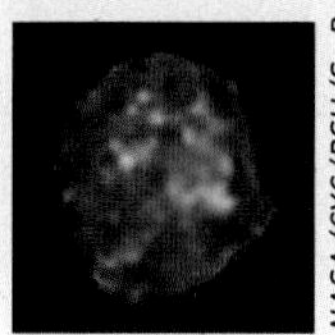
NASA/CXC/PSU/S. Park

# 7 Atoms and Starlight

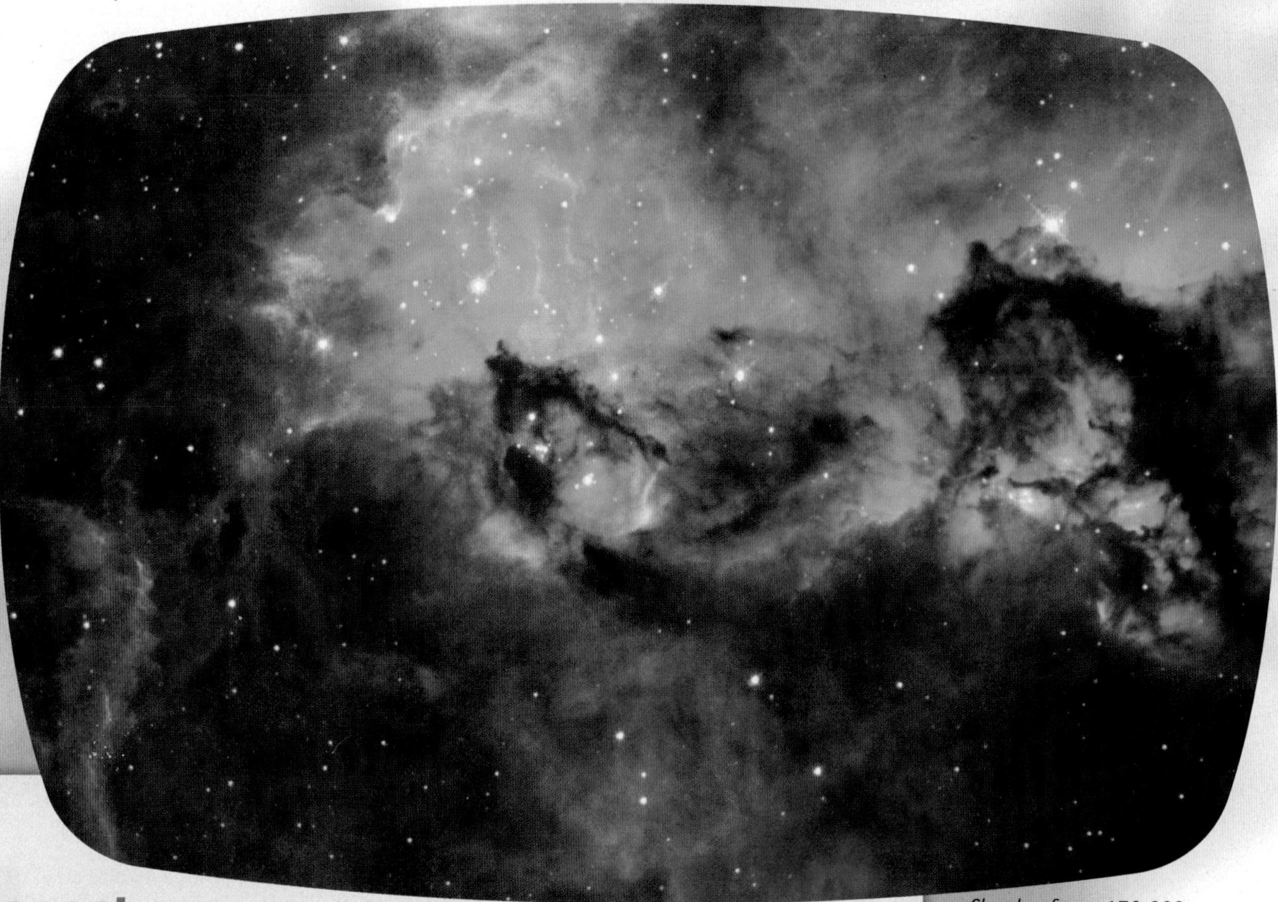

*Clouds of gas 170,000 light-years away glow in this false-color image. Red represents sulfur atoms, green hydrogen, and blue oxygen. Clues hidden in the starlight tell a story of star birth and star death.* (NASA, ESA, and M. Livio, STScI)

## Guidepost

In the last chapter, you read how telescopes gather starlight and how spectrographs spread out the light into spectra. Now you are ready to see what all the fuss is about. Here you will find answers to four essential questions:

- **What is an atom?**
- **How do atoms interact with light?**
- **What kinds of spectra do you see when you look at celestial objects?**
- **What can you learn from a star's spectrum?**

This chapter marks a change in the way you will look at nature. Up to this point, you have been thinking about what you can see with your eyes alone or aided by telescopes. In this chapter, you will begin using modern astrophysics to search out secrets that lie beyond what you can see.

The analysis of spectra is a powerful technique, and in the chapters that follow, you will use spectra to study stars, galaxies, and planets.

*Awake! for Morning in the Bowl of Night*

*Has flung the Stone that puts the Stars to Flight:*

*And Lo! the Hunter of the East has caught*

*The Sultan's Turret in a Noose of Light.*

— *THE RUBÁIYÁT OF OMAR KHAYYÁM*, TRANS. EDWARD FITZGERALD

THE UNIVERSE IS FILLED with fabulously beautiful clouds of glowing gas illuminated by brilliant stars, but they are all hopelessly beyond reach. No laboratory jar on Earth holds a sample labeled "star stuff," and no space probe has ever visited the inside of a star. The stars are far away, and the only information you can obtain about them comes hidden in starlight (■ Figure 7-1).

Earthbound humans knew almost nothing about stars until the early 19th century, when the German optician Joseph von Fraunhofer studied the solar spectrum and found it interrupted by some 600 dark lines representing colors that are missing from the sunlight Earth receives. When scientists realized that those spectral lines are related to the presence of various atoms in the sun's atmosphere, and found that the spectra of other stars have similar patterns of lines, a window finally opened to a scientific understanding of the sun and stars. In this chapter you will look through that window, seeing how the sun and other stars produce light, and how atoms interact with light to cause spectral lines. Once you understand that, you will know how astronomers determine the chemical composition of stars, as well as measure motions of gas at their surfaces and in their atmospheres.

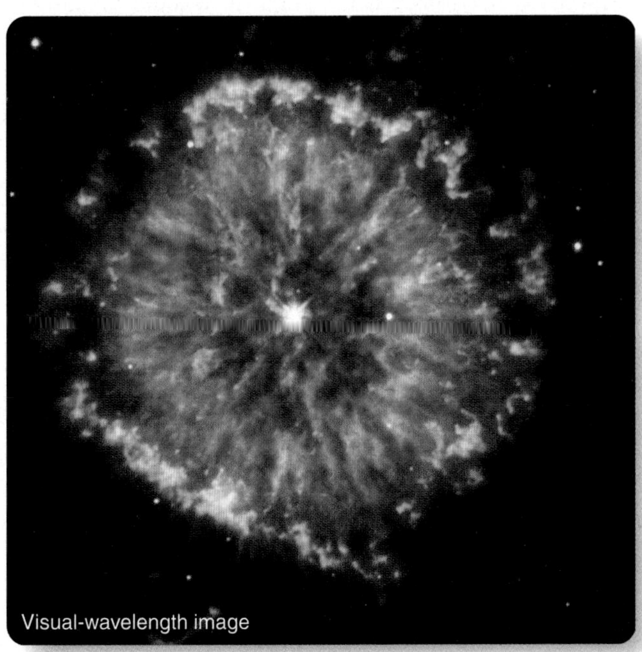


**■ Figure 7-1**

What's going on here? The sky is filled with beautiful and mysterious objects that lie far beyond your reach—in the case of the nebula NGC 6751, about 6500 ly beyond your reach. The only way to understand such objects is by analyzing their light. Such an analysis reveals that this object is a dying star surrounded by the expanding shell of gas it ejected a few thousand years ago. You will learn more about this phenomenon in a later chapter. (NASA Hubble Heritage Team/STScI/AURA)

## 7-1 Atoms

STARS ARE GREAT BALLS OF HOT GAS, and the atoms in that gas leave their marks on the light the stars emit. By understanding what atoms are and how they interact with light, you can decode the spectra of the stars and learn their secrets.

### A Model Atom

To think about atoms and how they interact with light, you need a working model of an atom. In Chapter 2, you used a model of the sky, the celestial sphere. In this chapter, you will begin your study of atoms by creating a model of an atom. Remember that a model can have practical value without being true. The stars are not actually attached to a sphere surrounding Earth, but to navigate a ship or point a telescope, it is useful to pretend they are. The electrons in an atom are not actually little beads orbiting the nucleus the way planets orbit the sun, but for some purposes it is useful to picture them as such.

Your model atom contains a positively charged **nucleus** at the center, which consists of two kinds of particles: **Protons** carry a positive electrical charge, and **neutrons** have no charge. This means the nucleus has a net positive charge.

The nucleus of this model atom is surrounded by a cloud of orbiting **electrons,** low-mass particles with negative charges. Normally the number of electrons equals the number of protons, and the positive and negative charges balance to produce a neutral atom. Protons and neutrons have masses about 1840 times greater than that of an electron, so most of the mass of an atom lies in the nucleus. Even so, a single atom is not a massive object. A hydrogen atom, for example, has a mass of only $1.67 \times 10^{-27}$ kg, about a trillionth of a trillionth of a gram.

An atom is mostly empty space. To see this, imagine constructing a simple scale model of a hydrogen atom. Its nucleus is a single proton with a diameter of about 0.0000016 nm, or $1.6 \times 10^{-15}$ m. If you multiply this by one trillion ($10^{12}$), you can represent the nucleus of your model atom with something about 0.16 cm in diameter—a grape seed would do. The region of a hydrogen atom that contains the electron has a diameter of about 0.4 nm, or $4 \times 10^{-10}$ m. Multiplying by a trillion increases the diameter to about 400 m, or about 4.5 football fields laid end to end (■ Figure 7-2). When you imagine a grape seed in the middle

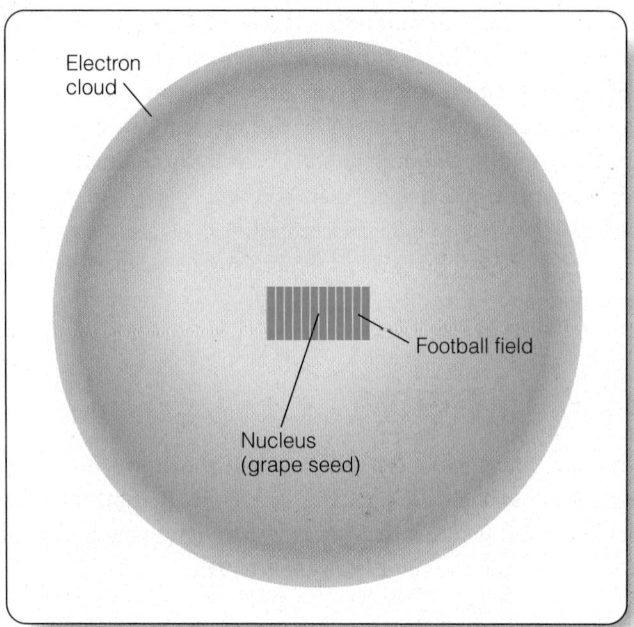


**Figure 7-2**

Magnifying a hydrogen atom by $10^{12}$ makes the nucleus the size of a grape seed and the diameter of the electron cloud about 4.5 times larger than a football field.

of a sphere 4.5 football fields in diameter, you can see that an atom is mostly empty space.

Now you can consider a **Common Misconception.** Most people, without thinking about it much, imagine that matter is solid, but you have seen that atoms are mostly empty space. The chair you sit on, the floor you walk on, are mostly not there. When you study the deaths of stars in a later chapter, you will see what happens to a star when most of the empty space gets squeezed out of its atoms.

## Different Kinds of Atoms

There are over a hundred chemical elements. The number of protons in the nucleus of an atom identifies which element it is. For example, a carbon atom has six protons in its nucleus. An atom with one more proton than this is nitrogen, and an atom with one fewer proton is boron.

Although an atom of a given element always has the same number of protons in its nucleus, the number of neutrons is less restricted. For instance, if you added a neutron to a carbon nucleus, it would still be carbon, but it would be slightly heavier. Atoms that have the same number of protons but a different number of neutrons are **isotopes**. Carbon has two stable isotopes. One contains six protons and six neutrons for a total of 12 particles and is thus called carbon-12. Carbon-13 has six protons and seven neutrons in its nucleus.

The number of electrons in an atom of a given element can vary. Protons and neutrons are bound tightly into the nucleus, but the electrons are held loosely in the electron cloud. Running a comb through your hair creates a static charge by removing a few electrons from their atoms. An atom that has lost or gained one or more electrons is said to be **ionized** and is called an **ion.** A neutral carbon atom has six electrons that balance the positive charge of the six protons in its nucleus. If you ionize the atom by removing one or more electrons, the atom is left with a net positive charge. Under some circumstances, an atom may capture one or more extra electrons, giving it more negative charges than positive. Such a negatively charged atom is also considered an ion.

Atoms that collide may form bonds with each other by exchanging or sharing electrons. Two or more atoms bonded together form a **molecule.** Atoms do collide in stars, but the high temperatures cause violent collisions that are unfavorable for chemical bonding. Only in the coolest stars are the collisions gentle enough to permit the formation of chemical bonds. You will see later that the presence of molecules such as titanium oxide (TiO) in a star is a clue that the star is very cool. In later chapters, you will see that molecules also can form in cool gas clouds in space and in the atmospheres of planets.

## Electron Shells

So far you have been thinking of the cloud of the whirling electrons in a general way, but now it is time to be more specific about how the electrons behave within the cloud.

Electrons are bound to the atom by the attraction between their negative charge and the positive charge on the nucleus. This attraction is known as the **Coulomb force,** after the French physicist Charles-Augustin de Coulomb (1736–1806). To ionize an atom, you need a certain amount of energy to pull an electron completely away from the nucleus. This energy is the electron's **binding energy,** the energy that holds it to the atom.

The size of an electron's orbit is related to the energy that binds it to the atom. If an electron orbits close to the nucleus, it is tightly bound, and a large amount of energy is needed to pull it away. Consequently, its binding energy is large. An electron orbiting farther from the nucleus is held more loosely, and less energy is needed to pull it away. That means it has less binding energy.

Nature permits atoms only certain amounts (quanta) of binding energy, and the laws that describe how atoms behave are called the laws of **quantum mechanics** (**How Do We Know? 7-1**). Much of this discussion of atoms is based on the laws of quantum mechanics.

Because atoms can have only certain amounts of binding energy, your model atom can have orbits only of certain sizes, called **permitted orbits.** These are like steps in a staircase: You can stand on the number-one step or the number-two step, but not on the number-one-and-one-quarter step–there isn't one. The electron can occupy any permitted orbit, but there are no orbits in between.

## Quantum Mechanics

**How can you understand nature if it depends on the atomic world you cannot see?** You can see objects such as stars, planets, aircraft carriers, and hummingbirds, but you can't see individual atoms. As scientists apply the principle of cause and effect, they study the natural effects they can see and work backward to find the causes. Invariably that quest for causes leads back to the invisible world of atoms.

Quantum mechanics is the set of rules that describe how atoms and subatomic particles behave. On the atomic scale, particles behave in ways that seem unfamiliar. One of the principles of quantum mechanics specifies that you cannot know simultaneously the exact location and motion of a particle. This is why physicists prefer to go one step beyond the simple atomic model that has electrons following orbits and instead describe the electrons in an atom as if they were a cloud of negative charge. That's a better model.

This raises some serious questions about reality. Is an electron really a particle at all? If you can't know simultaneously the position and motion of a specific particle, how can you know how it will react to a collision with a photon or another particle? The answer is that you can't know, and that seems to violate the principle of cause and effect.

Many of the phenomena you can see depend on the behavior of huge numbers of atoms, and quantum mechanical uncertainties average out. Nevertheless, the ultimate causes that scientists seek lie at the level of atoms, and modern physicists are trying to understand the nature of the particles that make up atoms. That is one of the most exciting frontiers of science.

*The world you see, including these neon signs, is animated by the properties of atoms and subatomic particles.* (Jeff Greenberg/PhotoEdit)

The arrangement of permitted orbits depends primarily on the charge of the nucleus, which in turn depends on the number of protons. Consequently, each kind of element has its own pattern of permitted orbits (■ Figure 7-3). Isotopes of the same elements have nearly the same pattern because they have the same number of protons. However, ionized atoms have orbital patterns that differ from their un-ionized forms. Thus, the arrangement of permitted orbits differs for every kind of atom and ion.

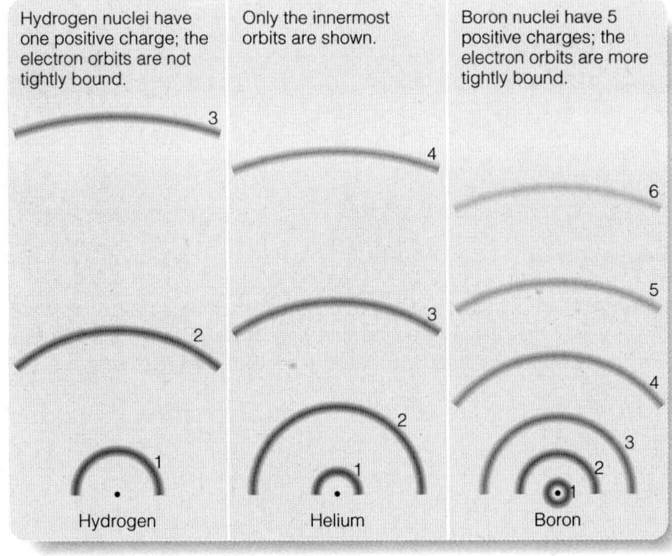

Hydrogen nuclei have one positive charge; the electron orbits are not tightly bound.
Only the innermost orbits are shown.
Boron nuclei have 5 positive charges; the electron orbits are more tightly bound.
3
2
1
4
3
2
1
6
5
4
3
2
1
Hydrogen
Helium
Boron

**■ Figure 7-3**

The electron in an atom may occupy only certain permitted orbits. Because different elements have different charges on their nuclei, the elements have different, unique patterns of permitted orbits.

### SCIENTIFIC ARGUMENT

***How many hydrogen atoms would it take to cross the head of a pin?***

This is not a frivolous question. In answering it, you will discover how small atoms really are, and you will see how powerful physics and mathematics can be as a way to understand nature. Many scientific arguments are convincing because they have the precision of mathematics. To begin, assume that the head of a pin is about 1 mm in diameter. That is 0.001 m. The size of a hydrogen atom is represented by the diameter of the electron cloud, roughly 0.4 nm. Because 1 nm equals $10^{-9}$ m, you can multiply and discover that 0.4 nm equals $4 \times 10^{-10}$ m. To find out how many atoms would stretch 0.001 m, you can divide the diameter of the pinhead by the diameter of an atom. That is, divide 0.001 m by $4 \times 10^{-10}$ m, and you get $2.5 \times 10^{6}$. It would take 2.5 million hydrogen atoms lined up side by side to cross the head of a pin.

Now you can see how tiny an atom is and also how powerful a bit of physics and mathematics can be. It reveals a view of nature beyond the capability of your eyes. Now build an argument using another bit of arithmetic. **How many hydrogen atoms would you need to add up to the mass of a paper clip (1 g)?**

## 7-2 The Interaction of Light and Matter

IF LIGHT DID NOT INTERACT WITH MATTER, you would not be able to see these words. In fact, you would not exist, because, among other problems, photosynthesis would be impossible, so there would be no grass, wheat, bread, beef, cheeseburgers, or any other kind of food. The interaction of light and matter makes life possible, and it also makes it possible for you to understand the universe.

You have already been considering a model hydrogen atom. Now you can use that model as you begin your study of light and matter. Hydrogen is both simple and common. Roughly 90 percent of all atoms in the universe are hydrogen.

### The Excitation of Atoms

Each electron orbit in an atom represents a specific amount of binding energy, so physicists commonly refer to the orbits as **energy levels.** Using this terminology, you can say that an electron in its smallest and most tightly bound orbit is in its lowest permitted energy level, which is called the atom's **ground state.** You could move the electron from one energy level to another by supplying enough energy to make up the difference between the two energy levels. It would be like moving a flowerpot from a low shelf to a high shelf; the greater the distance between the shelves, the more energy you would need to raise the pot. The amount of energy needed to move the electron is the energy difference between the two energy levels.

If you move an electron from a low energy level to a higher energy level, the atom becomes an **excited atom.** That is, you have added energy to the atom by moving its electron outward from the nucleus. An atom can become excited by collision. If two atoms collide, one or both may have electrons knocked into a higher energy level. This happens very commonly in hot gas, where atoms move rapidly and collide often.

Another way an atom can become excited is to absorb a photon. As you learned in the previous chapter, a photon is a bundle of electromagnetic waves with a specific energy. Only a photon with exactly the right amount of energy can move the electron from one level to another. If the photon has too much or too little energy, the atom cannot absorb it. Because the energy of a photon depends on its wavelength, only photons of certain wavelengths can be absorbed by a given kind of atom. ■ Figure 7-4 shows the lowest four energy levels of the hydrogen atom, along with three photons the atom could absorb. The longest-wavelength photon has only enough energy to excite the electron to the second energy level, but the shorter-wavelength photons can excite the electron to higher levels. Because the hydrogen atom has many more energy levels than shown in Figure 7-4, it can absorb photons of many different wavelengths.

Atoms, like humans, cannot exist in an excited state forever. An excited atom is unstable and must eventually (usually within $10^{-6}$ to $10^{-9}$ second) give up the energy it has absorbed and return its electron to a lower energy level. Thus the electron in an excited atom tends to tumble down to its lowest energy level, its ground state.

When an electron drops from a higher to a lower energy level, it moves from a loosely bound level to one that is more tightly bound. The atom then has a surplus of energy—the energy difference between the levels—that it can emit as a photon with a wavelength corresponding to that amount of energy (Chapter 6). Study the sequence of events shown in ■ Figure 7-5 to see how an atom can absorb and emit photons. Because each type of atom or ion has a unique set of energy levels, each type absorbs and

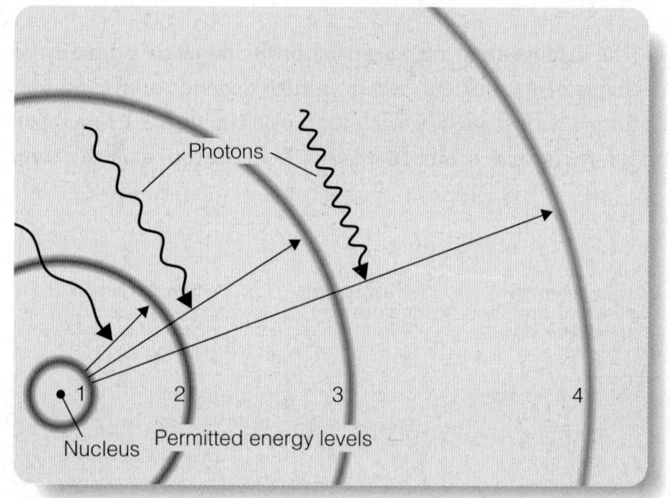


■ **Figure 7-4**

A hydrogen atom can absorb only those photons that move the atom's electron to one of the higher-energy orbits. Here three different photons are shown along with the change they would produce if they were absorbed.

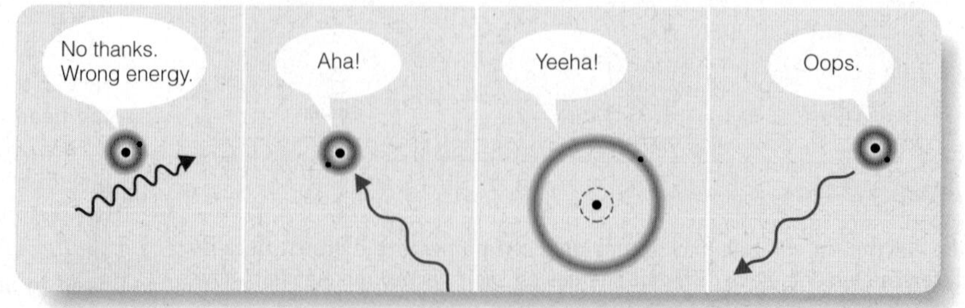


■ **Figure 7-5**

An atom can absorb a photon only if the photon has the correct amount of energy. The excited atom is unstable and within a fraction of a second returns to a lower energy level, re-radiating the photon in a random direction.

emits photons with a unique set of wavelengths. As a result, you can identify the elements in a gas by studying the characteristic wavelengths of light that are absorbed or emitted.

The process of excitation and emission is a common sight in urban areas at night. A neon sign glows when atoms of neon gas in a glass tube are excited by electricity flowing through the tube. As the electrons in the electric current flow through the gas, they collide with the neon atoms and excite them. Almost immediately after a neon atom is excited, its electron drops back to a lower energy level, emitting the surplus energy as a photon of a certain wavelength. The photons emitted by excited neon blend to produce a reddish-orange glow. Signs of other colors, generically called "neon signs," contain other gases or mixtures of gases instead of pure neon. Whenever you look at a neon sign, you are seeing atoms absorbing and emitting energy in the form of photons with specific colors determined by the structure of electron orbits in those atoms.

Neon signs are simple, but stars are complex. The colors of stars are not determined by the gases they contain. In the next section, you will discover why some stars are red and some are blue, and that will give you a new insight into how light interacts with matter.

## Radiation from a Heated Object

If you look closely at the stars in the constellation Orion, you will notice that they are not all the same color (see Figure 2-4). One of your Favorite Stars, Betelgeuse, in the upper left corner of Orion, is quite red; another Favorite Star, Rigel, in the lower right corner, is blue. These differences in color arise from differences in temperature.

The starlight that you see comes from gases that make up the visible surface of the star, its photosphere. (Recall that you met the photosphere of the sun in Chapter 3.) Layers of gas deeper inside the star also emit light, but that light is reabsorbed before it can reach the surface. The gas above the photosphere is too thin to emit much light. The photosphere is the visible surface of a star because it is dense enough to emit lots of light but transparent enough to allow that light to escape.

Stars produce their light for the same reason heated horseshoes glow in a blacksmith's forge—because they are hot. If a horseshoe is not too hot, it glows ruddy red, but as it heats up it grows brighter and yellower. Yellow-hot is hotter than red-hot but not as hot as white-hot.

The light from stars and from glowing horseshoes is produced by the acceleration of charged particles. Usually the accelerated particles are electrons because they are the least massive charged particles, and they are on the outsides of atoms, so they are the easiest to get moving. An electron is surrounded by an electric field; and, if you disturb an electron, the change in its electric field spreads outward at the speed of light as electromagnetic radiation. You learned in Chapter 5 that "acceleration" means any change in motion. Whenever you change the motion of an electron or other charged particle, you generate electromagnetic waves. If you run a comb through your hair, you disturb electrons in both hair and comb, producing static electricity. That produces electromagnetic radiation, which you can hear as snaps and crackles if you are standing near an AM radio. Stars don't comb their hair, of course, but they are hot, and they are made up of ionized gases, so there are plenty of electrons zipping around.

The molecules and atoms in any object are in constant motion, and in a hot object they are more agitated than in a cool object. You can refer to this agitation as **thermal energy.** If you touch an object that contains lots of thermal energy, it will feel hot as the thermal energy flows into your fingers. The flow of thermal energy is called **heat.** In contrast, **temperature** refers to the average speed of the particles. Hot cheese and hot green beans can have the same temperature, but the cheese can contain more thermal energy and can burn your tongue. Thus, *heat* refers to the flow of thermal energy, and *temperature* refers to the intensity of the agitation among the particles (**Focus on Fundamentals 3**).

When astronomers refer to the temperature of a star, they are talking about the temperature of the gases in the photosphere, and they express those temperatures on the **Kelvin temperature scale.** On this scale, zero degrees Kelvin (written 0 K) is **absolute zero** (–459.7°F), the temperature at which an object contains no thermal energy that can be extracted. Water freezes at 273 K and boils at 373 K. The Kelvin temperature scale is useful in astronomy because it is based on absolute zero and consequently is related directly to the motion of the particles in an object.

Now you can understand why a hot object glows, or to put it another way, why a hot object emits photon bundles of electromagnetic energy. The hotter an object is, the more motion there is among its particles. The agitated particles, including electrons, collide with each other, and when electrons accelerate—change their motion–part of the energy is carried away as electromagnetic radiation. The radiation emitted by a heated object is called **blackbody radiation,** a name translated from a German term that refers to the way a perfectly opaque object would behave. A perfectly opaque object would be both a perfect emitter and a perfect absorber of radiation. At room temperature, such a perfect absorber and emitter would look black, but at higher temperatures it would glow at wavelengths visible to a human eye. In astronomy and physics you will see the term *blackbody* referring to objects that glow brightly.

Blackbody radiation is quite common. In fact, it is responsible for the light emitted by an incandescent light bulb. Electricity flowing through the filament of the bulb heats it to high temperature, and it glows. You can also recognize the light emitted by a heated horseshoe as blackbody radiation. Many objects in the sky, including the sun and stars, primarily emit blackbody radiation because they are mostly opaque.

## Temperature, Heat, and Thermal Energy

One of the most **Common Misconceptions** in science involves temperature. People often say "temperature" when they really mean "heat," and sometimes they say "heat" when they mean something entirely different. This is a fundamental idea, so you need to understand the difference.

Even in an object that is solid, the atoms and molecules are continuously jiggling around bumping into each other. When something is hot, the particles are moving rapidly. Temperature is a measure of the average motion of the particles. (Mathematically, temperature is proportional to the square of the average velocity.) If you have your temperature taken, it will probably be 37.0°C (98.6°F), an indication that the atoms and molecules in your body are moving about at a normal pace. If you measure the temperature of a baby, the thermometer should register the same temperature, showing that the atoms and molecules in the baby's body are moving at the same average velocity as the atoms and molecules in your body.

The total energy of all of the moving particles in a body is called thermal energy. You have much more mass than the baby, so you must contain more thermal energy even though you have the same temperature. The thermal energy in your body and in the baby's body has the same intensity (temperature) but different amounts. People often confuse temperature and thermal energy, so you must be careful to distinguish between them. Temperature is like an intensity, and thermal energy is a total amount.

Many people say "heat" when they should say "thermal energy." Heat is the thermal energy that moves from a hot object to a cool object. If two objects have the same temperature, you and the infant for example, there is no transfer of thermal energy and no heat.

When you hear someone say "heat," check to see if he or she doesn't really mean thermal energy. You may have burned yourself on cheese pizza, but you probably haven't burned yourself on green beans. At the same temperature, cheese holds more thermal energy than green beans. It isn't the temperature that burns your tongue, but the flow of thermal energy, and that's heat.

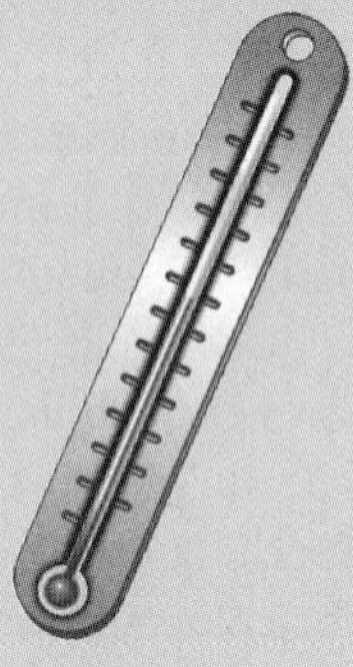

*What's the difference between temperature and heat?*

MASS | ENERGY | **TEMPERATURE AND HEAT** | DENSITY | PRESSURE

Hot objects emit blackbody radiation, but so do cold objects. Ice cubes are cold, but their temperature is higher than absolute zero, so they contain some thermal energy and must emit some blackbody radiation. The coldest gas drifting in space has a temperature only a few degrees above absolute zero, but it too emits blackbody radiation.

Two features of blackbody radiation are important. First, the hotter an object is, the more blackbody radiation it emits. Hot objects emit more radiation because their agitated particles collide more often and more violently with electrons. That's why a glowing coal from a fire emits more total energy than an ice cube of the same size.

The second feature is the relationship between the temperature of the object and the wavelengths of the photons it emits. The wavelength of the photon emitted when a particle collides with an electron depends on the violence of the collision. Only a violent collision can produce a short-wavelength (high-energy) photon. The electrons in an object have a distribution of speeds; a few travel very fast, and a few travel very slowly, but most travel at intermediate speeds. The hotter the object is, the faster, on average, the electrons travel. Because high-velocity electrons are rare, extremely violent collisions don't occur very often, and short-wavelength photons are rare. Similarly, most collisions are not extremely gentle, so long-wavelength (low-energy) photons are also rare. Consequently, blackbody radiation is made up of photons with a distribution of wavelengths with medium wavelengths most common. The **wavelength of maximum intensity ($\lambda_{max}$)** is the wavelength at which the object emits the most intense radiation and occurs at some intermediate wavelength. (Make special note that $\lambda_{max}$ does not refer to the maximum wavelength but to the wavelength of maximum.)

■ Figure 7-6 shows the intensity of radiation versus wavelength for three objects of different temperatures. The curves are high in the middle and low at either end because the objects emit most intensely at intermediate wavelengths. The total area under each curve is proportional to the total energy emitted, and you can see that the hotter object emits more total energy than the cooler objects. Look closely at the curves, and you will see that the wavelength of maximum intensity depends on temperature. The hotter an object, the shorter will be the wavelength of its maximum emitted intensity. The figure shows how temperature determines the color of a glowing blackbody. The hotter object emits more blue light than red and thus looks blue, and the cooler object emits more red than blue and consequently looks

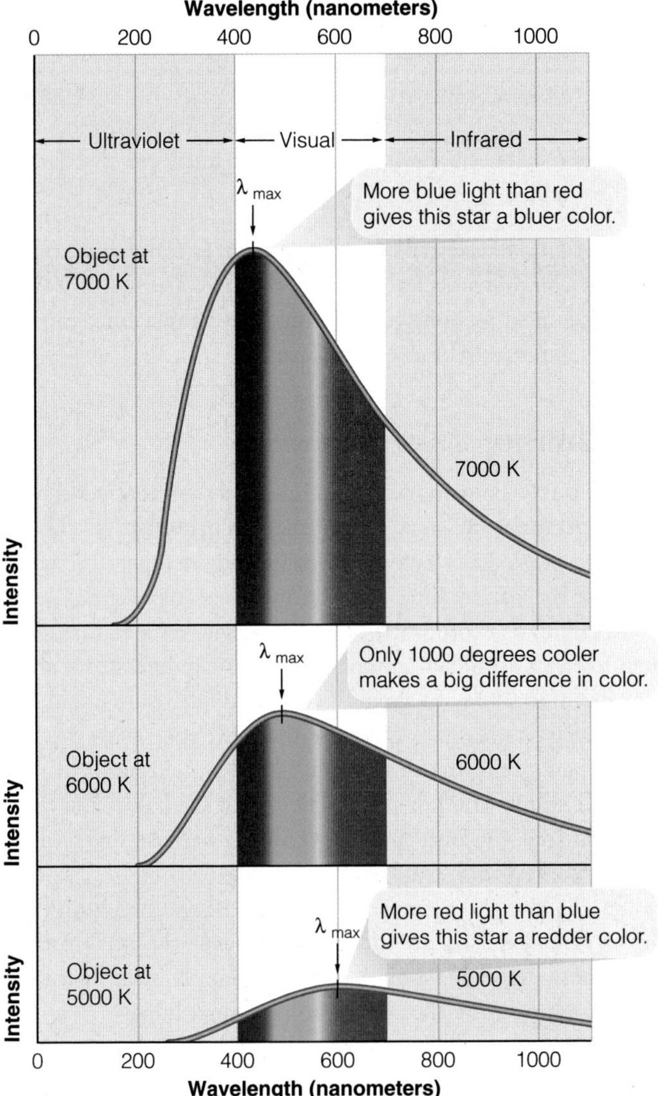


**■ Figure 7-6**

Graphs of blackbody radiation intensity versus wavelength from three objects at different temperatures demonstrate that a hot body radiates more total energy and that the wavelength of maximum intensity is shorter for hotter objects. The hotter object here would look blue to your eyes, while the cooler object would look red.

red. Now you can understand why two of your Favorite Stars, Betelgeuse and Rigel, have such different colors. Betelgeuse is cool and looks red, but Rigel is hot and looks blue.

Cool objects don't glow at visible wavelengths but still produce blackbody radiation. For example, the human body has a temperature of 310 K and emits blackbody radiation mostly in the infrared part of the spectrum. Infrared security cameras can detect burglars by the radiation they emit, and mosquitoes can track you down in total darkness by homing in on your infrared radiation. Although you emit lots of infrared radiation, you rarely emit higher-energy photons; and you almost never emit an X-ray or gamma-ray photon. Your wavelength of maximum intensity lies in the infrared part of the spectrum.

## Two Radiation Laws

The two features of blackbody radiation that you have just considered can be given precise mathematical form, and they have proven so dependable that they are known as laws. One law is related to energy and one to color.

As you saw in the previous section, a hot object emits more blackbody radiation than a cool object. That is, it emits more energy. Recall from Chapter 5 that energy is expressed in units called joules (J); 1 joule is about the energy of an apple falling from a table to the floor. The total radiation given off by 1 square meter of the surface of the object in joules per second equals a constant number, represented by $\sigma$ times the temperature raised to the fourth power.* This relationship is called the Stefan–Boltzmann law:

$$E = \sigma T^4 \ (\text{J/s/m}^2)$$

How does this help you understand stars? Suppose a star the same size as the sun had a surface temperature that was twice as hot as the sun's surface. Then each square meter of that star would radiate not twice as much energy, but $2^4$, or 16, times as much energy. From this law you can see that a small difference in temperature can produce a very large difference in the amount of energy a star's surface emits.

The second radiation law is related to the color of stars. In the previous section, you saw that hot stars look blue and cool stars look red. Wien's law written for conventional intensity units tells you that the wavelength at which a star radiates the most energy, its wavelength of maximum intensity ($\lambda_{max}$), depends only on the star's temperature:

$$\lambda_{max} = 2.90 \times 10^6 / T$$

That is, the wavelength of maximum radiation in nanometers equals 2.9 million divided by the temperature on the Kelvin scale.

This law is a powerful tool in astronomy, because it relates the temperature of a star to its wavelength of maximum intensity. For example, you might find a star emitting light with a maximum intensity at a wavelength of 1000 nm—in the near-infrared. Then the surface temperature of the star must be 2900 K. Later you will meet stars much hotter than the sun; such stars radiate most of their energy at very short wavelengths. The hottest stars, for instance, radiate most of their energy in the ultraviolet.

*For the sake of completeness, you can note that the constant $\sigma$ equals $5.67 \times 10^{-8}$ J/(s m$^2$ K$^4$) (units of joules per second per square meter per degree Kelvin to the fourth power).

## SCIENTIFIC ARGUMENT

***The infrared radiation coming out of your ear can tell a doctor your temperature. How does that work?***

You know two radiation laws, so your argument must use the right one. Doctors and nurses use a handheld device to measure body temperature by observing the infrared radiation emerging from a patient's ear. You might suspect the device depends on the Stefan–Boltzmann law and measures the intensity of the infrared radiation. A person with a fever will emit more energy than a healthy person. However, a healthy person with a large ear canal would emit more than a person with a small ear canal, so measuring intensity would not be helpful. The device actually depends on Wien's law, finding temperature by measuring the "color" of the infrared radiation. A patient with a fever will emit at a slightly shorter wavelength of maximum intensity, and the infrared radiation emerging from his or her ear will be a tiny bit "bluer" than that emitted by a person with normal temperature.

Astronomers can measure the temperatures of stars the same way. Adapt your argument for stars. **Use Figure 7-6 to explain how the colors of stars reveal their temperatures.**

# 7-3 Stellar Spectra

SCIENCE IS A WAY OF UNDERSTANDING NATURE, and the spectrum of a star can tell you a great deal about the star's temperature, motion, and composition. In later chapters, you will use spectra to study many more astronomical objects such as galaxies and planets, but you can begin by looking at the spectra of stars, including that of the sun.

## The Formation of a Spectrum

The spectrum of a star is formed as light passes outward through the gases near its surface. Read **Atomic Spectra** on pages 132–133 and notice that it describes three important properties of spectra and defines 12 new terms that will help you discuss astronomical spectra:

1. There are three kinds of spectra: (i) *continuous spectra;* (ii) *absorption* or *dark-line spectra,* which contain *absorption lines;* and (iii) *emission* or *bright-line spectra,* which contain *emission lines.* These spectra are described by *Kirchhoff's laws.* When you see one of these types of spectra, you can recognize the kind of matter that emitted the light.

2. Photons are emitted or absorbed when an electron in an atom makes a *transition* from one energy level to another. The wavelengths of the photons depend on the energy difference between the two levels. Hydrogen atoms can produce many spectral lines in series such as the *Lyman, Balmer,* and *Paschen* series. Only three lines in the Balmer series are visible to human eyes. The emitted photons coming from a hot cloud of hydrogen gas have the same wavelengths as the photons absorbed by hydrogen atoms in the gases of a star.

3. Most modern astronomy books display spectra as graphs of intensity versus wavelength. Be sure you recognize the connection between the dark absorption lines and the dips in the graphed spectrum.

Whatever kind of celestial object's spectrum astronomers look at, the most common spectral lines are the Balmer lines of hydrogen. In the next section, you will see how Balmer lines can tell you the temperature of a star's surface.

## The Balmer Thermometer

You can use the Balmer absorption lines as a thermometer to find the temperatures of stars. From the discussion of blackbody radiation, you already know how to estimate temperature using color, but the Balmer lines give you much greater accuracy.

Recall that astronomers use the Kelvin temperature scale when referring to stellar temperatures. These temperatures range from about 40,000 K to about 2500 K and refer to the temperature of the star's surface. Compare these extremes with the surface temperature of the sun, about 5800 K. The centers of stars are much hotter—millions of degrees—but the colors and spectra of stars tell you only about the surface because that's where the light comes from.

The Balmer thermometer works because the strength of the Balmer lines depends on the temperature of the star's surface layers. Both hot and cool stars have weak Balmer lines, but medium-temperature stars have strong Balmer lines.

The Balmer absorption lines are produced only by atoms with electrons in the second energy level. If a star is cool, there are few violent collisions between atoms to excite the electrons, so the electrons of most atoms are in the ground state, not the second level. Electrons in the ground state can't absorb photons in the Balmer series. As a result, you should expect to find weak Balmer absorption lines in the spectra of cool stars.

In the surface layers of hot stars, on the other hand, there are many violent collisions between atoms. These collisions can excite electrons to high energy levels or ionize some atoms by knocking electrons out of the atoms. Consequently, there are few hydrogen atoms with their electrons in the second orbit to form Balmer absorption lines. Hot stars, like cool stars, have weak Balmer absorption lines.

In stars of an intermediate temperature, roughly 10,000 K, the collisions are just right to excite large numbers of electrons into the second energy level. Hydrogen gas at that temperature absorbs Balmer wavelength photons very well and produces strong Balmer lines.

Theoretical calculations can predict exactly how strong the Balmer lines should be for stars of various temperatures. Such calculations are the key to finding temperatures from stellar

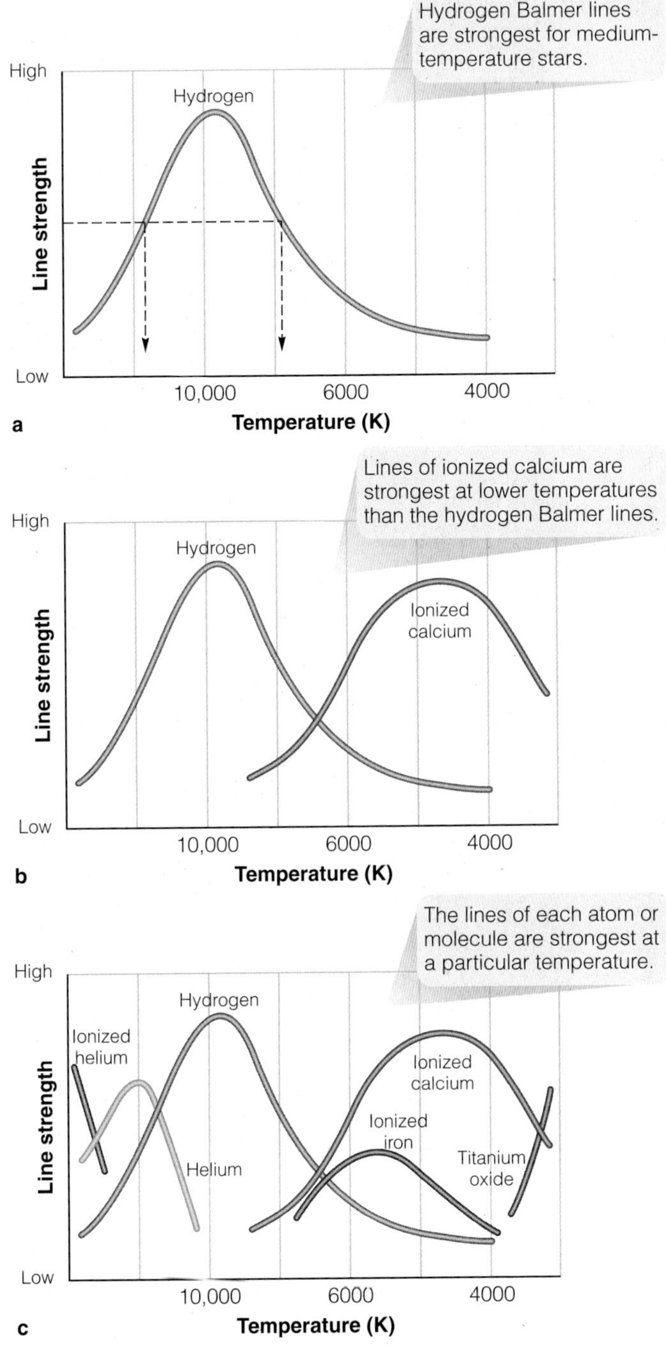


■ **Figure 7-7**

The strength of spectral lines can tell you the temperature of a star. (a) Balmer hydrogen lines alone are not enough because they give two answers. Balmer lines of a certain strength could be produced by a hotter star or a cooler star. (b) Adding another atom to the diagram helps, and (c) adding many atoms and molecules to the diagram creates a precise aid to find the temperatures of stars.

spectra. The curve in ■ Figure 7-7a shows the strength of the Balmer lines for various stellar temperatures. But you can see from the graph that a star with Balmer lines of a certain strength might have either of two temperatures, one high and one low. How do you know which is right? You must examine other spectral lines to choose the correct temperature.

Temperature has a similar effect on the spectral lines of other elements, but the temperature at which the lines reach their maximum strength differs for each element (Figure 7-7b). If you add a number of chemical elements to your graph, you will have a powerful aid for finding the stars' temperatures (Figure 7-7c).

Now you can determine a star's temperature by comparing the strengths of its spectral lines with your graph. For instance, if you found medium-strength Balmer lines and strong helium lines in a star's spectrum, you could conclude that it had a temperature of about 20,000 K. But if the star had weak hydrogen lines and strong lines of ionized iron, you would assign it a temperature of about 5800 K, similar to that of the sun.

The spectra of stars cooler than about 3500 K contain dark bands produced by molecules such as titanium oxide (TiO). Because of their structure, molecules can absorb photons at many wavelengths, producing numerous, closely spaced spectral lines that blend together to form bands. These molecular bands appear in the spectra of only the coolest stars because, as mentioned before, molecules in cool stars are not subject to the violent collisions that would break them apart in hotter stars.

## Spectral Classification

You have seen that the strengths of a star's various spectral lines depend on its surface temperature. From this you can conclude that all stars of a given temperature should have similar spectra. If you learn to recognize the pattern of spectral lines produced by a 6000 K star, for instance, you need not use Figure 7-7c every time you see that kind of spectrum. You can save time by classifying stellar spectra rather than analyzing each one individually.

The first widely used classification system was devised by astronomers at Harvard during the 1890s and 1900s. One of the astronomers, Annie J. Cannon, personally inspected and classified the spectra of over 250,000 stars. The spectra were first classified into groups labeled A through Q, but some groups were later dropped, merged with others, or reordered. The final classification includes the seven major **spectral classes,** or **types,** still used today: O, B, A, F, G, K, M.*

This sequence of spectral types, called the **spectral sequence,** is important because it is a temperature sequence. The O stars are the hottest, and the temperature decreases along the sequence to the M stars, the coolest. For increased precision, astronomers divide each spectral class into ten subclasses. For example, spectral class A consists of the subclasses A0, A1, A2, … A8, A9. Next come F0, F1, F2, and so on. This finer division gives a star's temperature to an accuracy of about 5 percent. The sun, for example, is not just a

*Generations of astronomy students have remembered the spectral sequence using the mnemonic "Oh, Be A Fine Girl (Guy), Kiss Me." More recent suggestions from students include "Oh Boy, An F Grade Kills Me" and "Only Bad Astronomers Forget Generally Known Mnemonics."

# Atomic Spectra

**1** To understand how to analyze a spectrum, begin with a simple incandescent lightbulb. The hot filament emits blackbody radiation, which forms a **continuous spectrum.**

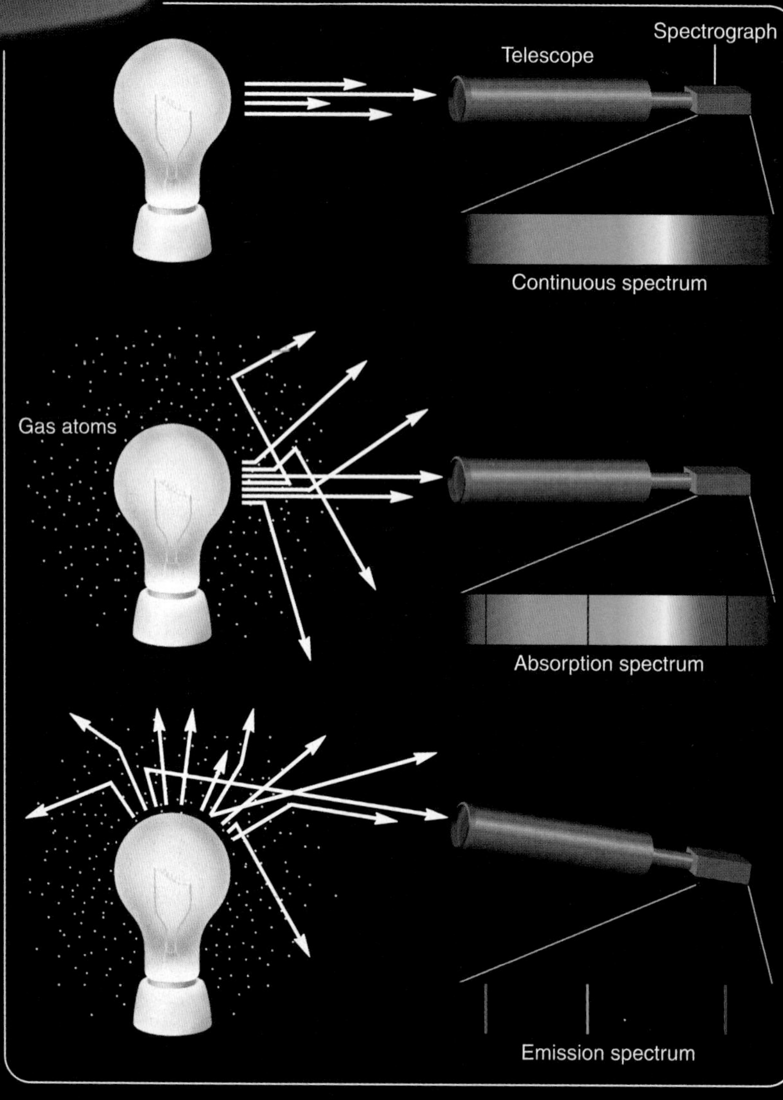


An **absorption spectrum** results when radiation passes through a cool gas. In this case you can imagine that the lightbulb is surrounded by a cool cloud of gas. Atoms in the gas absorb photons of certain wavelengths, which are missing from the spectrum, and you see their positions as dark **absorption lines.** Such spectra are sometimes called **dark-line spectra.**

An **emission spectrum** is produced by photons emitted by an excited gas. You could see **emission lines** by turning your telescope aside so that photons from the bright bulb did not enter the telescope. The photons you would see would be those emitted by the excited atoms near the bulb. Such spectra are also called **bright-line spectra.**

**1a** The spectrum of a star is an absorption spectrum. The denser layers of the photosphere emit blackbody radiation. Gases in the atmosphere of the star absorb their specific wavelengths and form dark absorption lines in the spectrum.

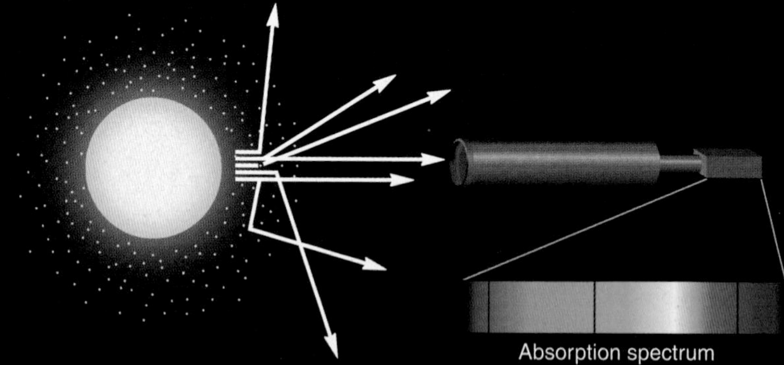

**1b** In 1859, long before scientists understood atoms and energy levels, the German scientist Gustav Kirchhoff formulated three rules, now known as **Kirchhoff's laws,** that describe the three types of spectra.

**KIRCHHOFF'S LAWS**

**Law I: The Continuous Spectrum**

A solid, liquid, or dense gas excited to emit light will radiate at all wavelengths and thus produce a continuous spectrum.

**Law II: The Emission Spectrum**

A low-density gas excited to emit light will do so at specific wavelengths and thus produce an emission spectrum.

**Law III: The Absorption Spectrum**

If light comprising a continuous spectrum passes through a cool, low-density gas, the result will be an absorption spectrum.

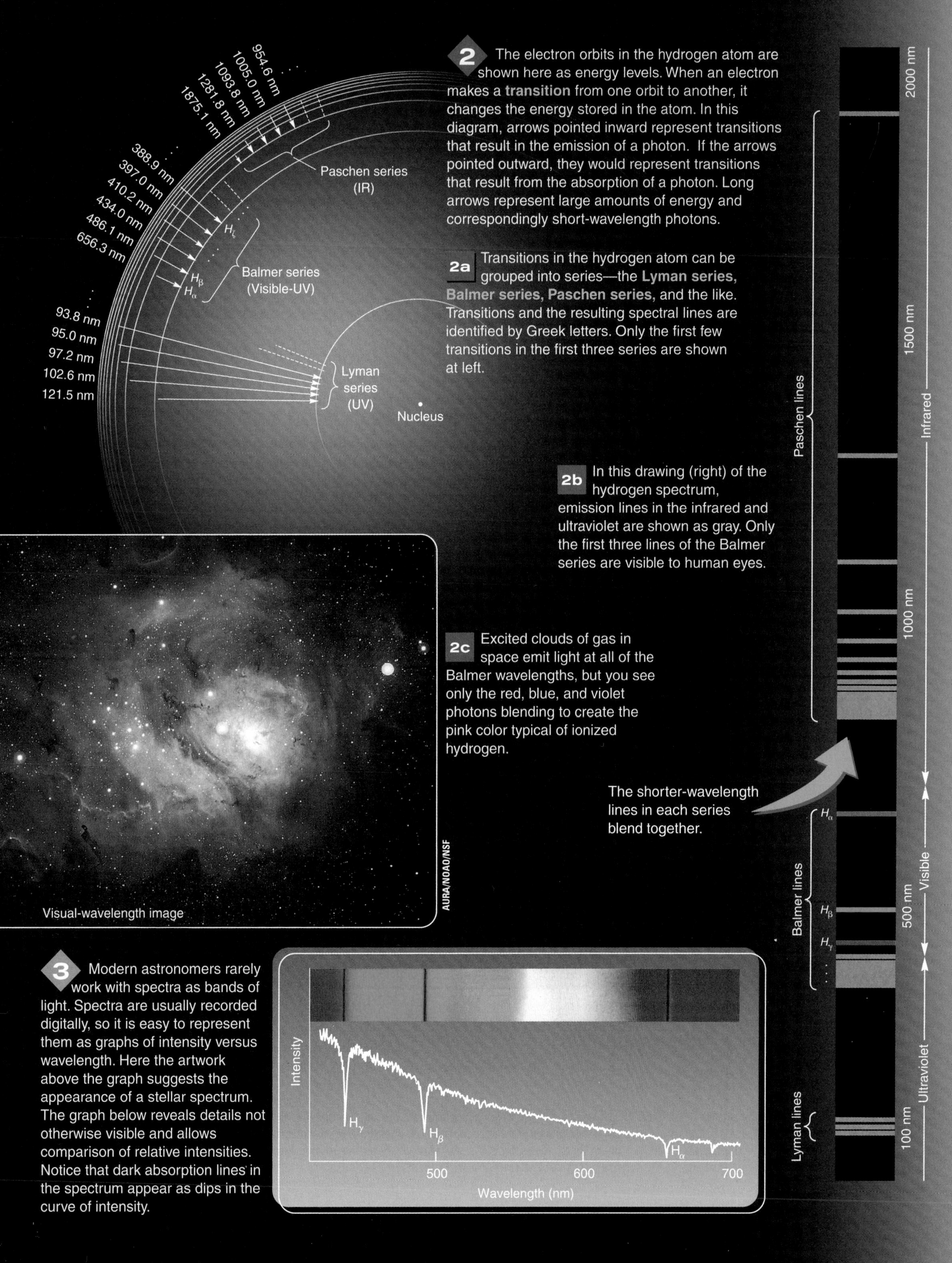


**2** The electron orbits in the hydrogen atom are shown here as energy levels. When an electron makes a **transition** from one orbit to another, it changes the energy stored in the atom. In this diagram, arrows pointed inward represent transitions that result in the emission of a photon. If the arrows pointed outward, they would represent transitions that result from the absorption of a photon. Long arrows represent large amounts of energy and correspondingly short-wavelength photons.

**2a** Transitions in the hydrogen atom can be grouped into series—the **Lyman series, Balmer series, Paschen series,** and the like. Transitions and the resulting spectral lines are identified by Greek letters. Only the first few transitions in the first three series are shown at left.

**2b** In this drawing (right) of the hydrogen spectrum, emission lines in the infrared and ultraviolet are shown as gray. Only the first three lines of the Balmer series are visible to human eyes.

**2c** Excited clouds of gas in space emit light at all of the Balmer wavelengths, but you see only the red, blue, and violet photons blending to create the pink color typical of ionized hydrogen.

**3** Modern astronomers rarely work with spectra as bands of light. Spectra are usually recorded digitally, so it is easy to represent them as graphs of intensity versus wavelength. Here the artwork above the graph suggests the appearance of a stellar spectrum. The graph below reveals details not otherwise visible and allows comparison of relative intensities. Notice that dark absorption lines in the spectrum appear as dips in the curve of intensity.

**Table 7-1 | Spectral Classes**

| Spectral Class | Approximate Temperature (K) | Hydrogen Balmer Lines | Other Spectral Features | Naked-Eye Example |
|---|---|---|---|---|
| O | 40,000 | Weak | Ionized helium | Meissa (O8) |
| B | 20,000 | Medium | Neutral helium | Achernar (B3) |
| A | 10,000 | Strong | Ionized calcium weak | Sirius (A1) |
| F | 7500 | Medium | Ionized calcium weak | Canopus (F0) |
| G | 5500 | Weak | Ionized calcium medium | Sun (G2) |
| K | 4500 | Very weak | Ionized calcium strong | Arcturus (K2) |
| M | 3000 | Very weak | TiO strong | Betelgeuse (M2) |

G star, but a G2 star. ■ Table 7-1 breaks down some of the information contained in Figure 7-7c and presents it in tabular form according to spectral class. For example, if a star has weak Balmer lines and lines of ionized helium, it must be an O star.

Thirteen stellar spectra are arranged in ■ Figure 7-8 from the hottest at the top to the coolest at the bottom. You can easily see how the strength of spectral lines depends on temperature. The Balmer lines are strongest in A stars, where the temperature is moderate but still high enough to excite the electrons in hydrogen atoms to the second energy level, where they can absorb Balmer wavelength photons. In the hotter stars (O and B), the Balmer lines are weaker because the higher temperature excites the electrons to energy levels above the second or ionizes the atoms. The Balmer lines in cooler stars (F through M) are also

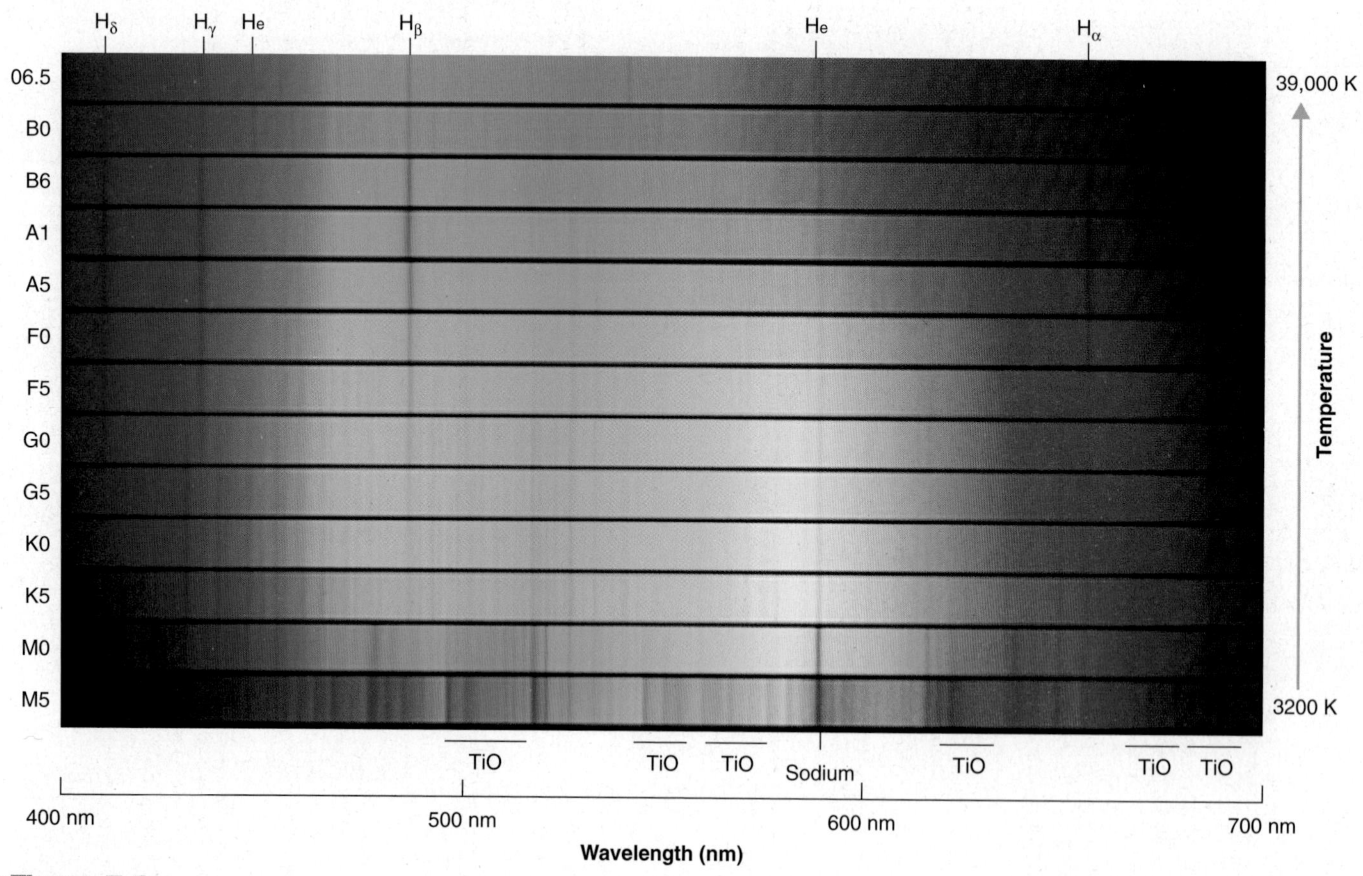


■ **Figure 7-8**

These spectra show stars ranging from hot O stars at the top to cool M stars at the bottom. The Balmer lines of hydrogen are strongest at spectral type A0, but the two closely spaced lines of sodium in the yellow are strongest for very cool stars. Helium lines appear only in the spectra of the hottest stars. Notice that the helium line visible in the top spectrum has nearly but not exactly the same wavelength as the sodium lines visible in cooler stars. Bands produced by the molecule titanium oxide are strong in the spectra of the coolest stars. (AURA/NOAO/NSF)

weaker but for a different reason. The lower temperature cannot excite many electrons to the second energy level, so few hydrogen atoms are capable of absorbing Balmer wavelength photons.

Although these spectra are attractive, astronomers rarely work with spectra as color images. Rather, they display spectra as graphs of intensity versus wavelength that show dark absorption lines as dips in the graph (■ Figure 7-9). Such graphs allow more detailed analysis than photographs. Notice, for example, that the overall curves are segments of blackbody curves with spectral lines superimposed. The wavelength of maximum intensity is in the infrared for the coolest stars and in the ultraviolet for the hottest stars. Look carefully at these graphs, and you can see that helium is visible only in the spectra of the hottest classes and titanium oxide bands only in the coolest. Two lines of ionized calcium increase in strength from A to K and then decrease from K through M. Because the strengths of these spectral lines depend on temperature, it requires only a few moments to study a star's spectrum and determine its temperature.

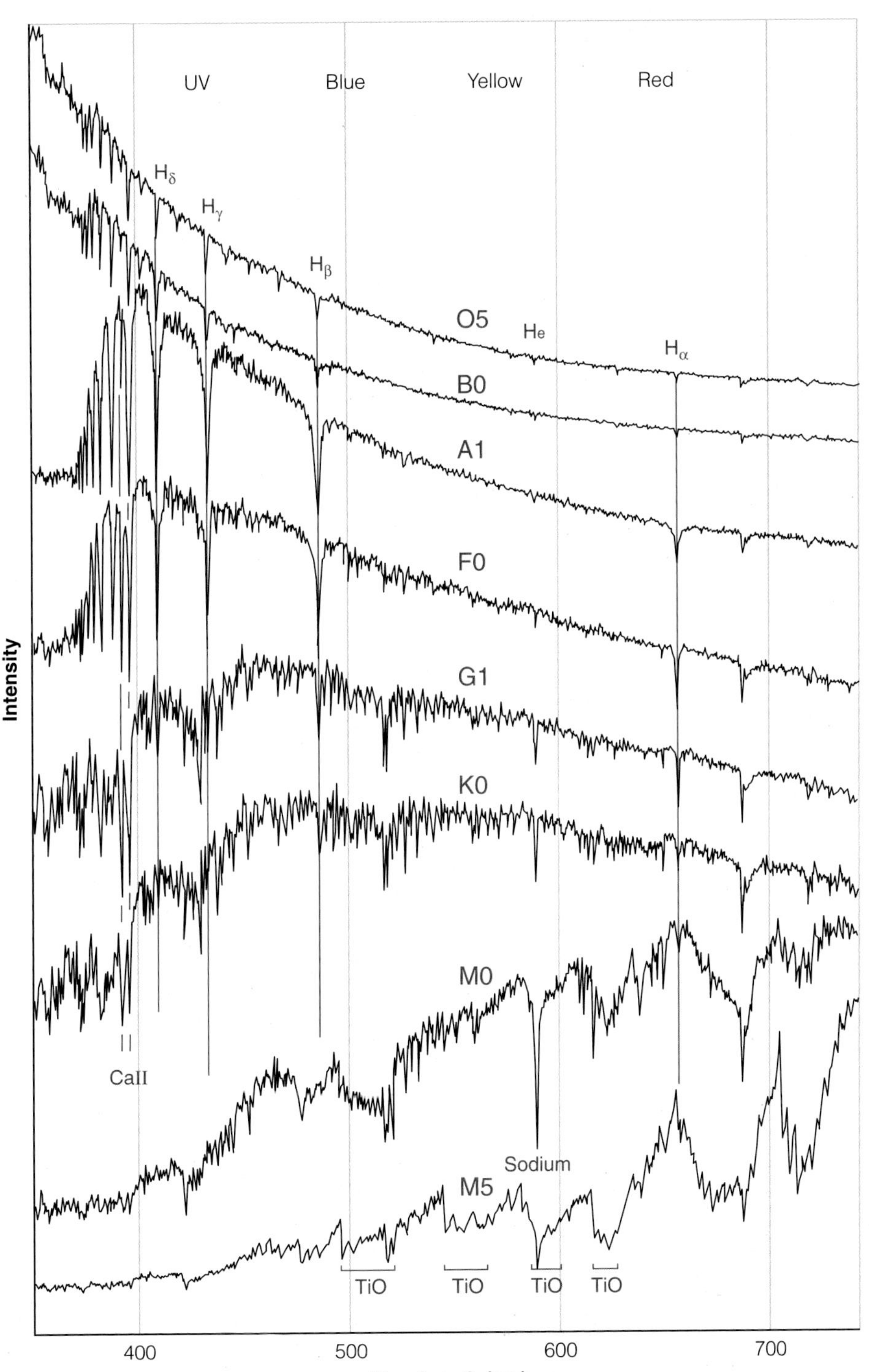


■ **Figure 7-9**

Modern digital spectra are often represented as graphs of intensity versus wavelength with dark absorption lines appearing as sharp dips in the curves. The hottest stars are at the top and the coolest at the bottom. Hydrogen Balmer lines are strongest at about A0, while lines of ionized calcium (Ca II) are strong in K stars. Titanium oxide (TiO) bands are strongest in the coolest stars. Compare these spectra with Figures 7-7c and 7-8. (Courtesy NOAO, G. Jacoby, D. Hunter, and C. Christian)

Now you can learn something new about your Favorite Stars. Sirius, brilliant in the winter sky, is an A1 star; and Vega, bright overhead in the summer sky, is an A0 star. They have nearly the same temperature and color, and both have strong Balmer lines in their spectra. The bright red star in Orion is Betelgeuse, a cool M2 star, but blue-white Rigel is a hot B8 star. Polaris, the North Star, is an F8 star a bit hotter than our sun, and Alpha Centauri, the closest star to the sun, is a G2 star just like the sun.

The study of spectral types is a century old, but astronomers continue to discover new types. The **L dwarfs,** found in 1998, are cooler and fainter than M stars, and are understood to be objects smaller than stars but larger than planets called "brown dwarfs" that you will learn about in a later chapter. The spectra of M stars contain bands produced by metal oxides such as titanium oxide (TiO), but L dwarf spectra contain bands produced by molecules such as iron hydride (FeH). The **T dwarfs,** discovered in 2000, are an even cooler and fainter type of brown dwarf than L dwarfs. Their spectra show absorption by methane ($CH_4$) and water vapor (■ Figure 7-10). The development of giant telescopes and highly sensitive infrared cameras and spectrographs is allowing astronomers to find and study these cool objects.

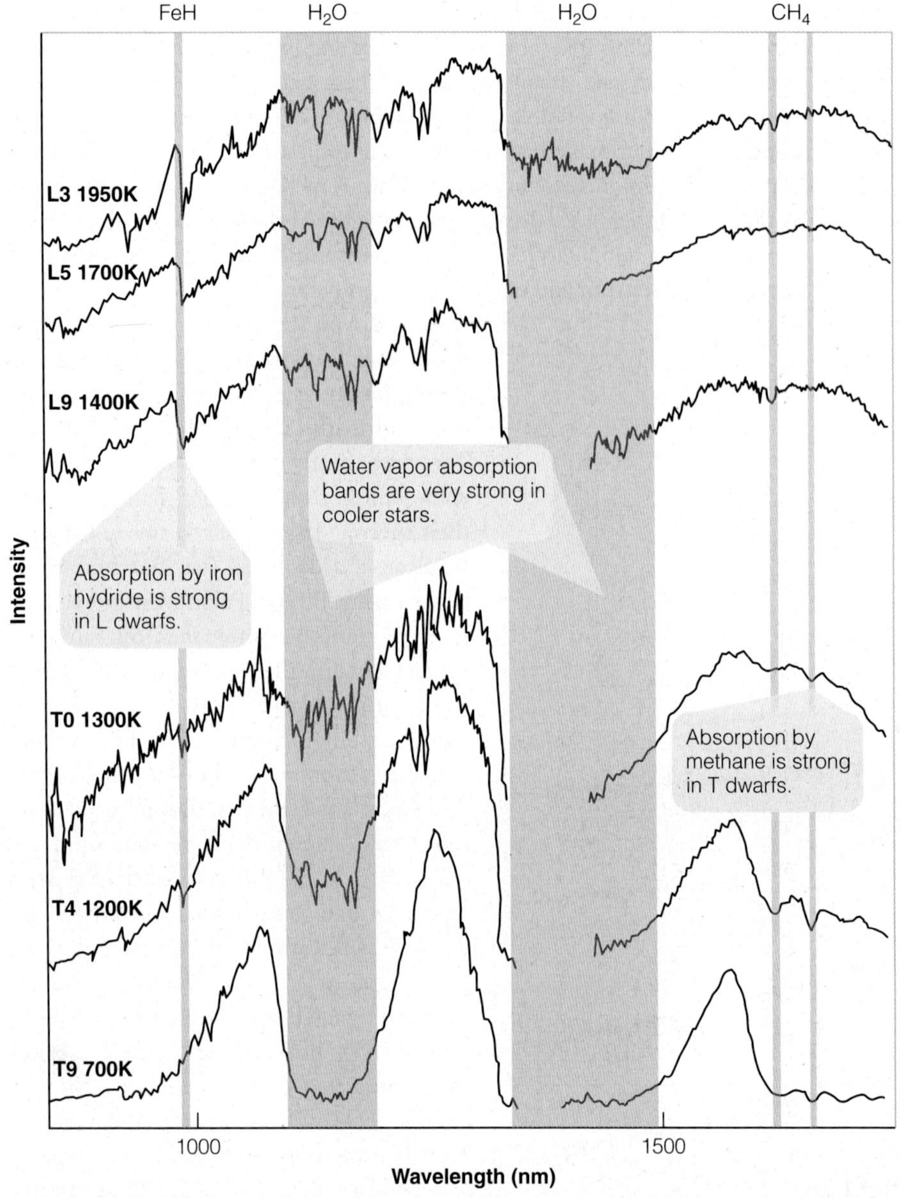


■ **Figure 7-10**

These six infrared spectra show the dramatic differences between L dwarfs and T dwarfs. Spectra of M stars show titanium oxide bands (TiO), but L and T dwarfs are so cool that TiO molecules do not form. Other molecules, such as iron hydride (FeH), water ($H_2O$), and methane ($CH_4$), can form in these very cool objects. (Adapted from Thomas R. Geballe, Gemini Observatory, from a graph that originally appeared in *Sky and Telescope Magazine,* February 2005, p. 37.)

## The Composition of the Stars

It seems as though it should be easy to learn the composition of the sun and stars just by looking at their spectra, but this is actually a difficult problem that wasn't well understood until the 1920s. The story is worth telling, not only because it illustrates the temperature dependence of stellar spectral types, it is also the story of an important American astronomer who waited decades to get proper credit for her work.

As a child in England, Cecilia Payne (1900–1979) excelled in classics, languages, mathematics, and literature, but her first love was astronomy. After finishing Newnham College in Cambridge, she left England, sensing that there were few opportunities there for a woman in science. In 1922, Payne arrived at Harvard, where she eventually earned her Ph.D. (although her degree was awarded by Radcliffe because Harvard did not then admit women).

In her thesis, Payne attempted to relate the strength of the absorption lines in stellar spectra to the physical conditions in the atmospheres of the stars. This was not easy; a given spectral line can be weak because the atom is rare or because the temperature is either too high or too low for it to absorb efficiently. If you see sodium lines in a star's spectrum, you can be sure that the star contains sodium atoms, but if you see no sodium lines, you must consider the possibility that sodium is present but the star is too hot or too cool for that type of atom to produce detectable spectral lines.

Density is another factor. If a gas is dense, the atoms are packed tighter together, collide more often, distort the electron orbits, and thereby change the strength of spectral lines (**Focus on Fundamentals 4**). If the gas is low density, the atoms collide less often and the spectral lines are less affected.

Payne's problem was to untangle these factors and find both the true temperatures of the stars and the true abundance of the atoms in their atmospheres. Solving this puzzle required some of the most recent advances in atomic physics and quantum mechanics. About the time Payne left Cambridge University, Indian physicist Meghnad Saha published his work on the ionization of atoms. Drawing from such theoretical work, Payne's calculations showed that over 90 percent of the atoms in stars (including the sun) must be hydrogen and most of the rest are helium (■ Table 7-2). The heavier atoms like calcium, sodium, and iron seem more abundant only because they are better at absorbing photons at the temperatures of stars.

At the time, astronomers found it hard to believe Payne's abundances of hydrogen and helium. They especially found such a high abundance of helium unacceptable. After all, hydrogen lines are at least visible in most stellar spectra, but helium lines are almost invisible in the spectra of all but the hottest stars. Nearly all astronomers assumed that the stars had roughly the same composition as Earth's surface; that is, they believed that the stars were composed mainly of heavier atoms such as carbon,

## Density

You are about as dense as an average star. What does that mean? As you study astronomy, you will use the term *density* often, so you should be sure to understand this fundamental concept. **Density** is a measure of the amount of matter in a given volume. Density is expressed as mass per volume, such as grams per cubic centimeter. The density of water, for example, is about 1 g/cm$^3$, and you are almost as dense as water.

To get a feel for density, imagine holding a brick in one hand and a similar-sized block of Styrofoam in the other hand. You can easily tell that the brick contains more matter than the Styrofoam block, even though both are the same size. The brick weighs more than the Styrofoam, but it isn't really the weight that you should consider. Rather, you should think about the mass of the two objects. In space, where they have no weight, the brick and the Styrofoam would still have mass, and you could tell just by moving them around that the brick contains more mass than the Styrofoam. For example, imagine tapping each object gently with your hand. The massive brick would be easy to distinguish from the low-mass Styrofoam block, even in weightlessness.

Density is a fundamental concept in science because it is a general property of materials. Metals tend to be dense; lead, for example, has a density of about 7 g/cm$^3$. Rock, in contrast, has a density of 3 to 4 g/cm$^3$. Water and ice have densities of about 1 g/cm$^3$. If you knew that a small moon orbiting Saturn had a density of 1.5 g/cm$^3$, you could immediately draw some conclusions about what kinds of materials the little moon might be made of—ice and a little rock, but not much metal. The density of an object is a basic clue to its composition.

Astronomical bodies can have dramatically different densities. The gas clouds between the stars can have very low densities, but the same kind of gas in a star can have a much higher density. The sun, for example, has an average density of about 1 g/cm$^3$, about the same as your body. As you study astronomical objects, pay special attention to their densities.

*A brick would be dense even in space where it had no weight.*



**Table 7-2 | The Most Abundant Elements in the Sun**

| Element | Percentage by Number of Atoms | Percentage by Mass |
|---|---|---|
| Hydrogen | 91.0 | 70.9 |
| Helium | 8.9 | 27.4 |
| Carbon | 0.03 | 0.3 |
| Nitrogen | 0.008 | 0.1 |
| Oxygen | 0.07 | 0.8 |
| Neon | 0.01 | 0.2 |
| Magnesium | 0.003 | 0.06 |
| Silicon | 0.003 | 0.07 |
| Sulfur | 0.002 | 0.04 |
| Iron | 0.003 | 0.1 |

silicon, iron, and aluminum. Even the most eminent astronomers dismissed Payne's result as illusory. Faced with this pressure and realizing that as a female scientist in the 1920s she faced an uphill battle, Payne could not press her discovery.

By 1929, astronomers generally understood the importance of temperature on measurements of composition derived from stellar spectra. At that point, they recognized that stars are mostly hydrogen and helium, but Payne received no credit.

Payne worked for many years as a staff astronomer at the Harvard College Observatory with no formal position on the faculty. She married Russian astronomer Sergei Gaposchkin in 1934 and was afterward known as Cecilia Payne-Gaposchkin. In 1956, Harvard accepted women to its faculty, and Payne-Gaposchkin was appointed a full professor and made chair of the Harvard astronomy department. By that time, the importance of her research 30 years earlier had come to be widely recognized.

Cecilia Payne-Gaposchkin's work on the chemical composition of the stars illustrates the importance of fully understanding the interaction between light and matter. It was her detailed understanding of the physics that led her to the correct composition. As you turn your attention to other information that can be derived from stellar spectra, you will again discover the importance of understanding light.

### The Doppler Effect

Surprisingly, one of the pieces of information hidden in a spectrum is the velocity of the light source. Astronomers can measure the wavelengths of the lines in a star's spectrum and find the velocity of the star. The **Doppler effect** is the apparent change in the wavelength of radiation caused by the motion of the source.

When astronomers talk about the Doppler effect, they are talking about a shift in the wavelength of electromagnetic radiation. But the Doppler shift can occur in any form of wave phenomena, including sound waves, so you probably hear the Doppler effect every day without noticing.

The pitch of a sound is determined by its wavelength. Sounds with long wavelengths have low pitches, and sounds with short wavelengths have higher pitches. You hear a Doppler shift every time a car or truck passes you and the pitch of its engine noise drops. Its sound is shifted to shorter wavelengths and higher pitches while it is approaching and is shifted to longer wavelengths and lower pitches after it passes.

To see why the sound waves are shifted in wavelength, consider a fire truck approaching you with a bell clanging once a second. When the bell clangs, the sound travels ahead of the truck to reach your ears. One second later, the bell clangs again, but, during that one second, the fire truck has moved closer to you, so the bell is closer at its second clang. Now the sound has a shorter distance to travel and reaches your ears a little sooner than it would have if the fire truck were not approaching. If you timed the clangs, you would find that you heard them slightly less than one second apart. After the fire truck passes you and is moving away, you hear the clangs sounding slightly more than one second apart because now each successive clang of the bell occurs farther from you and the sound travels farther to reach your ears.

■ Figure 7-11a shows a fire truck moving toward one observer and away from another observer. The position of the bell at each clang is shown by a small black bell. The sound of the clangs spreading outward is represented by black circles. You can see how the clangs are squeezed together ahead of the fire truck and stretched apart behind. If the fire truck has a siren instead of a bell, the sound coming from the siren will be a wave with a series of compressions and un-compressions. If the truck and siren are moving toward an observer, the compressions and un-compressions of the siren's sound wave will arrive more often—at a higher frequency—than if the truck were not moving, and the observer will hear the siren at a higher pitch than the same siren when it is stationary.

Now you can substitute a source of light for the clanging bell or wailing siren (Figure 7-11b). Imagine the light source emitting waves continuously as it approaches you. Each time the source emits the peak of a wave, it will be slightly closer to you than when it emitted the peak of the previous wave. From your vantage point, the successive peaks of the wave will seem closer together in the same way that the clangs of the bell seemed closer together. The light will appear to have a shorter wavelength, making it slightly bluer. Because the light is shifted slightly toward the blue end of the spectrum, this is called a **blueshift.** After the light source has passed you and is moving away, the peaks of successive waves seem farther apart, so the light has a longer wavelength and is redder. This is a **redshift.** The shifts are

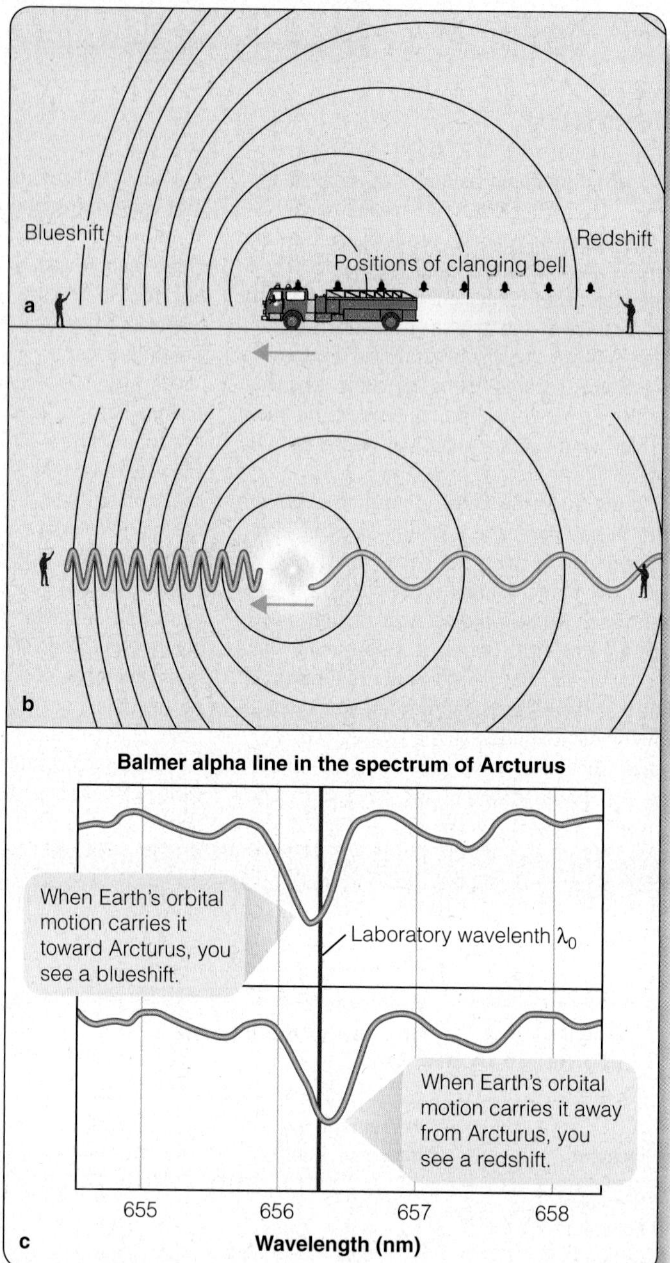


**■ Figure 7-11**

The Doppler effect. (a) The clanging bell on a moving fire truck produces sounds that move outward (black circles). An observer ahead of the truck hears the clangs closer together, while an observer behind the truck hears them farther apart. Similarly, the sound waves from a siren on an approaching truck will be received more often, and thus be heard with a higher tone, than a stationary truck, and the siren will have a lower tone if it is going away. (b) A moving source of light emits waves that move outward (black circles). An observer toward whom the light source is moving observes a shorter wavelength (a blueshift), and an observer for whom the light source is moving away observes a longer wavelength (a redshift). (c) Absorption lines in the spectrum of the bright star Arcturus are shifted to the blue in winter, when Earth's orbital motion carries it toward the star, and to the red in summer when Earth moves away from the star.

much too small to change the color of a star, but they are easily detected in spectra.

The terms *redshift* and *blueshift* are used to refer to any range of wavelengths. The light does not actually have to be red or blue, and the terms apply equally to wavelengths in other parts of the electromagnetic spectrum such as X-rays and radio waves. *Red* and *blue* refer to the direction of the shift, not to actual color.

The amount of change in wavelength, and thus the magnitude of the Doppler shift, depends on the velocity of the source. A moving car has a smaller Doppler shift than a jet plane, and a slow-moving star has a smaller Doppler shift than one that is moving more quickly. You can measure the velocity of a star by measuring the size of its Doppler shift. If a star is moving toward Earth, it has a blue shift and each of its spectral lines is shifted very slightly toward shorter wavelengths. If it is receding from Earth, it has a red shift. The next section will show how astronomers can convert Doppler shifts into velocities.

When you think about the Doppler effect, it is important to understand two things. Earth itself moves, so a measurement of a Doppler shift really measures the relative motion between Earth and the star. Figure 7-11c shows the Doppler effect in two spectra of the star Arcturus. Lines in the top spectrum are slightly blueshifted because the spectrum was recorded when Earth, in the course of its orbit, was moving toward Arcturus. Lines in the bottom spectrum are redshifted because it was recorded six months later, when Earth was moving away from Arcturus. To find the true motion of Arcturus, astronomers must subtract the motion of Earth.

The second point to remember is that the Doppler shift is sensitive only to the part of the velocity directed away from you or toward you—the **radial velocity ($V_r$)** (■ Figure 7-12a). You cannot use the Doppler effect to detect any part of the velocity that is perpendicular to your line of sight. A star moving to the left, for example, would have no blueshift or redshift because its distance from Earth would not be decreasing or increasing. This is why police using radar guns park right next to the highway (Figure 7-12b). They want to measure your full velocity as you drive past, not just part of your velocity.

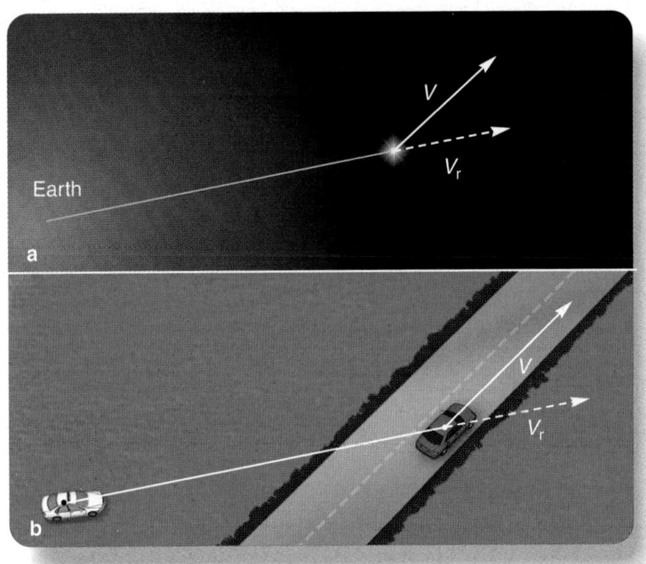


■ **Figure 7-12**

(a) From Earth, astronomers can use the Doppler effect to measure the radial velocity *($V_r$)* of a star, but they cannot measure its true velocity, *V*, through space. (b) Police radar can measure only the radial part of your velocity *($V_r$)* as you drive down the highway, not your true velocity along the pavement *(V)*. That is why police using radar should never park far from the highway. This police car is poorly placed to make a good measurement.

### Calculating the Doppler Velocity

It is easy to calculate the radial velocity of an object from its Doppler shift. The formula is a simple proportion relating the radial velocity $V_r$ divided by the speed of light *c,* to the change in wavelength, λ, divided by the un-shifted wavelength, $\lambda_0$:

$$\frac{V_r}{c} = \frac{\Delta\lambda}{\lambda_0}$$

For example, suppose you observed a line in a star's spectrum with a wavelength of 600.1 nm. Laboratory measurements show that the line should have a wavelength of 600 nm. That is, its un-shifted wavelength is 600 nm. What is the star's radial velocity? First note that the change in wavelength is 0.1 nm:

$$\frac{V_r}{c} = \frac{0.1}{600} = 0.000167$$

Multiplying by the speed of light, $3.00 \times 10^5$ km/s, gives the radial velocity, 50 km/s. Because the wavelength is shifted to the red (lengthened), the star must be receding.

## What Are We? Stargazers

Do you suppose chickens ever look at the sky and wonder what the stars are? Probably not. Chickens are very good at the chicken business, but they are not known for big brains and deep thought. Humans, in contrast, have highly evolved, sophisticated brains and are extremely curious. In fact, curiosity may be the most reliable characteristic of intelligence, and curiosity about the stars is a natural extension of our continual attempts to understand the world around us.

For early astronomers like Copernicus and Kepler, the stars were just points of light. There seemed to be no way to learn anything about them. Galileo's telescope revealed surprising details about the planets; but, even viewed through a large telescope, the stars are just points of light. Even when later astronomers began to assume that the stars were other suns, the stars seemed forever beyond human knowledge.

As you have seen, the key is understanding how light interacts with matter. In the last 150 years or so, scientists have discovered how atoms and light interact to form spectra, and astronomers have applied those discoveries to the ultimate object of human curiosity—the stars.

Chickens may never wonder what the stars are, or even wonder what chickens are, but humans are curious animals, and we do wonder about the stars and about ourselves. Our yearning to understand the stars is just part of our quest to understand what we are.

# Study and Review

## Summary

- An atom consists of a **nucleus (p. 123)** surrounded by a cloud of **electrons (p. 123)**. The nucleus is made up of positively charged **protons (p. 123)** and uncharged **neutrons (p. 123)**.
- The number of protons in an atom determines which element it is. Atoms of the same element (that is, having the same number of protons) with different numbers of neutrons are called **isotopes (p. 124)**.
- A neutral atom is surrounded by a number of negatively charged electrons equal to the number of protons in the nucleus. An atom that has lost or gained an electron is said to be **ionized (p. 124)** and is called an **ion (p. 124)**.
- Two or more atoms joined together form a **molecule (p. 124)**.
- The electrons in an atom are attracted to the nucleus by the **Coulomb force (p. 124)**. As described by **quantum mechanics (p. 124)**, the **binding energy (p. 124)** that holds electrons in an atom is limited to certain energies, and that means the electrons may occupy only certain **permitted orbits (p. 124)**.
- The size of an electron's orbit depends on its energy, so the orbits can be thought of as **energy levels (p. 126)** with the lowest possible energy level known as the **ground state (p. 126)**.
- An **excited atom (p. 126)** is one in which an electron is raised to a higher orbit by a collision between atoms or the absorption of a photon of the proper energy.
- The agitation among the atoms and molecules of an object is called **thermal energy (p. 127)**, and the flow of thermal energy is **heat (p. 127)**. In contrast, **temperature (p. 127)** refers to the intensity of the agitation and can be expressed on the **Kelvin temperature scale (p. 127)**, which gives temperature above **absolute zero (p. 127)**.
- Collisions among the particles in a body accelerate electrons and cause the emission of **blackbody radiation (p. 127)**. The hotter an object is, the more it radiates and the shorter is its **wavelength of maximum intensity,** $\lambda_{max}$ **(p. 128)**. This allows astronomers to estimate the temperatures of stars from their colors.
- **Density (p. 137)** is a measure of the amount of matter in a given volume.
- **Kirchhoff's laws (p. 132)** explain that a hot solid, liquid, or dense gas emits electromagnetic radiation at all wavelengths and produces a **continuous spectrum (p. 132)**. An excited low-density gas produces an **emission (bright-line) spectrum (p. 132)** containing **emission lines (p. 132)**. A light source viewed through a low-density gas produces an **absorption (dark-line) spectrum (p. 132)** containing **absorption lines (p. 132)**.
- An atom can emit or absorb a photon when an electron makes a **transition (p. 133)** between orbits.
- Because orbits of only certain energy differences are permitted in an atom, photons of only certain wavelengths can be absorbed or emitted. Each kind of atom has its own characteristic set of spectral lines. The hydrogen atom has the **Lyman (p. 133) series** of lines in the ultraviolet, the **Balmer series (p. 133)** partially in the visible, and the **Paschen series (p. 133)** (plus others) in the infrared.
- The strength of spectral lines depends on the temperature of the star. For example, in cool stars, the Balmer lines are weak because atoms are not excited out of the ground state. In hot stars, the Balmer lines are weak because atoms are excited to higher orbits or are ionized. Only at medium temperatures are the Balmer lines strong.
- A star's **spectral class** (or **type**) **(p. 131)** is determined by the absorption lines in its spectrum. The resulting **spectral sequence (p. 131)**, OBAFGKM, is important because it is a temperature

sequence. By classifying a star, the astronomer learns the temperature of the star's surface.

- Long after the spectral sequence was created, astronomers found the **L dwarfs (p. 135)** and **T dwarfs (p. 135)** at temperatures even cooler than the M stars.
- A spectrum can tell you the chemical composition of the stars. The presence of spectral lines of a certain element shows that that element must be present in the star, but you must proceed with care. Lines of a certain element may be weak or absent if the star is too hot or too cool even if that element is present in the star's atmosphere.
- The **Doppler effect (p. 137)** can provide clues to the motions of the stars. When a star is approaching, you observe slightly shorter wavelengths, a **blueshift (p. 138)**, and when it is receding, you observe slightly longer wavelengths, a **redshift (p. 138)**. This Doppler effect reveals a star's **radial velocity, $V_r$ (p. 139)**, that part of its velocity directed toward or away from Earth.

## Review Questions

1. Why might you say that atoms are mostly empty space?
2. What is the difference between an isotope and an ion?
3. Why is the binding energy of an electron related to the size of its orbit?
4. Explain why ionized calcium can form absorption lines, but ionized hydrogen cannot.
5. Describe two ways an atom can become excited.
6. Why do different atoms have different lines in their spectra?
7. Why does the amount of blackbody radiation emitted depend on the temperature of the object?
8. Why do hot stars look bluer than cool stars?
9. What kind of spectrum does a neon sign produce?
10. Why are Balmer lines strong in the spectra of medium-temperature stars and weak in the spectra of hot and cool stars?
11. Why are titanium oxide features visible in the spectra of only the coolest stars?
12. Explain the similarities among Table 7-1, Figure 7-7c, Figure 7-8, and Figure 7-9.
13. Explain why the presence of spectral lines of a given element in the solar spectrum tells you that element is present in the sun, but the absence of the lines would not necessarily mean the element is absent from the sun.
14. Why does the Doppler effect detect only radial velocity?
15. How can the Doppler effect explain shifts in both light and sound?
16. **How Do We Know?** How is the world you see around you determined by a world you cannot see?

## Discussion Questions

1. In what ways is the model of an atom a scientific model? In what ways is it incorrect?
2. Can you think of classification systems used to simplify what would otherwise be complex measurements? Consider foods, movies, cars, grades, and clothes.

## Problems

1. Human body temperature is about 310 K ($3.10 \times 10^2$ K, or 98.6°F). At what wavelength do humans radiate the most energy? What kind of radiation do we emit?
2. If a star has a surface temperature of 20,000 K ($2.00 \times 10^4$ K), at what wavelength will it radiate the most energy?
3. Infrared observations of a star show that it is most intense at a wavelength of 2000 nm ($2.00 \times 10^3$ nm). What is the temperature of the star's surface?
4. If you double the temperature of a blackbody, by what factor will the total energy radiated per second per square meter increase?
5. If one star has a temperature of 6000 K and another star has a temperature of 7000 K, how much more energy per second will the hotter star radiate from each square meter of its surface?
6. Electron orbit transition *A* produces light with a wavelength of 500 nm. Transition *B* involves twice as much energy as *A*. What wavelength light does it produce?
7. Determine the temperatures of the following stars based on their spectra. Use Figure 7-7c.
   a. medium-strength Balmer lines, strong helium lines
   b. medium-strength Balmer lines, weak ionized-calcium lines
   c. strong TiO bands
   d. very weak Balmer lines, strong ionized-calcium lines
8. To which spectral classes do the stars in Problem 7 belong?
9. In a laboratory, the Balmer beta line has a wavelength of 486.1 nm. If the line appears in a star's spectrum at 486.3 nm, what is the star's radial velocity? Is it approaching or receding?
10. The highest-velocity stars an astronomer might observe have velocities of about 400 km/s ($4.00 \times 10^2$ km/s). What change in wavelength would this cause in the Balmer gamma line? (*Hint:* Wavelengths are given on page 133.)

## Learning to Look

1. Consider Figure 7-3. When an electron in a hydrogen atom moves from the third orbit to the second orbit, the atom emits a Balmer-alpha photon in the red part of the spectrum. In what part of the spectrum would you look to find the photon emitted when an electron in a helium atom makes the same transition?
2. Where should the police car in Figure 7-12 have parked to make a good measurement?
3. The nebula shown below contains mostly hydrogen excited to emit photons. What kind of spectrum would you expect this nebula to produce?

T. Rector, University of Alaska, and WIYN/NURO/AURA/NSF

4. If the nebula in the image here crosses in front of the star, and the nebula and star have different radial velocities, what might the spectrum of the star look like?

# 8 The Sun

*This far-ultraviolet image of the sun made from space has been enhanced and rendered in false color to reveal complex structure on the surface and clouds of gas trapped in arched magnetic fields above the surface.* (NASA/TRACE)

## Guidepost

The sun is the source of light and warmth in our solar system, so it is a natural object of human curiosity. It is also the one star that is most clearly visible from Earth. The interaction of light and matter, which you studied in Chapter 7, can reveal the secrets of the sun and introduce you to the stars.

In this chapter, you will discover how the analysis of the solar spectrum can paint a detailed picture of the sun's atmosphere and how basic physics has solved the mystery of the sun's core. Here you will answer four essential questions:

**What do you see when you look at the sun?**

**How does the sun make its energy?**

**What are the dark sunspots?**

**Why does the sun go through a cycle of activity?**

Although this chapter is confined to the center of the solar system, it introduces you to a star and leads your thoughts onward among the stars and galaxies that fill the universe.

*All cannot live on the piazza, but everyone may enjoy the sun.*

— ITALIAN PROVERB

A WIT ONCE REMARKED that solar astronomers would know a lot more about the sun if it were farther away. The sun is so close that Earth's astronomers can see swirling currents of gas and arched bridges of magnetic force. The details seem overwhelming. But the sun is just an average star; and, in a sense, it is a simple object. It is made up almost entirely of the gases hydrogen and helium confined by its own gravity in a sphere 109 times Earth's diameter (**Celestial Profile 1**). The gases of the sun's surface are hot and radiate the light and heat that make life possible on Earth. That solar atmosphere is where you can begin your exploration.

## 8-1 The Solar Atmosphere

THE SUN'S ATMOSPHERE is made up of three layers. The visible surface is the *photosphere*, and above that are the *chromosphere* and the corona. (You first met these terms in Chapter 3 when you learned about solar eclipses.)

When you look at the sun you see a hot, glowing surface with a temperature of about 5800 K. At that temperature, every square millimeter of the sun's surface must be radiating more energy than a 60-watt light bulb. With all that energy radiating into space, the sun's surface would cool rapidly if energy did not flow up from the interior to keep the surface hot, so simple logic tells you that energy in the form of heat is flowing outward from the sun's interior. Not until the 1930s did astronomers understand that the sun makes its energy by nuclear reactions at the center. Those nuclear reactions are discussed in detail later in this chapter.

For now, you can consider the sun's atmosphere in its quiescent, average state. Later you can add the details of its continuous activity as heat flows up from the sun's interior and makes the outer layers churn like a pot of boiling soup.

### The Photosphere

The visible surface of the sun looks like a smooth layer of gas marked only by a few dark **sunspots** that come and go over a few weeks. Although the photosphere seems to be a distinct surface, it is not solid. In fact, the sun is gaseous from its outer atmosphere right down to its center. The photosphere is the thin layer of gas from which Earth receives most of the sun's light. It is less than 500 km deep, and if the sun magically shrank to the size of a bowling ball, the photosphere would be no thicker than a layer of tissue paper wrapped around the ball (■ Figure 8-1).

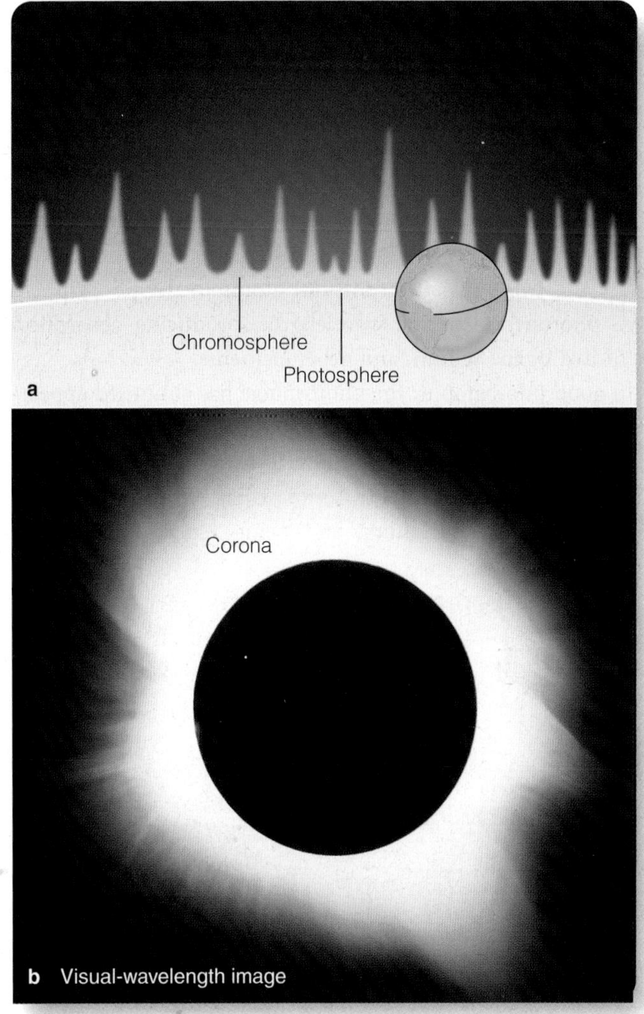


**■ Figure 8-1**

(a) A cross section at the edge of the sun shows the relative thickness of the photosphere and chromosphere. Earth is shown for scale. On this scale, the disk of the sun would be more than 1.5 m (5 ft) in diameter. The corona extends from the top of the chromosphere to great height above the photosphere. (b) This photograph, made during a total solar eclipse, shows only the inner part of the corona. (Daniel Good)

The photosphere is the layer in the sun's atmosphere that is dense enough to emit plenty of light but not so dense that the light can't escape. Below the photosphere, the gas is denser and hotter and therefore radiates plenty of light, but that light cannot escape from the sun because it is blocked by the outer layers of gas. So you cannot detect light from these deeper layers. Above the photosphere, the gas is less dense and so is unable to radiate much light.

Although the photosphere appears to be substantial, it is really a very-low-density gas. Even in its deepest and densest layers, the photosphere is less than 1/3000 as dense as the air you breathe. To find gases as dense as the air at Earth's surface, you would have to descend about 70,000 km below the photosphere, about 10 percent of the way to the sun's center. With fantastically efficient insulation, you could fly a spaceship right through the photosphere.

The spectrum of the sun is an absorption spectrum, and that can tell you a great deal about the photosphere. You know from Kirchhoff's third law that an absorption spectrum is produced when the source of a continuous spectrum is viewed through a gas. The deeper layers of the photosphere are dense enough to produce a continuous spectrum, but atoms in the photosphere absorb photons of specific wavelengths, producing absorption lines of hydrogen, helium, and other elements.

In good photographs, the photosphere has a mottled appearance because it is made up of dark-edged regions called granules. The overall pattern is called **granulation** (■ Figure 8-2a). Each granule is about the size of Texas and lasts for only 10 to 20 minutes before fading, shrinking, and being replaced by a new

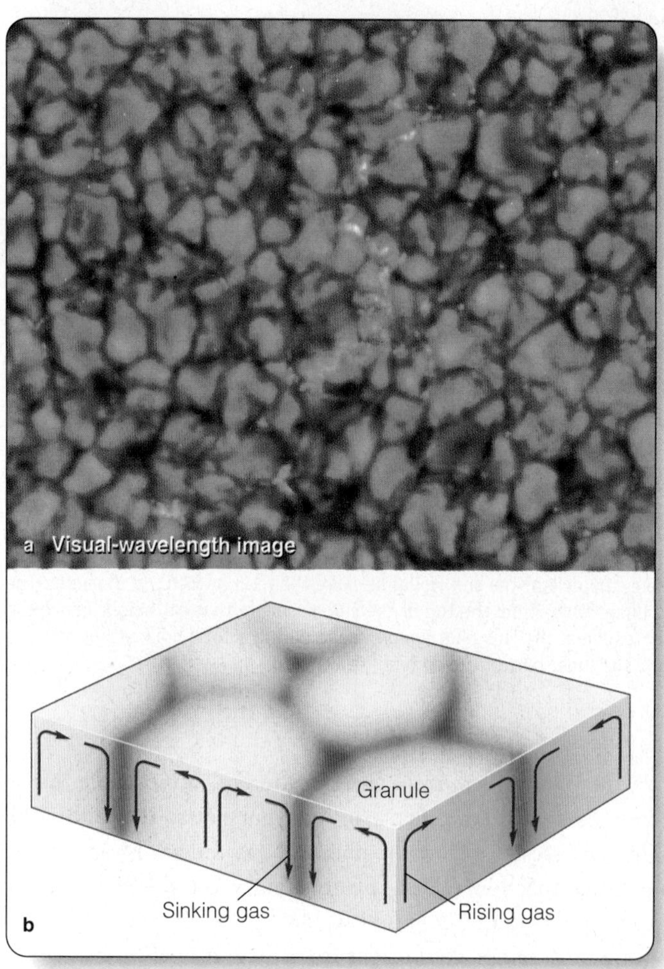


**■ Figure 8-2**

(a) This ultra-high-resolution image of the photosphere shows granulation. The largest granules here are about the size of Texas. (b) This model explains granulation as the tops of rising convection currents just below the photosphere. Heat flows upward as rising currents of hot gas and downward as sinking currents of cool gas. The rising currents heat the solar surface in small regions seen from Earth as granules.

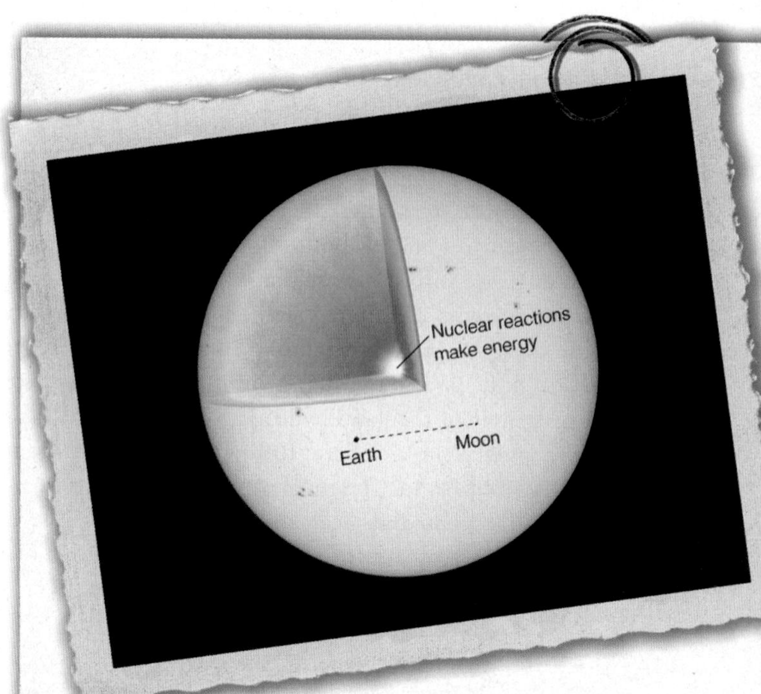


*This visible image of the sun shows a few sunspots and is cut away to show the location of energy generation at the sun's center. The Earth–moon system is shown for scale.* (Daniel Good)

## *Celestial Profile 1: The Sun*

### *From Earth:*

| | |
|---|---|
| Average distance from Earth | 1.000 AU ( $1.496 \times 10^8$ km) |
| Maximum distance from Earth | 1.017 AU ( $1.521 \times 10^8$ km) |
| Minimum distance from Earth | 0.983 AU ( $1.471 \times 10^8$ km) |
| Average angular diameter | 0.53° (32 arc minutes) |
| Period of rotation | 25.4 days at equator |
| Apparent visual magnitude | −26.74 |

### *Characteristics:*

| | |
|---|---|
| Radius | $6.96 \times 10^5$ km |
| Mass | $1.99 \times 10^{30}$ kg |
| Average density | 1.41 g/cm$^3$ |
| Escape velocity at surface | 618 km/s |
| Luminosity | $3.83 \times 10^{26}$ J/s |
| Surface temperature | 5800 K |
| Central temperature | $15 \times 10^6$ K |
| Spectral type | G2 V |
| Absolute visual magnitude | 4.83 |

### *Personality Profile:*

In Greek mythology, the sun was carried across the sky in a golden chariot pulled by powerful horses and guided by the sun god Helios. When Phaeton, the son of Helios, drove the chariot one day, he lost control of the horses, and Earth was nearly set ablaze before Zeus smote Phaeton from the sky. Even in classical times, people understood that life on Earth depends critically on the sun.

granule. Spectra of granules show that the centers are a few hundred degrees hotter than the edges, and Doppler shifts reveal that the centers are rising and the edges are sinking at speeds of about 0.4 km/s.

From this evidence, astronomers recognize granulation as the surface effects of convection just below the photosphere. **Convection** occurs when hot fluid rises and cool fluid sinks, as when, for example, a convection current of hot gas rises above a candle flame. You can watch convection in a liquid by adding a bit of cool non-dairy creamer to an unstirred cup of hot coffee. The cool creamer sinks, warms, expands, rises, cools, contracts, sinks again, and so on, creating small regions on the surface of the coffee that mark the tops of convection currents. Viewed from above, these regions look much like solar granules.

In the sun, rising currents of hot gas heat small regions of the photosphere, which, being slightly hotter, emit more blackbody radiation and look brighter. The cool sinking gas of the edges emits less light and thus looks darker (Figure 8-2b). The presence of granulation is clear evidence that energy is flowing upward through the photosphere.

Spectroscopic studies of the solar surface have revealed another less obvious kind of granulation. **Supergranules** are regions a little over twice Earth's diameter that include about 300 granules each. These supergranules are regions of very slowly rising currents that last a day or two. They appear to be produced by larger gas currents that lie deeper under the photosphere.

The edge, or **limb,** of the solar disk is dimmer than the center (see the figure in Celestial Profile 1). This **limb darkening** is caused by the absorption of light in the photosphere. When you look at the center of the solar disk, you are looking directly down into the sun, and you see deep, hot, bright layers in the photosphere. In contrast, when you look near the limb of the solar disk, you are looking at a steep angle to the surface and cannot see as deeply. The photons you see come from shallower, cooler, dimmer layers in the photosphere. Limb darkening proves that the temperature in the photosphere increases with depth, yet another confirmation that energy is flowing up from below.

## The Chromosphere

Above the photosphere lies the chromosphere. Solar astronomers define the lower edge of the chromosphere as lying just above the visible surface of the sun, with its upper regions blending gradually with the corona. You can think of the chromosphere as an irregular layer with a depth on average less than Earth's diameter (see Figure 8-1). Because the chromosphere is roughly 1000 times fainter than the photosphere, you can see it with your unaided eyes only during a total solar eclipse when the moon covers the brilliant photosphere. Then, the chromosphere flashes into view as a thin line of pink just above the photosphere. The word *chromosphere* comes from the Greek word *chroma,* meaning "color." The pink color is produced by the combined light of three bright emission lines—the red, blue, and violet Balmer lines of hydrogen.

Astronomers know a great deal about the chromosphere from its spectrum. The chromosphere produces an emission spectrum, and Kirchhoff's second law tells you it must be an excited, low-density gas. The chromosphere is about $10^8$ times less dense than the air you breathe.

Spectra reveal that atoms in the lower chromosphere are ionized, and atoms in the higher layers of the chromosphere are even more highly ionized. That is, they have lost more electrons. Hydrogen atoms contain only one electron, but atoms like calcium and iron contain many more, and at high enough temperatures they can lose a number of electrons and become highly ionized. From the ionization state of the gas, astronomers can find the temperature in different parts of the chromosphere. Just above the photosphere the temperature falls to a minimum of about 4500 K and then rises rapidly (■ Figure 8-3) to the extremely high temperatures of the corona. The upper chromosphere is hot enough to emit X-rays and can be studied by X-ray telescopes in space.

Solar astronomers can take advantage of some elegant physics to study the chromosphere. The gases of the chromosphere are transparent to nearly all visible light, but atoms in the gas are very good at absorbing photons of specific wavelengths. This produces certain dark absorption lines in the spectrum of the photosphere. A photon at one of those wavelengths is very

**■ Figure 8-3**

The chromosphere. If you could place thermometers in the sun's atmosphere, you would discover that the temperature increases from 5800 K at the photosphere to 1,000,000 K at the top of the chromosphere.

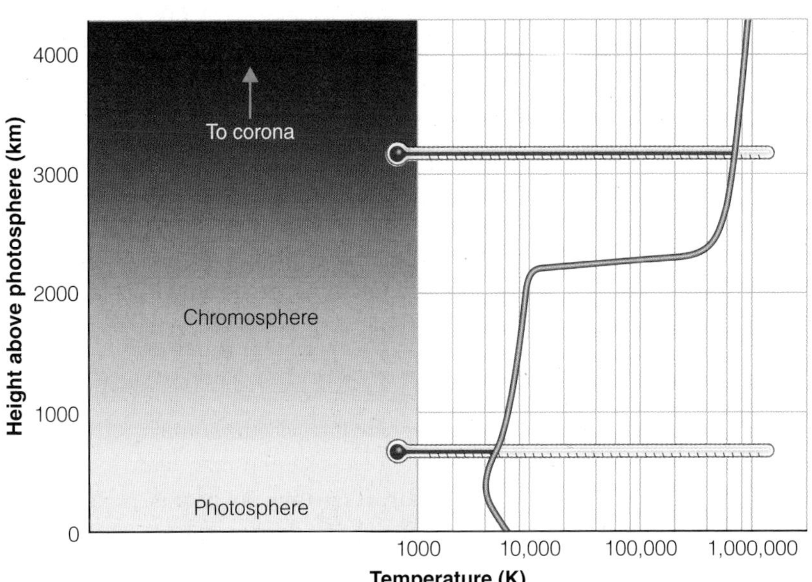

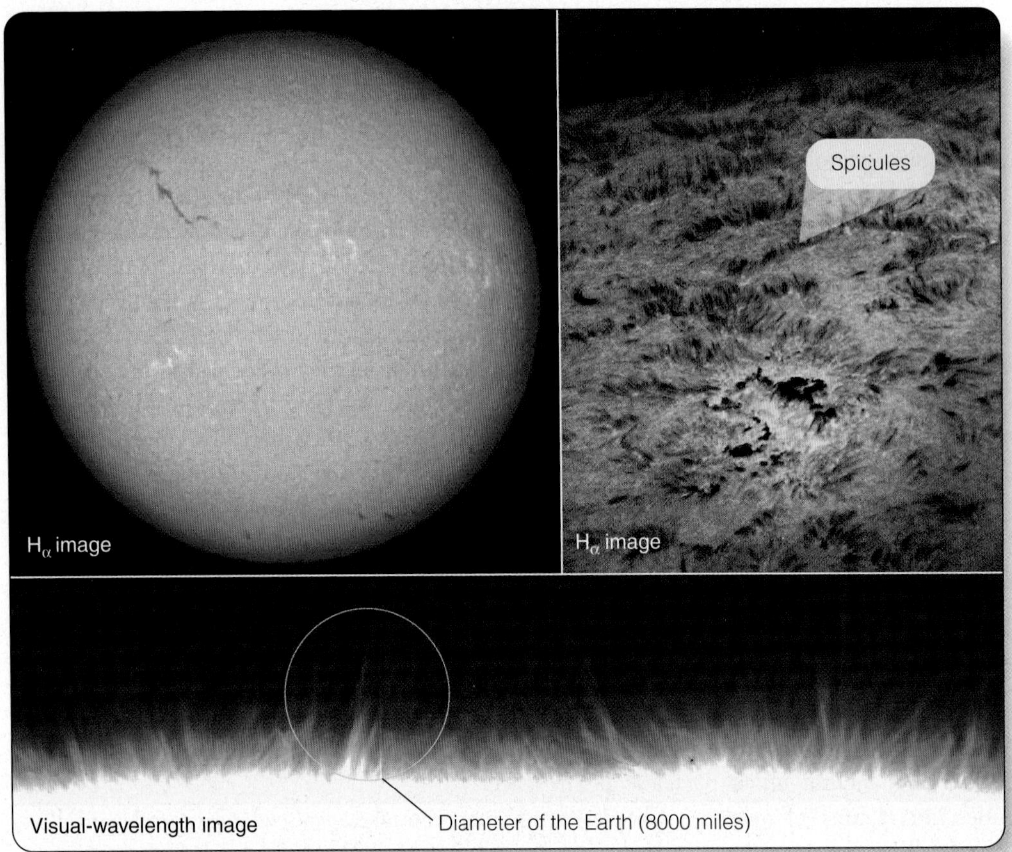


**Figure 8-4**

$H_\alpha$ filtergrams reveal complex structure in the chromosphere that cannot be seen at visual wavelengths, including spicules springing from the edges of supergranules over twice the diameter of Earth. Seen at the edge of the solar disk, spicules look like a burning prairie, but they are not at all related to burning. Compare with Figure 8-1. (BBSO; © 1971 NOAO/NSO; Hinode)

unlikely to escape from deeper layers. A **filtergram** is an image of the sun made using light in one of those dark absorption lines. Those photons can only have escaped from higher in the sun's atmosphere. In this way, filtergrams reveal detail in the upper layers of the chromosphere. Another way to study these high, hot layers of gas is to record solar images in the far-ultraviolet or in the X-ray part of the spectrum.

■ Figure 8-4 shows a filtergram made at the wavelength of the $H_\alpha$ Balmer line. This image shows complex structure in the chromosphere. **Spicules** are flamelike jets of gas extending upward into the chromosphere and lasting 5 to 15 minutes. Seen at the limb of the sun's disk, these spicules blend together and look like flames covering a burning prairie (Figure 8-1), but they are not flames at all. Spectra show that spicules are cooler gas from the lower chromosphere extending upward into hotter regions. Images at the center of the solar disk show that spicules spring up around the edge of supergranules like weeds around flagstones (Figure 8-4).

## The Solar Corona

The outermost part of the sun's atmosphere is called the corona, after the Greek word for crown. The corona is so dim that, like the chromosphere, it is not visible in Earth's daytime sky because of the glare of scattered light from the sun's brilliant photosphere. During a total solar eclipse, the innermost parts of the corona are visible to the unaided eye, as shown in Figure 8-1b. Observations made with specialized telescopes called **coronagraphs** can block the light of the photosphere and record the corona out beyond 20 solar radii, almost 10 percent of the way to Earth. Such images reveal streamers in the corona that follow magnetic lines of force in the sun's magnetic field (■ Figure 8-5).

The spectrum of the corona can tell you a great deal about the coronal gases and simultaneously illustrate how astronomers analyze a spectrum. Some of the light from the outer corona produces a spectrum with absorption lines that are the same as the photosphere's spectrum. This light is just sunlight reflected from dust particles in the corona. In contrast, some of the light from the corona produces a continuous spectrum that lacks absorption lines, and that happens when sunlight from the photosphere is scattered off free electrons in the ionized coronal gas. Because the coronal gas has a temperature over 1 million K and the electrons travel very fast, the reflected photons suffer large, random Doppler shifts that smear out absorption lines to produce a continuous spectrum.

Superimposed on the corona's continuous spectrum are emission lines of highly ionized gases. In the lower corona, the atoms are not as highly ionized as they are at higher altitudes, and this tells you that the temperature of the corona rises with altitude. Just above the chromosphere, the temperature is about 500,000 K, but in the outer corona the temperature can be 2 million K or more.

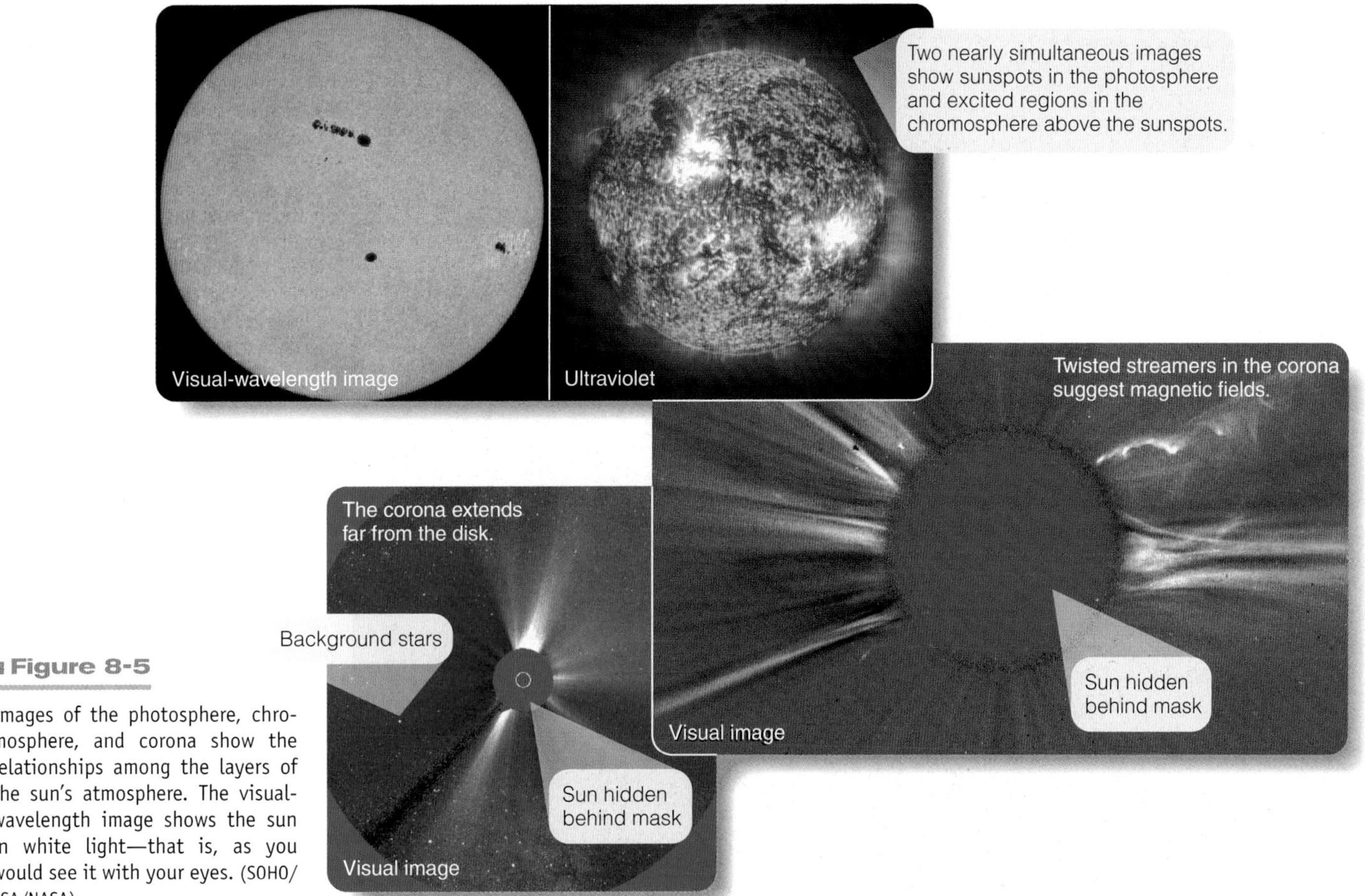
Two nearly simultaneous images show sunspots in the photosphere and excited regions in the chromosphere above the sunspots.
Visual-wavelength image
Ultraviolet
Twisted streamers in the corona suggest magnetic fields.
The corona extends far from the disk.
Background stars
Sun hidden behind mask
Visual image
Sun hidden behind mask
Visual image

**Figure 8-5**

Images of the photosphere, chromosphere, and corona show the relationships among the layers of the sun's atmosphere. The visual-wavelength image shows the sun in white light—that is, as you would see it with your eyes. (SOHO/ESA/NASA)

The corona is exceedingly hot gas. In fact, as you learned earlier, the gas in the corona and upper chromosphere is so hot it can emit X-rays. Nevertheless, the coronal gas is not very bright. Its density is very low, only $10^6$ atoms/cm$^3$ in its lower regions. That is about $10^{12}$ times less dense than the air you breathe. In its outer layers the corona contains only 1 to 10 atoms/cm$^3$, fewer than in the best vacuum in laboratories on Earth. Because of its low density, the hot gas does not emit much radiation.

Astronomers have wondered for years how the corona and chromosphere can be so hot. Heat flows from hot regions to cool regions, never from cool to hot. So how can the heat from the photosphere, with a temperature of only 5800 K, flow out into the much hotter chromosphere and corona? Observations made by the SOHO satellite have mapped a **magnetic carpet** of looped magnetic fields extending up through the photosphere (■ Figure 8-6). Remember that the gas of the chromosphere and corona has a very low density, so it can't resist movement of the magnetic fields. Turbulence below the photosphere seems to flick the magnetic loops back and forth and whip the gas about, heating the gas. Furthermore, observations with the Hinode spacecraft have revealed magnetic waves generated by turbulence below the photosphere traveling up into the chromosphere and corona and heating the gas. Because the gas in the corona has such a low density, it doesn't take much energy flowing outward as agitation in the magnetic fields to heat the corona to high temperature.

Ionized, low-density gas cannot cross magnetic fields, so where the sun's field loops back to the surface, the gas is trapped. However, some of the magnetic fields lead outward into space, and there the gas flows away from the sun in a breeze called the **solar wind.** Like an extension of the corona, the low-density gases of the solar wind blow past Earth at 300 to 800 km/s with gusts as high as 1000 km/s. Earth is bathed in the corona's hot breath.

Because of the solar wind, the sun is slowly losing mass, but this is only a minor loss for an object as massive as the sun. The sun loses about $10^7$ tons per second, but that is only $10^{-14}$ of a solar mass per year. Later in life, the sun, like many other stars, will lose mass rapidly in a more powerful wind. You will see in future chapters how this affects the evolution of stars.

Do other stars have chromospheres, coronae, and stellar winds like the sun? Stars are so far away they never look like more than points of light, but ultraviolet and X-ray observations suggest that the answer is yes. The spectra of many stars contain emission lines in the far-ultraviolet that could have formed only

# 8-2 Nuclear Fusion in the Sun

LIKE SOAP BUBBLES, stars are structures balanced between opposing forces that individually would destroy them. The sun is a ball of hot gas held together by its own gravity. If it were not for the sun's gravity, the hot, high-pressure gas in the sun's interior would explode outward. Likewise, if the sun were not so hot, its gravity would compress it into a small dense body. In this section, you will discover how the sun generates its heat.

The sun is powered by nuclear reactions that occur near its center.* The energy keeps the interior hot and keeps the gas totally ionized. That is, the electrons are not attached to atomic nuclei, so the gas is an atomic soup of rapidly moving particles colliding with each other at high velocities. Nuclear reactions inside stars involve atomic nuclei, not whole atoms.

How exactly can the nucleus of an atom yield energy? The answer lies in the force that holds the nuclei together.

## Nuclear Binding Energy

The sun generates its energy by breaking and reconnecting the bonds between the particles *inside* atomic nuclei. This is quite different from the way you would generate energy by burning wood in a fireplace. The process of burning wood extracts energy by breaking and rearranging chemical bonds among atoms in the wood. Chemical bonds are formed by the electrons in atoms, and you saw in Chapter 7 that the electrons are bound to the atoms by the electromagnetic force. So the chemical energy released when these bonds are broken and rearranged originates in the electromagnetic force.

There are only four forces in nature: the force of gravity, the electromagnetic force, the **weak force,** and the **strong force.** The weak force is involved in the radioactive decay of certain kinds of nuclear particles, and the strong force binds together atomic nuclei. The weak force and the strong force are both much stronger than either gravity or the electromagnetic force, but they are short-range forces that are effective only within the nuclei of atoms. Nuclear reactions break and reform the bonds that hold atomic nuclei together, so nuclear energy comes from the strong force. (In a later chapter you will learn that physicists have discovered the electromagnetic force and the weak force are actually different aspects of one force, and they have a goal of eventually finding a way to "unify" all four forces, including gravity, within one general mathematical description).

There are two ways to generate energy from atomic nuclei. Nuclear power plants on Earth generate energy through **nuclear fission** reactions that split uranium nuclei into less massive fragments. A uranium nucleus contains a total of 235 protons and neutrons, and when it decays, it splits into a range of possible fragments, each containing roughly half as many particles. Because the fragments that are produced are more tightly bound than the uranium nuclei, binding energy is released during uranium fission.

Stars don't use nuclear fission. They make energy in **nuclear fusion** reactions that combine light nuclei into heavier nuclei. The most common reaction inside stars, the one that occurs in the sun, fuses hydrogen nuclei (single protons) into helium nuclei, which contain two protons and two neutrons. Because the nuclei produced are more tightly bound than the original nuclei, energy is released.

■ Figure 8-8 shows how tightly different atomic nuclei are bound. The lower in the diagram, the more tightly the particles in a nucleus are held. Notice that both fusion and fission reactions move downward in the diagram toward more tightly bound nuclei. They both produce energy by releasing binding energy of atomic nuclei.

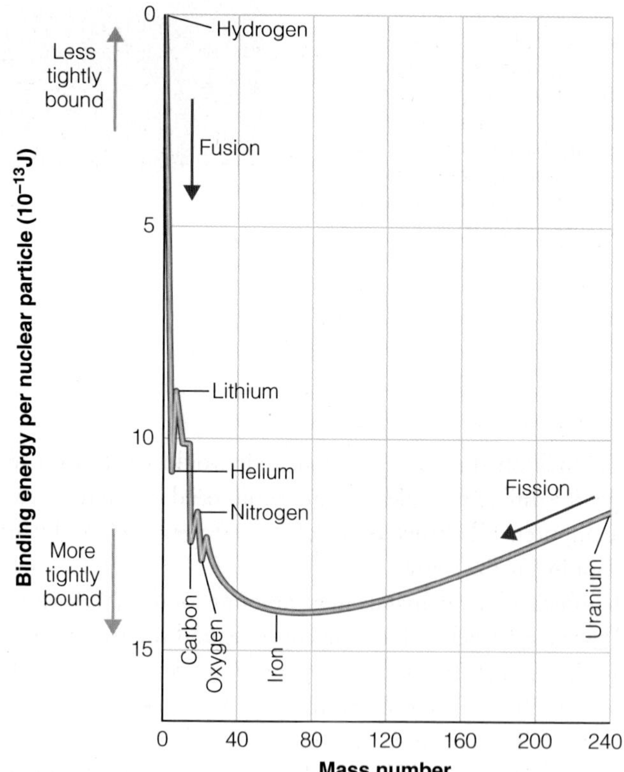


■ **Figure 8-8**

The red line in this graph shows the binding energy per particle, the energy that holds particles inside an atomic nucleus. The horizontal axis shows the atomic mass number of each element, the number of protons and neutrons in the nucleus. Both fission and fusion nuclear reactions move downward in the diagram (arrows), meaning the nucleus produced by a reaction is more tightly bound than the nuclei that went into the reaction. Iron has the most tightly bound nucleus, so no nuclear reactions can begin with iron and release energy.

*Astronomers sometimes use unhelpful words when they talk about nuclear reactions inside stars. They may use words like *burn* or *ignite.* What goes on inside stars is not related to simple burning but is comprised of nuclear reactions.

## Hydrogen Fusion

The sun fuses together four hydrogen nuclei to make one helium nucleus. Because one helium nucleus has 0.7 percent less mass than four hydrogen nuclei, it seems that some mass vanishes in the process. To see this, subtract the mass of a helium nucleus from the mass of four hydrogen nuclei:

$$\begin{array}{r l} \text{4 hydrogen nuclei} = & 6.693 \times 10^{-27}\text{ kg} \\ -1\text{ helium nucleus} = & 6.645 \times 10^{-27}\text{ kg} \\ \hline \text{Difference in mass} = & 0.048 \times 10^{-27}\text{ kg} \end{array}$$

This small amount of mass does not actually disappear but is converted to energy according to Einstein's famous equation:

$$\begin{aligned} E &= mc^2 \\ &= (0.048 \times 10^{-27}\text{ kg}) \times (3.0 \times 10^8\text{ m/s})^2 \\ &= 0.43 \times 10^{-11}\text{ J} \end{aligned}$$

You can symbolize the fusion reactions in the sun with a simple nuclear reaction:

$$4\,{}^1\text{H} \rightarrow {}^4\text{He} + \text{energy}$$

In this equation, ${}^1$H represents a proton, the nucleus of a hydrogen atom, and ${}^4$He represents the nucleus of a helium atom. The superscripts indicate the approximate weight of the nuclei (the number of protons plus the number of neutrons). The actual steps in the process are more complicated than this convenient summary suggests. Instead of waiting for four hydrogen nuclei to collide simultaneously, a highly unlikely event, the process normally proceeds step by step in a chain of reactions—the proton–proton chain.

The **proton–proton chain** is a series of three nuclear reactions that builds a helium nucleus by adding together protons. This process can only happen at temperatures above about 4 million K, and is efficient above 10 million K. The sun makes over 90 percent of its energy in this way.

The three steps in the proton–proton chain entail these reactions:

$$\begin{aligned} {}^1\text{H} + {}^1\text{H} &\rightarrow {}^2\text{H} + e^+ + \nu \\ {}^2\text{H} + {}^1\text{H} &\rightarrow {}^3\text{He} + \gamma \\ {}^3\text{He} + {}^3\text{He} &\rightarrow {}^4\text{He} + {}^1\text{H} + {}^1\text{H} \end{aligned}$$

In the first reaction, two hydrogen nuclei (two protons) combine and emit a particle called a **positron,** $e^+$ (a positively charged electron), and a **neutrino,** $\nu$ (a subatomic particle having an extremely low mass and a velocity nearly equal to the velocity of light). The emission of the positively charged positron results from the conversion of one of the protons into a neutron, and that forms a heavy hydrogen nucleus called **deuterium,** which contains a proton and a neutron. In the second reaction, the heavy hydrogen nucleus absorbs another proton and, with the emission of a gamma ray, $\gamma$, becomes a lightweight helium nucleus. Finally, two lightweight helium nuclei combine to form a nucleus of normal helium and two hydrogen nuclei. Because the last reaction needs two ${}^3$He nuclei, the first and second reactions must occur twice (■ Figure 8-9). The net result of this chain reaction is the transformation of four hydrogen nuclei into one helium nucleus plus energy.

The energy appears in the form of gamma rays, positrons, neutrinos and the energy of motion of all the particles. The gamma rays are photons that are absorbed by the surrounding gas before they can travel more than a fraction of a millimeter. This heats the gas. The positrons produced in the first reaction combine with free electrons, and both particles vanish, converting their mass into gamma rays, which are absorbed and also help keep the gas hot. In addition, when fusion produces new nuclei, they fly apart at high velocity and collide with other particles. This energy of motion helps raise the temperature of the gas. The neutrinos, on the other hand, don't heat the gas. Neutrinos are particles that almost never interact with other particles. The average neutrino could pass unhindered through a lead wall a light-year thick. Consequently, the neutrinos do not warm the gas but race out of the sun at nearly the speed of light, carrying away roughly 2 percent of the energy produced.

**■ Figure 8-9**

The proton–proton chain combines four protons (at far left) to produce one helium nucleus (at right). Energy is produced mostly as gamma rays and as positrons, which combine with electrons and convert their mass into energy. Neutrinos escape, carrying away about 2 percent of the energy produced.

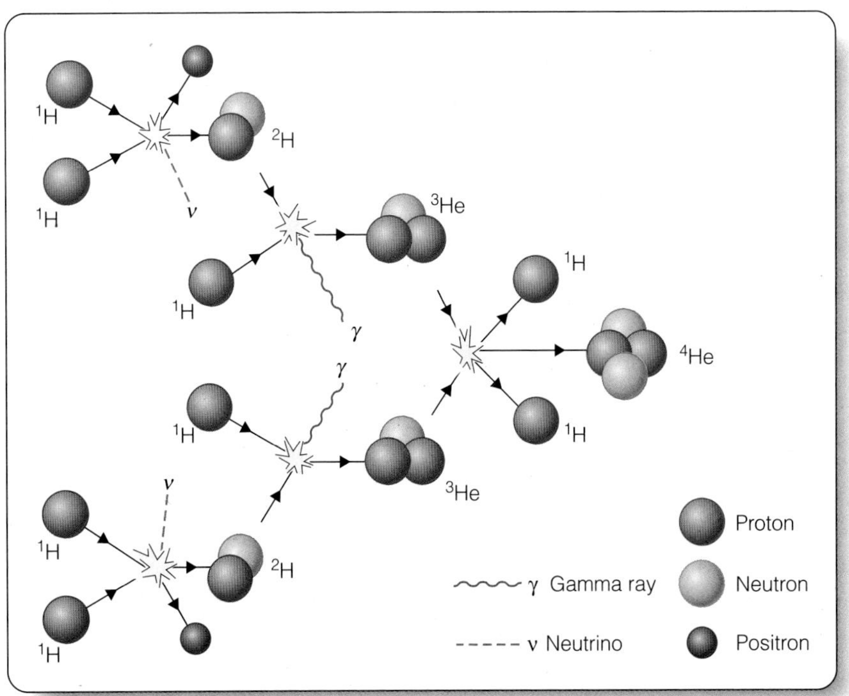

Creating one helium nucleus makes only a small amount of energy, hardly enough to raise a housefly one-thousandth of an inch. Because one reaction produces such a small amount of energy, it is obvious that many reactions are necessary to supply the energy needs of a star. The sun, for example, completes $10^{38}$ reactions per second, transforming 5 million tons of matter into energy every second. It might sound as if the sun is losing mass at a furious rate, but in its entire 10-billion-year lifetime, the sun will convert less than 0.07 percent of its mass into energy.

It is a **Common Misconception** that nuclear fusion in the sun is tremendously powerful. After all, the fusion of a milligram of hydrogen (roughly the mass of a match head) produces as much energy as burning 30 gallons of gasoline. However, at any one time, only a tiny fraction of the hydrogen atoms are fusing into helium, and the nuclear reactions in the sun are spread through a large volume in its core. Any single gram of matter produces only a little energy. A person of normal mass eating a normal diet produces about 4000 times more heat per gram than the matter in the core of the sun. Gram for gram, you are a much better heat producer than the sun. The sun produces a lot of energy because it contains a lot of grams of matter in its core.

Fusion reactions can occur only when the nuclei of two atoms get very close to each other. Because atomic nuclei carry positive charges, they repel each other with an electrostatic force called the Coulomb force. Physicists commonly refer to this repulsion between nuclei as the **Coulomb barrier.** To overcome this barrier and get close together, atomic nuclei must collide violently. Sufficiently violent collisions are rare unless the gas is very hot, in which case the nuclei move at high enough speeds. (Remember, an object's temperature is related to the speed with which its particles move.)

Nevertheless, the fusion of two protons is a highly unlikely process. If you could follow a single proton in the sun's core, you would see it encountering and bouncing off other protons millions of times a second, but you would have to follow it around for roughly a billion years before it happened to penetrate the Coulomb barrier and combine with another proton.

Because of the dependence of nuclear reactions on particle collisions, the reactions in the sun take place only near the center, where the gas is hot and dense. A high temperature ensures that collisions between nuclei are violent, and a high density ensures that there are enough collisions, and thus enough reactions, to make energy at the sun's rate.

## Energy Transport in the Sun

Now you are ready to follow the energy from the core of the sun to the surface. You will learn in a later chapter that astronomers have computed models indicating that the temperature at the center of the sun is about 15 million K. Compared with that, the sun's surface is very cool, only about 5800 K, so energy must flow from the high temperature core outward to the cooler surface where it is radiated into space.

Because the core is so hot, the photons bouncing around there are gamma rays. Each time a gamma ray encounters an electron, it is deflected or scattered in a random direction, and, as it bounces around, it slowly drifts outward toward the surface. That carries energy outward in the form of radiation, so astronomers refer to the inner parts of the sun as a **radiative zone.**

To examine this process, imagine picking a single gamma ray and following it to the surface. As your gamma ray is scattered over and over by the hot gas, it drifts outward into cooler layers, where the cooler gas tends to emit photons of longer wavelength. Your gamma ray will eventually be absorbed by the gas and reemitted as two X-rays. Now you must follow those two X-rays as they bounce around, and soon you will see them drifting outward into even cooler gas, where they will become a number of longer-wavelength photons.

The packet of energy you have been following, now represented by many photons, eventually reaches the outer layers of the sun, where the gas is so cool that it is not very transparent to radiation. There, energy backs up like water behind a dam, and the gas begins to churn in convection. Hot blobs of gas rise, and cool blobs sink. In this region, known as a **convective zone,** your packet of energy is carried outward not as photons but as circulating gas (■ Figure 8-10). Rising hot gas carries energy outward, of course, but sinking cool gas is a necessary part of the cycle that results in the net transport of energy outward.

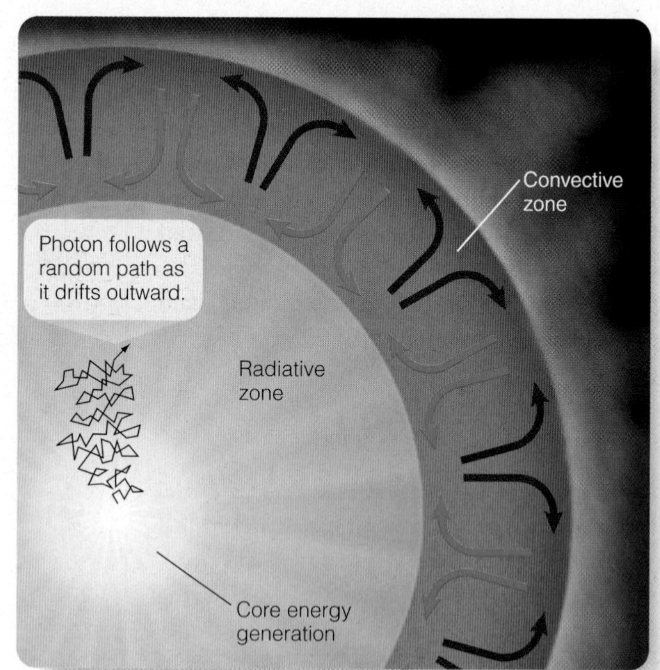


■ **Figure 8-10**

A cross-section of the sun. Near the center, nuclear fusion reactions generate high temperatures. Energy flows outward through the radiative zone as photons are randomly deflected over and over by electrons. In the cooler, more opaque outer layers, the energy is carried by rising convection currents of hot gas (red arrows) and sinking currents of cooler gas (blue arrows).

## How Do We Know? 8-1

### Scientific Confidence

**How can scientists be certain of something?** Sometimes scientists stick so firmly to their ideas in the face of contradictory claims that it sounds as if they are stubbornly refusing to consider alternatives. To see why they do this, consider the perpetual motion machine, a device that supposedly runs continuously with no source of energy. If you could invest in a real perpetual motion machine, you could sell cars that would run without any fuel. That's good mileage.

For centuries people have claimed to have invented perpetual motion machines, and for just as long scientists have been dismissing these claims as impossible. The problem with a perpetual motion machine is that it violates the law of conservation of energy, and scientists are not willing to accept that the law could be wrong. In fact, the Royal Academy of Sciences in Paris was so sure that a perpetual motion machine was impossible, and so tired of debunking hoaxes, that in 1775 they issued a formal statement refusing to deal with them. The U.S. Patent Office is so skeptical that they won't even consider granting a patent for one without seeing a working model first. Why do scientists seem so stubborn and close-minded on this issue?

Why isn't one person's belief in perpetual motion just as valid as another person's belief in the law of conservation of energy? In fact, the two positions are not equally valid. The confidence physicists have in their law is not a belief or even an opinion; it is an understanding founded on the fact that the law has been tested uncountable times and has never failed. The law is a fundamental truth about nature and can be used to understand what is possible and what is impossible. In contrast, no one has ever successfully demonstrated a perpetual motion machine.

When the first observations of solar neutrinos detected fewer than were predicted, some scientists speculated that astronomers misunderstood how the sun makes its energy or that they misunderstood the internal structure of the sun. But many astronomers stubbornly refused to reject their model because the nuclear physics of the proton–proton chain is well understood, and models of the sun's structure have been tested successfully many times. The confidence astronomers felt in their understanding of the sun was an example of scientific certainty, and that confidence in basic natural laws prevented them from abandoning decades of work in the face of a single contradictory observation.

*For centuries, people have tried to design a perpetual motion machine, but not a single one has ever worked. Scientists understand why.*

What seems to be stubbornness among scientists is really their confidence in basic principles that have been tested over and over. Those principles are the keel that keeps the ship of science from rocking before every little breeze. Without even looking at that perpetual motion machine, your physicist friends can warn you not to invest.

It can take a million years for the energy from a single gamma ray to work its way outward first as radiation and then as convection. When the energy finally reaches the photosphere, it is radiated into space as about 1800 photons of visible light.

It is time to ask the critical question that lies at the heart of science. What is the evidence to support this theoretical explanation of how the sun makes its energy?

### Counting Solar Neutrinos

Nuclear reactions in the sun's core produce floods of neutrinos that rush out of the sun and off into space. Over $10^{12}$ solar neutrinos flow through your body every second, but you never feel them because you are almost perfectly transparent to neutrinos. If you could detect these neutrinos, you could probe the sun's interior. You can't focus neutrinos with a lens or mirror, and they zip right through detectors used to count other atomic particles, but neutrinos of certain energies can trigger the radioactive decay of certain atoms. That gives astronomers a way to count solar neutrinos.

In the 1960s, chemist Raymond Davis Jr. devised a way to count neutrinos produced by hydrogen fusion in the sun. He buried a 100,000-gallon tank of cleaning fluid (perchloroethylene, $C_2Cl_4$) in the bottom of a South Dakota gold mine where cosmic rays could not reach it (■ Figure 8-11a) and counted the number of times a neutrino triggered a chlorine atom into becoming an argon atom. He expected to detect one neutrino a day, but he actually counted one-third as many as expected, only one every three days.

The Davis neutrino experiment created a huge controversy. Were scientists wrong about nuclear fusion in the sun? Did they misunderstand how neutrinos behave? Was the detector not working properly? Because astronomers had great confidence in their understanding of the solar interior, they didn't abandon their theories immediately (**How Do We Know? 8-1**). It took over 30 years, but eventually physicists were able to build better detectors, and they discovered that neutrinos oscillate among three different types, which physicists call flavors. Nuclear reactions in

**Figure 8-11**

(a) The Davis solar neutrino experiment used cleaning fluid and could detect only one of the three flavors of neutrinos. (Brookhaven National Laboratory) (b) The Sudbury Neutrino Observatory is a 12-meter-diameter globe containing water rich in deuterium (heavy hydrogen) in place of ordinary hydrogen. Buried 2100 m (6800 ft) deep in an Ontario mine, it can detect all three flavors of neutrinos and confirms that neutrinos oscillate. (Photo courtesy of SNO)

the sun produce only one flavor, and the Davis experiment was designed to detect (taste) that flavor. But during the 8-minute journey from the sun's core to Earth, the neutrinos oscillated so much that they were evenly distributed among the three different flavors by the time they arrived at Earth. That's why the Davis experiment detected only one-third of the number predicted.

The missing neutrinos have been detected by more sophisticated detectors (Figure 8-11b). In 2007, scientists announced that a supersensitive experiment in a tunnel under the Italian Alps had detected 50 neutrinos a day coming from the sun. The neutrinos have lower energies that those caught by the Davis experiment and are produced by a side reaction that produces beryllium-7. The number of neutrinos detected matches the prediction of models of nuclear fusion in the sun.

The center of the sun seems forever beyond human experience, but counting solar neutrinos provides evidence to confirm the theories. The sun makes its energy through nuclear fusion.

### SCIENTIFIC ARGUMENT

***Why does nuclear fusion require that the gas be very hot?***
This argument has to include the basic physics of atoms and thermal energy. Inside a star, the gas is so hot it is ionized, which means the electrons have been stripped off the atoms leaving bare, positively charged nuclei. In the case of hydrogen, the nuclei are single protons. These atomic nuclei repel each other because of their positive charges, so they must collide with each other at high velocity if they are to overcome that repulsion and get close enough together to fuse. If the atoms in a gas are moving rapidly, then the gas must have a high temperature, so nuclear fusion requires that the gas be very hot. If the gas is cooler than about 4 million K, hydrogen can't fuse because the protons don't collide violently enough to overcome the repulsion of their positive charges.

It is easy to see why nuclear fusion in the sun requires high temperature, but now expand your argument. **Why does it require high density?**

## 8-3 Solar Activity

THE SUN IS NOT QUIET. It is home to slowly changing spots larger than Earth and vast eruptions that dwarf human imagination. All of these seemingly different forms of solar activity have one thing in common—magnetic fields. The weather on the sun is magnetic.

### Observing the Sun

Solar activity is often visible with even a small telescope, but you should be very careful if you try to observe the sun. Sunlight is intense, and when it enters your eye it is absorbed and converted into thermal energy. The infrared radiation in sunlight is especially dangerous because your eyes can't detect it. You don't sense how intense the infrared is, but it is converted to thermal energy in your eyes and can burn and scar your retinas.

**Figure 8-12**

(a) Looking through a telescope at the sun is dangerous, but you can always view the sun safely with a small telescope by projecting its image on a white screen. (b) If you sketch the location and structure of sunspots on successive days, you will see the rotation of the sun and gradual changes in the size and structure of sunspots, just as Galileo did in 1610.

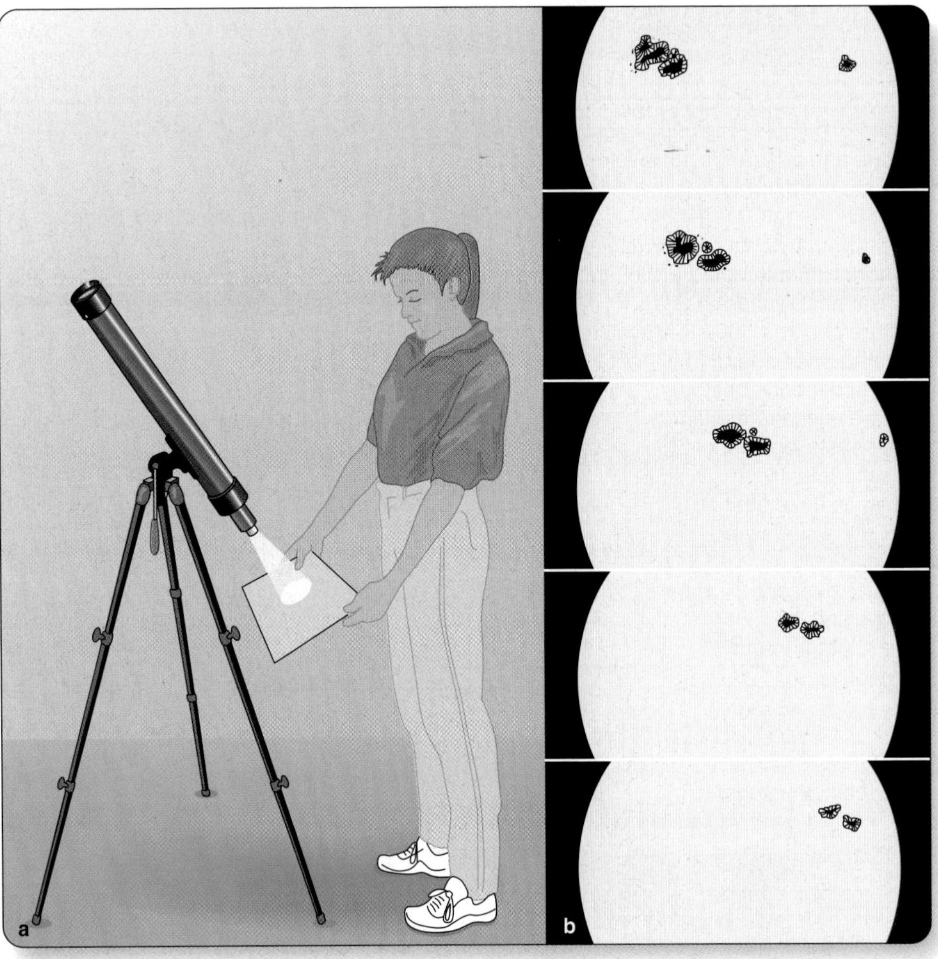


It is not safe to look directly at the sun, and it is even more dangerous to look at the sun through any optical instrument such as a telescope, binoculars, or even the viewfinder of a camera. The light-gathering power of such an optical system concentrates the sunlight and can cause severe injury. Never look at the sun with any optical instrument unless you are certain it is safe. ■ Figure 8-12a shows a safe way to observe the sun with a small telescope.

In the early 17th century, Galileo observed the sun and saw spots on its surface; day by day he saw the spots moving across the sun's disk. He rightly concluded that the sun was a sphere and was rotating. If you repeated his observations, you would probably also detect sunspots, a view that would look something like Figure 8-12b.

## Sunspots

The dark sunspots that you see at visible wavelengths only hint at the complex processes that go on in the sun's atmosphere. To explore those processes, you must analyze images and spectra at a wide range of wavelengths.

Study **Sunspots and the Sunspot Cycle** on pages 156–157 and notice five important points and four new terms:

1. Sunspots are cool spots on the sun's surface caused by strong magnetic fields.
2. Sunspots follow an 11-year cycle, becoming more numerous, reaching a maximum, and then becoming much less numerous. The *Maunder butterfly diagram* shows how the location of sunspots changes during a cycle.
3. The *Zeeman effect* gives astronomers a way to measure the strength of magnetic fields on the sun and provide evidence that sunspots contain strong magnetic fields.
4. The intensity of the sunspot cycle can vary from cycle to cycle and appears to have almost faded away during the *Maunder minimum* in the late 17th century. That seems to have affected Earth's climate.
5. The evidence is clear that sunspots are part of *active regions* dominated by magnetic fields that involve all layers of the sun's atmosphere.

The sunspot groups are merely the visible traces of magnetically active regions. But what causes this magnetic activity? The answer is linked to the growth and decay cycle of the sun's overall magnetic field.

# Sunspots and the Sunspot Cycle

**1** The dark spots that appear on the sun are only the visible traces of complex regions of activity. Observations over many years and at a range of wavelengths tell you that sunspots are clearly linked to the sun's magnetic field.

Spectra show that sunspots are cooler than the photosphere with a temperature of about 4200 K. The photosphere has a temperature of about 5800 K. Because the total amount of energy radiated by a surface depends on its temperature raised to the fourth power, sunspots look dark in comparison. Actually, a sunspot emits quite a bit of radiation. If the sun were removed and only an average-size sunspot were left behind, it would be brighter than the full moon.

Visual wavelength image

Umbra

Penumbra

A typical sunspot is about twice the size of Earth, but there is a wide range of sizes. They appear, last a few weeks to as long as 2 months, and then shrink away. Usually, sunspots occur in pairs or complex groups.

Earth to scale

NASA

Sunspots are not shadows, but astronomers refer to the dark core of a sunspot as its umbra and the outer, lighter region as the penumbra.

Hinode JAXA/NASA

Streamers above a sunspot suggest a magnetic field.

Hinode JAXA/NASA

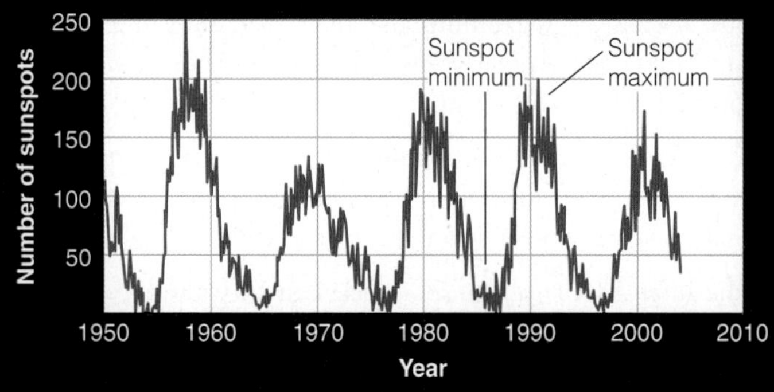


**2** The number of spots visible on the sun varies in a cycle with a period of 11 years. At maximum, there are often over 100 spots visible. At minimum, there are very few.

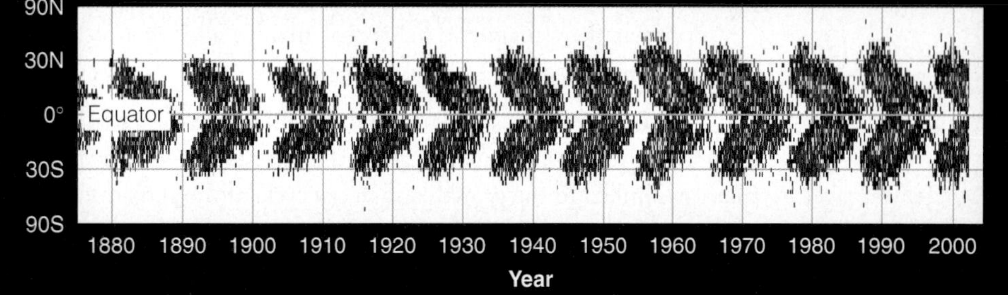


**2a** Early in the cycle, spots appear at high latitudes north and south of the sun's equator. Later in the cycle, the spots appear closer to the sun's equator. If you plot the latitude of sunspots versus time, the graph looks like butterfly wings, as shown in this **Maunder butterfly diagram,** named after E. Walter Maunder of Greenwich Observatory.

**3** Astronomers can measure magnetic fields on the sun using the **Zeeman effect** as shown below. When an atom is in a magnetic field, the electron orbits are altered, and the atom is able to absorb a number of different wavelength photons even though it was originally limited to a single wavelength. In the spectrum, you see single lines split into multiple components, with the separation between the components proportional to the strength of the magnetic field.

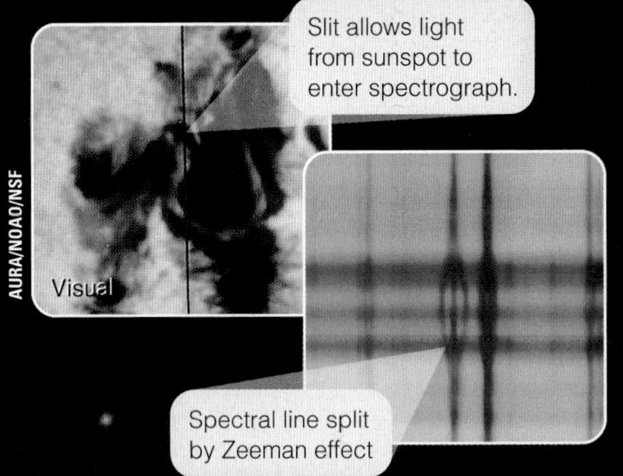


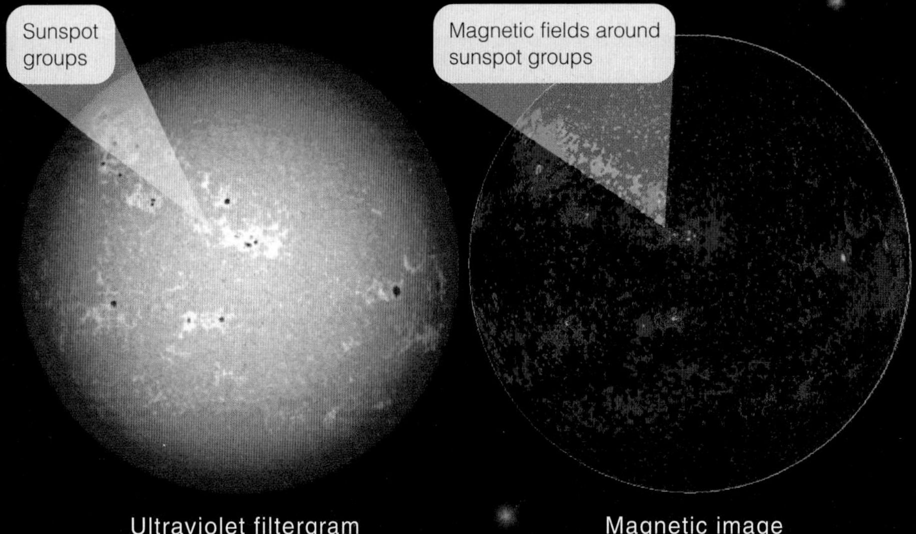


**Simultaneous images**

**3a** Images of the sun above show that sunspots contain magnetic fields a few thousand times stronger than Earth's. The strong fields are believed to inhibit gas motion below the photosphere; consequently, convection is reduced below the sunspot, and the surface there is cooler. Heat prevented from emerging through the sunspot is deflected and emerges around the sunspot, which can be detected in ultraviolet and infrared images.

Maunder minimum few spots colder winters

Winter severity in London and Paris

Warmer winters

↑Warm ↓Cold

Number of sunspots: 0, 50, 100, 150, 200, 250, 300, 350

Year: 1650, 1700, 1750, 1800, 1850, 1900, 1950, 2000

**4** Historical records show that there were very few sunspots from about 1645 to 1715, a phenomenon known as the **Maunder minimum.** This coincides with a period called the "little ice age," a period of unusually cool weather in Europe and North America from about 1500 to about 1850, as shown in the graph at left. Other such periods of cooler climate are known. The evidence suggests that there is a link between solar activity and the amount of solar energy Earth receives. This link has been confirmed by measurements made by spacecraft above Earth's atmosphere.

Magnetic fields can reveal themselves by their shape. For example, iron filings sprinkled over a bar magnet reveal an arched shape.

M. Seeds

Far -UV image

SOHO/EIT, ESA and NASA

**5** Observations at nonvisible wavelengths reveal that the chromosphere and corona above sunspots are violently disturbed in what astronomers call **active regions.** Spectrographic observations show that active regions contain powerful magnetic fields. Arched structures above an active region are evidence of gas trapped in magnetic fields.

The complexity of an active region becomes visible at short wavelengths.

Visual-wavelength image

**Simultaneous images**

Far-UV image

NASA/TRACE

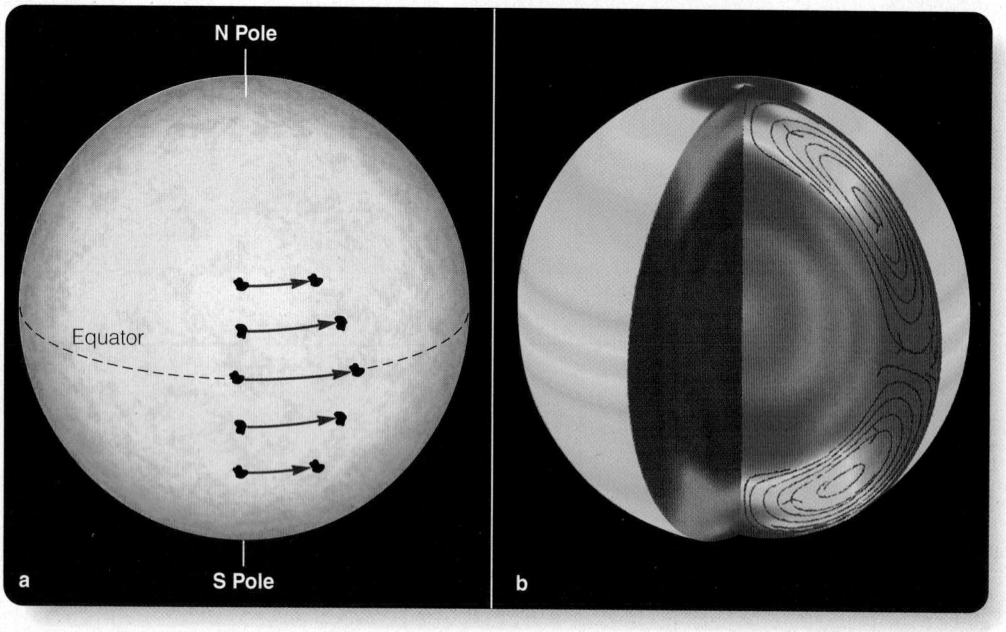

■ Figure 8-13

(a) In general, the photosphere of the sun rotates faster at the equator than at higher latitudes. If you started five sunspots in a row, they would not stay lined up as the sun rotates. (b) Detailed analysis of the sun's rotation from helioseismology reveals regions of slow rotation (blue) and rapid rotation (red). Such studies show that the interior of the sun rotates differentially and that currents similar to the trade winds in Earth's atmosphere flow through the sun. (NASA/SOI)

## The Sun's Magnetic Cycle

The sun's magnetic field is powered by the energy flowing outward through the moving currents of gas. The gas is highly ionized, so it is a very good conductor of electricity. When an electrical conductor rotates and is stirred by convection, it can convert some of the energy flowing outward as convection into a magnetic field. This process is called the **dynamo effect,** and it is understood also to operate in Earth's core and produce Earth's magnetic field. Helioseismologists have found evidence that the dynamo effect generates the sun's magnetic field at the bottom of the convection zone deep under the photosphere.

The sun's magnetic field cannot be as stable as Earth's. The sun does not rotate as a rigid body. It is a gas from its outermost layers down to its center, so some parts of the sun can rotate faster than other parts. The equatorial region of the photosphere rotates faster than do regions at higher latitudes (■ Figure 8-13a). At the equator, the photosphere rotates once every 25 days, but at latitude 45° one rotation takes 27.8 days. Helioseismology can map the rotation throughout the interior (Figure 8-13b) and has found that different levels in the sun rotate with different periods. This phenomenon is called **differential rotation,** and it is clearly linked with the magnetic cycle.

Although the magnetic cycle is not fully understood, the **Babcock model** (named for its inventor) explains the magnetic cycle as a progressive tangling of the solar magnetic field. Because the electrons in an ionized gas are free to move, the gas is a very good conductor of electricity, so any magnetic field in the gas is "frozen" into it. If the gas moves, the magnetic field must move with it. Differential rotation drags the magnetic field along and wraps it around the sun like a long string caught on a turning hubcap. Rising and sinking gas currents twist the field into rope-like tubes, which tend to float upward. The model predicts that sunspot pairs occur where these magnetic tubes burst through the sun's surface (■ Figure 8-14).

Sunspots tend to occur in groups or pairs, and the magnetic field around the pair resembles that around a bar magnet with one end magnetic north and the other end magnetic south, just as you would expect if a magnetic tube emerged through one sunspot in a pair and reentered through the other. At any one time, sunspot pairs south of the sun's equator have reversed polarity compared with those north of the sun's equator. ■ Figure 8-15 illustrates this by showing sunspot pairs south of the sun's equator moving with magnetic south poles leading, and sunspots north of the sun's equator moving with magnetic north poles leading. At the end of an 11-year sunspot cycle, the new spots appear with reversed magnetic polarity.

The Babcock model accounts for the reversal of the sun's magnetic field from cycle to cycle. As the magnetic field becomes tangled, adjacent regions of the sun are dominated by magnetic fields that point in different directions. After about 11 years of tangling, the field becomes so complex that adjacent regions of the sun begin changing their magnetic field to agree with neighboring regions. The entire field quickly rearranges itself into a simpler pattern, and differential rotation begins winding it up to start a new cycle. But the newly organized field is reversed, and the next sunspot cycle begins with magnetic north replaced by magnetic south. Consequently, the complete magnetic cycle is 22 years long, and the sunspot cycle is 11 years long.

This magnetic cycle also explains the Maunder butterfly diagram. As a sunspot cycle begins, the twisted tubes of magnetic force first begin to float upward and produce sunspot pairs at

The Solar Magnetic Cycle

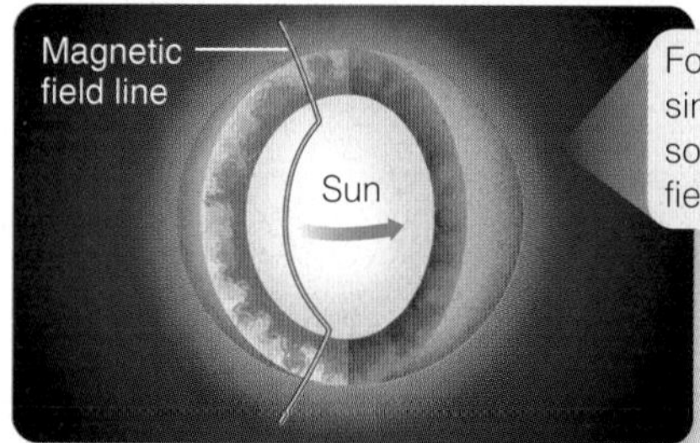
Magnetic field line
Sun
For simplicity, a single line of the solar magnetic field is shown.

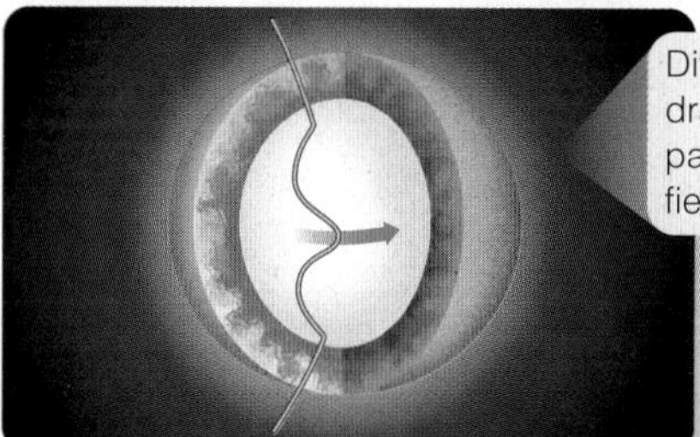
Differential rotation drags the equatorial part of the magnetic field ahead.

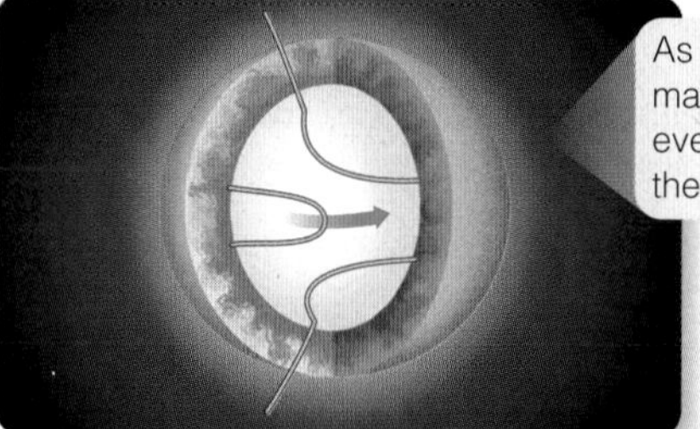
As the sun rotates, the magnetic field is eventually dragged all the way around.

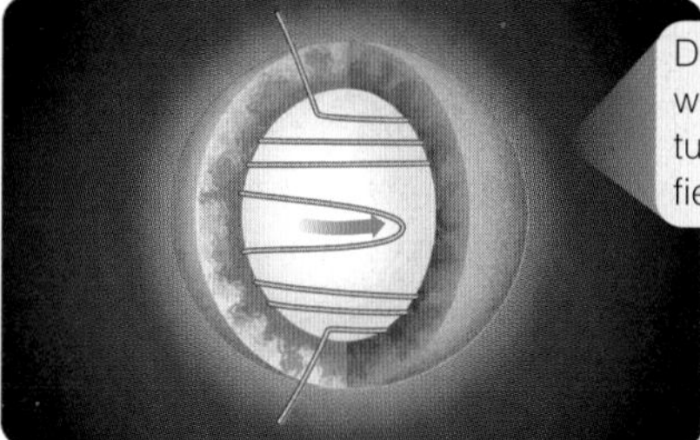
Differential rotation wraps the sun in many turns of its magnetic field.

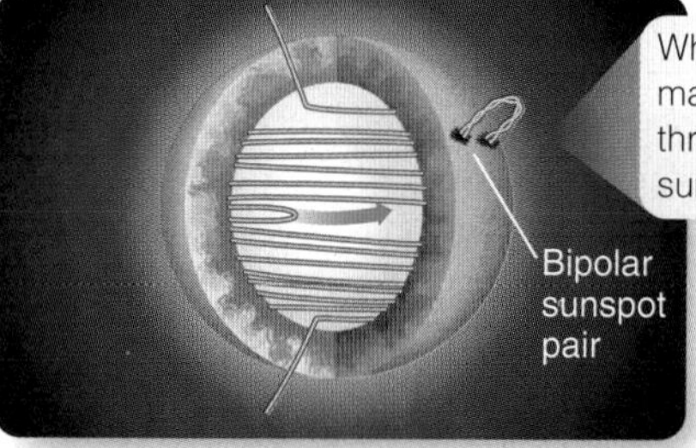
Where loops of tangled magnetic field rise through the surface, sunspots occur.
Bipolar sunspot pair

**Figure 8-14**

The Babcock model of the solar magnetic cycle explains the sunspot cycle as a consequence of the sun's differential rotation gradually winding up the magnetic field near the base of the sun's outer, convective layer.

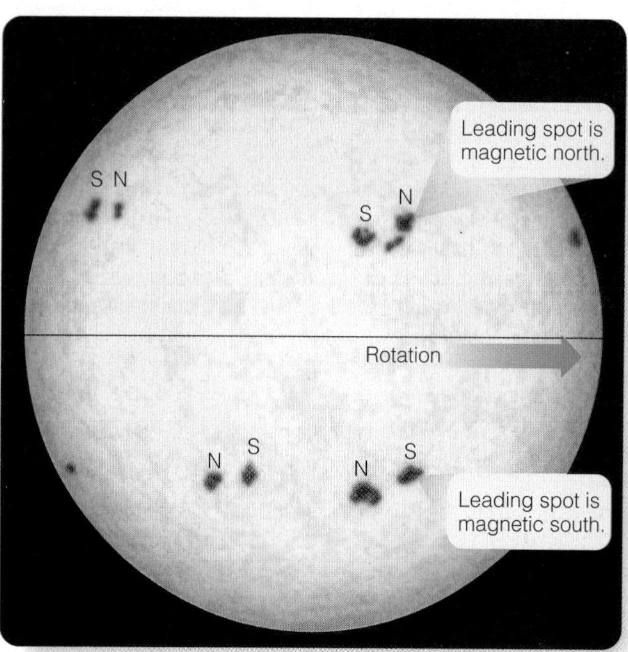
Leading spot is magnetic north.
S N
S
N
Rotation
N
S
N
S
Leading spot is magnetic south.

**Figure 8-15**

In sunspot groups, here simplified into pairs of major spots, the leading spot and the trailing spot have opposite magnetic polarity. Spot pairs in the southern hemisphere have reversed polarity from those in the northern hemisphere.

higher latitudes on the sun. Consequently the first sunspots in a cycle appear farther north and south of the equator. Later in the cycle, when the field is more tightly wound, the tubes of magnetic force arch up through the surface closer to the equator. As a result, the later sunspot pairs in a cycle appear closer to the equator.

Notice the power of a scientific model. The Babcock model may in fact be incorrect in some details, but it provides a framework around which to organize descriptions of complex solar activity. Even though the model of the sky in Chapter 2 and the model of the atom in Chapter 7 are only partially correct, they served as organizing themes to guide your thinking. Similarly, although the precise details of the solar magnetic cycle are not yet understood, the Babcock model gives you a general picture of the behavior of the sun's magnetic field (**How Do We Know? 8-2**).

## Spots and Magnetic Cycles of Other Stars

The sun seems to be a representative star, so you should expect other stars to have similar cycles of starspots and magnetic fields. This is a difficult topic, because, except for the sun, the stars are so far away that no surface detail is visible. Some stars, however, vary in brightness in ways that suggest they are mottled by dark spots. As these stars rotate, their total brightnesses change slightly, depending on the number of spots on the side facing Earth. High-precision spectroscopic analysis has even allowed astronomers to map the locations of spots on the surfaces of

## Confirmation and Consolidation

**What do scientists do all day?** The scientific method is sometimes portrayed as a kind of assembly line where scientists crank out new hypotheses and then test them through observation. In reality, scientists don't often generate entirely new hypotheses. It is rare that an astronomer makes an observation that disproves a long-held theory and triggers a revolution in science. Then what is the daily grind of science really about?

Many observations and experiments confirm already-tested hypotheses. The biologist knows that all worker bees in a hive are sisters because they are all female, and they all had the same mother, the queen bee. A biologist can study the DNA from many workers and confirm that hypothesis. By repeatedly confirming a hypothesis, scientists build confidence in the hypothesis and may be able to extend it. Do all of the workers in a hive have the same father, or did the queen mate with more than one male drone?

Another aspect of routine science is consolidation, the linking of a hypothesis to other well-studied phenomena. A biologist can study yellow jacket wasps from a single nest and discover that the wasps, too, are sisters. There must be a queen wasp who lays all of the eggs in a nest. But in a few nests, the scientist may find two sets of unrelated sister workers. Those nests must contain two queens sharing the nest for convenience and protection. From his study of wasps, the biologist consolidates what he knows about bees with what others have learned about wasps and reveals something new: That bees and wasps have evolved in similar ways for similar reasons.

Confirmation and consolidation allow scientists to build confidence in their understanding and extend it to explain more about nature.

*A yellow jacket is a wasp from a nest containing a queen wasp.* (Michael Dunham/ Getty Images)

certain stars (■ Figure 8-16a). Such results confirm that the sunspots you see on our sun are not unusual.

Certain features found in stellar spectra are associated with magnetic fields. Regions of strong magnetic fields on the solar surface emit strongly at the central wavelengths of the two strongest lines of ionized calcium. This calcium emission appears in the spectra of other sun-like stars and suggests that these stars, too, have strong magnetic fields on their surfaces. In some cases, the strength of this calcium emission varies over periods of days or weeks and suggests that the stars have active regions and are rotating with periods similar to that of the sun. These stars presumably have sunspots ("starspots") as well.

In 1966, astronomers began a long-term project that monitored the strengths of these calcium emission features in the spectra of 91 stars with temperatures ranging from 1000 K hotter than the sun to 3000 K cooler, considered most likely to have sun-like magnetic activity on their surfaces. The observations show that the strength of the calcium emission varies over periods of years. The calcium emission averaged over the sun's disk varies with the sunspot cycle, and similar periodic variations can be seen in the spectra of some of the stars studied (Figure 8-16b). The star 107 Piscium, for example, appears to have a starspot cycle lasting nine years. At least one star, tau Bootis, has been observed to reverse its magnetic field. This kind of evidence suggests that stars like the sun have similar magnetic cycles, and that the sun is normal in this respect.

## Chromospheric and Coronal Activity

The solar magnetic fields extend high into the chromosphere and corona, where they produce beautiful and powerful phenomena. Study **Magnetic Solar Phenomena** on pages 162–163 and notice three important points and seven new terms:

1. All solar activity is magnetic. The arched shapes of *prominences* are produced by magnetic fields, and *filaments* are prominences seen from above.
2. Tremendous energy can be stored in arches of magnetic field, and when two arches encounter each other a *reconnection* can release powerful eruptions called *flares.* Although these eruptions occur far from Earth, they can affect us in dramatic ways, and *coronal mass ejections (CMEs)* can trigger communications blackouts and *auroras.*
3. In some regions of the solar surface, the magnetic field does not loop back. High-energy gas from these *coronal holes* flows outward and produces much of the solar wind.

You may have heard the **Common Misconception** that an auroral display in the night sky is caused by sunlight reflecting off of the ice and snow at Earth's North Pole. It is fun to think of polar bears standing on sunlit slabs of ice, but that doesn't cause auroras. You know that auroras are produced by gases in Earth's upper atmosphere excited to glowing by energy from the solar wind.

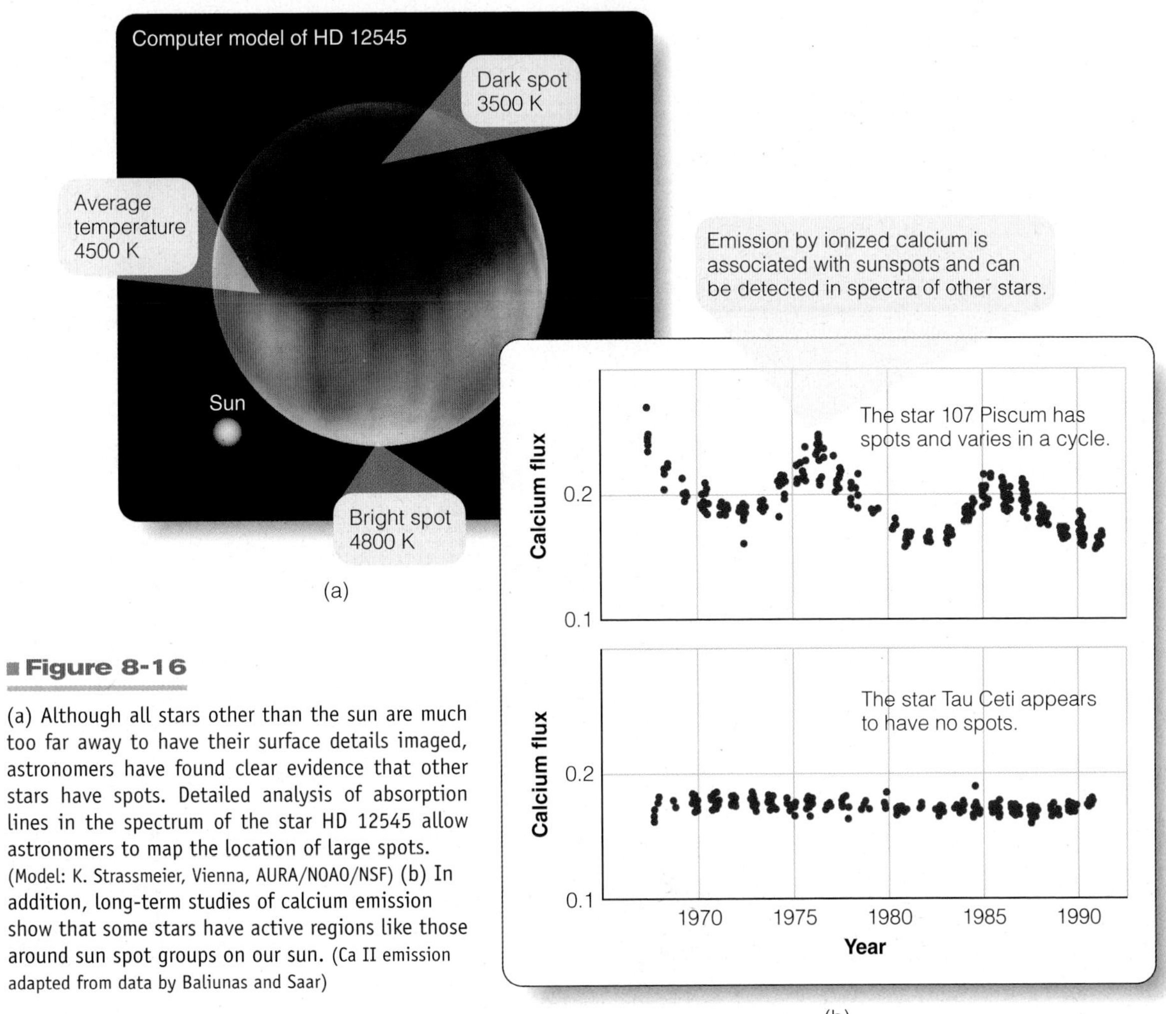


**Figure 8-16**

(a) Although all stars other than the sun are much too far away to have their surface details imaged, astronomers have found clear evidence that other stars have spots. Detailed analysis of absorption lines in the spectrum of the star HD 12545 allow astronomers to map the location of large spots. (Model: K. Strassmeier, Vienna, AURA/NOAO/NSF) (b) In addition, long-term studies of calcium emission show that some stars have active regions like those around sun spot groups on our sun. (Ca II emission adapted from data by Baliunas and Saar)

## The Solar Constant

Even a small change in the sun's energy output could produce dramatic changes in Earth's climate. The continued existence of the human species depends on the constancy of the sun, but we humans know very little about the variation of the sun's energy output.

The energy production of the sun can be monitored by adding up all of the energy falling on 1 square meter of Earth's surface during 1 second. Of course, some correction for the absorption of Earth's atmosphere is necessary, and you must count all wavelengths from X-rays to radio waves. The result, which is called the **solar constant,** amounts to about 1360 joules per square meter per second. Is the sun really constant? A change in the solar constant of only 1 percent could change Earth's average temperature by about 1°C (roughly 2°F). This may not seem like much, but during the last ice age Earth's average temperature was only about 5°C cooler than it is now.

Some of the best measurements of the solar constant were made by instruments aboard the Solar Maximum Mission satellite. These have shown variations in the energy received from the sun of about 0.1 percent that lasted for days or weeks. Superimposed on that random variation is a long-term decrease of about 0.018 percent per year that has been confirmed by observations made by sounding rockets, balloons, and satellites. This long-term decrease may be related to a cycle of activity on the sun with a period longer than the 22-year magnetic cycle.

Small, random fluctuations will not affect Earth's climate, but a long-term decrease over a decade or more could cause worldwide cooling. History contains some evidence that the solar constant may have varied in the past. As you saw on page 157, the "Little Ice Age" was a period of unusually cool weather in Europe and America that lasted from about 1500 to 1850.* The average temperature worldwide was about 1°C cooler than it is now. This period of cool weather corresponded to the Maunder minimum, a period of reduced solar activity—few sunspots, no auroral displays, and no solar coronas visible during solar eclipses.

*Ironically, the Maunder minimum coincides with the reign of Louis XIV of France, the "Sun King."

## Magnetic Solar Phenomena

**1** Magnetic phenomena in the chromosphere and corona, like magnetic weather, result as constantly changing magnetic fields on the sun trap ionized gas to produce beautiful arches and powerful outbursts. Some of this solar activity can affect Earth's magnetic field and atmosphere.

This ultraviolet image of the solar surface was made by the NASA TRACE spacecraft. It shows hot gas trapped in magnetic arches extending above active regions. At visual wavelengths, you would see sunspot groups in these active regions.

The gas in prominences may be 60,000 to 80,000 K, quite cold compared with the low-density gas in the corona, which may be as hot as a million Kelvin.

Trace/NASA

H-alpha filtergram

Sacramento Peak Observatory

**1a** A **prominence** is composed of ionized gas trapped in a magnetic arch rising up through the photosphere and chromosphere into the lower corona. Seen during total solar eclipses at the edge of the solar disk, prominences look pink because of the three Balmer emission lines. The image above shows the arch shape suggestive of magnetic fields. Seen from above against the sun's bright surface, prominences form dark **filaments.**

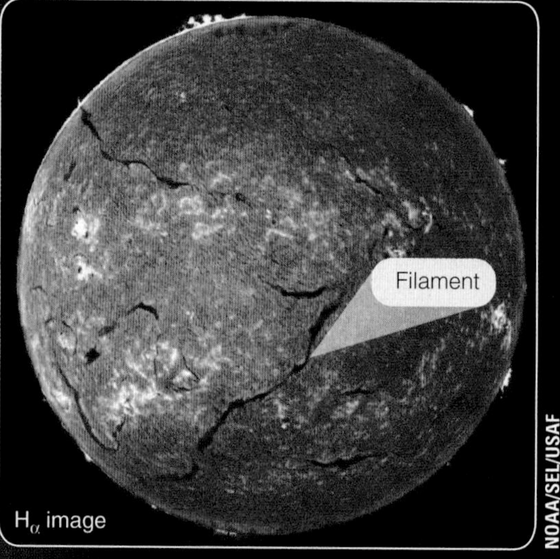


$H_\alpha$ image

NOAA/SEL/USAF

**1b** Quiescent prominences may hang in the lower corona for many days, whereas eruptive prominences burst upward in hours. The eruptive prominence below is many Earth diameters long.

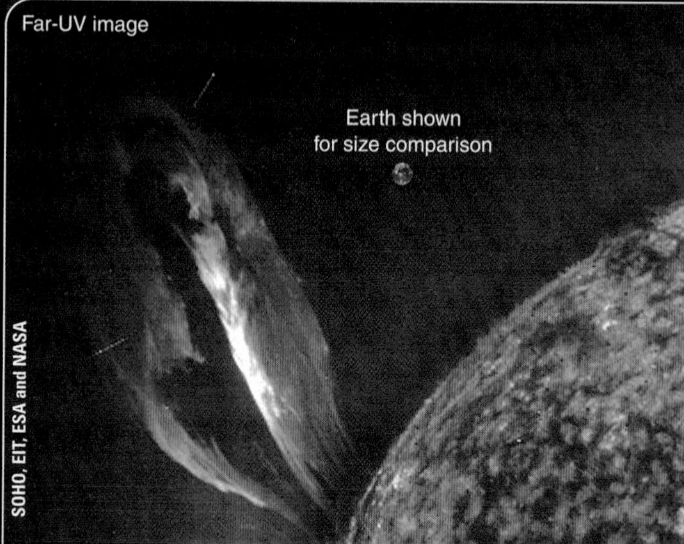


Far-UV image

SOHO, EIT, ESA and NASA

**2** Solar **flares** rise to maximum in minutes and decay in an hour. They occur in active regions where oppositely directed magnetic fields meet and cancel each other out in what astronomers call **reconnections.** Energy stored in the magnetic fields is released as short-wavelength photons and as high-energy protons and electrons. X-ray and ultraviolet photons reach Earth in 8 minutes and increase ionization in our atmosphere, which can interfere with radio communications. Particles from flares reach Earth hours or days later as gusts in the solar wind, which can distort Earth's magnetic field and disrupt navigation systems. Solar flares can also cause surges in electrical power lines and damage to Earth satellites.

This multiwavelength image shows a sunspot interacting with a neighboring magnetic field to produce a solar flare.

Hinode JAXA/NASA

**2a** At right, waves rush outward at 50 km/sec from the site of a solar flare 40,000 times stronger than the 1906 San Francisco earthquake. The biggest solar flares can be a billion times more powerful than a hydrogen bomb.

Helioseismology image

SOHO/MDI, ESA, and NASA

**2b** The solar wind, enhanced by eruptions on the sun, interacts with Earth's magnetic field and can create electrical currents up to a million megawatts. Those currents flowing down into a ring around Earth's magnetic poles excite atoms in Earth's upper atmosphere to emit photons as shown below. Seen from Earth's surface, the gas produces glowing clouds and curtains of **aurora.**

Auroras occur about 130 km above the Earth's surface.

Coronal mass ejection

Ring of aurora around the north magnetic pole

NSSDC, Holzworth and Meng

**2c** Magnetic reconnections can release enough energy to blow large amounts of ionized gas outward from the corona in **coronal mass ejections (CMEs).** If a CME strikes Earth, it can produce especially violent disturbances in Earth's magnetic field.

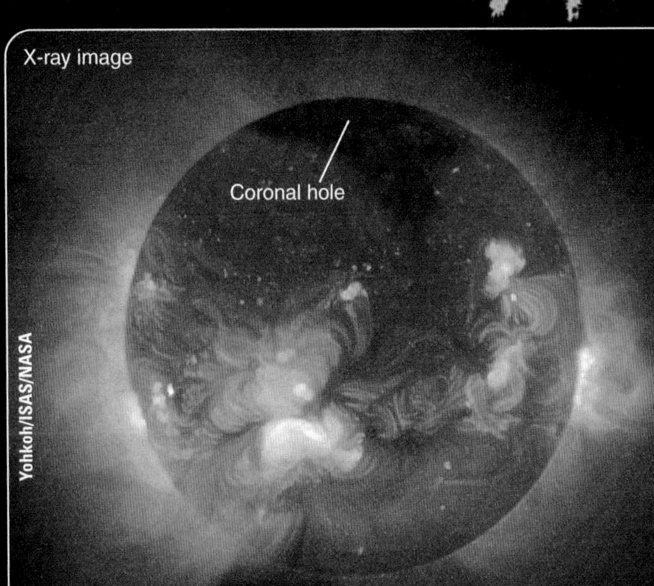


Yohkoh/ISAS/NASA

**3** Much of the solar wind comes from **coronal holes,** where the magnetic field does not loop back into the sun. These open magnetic fields allow ionized gas in the corona to flow away as the solar wind. The dark area in this X-ray image at right is a coronal hole.

In contrast, an earlier period called the Grand Maximum, lasting from about AD 1100 to about 1250, saw a warming of Earth's climate. The Vikings were able to explore and colonize Greenland, and native communities in parts of North America were forced to abandon their settlements because of long droughts. The Grand Maximum may have been caused by a small change in solar activity, but the evidence is not conclusive.

Other minima and maxima have been found in climate data taken from studies of the growth rings of trees. In good years, trees add a thicker growth ring than in poor years, so measuring tree rings can reveal the climate in the past. Evidently, solar activity can increase or decrease the solar constant very slightly and affect Earth's climate in dramatic ways. The future of our civilization on Earth may depend on our learning to understand the solar constant.

### SCIENTIFIC ARGUMENT

***What kind of activity would the sun have if it didn't rotate differentially?***

Once again, it can help you understand a concept by constructing an argument with one factor changed. Begin by thinking about the Babcock model. If the sun didn't rotate differentially, its equator traveling faster than its higher latitudes, then the magnetic field might not get wound up, and there might not be a solar cycle. Twisted tubes of magnetic field might not form and rise through the photosphere to produce prominences and flares, although convection might tangle the magnetic field slightly and produce some activity. Is the magnetic activity that heats the chromosphere and corona driven by differential rotation or by convection? It is hard to guess; but, without differential rotation, the sun might not have a strong magnetic field and high-temperature gas above its photosphere.

This is very speculative, but speculating within a scientific argument can be revealing. For example, redo the argument above. **What do you think the sun would be like if it had no convection inside?**

### What Are We? Sunlight

We live very close to a star and depend on it for survival. All of our food comes from sunlight that was captured by plants on land or by plankton in the oceans. We either eat those plants directly or eat the animals that feed on those plants. Whether you had salad, seafood, or a cheeseburger for supper last night, you dined on sunlight, thanks to photosynthesis.

Almost all of the energy that powers human civilization came from the sun through photosynthesis in ancient plants that were buried and converted to coal, oil, and natural gas. New technology is making energy from plant products like corn, soy beans, and sugar. It is all stored sunlight. Windmills generate electrical power, and the wind blows because of heat from the sun. Photocells make electricity directly from sunlight. Even our bodies have adapted to use sunlight to manufacture vitamin D.

Our planet is warmed by the sun, and without that warmth the oceans would be ice and much of the atmosphere would be a coating of frost. Books often refer to the sun as "our sun" or "our star." It is ours in the sense that we live beside it and by its light and warmth, but we can hardly say it belongs to us. It is better to say that we belong to the sun.

# Study and Review

## Summary

- The sun is very bright, and its light and infrared radiation can burn your eyes, so you must take great care in observing it. At sunset or sunrise when it is safe to look at the sun, you see the sun's photosphere, the level in the sun from which visible photons most easily escape. Dark **sunspots (p. 143)** come and go on the sun, but only rarely are they large enough to be visible to the unaided eye.
- The solar atmosphere consists of three layers of hot, low-density gas: the photosphere, chromosphere, and corona.
- The **granulation (p. 144)** of the photosphere is produced by **convection (p. 145)** currents of hot gas rising from below. Larger **supergranules (p. 145)** appear to be caused by larger convection currents deeper in the sun.
- The edge or **limb (p. 145)** of the solar disk is dimmer than the center. This **limb darkening (p. 145)** is evidence that the temperature in the solar atmosphere increases with depth.
- The chromosphere is most easily visible during total solar eclipses, when it flashes into view for a few seconds. It is a thin, hot layer of gas just above the photosphere, and its pink color is caused by the Balmer emission lines in its spectrum.
- **Filtergrams (p. 146)** of the chromosphere reveal **spicules (p. 146)**, flamelike structures extending upward into the lower corona.

- The corona is the sun's outermost atmospheric layer and can be imaged using a **coronagraph (p. 146)**. It is composed of a very-low-density, very hot gas extending many solar radii from the visible sun. Its high temperature—over 2,000,000 K—is believed to be maintained by the magnetic field extending up through the photosphere—the **magnetic carpet (p. 147)**—and by magnetic waves coming from below the photosphere.
- Parts of the corona give rise to the **solar wind (p. 147)**, a breeze of low-density ionized gas streaming away from the sun.
- Solar astronomers can study the motion, density, and temperature of gases inside the sun by analyzing the way the solar surface oscillates. Known as **helioseismology (p. 148)**, this field of study requires large amounts of data and extensive computer analysis.
- There are only four forces in nature: the electromagnetic force, the gravitational force, the **weak force (p. 150)**, and the **strong force (p. 150)**. In nuclear fission or nuclear fusion, the energy comes from the strong force. Physicists are working on unifying the four forces under one mathematical description.
- Nuclear reactors on Earth generate energy through **nuclear fission (p. 150)**, during which large nuclei such as uranium break into smaller fragments. The sun generates its energy through **nuclear fusion (p. 150)**, during which hydrogen nuclei fuse to produce helium nuclei.
- Hydrogen fusion in the sun proceeds in three steps known as the **proton–proton chain (p. 151)**. The first step in the chain combines two hydrogen nuclei to produce a heavy hydrogen nucleus called **deuterium (p. 151)**. The second step forms light helium, and the third step combines the light helium nuclei to form normal helium. Energy is released as **positrons (p. 151)**, **neutrinos (p. 151)**, gamma rays, and the rapid motion of particles flying away.
- Fusion can occur only at the center of the sun because charged particles repel each other, and high temperatures are needed to give particles high enough velocities to penetrate this **Coulomb barrier (p. 152)**. High densities are needed to provide large numbers of reactions.
- Neutrinos escape from the sun's core at nearly the speed of light, carrying away about 2 percent of the energy produced by fusion. Observations of fewer neutrinos than expected coming from the sun's core are now explained by the oscillation of neutrinos among three different types (flavors). The detection of solar neutrinos confirms the theory that the sun's energy comes from hydrogen fusion.
- Energy flows out of the sun's core as photons traveling through the **radiative zone (p. 152)** and closer to the surface as rising currents of hot gas and sinking currents of cooler gas in the **convective zone (p. 152)**.
- Sunspots seem dark because they are slightly cooler than the rest of the photosphere. The average sunspot is about twice the size of Earth. They appear for a month or so and then fade away, and the number of spots on the sun varies with an 11-year cycle.
- Early in a sunspot cycle, spots appear farther from the sun's equator, and later in the cycle they appear closer to the equator. This is shown in the **Maunder butterfly diagram (p. 155)**.
- Astronomers can use the **Zeeman effect (p. 155)** to measure magnetic fields on the sun. The average sunspot contains magnetic fields a few thousand times stronger than Earth's. This is part of the evidence that the sunspot cycle is produced by a solar magnetic cycle.
- The sunspot cycle does not repeat exactly each cycle, and the decades from 1645 to 1715, known as the **Maunder minimum (p. 155)**, seem to have been a time when solar activity was very low and Earth's climate was slightly colder.
- Sunspots are the visible consequences of **active regions (p. 155)** where the sun's magnetic field is strong. Arches of magnetic field can produce sunspots where the field passes through the photosphere.
- The sun's magnetic field is produced by the **dynamo effect (p. 158)** operating at the base of the convection zone.
- Alternate sunspot cycles have reversed magnetic polarity, which has been explained by the **Babcock model (p. 158)**, in which the **differential rotation (p. 158)** of the sun winds up the magnetic field. Tangles in the field arch above the surface and cause active regions visible to your eyes as sunspot pairs. When the field becomes strongly tangled, it reorders itself into a simpler but reversed field, and the cycle starts over.
- Arches of magnetic field are visible as **prominences (p. 162)** in the chromosphere and corona. Seen from above in filtergrams, prominences are visible as dark **filaments (p. 162)** silhouetted against the bright chromosphere.
- **Reconnections (p. 163)** of magnetic fields can produce powerful **flares (p. 163)**, sudden eruptions of X-ray, ultraviolet, and visible radiation plus high-energy atomic particles. Flares are important because they can have dramatic effects on Earth, such as communications blackouts.
- The solar wind originates in regions on the solar surface called **coronal holes (p. 163)**, where the sun's magnetic field leads out into space and does not loop back to the sun.
- **Coronal mass ejections, or CMEs (p. 163)**, occur when magnetic fields on the surface of the sun eject bursts of ionized gas that flow outward in the solar wind. Such bursts can produce **auroras (p. 163)** and other phenomena if they strike Earth.
- Other stars are too far away for starspots to be visible, but some stars vary in brightness in ways that show they have spots on their surfaces. Also, spectroscopic observations reveal that many other stars have spots and magnetic fields that follow long-term cycles like the sun's.
- Small changes in the **solar constant (p. 161)** over decades can affect Earth's climate and may be responsible or the Little Ice Age and other climate fluctuations in Earth's history.

## Review Questions

1. Why can't you see deeper into the sun than the photosphere?
2. What evidence can you give that granulation is caused by convection?
3. How are granules and supergranules related? How do they differ?
4. How can astronomers detect structure in the chromosphere?
5. What evidence can you give that the corona has a very high temperature?
6. What heats the chromosphere and corona to a high temperature?
7. How are astronomers able to explore the layers of the sun below the photosphere?
8. Why does nuclear fusion require high temperatures?
9. Why does nuclear fusion in the sun occur only near the center?
10. How can astronomers detect neutrinos from the sun?
11. How did neutrino oscillation affect the detection of solar neutrinos by the Davis experiment?
12. What evidence can you give that sunspots are magnetic?
13. How does the Babcock model explain the sunspot cycle?
14. What does the spectrum of a prominence reveal? What does its shape reveal?
15. How can solar flares affect Earth?
16. **How Do We Know?** What does it mean when scientists say they are certain? What does scientific certainty really mean?
17. **How Do We Know?** How does consolidation extend scientific understanding?

## Discussion Questions

1. What energy sources on Earth cannot be thought of as stored sunlight?
2. What would the spectrum of an auroral display look like? Why?

3. What observations would you make if you were ordered to set up a system that could warn astronauts in orbit of dangerous solar flares? Such a warning system exists.

## Problems

1. The radius of the sun is 0.7 million km. What percentage of the radius is taken up by the chromosphere?
2. The smallest detail visible with ground-based solar telescopes is about 1 arc second. How large a region does this represent on the sun? (*Hint:* Use the small-angle formula, Chapter 3.)
3. What is the angular diameter of a star like the sun located 5 ly from Earth? Is the Hubble Space Telescope able to detect detail on the surface of such a star?
4. How much energy is produced when the sun converts 1 kg of mass into energy?
5. How much energy is produced when the sun converts 1 kg of hydrogen into helium? (*Hint:* How does this problem differ from Problem 4?)
6. A 1-megaton nuclear weapon produces about $4 \times 10^{15}$ J of energy. How much mass must vanish when a 5-megaton weapon explodes?
7. A solar flare can release $10^{25}$ J. How many megatons of TNT would be equivalent?
8. The United States consumes about $2.5 \times 10^{19}$ J of energy in all forms in a year. How many years could you run the United States on the energy released by the solar flare in Problem 7?
9. Use the luminosity of the sun, the total amount of energy it emits each second, to calculate how much mass it converts to energy each second.
10. If a sunspot has a temperature of 4200 K and the solar surface has a temperature of 5800 K, how many times brighter is a square meter of the surface compared to a square meter of the sunspot? (*Hint:* Use the Stefan–Boltzmann law, Chapter 7.)
11. Neglecting energy absorbed or reflected by Earth's atmosphere, the solar energy hitting 1 square meter of Earth's surface is 1370 J/s. How long does it take a baseball diamond (90 ft on a side) to receive 1 megaton of solar energy?

## Learning to Look

1. Whenever there is a total solar eclipse, you can see something like the image shown at right. Explain why the shape and extent of the glowing gases is different for each eclipse.

NOAO

2. The two images at right show two solar phenomena. What are they, and how are they related? How do they differ?

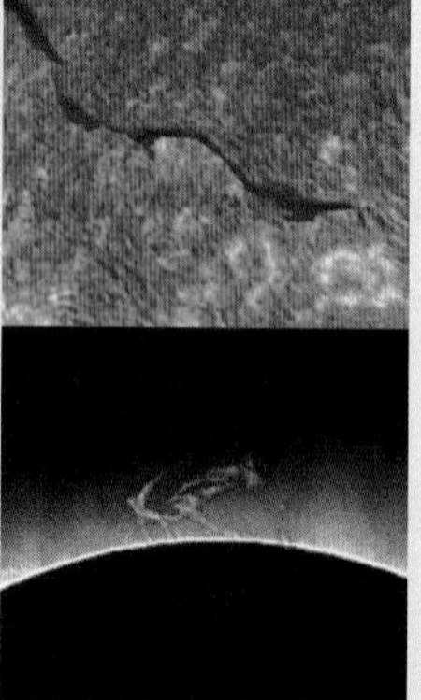

NOAO

3. This image of the sun was recorded in the extreme ultraviolet by the SOHO spacecraft. Explain the features you see.

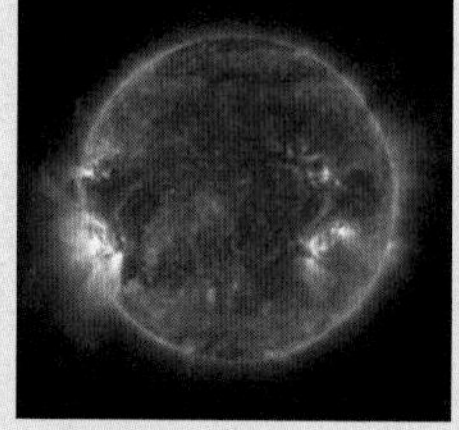

NASA/SOHO

# Perspective: Origins

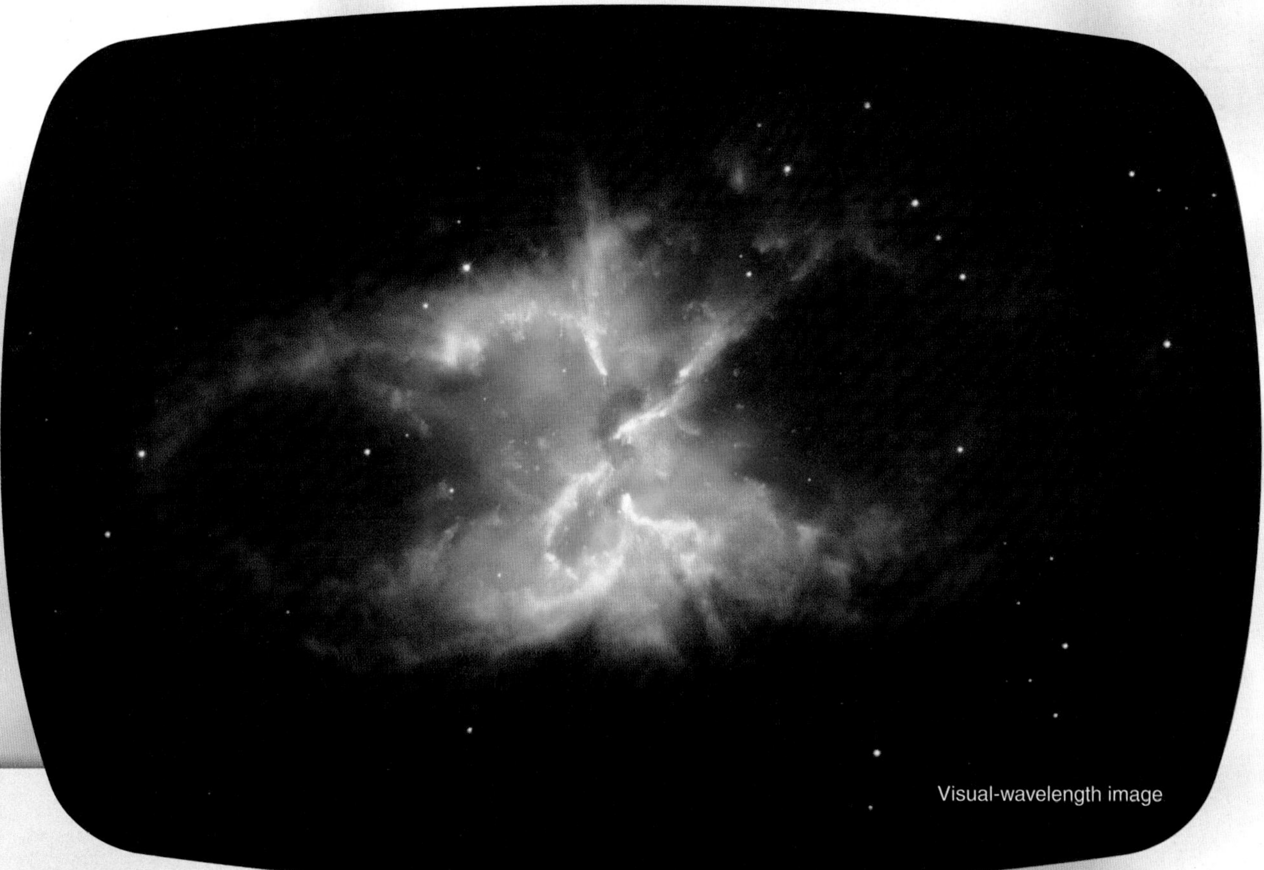


*This nebula cataloged as NGC 2438 was produced by an aging sunlike star that swelled to become a giant and then ejected its outer layers into space. Here red emission is produced by nitrogen atoms, green by hydrogen, and blue by oxygen.* (NASA, ESA, Hubble Heritage team, STScI/AURA)

## Guidepost

As you begin your study of Earth and its solar system, you can look around and get a perspective on where and what our solar system is, and that means you need to take a quick tour of the universe of stars and galaxies. That will help you put our solar system into perspective. Not only will you see how our solar system is located in the universe, but you will see how our solar system fits into the origin and evolution of the universe. Your tour of the cosmos will answer four essential questions:

- **How are stars born, and how do they die?**
- **What are galaxies?**
- **How did the universe begin and form galaxies?**
- **How were the atoms in our bodies formed?**

Most of all, your tour of the universe will show how the stars, our solar system, and planet Earth are made of the same stuff that is in your body and that those atoms had their birthdays long ago and far away.

*The Universe, as has been observed before, is an unsettlingly big place, a fact which for the sake of a quiet life most people tend to ignore.*

— DOUGLAS ADAMS, *THE RESTAURANT AT THE END OF THE UNIVERSE*

EVERYTHING HAS TO COME FROM SOMEWHERE. Look at your thumb. The atoms in your thumb are billions of years old. Only a hundred million years ago, they were inside dinosaurs. Five billion years ago, those atoms were part of a cloud of gas floating in space, and not long before that those atoms were inside stars. The atoms inside your body are old, but the matter the atoms are made of is even older. That matter had its beginning within minutes of the beginning of the universe 13.7 billion years ago. If your atoms could tell their stories, you would be amazed to know where they came from and where they have been.

As you study the solar system, remember the history of the matter it is made of. The iron inside Earth and in your blood, the oxygen, nitrogen and carbon in Earth's atmosphere that you breathe in and out, the calcium in your bones—all exist because stars have lived and died. You are no more isolated from the rest of the universe than a raindrop is isolated from the sea.

## P-1 The Birth of Stars

THE STARS ABOVE SEEM PERMANENT FIXTURES of the sky, but astronomers know that stars are born and stars die. Their lives are long compared with a human life; if you know where to look, you can see all of the stages of stellar birth, aging, and death represented in the sky. Astronomers have put those stages into the proper order and can tell the life story of the stars, a story that begins in the darkness of interstellar space.

Although space seems empty, it is actually filled with thinly spread gas and dust, the **interstellar medium.** The gas atoms are mostly hydrogen, a few centimeters apart, and the dust is made of microscopic grains of heavier atoms such as carbon and iron. The dust makes up only about one percent of the matter between the stars.

The interstellar medium is tenuous in the extreme, yet you can see clear evidence that it exists. In some places, the interstellar medium is collected into great dark clouds of dusty gas that obscure the stars beyond. In other cases, a nearby hot star can ionize the gas and create a glowing cloud (■ Figure P-1). Astronomers refer to both dark and glowing interstellar clouds as **nebulae** (singular, nebula), from the Latin word for cloud or mist.

Nebula N44
Roughly 40 young stars are inflating a bubble of hot gas inside the nebula from which they formed.
100 ly
Visual-wavelength image
Horsehead Nebula
Dusty foreground gas silhouetted against glowing gas illuminated by hot, young stars
Young star buried in the nebula
5 ly
Visual-wavelength image

**■ Figure P-1**

Young stars are found in clouds of gas and dust from which they have been born. The nebula N 44 is 170,000 ly from Earth in a nearby galaxy, and the Horsehead Nebula is only about 1500 ly distant in our own galaxy. Gas in both nebulae is excited to glow by hot, young stars, and dust is visible as dark, twisted clouds seen against the bright background gas. (N 44: ESO; Horsehead Nebula: NOAO and Nigel Sharp)

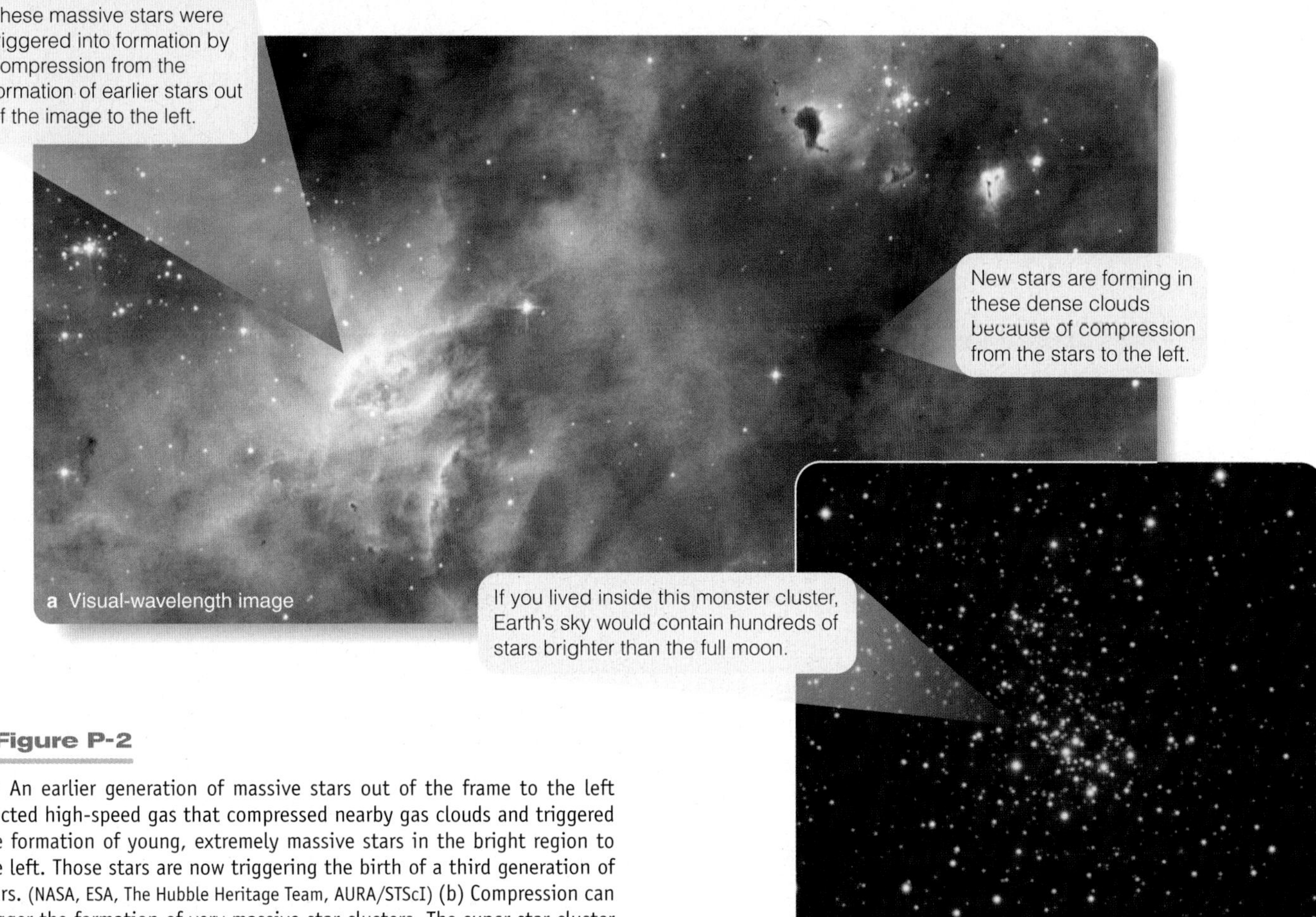

These massive stars were triggered into formation by compression from the formation of earlier stars out of the image to the left.
New stars are forming in these dense clouds because of compression from the stars to the left.
a Visual-wavelength image
If you lived inside this monster cluster, Earth's sky would contain hundreds of stars brighter than the full moon.
b Visual + infrared-wavelength image

**Figure P-2**

(a) An earlier generation of massive stars out of the frame to the left ejected high-speed gas that compressed nearby gas clouds and triggered the formation of young, extremely massive stars in the bright region to the left. Those stars are now triggering the birth of a third generation of stars. (NASA, ESA, The Hubble Heritage Team, AURA/STScI) (b) Compression can trigger the formation of very massive star clusters. The super star cluster Westerlund 1 contains roughly half a million stars, some extremely massive. It must have formed no more than 5 million years ago. (ESO)

These nebulae not only adorn the sky, they also mark the birthplace of stars. As the cold gases of a nebula grow denser, gravity can pull parts of it together to form warmer, denser bodies that eventually become **protostars**—objects destined to become stars.

Astronomers can't see these protostars easily because they are hidden deep inside the dusty gas clouds from which they form, but they are easily detected at infrared wavelengths. As the more massive protostars become hot, luminous stars, the light and gas flowing away from the newborn stars blow the nebula away to reveal the new stars (■ Figure P-2). In this way, a single gas cloud can give birth to a cluster of stars. Our sun was probably born in such a nebula about 5 billion years ago.

If you look at the constellation Orion, you can see one of these regions with your unaided eye. The patch of haze in Orion's sword is the Great Nebula in Orion. Read **Star Formation in the Orion Nebula** on pages 170–171 and notice four points and one new term:

1. The nebula you see is only a small part of a vast, dusty cloud. You see the nebula because the stars born within it have ionized the gas and driven it outward, breaking out of the cloud. In some cases, the turbulence has produced small, dense clouds of gas and dust called *Bok globules* that may be in the process of forming stars.

2. Also notice that a single very hot star (classified O6) is almost entirely responsible for producing the ultraviolet photons that ionize the gas and make the nebula glow.

3. Infrared observations reveal clear evidence of active star formation deeper in the cloud just to the northwest of the Trapezium, behind the visible nebula.

4. Finally, notice that many stars visible in the Orion Nebula are surrounded by disks of gas and dust. Planets form in such disks, but the disks do not last long and are clear evidence that the stars are very young.

As protostars contract, something happens that is quite important to planetwalkers like you—planets form. You will see in the next chapter how our solar system formed from the disk of gas and dust that orbited the protostar that became our sun.

You could identify the birth of a star as the moment when it begins fusing hydrogen into helium. That nuclear fusion reaction releases energy, supports the star, and stops its contraction. Stars that generate energy at their cores by hydrogen fusion are called **main-sequence stars.**

# Star Formation in the Orion Nebula

**Side view of Orion Nebula**

Hot Trapezium stars

Protostars

To Earth

Expanding ionized hydrogen

Molecular cloud

**1** The visible Orion Nebula shown below is a pocket of ionized gas on the near side of a vast, dusty molecular cloud that fills much of the southern part of the constellation Orion. The molecular cloud can be mapped by radio telescopes. To scale, the cloud would be many times larger than this page. As the stars of the Trapezium were born in the cloud, their radiation has ionized the gas and pushed it away. Where the expanding nebula pushes into the larger molecular cloud, it is compressing the gas (see diagram at right) and may be triggering the formation of the protostars that can be detected at infrared wavelengths within the molecular cloud.

Hundreds of stars lie within the nebula, but only the four brightest, those in the Trapezium, are easy to see with a small telescope. A fifth star, at the narrow end of the Trapezium, may be visible on nights of good seeing.

The cluster of stars in the nebula is less than 2 million years old. This must mean the nebula is similarly young.

Trapezium

Visual-wavelength image

**Credit: NASA, ESA, M. Robberto, STScI and the Hubble Space Telescope Orion Treasury Project Team**

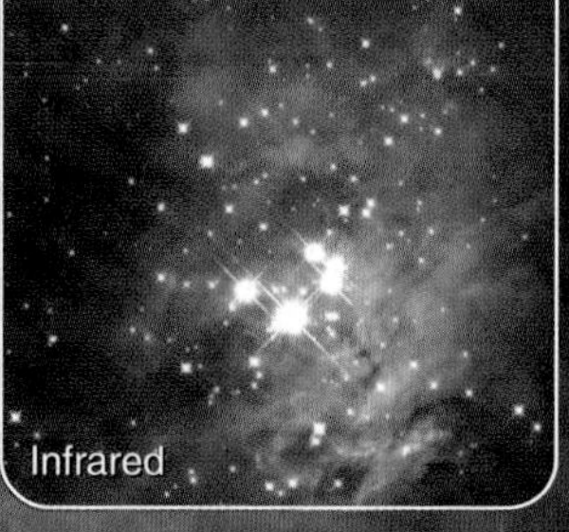


The near-infrared image above reveals more than 50 low-mass, cool protostars.

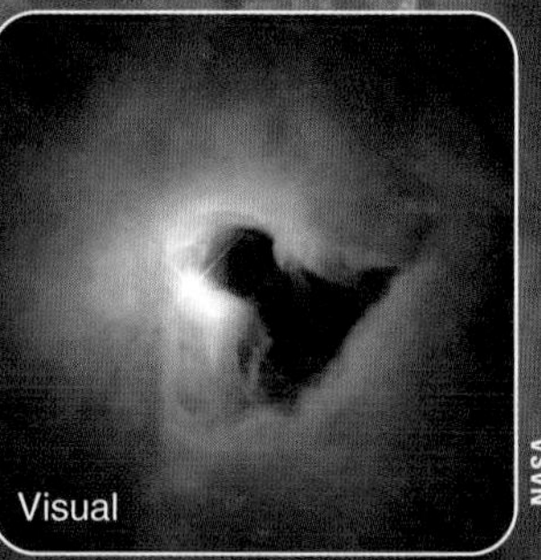


Small dark clouds called **Bok globules**, named after astronomer Bart Bok, are found in and near star-forming regions. The one pictured above is part of nebula NGC 1999 near the Orion Nebula. Typically about 1 light-year in diameter, they contain from 10 to 1000 solar masses.

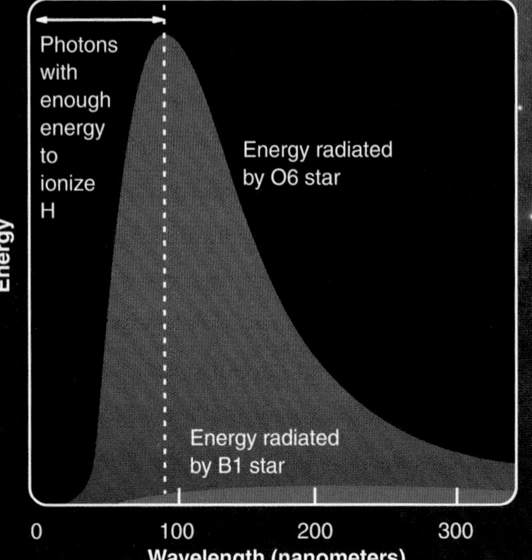


**2** Of all the stars in the Orion Nebula, only one is hot enough to ionize the gas. Only photons with wavelengths shorter than 91.2 nm can ionize hydrogen. The second-hottest stars in the nebula are B1 stars, and they emit little of this ionizing radiation. The hottest star, however, is an O6 star 30 times the mass of the sun. At a temperature of 40,000 K, it emits plenty of photons with wavelengths short enough to ionize hydrogen. Remove that one star, and the nebula would turn off its emission.

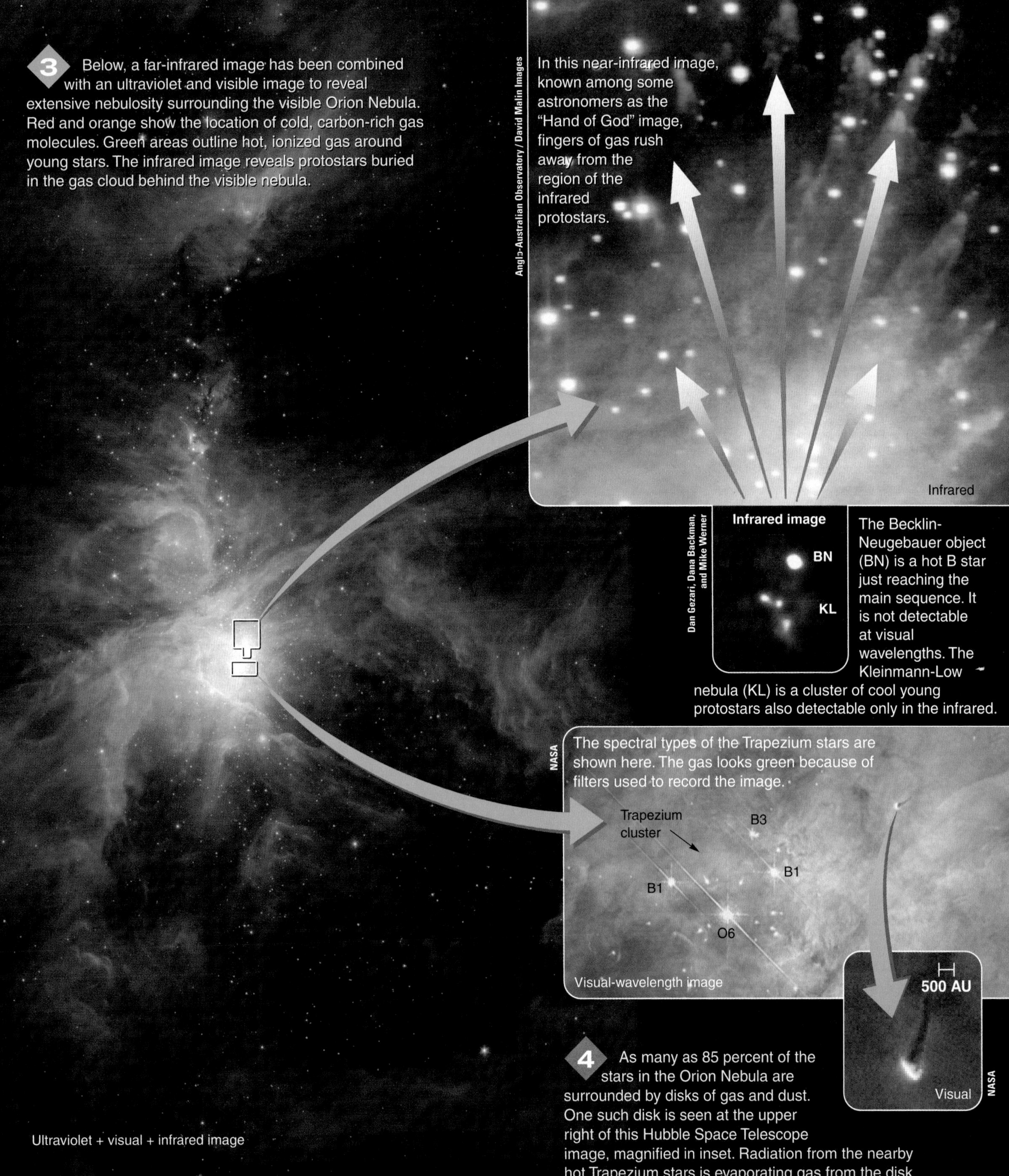
3
Below, a far-infrared image has been combined with an ultraviolet and visible image to reveal extensive nebulosity surrounding the visible Orion Nebula. Red and orange show the location of cold, carbon-rich gas molecules. Green areas outline hot, ionized gas around young stars. The infrared image reveals protostars buried in the gas cloud behind the visible nebula.
Anglo-Australian Observatory / David Malin Images
In this near-infrared image, known among some astronomers as the "Hand of God" image, fingers of gas rush away from the region of the infrared protostars.
Infrared
Dan Gezari, Dana Backman, and Mike Werner
Infrared image
BN
KL
The Becklin-Neugebauer object (BN) is a hot B star just reaching the main sequence. It is not detectable at visual wavelengths. The Kleinmann-Low nebula (KL) is a cluster of cool young protostars also detectable only in the infrared.
NASA
The spectral types of the Trapezium stars are shown here. The gas looks green because of filters used to record the image.
Trapezium cluster
B3
B1
B1
O6
Visual-wavelength image
500 AU
Visual
NASA
4
As many as 85 percent of the stars in the Orion Nebula are surrounded by disks of gas and dust. One such disk is seen at the upper right of this Hubble Space Telescope image, magnified in inset. Radiation from the nearby hot Trapezium stars is evaporating gas from the disk and driving it away to form an elongated nebula.
Ultraviolet + visual + infrared image

# How Do We Know? | P-1

## Mathematical Models

**How can scientists study aspects of nature that cannot be observed directly?** One of the most powerful methods in science is the mathematical model, a group of equations carefully designed to mimic the behavior of objects and processes that scientists want to study. Astronomers build mathematical models to study the structure hidden deep inside stars. Models can speed up the slow evolution of stars and slow down the rapid processes that generate energy. Stellar models are based on only four equations, but other models are much more complicated and may require many more equations.

For example, scientists and engineers designing a new airplane don't just build it, cross their fingers, and ask a test pilot to try it out. Long before any metal parts are made, mathematical models are created to test whether the wing design will generate enough lift, whether the fuselage can support the strain, and whether the rudder and ailerons can safely control the plane during takeoff, flight, and landing. Those mathematical models are put through all kinds of tests: Can a pilot fly with one engine shut down? Can the pilot recover from sudden turbulence? Can the pilot land in a crosswind? By the time the test pilot rolls the plane down the runway for the first time, the mathematical models have "flown" many thousands of miles.

Scientific models are only as good as the assumptions that go into them and must be compared with the real world at every opportunity. If you are an engineer designing a new airplane, you can test your mathematical models by making measurements in a wind tunnel. Models of stars are much harder to test against reality, but they do predict some observable things. Stellar models predict the existence of main sequence stars, the observed numbers of giant and supergiant stars, and the luminosities of stars of different masses. Without mathematical models, astronomers would know little about the lives of the stars, and designing new airplanes would be a very dangerous business.

*Before any new airplane flies, engineers build mathematical models to test its stability.* (The Boeing Company)

## P-2 The Deaths of Stars

STARS SPEND MOST OF THEIR LIVES fusing hydrogen into helium, and, when the hydrogen is exhausted, they fuse helium into carbon. The more massive stars can fuse carbon into even heavier atoms, but iron is the limit beyond which no star can go.

How long a star can live depends on its mass. The smallest stars are very common, but they are not very luminous. These cool **red dwarf** stars are hardly massive enough to fuse hydrogen, and they can survive for over 100 billion years. The sun, a medium-mass star, will live a total of about 10 billion years. In contrast, the most massive stars, up to almost 100 times the mass of the sun, are tremendously luminous and fuse their fuels so rapidly that they can live only a few million years.

This chapter tells the story of the birth and death of stars in a few paragraphs, but modern astronomers know a great deal about the lives of the stars. How could mere humans understand the formation, development, and deaths of objects that can survive for millions or even billions of years? The answer is the mathematical model (**How Do We Know? P-1**). Astronomers can express the physical laws that govern the gas and energy inside a star as equations that can be solved in computer programs. The resulting mathematical model describes the internal structure of a star and allows astronomers to follow its evolution as it ages (■ Figure P-3). Solving the mystery of the evolution of stars is one of the greatest accomplishments of modern astronomy.

How a star dies depends on its mass. A medium-mass star like the sun has enough hydrogen fuel to survive for billions of years, but it must eventually exhaust its fuel, swell to become a **giant** star, and then expel its outer layers in an expanding nebula like that shown in the photograph that opens this chapter. These **planetary nebulae** were named for their planet-like appearance in small telescopes, but they are, in fact, the remains of dying stars. Once the outer layers of a star are ejected into space, the hot core contracts to form a small, dense, cooling star—a **white dwarf.**

Study **The Formation of Planetary Nebulae** on pages 174–175 and notice four things:

1. First, you can understand what planetary nebulae are like by using simple observational methods such as Kirchhoff's laws and the Doppler effect.
2. Notice the model astronomers have developed to explain planetary nebulae. The real nebulae are more complex than the simple model of a slow wind and a fast wind, but the model provides a way to organize the observed phenomena.
3. Notice how oppositely directed jets and multiple shells produce many of the asymmetries seen in planetary nebulae.
4. Finally, notice the fate of the star itself; it must contract into a white dwarf.

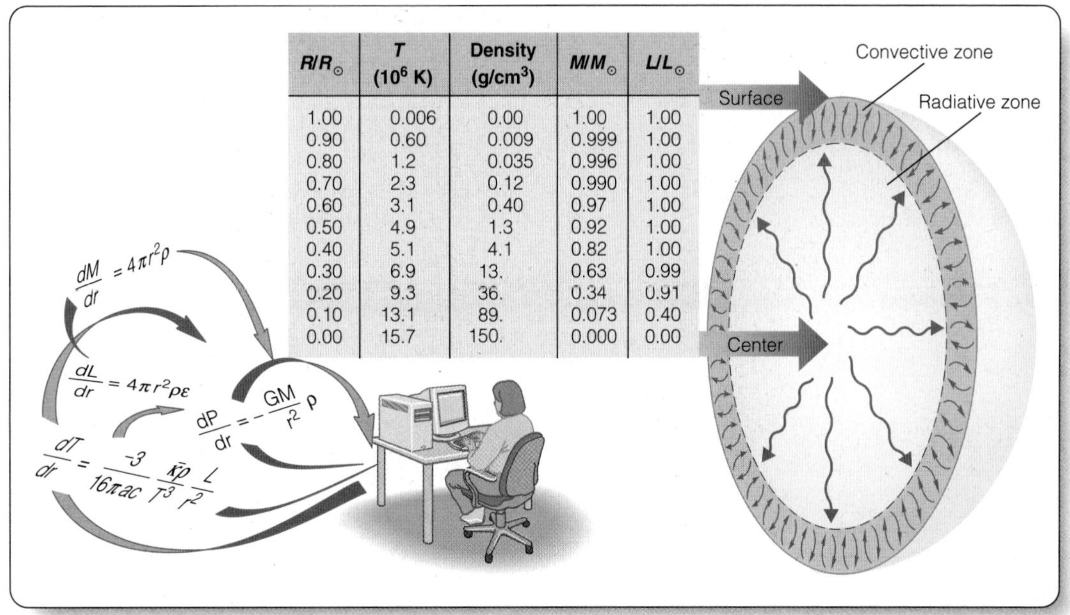

| $R/R_\odot$ | $T$ ($10^6$ K) | Density (g/cm$^3$) | $M/M_\odot$ | $L/L_\odot$ |
|---|---|---|---|---|
| 1.00 | 0.006 | 0.00 | 1.00 | 1.00 |
| 0.90 | 0.60 | 0.009 | 0.999 | 1.00 |
| 0.80 | 1.2 | 0.035 | 0.996 | 1.00 |
| 0.70 | 2.3 | 0.12 | 0.990 | 1.00 |
| 0.60 | 3.1 | 0.40 | 0.97 | 1.00 |
| 0.50 | 4.9 | 1.3 | 0.92 | 1.00 |
| 0.40 | 5.1 | 4.1 | 0.82 | 1.00 |
| 0.30 | 6.9 | 13. | 0.63 | 0.99 |
| 0.20 | 9.3 | 36. | 0.34 | 0.91 |
| 0.10 | 13.1 | 89. | 0.073 | 0.40 |
| 0.00 | 15.7 | 150. | 0.000 | 0.00 |



**■ Figure P-3**

A stellar model is a table of numbers that represent conditions inside a star. Such tables can be computed using basic laws of physics, shown here in mathematical form. The table in this figure describes the sun. (Illustration design by Mike Seeds)

When medium-mass stars like the sun die, they expel their outer layers back into space, and some of the atoms that have been cooked up inside the stars get mixed back into the interstellar medium.

In contrast with stars like the sun, massive stars become very large giants or even larger **supergiants.** These stars are so luminous that they exhale gas back into the interstellar medium (■ Figure P-4). They live very short lives, perhaps only millions of years, before they develop iron cores and explode as **supernovae** (singular, supernova) (■ Figure P-5). The core of such a dying massive star may form a **neutron star** or a **black hole,** but the outer parts of the star, newly enriched with the atoms

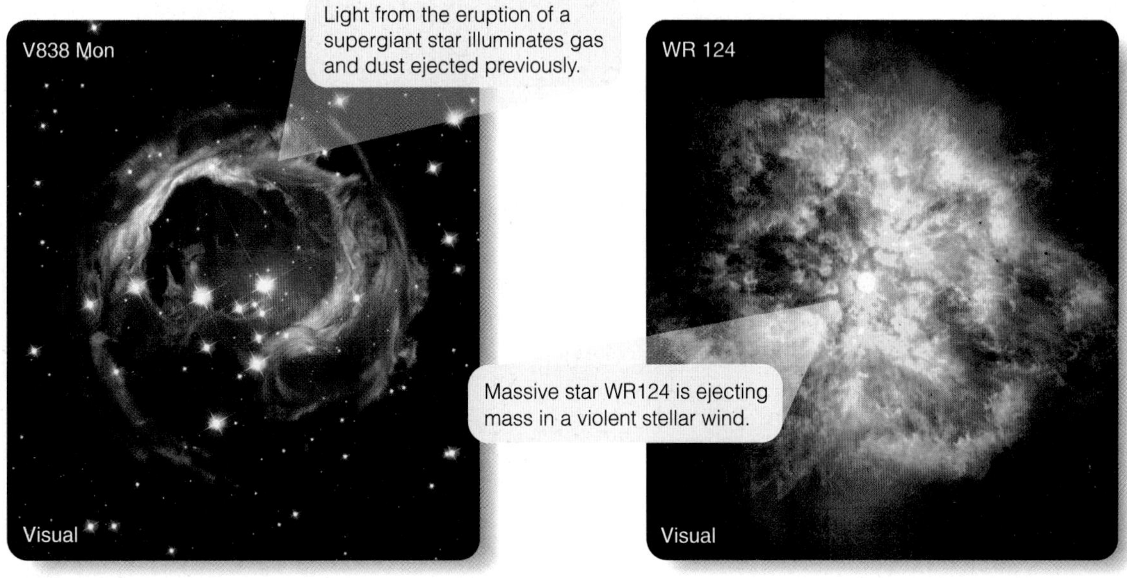



**■ Figure P-4**

Stars can lose mass if they are very hot, very large, or both. The massive red supergiant VY Canis Majoris is ejecting gas in loops, arcs, and knots as it ages. A young massive star such as WR124 constantly loses mass into space. Warmed dust in these gas clouds can make them glow in the infrared. (VY Canis Majoris: NASA, ESA, R. Humphreys; WR124: NASA)

# The Formation of Planetary Nebulae

**1** Simple observations tell astronomers what planetary nebulae are like. Their angular size and their distances indicate that their radii range from 0.2 to 3 ly. The presence of emission lines in their spectra assures us that they are excited, low-density gas. Doppler shifts show they are expanding at 10 to 20 km/s. If you divide radius b velocity, you find that planetary nebulae are no more than about 10,000 years old. Older nebulae evidently become mixed into the interstellar medium.

Astronomers find about 3000 planetary nebulae in the sky. Because planetary nebulae are short-lived formations, you can conclude that they must be a common part of stellar evolution Medium-mass stars up to a mas of about 8 solar masses are destined to die by forming planetary nebulae.

The Helix Nebula is 2.5 ly in diameter, and the radial texture shows how light and winds from the central star are pushing outward.

NASA/JPL-Caltech/ESA

Visual + Infrared

**2** The process that produces planetary nebulae involves two stellar winds. First, as an aging giant, the star gradually blows away its outer layers in a slow breeze of low-excitation gas that is not easily visible. Once the hot interior of the star is exposed, it ejects a high-speed wind that overtakes and compresses the gas of the slow wind like a snowplow, while ultraviolet radiation from the hot remains of the central star excites the gases to glow like a giant neon sign.

**Slow stellar wind from a red giant**

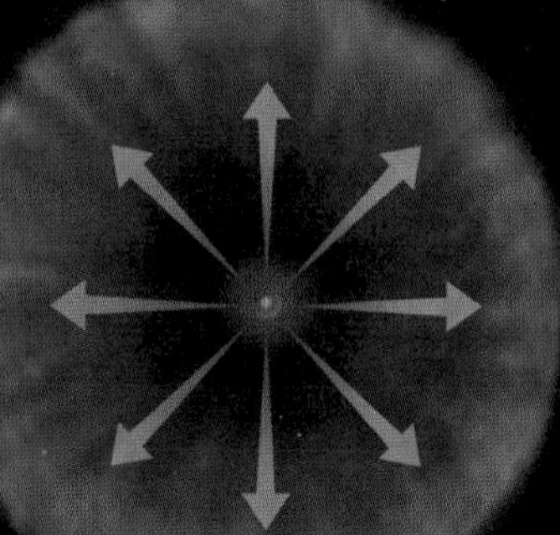

The gases of the slow wind are not easily detectable.

**Fast wind from exposed interior**

You see a planetary nebula where the fast wind compresses the slow wind.

**2a** The Cat's Eye, below, lies at the center of an extended nebula that must have been exhaled from the star long before the fast wind began forming the visible planetary nebula. See other images of the nebula on opposite page.

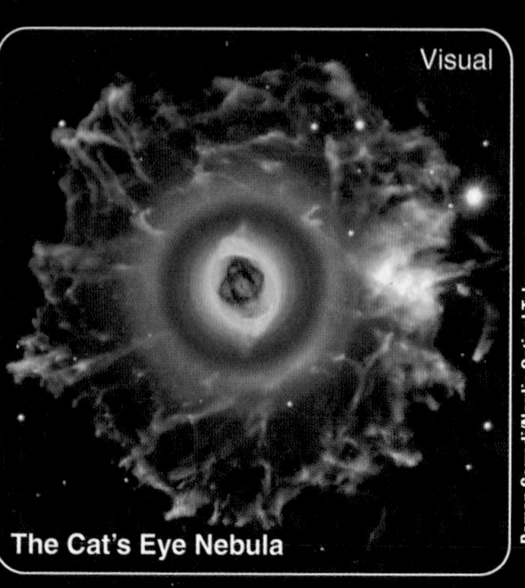


**The Cat's Eye Nebula**

Romano Corradi/Nordic Optical Telescope

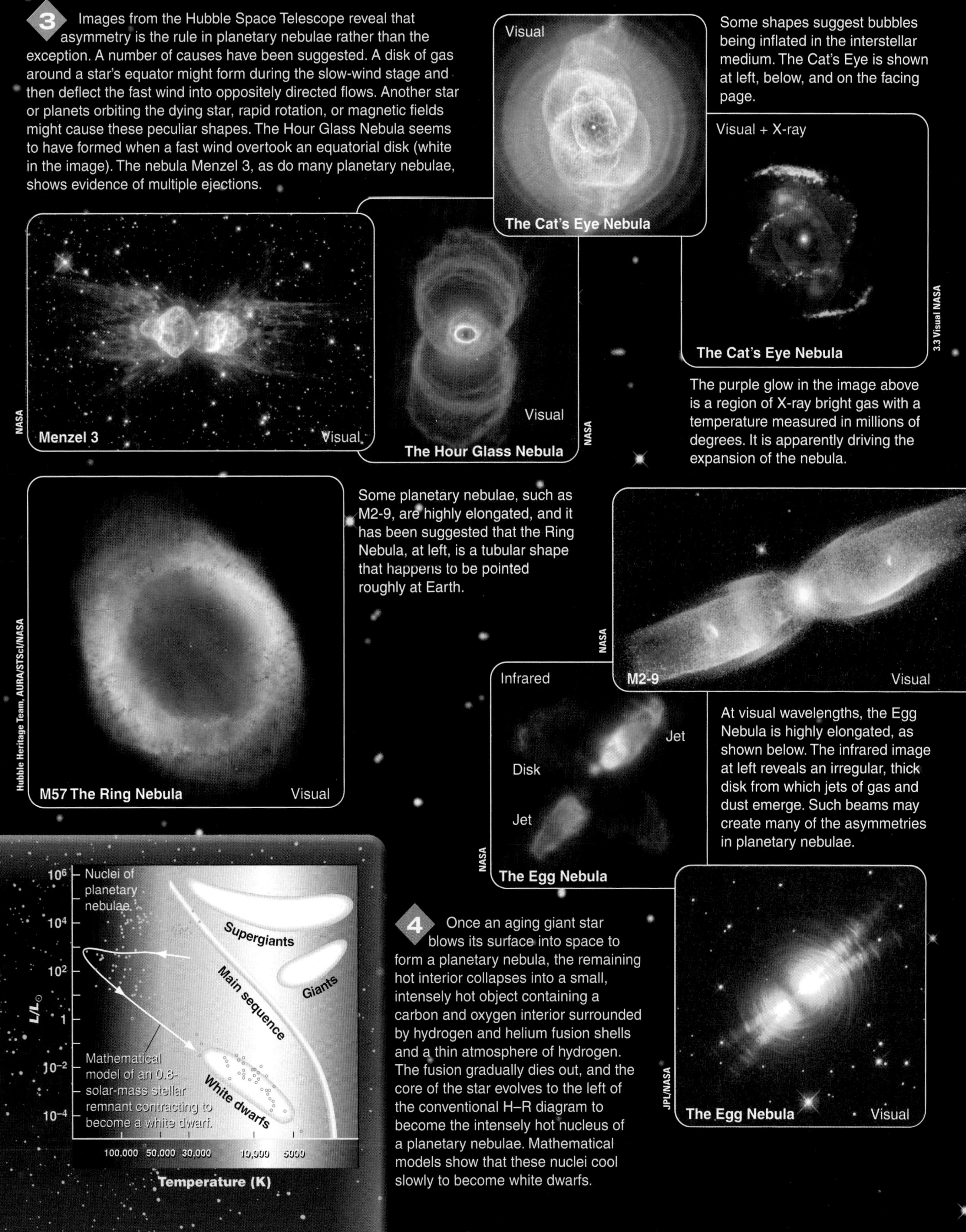


**3** Images from the Hubble Space Telescope reveal that asymmetry is the rule in planetary nebulae rather than the exception. A number of causes have been suggested. A disk of gas around a star's equator might form during the slow-wind stage and then deflect the fast wind into oppositely directed flows. Another star or planets orbiting the dying star, rapid rotation, or magnetic fields might cause these peculiar shapes. The Hour Glass Nebula seems to have formed when a fast wind overtook an equatorial disk (white in the image). The nebula Menzel 3, as do many planetary nebulae, shows evidence of multiple ejections.

Some shapes suggest bubbles being inflated in the interstellar medium. The Cat's Eye is shown at left, below, and on the facing page.

The purple glow in the image above is a region of X-ray bright gas with a temperature measured in millions of degrees. It is apparently driving the expansion of the nebula.

Some planetary nebulae, such as M2-9, are highly elongated, and it has been suggested that the Ring Nebula, at left, is a tubular shape that happens to be pointed roughly at Earth.

At visual wavelengths, the Egg Nebula is highly elongated, as shown below. The infrared image at left reveals an irregular, thick disk from which jets of gas and dust emerge. Such beams may create many of the asymmetries in planetary nebulae.

**4** Once an aging giant star blows its surface into space to form a planetary nebula, the remaining hot interior collapses into a small, intensely hot object containing a carbon and oxygen interior surrounded by hydrogen and helium fusion shells and a thin atmosphere of hydrogen. The fusion gradually dies out, and the core of the star evolves to the left of the conventional H–R diagram to become the intensely hot nucleus of a planetary nebulae. Mathematical models show that these nuclei cool slowly to become white dwarfs.

cooked up inside the star, are returned to the interstellar medium.

The violence of a supernova explosion can fuse atoms together to build elements heavier than iron. Gold, platinum, uranium, and other elements heavier than iron are rare and valuable because they are made only in the moments of a supernova explosion. The iodine atoms in your thyroid gland and the gold atoms in your class ring were made in supernova explosions.

The atoms of which you are made had their birth inside stars. That process is common in the universe because stars are common. Our galaxy contains billions of them.

## P-3 Our Home Galaxy

FROM A DARK LOCATION away from city lights, you can see the **Milky Way** stretching across the sky. The winter Milky Way is especially dramatic (■ Figure P-6a). It is actually the disk of the galaxy that we live in—the **Milky Way Galaxy** seen from the inside. Of course, no one has ever journeyed out into space to look back and take a picture of our galaxy, but astronomers have evidence that the nearby Andromeda Galaxy, shown in ■ Figure P-6b, looks much like our own. Our galaxy contains roughly 100 billion stars, and we live about two-thirds of the way from the center to the edge.

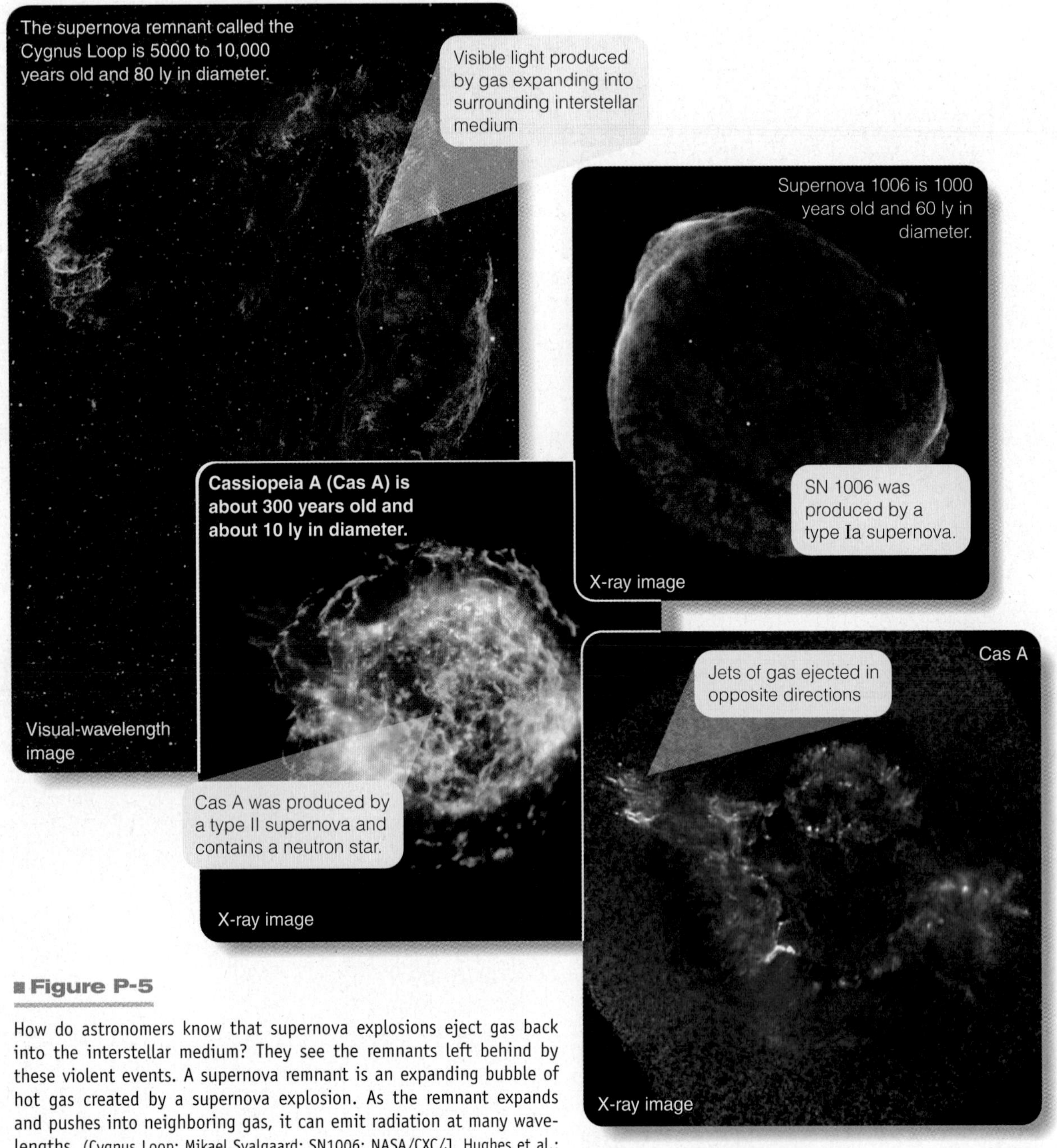
The supernova remnant called the Cygnus Loop is 5000 to 10,000 years old and 80 ly in diameter.
Visible light produced by gas expanding into surrounding interstellar medium
Visual-wavelength image
Supernova 1006 is 1000 years old and 60 ly in diameter.
SN 1006 was produced by a type Ia supernova.
X-ray image
Cassiopeia A (Cas A) is about 300 years old and about 10 ly in diameter.
Cas A was produced by a type II supernova and contains a neutron star.
X-ray image
Cas A
Jets of gas ejected in opposite directions
X-ray image

■ **Figure P-5**

How do astronomers know that supernova explosions eject gas back into the interstellar medium? They see the remnants left behind by these violent events. A supernova remnant is an expanding bubble of hot gas created by a supernova explosion. As the remnant expands and pushes into neighboring gas, it can emit radiation at many wavelengths. (Cygnus Loop: Mikael Svalgaard; SN1006: NASA/CXC/J. Hughes et al.; Cas A: NASA/CXC/GSFC/U. Hwang et al.)

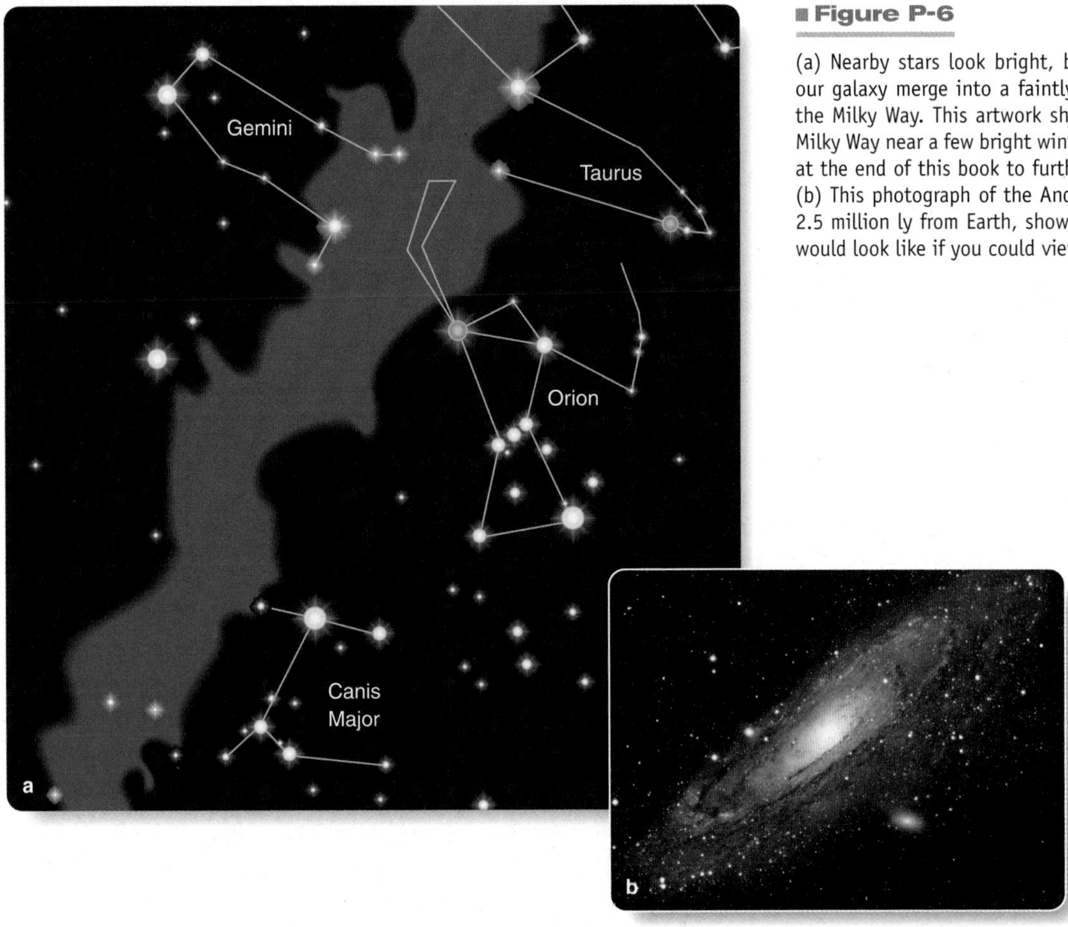

■ **Figure P-6**

(a) Nearby stars look bright, but the vast majority of the stars in our galaxy merge into a faintly luminous path that circles the sky, the Milky Way. This artwork shows the location of a portion of the Milky Way near a few bright winter constellations. (See the star charts at the end of this book to further locate the Milky Way in your sky.) (b) This photograph of the Andromeda Galaxy, a spiral galaxy about 2.5 million ly from Earth, shows approximately what our own galaxy would look like if you could view it from a distance. (AURA/NOAO/NSF)

Astronomers estimate that our galaxy is about 80,000 light-years in diameter. That is, light takes 80,000 years to travel from one edge to the other. If you had a photograph of the Milky Way Galaxy as big as North America, the entire solar system would be about the size of a small cookie, and the sun and planets would all be too small to see without a powerful microscope.

Large astronomical telescopes reveal other galaxies scattered across the sky. You have already met the Andromeda Galaxy in Figure P-6. It is so close you can see its nucleus with the unaided eye as a hazy patch in the constellation Andromeda. Other galaxies are all around us, and some are dramatically beautiful (■ Figure P-7).

## P-4 The Universe of Galaxies

Our galaxy is our home, but ours is only one of billions of galaxies. Some, like our Milky Way Galaxy, are disk shaped with graceful spiral arms marked by clouds of gas and bright newborn stars. But many galaxies are great swarms of stars with relatively little gas and dust.

Study **Galaxy Classification** on pages 178–179 and notice three important points and four new terms:

1. Many galaxies have no disk, no spiral arms, and almost no gas and dust. These *elliptical galaxies* range from huge giants to small dwarfs.

2. Notice that disk-shaped galaxies usually have spiral arms and contain gas and dust. Many of these *spiral galaxies* have a central region shaped like an elongated bar and are called *barred spiral galaxies.* The Milky Way Galaxy is a barred spiral. A few disk galaxies contain little gas and dust.

3. Finally, notice the *irregular galaxies*, which are generally shapeless and tend to be rich in gas and dust.

You might expect such titanic objects as galaxies to be rare, but large telescopes reveal that the sky is filled with galaxies. Like leaves on the forest floor, galaxies carpet the sky. They fill the universe in every direction as far as telescopes can see (■ Figure P-8). Grouped in clusters and superclusters, galaxies are the homes of the billions of stars that illuminate the universe and create the chemical elements.

When two galaxies collide, the stars swirl past each other without bumping, but the great clouds of gas and the magnetic fields inside the galaxies do collide and compress each other. The compression of the gas clouds can stimulate two colliding galaxies to form vast numbers of new stars and massive star clusters.

■ **Figure P-7**

The beautiful symmetry of the spiral pattern is clear in this image of the galaxy NGC 1300. Young star clusters containing hot, bright stars and ionized hydrogen are located along the spiral arms, while dark lanes of dust mark the inner edges of the arms. (NASA, ESA, and The Hubble Heritage Team; STScI/AURA)

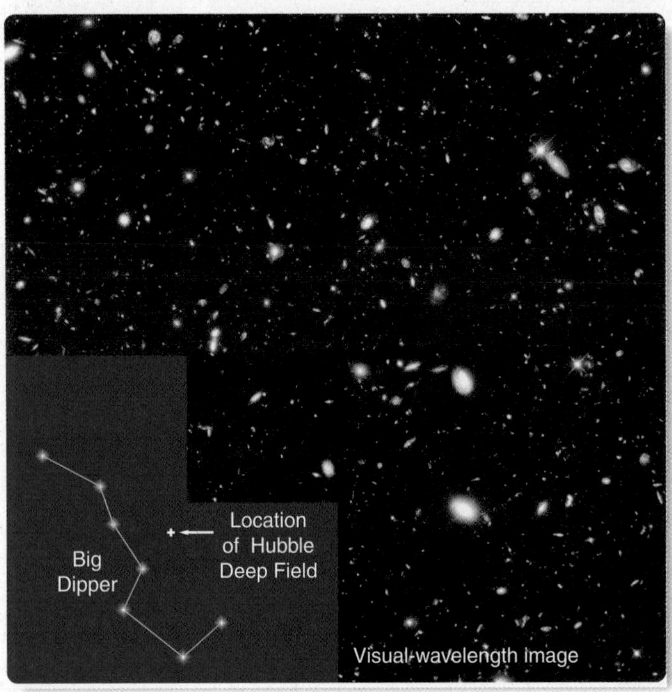


■ **Figure P-8**

An apparently empty spot on the sky only 1/30 the diameter of the full moon contains over 1500 galaxies in this extremely long time exposure known as the Northern Hubble Deep Field. Presumably the entire sky is similarly filled with galaxies. (R. Williams and the Hubble Deep Field Team, STScI/AURA, NASA, ESA)

Furthermore, tidal forces twist and distort the colliding galaxies. Images of such colliding galaxies show their twisted shapes, far-flung streamers of stars and gas, and extensive regions of star formation (■ Figure P-9). Such bursts of star formation salt the galaxies with newly formed heavy atoms. Our own Milky Way Galaxy has probably collided with more than one smaller galaxy, and some of the atoms of which you are made may have been formed because of those collisions.

## P-5 The Origin of the Universe

A LITTLE LESS THAN A CENTURY AGO, astronomers made an astonishing discovery. The galaxies in the universe are moving away from each other. The spectrum of a galaxy is the combined spectrum of billions of stars, and the spectral lines visible in galaxy spectra are shifted slightly toward the red end of the spectrum. This redshift is proportional to distance. That is, the farther away a galaxy is, the larger is the redshift that is visible in its spectrum (■ Figure P-10). This result, first discovered by the American astronomer Edwin Hubble in 1929, is clear evidence that the universe is expanding.

If the galaxies are all rushing away from each other, then you can imagine viewing a video running backward. You would see

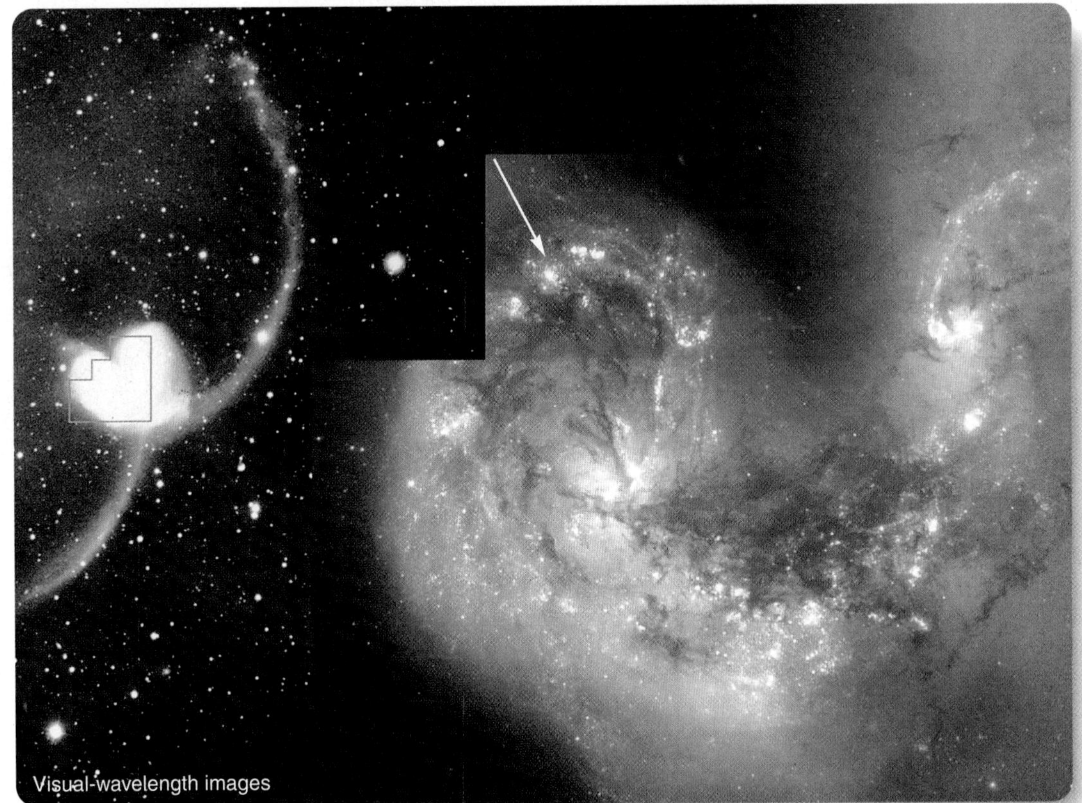


**Figure P-9**

The colliding galaxies NGC 4038 and NGC 4039 are known as the Antennae because their long curving tails resemble the antennae of an insect. Earth-based photos (left) show little detail, but a Hubble Space Telescope image (right) reveals the collision of the two galaxies producing thick clouds of dust and raging star formation, creating roughly a thousand massive star clusters such as the one at the top (arrow). Such collisions between galaxies are common. (Brad Whitmore, STScI, and NASA)

the galaxies moving toward each other, and eventually you would see the galaxies pushing into each other, compressing and heating the gas until your video screen was filled with the glare of an intensely hot, fantastically dense gas filling the universe. This, the beginning of the universe, is a state called the **big bang.**

How can anyone know the big bang really happened? The evidence is conclusive—astronomers can see it (**How Do We Know P-2**). To understand how you might see an event that occurred billions of years ago, you need to look once again at the galaxies. Nearby galaxies, such as the Andromeda Galaxy, are only a few million light-years away. The hazy light you see in the sky coming from that direction in the constellation Andromeda has been traveling through space for only 2.5 million years, and you see the Andromeda Galaxy as it was 2.5 million years ago when the light began its journey. If you used a telescope and looked at more distant galaxies, you would see them as they were a billion years ago or more; light

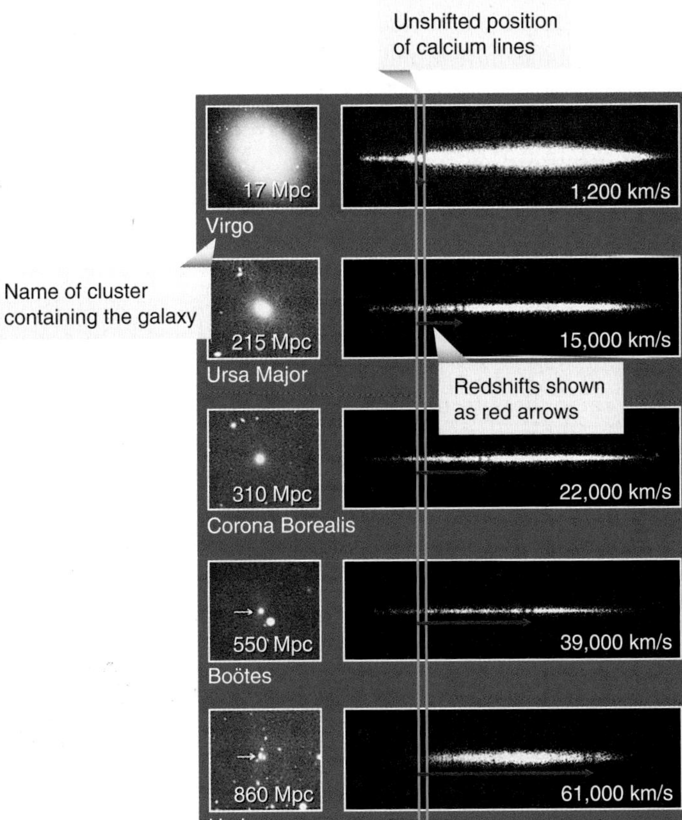


**Figure P-10**

These galaxy spectra extend from the near-ultraviolet at left to the blue part of the visible spectrum at right. The two dark absorption lines of once-ionized calcium are prominent in the near-ultraviolet. The redshifts in galaxy spectra are expressed here as velocities of recession. Note that the apparent velocity of recession is proportional to distance. (Caltech)

# How Do We Know? | P-2

## Science: A System of Knowledge

**What is the difference between believing in the big bang and understanding it?** If you ask a scientist, "Do you believe in the big bang?" she or he may hesitate before responding. The question implies something incorrect about the way science works.

Science is a system of knowledge and process but does not operate as a system of belief. Science is an attempt to understand logically how nature works and is based on observations and experiments used to test and confirm hypotheses and theories. A scientist does not really believe in even a well-confirmed theory in the way people normally use the word *believe.* Rather, the scientist understands the theory and recognizes how different pieces of evidence support or contradict the theory.

There are other ways to know things, and there are many systems of belief. Religions, for example, are systems of belief that are not entirely based on observation. In some cases, a political system is also a system of belief; many people believe that democracy is the best form of government and do not ask for, or expect, evidence supporting that belief. A system of belief can be powerful and lead to deep insights, but it is different from science.

Scientists try to be careful with words, so thoughtful scientists would not say they *believe* in the big bang. They would say that the evidence is overwhelming that the big bang really did occur and that they are compelled, by a logical analysis of both the observations and the theory, to conclude that the theory is very likely correct. In this way scientists try to be objective and reason without distortion by personal feelings and prejudices.

A scientist once referred to "the terrible rule of evidence." Sometimes the evidence forces a scientist to a conclusion she or he does not like, but science is not a system of belief, so the personal preferences of each scientist must take second place to the rule of evidence.

*Scientific knowledge is based objectively on evidence such as that gathered by spacecraft.* (NASA/WMAP Science Team)

Do you believe in the big bang? Or, instead, do you have confidence that the theory is right because of your analysis of the evidence? There is a big difference.

from those galaxies has been traveling that long. If you used a big enough telescope and looked at the most distant galaxies, you would be looking back in time and seeing them as they were over 10 billion years ago when the universe was young.

What would you see if you looked at the empty places on the sky between the most distant galaxies? You could detect a glow that was emitted by the hot, dense clouds of gas of the big bang, but you couldn't see that glow with your unaided eyes because the redshift is so great that the photons of light are shifted into the long-wavelength infrared and radio parts of the spectrum. Nevertheless, astronomers can "see" the radiation from the big bang with infrared and radio detectors. This is called the **cosmic microwave background radiation,** and it fills the universe, pouring in on Earth from all directions, and telling you that you are part of a universe that was very hot and very dense about 13.7 billion years ago (■ Figure P-11). Your atoms and Earth's atoms were part of the big bang.

Notice that the background radiation is visible in any direction in the sky. The big bang did not occur in a specific place, but it filled the entire volume of the universe. As the universe expanded, the total volume of space increased, and the hot gases of the big bang cooled and formed galaxies. Except for relatively small random motions, galaxies do not move as the universe expands, but are carried away from each other as the volume of space continues to increase. Galaxies are not fragments ejected from an explosion but rather are parts of a whole that fills all of the volume of the universe and expands continuously as the total volume increases.

The big bang theory seems fantastic, but it has been tested and confirmed over and over, and modern astronomers have great confidence that there really was a big bang (**How Do We Know? P-3**). Of course, there are more details to understand. How did the first galaxies form? Why did the galaxies form in giant clusters? You live in an exciting age when astronomers are able to ask these questions and expect to discover answers.

These ideas strain human imagination and challenge the most sophisticated mathematics, but the lesson is a simple one. The universe had a beginning not so long ago, and the matter you are made of had its birthday in that moment of cosmic beginnings. You are small, but you are part of something vast.

## P-6 The Story of Matter

ASTRONOMERS CAN TELL THE STORY of the matter in the universe. Mathematical models of the big bang show that the first few minutes in the history of the universe were unimaginably hot. Energy was so intense that it could form particles of matter, and

## Theories and Proof

**How do astronomers know the sun isn't made of burning coal?** People say dismissively of a theory they dislike, "That's only a theory," as if a theory were just a random guess. A hypothesis is something like a guess, but of course not a random one. What scientists mean by the word *theory* is a hypothesis that has "graduated" to being confidently considered a well-tested truth. You can think of a hypothesis as equivalent to having a suspect in a criminal case and a theory as equivalent to having finished the trial and convicted someone of the crime.

Of course, no matter how many tests and experiments you conduct, you can never prove that any scientific theory is absolutely true. It is always possible that the next observation you make will disprove the theory. And it is unfortunately sometimes true that innocent people go to jail and guilty people are free, but, occasionally, with further evidence, those legal mistakes can be fixed.

There have always been hypotheses about why the sun is hot. Some scientists once thought the sun was a ball of burning coal. Only a century ago, most astronomers accepted the hypothesis that the sun was hot because gravity was making it contract. In the late 19th century, geologists showed that Earth was much older than the sun could be if it was powered by gravity, so the gravity hypothesis had to be wrong. It wasn't until 1920 that a new hypothesis was proposed by Sir Arthur Eddington, who suggested that the sun is powered somehow by the energy in atomic nuclei. In 1938 the German-American astrophysicist Hans Bethe showed how nuclear fusion could power the sun. He won the Nobel Prize in 1967.

The fusion hypothesis is now so completely confirmed that it is fair to call it a theory. No one will ever go to the center of the sun, so you can't *prove* the fusion theory is right. Many observations and model calculations support this theory, and in Chapter 8 you saw further evidence in the neutrinos that have been detected coming from the sun's core. Nevertheless there remains some tiny possibility that all the observations and models are misunderstood and that the theory will be overturned by some future discovery. Astronomers have tremendous confidence that the sun is powered by fusion and not gravity or coal, but a scientific theory can never be proven conclusively correct.

There is a great difference between a theory in the colloquial sense of a far-fetched guess and a scientific theory that has undergone decades of testing and confirmation with observations, experiments, and models. But no theory can ever be proven absolutely true. It is up to you as a consumer of knowledge and a responsible citizen to distinguish between a flimsy guess and a well-tested theory that deserves to be treated like truth—at least pending further information.

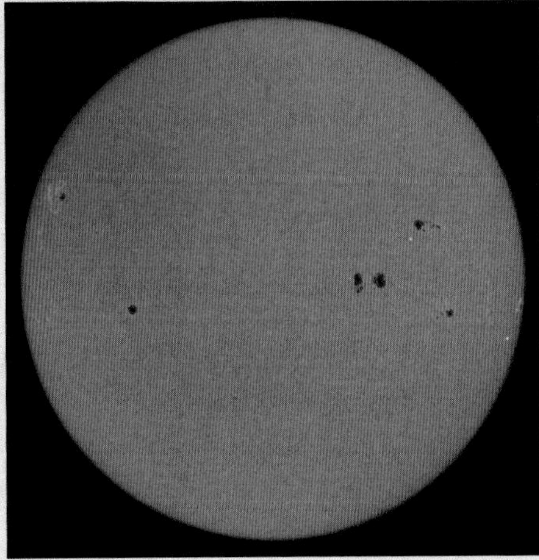

*It is only a theory, but astronomers have tremendous confidence that the sun generates its energy by hydrogen fusion and not by burning coal.* (SOHO/MDI)

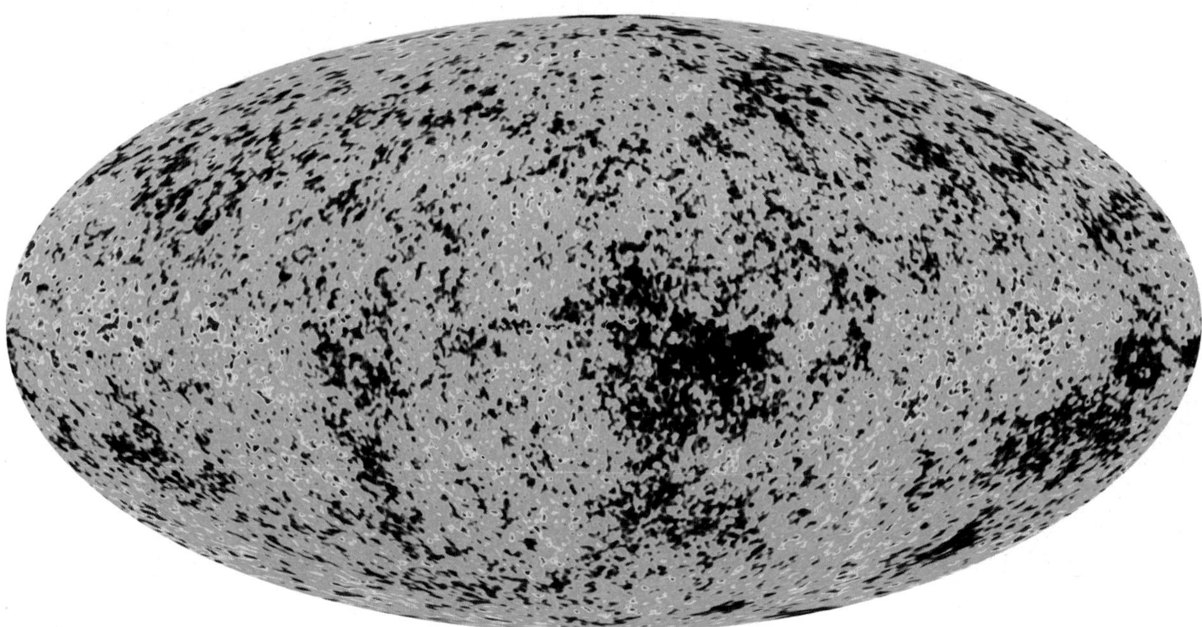

**Figure P-11**

Data from the WMAP spacecraft were used to make this far-infrared map of the sky. This infrared radiation was emitted when the universe was very young and filled with the hot gas of the big bang. The statistical distribution of the tiny irregularities in brightness allows astronomers to determine the age of the universe plus its rate of expansion. (NASA/WMAP Science Team)

that is when the first protons and electrons formed. Protons and electrons, even when they are not attached to each other, are hydrogen, so the first gas of the big bang was hydrogen. The temperature and density were so high that nuclear fusion reactions could occur, fusing some of the hydrogen into helium.

By the time the universe was a few minutes old, expansion had cooled its gases, and nuclear fusion stopped. In those first few minutes, about 25 percent of the mass had been converted into helium. But because there are no stable atomic nuclei with masses of five or eight times that of hydrogen, the fusion process could not go past helium. Few heavier atoms could be made.

As the universe expanded and cooled, it became dark. Gravity drew matter together to form great clouds that contracted to form the first galaxies. Eventually, stars began to form; and, as those first stars began to shine, they lit up the universe and ionized the hydrogen gas in great shells around the galaxies (■ Figure P-12).

Since the formation of the first stars, galaxies have collided and merged; generation after generation of stars have lived and died. Mass loss from aging stars, planetary nebulae, and supernova explosions has spread heavy elements out into the thin gas in space where they have become incorporated into new stars in a continuous gas-star-gas cycle (■ Figure P-13). Except for hydrogen, all of the atoms of which you are made were cooked up inside stars before the sun and Earth were born (■ Figure P-14). When the sun began to form from a cloud of gas, the atoms now inside your body were there. They were part of Earth as it formed, and now they are part of you. Every atom has a history that stretches back through the birth and death of stars to the first moment in time. You exist because stars have died.

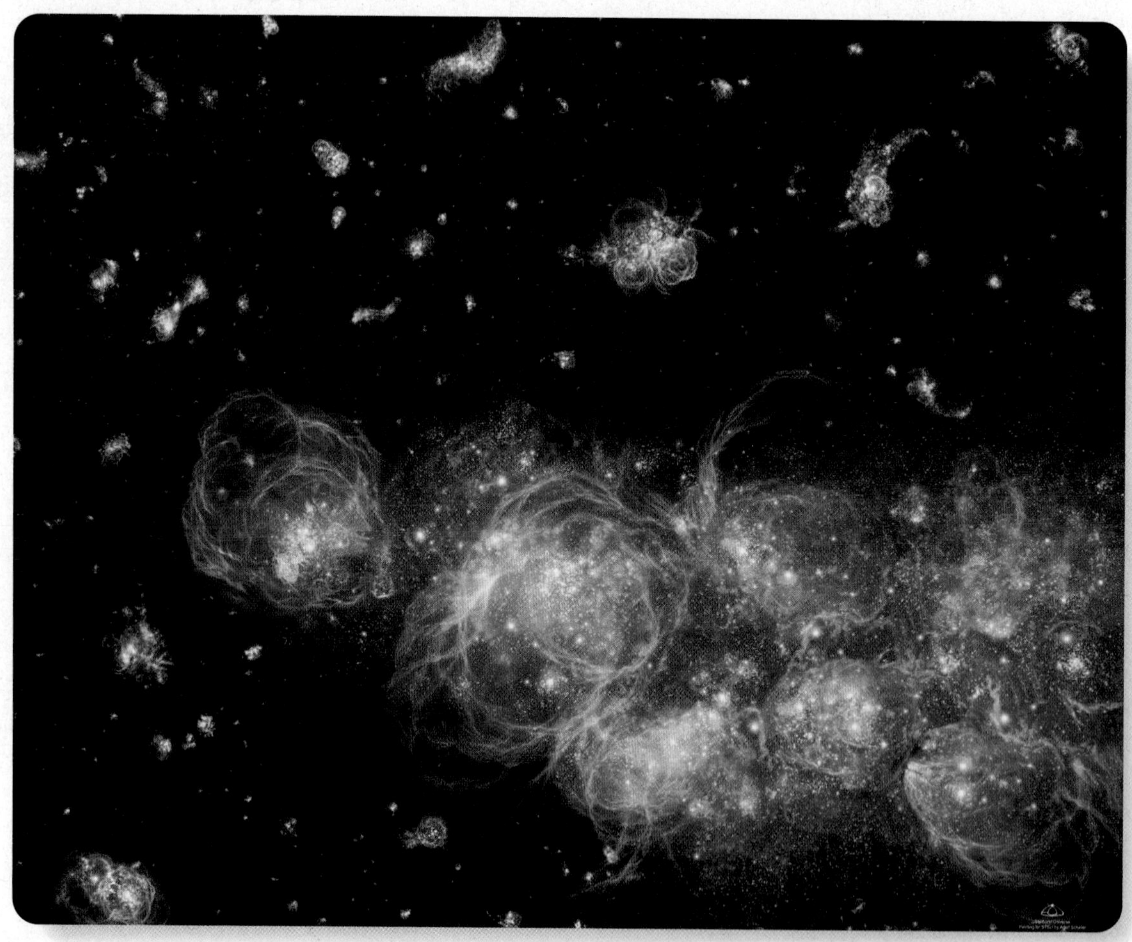

■ **Figure P-12**

In this artist's conception, the first stars produce floods of ultraviolet photons that ionize the gas in expanding bubbles. Such a storm of star formation ended an age when the universe had expanded in darkness. As star formation began in the universe, the production of the chemical elements heavier than helium also began. (K. Lanzetta, SUNY, A. Schaller for STScI, and NASA)

Aging stars exhale gas and dust into the interstellar medium.
Visual image
Stars forming from clouds of interstellar gas
Visual image
Gas expanding away from a supernova explosion
X-ray image
**The Gas-Stars-Gas Cycle**
Gas and dust in the interstellar medium.
Visual image

**■ Figure P-13**

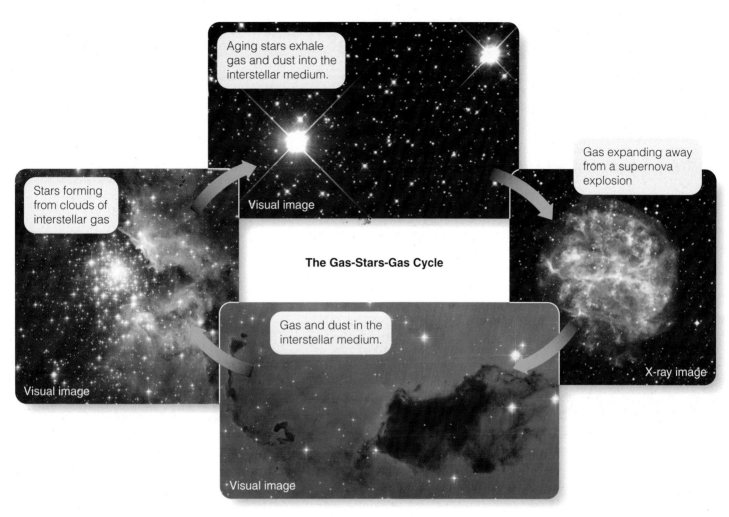

Matter cycles from the interstellar medium to form stars and back into the interstellar medium. Our galaxy is slowly converting hydrogen into elements heavier than helium and enriching the interstellar medium.

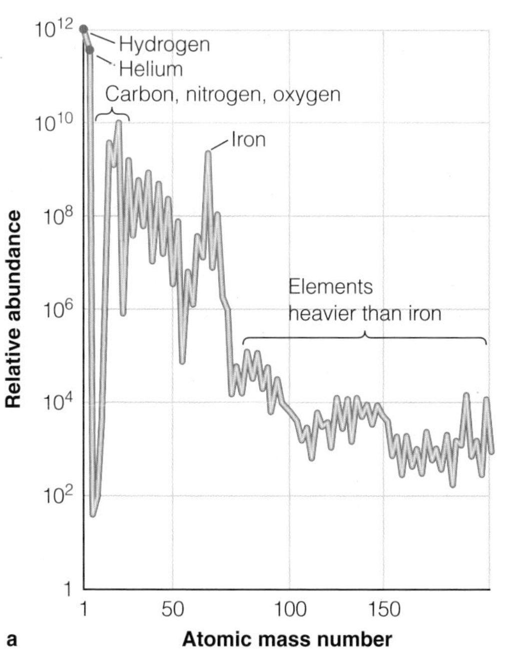


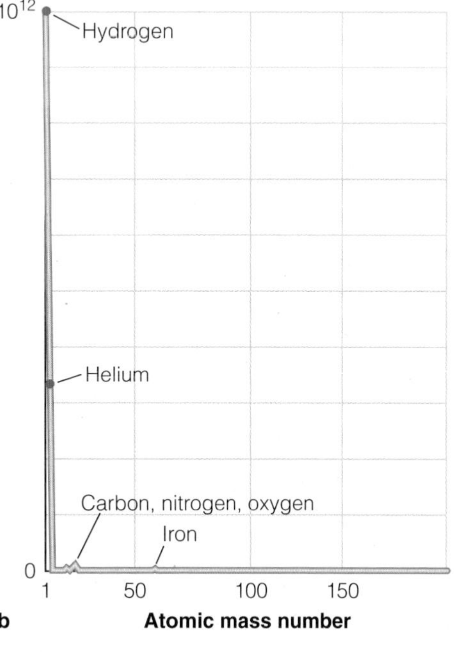


**■ Figure P-14**

The abundance of the chemical elements is usually plotted as in the graph in part (a). Hydrogen is most common, followed by helium. Carbon, nitrogen, oxygen, and iron also seem abundant, but heavier elements are quite rare because they are made only in supernova explosions. Plotting these abundances on a linear scale as in the graph in part (b) provides a more realistic impression. Most of the universe is hydrogen and helium. The atoms of which you are made are hardly more than a tiny impurity in the cosmos.

## What Are We? Children of the Milky Way

Hang on tight. The sun, with Earth in its clutch, is ripping along at about 220 km/sec (that's 490,000 mph) as it orbits the center of the Milky Way Galaxy. We live on a wildly moving ball of rock in a large galaxy that some people call our home galaxy, but the Milky Way is more than just our home. Perhaps "parent galaxy" would be a better name.

Except for hydrogen atoms, which have survived unchanged since the universe began, you and Earth are made of atoms heavier than helium. There is no helium in your body, but there are plenty of atoms such as calcium and iron. There is calcium in your bones and iron in your blood. All of those atoms and more were cooked up inside stars or in their supernova deaths.

Stars are born when clouds of gas are compressed and gravity pulls the gas together. That process has given birth to generations of stars, and each generation has produced elements heavier than helium and spread them back into space. The abundance of these atoms has grown slowly in the galaxy. About 4.6 billion years ago a cloud of gas enriched in those heavy atoms contracted and produced the sun, Earth, and you. Your atoms have been cooked up by the Milky Way Galaxy—your parent galaxy.

# Study and Review

## Summary

- The **interstellar medium (p. 168),** the gas and dust between the stars, can be seen as **nebulae (p. 168).** Small, dense, dark clouds are called **Bok globules (p. 170).** Stars are born when clouds of gas and dust contract to form **protostars (p. 169).**
- Protostars are cool and hidden inside clouds of gas and dust, so they are most easily observed in the infrared.
- Planets form in the disks of gas and dust around protostars.
- Stars that generate energy by hydrogen fusion, including the sun, are called **main-sequence stars (p. 169).**
- The lowest-mass stars, the **red dwarfs, (p. 172)** can survive for many billions of years.
- The sun is a medium-mass star and can live for about 10 billion years before it expands to become a **giant (p. 172),** expels its outer layers to form a **planetary nebula (p. 172),** and collapses to form a **white dwarf (p. 172).**
- The most massive stars live only a few million years and expand to become **supergiants (p. 173).** Giant and supergiant stars exhale gas back into the interstellar medium. The most massive stars die in explosions called **supernovae (p. 173)** that can form elements heavier than iron and blow them plus atoms of lighter elements made earlier in the star's life back into the interstellar medium. Supernova explosions can leave behind **neutron stars (p. 173)** or **black holes (p. 173)** as stellar remnants.
- Stars cook hydrogen atoms into atoms of heavier elements and return those atoms to the interstellar medium when the star dies.
- The **Milky Way (p. 176),** the hazy band of light across the night sky, is our galaxy, the **Milky Way Galaxy (p. 176)**, seen from the inside. It contains about 100 billion stars and is about 80,000 ly in diameter. We live about two-thirds of the way from the center to the edge.
- Billions of galaxies fill the sky, and some are **elliptical (p. 178),** some **spiral (p. 178),** and some **irregular (p. 179).** Some spiral galaxies, including the Milky Way Galaxy, contain a bar-shaped nucleus and are called **barred-spiral galaxies (p. 178).**
- Collisions between galaxies are common and can trigger star formation.
- The redshifts of the galaxies show that the universe is expanding, and imagining running the expansion backward in time leads to inference of the high-temperature, high-density beginning of the universe called the **big bang (p. 181).**
- The **cosmic microwave background radiation (p. 182)** is light released from matter soon after the big bang that has been redshifted by the expansion of the universe into the infrared and radio parts of the spectrum. It is direct evidence that the big bang really happened. Because Earth is part of the big bang, the radiation is detectable in every direction in the sky.
- The energy of the big bang produced protons and electrons (hydrogen gas), and some of that hydrogen fused into helium during the first few minutes, but no heavier elements could form.
- Gravity drew the gas into great clouds that formed galaxies, and stars fused hydrogen into heavier elements. Matter passes through a gas–star–gas cycle and is enriched in elements heavier than helium. A number of generations of stars have produced the chemical elements in Earth and in you.

## Review Questions

1. What evidence could you cite that there is an interstellar medium?
2. Why are protostars difficult to observe at visible wavelengths?
3. How are protostars related to planet formation?
4. How do stars like the sun die?
5. How do massive stars die?
6. What is the difference between the Milky Way and the Milky Way Galaxy?
7. Describe how star formation is related to gas and dust in galaxies such as ellipticals and spirals.
8. What evidence can you site to show that the universe is expanding? That there really was a big bang?
9. Why would it be accurate to say that the big bang formed the matter in your body, but the stars formed the atoms?
10. **How Do We Know?** How can scientists use mathematical models to understand processed that they cannot observe directly?
11. **How Do We Know?** Why do scientists hesitate to say they "believe" in a certain theory?
12. **How Do We Know?** How would you respond to someone who said, "Oh, that's only a theory"?

# 19 The Origin of the Solar System

*The human race lives on a planet in a planetary system that appears to have formed in a nebula around the protostar that became the sun. This artist's impression shows such a nebular disk around a forming star.* (NASA/JPL-Caltech)

## Guidepost

You have studied the appearance, origin, structure, and evolution of stars, galaxies, and the universe itself. So far, though, your studies have left out one important type of object—planets. Now it is time to correct that omission. In this chapter, once you learn the evidence for how the solar system formed, you can understand how the processes you have been studying produced Earth, your home planet.

As you explore our solar system in space and time, you will find answers to four essential questions:

- **What is the theory for the origin of the solar system?**
- **What are the observed properties of the solar system that the theory of its origin can explain?**
- **How do planets form?**
- **What do astronomers know about other planetary systems?**

In the next three chapters, you will explore in more detail each of the planets, plus asteroids, comets, and meteoroids. By studying the origin of the solar system before studying the individual objects in it, you give yourself a better framework for understanding these fascinating worlds.

*What place is this?*

*Where are we now?*

— CARL SANDBURG, "GRASS"

MICROSCOPIC CREATURES LIVE in the roots of your eyelashes. Don't worry. Everyone has them, and they are harmless.* They hatch, fight for survival, mate, lay eggs, and die in the tiny spaces around the roots of your eyelashes without doing any harm. Some live in renowned places—the eyelashes of a glamorous movie star, for example—but the tiny beasts are not self-aware; they never stop to say, "Where are we?"

You can study the solar system for many reasons. You can study Earth and its sibling planets because, as you are about to discover, there are almost certainly more planets in the universe than stars. Above all, you can study the solar system because it is your home in the universe. Because humans are an intelligent species, we have the ability and the responsibility to wonder where we are and what we are. Our kind has inhabited this solar system for several million years, but only within the last few hundred years have we begun to understand what the solar system is.

## 19-1 The Great Chain of Origins

YOU ARE LINKED through a great chain of origins that leads backward through time to the instant when the universe began, 13.7 billion years ago. The gradual discovery of the links in that chain is one of the most exciting adventures of the human intellect. In earlier chapters, you studied some of that story: the origin of the universe in the big bang, the formation of galaxies, the origin of stars, and the growth of the chemical elements. Now you will explore further to consider the origin of planets.

### The History of the Atoms in Your Body

The universe began in the big bang (see Chapter 18). By the time the universe was a few minutes old, the protons, neutrons, and electrons in your body had come into existence. You are made of very old matter.

Although those particles formed quickly, they were not linked together to form many of the atoms that are common today. Most of the matter in the early universe was hydrogen, and about 25 percent was helium. Although your body does not contain helium, it does contain many of those ancient hydrogen atoms unchanged since the universe began. Evidence indicates that almost no atoms heavier than helium were made in the big bang.

Within a few hundred million years after the big bang, matter began to collect to form galaxies containing billions of stars. You have learned that nuclear reactions inside stars are where low-mass atoms such as hydrogen are combined to make heavier atoms (see Chapters 12 and 13). Generation after generation of stars cooked the original particles, fusing them into atoms such as carbon, nitrogen, and oxygen that are common in your body. Even the calcium atoms in your bones were assembled inside stars.

Massive stars produce iron in their cores, but much of that iron is destroyed when the core collapses and the star explodes as a supernova. Most of the iron on Earth and in your body was produced instead by carbon fusion in type Ia supernova explosions and by the decay of radioactive atoms in the expanding matter ejected by type II supernovae. Atoms heavier than iron such as gold, silver, and iodine are created by rapid nuclear reactions that can occur only during supernova explosions. Iodine is critical to the function of your thyroid gland, and you probably have gold and silver jewelry or dental fillings. Realize that these types of atoms, which are part of your life on Earth, were made during the violent deaths of massive stars long ago.

Our galaxy contains at least 100 billion stars, of which the sun is one. Astronomers have a variety of evidence that the sun formed from a cloud of gas and dust almost 5 billion years ago, and the atoms in your body were part of that cloud. How the sun took shape, how the cloud gave birth to the planets, how the atoms in your body found their way onto Earth and into you, make up the story of this chapter. As you explore the origin of the solar system, keep in mind the great chain of origins that created the atoms. As the geologist Preston Cloud remarked, "Stars have died that we might live."

### Early Hypotheses for the Origin of Earth and the Solar System

The earliest descriptions of Earth's origin are myths and folktales that go back beyond the beginning of recorded history. In the time of Galileo, the telescope gave philosophers observational evidence on which to base rational explanations for natural phenomena. While people like Copernicus, Kepler, and Galileo tried to find logical explanations for the motions of Earth and the other planets, other philosophers began thinking about the origin of Earth and its sibling planets.

The first physical theory for the solar system's origin was proposed by the French philosopher and mathematician René Descartes (1596–1650). Because he lived and wrote before the time of Newton, Descartes did not recognize that gravitation is the dominant force in the universe. Rather, he believed that force was communicated by contact between bodies and that the entire

***Demodex folliculorum* has been found in 97 percent of individuals and is a characteristic of healthy skin.

universe was filled with vortices of whirling invisible particles. In 1644, he proposed that the sun and planets formed when a large vortex contracted and condensed. His hypothesis explained the general properties of the solar system known at the time.

A century later, in 1745, the French naturalist Georges-Louis de Buffon (1707–1788) proposed an alternative hypothesis that the planets were formed when a passing comet collided with or passed close to the sun and pulled matter out of the sun. He did not know that the solid parts of comets are small, insubstantial bodies, but later astronomers modified his hypothesis to propose that a star, rather than a comet, interacted with the sun. According to the modified hypothesis, matter ripped from the sun and the other star condensed to form the planets, which were driven into orbit around the sun by the motion of the two stars' collision. (■ Figure 19-1a). This **passing star hypothesis** was popular off and on for two centuries, but it contains serious flaws. First, stars are very small compared to the distances between them, so they collide very infrequently. Only a tiny fraction of stars in our galaxy have ever suffered a collision or close encounter with another star. More important, the gas pulled from the sun and the star would be much too hot to condense to make planets, and it would have dispersed instead. Furthermore, even if planets did form, they would not go into stable orbits around the sun.

The hypotheses of Descartes and Buffon fall into two broad categories. Descartes proposed an **evolutionary hypothesis** involving common, gradual processes to produce the sun and planets. If it were correct, stars with planets would be very common. Buffon's idea, on the other hand, is a **catastrophic hypothesis.** It involves unlikely, sudden events to produce the solar system, and thus implies that planetary systems are very rare. While your imagination may enjoy picturing the spectacle of colliding stars, modern scientists have observed that nature usually changes gradually in small steps rather than in sudden, dramatic events. The modern theory for the origin of the planets, based on abundant evidence, is evolutionary rather than catastrophic (**How Do We Know? 19-1**).

The modern theory of the origin of the solar system had its true beginning with Pierre-Simon de Laplace (1749–1827), a brilliant French astronomer and mathematician. In 1796, he combined Descartes's vortex idea with Newton's laws of gravity and motion to produce a model of a rotating cloud of matter that contracted under its own gravitation and flattened into a disk—the **nebular hypothesis.** As the disk grew smaller, it had to conserve angular momentum and spin faster and faster. Laplace reasoned that, when it was spinning as fast as it could, the disk would shed its outer edge to leave behind a ring of matter. Then the disk could contract further, speed up again, and leave another ring. In this way, he imagined, the contracting disk would leave behind a series of rings, each of

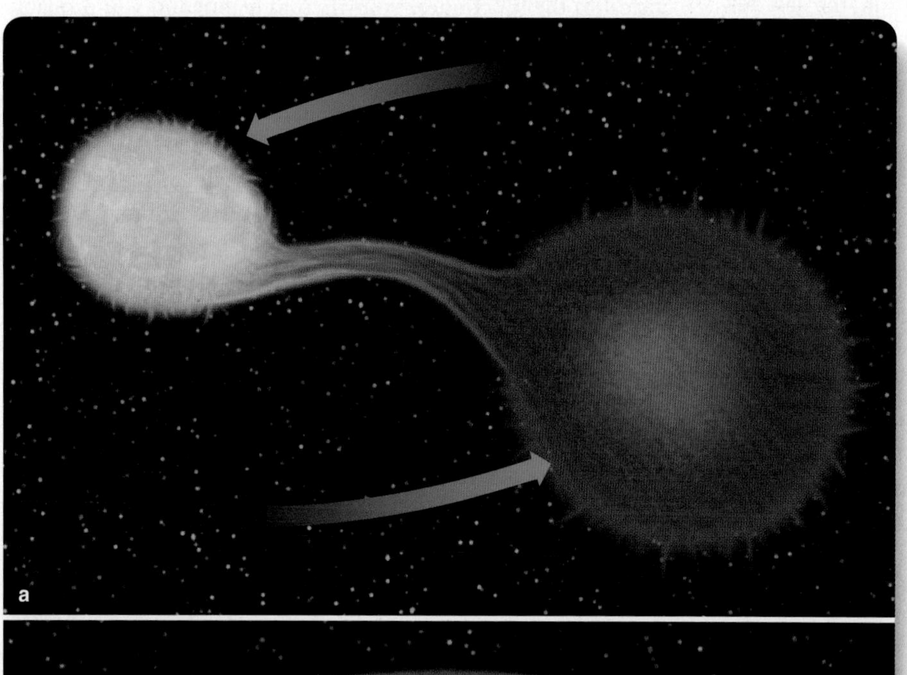

■ **Figure 19-1**

(a) The passing star hypothesis proposed that the sun was hit by, or had a very close encounter with, another star. Matter torn from the sun and the other star formed planets orbiting the sun and, perhaps, the other star. This is an example of a catastrophic hypothesis. (b) Originally proposed in the 18th century by Laplace, the nebular hypothesis proposed that a contracting disk of matter around the sun conserved angular momentum, spun faster, and shed rings of matter that then formed planets. This is an example of an evolutionary hypothesis.

which could become a planet circling the newborn sun at the center of the disk (Figure 19-1b).

According to the nebular hypothesis, the sun should be spinning very rapidly, or, to put it another way, the sun should have most of the angular momentum of the solar system. (Recall from Chapter 5 that angular momentum is the tendency of a rotating object to continue rotating.) As astronomers studied the planets and the sun, however, they found that the sun rotates relatively slowly and that the planets moving in their orbits actually have most of the angular momentum in the solar system. In fact, the rotation of the sun contains only about 0.3 percent of the angular momentum of the solar system. Because the nebular hypothesis could not explain this **angular momentum problem,** it was never fully successful, and astronomers instead considered various versions of the passing star hypothesis for over a century. Today astronomers have a consistent theory for the origin of our solar system that is a modified version of the nebular hypothesis.

## The Solar Nebula Theory

By 1940, astronomers were beginning to understand how stars form and how they generate their energy, and it became clear that the origin of the solar system was linked to that story.

The modern explanation of the origin of the planets is the **solar nebula theory,** which proposes that the planets, including Earth, formed in a rotating disk of gas and dust that surrounded the sun as it formed (■ Figure 19-2). Laplace's nebular hypothesis included a disk, but his process depended on rings of matter left behind as the disk contracted. Also, it was not based on a clear understanding of how gas and dust behave in such a disk. In the solar nebula theory, the planets grew within the disk by carefully described physical processes. This evolutionary hypothesis is so comprehensive and explains so many of the observations, both of our solar system and of other systems, that it can be considered to have "graduated" from being just a hypothesis to being properly called a theory. Astronomers are continuing to refine the details of that theory.

You have seen clear evidence that disks of gas and dust are common around young stars. Bipolar flows from protostars (see Chapter 11) were some of the first clues to the existence of such disks, but modern techniques can image the disks directly (■ Figure 19-3). The evidence is strong that our own planetary system formed in such a disk-shaped cloud around the sun. When the sun became luminous enough, the remaining gas and dust were blown away into space, leaving the planets orbiting the sun.

According to the solar nebula theory, our Earth and the other planets of the solar system formed billions of years ago as the sun condensed from a cloud of gas and dust. If planet formation is a natural part of star formation, most stars should have planets.

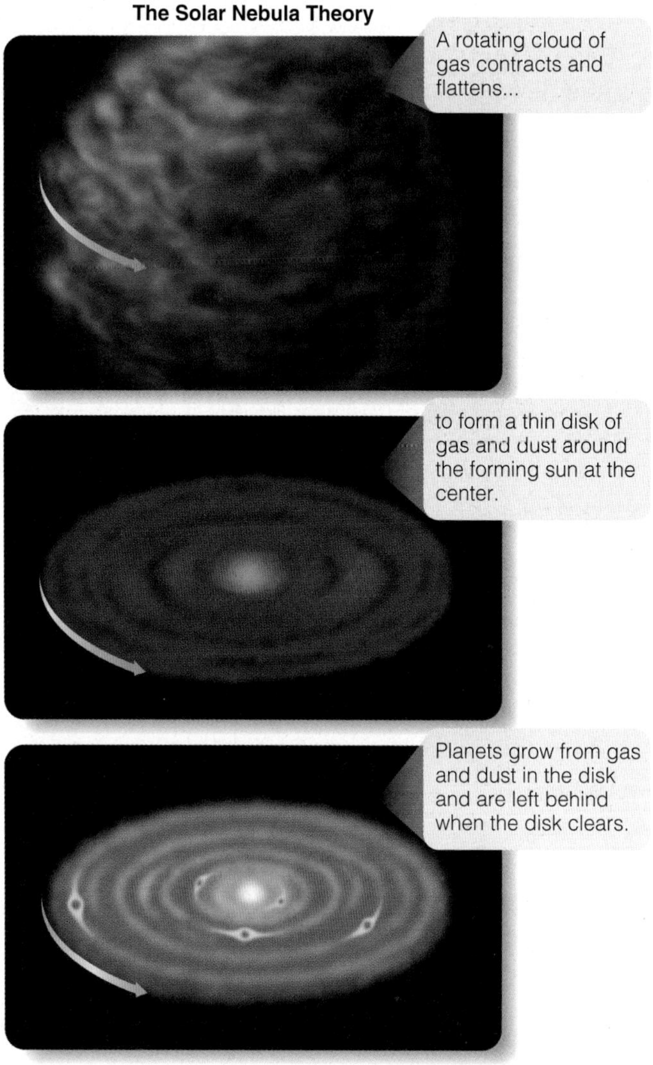

The Solar Nebula Theory

A rotating cloud of gas contracts and flattens...

to form a thin disk of gas and dust around the forming sun at the center.

Planets grow from gas and dust in the disk and are left behind when the disk clears.

**■ Figure 19-2**

The solar nebula theory implies that the planets formed along with the sun. The scale of the top panel is much larger than the lower two panels.

### SCIENTIFIC ARGUMENT

***Why does the solar nebula theory imply planets are common?***
Often, the implications of a theory are more important in building a scientific argument than the theory itself. The solar nebula theory is an evolutionary theory; and, if it is correct, the planets of our solar system formed from the disk of gas and dust that surrounded the sun as it condensed from the interstellar medium. That suggests it is a common process. Most stars form with disks of gas and dust around them, and planets should form in such disks. Planets should therefore be very common in the universe.

Now build a new scientific argument. **Why would a catastrophic hypothesis for the formation of the solar system suggest that planets are not common?**

## How Do We Know? 19-1

### Two Kinds of Theories: Catastrophic and Evolutionary

**How big a role have sudden, catastrophic events played in the history of the solar system?** Many theories in science can be classified as either evolutionary, in that they involve gradual processes, or catastrophic, in that they depend on sudden, unlikely events. Scientists have generally preferred evolutionary theories. Nevertheless, catastrophic events do occur.

Some people prefer catastrophic theories, perhaps because they like to see spectacular violence from a safe distance, which may explain the success of movies that include lots of car crashes and explosions. Also, catastrophic theories resonate with scriptural accounts of cataclysmic events and special acts of creation. Thus, many people have an interest in catastrophic theories.

Most scientific theories are evolutionary. Such theories do not depend on unlikely events or special acts. For example, geologists study theories of mountain building that are evolutionary and describe mountains being pushed up slowly as millions of years pass. The evidence of erosion and the folded rock layers show that the process is gradual. Because most such natural processes are evolutionary, scientists sometimes find it difficult to accept any theory that depends on catastrophic events.

You will see in this and later chapters that catastrophes do occur. For example, the planets are bombarded by debris from space, and some of those impacts are very large. As you study astronomy or any other natural science, notice that most theories are evolutionary but that you need to allow for the possibility of unpredictable catastrophic events.

*Mountains evolve to great heights by rising slowly, not catastrophically.* (Janet Seeds)

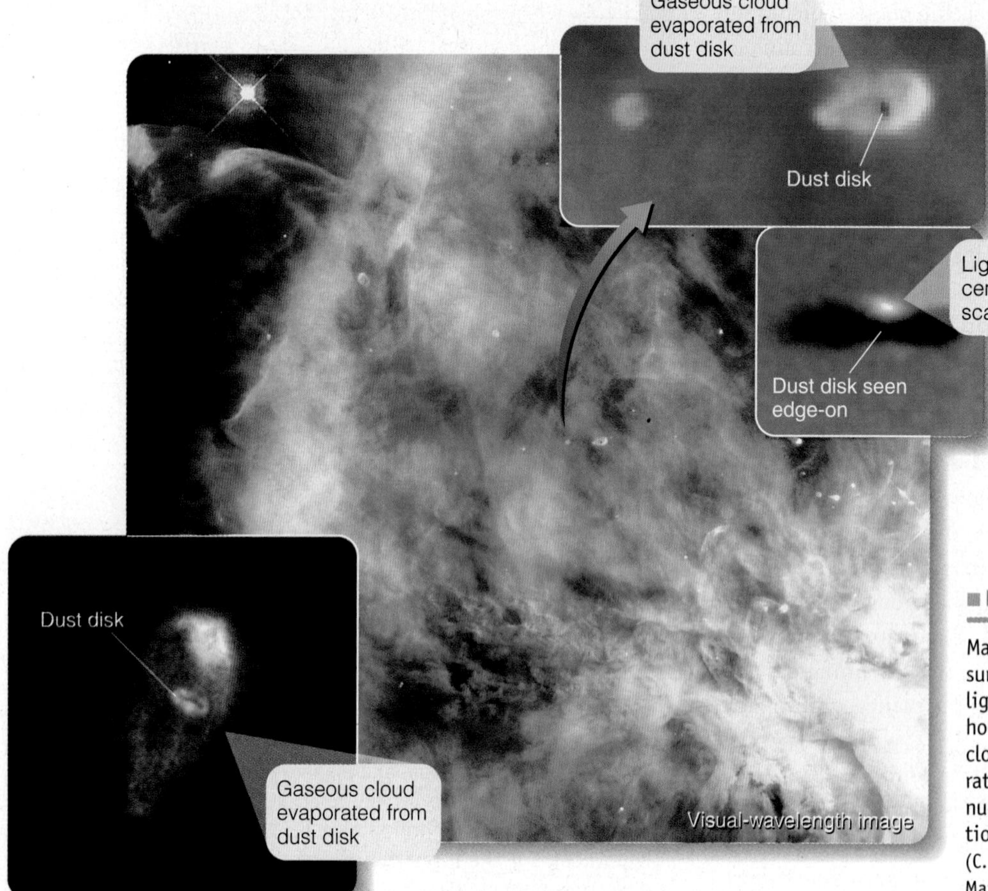


**Figure 19-3**

Many of the young stars in the Orion Nebula are surrounded by disks of gas and dust, but intense light from the brightest star in the neighborhood is evaporating the disks to form expanding clouds of gas. These particular disks may evaporate before they can form planets, but the large number of such disks shows that planet construction material around young stars is common. (C. R. O'Dell, Rice, NASA; Dark Disk: M. McCaughrean, Max-Planck Inst. für Astronomie, C. R. O'Dell, NASA; Lower left inset: J. Bally, H. Throop, C. R. O'Dell, NASA)

## 19-2 A Survey of the Solar System

To TEST THEIR HYPOTHESES about how the solar system was born, astronomers searched the present solar system for evidence of its past. In this section, you will survey the solar system and compile a list of its most significant characteristics that are potential clues to how it formed.

You can begin with the most general view of the solar system. It is, in fact, almost entirely empty space (look back to Figure 1-7). Imagine reducing the scale of the solar system until Earth is the size of a grain of table salt, about 0.3 mm (0.01 in.) in diameter. The moon is a speck of pepper about 1 cm (0.4 in.) away, and the sun is the size of a small plum located 4 m (13 ft) from Earth. Jupiter is an apple seed 20 m (66 ft) from the sun. Neptune, at the edge of the planetary zone, is a large grain of sand over 120 m (400 ft) from the central plum. Although your model solar system would be larger than a football field, you would need a powerful microscope to detect the asteroids orbiting between Mars and Jupiter. The planets are tiny specks of matter scattered around the sun—the last remains of the solar nebula.

### Revolution and Rotation

The planets revolve* around the sun in orbits that lie close to a common plane. The orbit of Mercury, the closest planet to the sun, is tipped 7.0° to Earth's orbit The rest of the planets' orbital planes are inclined by no more than 3.4°. As you can see, the solar system is basically flat and disk shaped.

The rotation of the sun and planets on their axes also seems related to the rotation of the disk. The sun rotates with its equator inclined only 7.2° to Earth's orbit, and most of the other planets' equators are tipped less than 30°. The rotations of Venus and Uranus are peculiar, however. Venus rotates backward compared with the other planets, whereas Uranus rotates on its side with its equator almost perpendicular to its orbit. You will explore these planets in detail in Chapters 21 and 23, but later in this chapter you will be able to understand how they might have acquired their peculiar rotations.

There is a preferred direction of motion in the solar system—counterclockwise as seen from the north, like the curl of the fingers of your right hand if you point your thumb toward your eyes. All the planets revolve counterclockwise around the sun; and with the exception of Venus and Uranus they rotate counterclockwise on their axes. Furthermore, nearly all of the moons in the solar system, including Earth's moon, orbit around their respective planets counterclockwise. With only a few exceptions, most of which are understood, revolution and rotation in the solar system follow a single theme. Apparently, these motions today are related to the original rotation of a disk of solar system construction material.

*Recall from Chapter 2 that the words *revolve* and *rotate* refer to different types of motion. A planet revolves around the sun but rotates on its axis. Cowboys in the Old West didn't carry revolvers. They actually carried rotators. And you don't rotate your tires every 6 months, you actually revolve them.

### Two Kinds of Planets

Perhaps the most striking clue to the origin of the solar system comes from the obvious division of the planets into two groups, the small Earth-like worlds and the giant Jupiter-like worlds. The difference is so dramatic that you are led to say, "Aha, this must mean something!" Study **Terrestrial and Jovian Planets** on pages 402–403, notice three important points, and learn two new terms:

1. The two kinds of planets are distinguished by their location. The four inner *Terrestrial planets* are quite different from the four outer *Jovian planets.*
2. Craters are common. Almost every solid surface in the solar system is covered with craters.
3. The two groups of planets are also distinguished by properties such as presence or absence of rings and numbers of moons. A theory of the origin of the planets needs to explain these properties.

The division of the planets into two groups is a clue to how our solar system formed. The present properties of individual planets, however, don't tell everything you need to know about their origins. The planets have all evolved since they formed. For further clues about the origin of the planets, you can look at smaller objects that have remained largely unchanged since soon after the birth of the solar system.

### Space Debris

The sun and planets are not the only products of the solar nebula. The solar system is littered with three kinds of space debris: asteroids, comets, and meteoroids. Although these objects represent a tiny fraction of the mass of the system, they are a rich source of information about the origin of the planets.

The **asteroids,** sometimes called minor planets, are small rocky worlds, most of which orbit the sun in a belt between the orbits of Mars and Jupiter. More than 100,000 asteroids have orbits that are charted, of which about 2000 follow paths that bring them into the inner solar system where they can potentially collide with a planet. Earth has been struck many times in its history. It is a **Common Misconception** that the asteroids are the remains of a planet that broke apart. In fact, planets are held together very tightly by their gravity and do not "break apart." Astronomers recognize the asteroids as debris left over from the failure of a planet to form at a distance of about 3 AU from the sun.

About 200 asteroids are more than 100 km (60 mi) in diameter, and tens of thousands are estimated to be more than 10 km

# Terrestrial and Jovian Planets

**1** The distinction between the Terrestrial planets and the Jovian planets is dramatic. The inner four planets, Mercury, Venus, Earth, and Mars, are **Terrestrial planets,** meaning they are small, dense, rocky worlds with little or no atmosphere. The outer four planets, Jupiter, Saturn, Uranus, and Neptune, are **Jovian planets,** meaning they are large, low-density worlds with thick atmospheres and liquid or ice interiors.

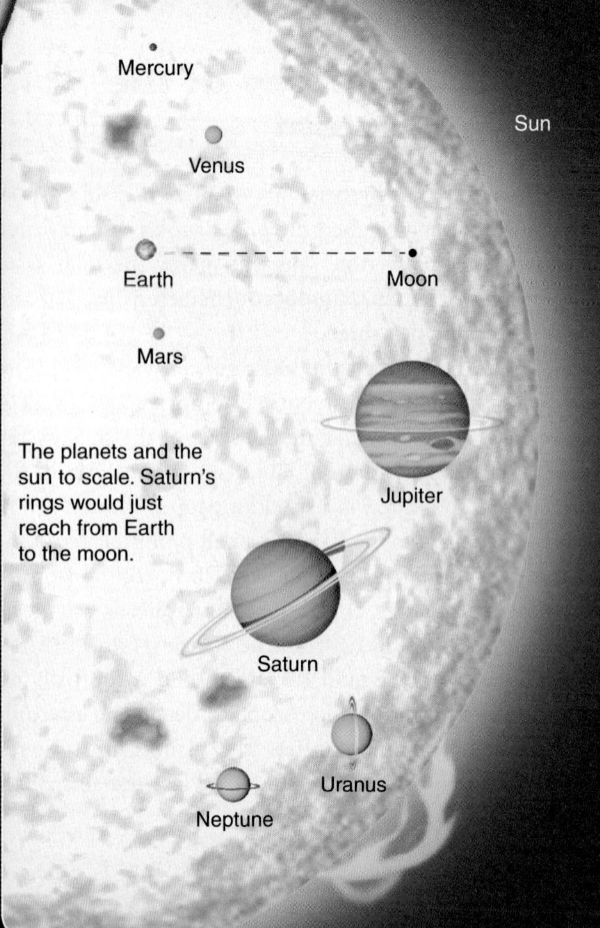


The planets and the sun to scale. Saturn's rings would just reach from Earth to the moon.

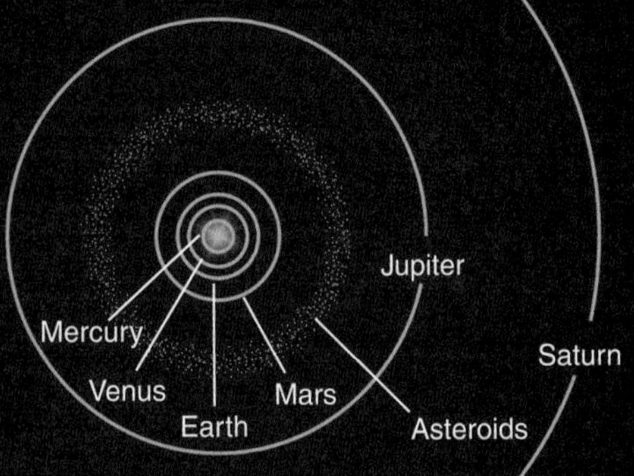


Planetary orbits to scale. The Terrestrial planets lie quite close to the sun, whereas the Jovian planets are spread far from the sun.

**1a** Of the Terrestrial planets, Earth is most massive, but the Jovian planets are much more massive. Jupiter is over 300 Earth masses, and Saturn is nearly 100 Earth masses. Uranus and Neptune are 15 and 17 Earth masses.

Mercury is only 40 percent larger than Earth's moon, and its weak gravity cannot retain a permanent atmosphere. Like the moon, it is covered with craters from meteorite impacts.

Mercury

Earth's moon

NASA

© UC Regents/Lick Observatory

**2** Craters are common on all of the surfaces in the solar system that are strong enough to retain them. Earth has about 150 impact craters, but many more have been erased by erosion. Besides the planets, the asteroids and nearly all of the moons in the solar system are scarred by craters. Ranging from microscopic to hundreds of kilometers in diameter, these craters have been produced over the ages by meteorite impacts. When astronomers see a rocky or icy surface that contains few craters, they know that the surface is young.

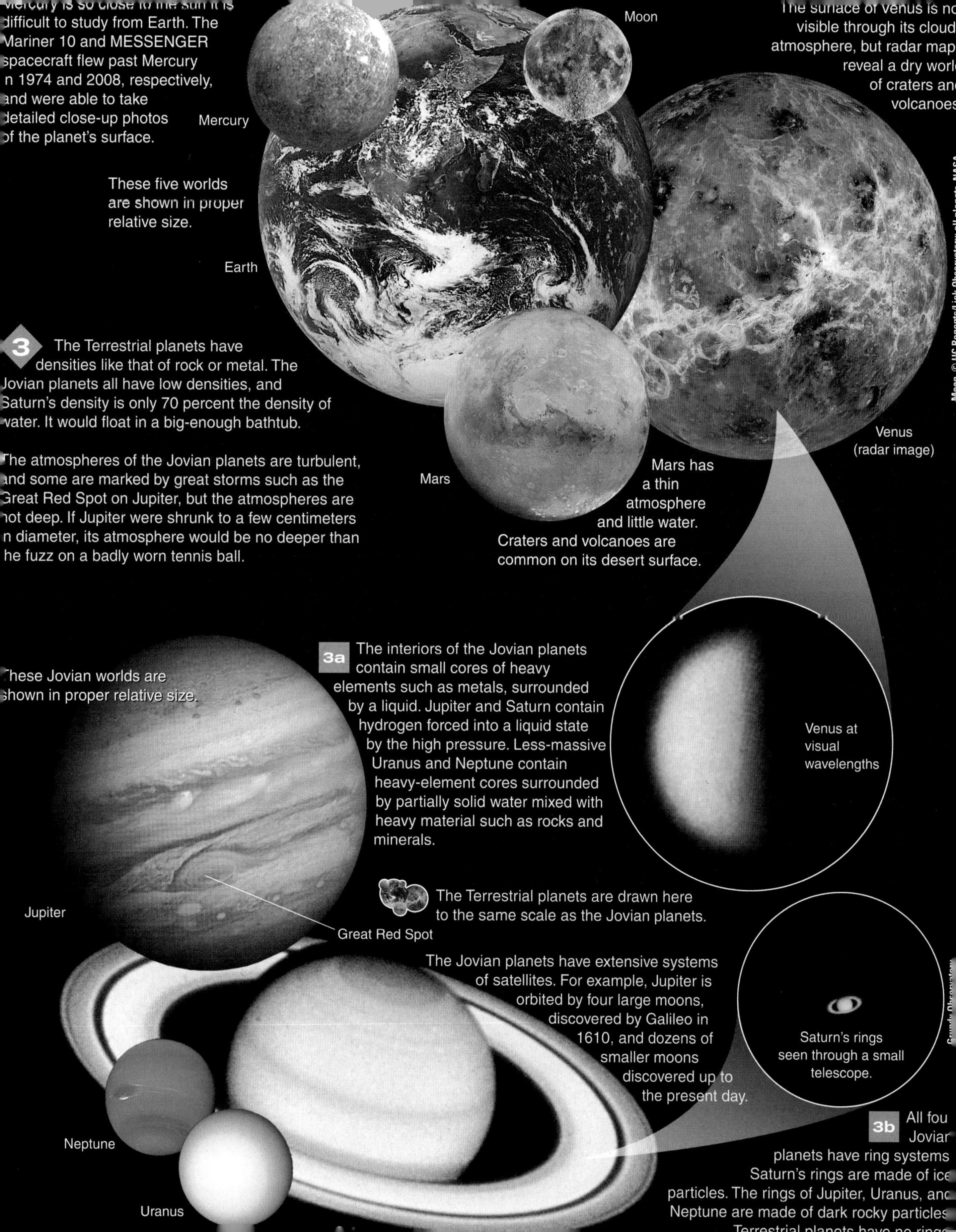
Mercury is so close to the sun it is difficult to study from Earth. The Mariner 10 and MESSENGER spacecraft flew past Mercury in 1974 and 2008, respectively, and were able to take detailed close-up photos of the planet's surface.
Moon
The surface of Venus is not visible through its cloudy atmosphere, but radar maps reveal a dry world of craters and volcanoes.
Mercury
These five worlds are shown in proper relative size.
Earth
3 The Terrestrial planets have densities like that of rock or metal. The Jovian planets all have low densities, and Saturn's density is only 70 percent the density of water. It would float in a big-enough bathtub.
The atmospheres of the Jovian planets are turbulent, and some are marked by great storms such as the Great Red Spot on Jupiter, but the atmospheres are not deep. If Jupiter were shrunk to a few centimeters in diameter, its atmosphere would be no deeper than the fuzz on a badly worn tennis ball.
Venus (radar image)
Mars
Mars has a thin atmosphere and little water. Craters and volcanoes are common on its desert surface.
3a The interiors of the Jovian planets contain small cores of heavy elements such as metals, surrounded by a liquid. Jupiter and Saturn contain hydrogen forced into a liquid state by the high pressure. Less-massive Uranus and Neptune contain heavy-element cores surrounded by partially solid water mixed with heavy material such as rocks and minerals.
These Jovian worlds are shown in proper relative size.
Venus at visual wavelengths
Jupiter
The Terrestrial planets are drawn here to the same scale as the Jovian planets.
Great Red Spot
The Jovian planets have extensive systems of satellites. For example, Jupiter is orbited by four large moons, discovered by Galileo in 1610, and dozens of smaller moons discovered up to the present day.
Saturn's rings seen through a small telescope.
Neptune
3b All four Jovian planets have ring systems. Saturn's rings are made of ice particles. The rings of Jupiter, Uranus, and Neptune are made of dark rocky particles. Terrestrial planets have no rings.
Uranus

(6 mi) in diameter. There are probably a million or more that are larger than 1 km (0.6 mi) and billions that are smaller. Because the largest are only a few hundred kilometers in size, Earth-based telescopes can detect no details on their surfaces, and even the Hubble Space Telescope can image only the largest features.

Spectroscopic observations indicate that asteroid surfaces are a variety of rocky and metallic materials. Photographs returned by robotic spacecraft show that asteroids are generally irregular in shape and covered with craters (■ Figure 19-4). These observations will be discussed in detail in Chapter 25, but in this quick survey of the solar system you have enough information to conclude that the solar nebula included elements that compose rock and metals and that collisions have played an important role in the solar system's history.

Since 1992, astronomers have discovered more than a thousand small, dark, icy bodies orbiting in the outer fringes of the solar system beyond Neptune. This collection of objects is called the **Kuiper belt** after astronomer Gerard Kuiper (KYE-per), who predicted their existence in the 1950s. There are probably 100 million bodies larger than 1 km in the Kuiper belt, many more than in the asteroid belt, and a successful theory should also explain how they came to be where they are. You will find out more about the origin of the Kuiper belt in Chapter 24.

In contrast to the rocky asteroids and dark Kuiper belt objects, the brightest **comets** can be seen with the naked eye and are impressively beautiful objects (■ Figure 19-5). Most comets are faint, however, and are difficult to locate even at their brightest. A comet may take months to sweep through the inner solar system, during which time it appears as a glowing ball with an extended tail of gas and dust.

The nuclei of comets are ice-rich bodies, a few kilometers or tens of kilometers in diameter (similar in size to asteroids), that are sometimes described as "dirty snowballs" or "icy mudballs," left over from the origin of the planets. From this you can conclude that at least some parts of the solar nebula had abundant icy material. You will see more evidence about the composition and history of comets in Chapter 25.

A comet nucleus remains frozen and inactive while it is far from the sun. As the nucleus moves along its elliptical orbit into the inner solar system, the sun's heat begins to vaporize the ices, releasing gas and dust. The pressure of sunlight and the solar wind push the gas and dust away, forming a long tail. The motion of the nucleus along its orbit, the effects of sunlight, and the outward flow of the solar wind can create comet tails that are long and straight or gently curved, but in any case the tail of a comet always points approximately away from the sun (Figure 19-5b), no matter what direction the comet itself is moving. The beautiful tail of a comet can be longer than an AU, but, again, note that it is produced by a relatively tiny nucleus only a few kilometers in diameter.

Unlike the stately comets, **meteors** flash across the sky in momentary streaks of light (■ Figure 19-6). They are commonly called "shooting stars." Of course, they are not stars but small bits of rock and metal colliding with Earth's atmosphere and bursting into incandescent vapor because of friction with the air about 80 km (50 mi) above the ground. This vapor condenses to form dust that settles slowly to Earth, adding about 40,000 tons per year to the planet's mass.

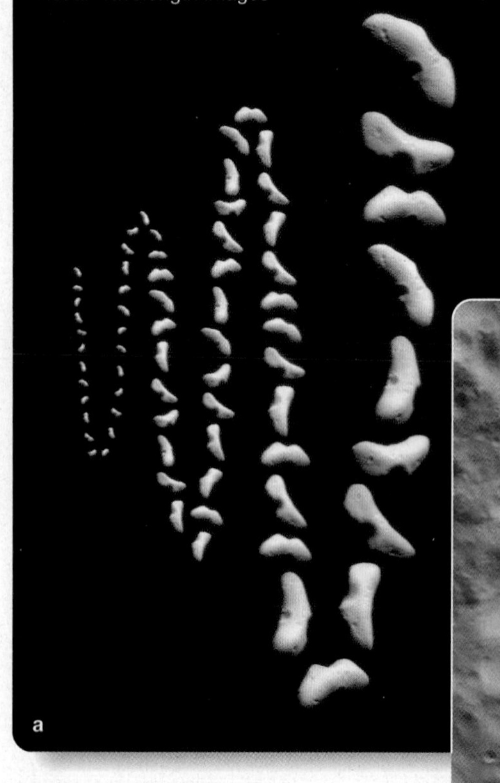


b

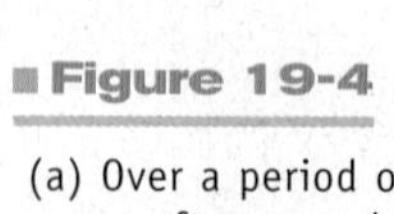

**■ Figure 19-4**

(a) Over a period of three weeks, the NEAR spacecraft approached the asteroid Eros and recorded a series of images arranged here in an entertaining pattern showing the irregular shape and 5-hour rotation period of the asteroid. Eros is 34 km (21 mi) long. (b) This close-up of the surface of Eros shows an area about 11 km (7 mi) from top to bottom. (Johns Hopkins University, Applied Physics Laboratory, NASA)

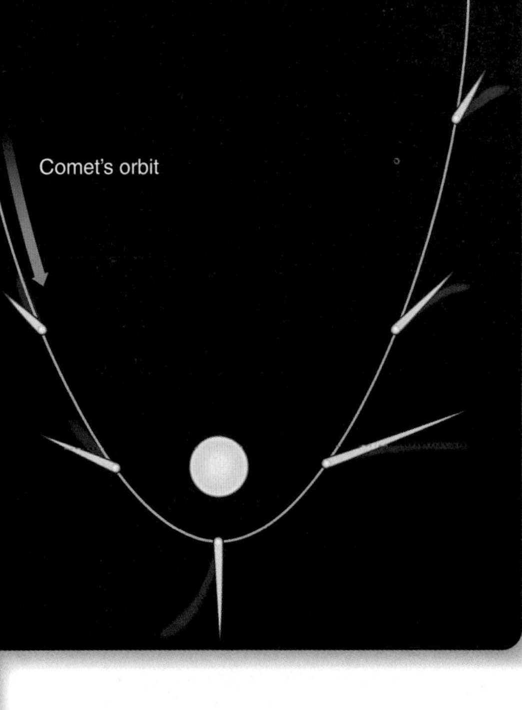


■ **Figure 19-5**

Comets orbit the sun in long, elliptical orbits and become visible when the sun's heat vaporizes its ices and pushes the gas and dust into a tail pointing away from the sun. (a) A comet may remain visible in the evening or morning sky for weeks as it moves through the inner solar system. Although comets are moving rapidly along their orbits, they are so distant that, on any particular evening, a comet seems to hang motionless in the sky. Comet Hyakutake is shown here near Polaris in 1996. (Mike Terenzoni) (b) A comet in a long, elliptical orbit becomes visible when the sun's heat vaporizes its ices and pushes the gas and dust away in a tail. (Celestron International)

Technically, the word *meteor* refers to the streak of light in the sky. In space, before its fiery plunge, the object is called a **meteoroid,** and any part of it that survives its fiery passage to Earth's surface is called a **meteorite.** Most meteoroids are specks of dust, grains of sand, or tiny pebbles. Almost all the meteors you see in the sky are produced by meteoroids that weigh less than 1 gram. Only rarely is a meteoroid massive and strong enough to survive its plunge and reach Earth's surface.

Thousands of meteorites have been found, and you will learn more about their various types in Chapter 25. Meteorites are mentioned here for one specific clue they can give you concerning the solar nebula: Meteorites can reveal the age of the solar system.

## The Age of the Solar System

According to the solar nebula theory, the planets should be about the same age as the sun. The most accurate way to find the age of a rocky body is to bring a sample into the laboratory and analyze the radioactive elements it contains.

When a rock solidifies, it incorporates known percentages of the chemical elements. A few of these elements have forms, called isotopes (see Chapter 7), that are radioactive, meaning they gradually decay into other isotopes. For example, potassium-40, called a parent isotope, decays into calcium-40 and argon-40, called daughter isotopes. The **half-life** of a radioactive substance is the time it takes for half of the parent isotope atoms to decay into daughter isotope atoms. The abundance of a radioactive substance gradually decreases as it decays, and the abundances of the daughter substances gradually increase (■ Figure 19-7). The half-life of potassium-40 is 1.3 billion years. If you also have information about the abundances of the elements in the original rock, you can measure the present abundances and find the age of the rock. For example, if you study a rock and find that only 50 percent of the potassium-40 remains and the rest has become a mixture of daughter isotopes, you could conclude that one half-life must have passed and that the rock is 1.3 billion years old.

Potassium isn't the only radioactive element used in radioactive dating. Uranium-238 decays with a half-life of 4.5 billion

■ **Figure 19-6**

A meteor is a sudden streak of glowing gases produced by a bit of material falling into Earth's atmosphere. Friction with the air vaporizes the material about 80 km (50 mi) above Earth's surface. (Daniel Good)

years to form lead-206 and other isotopes. Rubidium-87 decays into strontium-87 with a half-life of 47 billion years. Any of these substances can be used as a radioactive clock to find the age of mineral samples.

Of course, to find a radioactive age, you need to get a sample into the laboratory, and the only celestial bodies of which scientists have samples are Earth, the moon, Mars, and meteorites. The oldest Earth rocks so far discovered and dated are tiny zircon crystals from Australia that are 4.4 billion years old. That does not mean that Earth formed 4.4 billion years ago. As you will see in the next chapter, the surface of Earth is active, and the crust is continually destroyed and reformed with material welling up from beneath the crust. Those types of processes tend to dilute the daughter atoms and spread them away from the parent atoms, effectively causing the radioactive clocks to reset to zero. The radioactive age of a rock is actually the length of time since the material in that rock was last melted. Consequently, the dates of these oldest rocks tells you only a lower limit to the age of Earth, in other words, that Earth is *at least* 4.4 billion years old.

One of the most exciting goals of the Apollo lunar landings was to bring lunar rocks back to Earth's laboratories where their ages could be measured. Because the moon's

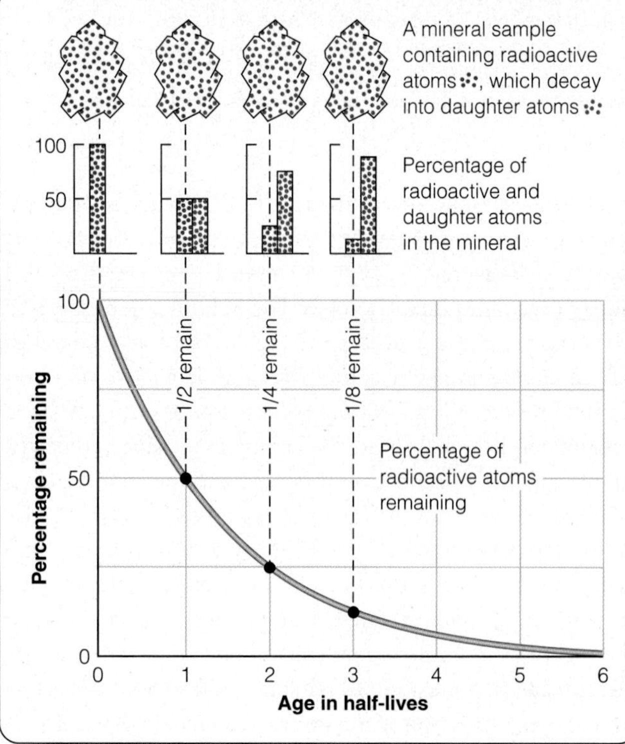


■ **Figure 19-7**

(a) The radioactive atoms (red) in a mineral sample decay into daughter atoms (blue). Half the radioactive atoms are left after one half-life, a fourth after two half-lives, an eighth after three half-lives, and so on. (b) Radioactive dating shows that this fragment of the Allende meteorite is 4.56 billion years old. It contains a few even older interstellar grains, which formed long before our solar system. (R. Kempton, New England Meteoritical Services)

surface is not geologically active like Earth's surface, some moon rocks might have survived unaltered since early in the history of the solar system. In fact, the oldest moon rocks are 4.5 billion years old. That means the moon must be *at least* 4.5 billion years old.

Although no one has yet been to Mars, over a dozen meteorites found on Earth have been identified by their chemical composition as having come from Mars. Most of these have ages of only a billion years or so, but one has an age of approximately 4.5 billion years. Mars must be at least that old.

The most important source for determining the age of the solar system is meteorites. Radioactive dating of meteorites yields a range of ages, but there is a fairly precise upper limit—many meteorite samples have ages of 4.56 billion years old, and none are older. That figure is widely accepted as the age of the solar system and is often rounded to 4.6 billion years. The true ages of Earth, the moon, and Mars are also assumed to be 4.6 billion years, although no rocks from those bodies have yet been found that have remained unaltered for that entire stretch of time.

One last celestial body deserves mention: the sun. Astronomers estimate the age of the sun to be about 5 billion years, but that is not a radioactive date because we have no samples of radioactive material from the sun. Instead, an independent estimate for the age of the sun can be made using helioseismological observations and mathematical models of the sun's interior (see Chapters 8 and 12). This yields a value of about 5 billion years, plus or minus 1.5 billion years, a number that is in agreement with the age of the solar system derived from the age of meteorites.

Apparently, all the bodies of the solar system formed at about the same time, some 4.6 billion years ago. You can add this as the final item to your list of characteristic properties of the solar system (■ Table 19-1).

**■ Table 19-1 | Characteristic Properties of the Solar System**

1. Disk shape of the solar system
   - Orbits in nearly the same plane
   - Common direction of rotation and revolution
2. Two planetary types
   - Terrestrial—inner planets; high density
   - Jovian—outer planets; low density
3. Planetary rings and large satellite systems
   - Yes for Jovian Planets (Jupiter, Saturn, Uranus, and Neptune)
   - No for Terrestrial Planets (Mercury, Venus, Earth, and Mars)
4. Space debris—asteroids, comets, and meteoroids
   - Composition, Orbits
     - Asteroids in inner solar system, composition like Terrestrial planets
     - Comets in outer solar system, composition like Jovian planets
5. Common age of about 4.6 billion years measured or inferred for Earth, the moon, Mars, meteorites, and the sun

### SCIENTIFIC ARGUMENT

***In what ways does the solar system resemble a disk?***

Notice that this argument is really a summary of pieces of evidence. First, the general shape of the solar system is that of a disk. The orbit of Mercury is inclined 7° to the plane of Earth's orbit, and the rest of the planets are in orbits inclined less than that. In other words, the planets are confined to a thin disk with the sun at its center.

Second, the motions of the sun and planets also follow this disk theme. The sun and most of the planets rotate in the same direction, counterclockwise as seen from the north, with their equators near the plane of the solar system. Also, all of the planets revolve around the sun in that same direction. The objects in our solar system mostly move in the same direction, which further reflects a disk theme.

One of the basic characteristics of our solar system is its disk shape, but another dramatic characteristic is the division of the planets into two groups. Build an argument to detail that evidence. **What are the distinguishing differences between the Terrestrial and Jovian planets?**

## 19-3 The Story of Planet Building

THE CHALLENGE FOR modern planetary scientists is to compare the characteristics of the solar system with predictions of the solar nebula theory, so they can work out details of how the planets formed.

### The Chemical Composition of the Solar Nebula

Everything astronomers know about the solar system and star formation suggests that the solar nebula was a fragment of an interstellar gas cloud. Such a cloud would have been mostly hydrogen with some helium and small amounts of the heavier elements.

That is precisely what you see in the composition of the sun (look back at Table 7-2). Analysis of the solar spectrum shows that the sun is mostly hydrogen, with a quarter of its mass being helium and only about 2 percent being heavier elements. Of course, nuclear reactions have fused some hydrogen into helium, but this happens in the sun's core and has not affected the composition of its surface and atmosphere, which are the parts you can observe directly. That means the composition revealed in the sun's spectrum is essentially the composition of the gases from which the sun formed.

This must have been the composition of the solar nebula, and you can also see that composition reflected in the chemical compositions of the planets. The inner planets are composed of rock and metal, and the outer planets are rich in low-density gases such as hydrogen and helium. The chemical composition of Jupiter resembles the composition of the sun. Furthermore, if you allowed low-density gases to escape from a blob of stuff with the same overall composition as the sun or Jupiter, the relative proportions of the remaining heavier elements would resemble Earth's chemical composition.

## The Condensation of Solids

An important clue to understanding the process that converted the nebular gas into solid matter is the variation in density among solar system objects. You have already noted that the four inner planets are small and have high density, resembling Earth, whereas the outermost planets are large and have low density, resembling Jupiter.

Even among the four Terrestrial planets, you will find a pattern of slight differences in density. Merely listing the observed densities of the Terrestrial planets does not reveal the pattern clearly because Earth and Venus, being more massive, have stronger gravity and have squeezed their interiors to higher densities. The **uncompressed densities**—the densities the planets would have if their gravity did not compress them, or to put it another way, the average densities of their original construction materials—can be calculated using the actual densities and masses of each planet (■ Table 19-2). In general, the closer a planet is to the sun, the higher its uncompressed density.

This density variation originated when the solar system first formed solid grains. The kind of matter that could condense in a particular region depended on the temperature of the gas there. In the inner regions, the temperature was evidently 1500 K or so. The only materials that can form grains at that temperature are compounds with high melting points, such as metal oxides and pure metals, which are very dense. Farther out in the nebula it was cooler, and silicates (rocky material) could also condense, in addition to metal. These are less dense than metal oxides and metals. Mercury, Venus, Earth, and Mars are evidently composed of a mixture of metals, metal oxides, and silicates, with proportionately more metals close to the sun and more silicates farther from the sun.

**■ Table 19-2 | Observed and Uncompressed Densities**

| Planet | Observed Density (g/cm³) | Uncompressed Density (g/cm³) |
|---|---|---|
| Mercury | 5.44 | 5.30 |
| Venus | 5.24 | 3.96 |
| Earth | 5.50 | 4.07 |
| Mars | 3.94 | 3.73 |

**■ Table 19-3 | The Condensation Sequence**

| Temperature (K) | Condensate | Planet (Estimated Temperature of Formation; K) |
|---|---|---|
| 1500 | Metal oxides | Mercury (1400) |
| 1300 | Metallic iron and nickel | |
| 1200 | Silicates | |
| 1000 | Feldspars | Venus (900) |
| 680 | Troilite (FeS) | Earth (600) |
| | | Mars (450) |
| 175 | $H_2O$ ice | Jovian (175) |
| 150 | Ammonia–water ice | |
| 120 | Methane–water ice | |
| 65 | Argon–neon ice | Pluto (65) |

Even farther from the sun there was a boundary called the **ice line** beyond which water vapor could freeze to form ice particles. Yet a little farther from the sun, compounds such as methane and ammonia could condense to form other types of ice. Water vapor, methane, and ammonia were abundant in the solar nebula, so beyond the ice line the nebula would have been filled with a blizzard of ice particles, mixed with small amounts of silicate and metal particles that could also condense there. Those ices are low-density materials. The compositions of Jupiter and the other outer planets are a mix of ices plus relatively small amounts of silicates and metal.

The sequence in which the different materials condense from the gas as you move away from the sun toward lower temperature is called the **condensation sequence** (■ Table 19-3). It suggests that the planets, forming at different distances from the sun, should have accumulated from different kinds of materials in a predictable way.

People who have read a little bit about the origin of the solar system may hold the **Common Misconception** that the matter in the solar nebula was sorted by density, with the heavy rock and metal sinking toward the sun and the low-density gases being blown outward. That is not the case. The chemical composition of the solar nebula was originally roughly the same throughout the disk. The important factor was temperature: The inner nebula was hot, and only metals and rock could condense there, whereas the cold outer nebula could form lots of ices along with metals and rock. The ice line seems to have been between Mars and Jupiter, and it separates the region for formation of the high-density Terrestrial planets from that of the low-density Jovian planets.

## The Formation of Planetesimals

In the development of a planet, three processes operate to collect solid bits of matter—rock, metal, ice—into larger bodies called

**planetesimals,** which eventually build the planets. The study of planet building is the study of these processes: condensation, accretion, and gravitational collapse, each of which will be described in detail in this section.

According to the solar nebula theory, planetary development in the solar nebula began with the growth of dust grains. These specks of matter, whatever their composition, grew from microscopic size first by condensation, then by accretion.

A particle grows by **condensation** when it adds matter one atom or molecule at a time from a surrounding gas. Snowflakes, for example, grow by condensation in Earth's atmosphere. In the solar nebula, dust grains were continuously bombarded by atoms of gas, and some of those stuck to the grains. A microscopic grain capturing a layer of gas molecules on its surface increases its mass by a much larger fraction than a gigantic boulder capturing a single layer of molecules. That is why condensation can increase the mass of a small grain rapidly, but, as the grain grows larger, condensation becomes less effective.

The second process is **accretion,** the sticking together of solid particles. You may have seen accretion in action if you have walked through a snowstorm with big, fluffy flakes. If you caught one of those "flakes" on your glove and looked closely, you saw that it was actually made up of many tiny, individual flakes that had collided as they fell and accreted to form larger particles. In the solar nebula, the dust grains were, on average, no more than a few centimeters apart, so they collided frequently and could accrete into larger particles.

When the particles grew to sizes larger than a centimeter, they would have been subject to new processes that tended to concentrate them. One important effect was that the growing solid objects would have collected into the plane of the solar nebula. Small dust grains could not fall into the plane because the turbulent motions of the gas kept them stirred up, but larger objects had more mass, and gas motions could not have prevented them from settling into the plane of the spinning nebula. Astronomers calculate this would have concentrated the larger solid particles into a relatively thin layer about 0.01 AU thick that would have made further growth more rapid. There is no clear distinction between a very large grain and a very small planetesimal, but you might consider an object to be a planetesimal when its diameter approaches a kilometer (0.6 mi) or so (■ Figure 19-8).

This concentration of large particles and planetesimals into the plane of the nebula is analogous to the flattening of a forming galaxy, and a process also found in galaxies may have become important once the plane of planetesimals formed. Computer models show that the rotating disk of particles should have been gravitationally unstable and would have been disturbed by spiral density waves resembling the much larger ones found in spiral galaxies. Those waves could have further concentrated the planetesimals and helped them coalesce into objects up to 100 km (60 mi) in diameter.

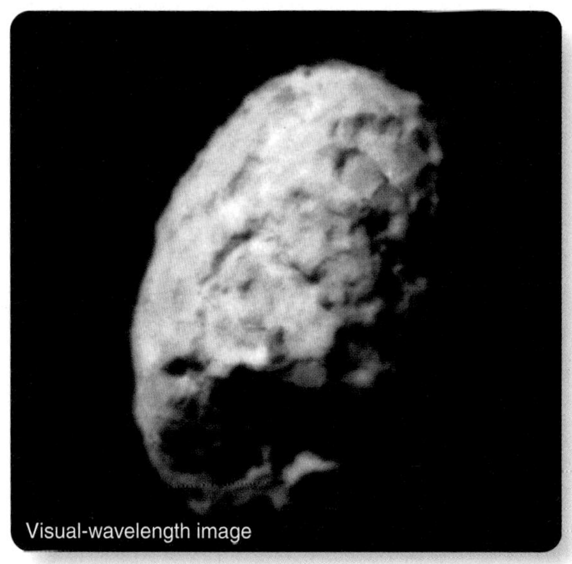


■ **Figure 19-8**

What did the planetesimals look like? You can get a clue from this photo of the 5-km-wide nucleus of Comet Wild 2 (pronounced *Vildt-two*). Whether rocky or icy, the planetesimals must have been small, irregular bodies, scarred by craters from collisions with other planetesimals. (NASA)

Through these processes, the theory proposes, the nebula became filled with trillions of solid particles ranging in size from pebbles to tiny planets. As the largest began to exceed 100 km in diameter, a third process began to affect them, and a new stage in planet building began, the formation of protoplanets.

## The Growth of Protoplanets

The coalescing of planetesimals eventually formed **protoplanets,** the name for massive objects destined to become planets. As these larger bodies grew, a new process helped them grow faster and altered their physical structure.

If planetesimals had collided at orbital velocities, it is unlikely that they would have stuck together. A typical orbital velocity in the solar system is about 10 km/s (22,000 mph). Head-on collisions at this velocity would vaporize the material. However, the planetesimals were all moving in the same direction in the nebular plane and didn't collide head on. Instead, they merely "rubbed shoulders," so to speak, at low relative velocities. Such gentle collisions would have been more likely to combine planetesimals than to shatter them.

Some adhesive effects probably helped. Sticky coatings and electrostatic charges on the surfaces of the smaller planetesimals probably aided formation of larger bodies. Collisions would have fragmented some of the surface rock; but, if the planetesimals were large enough, their gravity could have held on to some fragments to form a layer of soil composed of crushed rock. Such a relatively soft soil layer on the surfaces of larger planetesimals may have been effective in trapping smaller bodies.

The largest planetesimals would grow the fastest because they had the strongest gravitational field. Their stronger gravity could attract additional material, and they could also hold on to a cushioning layer to trap fragments. Astronomers calculate that the largest planetesimals would have grown quickly to protoplanetary dimensions, sweeping up more and more material.

Protoplanets had to begin growing by accumulating solid material because they did not have enough gravity to capture and hold large amounts of gas. In the warm solar nebula, the atoms and molecules of gas were traveling at velocities much larger than the escape velocities of modest-sized protoplanets. Therefore, in their early development, the protoplanets could grow only by attracting solid bits of rock, metal, and ice. Once a protoplanet approached a size of 15 Earth masses or so, however, it could begin to grow by **gravitational collapse,** the rapid accumulation of large amounts of in-falling gas from the nebula.

The theory of protoplanet growth into planets supposes that all the planetesimals had about the same chemical composition. The planetesimals accumulated to form a planet-sized ball of material with homogeneous composition throughout. Once the planet formed, heat would begin to accumulate in its interior from the decay of short-lived radioactive elements.

The violent impacts of in-falling particles would also have released energy called **heat of formation.** These two heating sources would eventually have melted the planet and allowed it to differentiate. **Differentiation** is the separation of material according to density. Once a planet melted, the heavy metals such as iron and nickel, plus elements chemically attracted to them, would settle to the core, while the lighter silicates and related materials floated to the surface to form a low-density crust. The scenario of planetesimals combining into planets that subsequently differentiated is shown in ■ Figure 19-9.

The process of differentiation depends partly on the presence of short-lived radioactive elements whose rapid decay would have released enough heat to melt the interior of planets. Astronomers know such radioactive elements were present because very old rock from meteorites contains daughter isotopes such as magnesium-26. That isotope is produced by the decay of aluminum-26 with a half-life of only 0.74 million years. The aluminum-26 and similar short-lived radioactive isotopes are gone now, but they must have been produced in a supernova explosion that occurred shortly before the formation of the solar nebula. In fact, some astronomers suspect that supernova explosions could have triggered the formation of the sun and other stars by compressing interstellar clouds (see Figure 11-2 and 11-3). Thus, our solar system may exist because of a supernova explosion that occurred about 4.6 billion years ago.

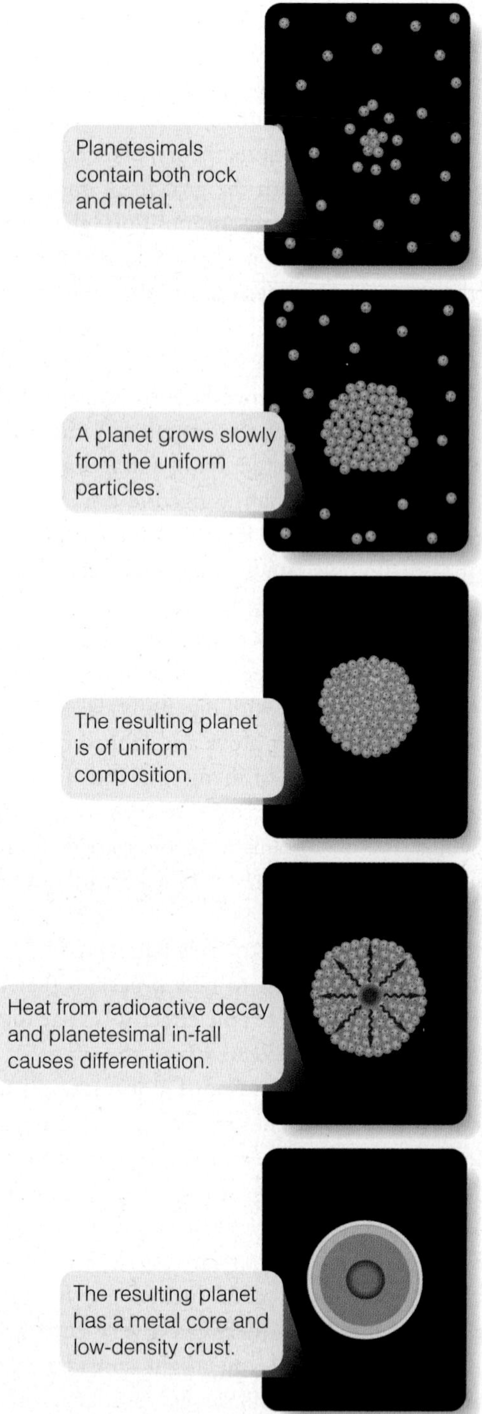

Planetesimals contain both rock and metal.
A planet grows slowly from the uniform particles.
The resulting planet is of uniform composition.
Heat from radioactive decay and planetesimal in-fall causes differentiation.
The resulting planet has a metal core and low-density crust.

■ **Figure 19-9**

This simple model of planet building assumes planets formed from accretion and collision of planetesimals that were of uniform composition, containing both metals and rocky material, and that the planets later differentiated, meaning they melted and separated into layers by density and composition.

If planets formed by accretion of planetesimals and were later melted by radioactive decay and heat of formation, then Earth's early atmosphere may have consisted of a combination of gases delivered by planetesimal impacts and released from the planet's interior during differentiation. The creation of a planetary atmosphere from a planet's interior is called **outgassing.** Given the location of Earth in the solar nebula, gases released from its interior during differentiation would not have included as much water as Earth now has. So, some astronomers think that Earth's water and even some of its present atmosphere and biosphere accumulated late in the formation of the planet as Earth swept up volatile-rich planetesimals. These icy planetesimals would have formed in the cool outer parts of the solar nebula and could have been scattered toward the Terrestrial planets by encounters with the Jovian planets, creating a comet bombardment.

According to the solar nebula theory, the Jovian planets could begin growing by the same processes that built the Terrestrial planets. However, in the inner solar nebula, only metals and silicates could form solids, so the Terrestrial planets grew slowly. In contrast, the outer solar nebula contained not just solid bits of metals and silicates but also plentiful ices. Astronomers calculate that the Jovian planets would have grown faster than the Terrestrial planets and quickly become massive enough to begin even faster growth by gravitational collapse, drawing in large amounts of gas from the solar nebula. The Terrestrial planet zone did not include ice particles, so those planets developed relatively slowly and never became massive enough to grow further by gravitational collapse.

The Jovian planets must have reached their present size in no more than about 10 million years, before the sun become hot and luminous enough to blow away the remaining gas in the solar nebula, removing the raw material for further Jovian growth. As you will learn in the next section, disturbances from outside the forming solar system may have reduced the time available for Jovian planet formation even more severely. The Terrestrial planets, in comparison, grew from solids and not from the gas, so they could have continued to grow by accretion from solid debris left behind after the gas was removed. Mathematical models indicate that the Terrestrial planets were at least half finished within 10 million years but could have continued to grow for another 20 million years or so.

The solar nebula theory has been very successful in explaining the formation of the solar system. But there are some problems, and the Jovian planets are the troublemakers.

## The Jovian Problem

New information about the star formation process makes it hard to explain the formation of the Jovian parents, and this has caused astronomers to expand and revise the theory of planet formation (■ Figure 19-10).

The new information is that gas and dust disks around newborn stars don't last long. Earlier you saw images of dusty gas disks around the young stars in the Orion star-forming region (Figure 19-2 and Chapter 11). Those disks are being evaporated by intense ultraviolet radiation from hot O and B stars forming nearby. Astronomers have calculated that nearly all stars form in clusters containing O and B stars, so this evaporation may happen to most disks. Even if a disk did not evaporate quickly, the gravitational influence of the crowded stars in a cluster could strip away the outer parts of the disk. Those are troublesome observations because they seem to indicate that disks can't last longer than about 7 million years at most, and many evaporate within the astronomically very short span of 100,000 years or so. That's not long enough to grow a Jovian planet by the combination of condensation, accretion, and gravitational collapse proposed in the standard solar nebular theory.

Yet, Jovian planets are common. In the final section of this chapter, you will see evidence that astronomers have found planets orbiting other stars, and almost all of those planets have the mass of Jovian planets. There may also be many Terrestrial planets orbiting those stars that are too small for astronomers to detect at present, but the important point is that there are lots of Jovian planets around.

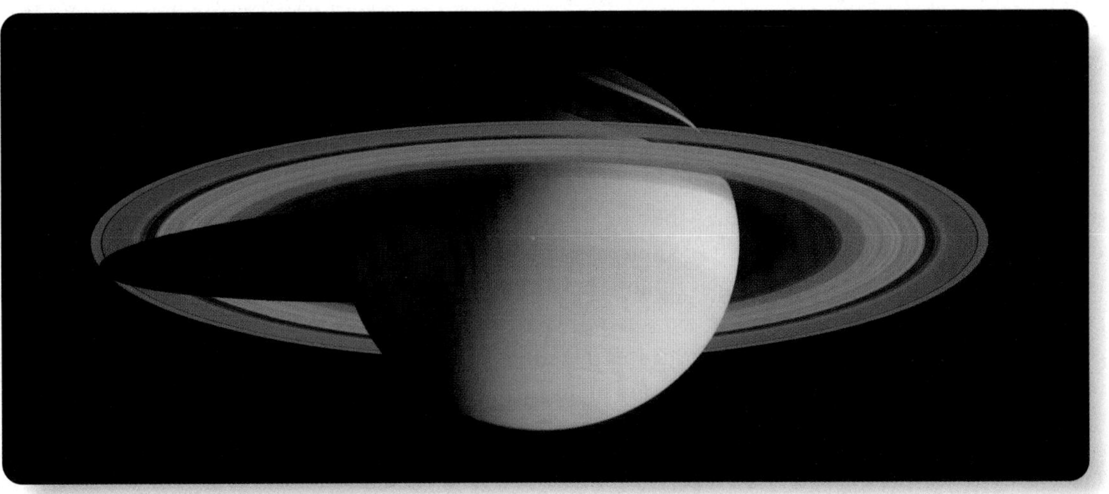

**■ Figure 19-10**

The Jovian worlds pose a problem for modern astronomers. Planet-forming nebulae are blown away in only a few million years by nearby luminous stars, so Jovian planets must form more quickly than initial calculations predicted. Newer research suggests that accretion followed by gravitational collapse can build Jovian planets in about a million years. Under certain conditions, direct gravitational collapse may form some large planets in just thousands of years.

Mathematical models of the solar nebula have been computed using specially built computers running programs that take weeks to finish a calculation. The results show that the rotating gas and dust of the solar nebula could have become unstable and formed Jovian planets by skipping straight to the step of gravitational collapse. That is, massive planets may have been able to form by **direct collapse** of gas without first forming a dense core by condensation and accretion of solid material. Jupiters and Saturns can form in these direct collapse models in only a few hundred years. If the Jovian planets formed in this way, they could have formed before the solar nebula disappeared, even if the nebula was eroded quickly by neighboring massive hot stars.

This new insight into the formation of the outer planets may also help explain a puzzle about the formation of Uranus and Neptune. Those planets are so far from the sun that accretion could not have built them rapidly. The gas and dust of the solar nebula must have been sparse out there, and Uranus and Neptune orbit so slowly they would not have swept up material very rapidly. The conventional view is that they grew by accretion so slowly that they never became quite massive enough to begin accelerated growth by gravitational collapse. In fact, it is hard to understand how they could have reached even their present sizes if they started growing by accretion so far from the sun. Theoretical calculations show that they might instead have formed closer to the sun, in the region of Jupiter and Saturn, and then could have been shifted outward by gravitational interactions with the bigger planets. In any case, the formation of Uranus and Neptune is part of the Jovian problem.

The traditional solar nebula theory proposes that the planets formed by accreting a core and then, if they became massive enough, accelerated growth by gravitational collapse. The proposed modification to the theory suggests that the outer planets could have skipped the core accretion phase.

## Explaining the Characteristics of the Solar System

Now you have learned enough to put all the pieces of the puzzle together and explain the distinguishing characteristics of the solar system in Table 19-1.

The disk shape of the solar system is inherited from the motion of material in the solar nebula. The sun and planets and moons mostly revolve and rotate in the same direction because they formed from the same rotating gas cloud. The orbits of the planets lie in the same plane because the rotating solar nebula collapsed into a disk, and the planets formed in that disk.

The solar nebula hypothesis is evolutionary in that it calls on continuing processes to gradually build the planets. To explain the odd rotations of Venus and Uranus, however, you may need to consider catastrophic events. Uranus rotates on its side. This might have been caused by an off-center collision with a massive planetesimal when the planet was nearly formed. Two hypotheses have been proposed to explain the backward rotation of Venus. Theoretical models suggest that the sun can produce tides in the thick atmosphere of Venus that could have eventually reversed the planet's rotation—an evolutionary hypothesis. It is also possible that the rotation of Venus was altered by an off-center impact late in the planet's formation, and that is a catastrophic hypothesis. Both may be true.

The second item in Table 19-1, the division of the planets into Terrestrial and Jovian worlds, can be understood through the condensation sequence. The Terrestrial planets formed in the inner part of the solar nebula, where the temperature was high and only compounds such as the metals and silicates could condense to form solid particles. That produced the small, dense Terrestrial planets. In contrast, the Jovian planets formed in the outer solar nebula, where the lower temperature allowed the gas to form large amounts of ices, perhaps three times more ices than silicates. That allowed the Jovian planets to grow rapidly and became massive, low-density worlds. Also, Jupiter and Saturn are so massive they were able to grow by drawing in cool gas by gravitational collapse from the solar nebula. The Terrestrial planets could not do this because they never became massive enough.

The heat of formation (the energy released by in-falling matter) was tremendous for these massive planets. Jupiter must have grown hot enough to glow with a luminosity of about 1 percent that of the present sun, although it never got hot enough to generate nuclear energy as a star would. Nevertheless, Jupiter is still hot inside. In fact, both Jupiter and Saturn radiate more heat than they absorb from the sun, so they are evidently still cooling.

A glance at the solar system suggests that you should expect to find a planet between Mars and Jupiter at the present location of the asteroid belt. Mathematical models indicate that the reason asteroids are there rather than a planet is that Jupiter grew into such a massive body that it was able to gravitationally disturb the motion of nearby planetesimals. The bodies that could have formed a planet just inward from Jupiter's orbit instead collided at high speeds and shattered rather than combining, were thrown into the sun, or were ejected from the solar system. The asteroids seen today are the last remains of those rocky planetesimals.

The comets, in contrast, are evidently the last of the icy planetesimals. Some may have formed in the outer solar nebula beyond Neptune and Pluto, but many probably formed among the Jovian planets where ices could condense easily. Mathematical models show that the massive Jovian planets could have ejected some of these icy planetesimals into the far outer solar system. In a later chapter, you will see evidence that some comets are icy bodies coming from those distant locations, falling back into the inner solar system.

The icy Kuiper belt objects appear to be ancient planetesimals that formed in the outer solar system but were never incorporated into a planet. They orbit slowly far from the light and warmth of the sun and, except for occasional collisions, have not changed much since the solar system was young. The gravitational influence of the planets deflects Kuiper belt objects into the inner solar system where they also are seen as comets.

The large satellite systems of the Jovian worlds may contain two kinds of moons. Some moons may have formed in orbit around forming planets in a miniature version of the solar nebula. Some of the smaller moons, in contrast, may be captured planetesimals, asteroids, and comets. The large masses of the Jovian planets would have made it easier for them to capture satellites.

In Table 19-1, you noted that all four Jovian worlds have ring systems, and you can understand this by considering the large mass of these worlds and their remote location in the solar system. A large mass makes it easier for a planet to hold onto orbiting ring particles; and, being farther from the sun, the ring particles are not as quickly swept away by the pressure of sunlight and the solar wind. It is hardly surprising, then, that the Terrestrial planets, low-mass worlds located near the sun, have no planetary rings.

The last entry in Table 19-1 is the common ages of solar system bodies, and the solar nebula theory has no difficulty explaining that characteristic. If the theory is correct, then the planets formed at the same time as the sun and should have roughly the same age.

The solar nebula theory can account for all of the distinguishing characteristics of the solar system, but there is yet another test you should apply to the theory. What about the problem that troubled Laplace and his nebular hypothesis—the angular momentum problem? If the sun and planets formed from a contracting nebula, the sun should have been left spinning very rapidly. That is, the sun should have most of the angular momentum in the solar system, and instead it has very little.

To study this problem, astronomers used the Spitzer Space Telescope to examine 500 young stars in the Orion Nebula. Those that rotate slowly are five times more likely to have a surrounding disk of gas and dust than the faster rotators. This confirms astronomer's expectations that the strong magnetic fields of young stars extend out into their disks. This allows the transfer of angular momentum, speeding the disk's rotation and slowing the star's rotation. Thus, the angular momentum problem is no longer an objection to the solar nebula theory.

Your general understanding of the origin of the solar system gives you a new way of thinking about asteroids, meteors, and comets. They are the last of the debris left behind by the solar nebula. These objects are such important sources of information about the history of our solar system that they will be discussed in detail in Chapter 25. But for now you can consider how the sun blew away the last of the gas and dust in the solar nebula.

## Clearing the Nebula

The sun probably formed along with many other stars in a swirling nebula. Observations of young stars in Orion (Figure 19-3) suggest that radiation from nearby hot stars would have evaporated the disk of gas and dust around the sun and that the gravitational influence of nearby stars would have pulled gas away. Even without the external effects, four internal processes would have gradually destroyed the solar nebula.

The most important of these internal processes was radiation pressure (see Chapter 11). When the sun became a luminous object, light streaming from its photosphere pushed against the particles of the solar nebula. Large bits of matter like planetesimals and planets were not affected, but low-mass specks of dust and individual atoms and molecules were pushed outward and eventually driven from the system.

The second effect that helped clear the nebula was the solar wind, the flow of ionized hydrogen and other atoms away from the sun's upper atmosphere. This flow is a steady breeze that rushes past Earth at about 400 km/s (250 mi/s). Young stars have even stronger winds than stars of the sun's age and irregular fluctuations in luminosity, like those observed in young stars such as T Tauri stars, which can accelerate the wind. The strong surging wind from the young sun may have helped push dust and gas out of the nebula.

The third effect that helped clear the nebula was the sweeping up of space debris by the planets. All of the old, solid surfaces in the solar system are heavily cratered by meteorite impacts (■ Figure 19-11). Earth's moon, Mercury, Venus, Mars, and most of the moons in the solar system are covered with craters. A few of these craters have been formed recently by the steady rain of meteorites that falls on all the planets in the solar system, but most of the craters appear to have been formed before roughly 4 billion years ago in what is called the **heavy bombardment,** as the last of the debris in the solar nebula were swept up by the planets.

The fourth effect was the ejection of material from the solar system by close encounters with planets. If a small object such as a planetesimal passes close to a planet, the small object's path will be affected by the planet's gravity field. In some cases, the small object can gain energy from the planet's motion and be thrown out of the solar system. Ejection is most probable in encounters with massive planets, so the Jovian planets were probably very efficient at ejecting the icy planetesimals that formed in their region of the nebula.

Attacked by the radiation and gravity of nearby stars and racked by internal processes, the solar nebula could not survive very long. Once the gas and dust were gone and most of the planetesimals were swept up, the planets could no longer gain significant mass, and the era of planet building ended.

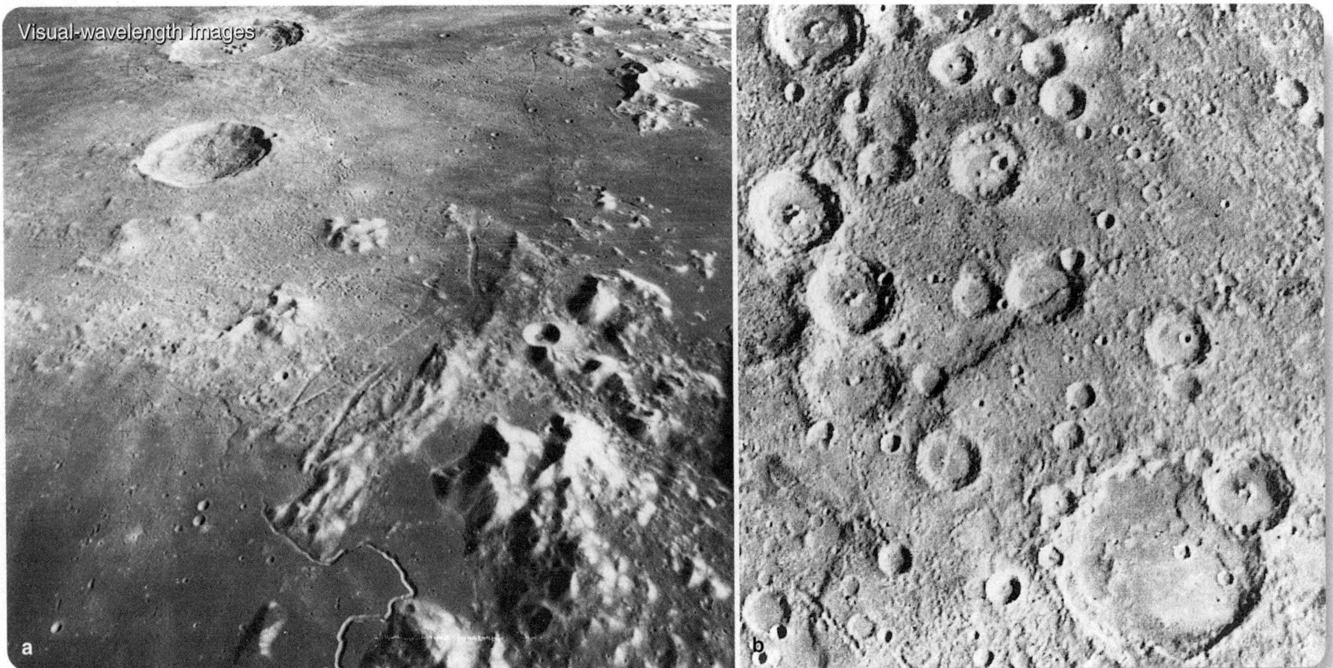


■ **Figure 19-11**

Every old, solid surface in the solar system is scarred by craters. (a) Earth's moon has craters ranging from basins hundreds of kilometers in diameter down to microscopic pits. (b) The surface of Mercury, as photographed by a passing spacecraft, shows vast numbers of overlapping craters. (NASA)

### SCIENTIFIC ARGUMENT

***Why are there two kinds of planets in our solar system?***

This is an opportunity for you to build an argument that closely analyzes the solar nebula theory. Planets begin forming from solid bits of matter, not from gas. Consequently, the kind of planet that forms at a given distance from the sun depends on the kind of substances that can condense out of the gas there to form solid particles. In the inner parts of the solar nebula, the temperature was so high that most of the gas could not condense to form solids. Only metals and silicates could form solid grains, and the innermost planets grew from this dense material. Much of the mass of the solar nebula consisted of hydrogen, helium, water vapor, and other gases, and they were present in the inner solar nebula but couldn't form solid grains. The Terrestrial planets could grow only from the solids in their zone, not from the gases, so the Terrestrial planets are small and dense.

In the outer solar nebula, the composition of the gas was the same, but it was cold enough for water vapor, and other simple molecules containing hydrogen, to condense to form ice grains. Because hydrogen was so abundant, there was lots of ice available. The outer planets grew from large amounts of ice combined with small amounts of metals and silicates. Eventually the outer planets grew massive enough that they could begin to capture gas directly from the nebula, and they became the hydrogen- and helium-rich Jovian worlds.

The condensation sequence combined with the solar nebula theory gives you a way to understand the difference between the Terrestrial and Jovian planets. Now expand your argument: **Why do some astronomers argue that the formation of the Jovian planets is a problem that needs further explanation?**

## 19-4 19-4 Planets Orbiting Other Stars

ARE THERE OTHER planetary systems? The evidence says yes. Do they contain planets like Earth? The evidence so far is incomplete.

### Planet-Forming Disks

Both visible- and radio-wavelength observations detect dense disks of gas and dust orbiting young stars. For example, at least 50 percent of the stars in the Orion nebula are surrounded by such disks (Figure 19-3). A young star is detectable at the center of most disks, and astronomers can measure that the disks contain many Earth masses of material in a region a few times larger in diameter than our solar system. The Orion star-forming region is only a few million years old, so planets may not have formed in these disks yet. Furthermore, the intense radiation from nearby hot stars is evaporating the disks so quickly that planets may never have a chance to grow large. The important point for astronomers is that so many young stars have disks. Evidently, disks of gas and dust are a common feature around stars that are forming.

The Hubble Space Telescope can detect dense disks around young stars in a slightly different way. The disks show up in silhouette against the nebulae that surround the newborn stars (■ Figure 19-12). These disks are related to the formation of

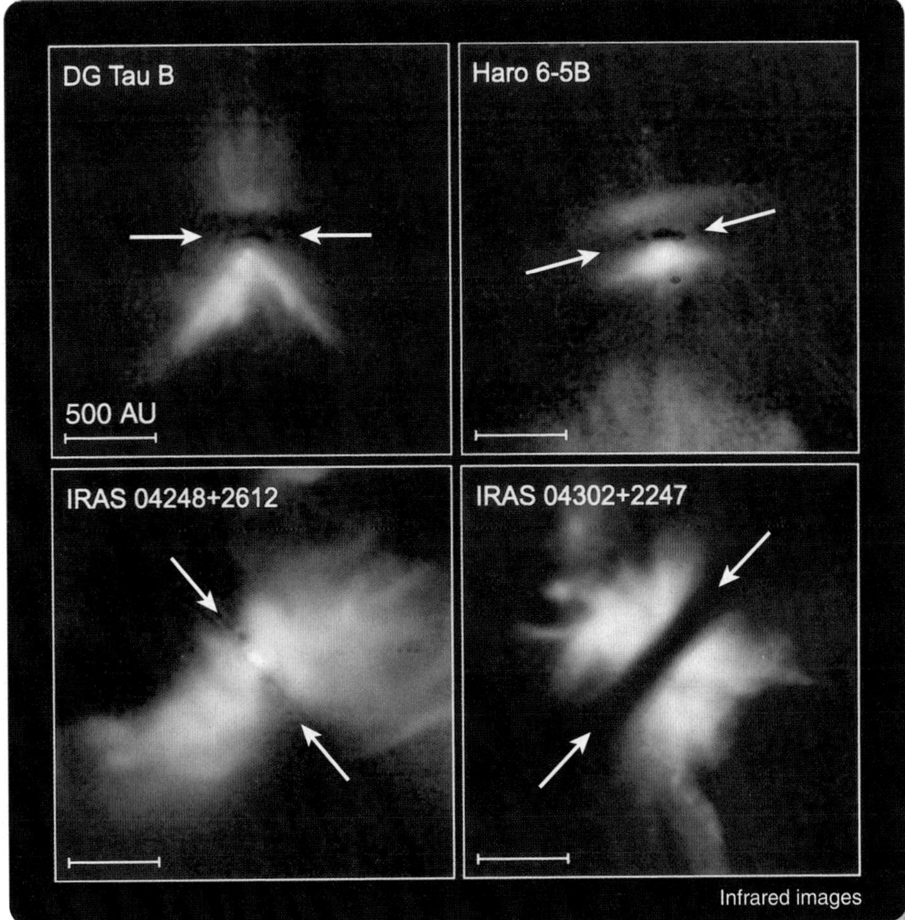


**■ Figure 19-12**

Dark bands (indicated by arrows) are edge-on disks of gas and dust around young stars seen in Hubble Space Telescope near-infrared images. Planets may eventually form in these disks. These systems are so young that material is still falling inward and being illuminated by light from the stars. (D. Padgett, IPAC/Caltech; W. Brandner, IPAC; K. Stapelfeldt, JPL; and NASA).

bipolar flows (page 223) in that they focus the gas flowing way from a young star into two jets shooting in opposite directions.

In addition to these dense, hot planet-forming disks around young stars, infrared astronomers have found cold, low-density dust disks around stars much older than the newborn stars in Orion, old enough to have finished forming. These tenuous dust disks are sometimes called **debris disks** because they are evidently made of dusty debris produced in collisions among small bodies such as comets, asteroids, and Kuiper belt objects, rather than dust left over from an original protostellar disk. That conclusion is based on calculations showing that the observed dust would be removed by radiation pressure in a much shorter time than the ages of those stars, meaning the dust there now must have been created relatively recently. Our own solar system contains such "second-generation" dust produced by asteroids, comets, and Kuiper belt objects, and astronomers have evidence that the solar system's Kuiper belt extending beyond the orbit of Neptune is an example of an old debris disk.

Some examples of debris disks are around the stars Beta Pictoris, HD 107146, and Epsilon Eridani (■ Figure 19-13). The dust disk around Beta Pictoris, an A-type star more massive and luminous than the sun, is about 20 times the diameter of our solar system. The dust disk around Epsilon Eridani, which is a K-type star somewhat smaller than the sun, is similar in size to the solar system's Kuiper belt. Like most of the other known low-density disks, both of these examples have central zones with even lower density. Those inner regions are understood to be places where planets have finished forming and swept up most of the construction material.

Few examples of planets orbiting stars with debris disks have been detected, but the presence of dust with short lifetimes around old stars assures you that small bodies such as asteroids and comets must be present as sources of the dust. If those small objects are there, then it is likely that there are also planets orbiting those stars.

Infrared observations reveal that the star Vega, easily visible in the northern hemisphere summer sky, also has a debris disk, and detailed studies show that much of the dust in that disk is tiny. The pressure of the light from Vega should blow away small dust particles quickly, so astronomers conclude that the dust being observed now must have been produced by a big event like the collision of two large planetesimals within the last million years (■ Figure 19-14). Fragments from that collision are still smashing into each other now and then and producing more dust, continuing to enhance the debris disk. This effect has also been found in the disk around the faint star known as HIP 8920. Such smashups probably happen rarely in a dust disk, but when they happen, they make the disk very easy to detect.

Notice the difference between the two kinds of planet-related disks that astronomers have found. The low-density dust disks such as the ones around Beta Pictoris, Epsilon Eridani, and Vega are produced by dust from collisions among comets, asteroids, and Kuiper belt objects. Such disks are evidence that planetary systems have already formed (■ Figure 19-15). In comparison, the dense disks of gas and dust such as those seen round the stars in Orion are sites where planets could be forming right now.

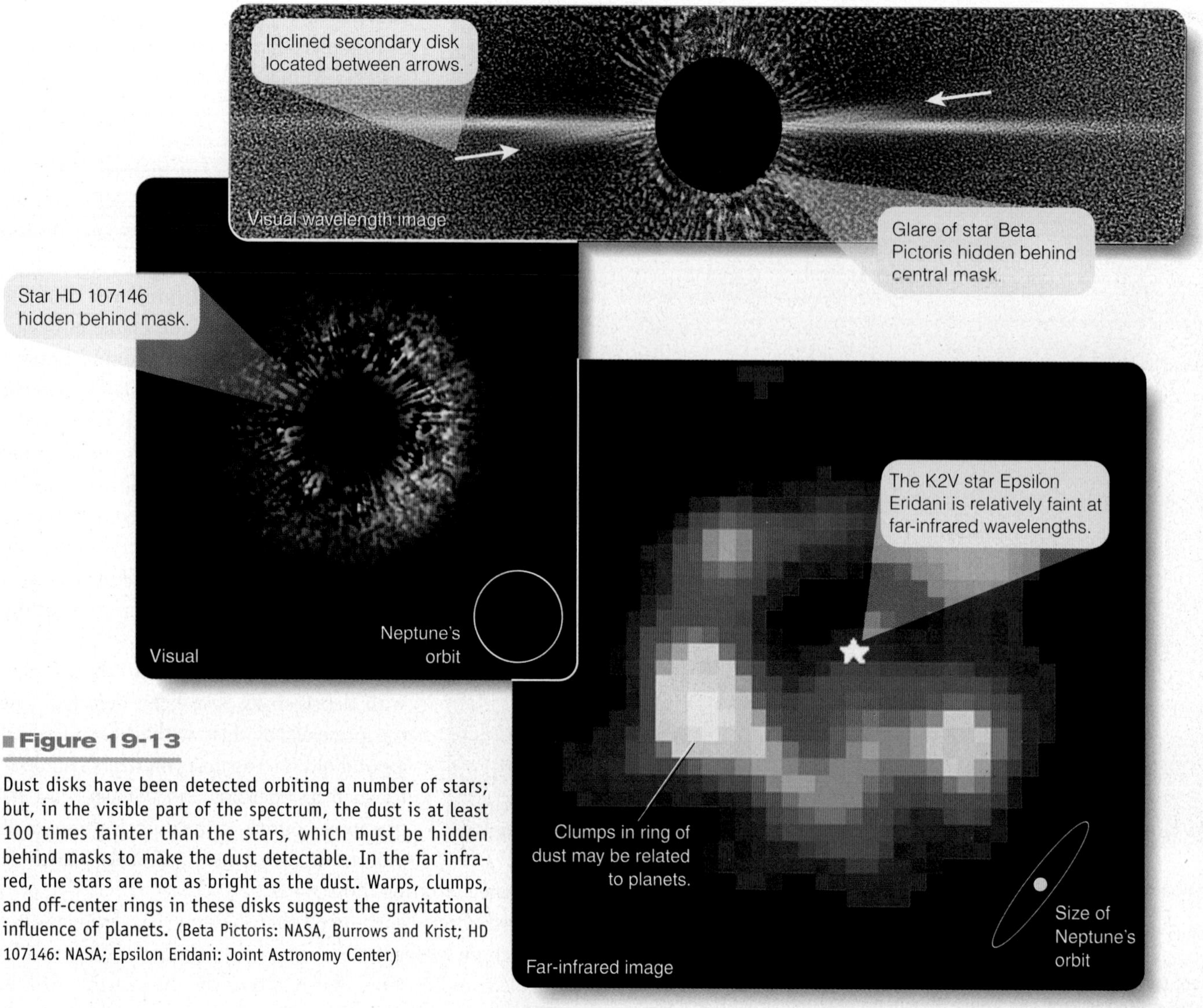

Inclined secondary disk located between arrows.

Visual wavelength image

Glare of star Beta Pictoris hidden behind central mask.

Star HD 107146 hidden behind mask.

Visual

Neptune's orbit

The K2V star Epsilon Eridani is relatively faint at far-infrared wavelengths.

Clumps in ring of dust may be related to planets.

Size of Neptune's orbit

Far-infrared image

**Figure 19-13**

Dust disks have been detected orbiting a number of stars; but, in the visible part of the spectrum, the dust is at least 100 times fainter than the stars, which must be hidden behind masks to make the dust detectable. In the far infrared, the stars are not as bright as the dust. Warps, clumps, and off-center rings in these disks suggest the gravitational influence of planets. (Beta Pictoris: NASA, Burrows and Krist; HD 107146: NASA; Epsilon Eridani: Joint Astronomy Center)

**Figure 19-14**

Collisions between asteroids are rare events, but they generate lots of dust and huge numbers of fragments, as in this artist's conception. Further collisions between fragments can continue to produce dust. Because such dust is blown away quickly, astronomers treat the presence of dust as evidence that objects of asteroid size are also present. (J. Lomberg/Gemini Observatory)

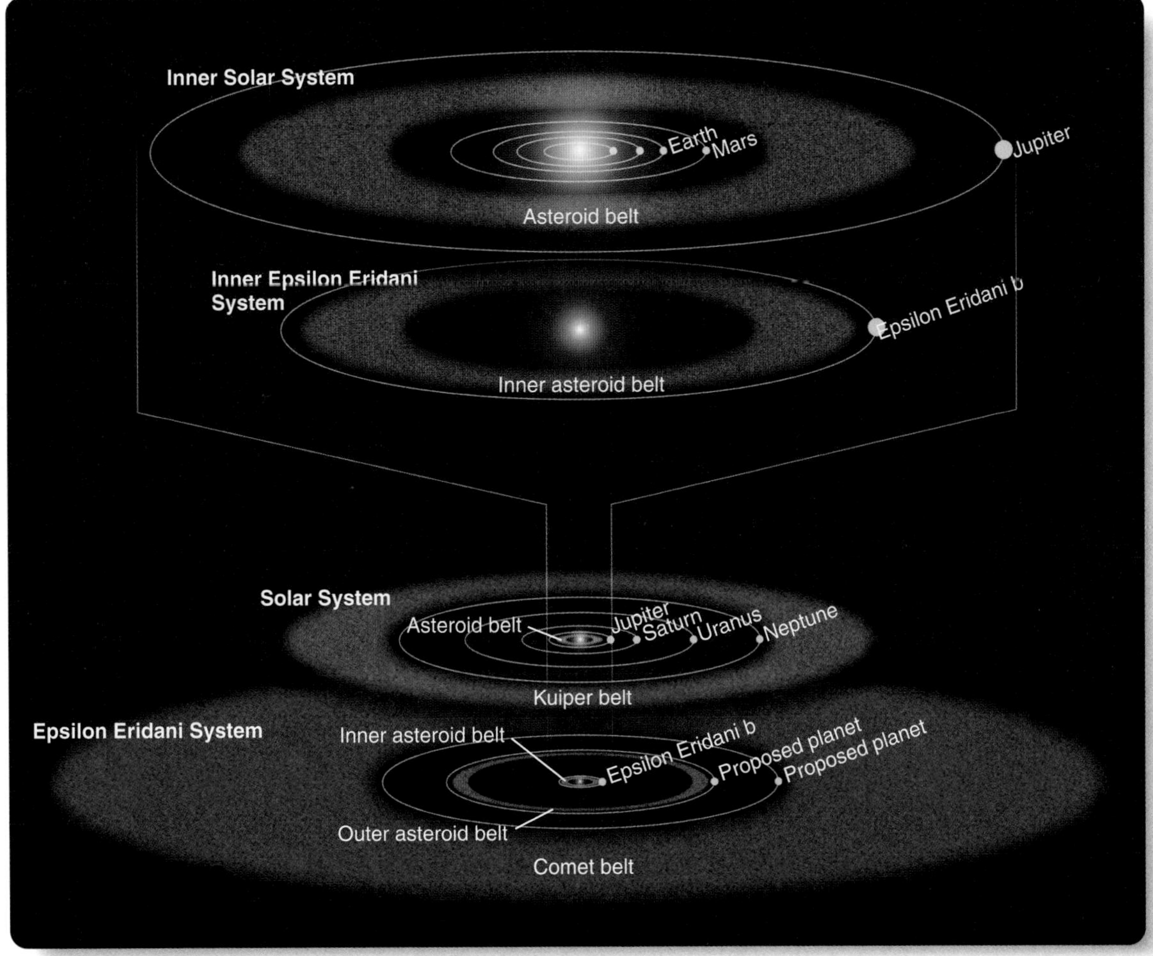


**Figure 19-15**

Dust in debris belts around older main-sequence stars indicates ongoing collisions of asteroids and comets. Such activity in our solar system is ultimately driven by the gravitational influence of planets. The locations of debris belt edges may be defined by orbits of adjacent planets. The inferred architecture of the Epsilon Eridani planetary system is shown in comparison with our solar system. (Artist's conception based on Spitzer Space Telescope and Caltech Submillimeter Observatory observations by D. Backman et al.; NASA/JPL-Caltech/R. Hurt)

## Observing Extrasolar Planets

A planet orbiting another star is called an **extrasolar planet.** Such a planet would be quite faint and difficult to see so close to the glare of its star. But there are ways to find these planets. To understand one important way, all you have to do is imagine walking a dog.

You will remember that Earth and the moon orbit around their common center of mass, and two stars in a binary system orbit around their center of mass. When a planet orbits a star, the star moves very slightly as it orbits the center of mass of the planet–star system. Think of someone walking a poorly trained dog on a leash; the dog runs around pulling on the leash, and even if it were an invisible dog, you could plot its path by watching how its owner was jerked back and forth. Astronomers can detect a planet orbiting another star by watching how the star moves as the planet tugs on it.

The first planet detected this way was discovered in 1995 orbiting the sun-like star 51 Pegasi. As the planet circles the star, the star wobbles slightly, and that very small motion of the star is detectable by Doppler shifts in the star's spectrum (■ Figure 19-16a) (see Chapters 7 and 9). From the motion of the star and estimates of the star's mass, astronomers can deduce that the planet has at least half the mass of Jupiter and orbits only 0.05 AU from the star. Half the mass of Jupiter amounts to 160 Earth masses, so this is a large planet. Note also that it orbits very close to its star, much closer than Mercury orbits around our sun.

Astronomers were not surprised by the announcement that a planet had been found orbiting a sun-like star; for years astronomers had assumed that many stars had planets. Nevertheless, they greeted the details of this discovery with professional skepticism (**How Do We Know? 19-2**). That skepticism led to careful tests of the data and further observations that confirmed

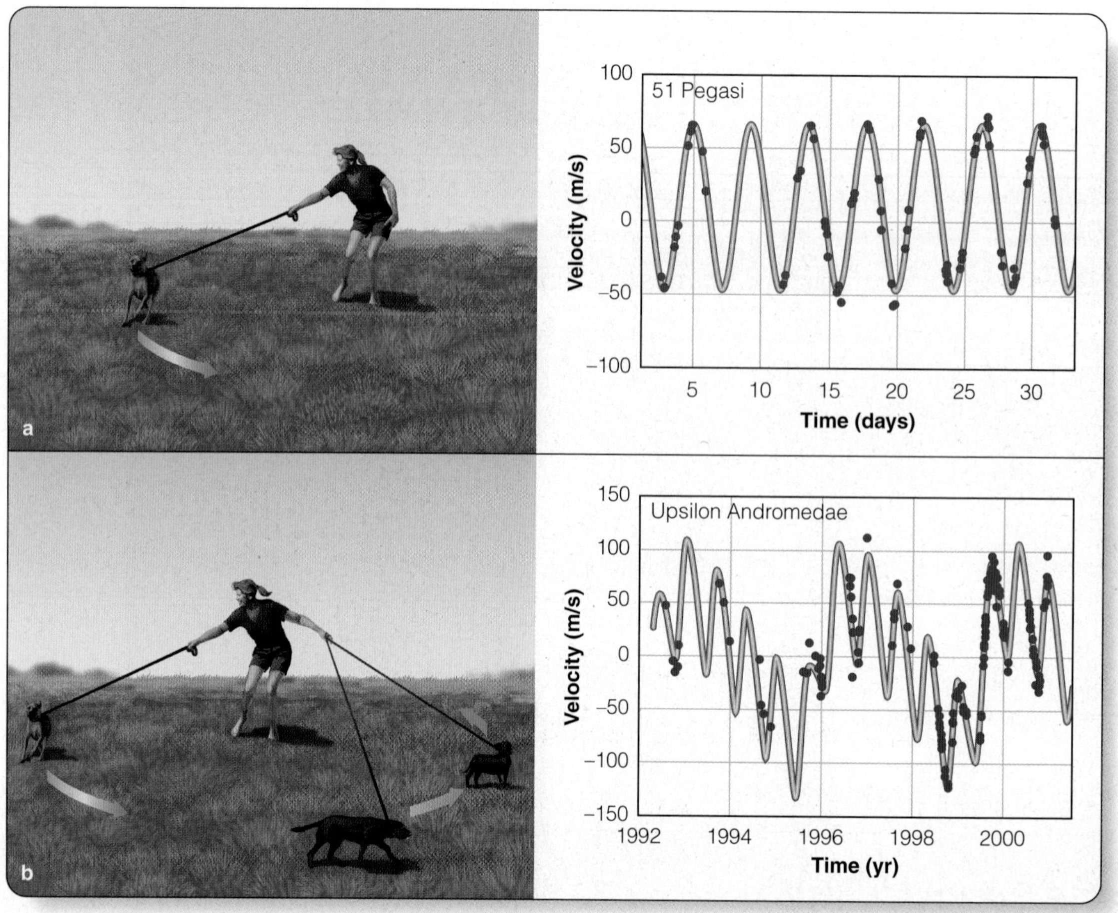


**Figure 19-16**

Just as someone walking a lively dog is tugged around, the star 51 Pegasi is pulled back and forth by the gravity of the planet that orbits it every 4.2 days. The wobble is detectable in precision observations of the star's Doppler shift. Someone walking three dogs is pulled about in a more complicated pattern, and you can see something similar in the Doppler shifts of Upsilon Andromedae, which is orbited by three planets detected so far. The influence of its shortest-period planet has been removed in this graph to reveal the orbital influences of the other two planets.

the discovery. In fact, almost 400 planets have been discovered in this way, including at least three planets orbiting the star Upsilon Andromedae (Figure 19-16b) and five orbiting 55 Cancri—true planetary systems. Over 25 such multiple-planet systems have been found.

Another way to search for planets is to look for changes in the brightness of the star when the orbiting planet crosses in front of the star, called a **transit.** The decrease in light during a planet transit is very small, but it is detectable, and astronomers have used this technique to find many planets. From the amount of light lost, astronomers can tell that these planets with Jovian masses have Jovian diameters, and thus Jovian densities and compositions.

The Spitzer Space Telescope has detected infrared radiation from two large hot planets already known from star wobble Doppler shifts. As these planets orbit their parent stars, the amount of infrared radiation from each system varies. When the planets pass behind their parent stars, the total infrared brightness of the systems noticeably decreases. These measurements confirm the existence of the planets and indicate their temperatures and sizes.

Notice how the techniques used to detect these planets resemble techniques used to study binary stars (see Chapter 9). Most of the planets were discovered using the same observational methods used to study spectroscopic binaries, but some were found by observing the stars as if they were eclipsing binaries. In contrast, a few extrasolar planets have been found by a technique called **microlensing**, in which an extrasolar planet passes precisely between Earth and a background star, briefly magnifying the distant star's brightness by gravitational lensing.

The planets discovered so far tend to be massive and have short orbital periods because lower-mass planets or longer-period

## How Do We Know? 19-2

### Scientists: Courteous Skeptics

**What does it mean to be skeptical yet also open to new ideas?** "Scientists are just a bunch of skeptics who don't believe in anything." That is a **Common Misconception** among people who don't understand the methods and goals of science. Yes, scientists are skeptical about new ideas and discoveries, but they do hold strong beliefs about how nature works. Scientists are skeptical not because they want to disprove everything but because they are searching for the truth and want to be sure that a new description of nature is reliable before it is accepted.

Another **Common Misconception** is that scientists automatically accept the work of other scientists. On the contrary, scientists skeptically question every aspect of a new discovery. They may wonder if another scientist's instruments were properly calibrated or whether the scientist's mathematical models are correct. Other scientists will want to repeat the work themselves using their own instruments to see if they can obtain the same results. Every observation is tested, every discovery is confirmed, and only an idea that survives many of these tests begins to be accepted as a scientific truth.

Scientists are prepared for this kind of treatment at the hands of other scientists. In fact, they expect it. Among scientists it is not bad manners to say, "Really, how do you know that?" or "Why do you think that?" or "Show me the evidence!" And it is not just new or surprising claims that are subject to such scrutiny. Even though astronomers had long expected to discover planets orbiting other stars, when a planet was finally discovered circling 51 Pegasi, astronomers were skeptical. This was not because they thought the observations were necessarily flawed but because that is how science works.

The goal of science is to tell stories about nature. Some people use the phrase "telling a story" to describe someone who is telling a fib. But the stories that scientists tell are exactly the opposite; perhaps you could call them "antifibs" because they are as true as scientists can make them. Skepticism eliminates stories with logical errors, flawed observations, or misunderstood evidence and eventually leaves only the stories that best describe nature.

Skepticism is not a refusal to hold beliefs. Rather, it is a way for scientists to find and keep those natural principles that are worthy of acceptance.

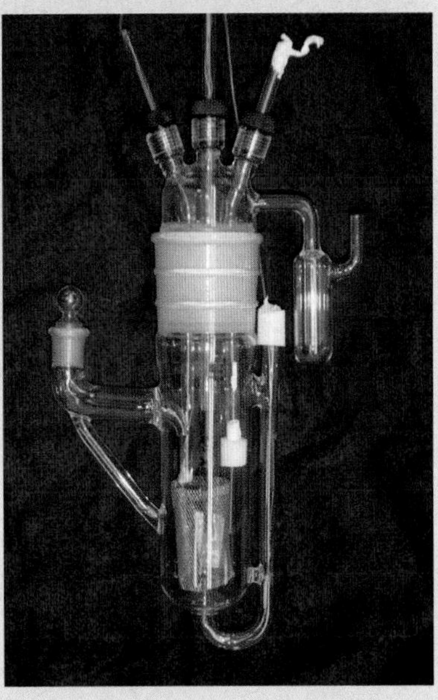

*A laboratory cell for the investigation of cold fusion claims that were thereby found to be false.* (Photo by Steven Krivit of a device at the U.S. Navy SPAWAR Systems Center in San Diego)

planets are harder to detect. Low-mass planets don't tug on their stars very much, and present-day spectrographs can't detect the very small velocity changes that these gentle tugs produce. Planets with longer periods are harder to detect because astronomers have not been making high-precision observations for a long enough time. Jupiter takes 11 years to circle the sun once, so it will take years for astronomers to see the longer-period wobbles produced by planets lying farther from their stars. You should not be surprised that the first planets discovered are massive and have short orbital periods.

The new planets may seem odd for another reason. As you have learned for our solar system, the large planets formed farther from the sun where the solar nebula was colder and ices could condense. How could big planets near their stars (called "**hot Jupiters**") have formed? Theoretical calculations indicate that planets forming in an especially dense disk of matter could spiral inward as they sweep up gas and planetesimals. That means it is possible for a few planets to become the massive, short-period planets that are detected most easily.

Many of the newly discovered extrasolar planets have elliptical orbits, and that seems odd when compared with our solar system, in which the planetary orbits are nearly circular. Theorists point out, however, that planets in some young planetary systems can interact with each other and can be thrown into elliptical orbits. This effect is probably rare in planetary systems, but astronomers find these extreme systems more easily because they tend to produce easily detected wobbles.

The preceding paragraphs should reassure you that massive planets in small or elliptical orbits do not fundamentally contradict the solar nebula theory. In fact, as astronomers continue to refine their instruments to detect smaller velocity shifts in stars, they find lower-mass planets. A planet with only about 5 times the mass of Earth, less than $\frac{1}{3}$ the mass of Uranus or Neptune, was found by the Doppler shift technique in 2007. This small extrasolar planet orbits its red dwarf parent star in only 13 Earth-days at a distance of 0.07 AU. With a diameter estimated to be 1.5 times larger than Earth and a

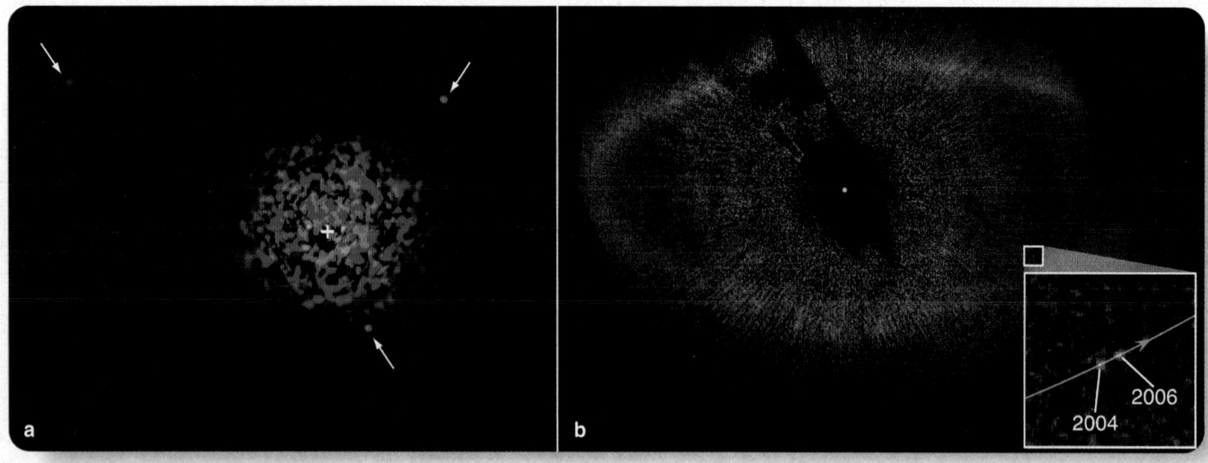


**Figure 19-17**

(a) Visible-wavelength images of 3 planets at apparent distances of 25, 40, and 70 AU from the star HR 8799. (C. Marois et al./ National Research Council of Canada/AURA/Keck Observatories) (b) Near-infrared image of a planet orbiting about 120 AU from the star Fomalhaut (alpha Piscis Austrinus), at the inside edge of that star's previously known debris ring. The inset shows motion of the object between 2004 and 2006, at the correct rate for a planet at that position orbiting a star of that mass. (P. Kalas et al./ STScI/NASA)

surface temperature that would permit liquid water, it is clearly not a Jovian planet.

Among the planets found so far, most are orbiting stars that are metal rich rather than metal poor. This supports the scenario of planet formation by the accretion of a core of solids and the later accumulation of gas. It is evidence against formation by direct collapse, which does not require the presence of solids such as metals and silicates to start planet formation.

Actually getting an image of a planet orbiting another star is about as easy as photographing a bug crawling on the bulb of a searchlight miles away. Planets are small and dim and get lost in the glare of the stars they orbit. Nevertheless, during 2008, astronomers managed to image three planets around the A-type star HR 8799, using adaptive optics and coronagraph instruments mounted on the Gemini and Keck telescopes atop Mauna Kea, and 1 planet around the A-type star Fomalhaut (alpha Piscis Austrinis), using the Hubble Space Telescope's near-infrared camera (■ Figure 19-17).

Searches for more extrasolar planets are being conducted. The Kepler space telescope mission, launched in 2009, will be sensitive enough to detect transits of planets as small as Earth. Space observatories will be able eventually to image Jovian and even Terrestrial planets directly around sunlike stars. The discovery of extrasolar planets gives astronomers added confidence in the solar nebula theory. The theory predicts that planets are common, and astronomers are finding them orbiting many stars.

## SCIENTIFIC ARGUMENT

***Why are debris disks evidence that planets have already formed?***

Sometimes a good scientific argument combines evidence, theory, and an astronomer's past experience, a kind of scientific common sense. Certainly the cold debris disks seen around stars like Vega are not places where planets are forming. They are not hot enough or dense enough to be young disks. Rather, the debris disks must be older, and the dust is being produced by collisions among comets, asteroids, and Kuiper belt objects. Small dust particles would be blown away or destroyed relatively quickly, so these collisions must be a continuing process. The successful solar nebula theory gives astronomers reason to believe that where you find comets, asteroids, and Kuiper belt objects, you should also find planets, so the dust disks seem to be evidence that planets have already formed in such systems.

Now build a new argument. **What direct evidence can you cite that planets orbit other stars?**

## What Are We? Planetwalkers

The matter you are made of came from the big bang, and it has been cooked into a wide variety of atoms inside stars. Now you can see how those atoms came to be part of Earth. Your atoms were in the cloud of gas that formed the solar system 4.6 billion years ago, and nearly all of that matter contracted to form the sun, but a small amount left behind in a disk formed planets. In the process, your atoms became part of Earth.

You are a planet-walker, and you have evolved to live on the surface of Earth. Are there other beings like you in the universe? Now you know that planets are common, and you can reasonably suppose that there are more planets in the universe than there are stars. However complicated the formation of the solar system was, it is a common process, so there may indeed be more planet-walkers living on other worlds.

But what are those distant planets like? Before you can go very far in your search for life beyond Earth, you need to explore the range of planetary types. It is time to pack your spacesuit and voyage out among the planets of our solar system, visit them one by one, and search for the natural principles that relate planets to each other. That journey begins in the next chapter.

# Study and Review

## Summary

- Descartes proposed that the solar system formed from a contracting vortex of matter—an **evolutionary hypothesis (p. 398)**. Buffon later suggested that a passing comet pulled matter out of the sun to form the planets—a **catastrophic hypothesis (p. 398)**. Later astronomers replaced the comet with a star to produce the **passing star hypothesis (p. 398)**.
- Laplace's **nebular hypothesis (p. 398)** required a contracting nebula to leave behind rings that formed each planet, but it could not explain the sun's low angular momentum, a puzzle known as the **angular momentum problem (p. 399)**.
- Hypotheses for the origin of the solar system have been either catastrophic or evolutionary. Catastrophic hypotheses depend on a rare event such as the collision of the sun with another star. Evolutionary hypotheses propose that the planets formed by gradual, natural processes. The evidence now strongly favors the **solar nebula theory (p. 399)**, an evolutionary scenario.
- Modern astronomy reveals that all the matter in the universe, including our solar system, was originally formed as hydrogen and helium in the big bang. Atoms heavier than helium were cooked up in nuclear reactions in later generations of stars. The sun and planets evidently formed from a cloud of gas and dust in the interstellar medium.
- The solar nebula theory proposes that the planets formed in a disk of gas and dust around the protostar that became the sun. Observations show that these disks are common.
- The solar system is disk shaped, with all the planets orbiting nearly in the same plane. The orbital revolution of all the planets, the rotation of most of the planets on their axes, and the revolution of most of their moons are all in the same direction, counterclockwise as seen from the north.
- The planets are divided into two groups. The inner four planets are **Terrestrial planets (p. 402)**—small, rocky, dense Earth-like worlds. The next four outward are Jupiter-like **Jovian planets (p. 402)** that are large and low density.
- All four of the Jovian worlds have ring systems and large families of moons. The Terrestrial planets have no rings and few moons.
- Most of the **asteroids (p. 401)**, small, irregular, rocky bodies, are located between the orbits of Mars and Jupiter.
- The **Kuiper belt (p. 404)** is composed of small, icy bodies that orbit the sun beyond the orbit of Neptune.
- **Comets (p. 404)** are icy bodies that pass through the inner solar system along long elliptical orbits. As the ices vaporize and release dust, the comet develops a tail that points approximately away from the sun.
- **Meteoroids (p. 405)** that fall into Earth's atmosphere are vaporized by friction and are visible as streaks of light called **meteors (p. 404)**. Larger and stronger meteoroids may survive to reach the ground, where they are called **meteorites (p. 405)**.
- The age of a rocky body can be found by radioactive dating, based on the decay **half-life (p. 405)** of radioactive atoms. The oldest rocks from Earth, the moon, and Mars have ages over 4 billion years. The oldest objects in our solar system are some meteorites that have ages of 4.6 billion years. This is taken to be the age of the solar system.
- **Condensation (p. 409)** in the solar nebula converted some of the gas into solid bits of matter, which grow by **accretion (p. 409)** to form billions of **planetesimals (p. 409)**.
- Planets begin growing by accretion of solid material into **protoplanets (p. 409)**. Once a protoplanet approaches about 15 Earth masses, it can begin growing by **gravitational collapse (p. 410)** as it pulls in gas from the solar nebula.

# 20 Earth: The Standard of Comparative Planetology

*The beauty of planet Earth can be deceptive. It was not always as it is now, and it will not survive unchanged forever. It was born in the solar nebula, and its end will come when the sun dies.* (Jean-Bernard Carillet/Getty Images)

## Guidepost

In the preceding chapter, you learned how our solar system formed as a by-product of the formation of the sun. You also saw how distance from the sun determined the general composition of each planet. In this chapter, you begin your study of individual planets with Earth. You will come to see Earth not only as your home, but as a planet among other planets. On the way, you will answer four essential questions:

- **How does Earth fit among the Terrestrial planets?**
- **How has Earth changed since it formed?**
- **What is the inside of Earth like?**
- **How has Earth's atmosphere formed and evolved?**

Like a mountain climber establishing a base camp before attempting the summit, you will, in this chapter, establish your basis of comparison on Earth. In the following chapters you will visit worlds that are un-Earthly but, in some ways, familiar.

size and tempera
complex. Where
that question in
cal history of the

*Why do you exp*
In Chapter 19,
inner parts of th
planets could gr
condense from
made mostly of
As you visit t
everywhere. **Wha**

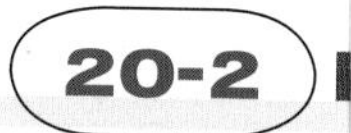

LIKE ALL THE TE
inner solar nebul
form, it began to

Four Stage

There is evidence
plus Earth's moo
(■ Figure 20-2).
The first stag
separation of mate
learned, Earth is d
different density.
occurred due to m
combination of ra
ing matter during
Earth melted, the
core.
The second st
surface formed. Th
tem made craters o
planets. As the deb
rate of cratering im
will learn in the n
crater counts and
large increase in th
heavy bombardme
affected all the plan
The third stage
ued to heat Earth's
mantle, where the
Some of that molte

*Nature evolves. The world was different yesterday.*

— PRESTON CLOUD, *COSMOS, EARTH AND MAN*

PLANETS, LIKE PEOPLE, ARE MORE ALIKE than they are different. They are described by the same basic principles, and their differences arise mostly because of small differences in background. To understand the planets, you can compare and contrast them to identify those principles, an approach called **comparative planetology.**

Earth is the ideal starting point for your study because it is the planet you know best. It is also a complex planet. Earth's interior is molten and generates a magnetic field. Its crust is active, with moving continents, earthquakes, volcanoes, and mountain building. Earth's oxygen-rich atmosphere is unique in the solar system. The properties of Earth will give you perspective on the other planets in our solar system.

## 20-1 A Travel Guide to the Terrestrial Planets

IF YOU VISIT THE CITY OF GRANADA in Spain, you will probably consult a travel guide. If it is a good guide, it will do more than tell you where to find museums and restrooms. It will give you a preview of what to expect. You are beginning a journey to the Earth-like worlds, so you should consult a travel guide and see what is in store.

### Five Worlds

You are about to visit Mercury, Venus, Earth, Earth's moon, and Mars. It may surprise you that the moon is included in your itinerary. It is, after all, just a natural satellite orbiting Earth and isn't one of the planets. But the moon is a fascinating world of its own, it makes a striking comparison with the other worlds on your list, and its history gives you important information about the history of Earth and the other planets.

■ Figure 20-1 compares the five worlds. The first feature you might notice is diameter. The moon is small, and Mercury is not much bigger. Earth and Venus are large and quite similar in size,

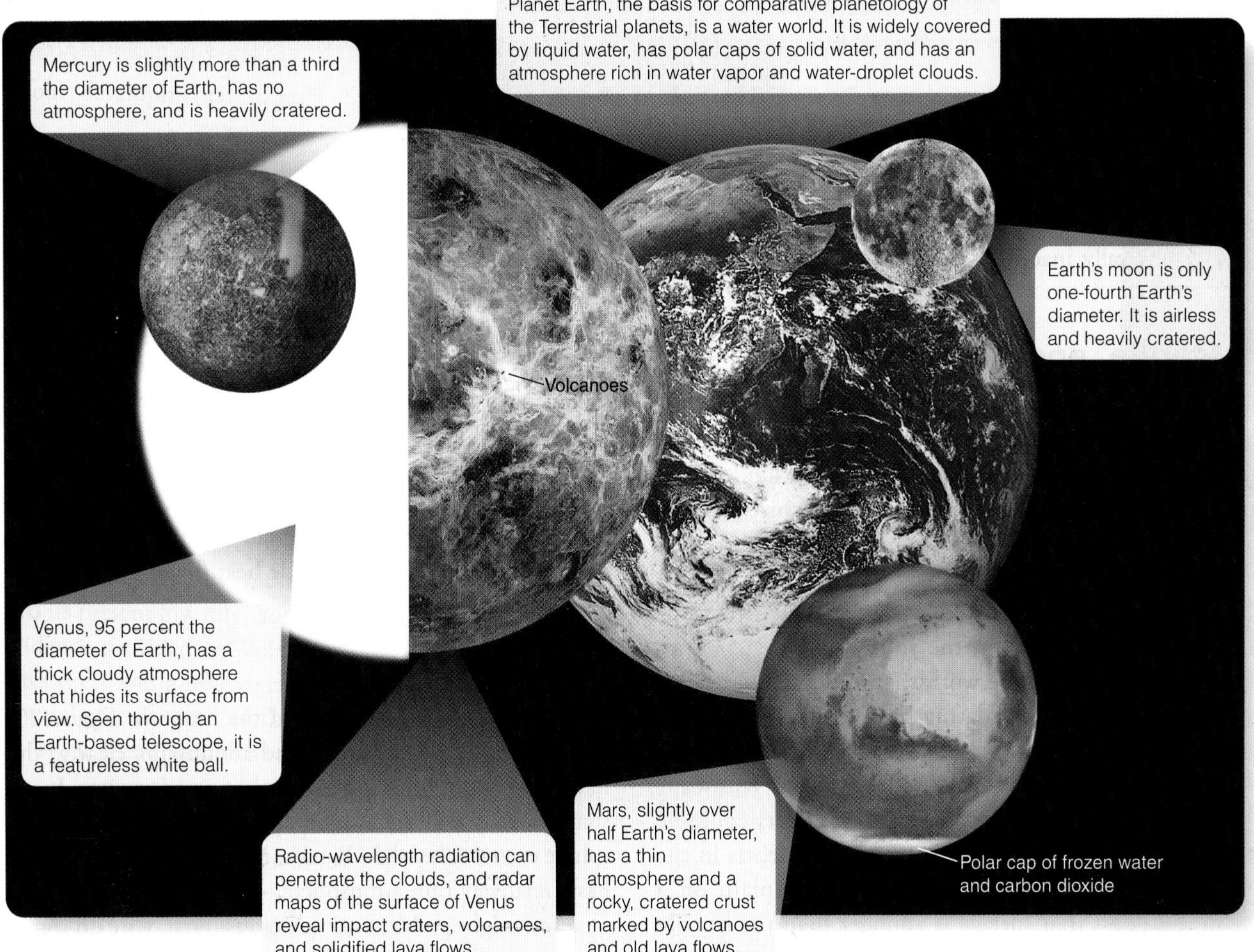

Planet Earth, the basis for comparative planetology of the Terrestrial planets, is a water world. It is widely covered by liquid water, has polar caps of solid water, and has an atmosphere rich in water vapor and water-droplet clouds.
Mercury is slightly more than a third the diameter of Earth, has no atmosphere, and is heavily cratered.
Earth's moon is only one-fourth Earth's diameter. It is airless and heavily cratered.
Volcanoes
Venus, 95 percent the diameter of Earth, has a thick cloudy atmosphere that hides its surface from view. Seen through an Earth-based telescope, it is a featureless white ball.
Radio-wavelength radiation can penetrate the clouds, and radar maps of the surface of Venus reveal impact craters, volcanoes, and solidified lava flows.
Mars, slightly over half Earth's diameter, has a thin atmosphere and a rocky, cratered crust marked by volcanoes and old lava flows.
Polar cap of frozen water and carbon dioxide

**■ Figure 20-1**

Planets in comparison. Earth and Venus are similar in size, but their atmospheres and surfaces are very different. The moon and Mercury are much smaller, and Mars is intermediate in size. (Moon: UCO/Lick Observatory: All planets: NASA)

can emphasize some of those stages over others and produce surprisingly different worlds.

## Exceptional Earth

Earth is a good standard for comparative planetology (**Celestial Profile 2**). Every major process on any rocky world in our solar system is represented in some form on Earth. Nevertheless, Earth is unusual, if not unique, among the planets in our solar system, in two ways: the presence of abundant liquid water and the presence of life.

First, 75 percent of Earth's surface is covered by liquid water. No other planet in our solar system has liquid water on its surface, although, as you will learn in later chapters, Mars had surface water long ago, Venus probably did, and some moons in the outer solar system show evidence of having liquid water under their surfaces. Water fills Earth's oceans, evaporates into the atmosphere, forms clouds, and then falls as rain. Water falling on the continents flows downhill to form rivers that flow back to the sea and, in doing so, produces intense erosion. Entire mountain ranges can literally dissolve and wash away in only a few tens of millions of years, less than 1 percent of Earth's total age. You will not see such rapid erosion on most worlds. Liquid water is, in fact, a rare material on most planets.

Your home planet is special in a second way. Some of the matter on the surface of this world is alive, and a small part of that living matter, including you, is aware. No one is sure how the presence of living matter has affected the evolution of Earth, but this process seems to be totally missing from other worlds in our solar system. Furthermore, as you will learn later in this chapter, the thinking part of the life on Earth, humankind, is actively altering our planet.

### SCIENTIFIC ARGUMENT

***Why should you think Earth went through an early stage of cratering?***

When you build a scientific argument, take great care to distinguish between theory and evidence. Recall from the previous chapter that the planets formed by the accretion of planetesimals from the solar nebula. The proto-Earth may have been molten as it formed, but as soon as it grew cool enough to form a solid crust, the remaining planetesimal impacts would have formed craters. So you can reason from the solar nebula hypothesis that Earth should have been cratered. But you can't use a theory as evidence to support some other theory. To find real observational evidence, you need only look at the moon. The moon has craters, and so does every other old surface in our solar system. There must have been a time, when the solar system was young, during which there were large numbers of objects striking all the planets and moons and blasting out craters. If it happened to other worlds in our solar system, it must have happened to Earth, too.

The best evidence to support your argument would be lots of craters on Earth, but, of course, there are few craters on Earth. Extend your argument. **How has the presence of liquid water on Earth affected the appearance of Earth's crust?**

*Earth's surface is marked by high continents and low seafloors, but the crust is only 10 to 60 km thick. Below that lie a deep mantle and an iron core.* (NGDC)

## *Celestial Profile 2: Earth*

### *Motion:*

| | |
|---|---|
| Average distance from the sun | 1.00 AU (1.50 × $10^8$ km) |
| Eccentricity of orbit | 0.017 |
| Maximum distance from the sun | 1.017 AU (1.52 × $10^8$ km) |
| Minimum distance from the sun | 0.983 AU (1.47 × $10^8$ km) |
| Inclination of orbit to ecliptic | 0° (by definition) |
| Average orbital velocity | 29.8 km/s |
| Orbital period | 1.0000 y (365.25 days) |
| Period of rotation | 24.00 h (with respect to the sun) |
| Period of rotation | 23.93 h (with respect to the stars) |
| Inclination of equator to orbit | 23.4° |

### *Characteristics:*

| | |
|---|---|
| Equatorial diameter | 1.28 × $10^8$ km |
| Mass | 5.97 × $10^{24}$ kg |
| Average density | 5.50 g/cm$^3$ (4.07 g/cm$^3$ uncompressed) |
| Surface gravity | 1.00 Earth gravity |
| Escape velocity | 11.2 km/s |
| Surface temperature | −90° to 60°C (−130° to 140°F) |
| Average albedo | 0.39 |
| Oblateness | 0.0034 |

### *Personality Point:*

Earth comes, through Old English *eorthe* and Greek *Eraze,* from the Hebrew *erez,* which means ground. *Terra* comes from the Roman goddess of fertility and growth; thus, *Terra Mater,* Mother Earth.

size and temperature are important. The second question is more complex. Where did those atmospheres come from? To answer that question in later chapters, you will have to study the geological history of these worlds.

### SCIENTIFIC ARGUMENT

***Why do you expect the inner planets to be high-density worlds?*** In Chapter 19, you saw how the inner planets formed from hot inner parts of the solar nebula. No ice solidified there, so the inner planets could grow only from particles of rock and metal able to condense from hot gas. So, you expect the inner planets to be made mostly of rock and metal, which are dense materials.

As you visit the Terrestrial planets, you will find craters almost everywhere. **What made all of those craters?**

## 20-2 Earth as a Planet

LIKE ALL THE TERRESTRIAL PLANETS, Earth formed from the inner solar nebula about 4.6 billion years ago. Even as it took form, it began to change.

### Four Stages of Planetary Development

There is evidence that Earth and the other Terrestrial planets, plus Earth's moon, passed through four developmental stages (■ Figure 20-2).

The first stage of planetary evolution is *differentiation,* the separation of material according to density. As you have already learned, Earth is differentiated, meaning, separated into layers of different density. That differentiation is understood to have occurred due to melting of Earth's interior caused by heat from a combination of radioactive decay plus energy released by infalling matter during the planet's formation. Once the interior of Earth melted, the densest materials were able to sink to the core.

The second stage, *cratering,* could not begin until a solid surface formed. The heavy bombardment of the early solar system made craters on Earth just as it did on the moon and other planets. As the debris in the young solar system cleared away, the rate of cratering impacts fell gradually to its present low rate. You will learn in the next chapter that evidence provided by lunar crater counts and rock samples indicates there was a temporary large increase in the impact cratering rate near the end of the heavy bombardment, and this violent event likely would have affected all the planets.

The third stage, *flooding,* began as radioactive decay continued to heat Earth's interior and caused rock to melt in the upper mantle, where the pressure was lower than in the deep interior. Some of that molten rock welled up through cracks in the crust and flooded the deeper impact basins. Later, as the environment cooled, water fell as rain and flooded the basins to form the first oceans. Note that on Earth, basin flooding was first by lava and later by water.

The fourth stage, *slow surface evolution,* has continued for at least the past 3.5 billion years. Earth's surface is constantly changing as sections of crust slide over and against each other, push up mountains, and shift continents. In addition, moving air and water erode the surface and wear away geological features. Almost all traces of the first billion years of Earth's history have been destroyed by the active crust and erosion.

Terrestrial planets pass through these four stages, but differences in mass, temperature, and composition among the planets

**Four Stages of Planetary Development**

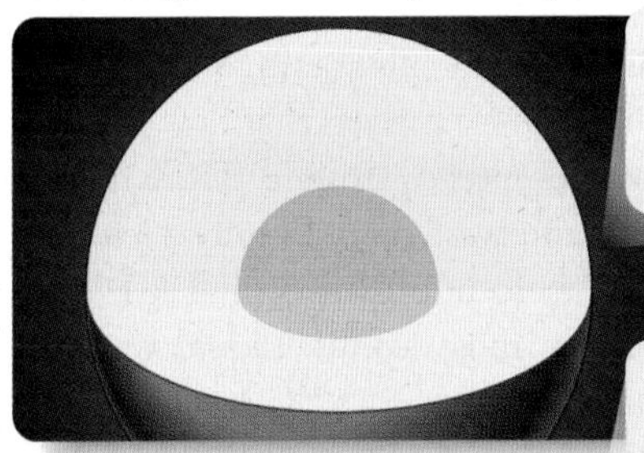

Differentiation produces a dense core, thick mantle, and low-density crust.


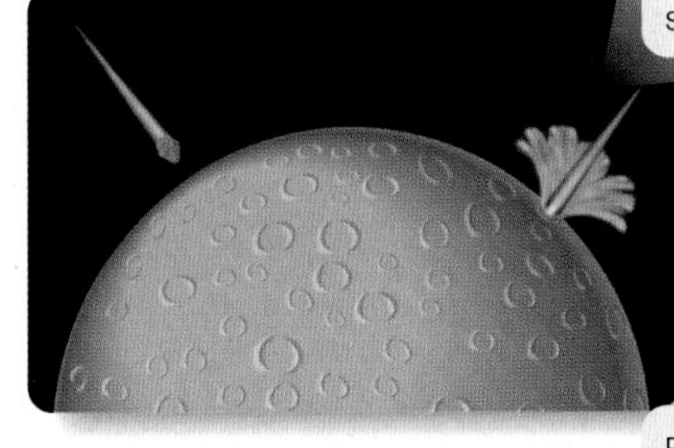

The young Earth was heavily bombarded in the debris-filled early solar system.

Flooding by molten rock and later by water can fill lowlands.

Slow surface evolution continues due to geological processes, including erosion.


■ **Figure 20-2**

The four stages of Terrestrial planetary development are illustrated for Earth.

can emphasize some of those stages over others and produce surprisingly different worlds.

## Exceptional Earth

Earth is a good standard for comparative planetology (**Celestial Profile 2**). Every major process on any rocky world in our solar system is represented in some form on Earth. Nevertheless, Earth is unusual, if not unique, among the planets in our solar system, in two ways: the presence of abundant liquid water and the presence of life.

First, 75 percent of Earth's surface is covered by liquid water. No other planet in our solar system has liquid water on its surface, although, as you will learn in later chapters, Mars had surface water long ago, Venus probably did, and some moons in the outer solar system show evidence of having liquid water under their surfaces. Water fills Earth's oceans, evaporates into the atmosphere, forms clouds, and then falls as rain. Water falling on the continents flows downhill to form rivers that flow back to the sea and, in doing so, produces intense erosion. Entire mountain ranges can literally dissolve and wash away in only a few tens of millions of years, less than 1 percent of Earth's total age. You will not see such rapid erosion on most worlds. Liquid water is, in fact, a rare material on most planets.

Your home planet is special in a second way. Some of the matter on the surface of this world is alive, and a small part of that living matter, including you, is aware. No one is sure how the presence of living matter has affected the evolution of Earth, but this process seems to be totally missing from other worlds in our solar system. Furthermore, as you will learn later in this chapter, the thinking part of the life on Earth, humankind, is actively altering our planet.

### SCIENTIFIC ARGUMENT

***Why should you think Earth went through an early stage of cratering?***

When you build a scientific argument, take great care to distinguish between theory and evidence. Recall from the previous chapter that the planets formed by the accretion of planetesimals from the solar nebula. The proto-Earth may have been molten as it formed, but as soon as it grew cool enough to form a solid crust, the remaining planetesimal impacts would have formed craters. So you can reason from the solar nebula hypothesis that Earth should have been cratered. But you can't use a theory as evidence to support some other theory. To find real observational evidence, you need only look at the moon. The moon has craters, and so does every other old surface in our solar system. There must have been a time, when the solar system was young, during which there were large numbers of objects striking all the planets and moons and blasting out craters. If it happened to other worlds in our solar system, it must have happened to Earth, too.

The best evidence to support your argument would be lots of craters on Earth, but, of course, there are few craters on Earth. Extend your argument. **How has the presence of liquid water on Earth affected the appearance of Earth's crust?**

*Earth's surface is marked by high continents and low seafloors, but the crust is only 10 to 60 km thick. Below that lie a deep mantle and an iron core.* (NGDC)

## *Celestial Profile 2: Earth*

### *Motion:*

| | |
|---|---|
| Average distance from the sun | 1.00 AU ($1.50 \times 10^8$ km) |
| Eccentricity of orbit | 0.017 |
| Maximum distance from the sun | 1.017 AU ($1.52 \times 10^8$ km) |
| Minimum distance from the sun | 0.983 AU ($1.47 \times 10^8$ km) |
| Inclination of orbit to ecliptic | 0° (by definition) |
| Average orbital velocity | 29.8 km/s |
| Orbital period | 1.0000 y (365.25 days) |
| Period of rotation | 24.00 h (with respect to the sun) |
| Period of rotation | 23.93 h (with respect to the stars) |
| Inclination of equator to orbit | 23.4° |

### *Characteristics:*

| | |
|---|---|
| Equatorial diameter | $1.28 \times 10^8$ km |
| Mass | $5.97 \times 10^{24}$ kg |
| Average density | 5.50 g/cm$^3$ (4.07 g/cm$^3$ uncompressed) |
| Surface gravity | 1.00 Earth gravity |
| Escape velocity | 11.2 km/s |
| Surface temperature | −90° to 60°C (−130° to 140°F) |
| Average albedo | 0.39 |
| Oblateness | 0.0034 |

### *Personality Point:*

*Earth* comes, through Old English *eorthe* and Greek *Eraze,* from the Hebrew *erez,* which means ground. *Terra* comes from the Roman goddess of fertility and growth; thus, *Terra Mater,* Mother Earth.

## 20-3 The Solid Earth

ALTHOUGH YOU MIGHT THINK of earth as solid rock, it is in fact neither entirely solid nor entirely rock. The thin crust seems solid, but it floats and shifts on a semiliquid layer of molten rock just below the crust. Below that lies a deep, rocky mantle surrounding a core of liquid metal. Much of what you see on Earth's surface is determined by conditions and processes in its interior.

### Earth's Interior

The theory of the origin of planets from the solar nebula predicts that Earth should have melted and differentiated into a dense metallic core and a dense mantle with a low-density silicate crust. But did it? Where's the evidence? Earth's average density can be calculated easily from its known mass and density. Clearly, the silicate rocks on Earth's surface have lower density than material inside the planet. But, what more can be determined about Earth's interior?

High temperature and tremendous pressure in Earth's interior make any direct exploration impossible. Even the deepest oil wells extend only a few kilometers down and don't reach through the crust. It is impossible to drill far enough to sample Earth's core. Yet Earth scientists have studied the interior and found clear evidence that Earth did differentiate (**How Do We Know? 20-2**).

This exploration of Earth's interior is possible because earthquakes produce vibrations called **seismic waves** that travel through the crust and interior and eventually register on sensitive detectors called **seismographs** all over the world (■ Figure 20-3). Two kinds of seismic waves are important to this discussion. The **pressure (*P*) waves** are much like sound waves in that they travel as a sequence of compressions and decompressions. As a *P* wave passes, particles of matter vibrate back and forth parallel to the direction of wave travel (■ Figure 20-4a). In contrast, the **shear (*S*) waves** move as displacements of particles perpendicular to the waves' direction of travel (Figure 20-4b). That means that *S* waves distort the material but do not compress it. Normal sound waves are pressure waves, whereas the water waves you can surf on, and vibrations you see in a bowl of jelly, are shear waves. Because *P* waves are compression waves, they can move through a liquid. *S* waves can move along the surface of a liquid, but not through it. A glass of water can't shimmy like jelly because a liquid does not have the rigidity required to transmit *S* waves.

The *P* and *S* waves caused by an earthquake do not travel in straight lines or at constant speed within Earth. The waves may reflect off boundaries between layers of different density, or they may be refracted as they pass through a boundary. In addition, the gradual increase in temperature and density toward Earth's center means the speed of sound increases as

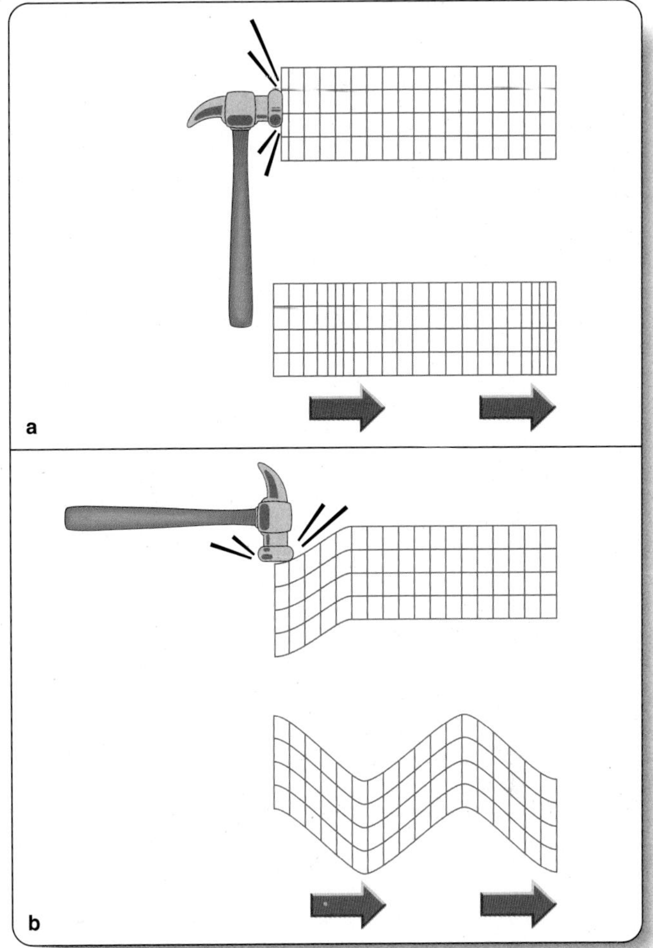


■ **Figure 20-4**

(a) *P*, or pressure, waves, like sound waves in air, travel as a region of compression. (b) *S*, or shear, waves, like vibrations in a bowl of jelly, travel as displacements perpendicular to the direction of travel. *S* waves tend to travel more slowly than *P* waves and cannot travel through liquids.

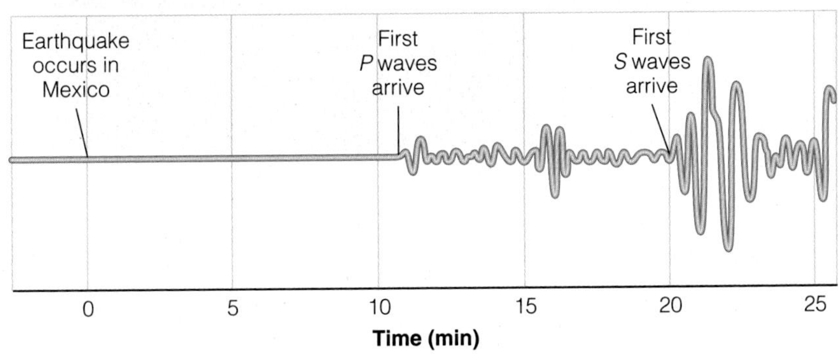


■ **Figure 20-3**

A seismograph in northern Canada made this record of seismic waves from an earthquake in Mexico. The first vibrations, *P* waves, arrived 11 minutes after the quake, but the slower *S* waves took 20 minutes to make the journey.

# How Do We Know? | 20-2

## Studying an Unseen World

**How can studying what can't be seen save your life?** Science tells us how nature works, and the basis for scientific knowledge is evidence gathered through observation. But much of the natural world can't be observed directly because it is too small, or far away, or deep underground. Yet geologists describe molten rock deep inside Earth, and biologists discuss the structure of genetic molecules. So how can these scientists know about things they can't observe directly?

A virus, which can be as common as a cold or as deadly as Ebola, contains a tiny bit of genetic information in the form of a DNA molecule. You have surely had a virus, but you've never seen one. Even under the best electron microscopes, a virus can be seen only as a hazy pattern of shadows. Nevertheless scientists know enough about them to devise ingenious ways to protect us from viral disease.

A virus is DNA hidden inside a protective coat of protein molecules, which is a rigid molecular lattice almost like a mineral. In fact, a culture of viruses can be crystallized, and the shapes of the crystals reveal the shape and structure of the virus. Unlike a crystal of calcite, however, a crystal of viruses also contains genetic information.

Scientists can make a vaccine to protect against a certain virus if they can identify a unique molecular pattern on the protein coat. The vaccine is harmless but contains that same pattern and trains your body's immune system to recognize the pattern and attack it. Vaccines significantly reduce the danger of common illness such as chicken pox and influenza, and they have virtually wiped out devastating diseases like polio and smallpox in the developed world. Researchers are currently working on a vaccine for HIV/AIDS that would potentially save millions of lives.

Even though viruses are too small to see, scientists can use chains of inference and the interaction of theory and evidence to deduce the structure of a virus. Whether it is a virus or the roots of a volcano, science takes us into realms beyond human experience and allows us to see the unseen.

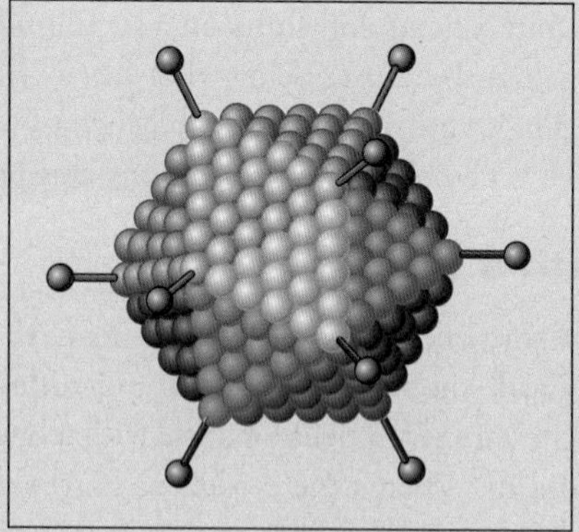

*The electron microscope allows biologists to deduce the elegant structure of the virus that causes the common infection called pink eye.*

well. These changes cause sound waves to be refracted as they travel through Earth's interior, meaning that, instead of following straight lines, seismic waves curve away from the denser, hotter central regions. Geoscientists can use the arrival times of reflected and refracted seismic waves from distant earthquakes to construct a model of Earth's interior.

Such studies confirm that the interior consists of three parts: a central core, a thick mantle, and a thin crust. *S* waves provide an important clue to the nature of the core. When an earthquake occurs, no direct *S* waves pass through the core to register on seismographs on the opposite side of Earth, as if the core were casting a shadow (■ Figure 20-5). The absence of S waves shows that the core is mostly liquid, and the size of the *S*-wave shadow fixes the size of the core at about 55 percent of Earth's radius. Mathematical models predict that the core is also hot (about 6000 K), dense (about 14 g/cm$^3$), and composed of iron and nickel.

Earth's core is as hot as the surface of the sun, but it is under such tremendous pressure that the material cannot vaporize. Because of its high temperature, most of the core is a liquid. Nearer the center the material is under even higher pressure,

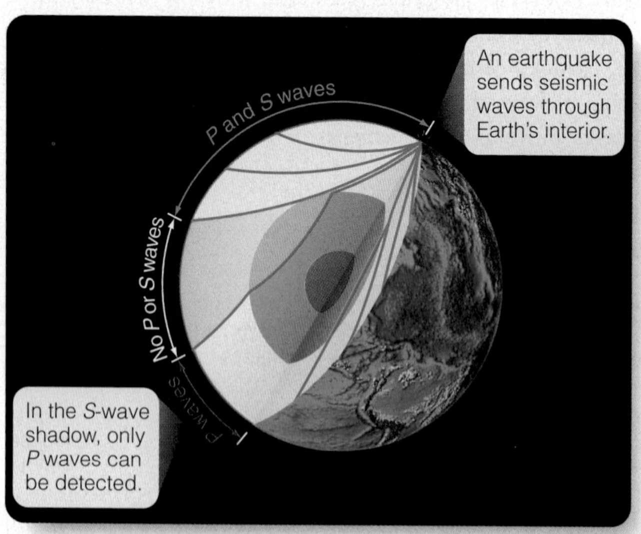
An earthquake sends seismic waves through Earth's interior.
P and S waves
No P or S waves
P waves
In the S-wave shadow, only P waves can be detected.

■ **Figure 20-5**

*P* and *S* waves give you clues to the structure of Earth's interior. No direct *S* waves from an earthquake reach the side of Earth opposite their source, indicating that Earth's core is liquid. The size of the *S* wave "shadow" tells you the size of the liquid outer part of the core.

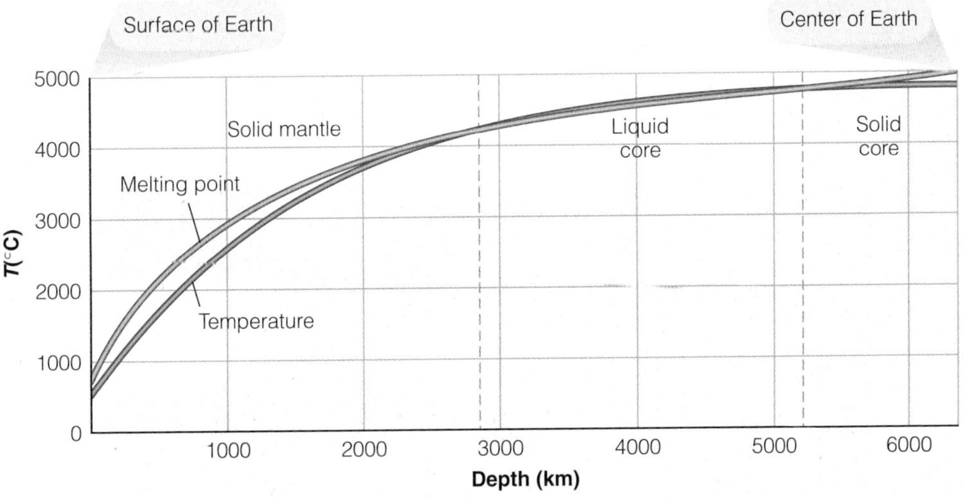


■ **Figure 20-6**

Theoretical models combined with observations of the velocity of seismic waves reveal the temperature inside Earth (blue line). The melting point of the material (red line) is determined by its composition and by the pressure. In the mantle and in the inner core, the melting point is higher than the existing temperature, and the material is not molten.

which in turn raises the melting point so high that the material cannot melt (■ Figure 20-6). That is why there is an inner core of solid iron and nickel. Estimates suggest the inner core's radius is about 22 percent that of Earth.

The paths of seismic waves in the mantle, the layer of dense rock that lies between the molten core and the crust, show that it is not molten, but it is not precisely solid either. Mantle material behaves like a **plastic,** a material with the properties of a solid but capable of flowing under pressure. The asphalt used in paving roads is a common example of a plastic. It shatters if struck with a sledgehammer, but it bends under the steady weight of a heavy truck. Just below Earth's crust, where the pressure is less than at greater depths, the mantle is most plastic.

Earth's rocky crust is made up of low-density rocks and floats on the denser mantle. The crust is thickest under the continents, up to 60 km thick, and thinnest under the oceans, where it is only about 10 km thick. Unlike the mantle, the crust is brittle and can break when it is stressed.

Although Earth's core is only 2000 miles from you, it is completely inaccessible. Earth's seismic activity reveals some of Earth's innermost secrets. But there is another source of evidence about Earth's interior—its magnetic field.

## The Magnetic Field

Apparently, Earth's magnetic field is a direct result of its rapid rotation and its molten metallic core. Internal heat forces the liquid core to circulate with convection while Earth's rotation turns it about an axis. The core is a highly conductive iron–nickel alloy, an even better electrical conductor than copper, the material commonly used for electrical wiring. The rotation of this convecting, conducting liquid generates Earth's magnetic field in a process called the dynamo effect (■ Figure 20-7). That is the same process that generates the solar magnetic field in the convective layers of the sun (see Chapter 8), and you will see it again when you explore other planets.

Earth's magnetic field protects it from the solar wind. Blowing outward from the sun at about 400 km/s, the solar wind consists of ionized gases carrying a small part of the sun's magnetic field. When the solar wind encounters Earth's magnetic field, it is deflected like water flowing around a boulder in a stream. The surface where the solar wind is first deflected is called the **bow shock,** and the cavity dominated by Earth's magnetic field is called the **magnetosphere** (■ Figure 20-8a). High-energy particles from the solar wind leak into the magnetosphere and become trapped within Earth's magnetic field to produce the **Van Allen belts** of radiation. You will see in later chapters that all planets that have magnetic fields have bow shocks, magnetospheres, and radiation belts.

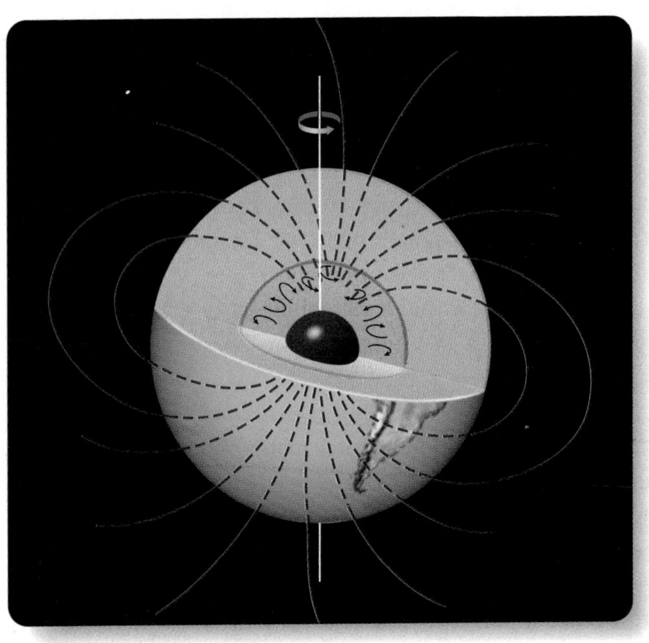

■ **Figure 20-7**

The dynamo effect couples convection in the liquid core with Earth's rotation to produce electric currents that are believed to be responsible for Earth's magnetic field.

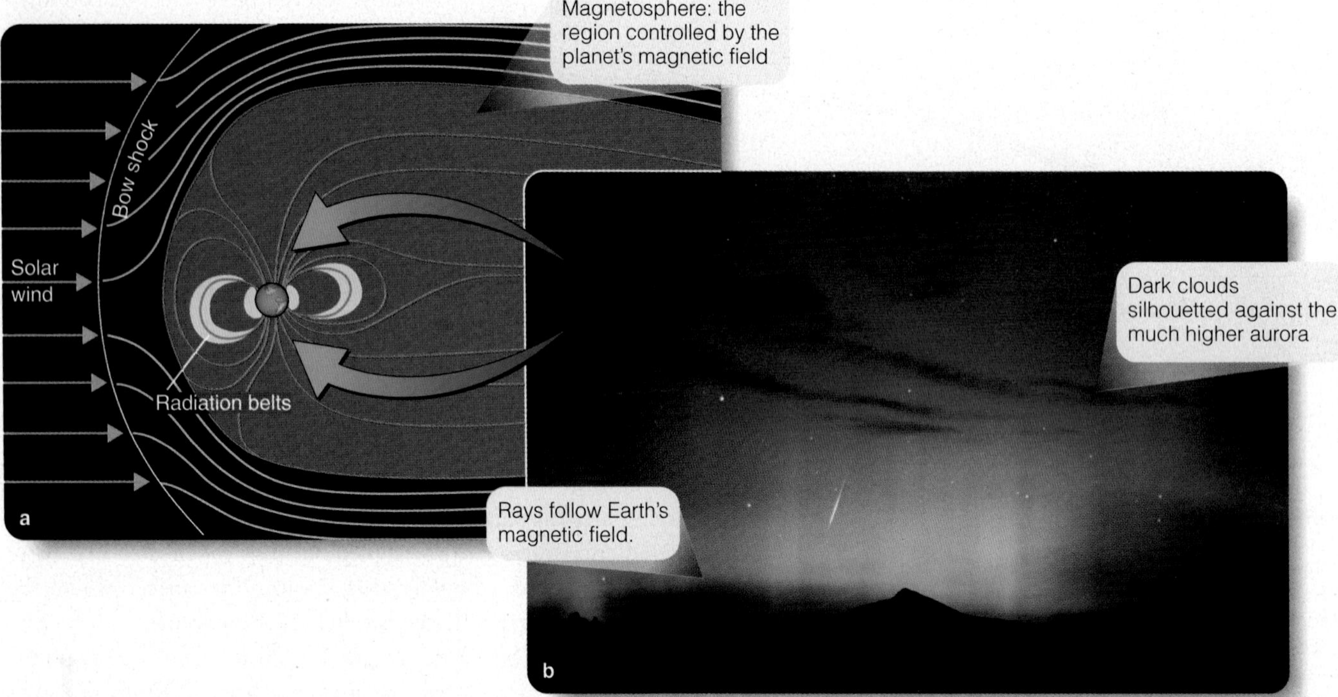
Magnetosphere: the region controlled by the planet's magnetic field
Bow shock
Solar wind
Radiation belts
a
Dark clouds silhouetted against the much higher aurora
Rays follow Earth's magnetic field.
b

**Figure 20-8**

(a) Earth's magnetic field dominates space around Earth by deflecting the solar wind and trapping high-energy particles in radiation belts. The magnetic field lines enter Earth's atmosphere around the north and south magnetic poles. (b) Powerful currents flow down along the magnetic field lines near the poles and excite gas atoms to emit photons, creating auroras. Colors are produced as different atoms are excited. Note the meteor (shooting star). (Jimmy Westlake)

Earth's magnetic field produces the dramatic and beautiful auroras, glowing rays and curtains of light in the upper atmosphere (Figure 20-8b). The solar wind carries charged particles past Earth's extended magnetic field, and this generates tremendous electrical currents that flow into Earth's atmosphere near the north and south magnetic poles (see Chapter 8). The currents ionize gas atoms in Earth's atmosphere, and when the ionized atoms capture electrons and recombine, they produce an emission spectrum as if they were part of a vast "neon" sign (see Chapter 7).

Although you can be confident that Earth's magnetic field is generated within its molten core, many mysteries remain. For example, rocks retain traces of the magnetic field in which they solidify, and some contain fields that point backward. That is, they imply that Earth's magnetic field was reversed at the time they solidified. Careful analysis of such rocks indicates that Earth's field has reversed itself in a very irregular pattern, on average every 700,000 years or so during the past 30 million years, with the north magnetic pole becoming the south magnetic pole and vice versa. The reversals are poorly understood, but they may be related to changes in the core's convection.

Convection in Earth's core is important because it generates the magnetic field. As you will see in the next section, convection in the mantle constantly remakes Earth's surface.

## Earth's Active Crust

Earth's crust is composed of low-density rock that floats on the mantle. The image of a rock floating may seem odd, but recall that the rock of the mantle is very dense. Also, just below the crust, the mantle rock tends to be highly plastic, so great sections of low-density crust do indeed float on the semi-liquid mantle like great lily pads floating on a pond.

The motion of the crust and the erosive action of water make Earth's crust highly active. Read **The Active Earth** on pages 434–435 and notice three important points and six new terms:

1. *Plate tectonics*, the motion of crustal plates, produces much of the geological activity on Earth. Plates spreading apart can form *rift valleys*, or, on the ocean floor, *midocean rises* where molten rock solidifies to form *basalt.* A plate sliding into a *subduction zone* can trigger volcanism, and the collision of plates can produce *folded mountain ranges.* Chains of volcanoes such as the Hawaiian islands can result when a plate moves horizontally across a hot spot.

2. Notice how the continents on Earth's surface have moved and changed over periods of hundreds of millions of years. A hundred million years is only 0.1 billion years, so

sections of Earth's crust are in rapid motion relative to the 4.6 billion-year age of the planet.

3 Most of the geological features you know—mountain ranges, the Grand Canyon, and even the familiar outline of the continents—are recent products of Earth's active surface. Earth's surface is constantly renewed. The oldest rocks on Earth, small crystals of the mineral zircon from western Australia, are 4.4 billion years old. Most of the crust is much younger than that. Most of the mountains and valleys you see around you are no more than a few tens of millions of years old.

The average speed of plate movement is slow, but sudden movements do occur. Plate margins can stick, accumulate stress, and then release it suddenly. For example, that's what happened in 2004 along a major subduction zone in the Indian Ocean. The total motion was as much as 15 meters, and the resulting earthquake caused devastating tsunamis (tidal waves). Every day, minor earthquakes occur on moving faults, and the stress that builds in those faults that are sticking will eventually be released in major earthquakes.

Earth's active crust explains why Earth contains so few impact craters. The moon is richly cratered, but Earth's surface has only about 150 impact craters. Plate tectonics and erosion have destroyed all but the most recent craters on Earth.

You can see that Earth's geology is dominated by two dramatic forces. Heat rising from the interior drives plate tectonics. Just below the thin crust of solid rock lies a churning molten layer that rips the crust to fragments and pushes the pieces about like rafts of wood on a pond. The second force modifying the crust is water. It falls as rain and snow and tears down mountains, erodes river valleys, and washes any raised ground into the sea. Tectonics builds up mountains and continents, and then erosion rips them down.

### SCIENTIFIC ARGUMENT

***What evidence indicates that Earth has a liquid metal core?***

A good scientific argument focuses on evidence. In this case, the evidence is indirect because you can never visit Earth's core. Seismic waves from distant earthquakes pass through Earth, but a certain kind of wave, the *S* type, does not pass through the core. Because the *S* waves cannot move through a liquid, scientists conclude that Earth's core is partly liquid. Earth's magnetic field is further evidence of a liquid metallic core. The theory for the generation of magnetic fields, the dynamo effect, requires a moving, conducting liquid (for a planet) or gas (for a star) in the interior. If Earth's core were not partly a liquid metal, it would not be able to generate a magnetic field.

Two different kinds of evidence tell you that our planet has a liquid core. Can you build a new argument? **What evidence can you cite to support the theory of plate tectonics?**

## 20-4 Earth's Atmosphere

You can't tell the story of Earth without mentioning its atmosphere. Not only is it necessary for life, but it is also intimately related to the crust. It affects the surface through erosion by wind and water, and in turn the chemistry of Earth's surface affects the composition of the atmosphere.

### Origin of the Atmosphere

Until a few decades ago, planetary scientists thought the early Earth might have attracted small amounts of gases such as hydrogen, helium, and methane from the solar nebula to form a **primeval atmosphere.** According to that old hypothesis, slow decay of radioactive elements eventually heated Earth's interior; melted it; caused it to differentiate into core, mantle, and crust; and triggered widespread volcanism.

When a volcano erupts, 50 to 80 percent of the gas released is water vapor. The rest is mostly carbon dioxide, nitrogen, and smaller amounts of sulfur gases such as hydrogen sulfide—the rotten-egg gas that you smell if you visit geothermal pools and geysers such as those at Yellowstone National Park. These gases could have diluted the primeval atmosphere and eventually produced a **secondary atmosphere** rich in carbon dioxide, nitrogen, and water vapor.

In contrast, a modern understanding of planet building shows that Earth formed so rapidly that it was substantially heated by the impacts of infalling material, as well as by radioactive decay. If Earth's surface was molten as it formed, then outgassing would have been continuous, and the early atmosphere would have been rich in carbon dioxide, nitrogen, and water vapor from the beginning. In other words, planetary scientists now think Earth went straight to the volcanic "secondary atmosphere" and never had a hydrogen-rich primeval atmosphere. You will find in Chapter 26 that the lack of hydrogen in Earth's original atmosphere has important implications for how life began on our planet.

Astronomers also have suspected that some of the abundant water on Earth arrived late in the formation process as a bombardment of volatile-rich planetesimals. These icy bodies, the theory goes, were scattered by the growing mass of the outer planets and by the outward migration of Saturn, Uranus, and Neptune. The inner solar system, including Earth, would then have been bombarded by a storm of comets, some of which could have supplied some or all of Earth's water. That hypothesis once faced a serious objection. Spectroscopic studies of a few comets revealed that the ratio of deuterium to hydrogen in comets does not match the ratio in the water on Earth. Some astronomers thought this meant that the water now on Earth could not have arrived in cometlike planetesimals. However, studies of Comet LINEAR, which broke up in 1999 as it passed near the sun, show that the water in that comet had a ratio of isotopes

# The Active Earth

**1** Our world is an astonishingly active planet. Not only is it rich in water and therefore subject to rapid erosion, but its crust is divided into moving sections called plates. Where plates spread apart, lava wells up to form new crust; where plates push against each other, they crumple the crust to form mountains. Where one plate slides over another, you see volcanism. This process is called **plate tectonics**, referring to the Greek word for "builder." (An architect is literally an arch builder.)

A **rift valley** forms where continental plates begin to pull apart. The Red Sea has formed where Africa has begun to pull away from the Arabian peninsula.

Midocean rise

Red Sea

Midocean rise

National Geophysical Data Center

A typical view of planet Earth

William K. Hartmann

Janet Seeds

Mountains are common on Earth, but they erode away rapidly because of the abundant water.

**1a** Evidence of plate tectonics was first found in ocean floors, where plates spread apart and magma rises to form **midocean rises** made of rock called **basalt,** a rock typical of solidified lava. Radioactive dating shows that the basalt is younger near the midocean rise. Also, the ocean floor carries less sediment near the midocean rise. As Earth's magnetic field reverses back and forth, it is recorded in the magnetic fields frozen into the basalt. This produces a magnetic pattern in the basalt that shows that the seafloor is spreading away from the midocean rise.

**1b** A **subduction zone** is a deep trench where one plate slides under another. Melting releases low-density magma that rises to form volcanoes such as those along the northwest coast of North America, including Mount St. Helens.

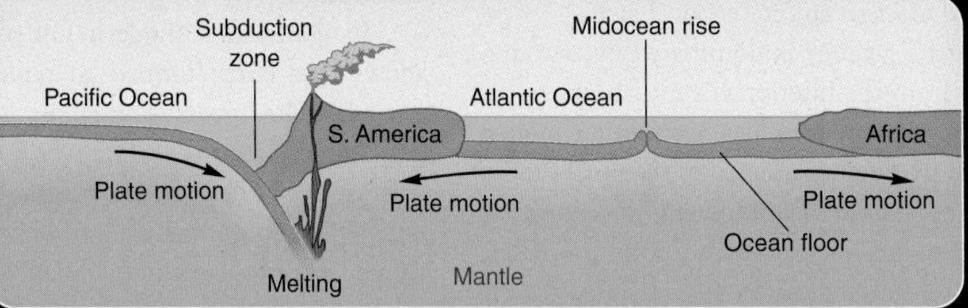


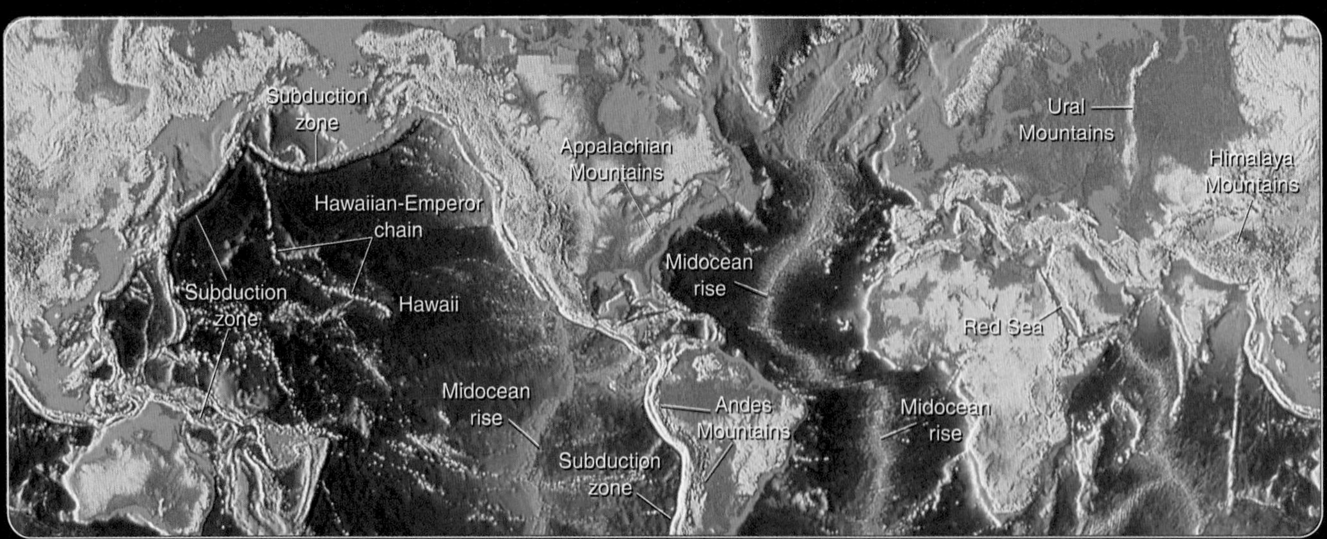


**1c** Hot spots caused by rising magma in the mantle can poke through a plate and cause volcanism such as that in Hawaii. As the Pacific plate has moved northwestward, the hot spot has punched through to form a chain of volcanic islands, now mostly worn below sea level. **Folded mountain ranges** can form where plates push against each other. For example, the Ural Mountains lie between Europe and Asia, and the Himalaya Mountains are formed by India pushing north into Asia. The Appalachian Mountains are the remains of a mountain range thrust up when North America was pushed against Africa.

1d The floor of the Pacific Ocean is sliding into subduction zones in many places around its perimeter. This pushes up mountains uch as the Andes and triggers earthquakes and active volcanism all round the Pacific in what is called the Ring of Fire. In places such as outhern California, the plates slide past each other, causing frequent arthquakes.

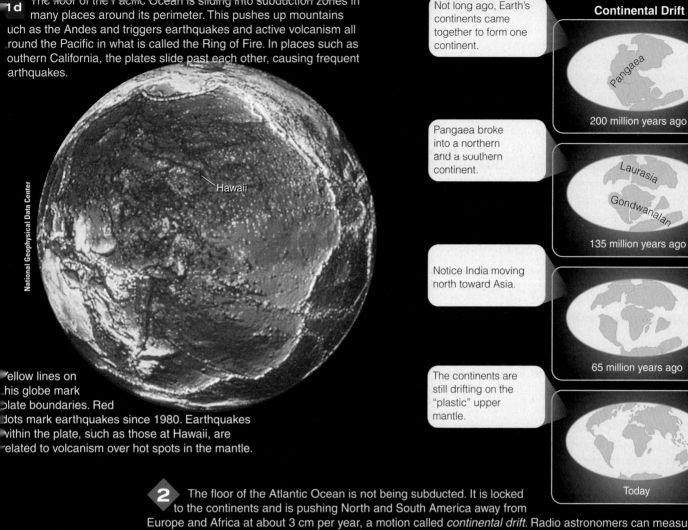

Hawaii

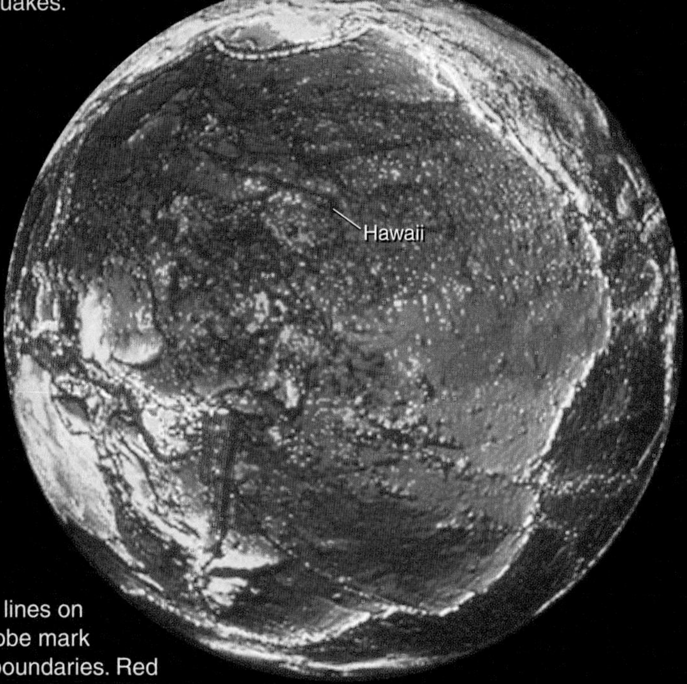

National Geophysical Data Center

ellow lines on his globe mark late boundaries. Red lots mark earthquakes since 1980. Earthquakes vithin the plate, such as those at Hawaii, are elated to volcanism over hot spots in the mantle.

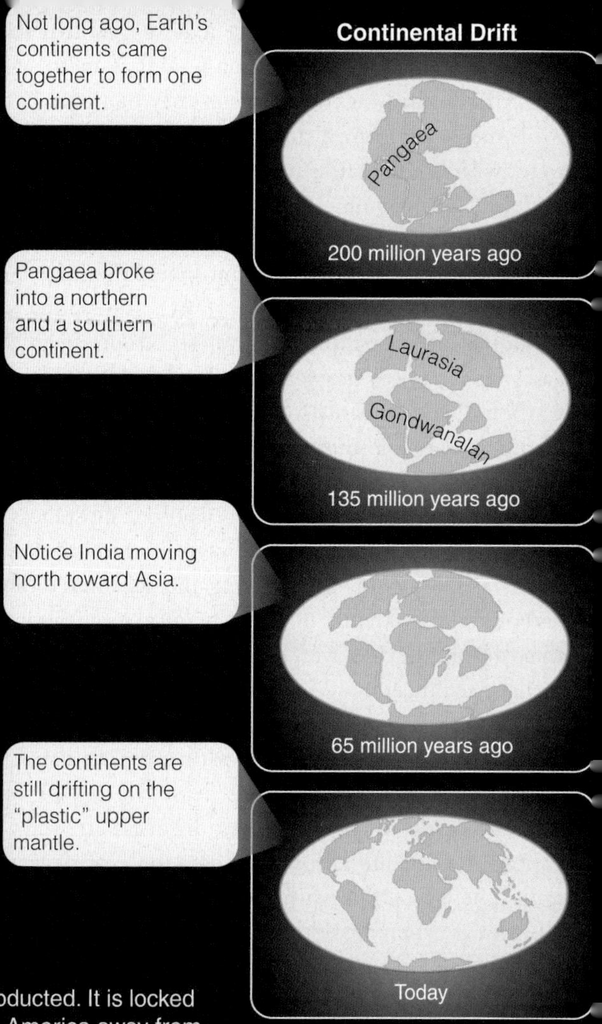

**Continental Drift**

Not long ago, Earth's continents came together to form one continent.

Pangaea

200 million years ago

Pangaea broke into a northern and a southern continent.

Laurasia

Gondwanalan

135 million years ago

Notice India moving north toward Asia.

65 million years ago

The continents are still drifting on the "plastic" upper mantle.

Today

2 The floor of the Atlantic Ocean is not being subducted. It is locked to the continents and is pushing North and South America away from Europe and Africa at about 3 cm per year, a motion called *continental drift*. Radio astronomers can measure this motion by timing and comparing radio signals from pulsars using European and American radio telescopes. Roughly 200 million years ago, North and South America were joined to Europe and Africa. Evidence of that lies in similar fossils and similar rocks and minerals found in the matching parts of the continents. Notice how North and South America fit against Europe and Africa like a puzzle.

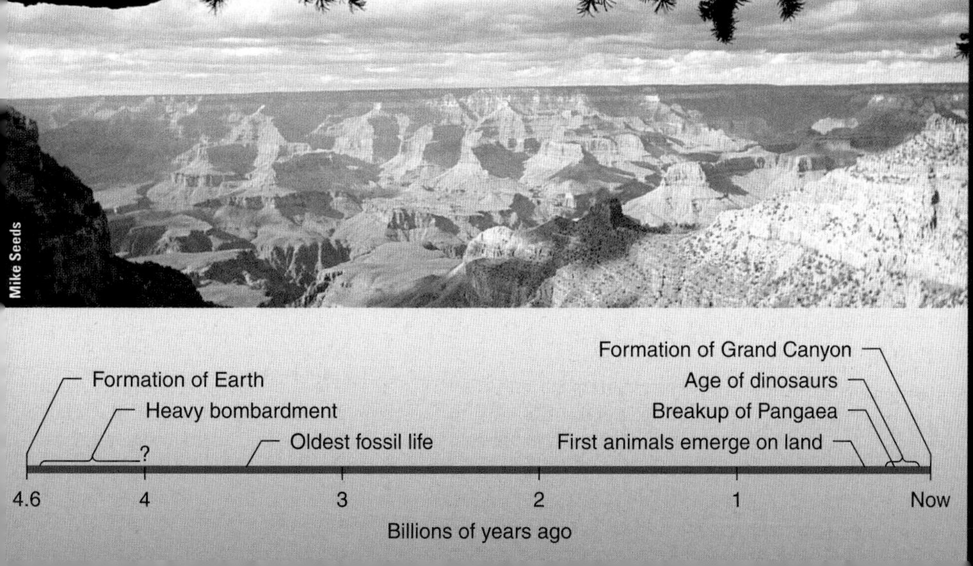


3 Plate tectonics pushes up mountain ranges and causes bulges in the crust, and water erosion wears the rock away. The Colorado River began cutting the Grand Canyon only about 10 million years ago when the Colorado plateau warped upward under the pressure of moving plates. That sounds like a long time ago, but it is only 0.01 billion years. A mile down, at the bottom of the canyon, lie rocks 0.57 billion years old, the roots of an earlier mountain range that stood as high as the Himalayas. It was pushed up, worn away to nothing, and covered with sediment long ago. Many of the geological features we know on Earth have been produced by relatively recent events.

similar to water on Earth. Those data suggest that there may be major differences in composition among comets. Icy planetesimals that formed far from the sun may be richer in deuterium, while those that formed closer to the orbit of Jupiter may contain water with isotope ratios more like those in Earth's water. This is a subject of continuing research, and it shows that the origin of Earth's atmosphere and oceans is yet to be fully understood.

In whatever way Earth's atmosphere originated, the mix of gases must have changed over time. The young atmosphere would have been rich in water vapor, carbon dioxide, and other gases. As it cooled, the water condensed to form the first oceans. Carbon dioxide is easily soluble in water—which is why carbonated beverages are so easy to manufacture—and the first oceans began to absorb atmospheric carbon dioxide. Once in solution, the carbon dioxide reacted with dissolved substances in the seawater to form silicon dioxide, limestone, and other mineral sediments in the ocean floor, freeing the seawater to absorb more carbon dioxide. Thanks to those chemical reactions in the oceans, the carbon dioxide was transferred from the atmosphere to seafloor sediments.

When Earth was young, its atmosphere had no free oxygen, that is, oxygen not combined with other elements. Oxygen is very reactive and quickly forms oxides in the soil or combines with iron and other substances dissolved in water. Only the action of plant life keeps a steady supply of oxygen in Earth's atmosphere via photosynthesis, which makes energy for plants by absorbing carbon dioxide and releasing oxygen. Beginning about 2 to 2.5 billion years ago, photosynthetic plants in the oceans had multiplied to the point where they made oxygen at a rate faster than chemical reactions could remove it from the atmosphere. After that time, atmospheric oxygen increased rapidly. (This topic will be discussed again in Chapter 26.) It is a **Common Misconception** that there is life on Earth because of oxygen. The truth is exactly the opposite: There is oxygen in Earth's atmosphere because of life. Most life forms on Earth do not need oxygen (except the minority of creatures that are animals, including us), and some are even poisoned by it.

An ozone molecule consists of three oxygen atoms linked together ($O_3$). Ozone molecules are very good at absorbing ultraviolet photons. Earth's lower atmosphere is now protected from ultraviolet radiation by an **ozone layer** about 15 to 30 km above the surface that exists because the atmosphere contains abundant ordinary oxygen ($O_2$) from which ozone is made. Because the atmosphere of the young Earth did not contain oxygen, an ozone layer could not form, and the sun's ultraviolet radiation was able to penetrate deep into the atmosphere. There the energetic ultraviolet photons would have broken up weaker molecules such as water ($H_2O$). The hydrogen from the water then escaped to space, and the oxygen formed oxides in the crust. Earth's atmosphere could not reach its present composition (■ Table 20-1) until it was protected by an ozone layer, and that required oxygen.

**■ Table 20-1 | Earth's Atmosphere**

| Gas | Percent by Weight |
|---|---|
| $N_2$ | 75.5 |
| $O_2$ | 23.1 |
| Ar | 1.29 |
| $CO_2$ | 0.05 |
| Ne | 0.0013 |
| He | 0.00007 |
| $CH_4$ | 0.0001 |
| Kr | 0.0003 |
| $H_2O$ (vapor) | 1.7–0.06 |

## Climate and Human Effects on the Atmosphere

If you climb to the top of a high mountain, you will find the temperature to be much lower than at sea level (■ Figure 20-9). Most clouds form at such altitudes. Higher still, you would find the air much colder, and so thin it could not protect you from the intense ultraviolet radiation in sunlight.

You can live on Earth's surface in safety because of Earth's atmosphere, but modern civilization is altering Earth's atmosphere in at least two serious ways, by adding carbon dioxide ($CO_2$) and by destroying ozone.

The concentration of $CO_2$ in Earth's atmosphere is important because $CO_2$ can trap heat in a process called the

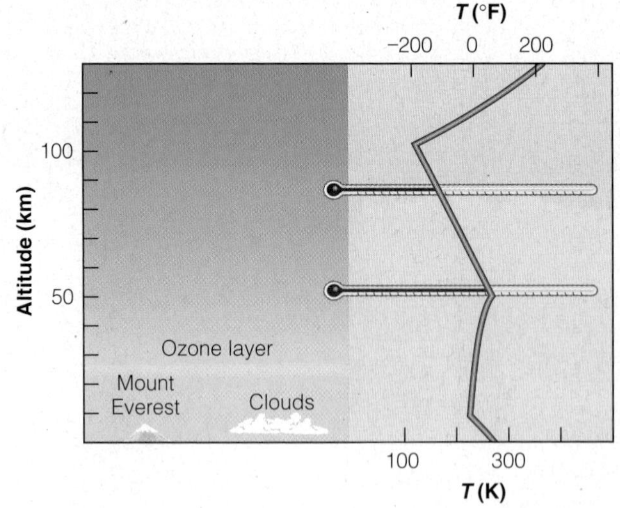


**■ Figure 20-9**

Thermometers placed in Earth's atmosphere at different levels would register the temperatures shown in the graph at the right. The lower few kilometers where you live are comfortable, but higher in the atmosphere the temperature is quite low. The ozone layer lies about 15 to 30 km above Earth's surface.

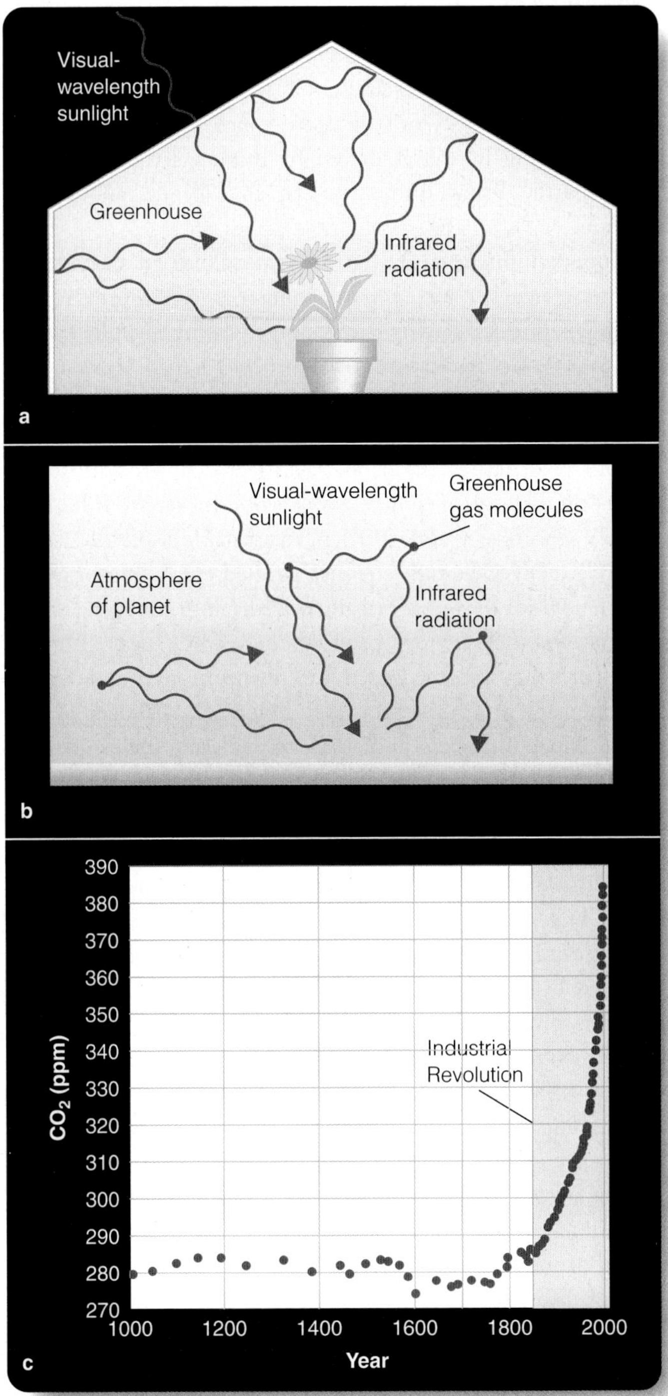


**Figure 20-10**

The greenhouse effect: (a) Visual-wavelength sunlight can enter a greenhouse and heat its contents, but the longer-wavelength infrared radiation cannot get out. (b) The same process can heat a planet's surface if its atmosphere contains greenhouse gases such as $CO_2$. (c) The concentration of $CO_2$ in Earth's atmosphere as measured in Antarctic ice cores remained roughly constant for thousands of years until the beginning of the Industrial Revolution around the year 1800. Since then it has increased by more than 30 percent. Evidence from proportions of carbon isotopes and oxygen in the atmosphere proves that most of the added $CO_2$ is the result of burning fossil fuels. (Graph adapted from a figure by Etheridge, Steele, Langenfelds, Francey, Barnola, and Morgan)

**greenhouse effect** (■ Figure 20-10a). When sunlight shines through the glass roof of a greenhouse, it heats the benches and plants inside. The warmed interior radiates infrared radiation, but the glass is opaque to infrared. Warm air in the greenhouse cannot mix with cooler air outside, so heat is trapped within the greenhouse, and the temperature climbs until the glass itself grows warm enough to radiate heat away as fast as the sunlight enters. This is the same process that heats a car when it is parked in the sun with the windows rolled up.

Earth's atmosphere is transparent to sunlight, and when the ground absorbs the sunlight, it grows warmer and radiates at infrared wavelengths. However, $CO_2$ makes the atmosphere less transparent to infrared radiation, so infrared radiation from the warm surface is absorbed by the atmosphere and cannot escape back into space. That traps heat and makes Earth warmer (Figure 20-10b).

It is a **Common Misconception** that the greenhouse effect is only bad. Evidence indicates that Earth has had a greenhouse effect for its entire history. Without the greenhouse effect, Earth presently would be at least 30°C (54°F) colder and uninhabitable for humans. The problem is that human civilization is adding $CO_2$ to the atmosphere more rapidly than it can be naturally removed, thereby increasing the intensity of the greenhouse effect.

$CO_2$ is not the only greenhouse gas. Water vapor, methane, and other gases also help warm Earth, but $CO_2$ is the most important. For 4 billion years, natural processes on Earth have removed $CO_2$ from the atmosphere and buried the carbon in the form of limestone, coal, oil, and natural gas. Since the beginning of the Industrial Revolution in the late 18th century, humans have been digging up lots of carbon-rich fuels, burning them to get energy, and releasing $CO_2$ back into the atmosphere. It is a **Common Misconception** that human output of $CO_2$ is minor compared to natural sources such as volcanoes. Careful measurements of carbon isotope ratios and relative amounts of $CO_2$ versus $O_2$ in the atmosphere show that the $CO_2$ added to the atmosphere since the year 1800 is mostly or entirely due to human burning of fossil fuels. Estimates are that the amount of $CO_2$ in Earth's atmosphere could double during the 21st century. The increased concentration of $CO_2$ is increasing the greenhouse effect and warming Earth in what is known as **global warming.** Studies of the growth rings in very old trees show that the average Earth climate had been cooling for most of the last 1000 years, but the 20th century reversed that trend with a rise of 0.56 to 0.92°C (1.01 to 1.66°F). The amount of warming to expect in the future is difficult to predict because Earth's climate is critically sensitive to a number of different factors, not just the abundance of greenhouse gases. For example, a planet's **albedo** is the fraction of the sunlight hitting it that gets reflected away. A planet with an albedo of 1 would be perfectly white, and a planet with an albedo of 0 would be perfectly black. Earth's overall albedo is 0.39, meaning it reflects back into space 39 percent of the sunlight that hits it. Much of that reflection is caused by clouds, and the formation of clouds depends critically on the presence of water vapor in the upper atmosphere, the temperature of the upper atmosphere, and the patterns of atmospheric circulation. Even a small change in any of these factors could change Earth's albedo and thus its climate. For example, a slight warming should increase water vapor in the atmosphere, and water vapor is another greenhouse gas that would enhance the warming. But increased water vapor could increase cloud cover, increase Earth's albedo, and partially reduce the warming. On the other hand, high icy clouds tend to enhance the greenhouse effect. The situation is complex, and therefore precise calculations of future warming are not easy to make. Also, even small changes in temperature can alter circulation patterns in the atmosphere and in the oceans, and the consequences of such changes are very difficult to model.

Even though the future is uncertain, general trends now point to substantial continued warming. Mountain glaciers have melted back dramatically since the 19th century. Measurements show that polar ice in the form of permafrost, ice shelves, and ice on the open Arctic Ocean is melting. It is a **Common Misconception** that the observed warming of the Earth is due to natural causes rather than the greenhouse effect. Regular and predictable changes in Earth's axis inclination and orientation and in the shape of its orbit, called Milankovitch cycles (see Chapter 3), currently would be driving Earth's climate toward lower, not higher, temperatures. Also, careful observations during recent decades by space probes indicate the sun has not been increasing measurably in luminosity. The observed warming must be strong to be occurring in the face of opposing astronomical effects.

Although changes are small now, it is a serious issue for the future. Even a small rise in temperatures will dramatically affect agriculture, not only through rising temperatures but also through changes in rainfall. It is a **Common Misconception** that all of Earth will warm at the same rate if there is global warming. Models predict that although most of North American will grow warmer and dryer, Europe initially will become cooler and wetter. Also, the melting of ice on polar landmasses such as Greenland can cause a rise in sea levels that will flood coastal regions and alter shore environments. A modest rise will cover huge low-lying areas such as the entire state of Florida.

There is no doubt that civilization is warming Earth through an enhanced greenhouse effect, but a remedy is difficult to imagine. Reducing the amount of $CO_2$ and other greenhouse gases released to the atmosphere is difficult because modern society depends on burning fossil fuels for energy. Conserving forests is difficult because growing populations, especially in developing countries with large forest reserves, demand the wood and the agricultural land produced when forests are cut. Political, business, and economic leaders argue that the issue is uncertain, but all around the world scientists of stature have reached agreement: Global warming is real, is driven by human activity, and will change Earth. What humanity can or will do about it is uncertain.

Human influences on Earth's atmosphere go beyond the greenhouse effect. Our modern industrial civilization is also reducing ozone in Earth's atmosphere. Many people have a **Common Misconception** that ozone is bad because they hear it mentioned as a pollutant of city air, produced by auto emissions. Breathing ozone is bad for you, but, as you learned earlier in this chapter, the ozone layer in the upper atmosphere protects the lower atmosphere and Earth's surface from harmful solar UV photons. Ozone ($O_3$) is an unstable molecule and is chemically active. Certain chemicals called chlorofluorocarbons (CFCs), used for refrigeration, air conditioning, and some industrial processes, can destroy ozone. As these CFCs escape into the atmosphere, they become mixed into the ozone layer and convert the ozone back into normal oxygen molecules. Ordinary oxygen does not block ultraviolet radiation, so depleting the ozone layer causes an increase in ultraviolet radiation at Earth's surface. In small doses, ultraviolet radiation can produce a suntan, but in larger doses it can cause skin cancers.

The ozone layer over the Antarctic may be especially sensitive to CFCs. Starting in the late 1970s, the ozone concentration fell significantly over the Antarctic, and a hole in the ozone layer developed over the continent each October at the time of the Antarctic spring (■ Figure 20-11). Satellite and ground-based measurements showed the same thing beginning to happen at higher northern latitudes, with the amount of ultraviolet radiation reaching the ground increasing. This was an early warning that human activity is modifying Earth's atmosphere in a potentially dangerous way. Fortunately, as a result of that warning, international agreements banned most uses of CFCs, and the trend of ozone hole expansion seems to have slowed and may be reversing.

There is yet another **Common Misconception** that global warming and ozone depletion are two names for the same thing. Take careful note that the ozone hole is a second Earth environmental issue that is basically separate from global warming. The $CO_2$ and ozone problems in Earth's atmosphere are paralleled on Venus and Mars. When you study Venus in Chapter 22, you will discover a runaway greenhouse effect that has made the surface of the planet hot enough to melt lead. On Mars you will discover an atmosphere without an ozone layer. A few minutes of sunbathing on Mars would kill you. Once again, you can learn more about your own planet by studying the extreme conditions on other planets.

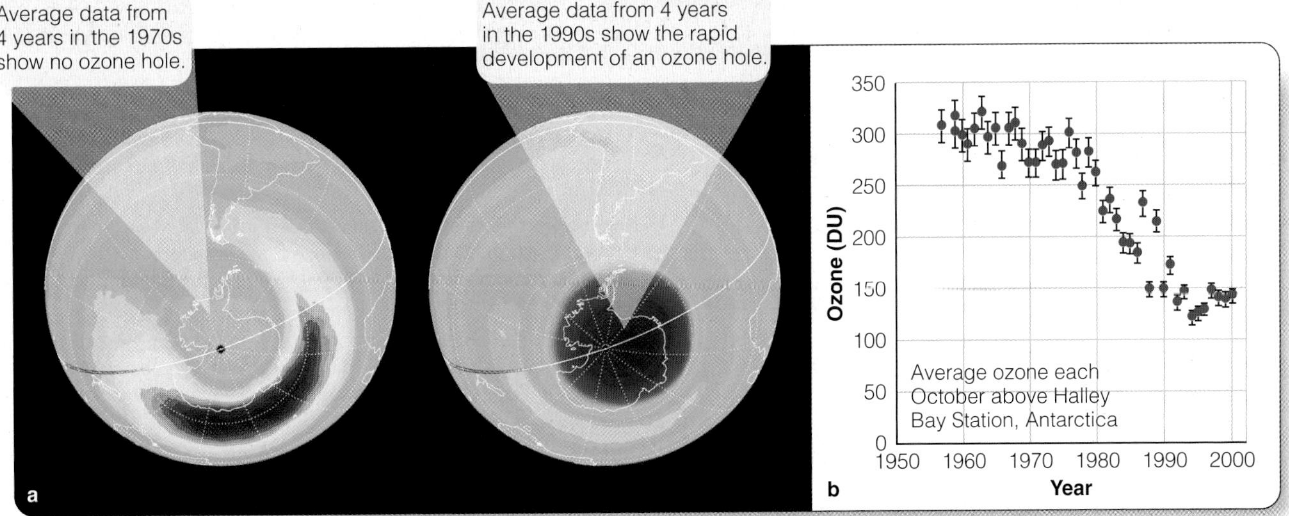


**Figure 20-11**

(a) Satellite observations of ozone concentrations over Antarctica are shown here as red for highest concentration and violet for lowest. Since the 1970s, a hole in the ozone layer has developed over the South Pole. (b) Although ozone depletion is most dramatic above the South Pole, ozone concentrations have declined at all latitudes. (NASA/GSFC/TOMS and Glenn Carver)

## SCIENTIFIC ARGUMENT

***Why does Earth's atmosphere contain little carbon dioxide and lots of oxygen?***

Sometimes as you build a scientific argument, you must contradict what seems, at first glance, a simple truth. In this case, because volcanic outgassing releases mostly $CO_2$, $N_2$, and water vapor, you might expect Earth's atmosphere to be very rich in $CO_2$. Luckily for the human race, $CO_2$ is highly soluble in water, and Earth's surface temperature allows it to be covered with liquid water. The $CO_2$ dissolves in the oceans and combines with minerals in seawater to form deposits of silicon dioxide, limestone, and other mineral deposits. In this way, the $CO_2$ is removed from the atmosphere and buried in Earth's crust. Oxygen, in contrast, is highly reactive and forms oxides so easily you might expect it to be rare in the atmosphere. Happily for us animals, it is continually replenished as green plants release oxygen into Earth's atmosphere faster than chemical reactions can remove it. Were it not for liquid-water oceans and plant life, Earth would have a thick $CO_2$ atmosphere with no free oxygen.

Now follow up on your argument. **Why would an excess of $CO_2$ and a deficiency of free oxygen be harmful to all life on Earth in ways that go beyond mere respiration?**

## What Are We? Scientific Imagineers

One of the most fascinating aspects of science is its power to reveal the unseen. That is, it reveals regions you can never visit. You saw this in earlier chapters when you studied the inside of the sun and stars, the surface of neutron stars, the event horizon around black holes, the cores of active galaxies, and more. In this chapter, you have "seen" Earth's core.

An engineer is a person who builds things, so you can call a person who imagines things an imagineer. Most creatures on Earth cannot imagine situations that do not exist, but humans have evolved the ability to say, "What if?" Our ancient ancestors could imagine what would happen if a tiger was hiding in the grass, and we can imagine the inside of Earth.

A poet can imagine the heart of Earth, and a great writer can imagine a journey to the center of Earth. In contrast, scientists use their imagination in a carefully controlled way. Guided by evidence and theory, they can imagine the molten core of our planet. As you read this chapter you saw the yellow-orange glow and felt the heat of the liquid iron, and you were a scientific imagineer.

Human imagination makes science possible and provides one of the great thrills of science—exploring beyond the limits of normal human experience.

# Study and Review

## Summary

- Earth is the standard of **comparative planetology (p. 425)** in the study of the Terrestrial planets because we know it best and because it contains all of the phenomena found on the other Terrestrial planets.
- Our discussion of the Terrestrial planets considers Earth, the moon, Mercury, Venus, and Mars. Earth's moon is included because it is a complex world and makes a striking comparison with Earth.
- The Terrestrial worlds differ mainly in size, but they all have low-density crusts, **mantles (p. 426)** of dense rock, and metallic cores.
- Comparative planetology warns you to expect that cratered surfaces are old, that heat flowing out of a planet drives geological activity, and that the nature of a planet's atmosphere depends on the size of the planet and its temperature.
- At some point early in its history Earth was hot enough to be completely molten, which caused it to differentiate into layers of different density.
- Earth has passed through four stages as it has evolved: (1) differentiation, (2) cratering, (3) flooding by lava and water, and (4) slow surface evolution. The other Terrestrial planets and the moon also passed through the same stages, which had different effects and durations depending on the specific properties of each body.
- Earth is peculiar in that it has large amounts of liquid water on its surface, and that water drives strong erosion that alters the surface geology.
- Earth is also peculiar in that it is the only known home for life.
- **Seismic waves (p. 429)** generated by earthquakes can be detected by **seismographs (p. 429)** all over the world and can reveal Earth's internal structure.
- **Pressure *(P)* waves (p. 429)** can travel through a liquid, but **shear *(S)* waves (p. 429)** cannot. Observations show that *S* waves cannot pass through Earth's core, and that is evidence that the core is liquid. Measurements of heat flowing outward from the interior, combined with mathematical models, reveal that the core is very hot and composed of iron and nickel.
- Although Earth's crust is brittle and breaks under stress, the mantle is **plastic (p. 431)** and can deform and flow under pressure.
- Earth's magnetic field is generated by the dynamo effect in the liquid, convecting, rotating, conducting core. The magnetic field shields Earth from the solar wind by producing a **bow shock (p. 431)** and a **magnetosphere (p. 431)** around the planet. Radiation belts called the **Van Allen belts (p. 431)**, as well as aurora displays, are also produced by the magnetic field.
- Earth is dominated by **plate tectonics (p. 434)**, which breaks the crust into moving sections. Plate tectonics is driven by heat flowing upward from the interior.
- Tectonic plates are made of low-density, brittle rock that floats on the hotter plastic upper layers of the mantle. **Rift valleys (p. 434)** can be produced where plates begin pulling away from each other.
- New crust is formed along **midocean rises (p. 434)** where molten rock solidifies to form **basalt (p. 434)**. Crust is destroyed where it sinks into the mantle along **subduction zones (p. 434)**. Volcanism and earthquakes are common along the edge of the plates.
- The motion of a plate across a hot spot can produce a chain of volcanic islands such as the Hawaiian island chain. Hot-spot volcanism is not related to subduction zones.
- The continents are drifting slowly on the plastic mantle, and their arrangement changes with time. Where they collide, they can form **folded mountain ranges (p. 434)**.
- Most geological features on Earth, such as mountain ranges and the Grand Canyon, have been formed recently. The first billion years of Earth's geology are almost entirely erased by plate tectonics and erosion.
- Because Earth formed hot, it never had a **primeval atmosphere (p. 433)** rich in hydrogen and helium that was later replaced by a **secondary atmosphere (p. 433)** baked out of the interior.
- Because Earth formed in a molten state, its first atmosphere was probably mostly carbon dioxide, nitrogen, and water vapor. Most of the carbon dioxide eventually dissolved in seawater and was added to ocean sediments. Plant life has added oxygen to the atmosphere.
- Ultraviolet photons can break up water molecules in a planet's atmosphere, but as soon as Earth had enough oxygen, an **ozone layer (p. 436)** could form high in Earth's atmosphere. The ozone absorbs ultraviolet photons and protects water molecules.
- The **albedo (p. 437)** of a planet is the fraction of sunlight hitting it that it reflects into space. Small changes in the albedo of Earth caused by changes in clouds and atmospheric currents can have a dramatic effect on climate.
- The **greenhouse effect (p. 437)** can warm a planet if gases such as carbon dioxide in the atmosphere are transparent to light but opaque to infrared. The natural greenhouse effect warms Earth and makes it comfortable for life, but greenhouse gases added by industrial civilization are responsible for **global warming (p. 437)**.
- Measurement of carbon isotope ratios and carbon dioxide versus oxygen abundances make it clear that the $CO_2$ added to the atmosphere since 1800 is predominantly from burning of fossil fuels. Observations and model calculations have eliminated other candidate causes for the current warming such as natural climate cycles or variations in the sun's output.
- The ozone layer high in Earth's atmosphere protects the surface from ultraviolet radiation, but certain chemicals called chlorofluorocarbons released in industrial processes attack the ozone layer and thin it. This is allowing more harmful ultraviolet radiation to reach Earth's surface.

## Review Questions

1. Why would you include the moon in a comparison of the Terrestrial planets?
2. In what ways is Earth peculiar among the Terrestrial planets?
3. What are the four stages of planetary development?
4. How do you know that Earth differentiated?
5. What evidence can you cite that Earth's metallic core is liquid?
6. How are earthquakes in Hawaii different from those in Southern California?
7. What characteristics must Earth's core have in order to generate a magnetic field?
8. How do island chains located in the centers of tectonic plates such as the Hawaiian islands help you understand plate tectonics?
9. What evidence can you cite that the Atlantic Ocean is growing wider?
10. How is your concept of Earth's first atmosphere related to the speed with which Earth formed from the solar nebula?
11. What has produced the oxygen in Earth's atmosphere?
12. How does the increasing abundance of $CO_2$ in Earth's atmosphere cause a rise in Earth's temperature?
13. Why would a decrease in the density of the ozone layer cause public health problems?
14. **How Do We Know?** How is deducing the structure of a virus like finding the composition of Earth's core?

## Discussion Questions

1. If you orbited a planet in another solar system and discovered oxygen in its atmosphere, what might you expect to find on its surface?
2. If liquid water is rare on the surface of planets, then most Terrestrial planets must have $CO_2$-rich atmospheres. Correct? Why?

## Problems

1. In Figure 20-3, the earthquake occurred 7440 km from the seismograph. How fast did the *P* waves travel in km/s? How fast did the *S* waves travel?
2. What percentage of Earth's volume is taken up by its metallic core?
3. If the Atlantic seafloor is spreading at 3 cm/year and is now 6400 km wide, how long ago were the continents in contact?
4. The Hawaiian-Emperor chain of undersea volcanoes is about 7500 km long, and the Pacific plate is moving 9.2 cm a year. How old is the oldest detectable volcano in the chain? What has happened to older volcanoes in the chain?
5. From Hawaii to the bend in the Hawaiian-Emperor chain is about 4000 km. Use the speed of plate motion given in Problem 4 to estimate how long ago the direction of plate motion changed. (Note that the San Andreas fault in southern California became active at about the same time!)
6. Calculate the age of the Grand Canyon as a percent of Earth's age.

## Learning to Look

1. Look at the globe of Earth shown on page 435 and look for volcanoes scattered over the Pacific Ocean. What is producing these volcanoes?
2. In what ways is the photo at the right a typical view of the surface of planet Earth? How is it unusual among planets in general?

William K. Hartmann

3. What do you see in this photo that suggests heat is flowing out of Earth's interior?

USGS

# 21 The Moon and Mercury: Comparing Airless Worlds

*When astronauts stepped onto the surface of the moon, they found an unearthly world with no air, no water, weak gravity, and a dusty, cratered surface. Through comparative planetology, the moon reveals a great deal about our own beautiful Earth.* (JSC/NASA)

## Guidepost

Want to fly to the moon? You will need to pack more than your lunch. There is no air and no water, and the sunlight is strong enough to kill you. Mercury is the same kind of world. Take shelter in the shade, and you will freeze to death in moments. Earth seems normal to you, and other worlds are, well, unearthly, but they are related to Earth in surprising ways. Exploring these two airless worlds will answer five essential questions:

- **Why is the moon airless and cratered?**
- **How did the moon form and evolve?**
- **In what ways is Mercury similar to, and different from, the moon?**
- **How did Mercury form and evolve?**
- **How are the histories of the moon and Mercury connected to Earth?**

You are beginning your detailed study of planets by exploring airless worlds; in the next chapter you will move on to bigger planets with atmospheres. They are not necessarily more interesting places, but they are just a bit less unearthly.

*That's one small step for [a] man . . .*
*one giant leap for mankind.*

— NEIL ARMSTRONG, ON THE MOON

*Beautiful, beautiful. Magnificent desolation.*

— EDWIN E. (BUZZ) ALDRIN JR., ON THE MOON

IF YOU HAD BEEN THE FIRST PERSON to step onto the surface of the moon, what would you have said? Neil Armstrong responded to the historic significance of being the first human to step onto the surface of another world. Buzz Aldrin was second, and he responded to the moon itself. It *is* desolate, and it *is* magnificent. But it is not unusual. Many planets in the universe probably look like Earth's moon, and astronauts may someday walk on such worlds and compare them with our moon.

In this chapter, you will use comparative planetology to study the moon and Mercury, and you will continue following three important themes of planetary astronomy: impact cratering, internal heat flow, and giant impact-induced volcanism. These three themes will help you organize the flood of details astronomers have learned about the moon and Mercury.

## 21-1 The Moon

ONLY 12 PEOPLE HAVE STOOD on the moon, but planetary astronomers know it well. The photographs, measurements, and samples brought back to Earth paint a picture of an airless, ancient, battered crust, and a world created by a planetary catastrophe.

### The View from Earth

A few billion years ago, the moon must have rotated faster than it does today, but Earth is over 80 times more massive than the moon (**Celestial Profile 3**), and its tidal forces on the moon are strong. Earth's gravity raised tidal bulges on the moon, and friction in the bulges has slowed the moon until it now rotates once each orbit, keeping the same side facing Earth. A moon whose rotation is locked to its planet is said to be **tidally coupled.** That is why we always see the same side of the moon; the back of the moon is never visible from Earth. The moon's familiar face has shone down on Earth since long before there were humans (■ Figure 21-1).

Based on what you already know, you can predict that the moon should have no atmosphere. It is a small world with an escape velocity too low to keep gas atoms and molecules from escaping into space. You can confirm your theory with even a small telescope. You see no clouds or other obvious traces of an atmosphere, and shadows near the **terminator,** the dividing line between

*Earth's moon is about one-fourth the diameter of Earth. Its low density indicates that it contains little iron, but the size of its iron core and the amount of remaining heat are unknown.* (NASA)

## *Celestial Profile 3: The Moon*

### *Motion:*

| | |
|---|---|
| Average distance from Earth | $3.84 \times 10^5$ km (center to center) |
| Eccentricity of orbit | 0.055 |
| Inclination of orbit to ecliptic | 5.1° |
| Average orbital velocity | 1.02 km/s |
| Orbital period (sidereal) | 27.3 d |
| Orbital period (synodic) | 29.5 d |
| Inclination of equator to orbit | 6.7° |

### *Characteristics:*

| | |
|---|---|
| Equatorial diameter | $3.48 \times 10^3$ km |
| Mass | $7.35 \times 10^{22}$ kg (0.0123 $M_\oplus$) |
| Average density | 3.36 g/cm$^3$ (3.3 g/cm$^3$ uncompressed) |
| Surface gravity | 0.17 Earth gravity |
| Escape velocity | 2.4 km/s (0.21 $V_\oplus$) |
| Surface temperature | −170° to 130°C (−275° to 265°F) |
| Average albedo | 0.07 |

### *Personality Point:*

Lunar superstitions are common. The words *lunatic* and *lunacy* come from *luna*, the moon. Someone who is *moonstruck* is supposed to be a bit nutty. Because the moon affects the ocean tides, many superstitions link the moon to water, to weather, and to women's cycle of fertility. According to legend, moonlight is supposed to be harmful to unborn children; but, on the plus side, moonlight rituals are said to remove warts.

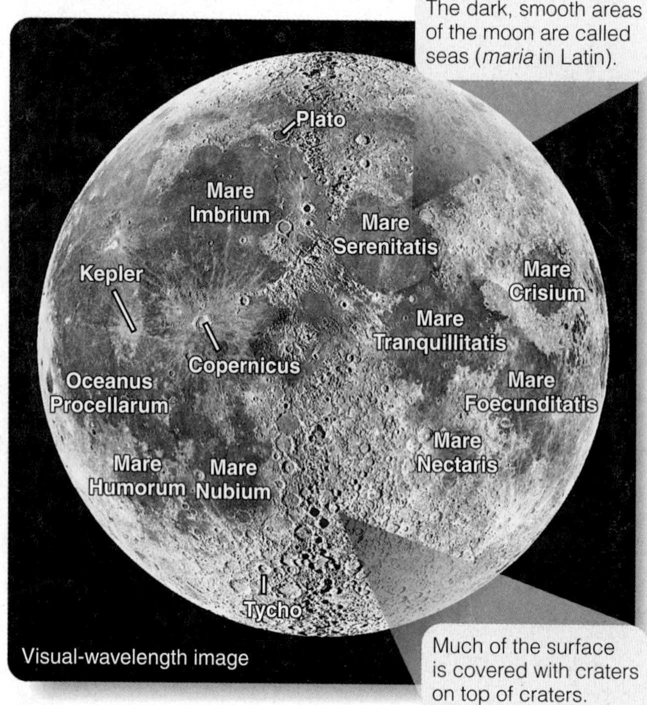


**■ Figure 21-1**

The side of the moon that faces Earth is a familiar sight. Craters have been named for famous scientists and philosophers, and the so-called seas have been given romantic names. Mare Imbrium is the Sea of Rains, and Mare Tranquillitatis is the Sea of Tranquillity. There is, in fact, no water on the moon. (Photo © UC Regents/Lick Observatory)

daylight and darkness, are sharp and black. There is no air on the moon to scatter light and soften shadows. Also, with even a small telescope you could watch stars disappear behind the **limb** of the moon—the edge of its disk—without dimming. Clearly, the moon is an airless (and, therefore, soundless) world.

The surface of the moon is divided into two dramatically different kinds of terrain. The lunar highlands are filled with jumbled mountains, but there are no folded mountain ranges like the ones on Earth. This shows that the moon has no plate tectonics. The mountains are pushed up by millions of impact craters one on top of the other. In fact, the highlands are saturated with craters, meaning that it would be impossible to form a new crater without destroying the equivalent of one old crater. In contrast the lowlands, about 3 km (2 mi) lower than the highlands, are smooth, dark plains called **maria,** Latin for "seas." (The singular of *maria* is **mare,** pronounced MAH-ray.) The first telescope observers thought these were bodies of water, but further examination showed that the maria are marked by ridges, faults, smudges, and scattered craters and can't be water. Rather, the maria are ancient lava flows that have apparently covered the older, cratered lowlands.

Those lava flows suggest volcanism, but only small traces of past volcanic activity are visible from Earth. No major volcanic peaks are visible on the moon, and no active volcanism has ever been detected. The lava flows that created the maria happened long ago and were much too fluid to build peaks. With a good telescope and some diligent searching you could see a few small domes pushed up by lava below the surface, and you could see long, winding channels called **sinuous rilles** (■ Figure 21-2). These channels are often found near the edges of the maria and were evidently cut by flowing lava. In some cases, such a channel may once have had a roof of solid rock, forming a lava tube. After the lava drained away, meteorite impacts collapsed the roof to form a sinuous rille. The view from Earth provides only hints of ancient volcanic activity associated with the maria.

Lava flows and impact cratering have dominated the history of the moon. Study **Impact Cratering** on pages 446–447 and notice three important points and five new terms:

1. Impact craters have certain distinguishing characteristics, such as their shape and the *ejecta, rays,* and *secondary craters* around them.
2. Lunar impact craters range from tiny pits formed by *micrometeorites* to giant *multiringed basins.*
3. Most of the craters on the moon are old; they were formed long ago when the solar system was young.

Meteorites strike the moon all the time, but large impacts are rare today. Astronomers estimate that meteorites with diameters of tens of meters strike the moon every few decades, but no one has ever seen such an impact with certainty. Small flashes of light have been seen on the dark side of the moon during showers of meteors, but those impacts must have been made by very small objects, perhaps only centimeters in size. No significant change has been seen on the moon since the invention of the telescope. As you will learn in Chapter 25, large impacts do happen on the moon and Earth, but they are quite rare, and nearly all of the lunar craters seen through telescopes date from the solar system's youth.

The lunar features visible from Earth allow you to begin constructing a tentative hypothesis to explain the history of the moon. Such a hypothesis provides a framework that will help you organize all of the details and observations (**How Do We Know? 21-1**). As the moon formed, its crust would have been heavily cratered by debris left over from the formation of the planets. Sometime after the cratering subsided, lava welled up from below the crust and flooded the lowlands, covering the craters there and forming the smooth maria. You can locate a few large craters on the maria such as Kepler and Copernicus in Figure 21-1, but note that the bright rays around them show that they are relatively young, because rays are darkened and erased by exposure to sunlight, solar wind, and micrometeorite bombardment. The maria are only

Hadley Rille is a sinuous rille near the edge of Mare Imbrium.
Faults where the crust has broken
Apollo 15 found Hadley Rille to be a 1200-foot-deep lava channel.
The lunar highlands are saturated with craters on top of craters.
The smooth maria are generally circular with faults and wrinkles in the old lava surface.
Visual-wavelength images
Ancient craters partly flooded by lava

**■ Figure 21-2**

Details visible in photographs show that meteorite impacts long ago covered the moon with craters, but that lava flooded out and filled the largest basins covering the craters there with smooth plains. (Hadley: NASA; Moon disk, highlands, and mare: © UC Regents/Lick Observatory)

lightly scarred by impacts and must be younger than the cratered highlands.

It is difficult to estimate the true age of any specific crater. In some cases, you can find **relative ages** by noting that a crater or its rays partially covers other craters. Clearly the crater on top must be younger than the craters on the bottom. Relative ages can be calibrated using radioactive ages of lunar samples, which indicate how the cratering rate decreased when the moon was young. Combining all this information, astronomers can study the size and number of craters on a section of the moon's surface and estimate that section's **absolute age** in years. The maria, containing few craters, appear to be 3 to 4 billion years old, and the highlands are older.

Earth-based observations allowed astronomers to begin telling the story of the moon's surface, but the view from Earth does not provide enough evidence. To really understand the history of the lunar surface, humans had to go there and bring back samples.

## The Apollo Missions

On May 25, 1961, President John Kennedy committed the United States to landing a human being on the moon by 1970. Although the reasons for that decision related more to economics, international politics, and the stimulation of technology than to science, the Apollo program became a fantastic scientific adventure, including six expeditions to the surface of the moon that changed how we all think about Earth.

Flying to the moon is not particularly difficult. With powerful enough rockets and enough food, water, and air, it is a straightforward trip. Landing on the moon is more difficult but not impossible. The moon's gravity is only one-sixth that of Earth, and there is no atmosphere to disturb the trajectory of the spaceship. Getting to the moon isn't too hard, and landing is possible; the difficulty is doing both on one trip. The spaceship must carry food, water, and air for a number of days in space plus fuel and rockets for mid-course corrections and for a return to Earth. All of this adds up to a ship that is too massive to make a safe landing on the lunar surface. The solution was to take two spaceships to the moon, one to make the round-trip in and one to land in (■ Figure 21-3).

The command module was the long-term living space and command center for the trip. Three astronauts had to live in it for a week, and it had to carry all the life-support equipment, navigation instruments, computers, power packs, and so on for a week's jaunt in space. The lunar landing module (LM for short)

## Impact Cratering

**1** The craters that cover the moon and many other bodies in the solar system were produced by the high-speed impact of meteorites of all sizes. Meteorites striking the moon travel 10 to 70 km/s and can hit with the energy of many nuclear bombs.

A meteorite striking the moon's surface can deliver tremendous energy and can produce an impact crater 10 or more times larger in diameter than the meteorite. The vertical scale is exaggerated at right for clarity.

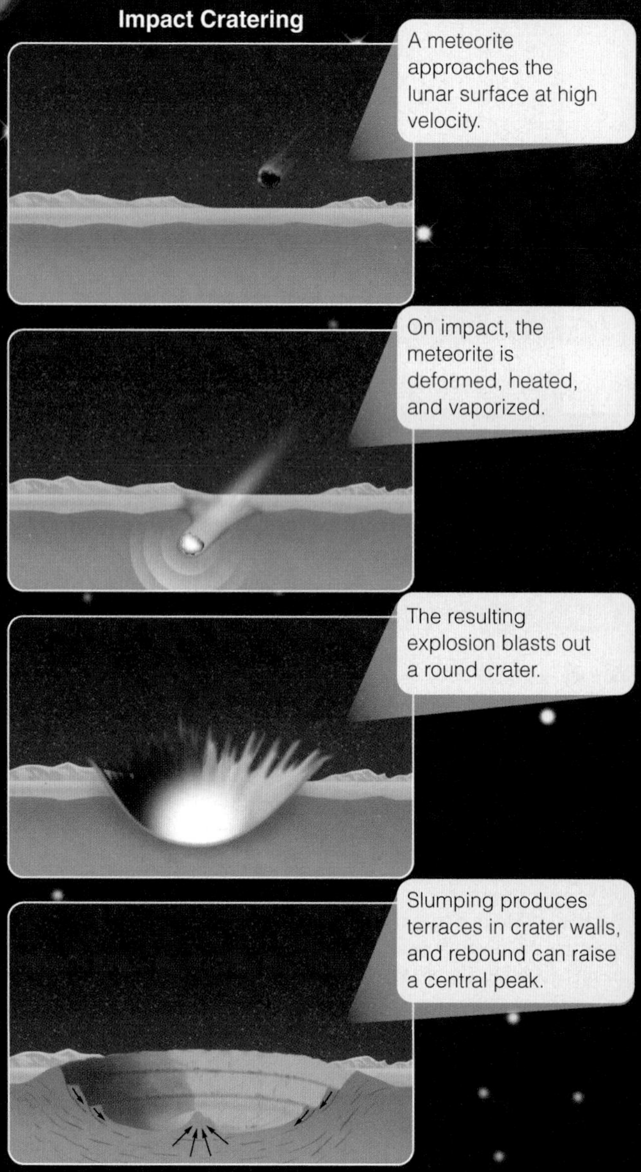

**Impact Cratering**

A meteorite approaches the lunar surface at high velocity.

On impact, the meteorite is deformed, heated, and vaporized.

The resulting explosion blasts out a round crater.

Slumping produces terraces in crater walls, and rebound can raise a central peak.

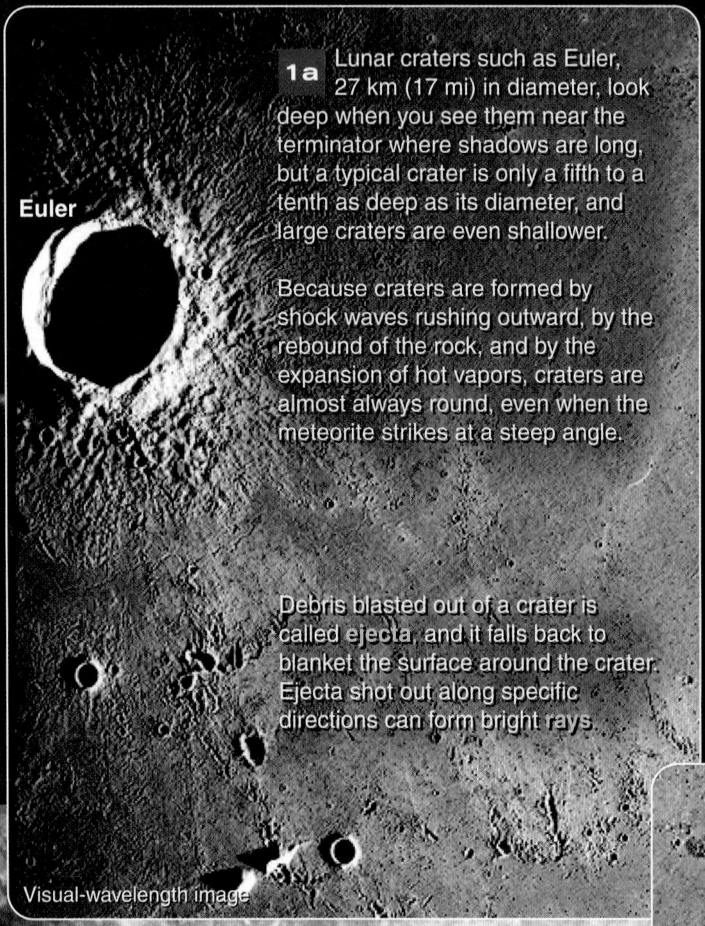

**1a** Lunar craters such as Euler, 27 km (17 mi) in diameter, look deep when you see them near the terminator where shadows are long, but a typical crater is only a fifth to a tenth as deep as its diameter, and large craters are even shallower.

Because craters are formed by shock waves rushing outward, by the rebound of the rock, and by the expansion of hot vapors, craters are almost always round, even when the meteorite strikes at a steep angle.

Debris blasted out of a crater is called ejecta, and it falls back to blanket the surface around the crater. Ejecta shot out along specific directions can form bright rays.

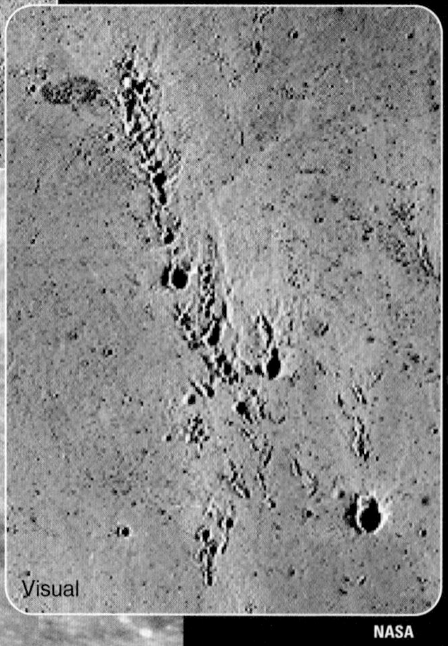

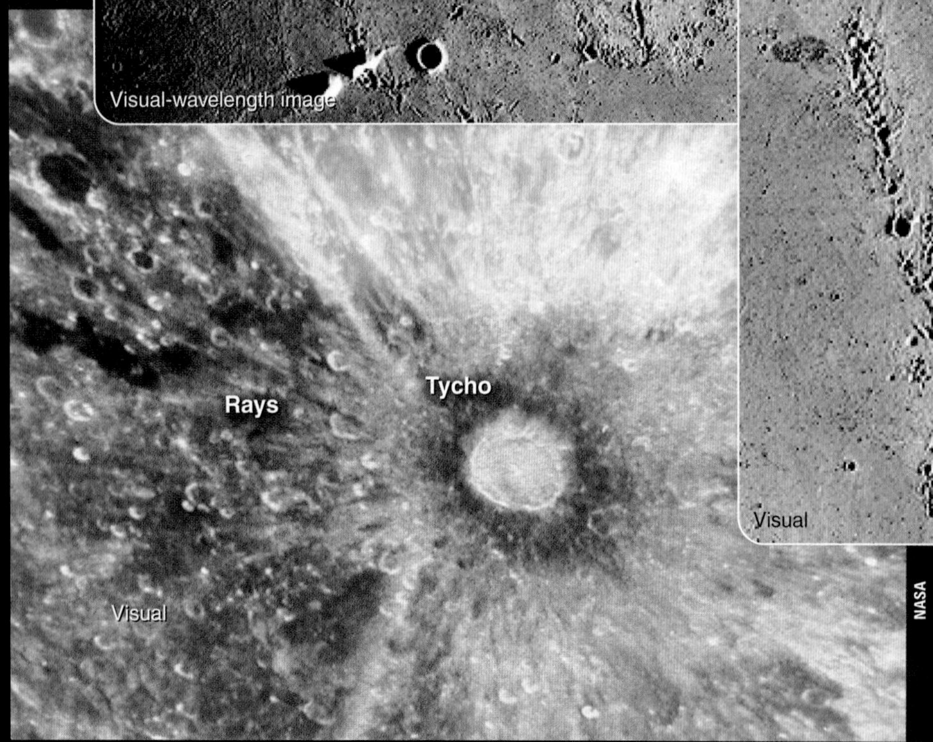

Euler

Visual-wavelength image

NASA

Rays

Tycho

Visual

Visual

NASA

NASA

**1b** Rock ejected from distant impacts can fall back to the surface and form smaller craters called **secondary craters.** The chain of craters here is a 45-km-long chain of secondary craters produced by ejecta from the large crater Copernicus 200 km out of the frame to the lower right.

Bright ejecta blankets and rays gradually darken as sunlight alters minerals and small meteorites stir the dusty surface. Bright rays are signs of youth. Rays from the crater Tycho, perhaps only 100 million years old, extend halfway around the moon.

**2** Plum Crater (right), 40 m (130 ft) in diameter, was visited by Apollo 16 astronauts. Note the many smaller craters visible. Lunar craters range from giant impact basins to tiny pits in rocks struck by **micrometeorites,** meteorites of microscopic size.

Lunar rover
Sun glare in camera lens
NASA

Visual-wavelength images

Mare Orientale

Solidified lava

NASA

**2a** In larger craters, the deformation of the rock can form one or more inner rings concentric with the outer rim. The largest of these craters are called **multiringed basins.** In Mare Orientale on the west edge of the visible moon, the outermost ring is almost 900 km (550 mi) in diameter.

**2b** The energy of an impact can melt rock, some of which falls back into the crater and solidifies. When the moon was young, craters could also be flooded by lava welling up from below the crust.

A few meteorites found on Earth have been identified chemically as fragments of the moon's surface blasted into space by cratering impacts. The fragmented nature of these meteorites indicates that the moon's surface has been battered by impact craters.

Meteorite from moon
NASA

**3** Most of the craters on the moon were produced long ago when the solar system was filled with debris from planet building. As that debris was swept up, the cratering rate fell rapidly, as shown schematically below.

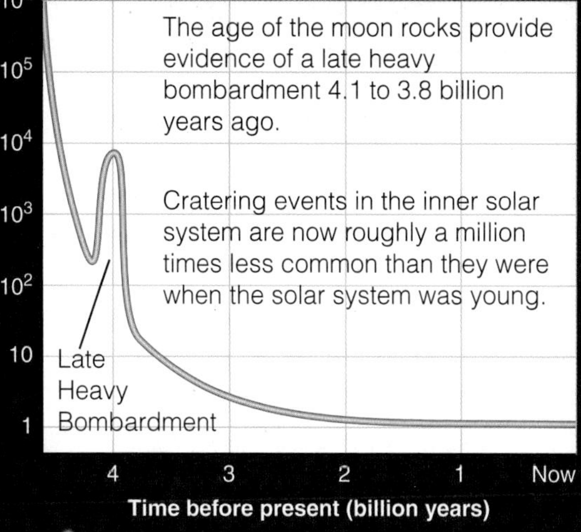

Rate of Crater Formation
The age of the moon rocks provide evidence of a late heavy bombardment 4.1 to 3.8 billion years ago.
Cratering events in the inner solar system are now roughly a million times less common than they were when the solar system was young.
Late Heavy Bombardment
Cratering rate relative to present
$10^6$
$10^5$
$10^4$
$10^3$
$10^2$
10
1
4
3
2
1
Now
Time before present (billion years)

## How Do We Know? 21-1

### How Hypotheses and Theories Unify the Details

**Why is playing catch more than just looking at the ball?** Like any technical subject, science includes a mass of details, facts, figures, measurements, and observations. The flood of details can be overwhelming, but one of the most important characteristics of science comes to your rescue. The goal of science is not to discover more details but to explain the details with a unifying hypothesis or theory. A good theory is like a basket that makes it easier for you to carry a large assortment of details.

For example, when a psychologist begins studying the way the human eye and brain respond to moving objects, the data are a sea of detailed measurements and observations. Infants look at a moving ball for only moments, but older children look longer. Adults can concentrate longer on the moving ball, but their eyes move differently if they are given a stick to point with. Scans of brain activity show that different areas of the brain are active in different age subjects and under different circumstances.

From the data, the psychologist might form the hypothesis that the human brain processes visual information differently depending on its intended use. If you look at a baseball being rubbed in the hands of a pitcher, your brain processes the visual information one way. If you see a baseball flying at you and you have to catch it, your brain processes the information in a different way. The psychologist's theory brings all of the details into place as parts of a logical theory about the ability and necessity of action. Babies don't catch balls. Sometimes a ball is an object that might be rough or smooth, but sometimes it is an object to be caught. The brain responds appropriately.

The goal of science is to understand nature, not to memorize details. Whether scientists are psychologists studying brain functions or astronomers studying the formation of other worlds, they are trying to unify their data and explain it with a single hypothesis or theory.

*When scientists create a hypothesis, it draws together a great many observations and measurements.* (Phyllis Leber)

**Figure 21-3**

Inside the lunar module, the two astronauts stood in a space hardly bigger than two telephone booths. The metal skin was so thin it was easily flexible, like metal foil, and the legs of the module, designed specifically for the moon's weak gravity, could not support the lander's weight on Earth. Only the upper half of the lander blasted off from the surface to return the astronauts to the command module in orbit around the moon. (NASA)

was tacked to the front of the command module like a bicycle strapped to the front of the family camper. It carried only enough fuel and supplies for the short trip to the lunar surface, and it was built to minimize weight and maximize maneuverability.

The weaker gravity of the moon made the design of the LM simpler. Landing on Earth requires reclining couches for the astronauts, but the trip to the lunar surface involved smaller accelerations. In an early version of the LM, the astronauts sat on what looked like bicycle seats, but these were later scrapped to save weight. The astronauts had no seats at all in the LM, and once they began their descent and acquired weight, they stood at the controls held by straps, riding the LM like daredevils on a rocket surfboard.

When it lifted off from the lunar surface, the LM saved weight by leaving the larger descent rocket and support stage behind. Only the compartment containing the two astronauts, their instruments, and their cargo of rocks returned to the command module orbiting above. The astronauts in the LM blasting up from the lunar surface were again standing at the controls. The rocket engine that lifted them back into orbit around the moon was not much bigger than a dishwasher.

The most complicated part of the trip was the rendezvous and docking between the tiny remains of the LM and the command module. Aided by radar systems and computers, the two astronauts docked with the command module, transferred their moon rocks, and jettisoned the remains of the LM. Only the command module returned to Earth.

The first human-piloted lunar landing was made on July 20, 1969. While Michael Collins waited in orbit around the moon, Neil Armstrong and Edwin Aldrin Jr. took the LM down to the surface. Although much of the descent was controlled by computers, the astronauts had to override a number of computer alarms and take control of the LM to avoid a boulder-strewn crater bigger than a football field.

Between July 1969 and December 1972, 12 people reached the lunar surface and collected 380 kg (840 lb) of rocks and soil (■ Table 21-1). The flights were carefully planned to visit different regions and develop a comprehensive understanding of the lunar surface.

The first flights went to relatively safe landing sites (■ Figure 21-4)—Mare Tranquillitatis for Apollo 11 and Oceanus Procellarum for Apollo 12. Apollo 13 was aimed at a more complicated site, but an explosion in an oxygen tank on the way to the moon ended all chances of a landing and nearly cost the astronauts their lives. They succeeded in using the life support in the LM to survive the trip around the moon, and they eventually returned to Earth safely in the crippled command module.

The last four Apollo missions, 14 through 17, sampled geologically important places on the moon. Apollo 14 visited the Fra Mauro region, which is covered by ejecta from the impact that dug the multiringed basin now filled by Mare Imbrium. Apollo 15 visited the edge of Mare Imbrium at the foot of the Apennine

**■ Table 21-1 | Apollo Lunar Landings**

| Apollo Mission* | Astronauts: Commander LM Pilot CM Pilot | Date | Mission Goals | Sample Weight (kg) | Typical Samples | Ages ($10^9$ y) |
|---|---|---|---|---|---|---|
| 11 | Armstrong<br>Aldrin<br>Collins | July 1969 | First human landing; Mare Tranquillitatis | 21.7 | Mare basalts | 3.48–3.72 |
| 12 | Conrad<br>Bean<br>Gordon | Nov. 1969 | Visit Surveyor 3; sample Oceanus Procellarum (mare) | 34.4 | Mare basalts | 3.15–3.37 |
| 14 | Shepard<br>Mitchell<br>Roosa | Feb. 1971 | Fra Mauro, Imbrium ejecta sheet | 42.9 | Breccia | 3.85–3.96 |
| 15 | Scott<br>Irwin<br>Worden | July 1971 | Edge of Mare Imbrium and Apennine Mountains, Hadley Rille | 76.8 | Mare basalts;<br>highland anorthosite | 3.28–3.44<br>4.09 |
| 16 | Young<br>Duke<br>Mattingly | April 1972 | Sample highland crust; Cayley formation (ejecta); Descartes | 94.7 | Highland basalt;<br>breccia | 3.84<br>3.92 |
| 17 | Cernan<br>Schmitt<br>Evans | Dec, 1972 | Sample highland crust; dark halo craters; Taurus–Littrow | 110.5 | Mare basalt;<br>highland breccia<br>fractured dunite | 3.77<br>3.86<br>4.48 |

*The Apollo 13 mission suffered an explosion on the way to the moon and did not land.

**■ Figure 21-4**

Apollo 11, the first mission to the moon, landed on the smooth surface of Mare Tranquillitatis in the lunar lowlands, and the horizon was straight and level. When Apollo 17 landed at Taurus–Littrow in the lunar highlands, the astronauts found the horizon mountainous and the terrain rugged. Landing sites for the other Apollo missions are shown on the photo of the moon. (Moon: © UC Regents/Lick Observatory; Astronauts: NASA)

Mountains and examined Hadley Rille (see Figure 21-2). Apollo missions 16 and 17 visited highland regions to sample older parts of the lunar crust (look again at Figure 21-4). Almost all of the lunar samples from these six landings are now held at the Planetary Materials Laboratory at the Johnson Space Center in Houston, although one has been embedded in a stained glass window in the National Cathedral in Washington, DC. These lunar samples are a national treasure containing clues to the beginnings of our solar system.

## Moon Rocks

Many scientific measurements were made on the moon, but the most exciting prospect was the return of moon rocks to Earth. Analysis could reveal clues to the chemical and physical history of the moon, the origin and evolution of Earth, and the conditions in the solar nebula under which the planets formed.

Of the many rock samples that the Apollo astronauts carried back to Earth, every one is igneous. That is, they formed by the cooling and solidification of molten rock. No sedimentary rocks were found, consistent with the understanding that the moon has never had liquid water on its surface. In addition, the rocks were extremely dry. Almost all Earth rocks contain 1 to 2 percent water, either as free water trapped in the rock or as water molecules chemically bonded with certain minerals. But moon rocks contain no water at all.

Rocks from the lunar maria are dark-colored, dense basalts much like the solidified lava produced by the Hawaiian volcanoes (■ Figure 21-5). These rocks are rich in heavy elements such as iron, manganese, and titanium, which give them their dark color. Some of the basalts are **vesicular,** meaning that they contain holes caused by bubbles of gas in the molten rock. Like bubbles in a carbonated beverage, these bubbles do not form while the magma is under pressure. Only when the molten rock flows out onto the surface, where the pressure is low, do bubbles appear. The vesicular nature of some of the basalts shows that these rocks formed in lava flows that reached the surface and did not solidify underground.

Absolute ages of the mare basalts can be found from the radioactive atoms they contain (see Chapter 19), and these ages range from about 2 to 4 billion years. These ages confirm that the lava flows happened after the end of the heavy bombardment.

The highlands are composed of low-density rock containing calcium-, aluminum-, and oxygen-rich minerals that would have been among the first to solidify and float to the top of molten rock.

The Apollo astronauts found that all moon rocks are igneous, meaning they solidified from molten rock.
Rocks exposed on the lunar surface become pitted by micrometeorites.
Vesicular basalt contains bubbles frozen into the rock when it was molten.
A breccia is formed by rock fragments bonded together by heat and pressure.


**Figure 21-5**

Rocks returned from the moon show that the moon formed in a molten state, that it was heavily fractured by cratering when it was young, and that it is now affected mainly by micrometeorites grinding away at surface rock. (NASA)

Some of this rock is **anorthosite,** a light-colored rock that contributes to the highlands' bright contrast with the dark, iron-rich basalts of the lowlands. The rocks of the highlands, although badly shattered by impacts, represent the moon's original low-density crust, whereas the mare basalts rose as molten rock from the deep crust and upper mantle. The highland crustal rocks range in age from 4.0 to 4.5 billion years old, significantly older than the mare basalts.

Moon rocks are igneous, but many are classified as **breccias,** rocks that are made up of fragments of earlier rocks cemented together by heat and pressure. Evidently, after the molten rock solidified, meteorite impacts broke up the rocks and fused them together time after time.

If you went to the moon, you would get your spacesuit dirty. Both the highlands and the lowlands of the moon are covered by a layer of powdered rock and crushed fragments called the **regolith.** It is about 10 m deep on the maria but over 100 m deep in certain places in the highlands. About 1 percent of the regolith is meteoric fragments; the rest is the smashed remains of moon rocks that have been pulverized by the constant rain of meteorites. The smallest meteorites, the micrometeorites, do the most damage by constantly sandblasting the lunar surface, grinding the rock down to fine dust. The Apollo astronauts found that the dust coated their spacesuits and equipment, and then the interior of the Lunar Module after they climbed back inside. Impact cratering, a theme of this chapter, dominates the lunar surface and is responsible for the lunar regolith.

The moon rocks are old, dry, igneous, and badly shattered by impacts. You can use these facts, combined with what you know about lunar features, to tell the story of the moon.

## The History of the Moon

Evidence preserved in the Apollo moon rocks shows that the moon must have formed in a molten state. Planetary geologists now refer to the newborn moon as a magma ocean. Evidently, denser materials sank to form a small core, and, as the magma ocean cooled, low-density minerals crystallized and floated to the top to form a low-density crust. In this way the moon differentiated into core, mantle, and crust. This corresponds to the first of the four stages of Terrestrial planet development displayed in Figure 20-2. The radioactive ages of the moon rocks show that the surface solidified between 4.6 and 4.1 billion years ago. The moon has a low average density and no magnetic field, so its dense core must be small. The core may still retain enough heat to be partially molten, but it can't contain much molten iron, or the dynamo effect would produce a magnetic field.

The second stage, cratering, began as soon as the crust solidified, and the older highlands show that the cratering was intense during the first 0.5 billion years or so—during the heavy bombardment at the end of planet building. The cratering rate should have fallen rapidly as the solar system was cleared of debris. However, there is evidence from lunar crater counts and rock sample ages that there was a temporary surge in impact rate, called the **late heavy bombardment,** about 4 billion years ago, near the end of the heavy bombardment era. Models of the solar system's evolution mentioned in Chapter 19 and 23 indicate that the late heavy bombardment could have been caused by an episode of Jovian planet migration that would have scattered remnant planetesimals into collisions with all the solar system's planets and moons.

The moon's crust was shattered to a depth of 10 kilometers or so, and the largest impacts during the heavy bombardment and late heavy bombardment formed giant multiringed crater basins hundreds of kilometers in diameter, such as Mare Imbrium and Mare Orientale. This led to the third stage—flooding. Although Earth's moon cooled rapidly after its formation, radioactive decay heated material deep in the crust, and part of it melted. Molten rock followed the cracks up to the surface and flooded the giant basins with successive lava flows of dark basalts from about 4 to about 2 billion years ago. This formed the maria (■ Figure 21-6).

It is a **Common Misconception** that the lava that floods out on the surfaces of Earth and other planets comes from the molten core. The lava comes from the lower crust and upper mantle. The pressure is lower there, and that lowers the melting point of the rock enough that radioactive decay can melt portions of the rock. If there are faults and cracks, the magma can reach the surface and form volcanoes and lava flows. Whenever you see lava flows on a planet, you can be sure heat is flowing out of the interior, but the lava did not come all the way from the core.

Some maria on the moon, such as Mare Imbrium, Mare Serenitatis, Mare Humorum, and Mare Crisium, retain their round shapes, but others are irregular because the lava overflowed the edges of the basin or because the shape of the basin was modified by further cratering. The floods of lava left other characteristic features frozen into the maria. In places, the lava formed channels that are seen from Earth as sinuous rilles. Also, the weight of the maria pressed the crater basins downward, and the solidified lava was compressed and formed wrinkle ridges visible even in small telescopes. The tension at the edges of the maria broke the hard lava to produce straight fractures and faults. All of these features are visible in Figure 21-2. As time passed, further cratering and overlapping lava floods modified the maria. Consequently, you should think of the maria as accumulations of features reflecting multiple events during the moon's complex history.

Mare Imbrium is a dramatic example of how the great basins became the maria. Its story can be told in detail partly because of evidence gathered by Apollo 14 astronauts, who landed on ejecta

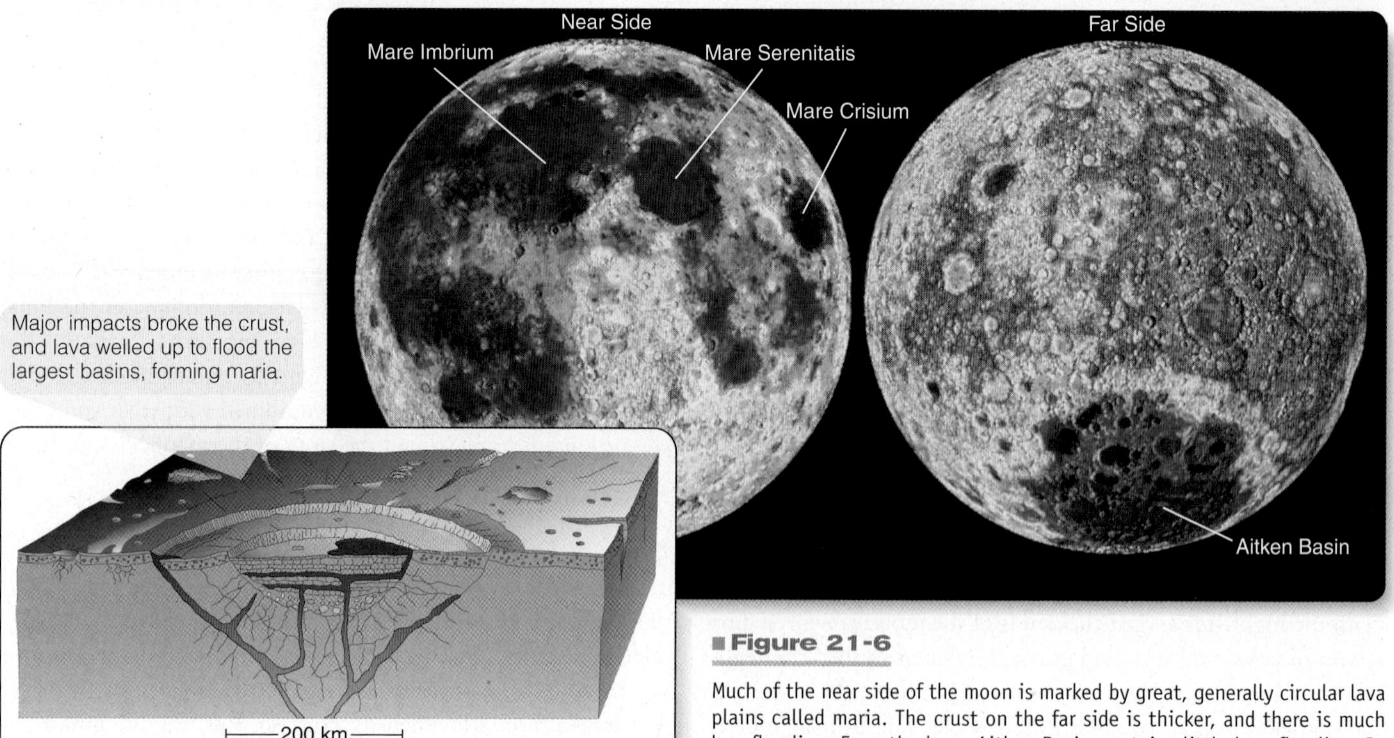


■ **Figure 21-6**

Much of the near side of the moon is marked by great, generally circular lava plains called maria. The crust on the far side is thicker, and there is much less flooding. Even the huge Aitken Basin contains little lava flooding. In these maps, color marks elevation, with red the highest regions and purple the lowest. (Adapted from a diagram by William Hartmann; NASA/Clementine)

from the Imbrium impact (■ Figure 21-7), and by the Apollo 15 astronauts, who landed at the edge of the mare itself. Near the end of the heavy bombardment, roughly 4 billion years ago, a planetesimal estimated to have been at least 60 km (40 mi) across (about the size of Rhode Island) struck the moon and blasted out a giant multiringed basin. The impact was so violent the ejecta blanketed 16 percent of the moon's surface. After the cratering rate fell at the end of the heavy bombardment, lava flows welled up time after time and flooded the Imbrium Basin, burying all but the highest parts of the giant multiringed basin. The Imbrium Basin is now a large, generally round mare marked by only a few craters that have formed since the last of the lava flows (■ Figure 21-8).

This story of the moon might suggest that it was a violent place during the cratering phase, but large impacts were in fact rare; the moon was, for the most part, a peaceful place even during the heavy bombardment. Had you stood on the moon at that time you would have experienced a continuous rain of micrometeorites and much less common pebble-size impacts. Centuries might pass between major impacts. Of course, when a large impact did occur far beyond the horizon, it might have buried you under ejecta or jolted you by seismic shocks. You could have felt the Imbrium impact anywhere on the moon, but had you been standing on the side of the moon directly opposite that impact, you would have been at the focus of seismic waves traveling around the moon from different directions. When the waves met under your feet, the surface would have jerked up and down by as much as 10 m. The place on the moon opposite the Imbrium Basin is a strangely disturbed landscape called **jumbled terrain.** You will see similar effects of large impacts on other worlds.

Studies of our moon show that its crust is thinner on the side facing Earth, perhaps due to tidal effects. Consequently, while lava flooded the basins on the Earth-ward side, it was unable to rise through the thicker crust to flood the lowlands on the far side. The largest known impact basin in the solar system is the Aitken Basin near the moon's south pole (Figure 21-6b). It is about 2500 ki (1500 mi) in diameter and as deep as 13 kilometers (8 mi) in places, but flooding has never filled it with smooth lava flows.

**■ Figure 21-7**

Apollo 14 landed on rugged terrain suspected of being ejecta from the Imbrium impact. The large boulder here is ejecta that, at some time in the past, fell here from an impact far beyond the horizon. (NASA)

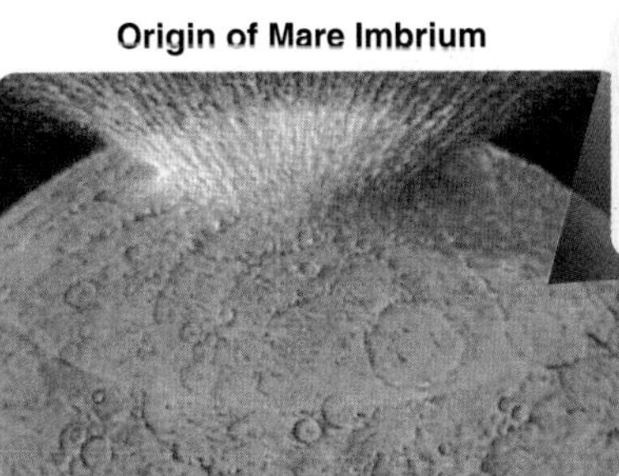

Origin of Mare Imbrium
Four billion years ago, an impact forms a multiringed basin over 1000 km (700 mi) in diameter.

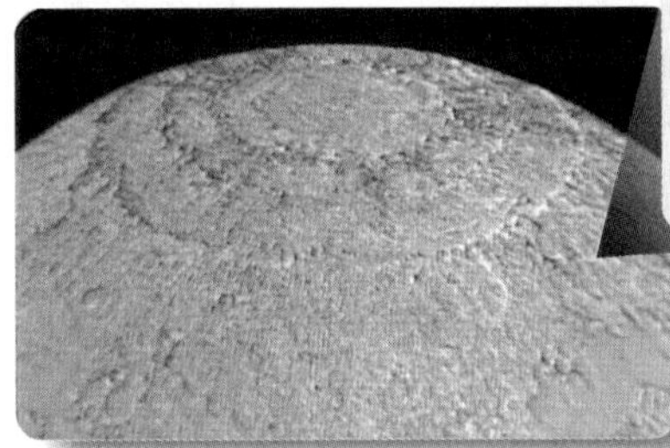

Continuing impacts crater the surface but do not erase the high walls of the multiringed basin.

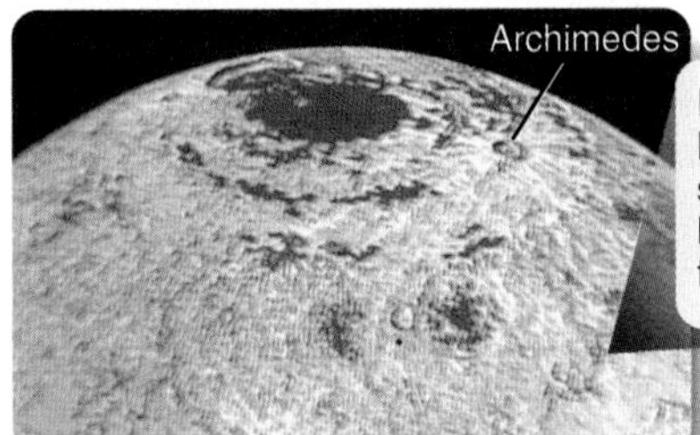

Archimedes
Impacts form a few large craters, and, starting about 3.8 billion years ago, lava floods low regions.

Repeated lava flows cover most of the inner rings and overflow the basin to merge with other flows.

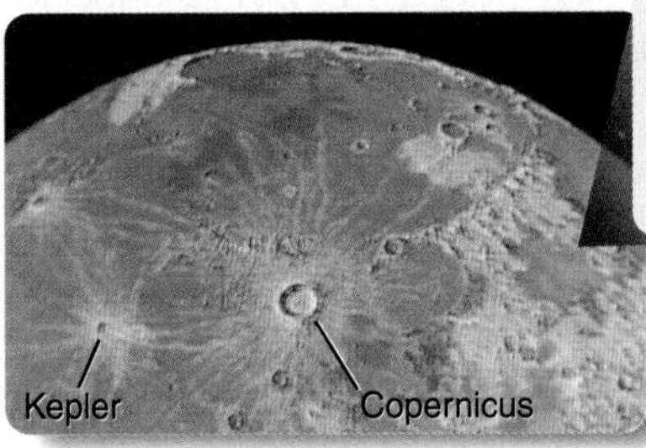

Impacts continue, including those that formed the relatively young craters Copernicus and Kepler.
Kepler
Copernicus

**■ Figure 21-8**

Lava flooding after the end of the heavy bombardment filled a giant multiringed basin and formed Mare Imbrium. (Don Davis)

The moon is small, and small worlds cool rapidly because they have a large ratio of surface area to volume. The rate of heat loss is proportional to the surface area, and the amount of heat in a world is proportional to the volume. The smaller a world is, the easier it is for the heat to escape. That is why a small cupcake fresh from the oven cools more rapidly than a large cake. The moon lost much of its internal heat when it was young, and it is the outward flow of heat that drives geological activity, so the moon is mostly inactive today. The crust of the moon rapidly grew thick and never divided into moving plates. There are no rift valleys or folded mountain chains on the moon. The last lava flows on the moon ended about 2 billion years ago when the moon's internal temperature fell too low to maintain subsurface lava.

The overall terrain on the moon is almost fixed. On Earth a billion years from now, plate tectonics will have totally changed the shapes of the continents, and erosion will have long ago worn away the mountain ranges you see today. On the moon, with no atmosphere and no water, there is no Earth-like erosion. Over the next billion years, impacts will have formed only a few more large craters, and nearly all of the lunar scenery will be unchanged. Micrometeorites are the biggest influence; they will have blasted the soil, erasing the footprints left by the Apollo astronauts and reducing the equipment they left behind to peculiar chemical contamination in the regolith at the six Apollo landing sites.

You have studied the story of the moon's evolution in detail for later comparison with other planets and moons in our solar system, but the story has skipped one important question: Where did Earth get such a large satellite?

## The Origin of Earth's Moon

Over the last two centuries, astronomers developed three different hypotheses for the origin of Earth's moon. The *fission hypothesis* proposed that the moon broke from a rapidly spinning young Earth. The *condensation hypothesis* suggested that Earth and its moon condensed together from the same cloud of matter in the solar nebula. The *capture hypothesis* suggested that the moon formed elsewhere in the solar nebula and was later captured by Earth. Each of these older ideas had problems and failed to survive comparison with all the evidence.

In the 1970s, after moon rocks were returned to Earth and studied in detail, a new hypothesis originated that combined some aspects of the three older hypotheses. The **large-impact hypothesis** proposes that the moon formed when a very large planetesimal, estimated to have been at least as massive as Mars, smashed into the proto-Earth. Model calculations indicate that this collision would have ejected a disk of debris into orbit around Earth that would have quickly formed the moon (■ Figure 21-9).

This hypothesis explains several phenomena. If the proto-Earth and impactor had each already differentiated, the ejected material would have been mostly iron-poor mantle and crust, which would explain the moon's low overall density and iron-poor composition. Furthermore, the material would have lost its volatile components while it was in space, so the moon would have formed lacking volatiles. Such an impact would have melted the proto-Earth, and the material falling together to form the moon would also have been heated hot enough to melt completely. This fits the evidence that the highland anorthosite in the moon's oldest rocks formed by differentiation of large quantities of molten material.

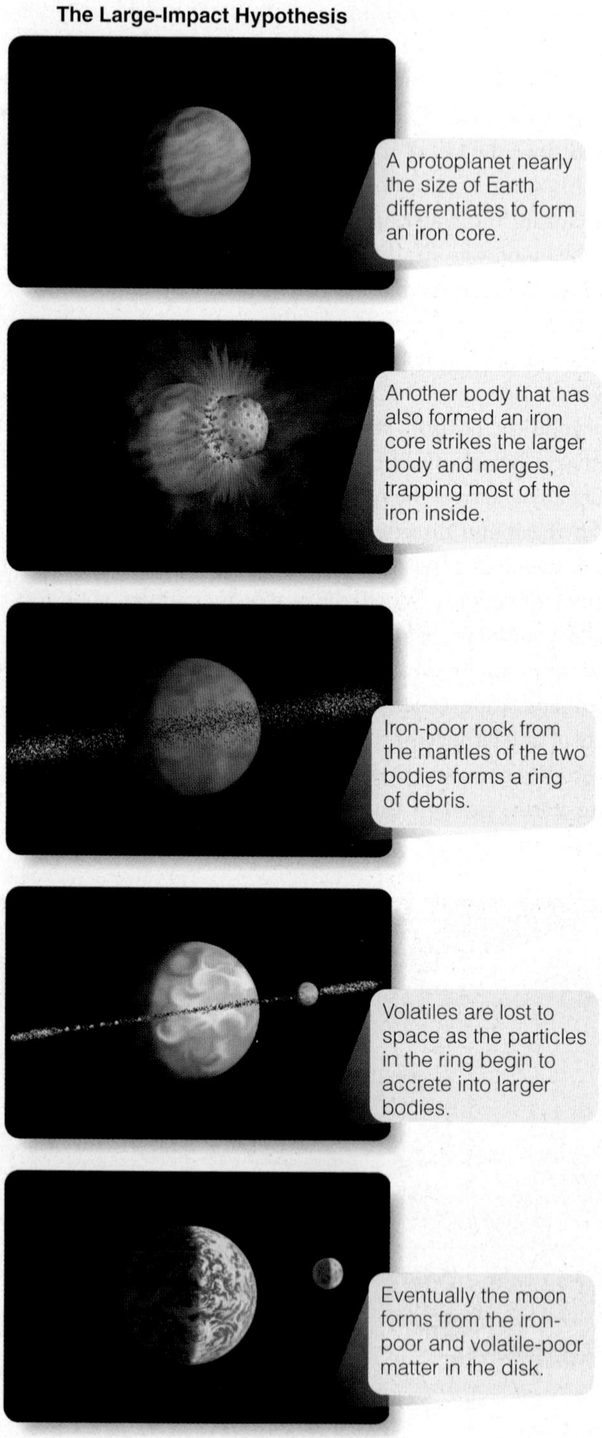

**The Large-Impact Hypothesis**

A protoplanet nearly the size of Earth differentiates to form an iron core.

Another body that has also formed an iron core strikes the larger body and merges, trapping most of the iron inside.

Iron-poor rock from the mantles of the two bodies forms a ring of debris.

Volatiles are lost to space as the particles in the ring begin to accrete into larger bodies.

Eventually the moon forms from the iron-poor and volatile-poor matter in the disk.

**■ Figure 21-9**

Sometime before the solar system was 50 million years old, a collision produced Earth and the moon in its orbit inclined to Earth's equator.

The large-impact hypothesis is consistent with the known evidence and is now considered likely to be correct.

This hypothesis would explain other things. The collision must have occurred at a steep angle to eject enough matter to make the moon, so the objects could not have collided head-on. A glancing collision would have spun the material rapidly enough to explain the observed angular momentum in the Earth–moon system. And, as mentioned above, if the two colliding planetesimals had already differentiated, the ejected material would be mostly iron-poor mantle and crust. Calculations indicate that the iron core of the impacting body would have fallen into the larger body that became Earth. This would explain why the moon is so poor in iron and why the abundances of other elements are so similar to those in Earth's mantle. Finally, the collision-heated material that eventually became the moon would have remained in a disk long enough for water and other volatile elements, which the moon lacks, to be outgassed and lost to space.

The moon is evidently the result of a giant impact. Until recently, astronomers have been reluctant to consider such catastrophic events, but a number of lines of evidence suggest that other planets also may have been affected by giant impacts. Consequently, the third theme identified in the introduction to this chapter, giant impacts, has the potential to help you understand other worlds. Catastrophic events are rare, but they can occur.

### SCIENTIFIC ARGUMENT

***If the moon was intensely cratered by the heavy bombardment and then formed great lava plains, why didn't the same thing happen on Earth?***

Is this argument obvious? It is still worth reviewing as a way to test your understanding. In fact, the same thing did happen on Earth. Although the moon has more craters than Earth, the moon and Earth are the same age, and both were battered by meteorites during the heavy bombardment. Some of those impacts on Earth must have been large and dug giant multiringed basins. Lava flows must have welled up through Earth's crust and flooded the lowlands to form great lava plains much like the lunar maria.

Earth, however, is a larger world and has more internal heat, which escapes more slowly than the moon's heat did. The moon is now geologically dead, but Earth is very active, with heat flowing outward from the interior to drive plate tectonics. The moving plates long ago erased all evidence of the cratering and lava flows dating from Earth's youth.

Comparative planetology is a powerful tool in that it allows you to see similar processes occurring under different circumstances. For example, expand your argument to explain a different phenomenon. **Why doesn't the moon have a magnetic field?**

## 21-2 Mercury

EARTH'S MOON AND MERCURY are good subjects for comparative planetology. They are similar in a number of ways. Most important, they are small worlds (**Celestial Profile 4**); the moon

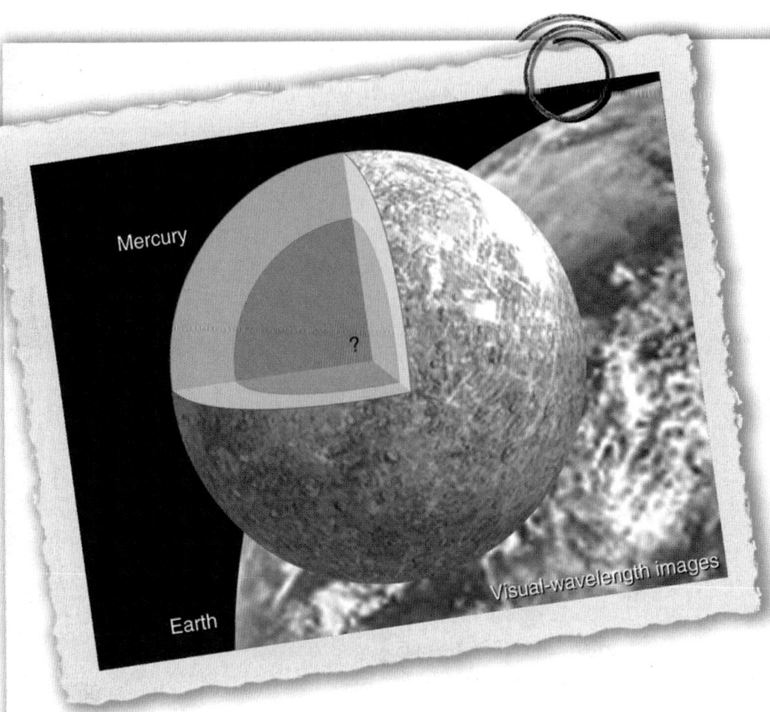


*Mercury is a bit over one-third the diameter of Earth, but its high density must mean it has a large iron core. The amount of heat it retains is unknown.* (NASA)

## Celestial Profile 4: Mercury

### Motion:

| | |
|---|---|
| Average distance from the sun | 0.387 AU ($5.79 \times 10^7$ km) |
| Eccentricity of orbit | 0.206 |
| Inclination of orbit to ecliptic | 7.0° |
| Average orbital velocity | 47.9 km/s |
| Orbital period | 0.241 y (88.0 days) |
| Period of rotation | 58.6 d (direct) |
| Inclination of equator to orbit | 0° |

### Characteristics:

| | |
|---|---|
| Equatorial diameter | $4.89 \times 10^3$ km (0.382 $D_\oplus$) |
| Mass | $3.31 \times 10^{23}$ kg (0.0558 $M_\oplus$) |
| Average density | 5.44 g/cm$^3$ (5.4 g/cm$^3$ uncompressed) |
| Surface gravity | 0.38 Earth gravity |
| Escape velocity | 4.3 km/s (0.38 $V_\oplus$) |
| Surface temperature | −170° to 430°C (−275° to 805°F) |
| Average albedo | 0.1 |
| Oblateness | 0 |

### Personality Point:

Mercury lies very close to the sun and completes an orbit in only 88 days. For this reason, the ancients named the planet after Mercury, the fleet-footed messenger of the gods. The name is also applied to the element mercury, which is also known as quicksilver because it is a heavy, quickly flowing silvery liquid at room temperatures.

is only a fourth of Earth's diameter, and Mercury is just over a third of Earth's diameter. Their rotation has been altered by tides, their surfaces are heavily cratered, their lowlands are flooded in places by ancient lava flows, and both are airless and have ancient, inactive surfaces. The impressive differences between them also will help you understand the nature of these airless worlds.

Mercury is the innermost planet in the solar system, and thus its orbit keeps it near the sun in the sky, viewed from Earth. It is sometimes visible near the horizon in the evening sky after sunset or in the dawn sky just before sunrise. Earth-based telescopes show the small disk of Mercury passing through phases like the moon and Venus. The Mariner 10 spacecraft looped through the inner solar system in 1974 and 1975, taking photographs and other measurements during three flybys of Mercury. Astronomers managed in 2007 to make impressively high-resolution images from Earth of the parts of Mercury not covered by Mariner 10's images (■ Figure 21-10). A new spacecraft called MESSENGER (MErcury Surface, Space ENvironment, GEochemistry, and Ranging mission) has made two flybys of Mercury at the time of this writing and will make one more before it goes into orbit around the planet in 2011 and begins a yearlong close-up study (■ Figure 21-11).

**■ Figure 21-11**

The MESSENGER spacecraft will make three flybys and then spend a year orbiting Mercury, observing from behind a fabric sunscreen. (NASA/Johns Hopkins University Applied Physics Lab/Carnegie Institute of Washington)

## Rotation and Revolution

During the 1880s, the Italian astronomer Giovanni Schiaparelli sketched the faint features he thought he saw on the disk of Mercury and concluded that the planet was tidally locked to the sun and kept the same side facing the sun throughout its orbit. This was actually a very good guess because, as you will see, tidal coupling between rotation and revolution is common in the solar system. You have already seen that the moon is tidally locked to Earth. But the rotation of Mercury is more complex than Schiaparelli thought.

In 1962, radio astronomers detected blackbody emissions from the planet and concluded that the dark side was not as cold as it should have been if the planet kept one side in perpetual darkness. In 1965, radio astronomers made radar contact with Mercury by using the 305-m Arecibo dish (see Figure 6-21) to transmit a pulse of radio energy at Mercury and then waiting for the reflected signal to return. Doppler shifts in the reflected radio pulse showed that the planet was rotating with a period of only about 59 days, noticeably shorter than the orbital period of 88 days.

Mercury is tidally coupled to the sun but in a different way than the moon is coupled to Earth. Mercury rotates not once per orbit but 1.5 times per orbit. That is, its period of rotation is two-thirds its orbital period. This means that a mountain on Mercury directly facing the sun at one place in its orbit will point away from the sun one orbit later and toward the sun after the next orbit (■ Figure 21-12).

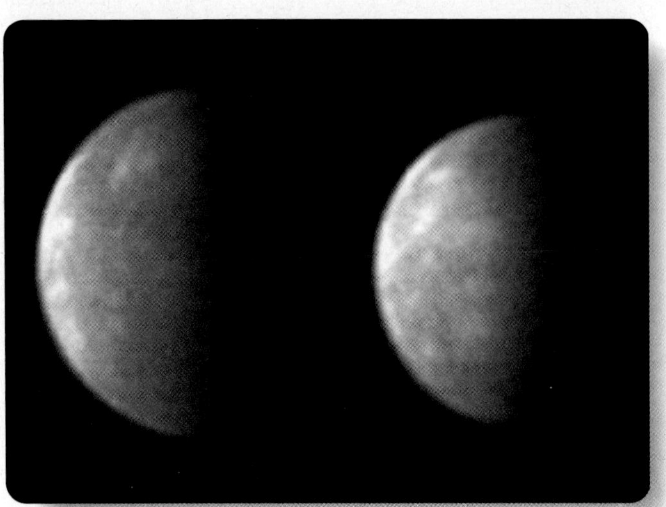

**■ Figure 21-10**

Using the 4.1-meter SOAR telescope in Chile by remote control from their offices in North Carolina, astronomer Gerald Cecil and undergraduate student Dmitry Rashkeev resolved details on Mercury as small as 150 km during March and April 2007, the highest-quality images of that planet ever made from Earth. The observations were timed to view mostly the side of Mercury that had not been previously imaged by the Mariner 10 probe. The telescope aperture was masked down to 1.35 meters in diameter to reduce the effect of Earth atmosphere "seeing." A series of very rapid exposures were taken with a digital camera, with most images being discarded, saving only the highest-resolution data. (G. Cecil/University of North Carolina)

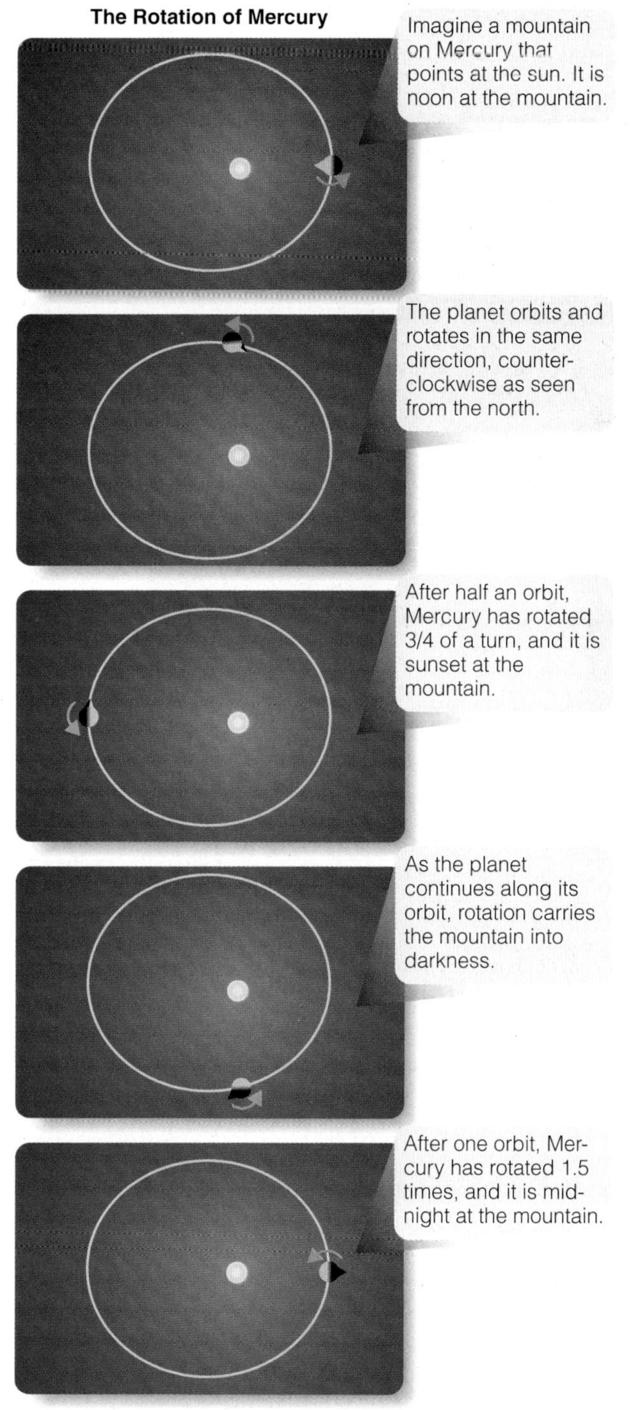

The Rotation of Mercury
Imagine a mountain on Mercury that points at the sun. It is noon at the mountain.
The planet orbits and rotates in the same direction, counter-clockwise as seen from the north.
After half an orbit, Mercury has rotated 3/4 of a turn, and it is sunset at the mountain.
As the planet continues along its orbit, rotation carries the mountain into darkness.
After one orbit, Mercury has rotated 1.5 times, and it is midnight at the mountain.


■ **Figure 21-12**

Mercury's rotation is in resonance with its orbital motion. It orbits the sun in 88 days and rotates on its axis in two-thirds of that time. One full day on Mercury from noon to noon takes two orbits.

If you flew to Mercury and landed your spaceship in the middle of the day side, the sun would be high overhead, and it would be noon. Your watch would show almost 44 Earth days passing before the sun set in the west, and a total of 88 Earth days would pass before the sun reached the midnight position. In those 88 Earth days, Mercury would have completed one orbit around the sun (Figure 21-12). It would require another entire orbit of Mercury for the sun to return to the noon position overhead. So a full day on Mercury is two Mercury years long!

The complex tidal coupling between the rotation and revolution of Mercury is an important illustration of the power of tides. Just as the tides in the Earth–moon system have slowed the moon's rotation and locked it to Earth, so have the sun–Mercury tides slowed the rotation of Mercury and coupled its rotation to its revolution. Astronomers refer to such a relationship as a **resonance.** You will see other such resonances as you continue to explore the solar system.

Like its rotation, Mercury's orbital motion is complex. Recall from Chapter 5 that Mercury's orbit is modestly eccentric and precesses faster than can be explained by Newton's laws but at just the rate predicted by Einstein's theory of general relativity. The orbital motion of Mercury is taken as strong confirmation of the curvature of space-time as predicted by general relativity.

## The Surface of Mercury

Because Mercury is close to the sun, the temperatures on Mercury are extreme. If you stood in direct sunlight on Mercury, you would hear your spacesuit's cooling system cranking up to top speed as it tried to keep you cool. Daytime temperatures can exceed 700 K (800°F), although about 500 K is a more usual high temperature. If you stepped into shadow on Mercury or took a walk at night, with no atmosphere to distribute heat, your spacesuit heaters would struggle to keep you warm. The surface can cool to 100 K (-280°F). Nights on Mercury are bitter cold. Don't go to Mercury in a cheap spacesuit.

Nights are cold on Mercury because it has almost no atmosphere. It borrows hydrogen and helium atoms from the solar wind, and atoms such as oxygen, sodium, potassium, and calcium have been detected in a cloud above the planet's surface that has such a low density that the atoms do not collide with each other. They just bounce from place to place on the surface and, because of the low escape velocity, eventually disappear into space. Some of these atoms are probably baked out of the crust, or possibly produced by very low-level remnant volcanic venting.

In photographs, Mercury looks much like Earth's moon (■ Figure 21-13). It is heavily battered, with craters of all sizes, including some large basins. Some craters are obviously old and degraded; others seem quite young and have bright rays of ejecta. However, a quick glance at photos of Mercury shows no large, dark maria like the moon's flooded basins.

When planetary scientists began looking at Mercury flyby photographs in detail, they discovered something not seen on the moon. Mercury is marked by great curved cliffs called **lobate scarps** (■ Figure 21-14). These seem to have formed when the planet cooled and shrank in diameter by a few kilometers, wrinkling its crust as a drying apple wrinkles its skin. Some of these scarps are as high as 3 km and reach hundreds of kilometers across the surface. Other faults in Mercury's crust are straight and

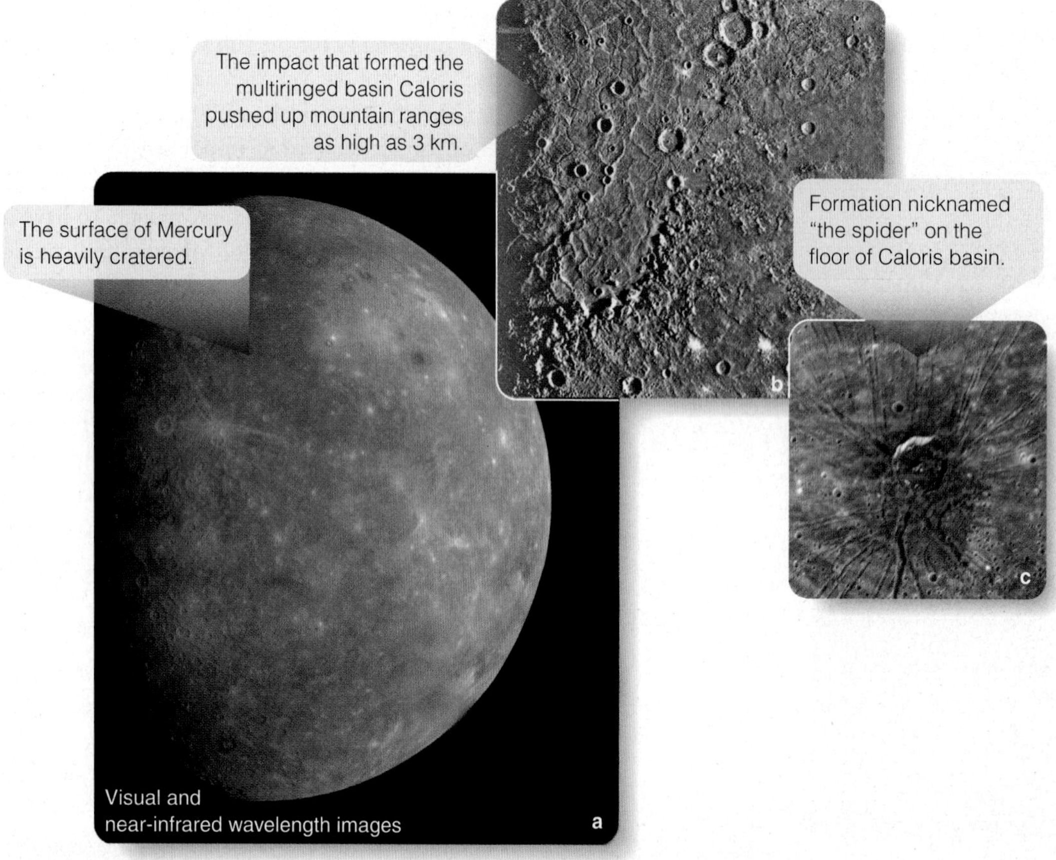
The impact that formed the multiringed basin Caloris pushed up mountain ranges as high as 3 km.
The surface of Mercury is heavily cratered.
Formation nicknamed "the spider" on the floor of Caloris basin.
Visual and near-infrared wavelength images
a
b
c

**■ Figure 21-13**

(a) This 2008 MESSENGER spacecraft flyby image in false-color highlights differences in composition between different parts of Mercury's surface. (NASA/JHU/APL) (b) The Caloris ringed basin, photographed by the Mariner 10 spacecraft as it flew past the planet in 1975. (NASA) (c) The origin of the spider formation photographed by MESSENGER is a puzzle. (NASA/JHU/APL)

may have been produced by tidal stresses generated when the sun slowed Mercury's rotation.

The largest basin on Mercury is called Caloris Basin after the Latin word for "heat," recognition of its location at one of the two "hot poles" that face the sun at alternate perihelions. At the times of the Mariner encounters, the Caloris Basin was half in shadow (■ Figure 21-15a). Although half cannot be seen, the low angle of illumination was ideal for the study of the lighted half because it produced dramatic shadows.

Caloris is a gigantic multiringed impact basin 1300 km (800 mi) in diameter with concentric mountain rings up to 3 km high. The impact threw ejecta 600 to 800 km across the planet, and the focusing of seismic waves on the far side produced peculiar terrain that looks much like the jumbled area on the moon's surface that lies opposite the Imbrium basin (Figure 21-15b and c). The Caloris Basin is partially filled with lava flows. Some of this lava may be material melted by the energy of the impact, but some may be lava from below the crust that leaked up through cracks. The weight of this lava and the sagging of the crust have produced deep cracks in the central lava plains. The geophysics of such large, multiringed crater basins is not well understood at present, but Caloris Basin seems to be the same kind of structure as the Imbrium Basin on the moon, although it has not been as deeply flooded with lava.

When the MESSENGER spacecraft begins orbiting Mercury in 2011, it will photograph nearly all the planet's surface at a much higher resolution than did Mariner 10. Those new photographs will help planetary scientists build a much more complete understanding of Mercury's surface.

## The Plains of Mercury

The most striking difference between Mercury and the moon is that Mercury lacks the great dark lava plains so obvious on the moon. Under careful examination, the Mariner 10 photographs show that Mercury has plains, two different kinds, in fact, but they are different from the moon's. Understanding these differences is the key to understanding the history of Mercury.

■ **Figure 21-14**

A lobate scarp (arrow) crosses craters, indicating that Mercury cooled and shrank, wrinkling its crust, after many of its craters had formed. (NASA)

Much of Mercury's surface is old, cratered terrain (■ Figure 21-16), but other areas called **intercrater plains** are less heavily cratered. Those plains are marked by meteorite craters less than 15 km in diameter and secondary craters produced by chunks of ejecta from larger impacts. Unlike the heavily cratered regions, the intercrater plains are not totally saturated with craters. As an expert in comparative planetology, you can recognize that this means that the intercrater plains were produced by later lava flows that buried older terrain.

Smaller regions called **smooth plains** appear to be even younger than the intercrater plains. They have fewer craters and appear to be lava flows that occurred after most cratering had ended. Much of the region around the Caloris Basin is composed of these smooth plains (Figure 21-16), and they appear to have formed soon after the Caloris impact.

Given the available evidence, planetary astronomers conclude that the plains of Mercury are solidified lava flows much like the maria on the moon. Unlike the maria, Mercury's lava plains are not significantly darker than the rest of the crust. This may be due to a compositional difference between Mercury's lava flows and the moon's. Except for a few bright crater rays, Mercury's surface is a uniform gray with an albedo of only about 0.1. That means Mercury's lava plains are not as dramatically obvious on photographs as the much darker maria on our own moon, which show up in contrast to the lighter highlands.

## The Interior of Mercury

One of the most striking differences between Mercury and the moon is the composition of their interiors. You have seen that the moon is a low-density world that contains no more than a small core of metals. In contrast, evidence suggests that Mercury has a large metallic core.

Mercury is over 60 percent denser than the moon. Yet Mercury's surface appears to be normal rock, so you can conclude that its interior contains a large core of dense metals, mostly iron. In proportion to its size, Mercury must have a larger metallic core than Earth (see the diagram in Celestial Profile 4).

If Mercury had a large metallic core that remained molten, then the dynamo effect would generate a magnetic field (see Chapter 20). The Mariner 10 spacecraft found a magnetic field only about 0.5 percent as strong as Earth's, and this weak field made it difficult to understand the interior. Because Mercury is a small world, it should have lost most of its internal heat long ago and should not have a molten core. Nevertheless, radar observations of Mercury's rotation show that the surface of the planet is shifting back and forth slightly in response to the sun's gravity. That must mean that at least the outer core is molten. If Mercury's iron core contains a higher-than-Earthly concentration of sulfur, the melting point would be lowered, and the outer core, where the pressure is lower, could be molten. It is not clear, however, why a planet that formed so close to the sun could contain so much sulfur, which is a volatile material. The

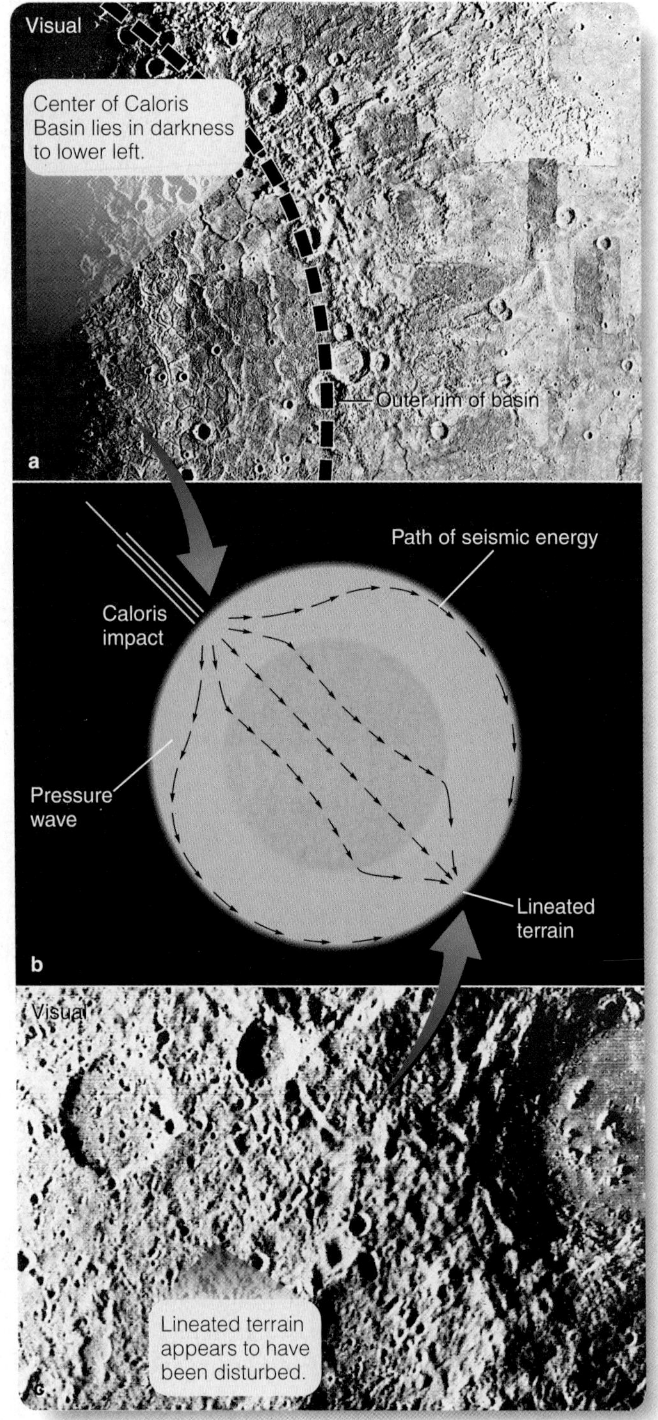


**Figure 21-15**

The huge impact that formed the Caloris Basin on Mercury sent seismic waves through the planet. Where the waves came together on the far side, they produced this lineated terrain, which resembles the jumbled terrain on Earth's moon opposite the Imbrium impact. (NASA)

MESSENGER spacecraft will make measurements of Mercury's gravitational and magnetic fields that will help planetary astronomers understand its core.

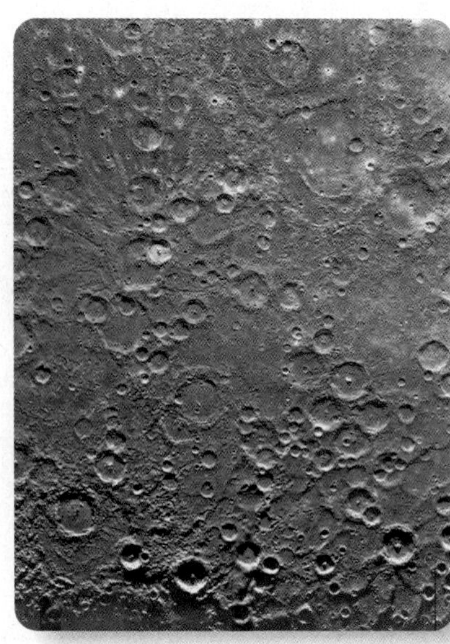

**Figure 21-16**

Study this region of Mercury carefully, and you will notice the plains between the craters. This surface is not saturated as are the lunar highlands, but it contains more craters than the surface of the maria on Earth's moon. This shows that these plains formed after the heaviest of the cratering in the early solar system. (NASA)

It is also difficult to explain the proportionally large size of the metallic core inside Mercury. In Chapter 19, you saw that planets forming near the sun should incorporate more metals, but some models suggest that this effect alone cannot account for the size of Mercury's very large metallic core. One hypothesis is that heat from the sun vaporized and drove away some of the rock-forming elements in the inner solar nebula. Another theory involves a giant impact when Mercury was young, an impact much like the planet-shattering impact proposed to explain the origin of Earth's moon. If the forming planet had differentiated and was then struck by a large planetesimal, the impact could have shattered the crust and mantle and blasted much of the lower-density material into space. The denser core could have survived, re-formed, and then swept up some of the lower-density debris to form a thin mantle and crust. This scenario would leave Mercury with a deficiency of low-density crustal rock. It is possible that both Mercury and the moon may be products of giant impacts.

The MESSENGER spacecraft will be able to test this hypothesis when it goes into orbit around Mercury in 2011. The probe will analyze the composition of the crust by remote measurements from orbit, and the chemical composition of the rock will reveal its history.

## A History of Mercury

Can you combine evidence and theory to tell the story of Mercury? It formed in the innermost part of the solar nebula, and, as you have seen, a giant impact may have robbed it of some of its lower-density rock and left it a small, dense world with a surprisingly large metallic core.

Like the moon, Mercury suffered heavy cratering by debris in the young solar system. Planetary scientists don't know accurate ages for features on Mercury, but you can assume that cratering, the second stage of planetary development, occurred over the same period as the cratering on the moon. This intense cratering declined rapidly as the planets swept up the last of the debris left over from planet building.

The cratered surface of Mercury is not exactly like that of the moon. Because of Mercury's stronger gravity, the ejecta from an impact on Mercury is thrown only about 65 percent as far as on the moon, and that means the ejecta from an impact on Mercury does not blanket as much of the surface. Also, the intercrater plains appear to have formed when lava flows occurred during the heavy bombardment, burying the older surface, and then accumulated more craters. Sometime near the end of cratering, a planetesimal over 100 km in diameter smashed into the planet and blasted out the great multiringed Caloris Basin. Only parts of that basin have been flooded by lava flows.

The smooth plains contain fewer craters and may date from the time of the Caloris impact. The impact may have been so big it fractured the crust and allowed lava flows to resurface wide areas. Because this happened near the end of cratering, the smooth plains have few craters.

Finally, the cooling interior contracted, and the crust broke to form the lobate scarps. Lava flooding ended quickly, perhaps because the shrinking planet squeezed off the lava channels to the surface. Mercury lacks a true atmosphere, so you would not expect flooding by water, but radar images show a bright spot at the planet's north pole that may be caused by ice trapped in perpetually shaded crater floors where the temperature never exceeds 60 K (–350°F). This may be water from comets that occasionally collide with Mercury and deliver a burst of water vapor. Similar water deposits may exist at the poles of Earth's moon.

The fourth stage in the story of Mercury, slow surface evolution, is now limited to micrometeorites, which grind the surface to dust; rare larger meteorites, which leave bright-rayed craters; and the slow but intense cycle of heat and cold, which weakens the rock at the surface. The planet's crust is now thick, and although its core may be partially molten, the heat flowing outward is unable to drive plate tectonics that would actively erase craters and build folded mountain ranges.

## SCIENTIFIC ARGUMENT

***Why don't Earth and the moon have lobate scarps?***

Of course, this calls for a scientific argument based on comparative planetology. You might expect that any world with a large metallic interior should have lobate scarps. When the metallic core cools and contracts, the world should shrink, and the contraction should wrinkle and fracture the brittle crust to form lobate scarps. But there are other factors to consider. Earth has a fairly large metallic core; but it has not cooled very much, and the crust is thin, flexible, and active. If any lobate scarps ever did form on Earth, they would have been quickly destroyed by plate tectonics.

The moon does not have a large metal core. It may be that the rocky interior did contract slightly as the moon lost its internal heat, but that smaller contraction may not have produced major lobate scarps. Also, any surface features that formed early in the moon's history would have been destroyed by the heavy bombardment.

You can, in a general way, understand lobate scarps, but now add timing to your argument. **How do you know the lobate scarps on Mercury formed after most of the heavy bombardment was over?**

## What Are We? Comfortable

Many planets in the universe probably look like the moon and Mercury. They are small, airless, and cratered. Some are made of stone; and some, because they formed farther from their star, are made mostly of ices. If you randomly visited a planet anywhere in the universe, you might find yourself standing on a cratered moonscape.

Earth-like worlds are unusual, but perhaps not rare. The Milky Way Galaxy contains over 100 billion stars, and over 100 billion galaxies are visible with existing telescopes. Most of those stars probably have planets, and although many planets look like Earth's moon and Mercury, there are also probably plenty of Earth-like worlds.

As you look around your planet, you should feel comfortable living on such a beautiful planet, but it was not always such a nice place. The craters on the moon and the moon rocks returned by astronauts show that the moon formed as a sea of magma. Mercury seems to have had a similar history, so the Earth likely formed the same way. It was once a seething ocean of liquid rock swathed in a hot, thick atmosphere, torn by eruptions of more rock, explosions of gas from the interior, and occasional impacts from space. The moon and Mercury show that that is the way Terrestrial planets begin. Earth has evolved to become your home world, but mother Earth has had a violent past.

# Study and Review

## Summary

- The moon is **tidally coupled (p. 443)** to Earth and rotates on its axis once each orbit, keeping the same side facing Earth.
- The moon is small and has only one-sixth the gravity of Earth. It has such a low escape velocity that it is unable to retain an atmosphere. Observers on Earth see sharp shadows on the moon's surface, especially near the **terminator (p. 443)**, and stars disappear behind the **limb (p. 444)** of the moon without dimming. Both observations are evidence of the lack of an atmosphere. Astronauts visiting the moon verified that the moon has no measurable atmosphere.
- Large, smooth, dark plains on the moon called **maria** (singular, **mare**) **(p. 444)** are old lava flows that fill the lowlands. Evidence of lava is seen as **sinuous rilles (p. 444)** that once carried flowing lava, faults where the lava plain cracked, and wrinkles in the surface.
- When a meteorite strikes the moon, it digs a large round impact crater and throws out debris that falls back as **ejecta (p. 446)** and can form **rays (p. 446)** and **secondary craters (p. 446)**.
- The largest impacts on the moon dug huge **multiringed basins (p. 447)**.
- The highlands on the moon are saturated with craters. Rare meteorite impacts continue to form new craters, and **micrometeorites (p. 447)** constantly grind the surface down to dust, but most of the craters on the moon are very old.
- Astronomers can find **relative ages (p. 445)** for lunar features by looking to see which lies on top and which lies below. **Absolute ages (p. 445)** in years can be found by counting craters on surfaces and comparing the results with an estimate of the rate of crater formation calibrated using radioactive ages of lunar rock samples.
- Most of the craters on the moon were formed during the heavy bombardment at the end of planet building about 4 billion years ago. No crater is known with certainty to have been formed on the moon in historic times.
- Between 1969 and 1972, 12 astronauts set foot on the moon and returned specimens to Earth.
- The moon rocks are all igneous, showing that they solidified from molten rock. Some are **vesicular (p. 450)** basalts, showing that they formed in lava flows on the surface. Light-colored **anorthosite (p. 451)** is part of the old crust and helps make the highlands brighter than the maria in the lowlands. Many of the rocks are **breccias (p. 451)**, which shows that much of the lunar crust was fractured by meteorite impacts.
- The surface of the moon is covered by rock crushed and powdered by meteorite impacts to form a soil called **regolith (p. 451)**.
- There is evidence from lunar crater counts and rock sample ages of an episode of increased impact rate, labeled the **late heavy bombardment (p. 452)**, about 4 billion years ago near the end of the heavy bombardment era. Many of the giant impact basins containing mare were formed during this time. That event may have been caused by Jovian planets migrating and scattering remnant planetesimals across the solar system.
- The Imbrium basin was formed about 4 billion years ago by the impact of a planetesimal at least 60 km (40 mi) in diameter. Seismic waves traveling through the moon were focused to the far side and produced **jumbled terrain (p. 453)**. Later flooding has nearly buried the original basin.
- The fission, condensation, and capture hypotheses for the origin of the moon have all been abandoned. The commonly accepted explanation is called the **large-impact hypothesis (p. 454)**. The moon appears to have formed from a ring of debris ejected into space when a large planetesimal struck the proto-Earth after it had differentiated. This would explain, for example, the moon's low density and lack of volatiles.
- The moon rocks show that the moon formed in a molten state.
- The moon differentiated but contains little iron. Its low-density crust was heavily cratered and shattered to great depth.
- Lava, flowing up from below the crust, filled the lowlands to form the smooth maria plains. The maria formed after the end of the heavy bombardment and contain few craters.
- The near side of the moon has a thin crust, possibly due to tidal forces. The back side has a thicker crust and little lava flooding.
- Because the moon is small, it has lost its internal heat and is geologically dead. The only slow surface evolution occurring now is the blasting of micrometeorites.
- Mercury is tidally locked to the sun and rotates 1.5 times per orbit in a **resonance (p. 457)** relationship between its rotation and its revolution.
- Mercury is a small world that has been unable to retain a true atmosphere and has lost most of its internal heat. It is cratered and geologically inactive.
- As the metallic core cooled and contracted, the brittle crust broke to form **lobate scarps (p. 457)**, like wrinkles in the skin of a drying apple.
- The Caloris Basin is a large, multiringed basin on Mercury that has been partially flooded by lava flows.
- The **intercrater plains (p. 459)** on Mercury appear to be later lava flows that covered older craters and then accumulated more craters. The **smooth plains (p. 459)** seem to be more recent lava flows that contain few craters. All of these plains are a similar shade of gray, so dark lava flows are not visible on Mercury as they are in the moon's maria.
- Mercury formed in the inner solar system and contains a larger proportion of dense metals. It is possible that a large impact shattered and drove off some of the planet's crust. This could explain why it has a larger metallic core even than would be predicted by the condensation sequence.
- It is not clear how much heat remains in Mercury. It is not geologically active, and it does not have a strong magnetic field. Nevertheless it has a weak magnetic field, and radar observations of its rotation suggest that the outer layers of its large iron core remain molten.
- Mercury was heavily cratered during the heavy bombardment, but lava flows covered some of those craters, and new craters formed the intercrater plains. Fractures produced by the Caloris impact may have triggered later lava flows that formed the smooth plains.
- The MESSENGER spacecraft will go into orbit around Mercury in 2011 and provide more extensive photography of its surface and detailed measurements of its physical properties.

## Review Questions

1. How could you find the relative ages of the moon's maria and highlands?
2. How can you tell that Copernicus is a young crater?
3. Why did the first Apollo missions land on the maria? Why were the other areas of more scientific interest?
4. Why do astronomers suppose that the moon formed with a molten surface?

5. Why are so many lunar samples breccias?
6. What do the vesicular basalts tell you about the evolution of the lunar surface?
7. What evidence would you expect to find on the moon if it had been subjected to plate tectonics? Do you find such evidence?
8. How does the large-impact hypothesis explain the moon's lack of iron? Of volatiles?
9. How does the tidal coupling between Mercury and the sun differ from that between the moon and Earth?
10. What is the difference between the intercrater plains and the smooth plains in terms of time of formation?
11. What evidence can you cite that Mercury has a partially molten, metallic core?
12. How were the histories of the moon and Mercury similar? How were they different?
13. What is the main reason that the surface evolution of both the moon and Mercury has essentially come to a stop, but the Earth's surface evolution continues?
14. **How Do We Know?** How is a hypothesis or theory like a container in which you can carry an assortment of ideas, observations, facts, and measurements?

## Discussion Questions

1. Old science-fiction paintings and drawings of colonies on the moon often showed very steep, jagged mountains. Why did the artists assume that the mountains would be more jagged than mountains on Earth? Why are lunar mountains actually less jagged than mountains on Earth?
2. From your knowledge of comparative planetology, propose a description of the view that astronauts would have if they landed on the surface of Mercury.

## Problems

1. Calculate the escape velocity of the moon from its mass and diameter. (*Hint:* See Chapter 5.)
2. Why do small planets cool faster than large planets? Compare surface area to volume. Why is that comparison important?
3. The smallest detail visible through Earth-based telescopes is about 1 arc second in diameter. What linear size is this on the moon? (*Hint:* Use the small-angle formula, Chapter 3.)
4. The trenches where Earth's seafloor slips downward are 1 km or less wide. Could Earth-based telescopes resolve such features on the moon? Why can you be sure that such features are not present on the moon?
5. The Apollo command module orbited the moon about 100 km above the surface. What was its orbital period? (*Hint:* See Chapter 5.)
6. From a distance of 100 km above the surface of the moon, what is the angular diameter of an astronaut in a spacesuit? Could someone have seen the astronauts from the command module? (*Hint:* Use the small-angle formula, Chapter 3.)
7. If you transmitted radio signals to Mercury when it was closest to Earth and waited to hear the radar echo, how long would you wait?
8. Suppose you sent a spacecraft to land on Mercury, and it transmitted radio signals to Earth at a wavelength of 10 cm. If you saw Mercury at its greatest angular distance west of the sun, to what wavelength would you have to tune your radio telescope to detect the signals? (*Hints:* See Celestial Profile 4 and the section on the Doppler shift in Chapter 7.)
9. What would the wavelength of maximum output be for infrared radiation from the surface of Mercury? How would that differ for the moon?
10. Calculate the escape velocity from Mercury. How does that compare with the escape velocity from the moon and from Earth?

## Learning to Look

1. Examine the mountains at the Apollo 17 landing site (Figure 21-4). What processes shape mountains on Earth that have not affected mountains on the moon?
2. The rock shown in Figure 21-7 was thrown from beyond the horizon and landed where the astronauts found it. It must have formed a small impact crater, but the astronauts didn't find traces from that impact. Why not? (*Hint:* When did the rock fall?)
3. In this photo, astronaut Alan Bean works at the Apollo 12 lander. Describe the surface you see. What kind of terrain did they land on for this, the second human moon landing?

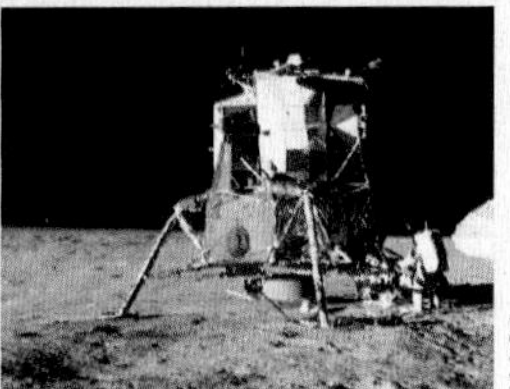
NASA

# 22 Comparative Planetology of Venus and Mars

*Mars rover Spirit, about the size of a riding lawnmower, could not take its own picture. It has been digitally added to this photo of the Martian surface. Rovers carry instruments to analyze rocks and are controlled from Earth.* (NASA/JPL-Caltech/Cornell)

## Guidepost

You have been to the moon and to Mercury, and now you are going to find Venus and Mars dramatically different from those small, inactive and airless worlds. Venus and Mars have internal heat and atmospheres. The internal heat means they are geologically active, and the atmospheres mean they have weather. As you explore, you will discover answers to six essential questions:

- **Why is the atmosphere of Venus so thick?**
- **What is the hidden surface of Venus really like?**
- **How did Venus form and evolve?**
- **Why is the atmosphere of Mars so thin?**
- **What is the evidence that Mars once had water on its surface?**
- **How did Mars form and evolve?**

The comparative planetology questions that need to always be on your mind when you explore another world are these: How and why is this world similar to Earth? How and why is this world different from Earth? You will see that small initial differences can have big effects.

You are a planet-walker, and you are becoming an expert on the kind of planets you can imagine walking on. But there are other worlds beyond Mars in our solar system so peculiar they have no surfaces to walk on, even in your imagination. You will explore them in the next two chapters.

*There wasn't a breath in that land of death . . .*

— ROBERT SERVICE, "THE CREMATION OF SAM McGEE"

THE TEMPERATURE ON MARS on a hot summer day at noon might feel pretty pleasant at around 20°C (68°F). But without a spacesuit, you could live only about 30 seconds because the air is mostly carbon dioxide with almost no oxygen and is deadly dry. Even more important, the air pressure is less than 1 percent that at the surface of Earth, so your exposed body fluids such as tears and saliva would boil if you stepped outside your spaceship unprotected.

Not even a spacesuit would save you on Venus. The air pressure there is 90 times Earth's, and the air is almost entirely carbon dioxide, with traces of various acids. Worse yet, the surface is hot enough to melt lead.

Venus and Mars resemble Earth in some ways, so why are they such unfriendly places to visit? Comparative planetology will give you some clues.

## 22-1 Venus

AN ASTRONOMER ONCE BECAME ANNOYED when someone referred to Venus as a planet gone wrong. "No," she argued, "Venus is probably a fairly normal planet. It is Earth that is peculiar. The universe probably contains more planets like Venus than like Earth." To understand how unusual Earth is, you need only to compare it with Venus.

In many ways, Venus is a twin of Earth, and you might expect surface conditions on the two planets to be quite similar. Venus and Earth are almost exactly the same size and mass, with Venus having 95 percent the diameter of Earth (**Celestial Profile 5**). Venus and Earth have similar average densities, and they formed in the same part of the solar nebula—Venus's orbit is the closest to Earth's of all the planets, so Venus's overall composition is similar to Earth's (see Chapter 19). Also, planets the size of Earth and Venus cool slowly, so you might expect Venus, like Earth, to have a molten metallic core and an active crust with plate tectonics.

Unfortunately, the surface of Venus is perpetually hidden by thick clouds that completely envelop the planet, preventing us from easily observing conditions there. From the time of Galileo until the early 1960s, astronomers could only speculate about Earth's twin. Some science fiction writers imagined that Venus was a steamy swamp planet inhabited by strange creatures, while others imagined that it was completely covered by an ocean or by a planetwide dusty desert.

Starting in the 1960s, astronomers have used microwave blackbody emission from Venus to measure its surface temperature, and radar to penetrate the clouds, image the surface, and measure the planet's rotation. At least 23 spacecraft have flown

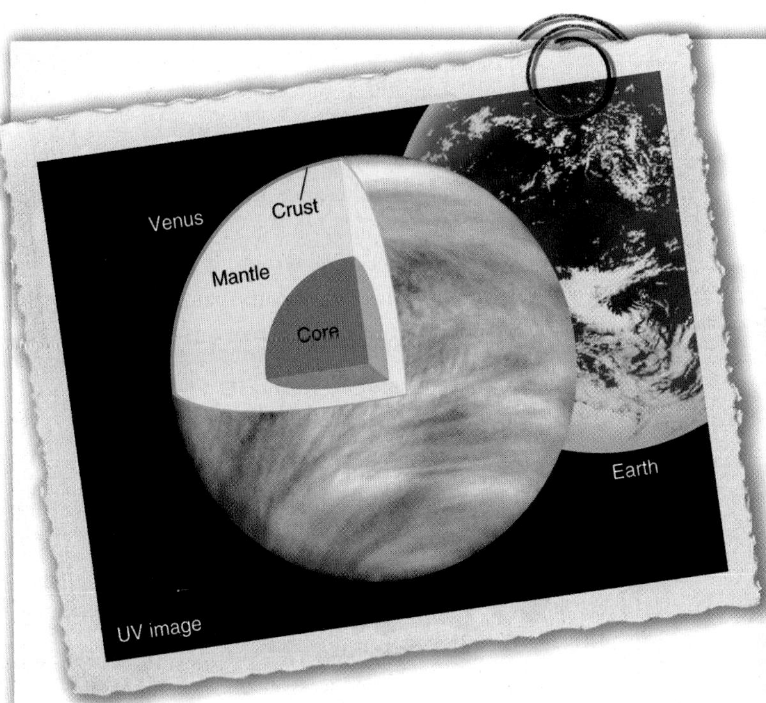


*Venus is only 5 percent smaller than Earth, but its atmosphere is perpetually cloudy, and its surface is hot enough to melt lead. It may have a hot core about the size of Earth's.*

## *Celestial Profile 5: Venus*

### *Motion:*

| | |
|---|---|
| Average distance from the sun | 0.723 AU (1.08 × $10^8$ km) |
| Eccentricity of orbit | 0.007 |
| Inclination of orbit to ecliptic | 3.4° |
| Average orbital velocity | 35.0 km/s |
| Orbital period | 0.6152 y (224.7days) |
| Period of rotation | 243.0 days (retrograde) |
| Inclination of equator to orbit | 177° |

### *Characteristics:*

| | |
|---|---|
| Equatorial diameter | 1.21 × $10^4$ km (0.949 $D_\oplus$) |
| Mass | 4.87 × $10^{24}$ kg (0.815 $M_\oplus$) |
| Average density | 5.24 g/cm$^3$ (4.2 g/cm$^3$ uncompressed) |
| Surface gravity | 0.90 Earth gravity |
| Escape velocity | 10.3 km/s (0.92 $V_\oplus$) |
| Surface temperature | 470°C (880°F) |
| Albedo (cloud tops) | 0.76 |
| Oblateness | 0 |

### *Personality Point:*

Venus is named for the Roman goddess of love, perhaps because the planet often shines so beautifully in the evening or dawn sky. In contrast, the ancient Maya identified Venus as their war god Kukulkan and sacrificed human victims to the planet when it rose in the dawn sky.

past or orbited Venus, and over a dozen have landed on its surface. The resulting picture of Venus is dramatically different from the murky swamps of fiction. In fact, the surface of Venus is drier than any desert on Earth and twice as hot as a kitchen oven set to its highest temperature. Another startling contrast to Earth is that Venus rotates very slowly, once in 243 Earth days, in the retrograde (backward) direction.

If Venus is not a planet gone wrong, it is certainly a planet gone down a different evolutionary path than your home planet. How did Earth's twin become so different?

## The Atmosphere of Venus

Although Venus is Earth's twin in size, its atmosphere is truly unearthly. The composition, temperature, and density of Venus' atmosphere make the planet's surface entirely inhospitable. About 96 percent of its atmosphere is carbon dioxide, and 3.5 percent is nitrogen. The remaining 0.5 percent is water vapor, sulfuric acid ($H_2SO_4$), hydrochloric acid (HCl), and hydrofluoric acid (HF). In fact, the thick clouds that hide the surface are composed of sulfuric acid droplets and microscopic sulfur crystals.

Soviet and U.S. spacecraft dropped probes into the atmosphere of Venus, and those probes radioed data back to Earth as they fell toward the surface. These studies show that Venus's cloud layers are much higher and more stable than those on Earth. The highest layer of clouds, the layer visible from Earth, extends from about 60 to 70 km (about 40 to 45 mi) above the surface (■ Figure 22-1). For comparison, the clouds on Earth normally do not extend higher than about 16 km (10 mi).

These cloud layers are highly stable because the atmospheric circulation on Venus is much more regular than that on Earth. The heated atmosphere at the **subsolar point,** the point on the planet where the sun is directly overhead, rises and spreads out in the upper atmosphere. Convection circulates this gas toward the dark side of the planet and the poles, where it cools and sinks. This circulation produces 300 km/hour jet streams in the upper atmosphere, which move from east to west (the same direction the planet rotates) so rapidly that the entire atmosphere rotates with a period of only four days.

The details of this atmospheric circulation are not well understood, but it seems that the slow rotation of the planet is an important factor. On Earth, large-scale circulation patterns are broken into smaller cyclonic disturbances by Earth's rapid rotation. Because Venus rotates more slowly, its atmospheric circulation is not broken up into cyclonic storms but instead is organized as a planetwide wind pattern.

Although Venus's upper atmosphere is cool, the lower atmosphere is quite hot (Figure 22-1b). Instrumented probes that have reached the surface report that the temperature is 470°C (880°F), and the atmospheric pressure is 90 times that of Earth. Earth's atmosphere is 1000 times less dense than water, but on Venus the air is only 10 times less dense than water. If you could survive the unpleasant atmospheric composition, intense heat, and high pressure, you could strap wings to your arms and fly.

The present atmosphere of Venus is extremely dry, but there is evidence that it once had significant amounts of water. As one of the Pioneer Venus probes descended through the

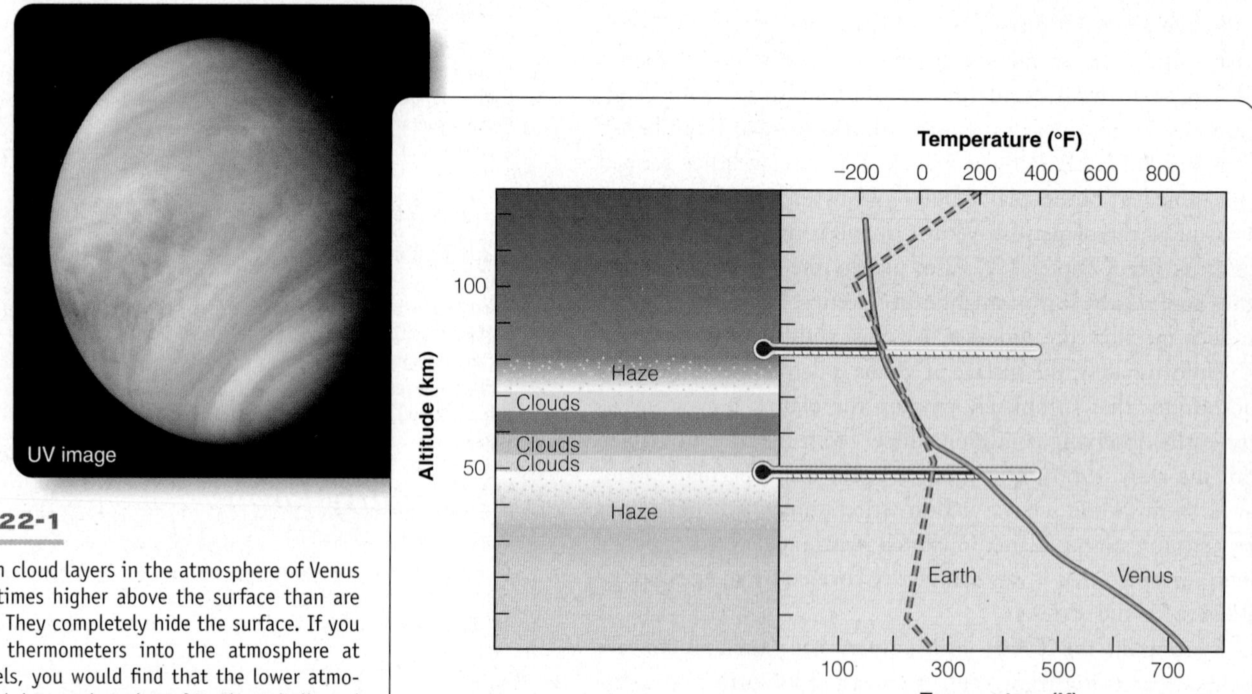


■ **Figure 22-1**

The four main cloud layers in the atmosphere of Venus are over 10 times higher above the surface than are Earth clouds. They completely hide the surface. If you could insert thermometers into the atmosphere at different levels, you would find that the lower atmosphere is much hotter than that of Earth, as indicated by the red line in the graph. (NASA)

atmosphere of Venus in 1978, it discovered that deuterium is about 150 times more abundant compared to normal hydrogen atoms than it is on Earth. This abundance of deuterium, the heavy isotope of hydrogen, could have developed because Venus has no ozone layer to absorb the ultraviolet radiation in sunlight. These UV photons broke water molecules into hydrogen and oxygen. The oxygen would have formed oxides in the soil, and the hydrogen would have leaked away into space. The heavier deuterium atoms would leak away more slowly than normal hydrogen atoms, which would increase the ratio of deuterium to normal hydrogen.

Venus has essentially no water now, but the amount of deuterium in the atmosphere suggests that it may have once had enough water to make a planetwide ocean at least 25 meters deep. (For comparison, the water on Earth would make a uniform ocean about 3000 meters deep.) Venus is now a deadly dry world with only enough water vapor in its atmosphere to make a planet-wide ocean 0.3 meter (1 foot) deep. Venus's present lack of water is one of the biggest differences between that planet and Earth.

## The Venusian Greenhouse

You saw in Chapter 20 how the greenhouse effect warms Earth. Carbon dioxide ($CO_2$) is transparent to light but opaque to infrared (heat) radiation. That means energy can enter the atmosphere as light and warm the surface, but the surface cannot radiate the energy back to space easily because the atmospheric $CO_2$ is opaque to infrared radiation. Venus also has a greenhouse effect, but on Venus the effect is fearsomely strong; that is the explanation for why Venus, although it is farther from the sun than Mercury, is actually hotter. Whereas Earth's atmosphere contains only about 0.04 percent $CO_2$, the atmosphere of Venus contains 96 percent $CO_2$, and as a result temperatures on the surface of Venus are more than hot enough to melt lead. The thick atmosphere and its high winds carry heat around the planet efficiently enough to make surface temperatures on Venus nearly the same everywhere. This evidently offsets the effect of the planet's slow rotation that should cause a large temperature difference between the day side and the night side.

Planetary astronomers think they know how Venus got into such a jam. When Venus was young, it may have been cooler than it is now; but, because it formed 30 percent closer to the sun than did Earth, it was always warmer than Earth, and that unleashed processes that made it even hotter. Calculations indicate that Venus and Earth should have outgassed about the same amount of $CO_2$, but Earth's oceans have dissolved most of Earth's $CO_2$ and converted it to sediments such as limestone. If all of Earth's carbon were dug up and converted back to $CO_2$, our atmosphere would be about as dense as the air on Venus and composed mostly of $CO_2$, like Venus's atmosphere.

As you saw in the previous section, Venus may once have had water on its surface, but that water would have begun to evaporate at the temperatures of early Venus. Carbon dioxide is highly soluble in water, but as surface water disappeared on Venus, the ability to dissolve $CO_2$ and remove it from the atmosphere also would have disappeared. The surface of Venus is now so hot that even sulfur, chlorine, and fluorine have baked out of the rock and formed sulfuric, hydrochloric, and hydrofluoric acid vapors.

## The Surface of Venus

Given that the surface of Venus is perpetually hidden by clouds, is hot enough to melt lead, and suffers under crushing atmospheric pressure, it is surprising how much planetary scientists know about the geology of Venus. Early radar maps made from Earth penetrated the clouds and showed that it had mountains, plains, and some craters. The Soviet Union launched a number of spacecraft that landed on Venus, and although the surface conditions caused the spacecraft to fail within an hour or so of landing, they did analyze the rock and transmit a few images back to Earth. The rock seems to be basalt, a typical product of volcanism. The images revealed dark-gray rocky plains bathed in a deep-orange glow caused by sunlight filtering down through the thick atmosphere (■ Figure 22-2).

The U.S. Pioneer Venus main probe orbited Venus in 1978 and made radar maps showing features as small as 25 km in diameter. Later, two Soviet Venera spacecraft mapped the north polar region with a resolution of 2 km. From 1992 to 1994, the U.S. Magellan spacecraft orbited Venus and created radar maps showing details as small as 100 m. These radar maps provide a comprehensive look below the clouds.

The color of Venus radar maps is mostly arbitrary. Human eyes can't see radio waves, so the radar maps must be given some false colors to distinguish height or roughness or composition. Some maps use blues and greens for lowlands and yellows and reds for highlands. Remember when you look at these maps that there is no liquid water on Venus. Magellan scientists chose to use yellows and oranges for their radar maps in an effort to mimic the orange color of daylight caused by the thick atmosphere (■ Figure 22-3). When you look at these orange images, you need to remind yourself that the true color of the rock is dark gray (**How Do We Know? 22-1**).

Radar maps of Venus reveal a number of things about the surface. If you transmit a radio signal down through the clouds and measure the time until you hear the echo coming back up, you can measure the altitude of the surface. As a result, part of the Magellan data is a detailed altitude map of the surface. You can also measure the amount of the signal that is reflected from each spot on the surface. Much of the surface of Venus is made up of old smooth lava flows that do not look bright in radar maps, but faults and uneven terrain look brighter. Some young, rough lava flows are very rough and contain billions of tiny crevices

**■ Figure 22-2**

The Venera lander touched down on Venus in 1982 and carried a camera that swiveled from side to side to photograph the surface. The orange glow is produced by the thick atmosphere; when that is corrected to produce the view as it would be under white light, you can see that the rocks are dark gray. Isotopic analysis suggests they are basalts.

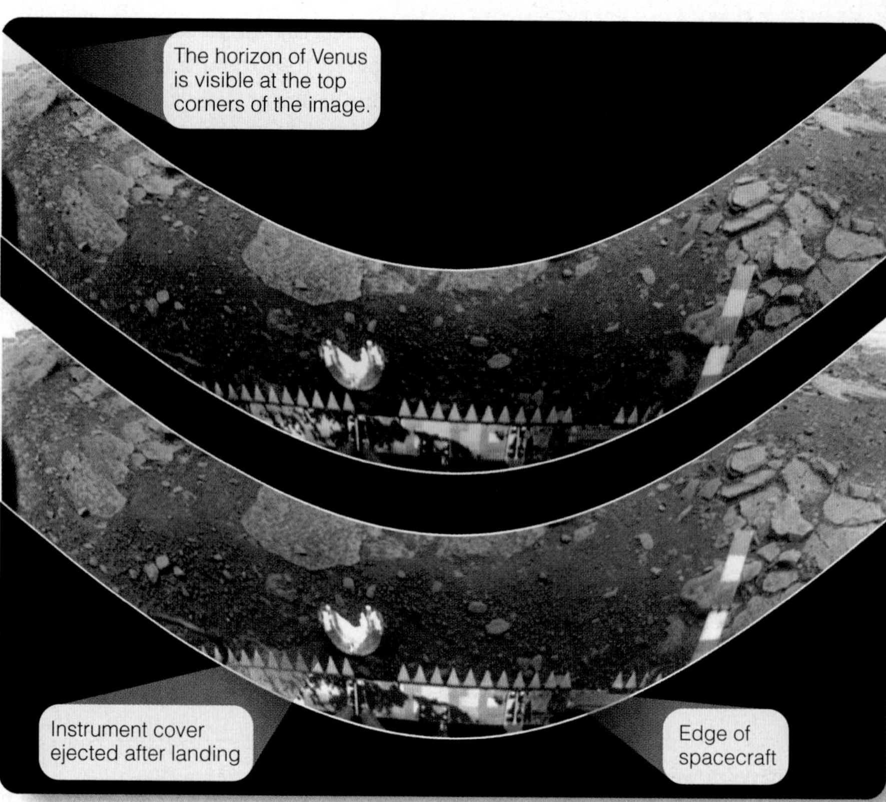


(■ Figure 22-4) that bounce the radar signal around and shoot it back the way it came. These rough lava flows look very bright in radar maps. Certain mineral deposits are also bright. Radar maps do not show how the surface would look to human eyes but rather provide information about altitude, roughness, and, in some cases, chemical composition.

The big map in ■ Figure 22-5 is a map of all of Venus except the polar regions. (Note that this map has been color coded by altitude.) By international agreement, the names of celestial bodies and features on celestial bodies are assigned by the International Astronomical Union, which has decided that all names on Venus should be feminine. There are only a few exceptions, such as the mountain Maxwell Montes, 50 percent higher than Mount Everest. It was discovered during early Earth-based radar mapping and named for James Clark Maxwell, the 19th-century physicist who first described electromagnetic radiation. Alpha Regio and Beta Regio, later found to be volcanic peaks, were also discovered by radar before the naming convention was adopted. Other names on Venus are feminine, such as the highland regions Ishtar Terra and Aphrodite Terra, named for the Babylonian and Greek goddesses of love. The insets in Figure 22-5 show roughness and composition of the surface in maps made by different satellites and color coded in different ways. All of these radar maps paint a picture of a hot, violent, desert world.

Radar maps show that the surface of Venus consists of low, rolling plains and highland regions. The rolling plains appear to be large-scale smooth lava flows, and the highlands are regions of deformed crust.

Just as in the case of the lunar landscape, craters are the key to figuring out the age of the surface. With nearly 1000 impact craters on its surface, Venus has more craters than

**■ Figure 22-3**

Venus without its clouds: This mosaic of Magellan radar maps has been given an orange color to mimic the sunset coloration of daylight at the surface of the planet. The image shows scattered impact craters and volcanic regions such as Beta Regio and Atla Regio. (NASA)

## How Do We Know? 22-1

### Data Manipulation

**Why do scientists think it is OK to visually enhance their data?** Planetary astronomers studying Venus change the colors of radar maps and stretch the height of mountains. If they were making political TV commercials and were caught digitally enhancing a politician's voice, they would be called dishonest, but scientists often manipulate and enhance their data. It's not dishonest because the scientists are their own audience.

Research physiologists studying knee injuries, for instance, can use magnetic resonance imaging (MRI) data to study both healthy and damaged knees. By placing a patient in a powerful magnetic field and irradiating his or her knee with precisely tuned radio frequency pulses, the MRI machine can force one in a million hydrogen atoms to emit radio frequency photons. The intensity and frequency of the emitted photons depends on how the hydrogen atoms are bonded to other atoms, so bone, muscle, and cartilage emit different signals. An antenna in the machine picks up the emitted signals and stores huge masses of data in computer memory as tables of numbers.

The tables of numbers are meaningless in that form to the physiologists, but by manipulating the data, they can produce images that reveal the anatomy of a knee. By enhancing the data, they can distinguish between bone and cartilage and see how tendons are attached. They can filter the data to see fine detail or smooth the data to eliminate distracting textures. Because the physiologists are their own audience, they know how they are manipulating the data and can use it to devise better ways to treat knee injuries.

When scientists say they are "massaging the data," they mean they are filtering, enhancing, and manipulating it to bring out the features they need to study. If they were presenting that data to a television audience to promote a cause or sell a product, it would be dishonest, but scientists' manipulation of the data allows them to better understand how nature works.

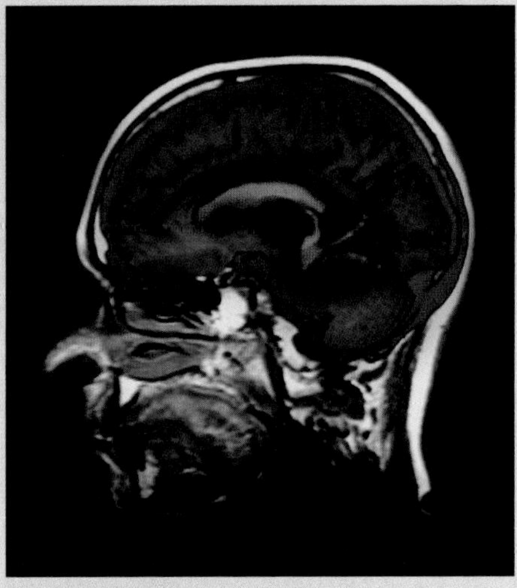

*You are accustomed to seeing data manipulated and presented in convenient ways.* (PhotoDisc/Getty Images)

**■ Figure 22-4**

Although it is nearly 1000 years old, this lava flow near Flagstaff, Arizona, is still such a rough jumble of sharp rock that it is dangerous to venture onto its surface. Rough surfaces are very good reflectors of radio waves and look bright in radar maps. Solidified lava flows on Venus show up as bright regions in the radar maps because they are rough. (Mike Seeds)

Earth but not nearly as many as the moon. The craters are uniformly scattered over the surface and look sharp and fresh (■ Figure 22-6). With no water and a thick, slow-moving lower atmosphere, there is little erosion on Venus, and the thick atmosphere protects the surface from small meteorites. Consequently, there are no small craters. Planetary scientists conclude that the surface of Venus is older than Earth's surface but not as old as the moon's. Unlike the moon, there are no very old, cratered highlands. Lava flows seem to have covered the entire surface of Venus within approximately the last half-billion years.

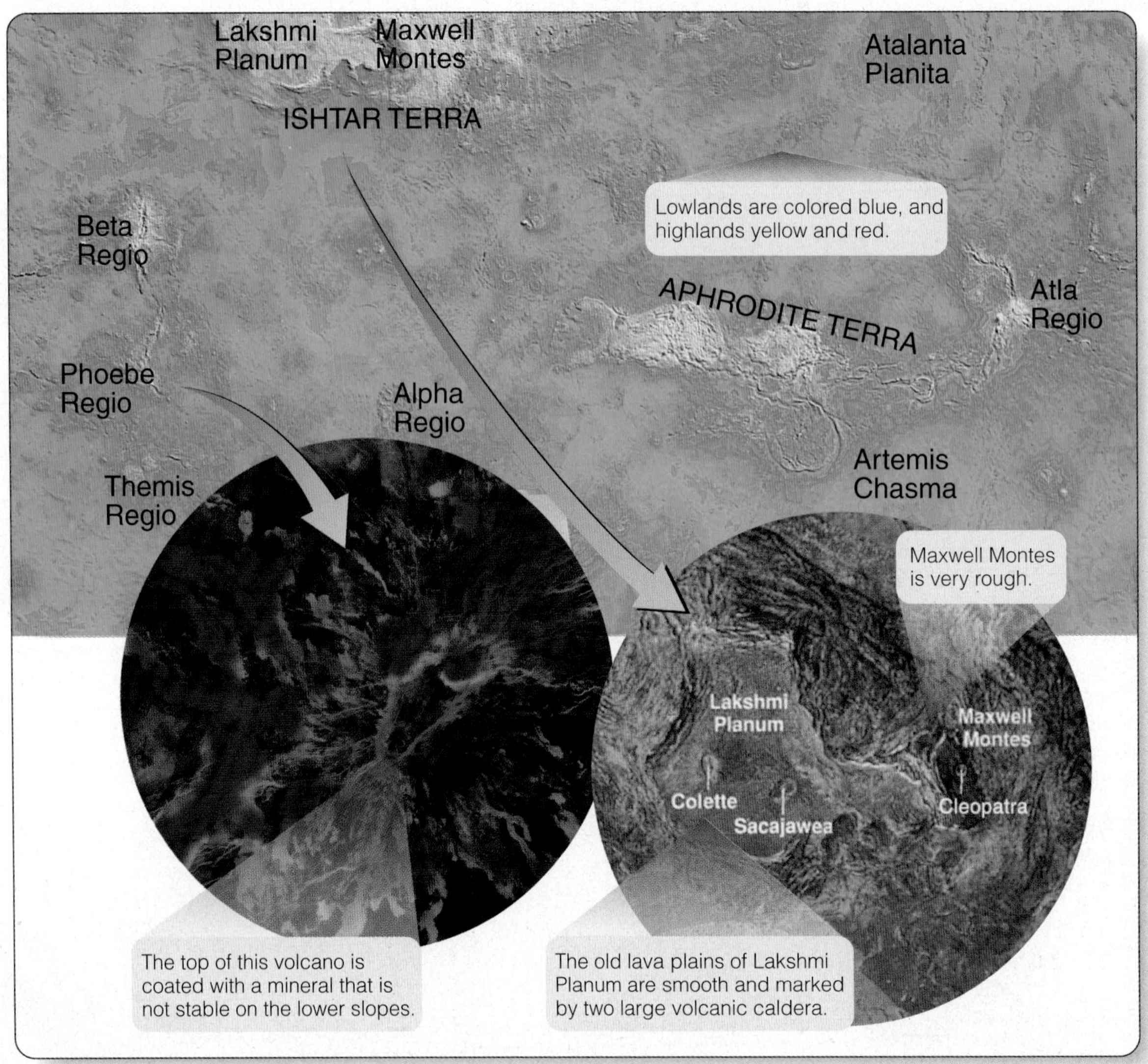
Lakshmi Planum
Maxwell Montes
Atalanta Planita
ISHTAR TERRA
Lowlands are colored blue, and highlands yellow and red.
Beta Regio
APHRODITE TERRA
Atla Regio
Phoebe Regio
Alpha Regio
Artemis Chasma
Themis Regio
Maxwell Montes is very rough.
Lakshmi Planum
Maxwell Montes
Colette
Sacajawea
Cleopatra
The top of this volcano is coated with a mineral that is not stable on the lower slopes.
The old lava plains of Lakshmi Planum are smooth and marked by two large volcanic caldera.

**■ Figure 22-5**

Notice how these three radar maps show different things. The main map here shows elevation over most of the surface of Venus. Only the polar areas are not shown. The inset map at left shows an electrical property of surface minerals related to chemical composition. The detailed map of Maxwell Montes and Lakshmi Planum at right is color coded to show degree of roughness, with purple smooth and orange rough. (Maxwell and Lakshmi Planum map: USGS; other maps: NASA)

## Volcanism on Venus

Volcanism seems to dominate the surface of Venus. Much of Venus is covered by lava flows such as those photographed by the Venera landers (Figure 22-2). Also, volcanic peaks and other volcanic features are evident in radar maps.

As usual, you can learn more about other worlds by comparing them with each other and with Earth. Read **Volcanoes** on pages 472–473 and notice three important ideas and two new terms:

1. There are two main types of volcanoes found on Earth. *Composite volcanoes* tend to be associated with plate motion and located near the edges of plates, whereas *shield volcanoes* are associated with hot spots caused by columns of magma rising from deep in the mantle, and not by plate motion.

2. Notice that you can recognize the volcanoes on Venus and Mars by their shapes, even when the images are manipulated in computer mapping. The volcanoes found on Venus and Mars are shield volcanoes, produced by hot-spot volcanism and not by plate tectonics.

3. Also notice the large size of volcanoes on Venus and Mars. They have grown very large because of repeated eruptions at the same place in the crust. This is also evidence that, unlike Earth, neither Venus nor Mars has been dominated by plate tectonics and horizontal crust motions.

**■ Figure 22-6**

Impact crater Howe in the foreground of this Magellan radar image is 37 km (23 mi) in diameter. Craters in the background are 47 km (29 mi) and 63 km (39 mi) in diameter. This radar map has been digitally modified to represent the view as if from a spacecraft flying over the craters. (NASA)

The radar image of Sapas Mons in ■ Figure 22-7 shows a dramatic overhead view of this volcano, which is 400 km (250 mi) in diameter at its base and 1.5 km (0.9 mi) high. Many bright, young lava flows radiate outward, covering older, darker flows. Remember that the colors in this image are artificial; if you could walk across these lava flows and shine your spacesuit's white headlight on them, you would find them solid, dark gray stone. Sapas Mons and other nearby volcanoes are located along a system of faults where rising magma evidently broke through the crust.

In addition to the volcanoes, radar images reveal other volcanic features on the surface of Venus. Lava channels are common, and they appear similar to the sinuous rilles visible on Earth's moon. The longest channel on Venus is also the longest known lava channel in the solar system. It stretches 6800 km (4200 mi), roughly twice the distance from Chicago to Los Angeles. These channels are 1 to 2 km wide and can sometimes be traced back to collapsed areas where lava appears to have drained from beneath the crust.

For further evidence of volcanism on Venus, you can look at features called **coronae,** circular bulges up to 2100 km in diameter containing volcanic peaks and lava flows. The coronae appear to be caused by rising currents of molten magma below the crust that create an uplifted dome and then withdraw to allow the surface to subside and fracture. Coronae are sometimes accompanied by circular outpourings of viscous lava called pancake domes, and by domes and hills pushed up by molten

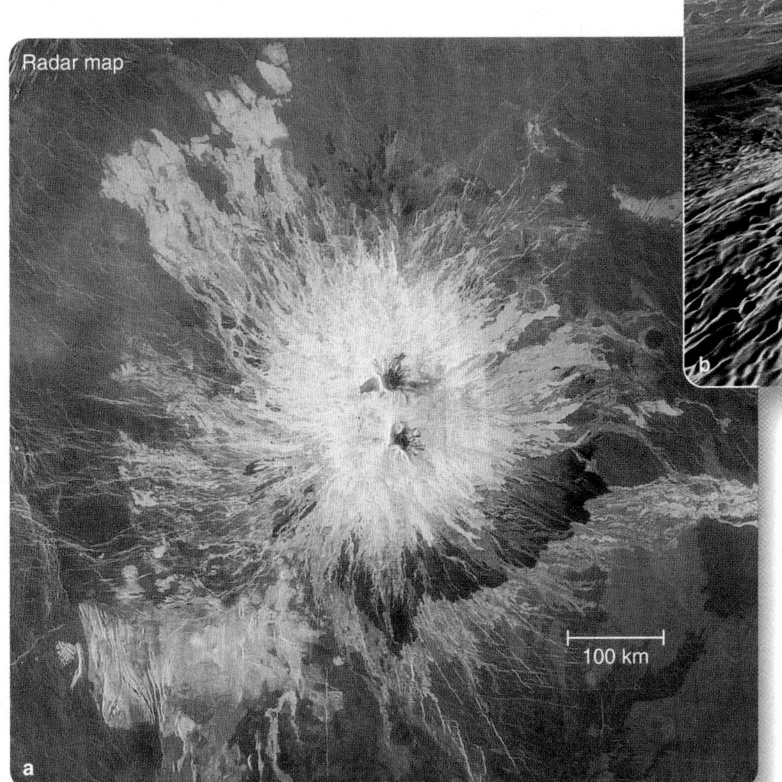


**■ Figure 22-7**

(a) Volcano Sapas Mons, lying along a major fracture zone, is topped by two lava-filled calderas and flanked by rough lava flows. The orange color of this radar map mimics the orange light that filters through the thick atmosphere. (NASA) (b) Seen by light typical of Earth's surface, Sapas Mons might look more like this computer-generated landscape. Volcano Maat Mons rises in the background. The vertical scale has been exaggerated by a factor of 15 to reveal the shape of the volcanoes and lava flows. (Copyright © 1992, David P. Anderson, Southern Methodist University)

# Volcanoes

**1** Molten rock (magma) is less dense than the surrounding rock and tends to rise. Where it bursts through Earth's crust, you see volcanism. The two main types of volcanoes on Earth provide good examples for comparison with those on Venus and Mars.

On Earth, **composite volcanoes** form above subduction zones where the descending crust melts and the magma rises to the surface. This forms chains of volcanoes along the subduction zone, such as the Andes along the west coast of South America.

Magma rising above subduction zones is not very fluid, and it produces explosive volcanoes with sides as steep as 30°.

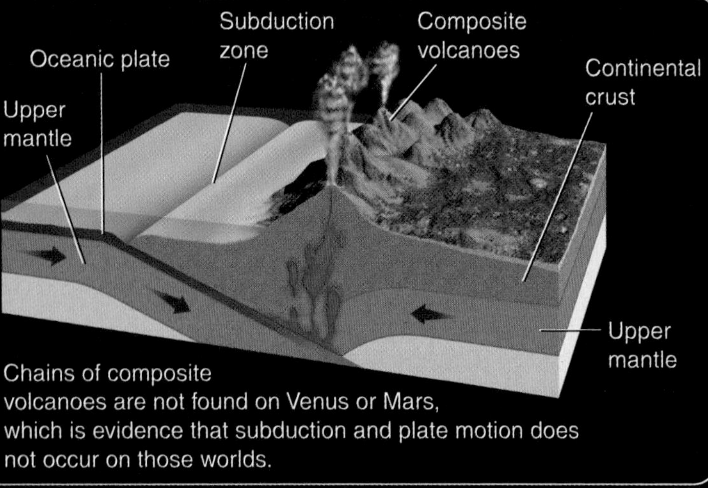


Chains of composite volcanoes are not found on Venus or Mars, which is evidence that subduction and plate motion does not occur on those worlds.

Based on *Physical Geology,* 4th edition, James S. Monroe and Reed Wicander, Wadsworth Publishing Company. Used with permission.

Mount St. Helens exploded northward on May 18, 1980, killing 63 people and destroying 600 km$^2$ (230 mi$^2$) of forest with a blast of winds and suspended rock fragments that moved as fast as 480 km/hr (300 mph) and had temperatures as hot as 350°C (660°F). Note the steep slope of this composite volcano.

USGS

Shield volcano

Lava flow

Magma chamber

Oceanic crust

**1a** A **shield volcano** is formed by highly fluid lava (basalt) that flows easily and creates low-profile volcanic peaks with slopes of 3° to 10°. The volcanoes of Hawaii are shield volcanoes that occur over a hot spot in the middle of the Pacific plate.

Magma collects in a chamber in the crust and finds its way to the surface through cracks.

Magma forces its way upward through cracks in the upper mantle and causes small, deep earthquakes.

A hot spot is formed by a rising convection current of magma moving upward through the hot, deformable (plastic) rock of the mantle.

The Cascade Range composite volcanoes are produced by an oceanic plate being subducted below North America and partially melting.

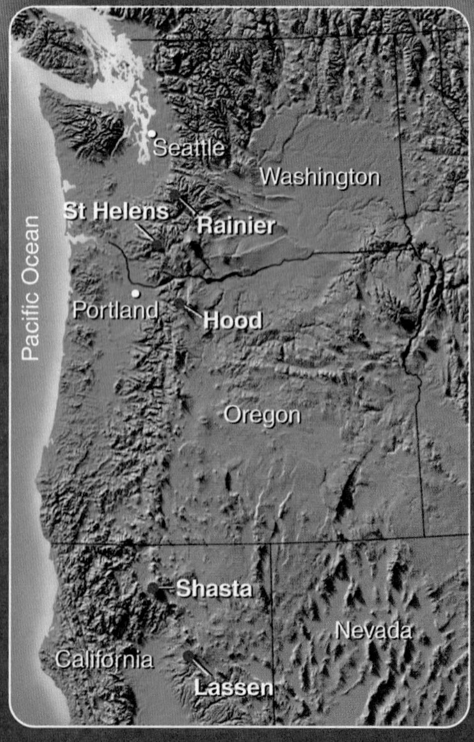

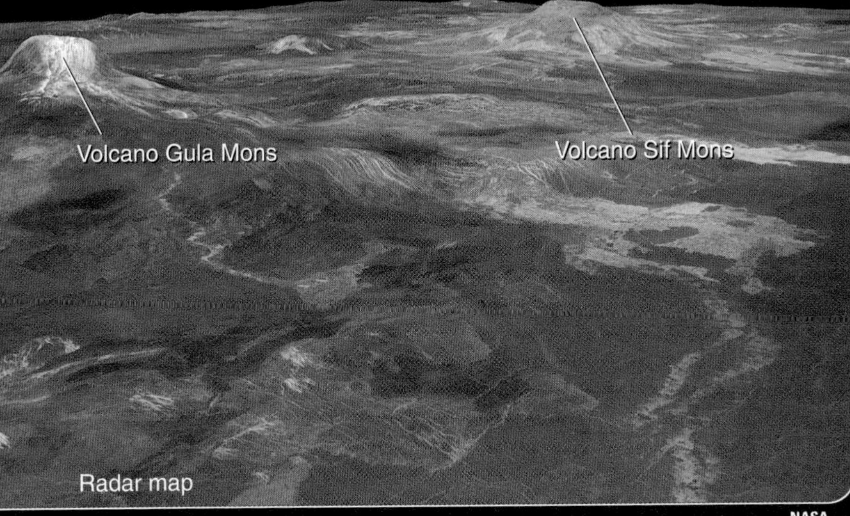

NASA

**2** Volcanoes on Venus are shield volcanoes. They appear to be steep sided in some images created from Magellan radar maps, but that is because the vertical scale has been exaggerated to enhance detail. The volcanoes of Venus are actually shallow-sloped shield volcanoes.

**2a** This computer model of a mountain with the vertical scale magnified 10 times appears to have steep slopes such as those of a composite volcano.

A true profile of the computer model shows the mountain has very shallow slopes typical of shield volcanoes.

Mike Seeds

**3** Volcanism over a hot spot results in repeated eruptions that build up a shield volcano of many layers. Such volcanoes can grow very large.

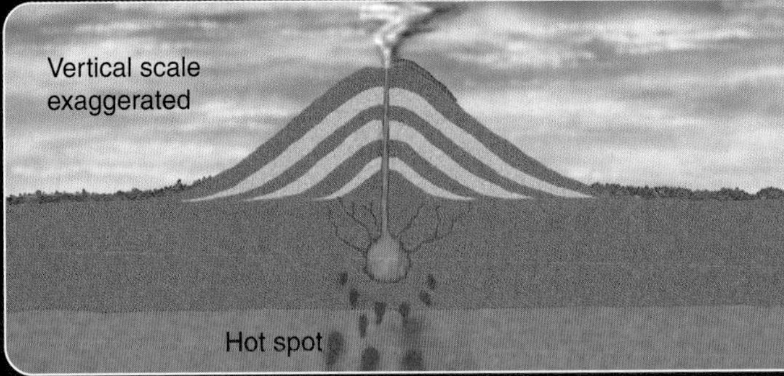


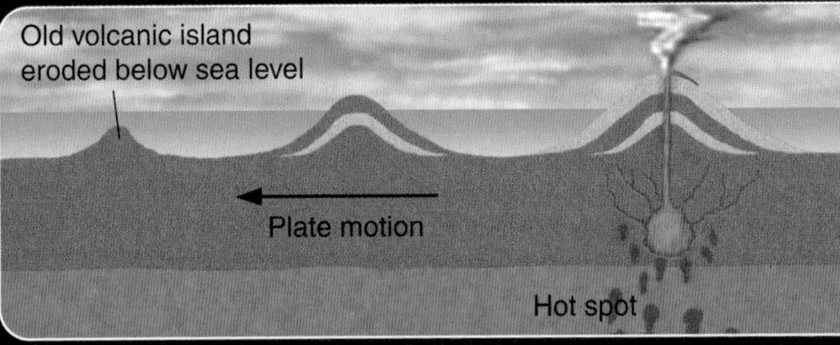


If the crustal plate is moving, magma generated by the hot spot can repeatedly penetrate the crust to build a chain of volcanoes. Only the volcanoes over the hot spot are active. Older volcanoes slowly erode away. Such volcanoes cannot grow large because the moving plate carries them away from the hot spot.

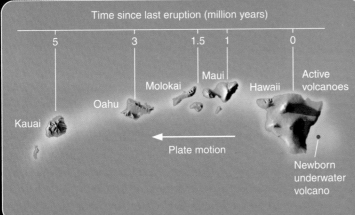

Time since last eruption (million years)
5 3 1.5 1 0
Kauai
Oahu
Molokai
Maui
Hawaii
Active volcanoes
Plate motion
Newborn underwater volcano

**3a** The volcanoes that make up the Hawaiian Islands as shown at left have been produced by a hot spot poking upward through the middle of the moving Pacific plate.

NASA

**3b** The plate moves about 9 cm/yr and carries older volcanic islands northwest, away from the hot spot. The volcanoes cannot grow extremely large because they are carried away from the hot spot. New islands form to the southeast over the hot spot.

Olympus Mons at right is the largest volcano on Mars. It is a shield volcano 25 km (16 mi) high and 700 km (440 mi) in diameter at its base. Its vast size is evidence that the crustal plate must have remained stationary over the hot spot. This is evidence that Mars has not had plate tectonics.

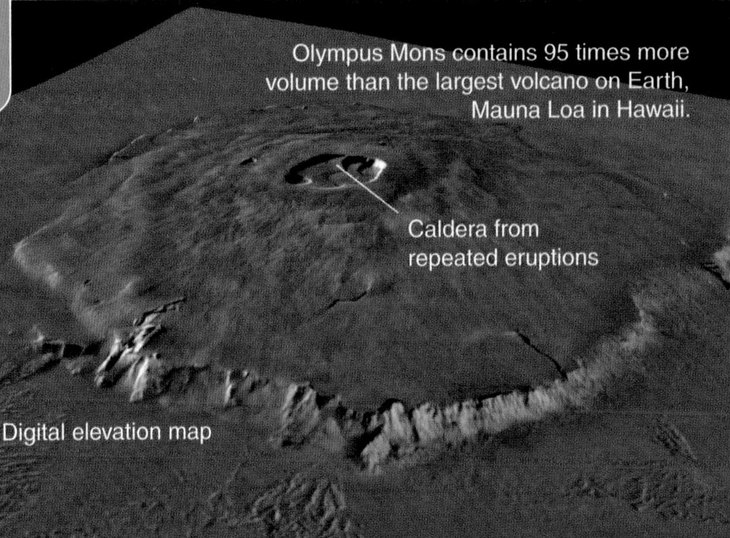

Olympus Mons contains 95 times more volume than the largest volcano on Earth, Mauna Loa in Hawaii.
Caldera from repeated eruptions
Digital elevation map

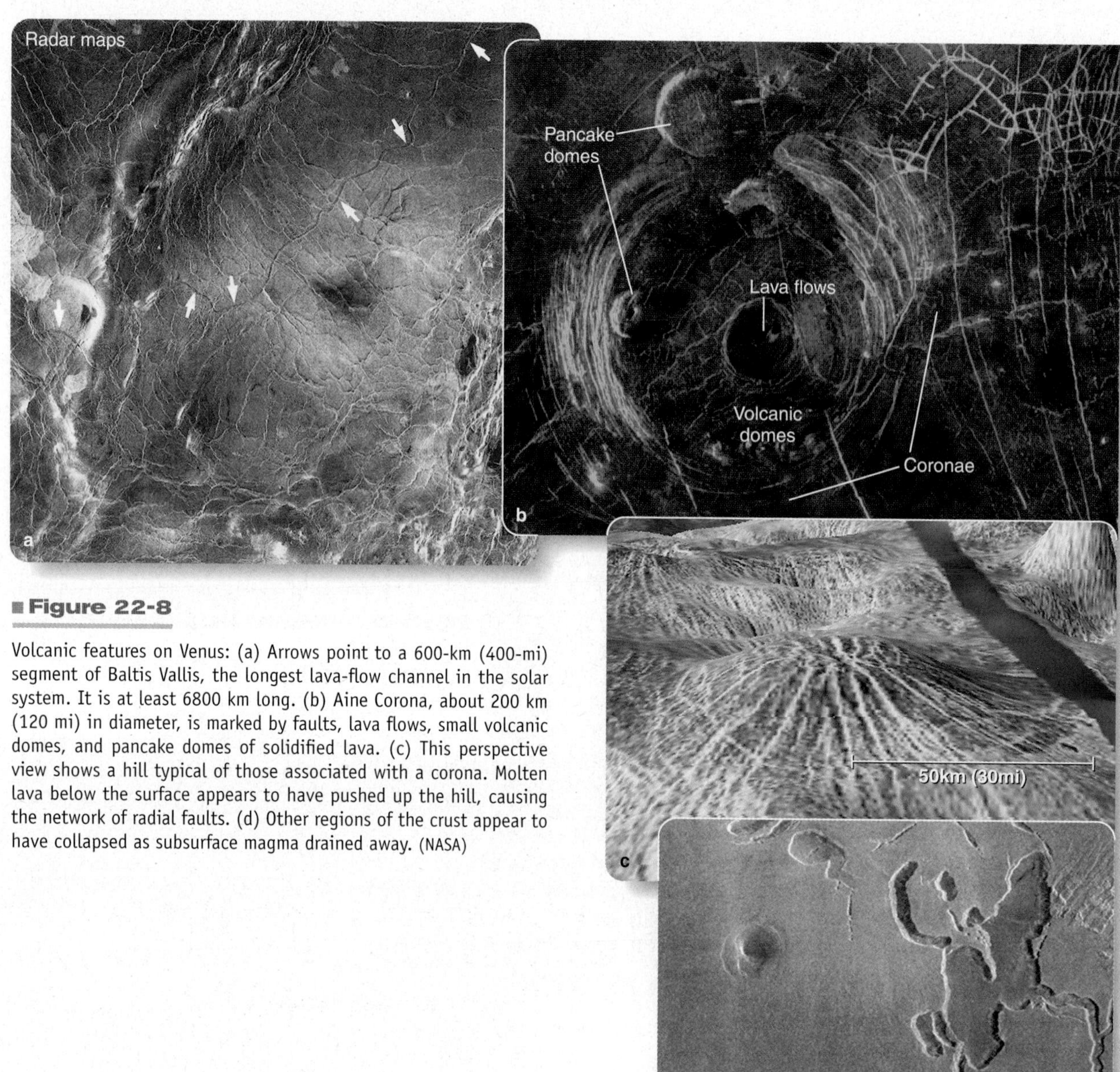


**■ Figure 22-8**

Volcanic features on Venus: (a) Arrows point to a 600-km (400-mi) segment of Baltis Vallis, the longest lava-flow channel in the solar system. It is at least 6800 km long. (b) Aine Corona, about 200 km (120 mi) in diameter, is marked by faults, lava flows, small volcanic domes, and pancake domes of solidified lava. (c) This perspective view shows a hill typical of those associated with a corona. Molten lava below the surface appears to have pushed up the hill, causing the network of radial faults. (d) Other regions of the crust appear to have collapsed as subsurface magma drained away. (NASA)

rock below the surface. All of these volcanic features are shown in ■ Figure 22-8.

There is no reason to suppose that all of the volcanoes found on Venus are extinct, so volcanoes may be erupting on Venus right now. However, the radar maps caught no evidence of actual eruptions in progress.

If you want to learn more of the secrets of Venus, you will want to visit the big land mass called Ishtar Terra. Western Ishtar consists of a high volcanic plateau called Lakshmi Planum. It rises 4 km (2.5 mi) above the rolling plains and appears to have formed from lava flows originating in volcanic vents such as Collette and Sacajawea (see Figure 22-5). Although folded mountain ranges showing widespread plate tectonics do not occur on Venus as they do on Earth, the areas north and west of Lakshmi Planum appear to be ranges of wrinkled mountains where some type of horizontal crust motion has pushed up against Ishtar Terra. Furthermore, faults and deep chasms are widespread over the surface of Venus, and that suggests stretching of the crust in those areas. So, although no direct evidence of plate motion is visible, there has certainly been compression and wrinkling of the crust in some areas and stretching and faulting in other areas, suggesting some limited crustal motion.

While Earth's crust has broken into rigid moving plates, the surface of Venus seems more pliable and does not break easily into plates. The history of Venus has been a fiery tale, dominated by volcanism.

## The Rotation of Venus

Nearly all of the planets in our solar system rotate counterclockwise as seen from the north. Uranus is an exception, and so is Venus.

In 1962, radio astronomers were able to transmit a radio pulse of precise wavelength toward Venus and detect the echo returning some minutes later. That is, they detected Venus by radar. But the echo was not at one precise wavelength. Part of the reflected signal had a longer wavelength, and part had a shorter wavelength. Evidently the planet was rotating; radio energy reflected from the receding edge was redshifted, and radio energy reflected from the approaching edge was blueshifted. From this Doppler effect, the radio astronomers could tell that Venus was rotating once every 243.0 Earth days. Furthermore, because the western edge of Venus produced the blueshift, the planet had to be rotating in the backward direction.

Why does Venus rotate backward, and very slowly? For decades, textbooks have suggested that proto-Venus was set spinning backward when it was struck off-center by a large planetesimal. That is a reasonable possibility; you have seen that a similar collision probably gave birth to Earth's moon. But there is an alternative. Mathematical models suggest that the rotation of a Terrestrial planet with a molten core and a dense atmosphere can be gradually reversed by solar tides in its atmosphere. Notice the contrast between the catastrophic theory of a giant impact and the evolutionary theory of atmospheric tides. It is possible that both mechanisms played a role in causing Venus's peculiar rotation. Perhaps Venus's very slow rotation is the explanation for why there is no dynamo effect to produce a magnetic field.

## A History of Venus

Earth passed through four major stages in its history (see Chapter 20), and you have seen how the moon and Mercury were affected by their own versions of the same stages. Venus, however, has had a peculiar passage through planetary development, and its history is difficult to understand. Planetary scientists are not sure of all the details about how the planet formed and differentiated, how it was cratered and flooded, or how its surface has evolved.

Venus is only slightly closer to the sun than Earth, so you might expect from the condensation sequence that it should have a similar over-all composition with perhaps slightly higher metal content than Earth. Instead, Venus's uncompressed density is slightly *less* than Earth's (look back to Table 19-2). The density and size of Venus still require that it have a dense metallic interior much like Earth's. However, if the metal in Venus's core is molten, then you would expect the dynamo effect to generate a magnetic field. But, no spacecraft has detected a magnetic field around Venus; its magnetic field must be at least 25,000 times weaker than Earth's. Some theorists wonder if the core of the planet is solid. If it is solid, scientists are puzzled by how Venus got rid of its internal heat faster than Earth did.

Because the planet lacks a magnetic field, it is not protected from the solar wind. The solar wind slams into the uppermost layers of Venus' atmosphere, forming a bow shock where the wind is slowed and deflected (■ Figure 22-9). Planetary scientists know little about the differentiation of the planet into core and mantle, so the size of the core shown in Figure 22-9 is estimated by analogy with Earth's. The magnetic field carried by the solar wind drapes over Venus like seaweed over a fishhook, forming a long tail within which ions flow away from the planet. You will see in Chapter 25 that comets, which also lack magnetic fields, interact with the solar wind in the same way.

Studies of moon rocks show that the moon formed as a sea of magma; and, presumably, Venus and Earth formed in the same way and never had primeval atmospheres rich in hydrogen. Instead, they outgassed carbon dioxide atmospheres as they formed. Calculations show that Venus and Earth have outgassed about the same amount of $CO_2$, but Earth's oceans have dissolved that $CO_2$ and converted it to sediments such as limestone. The main cause of the difference between surface conditions on Earth and Venus is the lack of water on Venus. Venus may have had oceans when it was young; but, because Venus is closer to the sun than Earth, it was initially warmer, and the $CO_2$ in the atmosphere created a greenhouse effect that made the planet even warmer. That process could have dried up any oceans that did exist and reduced the ability of the planet to clear its atmosphere of $CO_2$. As more $CO_2$ was outgassed, the greenhouse effect grew even more severe. Venus became trapped in a runaway greenhouse effect.

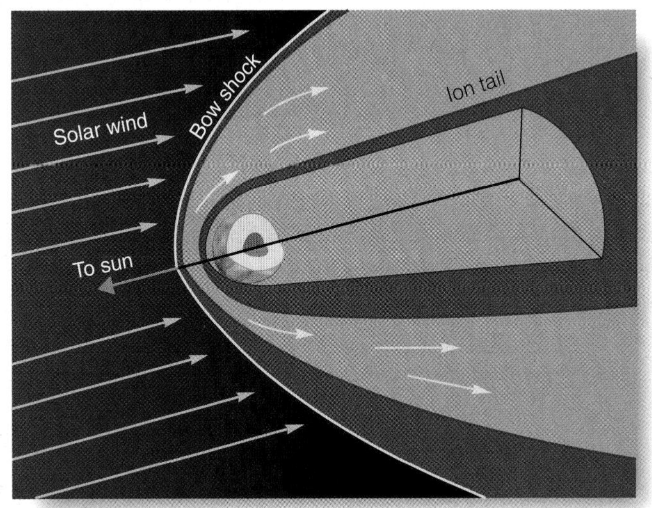


**■ Figure 22-9**

By analogy with Earth, the interior of Venus should contain a molten core (size estimated here), but no spacecraft has detected a planetary magnetic field. Thus, Venus is unprotected from the solar wind, which strikes the planet's upper atmosphere and is deflected into an ion tail.

In comparison, Earth avoided a runaway greenhouse effect because it was farther from the sun and always cooler than Venus. Consequently, it could form and preserve liquid-water oceans, which absorbed the $CO_2$ and left an atmosphere of nitrogen that was relatively transparent in some parts of the infrared. As you learned earlier, if all of the carbon in Earth's sediments was put back into the atmosphere as $CO_2$, our air would be as dense as that of Venus, and Earth would suffer from a tremendous greenhouse effect. Recall also from Chapter 20 how the use of fossil fuels and the destruction of forests are increasing the $CO_2$ concentration in our atmosphere and warming the planet. Venus warns us of what a greenhouse effect can do.

Fully 70 percent of the heat from Earth's interior flows outward through volcanism along midocean ridges. But Venus lacks crustal rifts, and even its numerous volcanoes cannot carry much heat out of the interior. Rather, Venus seems to get rid of its interior heat through large currents of hot magma that rise beneath the crust. Coronae, lava flows, and volcanism occur above such currents. The surface rock on Venus is the same kind of dark-gray basalt found in ocean crust on Earth.

True plate tectonics is not important on Venus. For one thing, the crust is very dry and is consequently about 12 percent less dense than Earth's crust. This low-density crust is more buoyant than Earth's crust and resists being pushed into the interior. Also, the crust is so hot that it is halfway to its melting point. Such hot rock is not very stiff, so it cannot form the rigid plates typical of plate tectonics on Earth.

There is no sign of plate tectonics on Venus, but there is evidence that convection currents below the crust are deforming the crust to create coronae and push up mountains such as Maxwell. Detailed measurements of the strength of gravity over Venus's mountains show that some must be held up not by deep roots like mountains on Earth but by rising currents of magma. Other mountains, like those around Ishtar Terra, appear to be folded mountains caused by limited horizontal motions in the crust, driven perhaps by convection currents in the mantle.

The small number of craters on the surface of Venus hints that the entire crust has been replaced within the last half-billion years or so. This may have occurred in a planetwide overturning as the old crust broke up and sank and lava flows created a new crust. This could happen periodically on Venus, or the planet may have had geological processes more like Earth's until a single resurfacing geologically recently. In either case, unearthly Venus may eventually reveal more about how our own world works.

### SCIENTIFIC ARGUMENT

***What evidence can you point to that Venus does not have plate tectonics?***

Sometimes a scientific argument can be helpful by eliminating a possibility. On Earth, plate tectonics is identifiable by the worldwide network of faults, subduction zones, volcanism, and folded mountain chains that outline the plates. Although some of these features are visible on Venus, they do not occur in a planetwide network of plate boundaries. Volcanism is widespread, but folded mountain ranges occur in only a few places, such as near Lakshmi Planum and Maxwell Montes, and, unlike on Earth, they do not make up long mountain chains. Also, the large size of the shield volcanoes on Venus shows that the crust is not moving over the hot spots in the way the Pacific seafloor is moving over the Hawaiian hot spot.

At first glance, you might think that Earth and Venus should be as similar as siblings, but comparative planetology reveals that they are more like cousins. You can blame the thick atmosphere of Venus for altering its geology, but that calls for a new scientific argument: **Why isn't Earth's atmosphere similar to that of Venus?**

## 22-2 Mars

MERCURY AND THE MOON ARE SMALL. Venus and Earth are the largest of the Terrestrial planets. Mars has an intermediate size. It is twice the diameter of the moon but only a little more than half of Earth's diameter (**Celestial Profile 6**). Mars's small size has allowed it to cool faster than Earth, and much of its atmosphere has leaked away. Its present carbon-dioxide atmosphere is a bit less than 1 percent as dense as Earth's.

### No Canals on Mars

Long before the space age, the planet Mars was a mysterious landscape in the public mind. In the century following Galileo's first astronomical use of the telescope, astronomers discovered dark markings on Mars as well as bright polar caps. Timing the motions of the markings, they concluded that a Martian day was about 24 hours 40 minutes long. Its axis is tipped 25.2° to its orbit, almost exactly the same as Earth's 23.4° tilt, so Mars has seasons with about the same winter/summer contrast as Earth. Mars's year is about 1.88 Earth years long. These similarities with Earth encouraged the belief that Mars might be inhabited.

In 1858, the Jesuit astronomer Angelo Secchi referred to a region on Mars as *Atlantic Canale.* This is the first use of the Italian word *canale* (channel) to refer to a feature on Mars. Then, in the late summer of 1877, the Italian astronomer Giovanni Schiaparelli, using a telescope only 8.75 in. in diameter, thought he glimpsed fine, straight lines on Mars. He too used the Italian word *canali* (plural) for these lines, and the word was translated into English not as "channel," a narrow body of water that is a natural geological feature, but as "canal," an artificially dug channel. Thus the "canals of Mars" were born. Many astronomers could not see the canals at all, but others drew maps showing hundreds (■ Figure 22-10).

In the decades that followed Schiaparelli's "discovery," many people assumed that the canals were watercourses built by an intelligent race to carry water from the polar caps to the lower latitudes. Much of this excitement was generated by Percival Lowell, a wealthy Bostonian who, in 1894, founded Lowell Observatory in Flagstaff, Arizona, principally for the study of Mars. He not only mapped hundreds of canals but also popularized his results in books and lectures. Although some astronomers claimed the canals were merely illusions, by 1907 the general public was so sure that life existed on Mars that the *Wall Street Journal* suggested that the most extraordinary event of the previous year had been "the proof by astronomical observations... that conscious, intelligent human life exists upon the planet Mars." Further sightings of bright clouds and flashes of light on Mars strengthened this belief, and some urged that gigantic geometrical diagrams be traced in the Sahara Desert to signal to the Martians that Earth, too, is inhabited. All seemed to agree that the Martians were older and wiser than humans.

This fascination with men from Mars was not a passing fancy. Beginning in 1912, Edgar Rice Burroughs (author of the Tarzan stories) wrote a series of 11 novels about the adventures of the Earthman John Carter, lost on Mars. Burroughs made the geography of Mars, named by Schiaparelli after Mediterranean lands both real and mythical, into household words. He also made his Martians small and gave them green skin.

By Halloween night of 1938, people were so familiar with life on Mars that they were ready to believe that Earth could be invaded. When a radio announcer repeatedly interrupted dance music to report the landing of a spaceship in New Jersey, the emergence of monstrous creatures, and their destruction of whole cities, thousands of otherwise sensible people fled in panic, not knowing that Orson Welles and other actors were dramatizing H. G. Wells's book *The War of the Worlds*.

Public fascination with Mars, its canals, and its little green men lasted right up until July 15, 1965, when Mariner 4, the first spacecraft to fly past Mars, radioed back photos of a dry, cratered surface and proved that there are no canals and no Martians. The canals are optical illusions produced by the human brain's powerful ability to assemble a field of disconnected marks into a coherent image. If your brain could not do this, the photos on these pages would be nothing but swarms of dots, and the images on a TV screen would never make sense. The downside of this is that the brain of an astronomer looking for something at the edge of visibility is capable of connecting faint, random markings on Mars into the straight lines of canals.

Even today, Mars holds some fascination for the general public. Grocery store tabloids regularly run stories about a giant face carved on Mars by an ancient race. Although planetary scientists recognize it as nothing more than chance shadows in a photograph and dismiss the issue as a silly hoax, the stories persist.

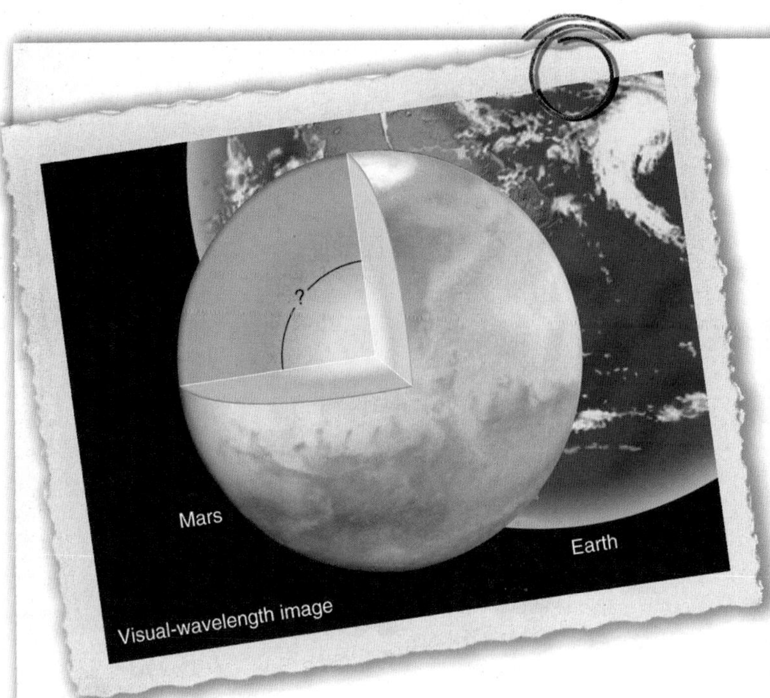


*Mars is only half the diameter of Earth and probably retains some internal heat, but the size and composition of its core are not well known.* (NASA)

## Celestial Profile 6: Mars

### Motion:

| | |
|---|---|
| Average distance from the sun | 1.52 AU ($2.28 \times 10^8$ km) |
| Eccentricity of orbit | 0.093 |
| Inclination of orbit to ecliptic | 1.9° |
| Average orbital velocity | 24.1 km/s |
| Orbital period | 1.881 y (687.0 days) |
| Period of rotation | 24.62 h |
| Inclination of equator to orbit | 25.2° |

### Characteristics:

| | |
|---|---|
| Equatorial diameter | $6.79 \times 10^3$ km (0.531 $D_\oplus$) |
| Mass | $6.42 \times 10^{25}$ kg (0.108 $M_\oplus$) |
| Average density | 3.94 g/cm$^3$ (3.3 g/cm$^3$ uncompressed) |
| Surface gravity | 0.38 Earth gravity |
| Escape velocity | 5.0 km/s (0.45 $V_\oplus$) |
| Surface temperature | –140° to 15°C (–220° to 60°F) |
| Average albedo | 0.16 |
| Oblateness | 0.009 |

### Personality Point:

Mars is named for the god of war. Minerva was the goddess of defensive war, but Bullfinch's *Mythology* refers to Mars's "savage love of violence and bloodshed." You can see how the planet glows blood red in the evening sky because of iron oxides in its soil.

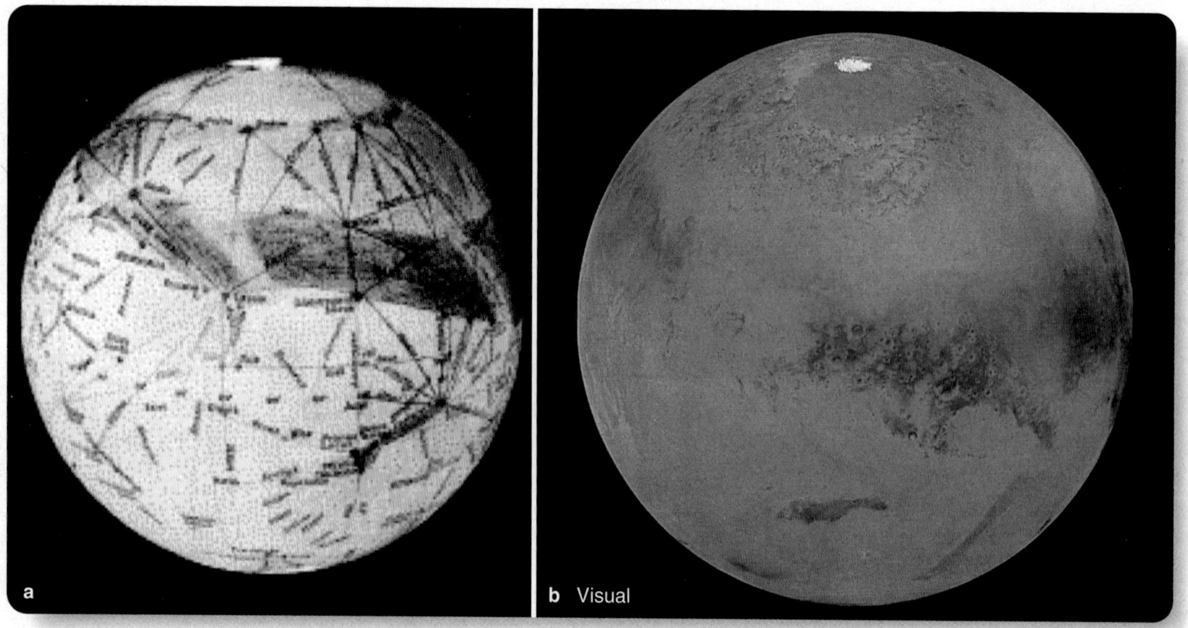


**■ Figure 22-10**

(a) Early in the 20th century, Percival Lowell mapped canals over the face of Mars and concluded that intelligent life resided there. (Lowell Observatory) (b) Modern images recorded by spacecraft reveal a globe of Mars with no canals. Instead, the planet is marked by craters and, in some places, volcanoes. Both of these images are reproduced with south at the top, as they appear in telescopes. Lowell's globe is inclined more nearly vertically and is rotated slightly to the right compared with the modern globe. (U.S. Geological Survey)

A hundred years of speculation have raised high expectations for Mars. If there were intelligent life on Mars and its representatives came to Earth, they would probably be a big disappointment to the readers of the tabloids.

## The Atmosphere of Mars

If you visited Mars, your first concern, even before you opened the door of your spaceship, would be the atmosphere. Is it breathable? Even for the astronomer observing safely from Earth, the atmosphere of Mars is of major interest. The gases that cloak Mars are critical to understanding the history of the planet.

The air on Mars is 95 percent carbon dioxide, with a few percent each of nitrogen and argon. You probably noticed that this is quite similar to the composition of Venus's atmosphere. The reddish color of the Martian soil is caused by iron oxides (rusts), and this is a warning that the oxygen humans would prefer to find in the atmosphere is locked in chemical compounds in the soil. The Martian atmosphere contains almost no water vapor or oxygen, and its density at the surface of the planet is only about 1 percent that of Earth's atmosphere. This does not provide enough pressure to prevent liquid water from boiling into vapor. Water can exist at the Martian surface only as ice or vapor.

Although the air is thin, it is dense enough to be visible in photographs (■ Figure 22-11). Haze and clouds come and go, and occasional weather patterns are visible. Winds on Mars can be strong enough to produce dust storms that envelop the entire planet. The polar caps visible in photos are also related to the Martian atmosphere. The ices in the polar caps are frozen carbon dioxide ("dry ice") with frozen water underneath.

**■ Figure 22-11**

The atmosphere of Mars is evident in this image made by the Hubble Space Telescope. The haze is made up of high, water-ice crystals in the thin $CO_2$ atmosphere. The spot at extreme left is the volcano Ascraeus Mons, 25 km (16 mi) high, poking up through the morning clouds. Note the north polar cap at the top. (Philip James, University of Toledo; Steven Lee, University of Colorado, Boulder; and NASA)

**■ Figure 22-12**

Mars is a red desert planet, as shown in this true-color photo made by the Rover Opportunity. The rock outcrop is a meter-high crater wall. After taking this photo, Opportunity descended into the crater to study the wall. Dust suspended in the atmosphere colors the sky red-orange. (NASA © 1995–2007 by Calvin J. Hamilton)

If you could visit Mars you would find it a reddish, airless, bone-dry desert (■ Figure 22-12). To understand Mars, you can ask why its atmosphere is so thin and dry and why the surface is rich in oxides. To find those answers you need to consider the origin and evolution of the Martian atmosphere.

Presumably, the gases in the Martian atmosphere were mostly outgassed from its interior. Volcanism on Terrestrial planets typically releases carbon dioxide and water vapor, plus other gases. Because Mars formed farther from the sun, you might expect that it would have incorporated more volatiles when it formed. But Mars is smaller than Earth, so it has had less internal heat to drive geological activity, and that would lead you to suspect that it has not outgassed as much as Earth. In any case, whatever outgassing took place occurred early in the planet's history, and Mars, being small, cooled rapidly and now releases little gas.

How much atmosphere a planet has depends on how rapidly it loses gas to space, and that depends on the planet's mass and temperature. The more massive the planet, the higher its escape velocity (see Chapter 5), and the more difficult it is for gas atoms to leak into space. Mars has a mass less than 11 percent that of Earth, and its escape velocity is only 5 km/s, less than half Earth's. Consequently, gas atoms can escape from it much more easily than they can escape from Earth.

The temperature of a planet's atmosphere is also important. If a gas is hot, its molecules have a higher average velocity and are more likely to exceed escape velocity. That means a planet near the sun is less likely to retain an atmosphere than a more distant, cooler planet. The velocity of a gas molecule, however, also depends on the mass of the molecule. On average, a low-mass molecule travels faster than a massive molecule. For that reason, a planet loses its lowest-mass gases more easily because those molecules travel fastest.

You can see this principle of comparative planetology if you plot a diagram such as that in ■ Figure 22-13. The data points show the escape velocity versus temperature for the larger objects in our solar system. The temperature used in the diagram is the temperature of the gas that is in a position to escape. For the moon, which has essentially no atmosphere, this is the temperature of the sunlit surface. For Mars, the temperature that is important is that at the top of the atmosphere.

The lines in Figure 22-13 show the typical velocities of the fastest-traveling examples of various molecules. At any given temperature, some water molecules, for example, travel faster than others, and it is the highest-velocity molecules that escape from a planet. The diagram shows that the Jovian planets have escape velocities so high that very few molecules can escape. Earth and Venus can't hold hydrogen, and Mars can hold only the more massive molecules. Earth's moon is too small to keep any gases from leaking away. You can refer back to this diagram when you study the atmospheres of other worlds in later chapters.

Over the 4.6 billion years since Mars formed, it has lost some of its lower-mass gases. Water molecules are massive enough for Mars to keep, but ultraviolet radiation can break them up. The hydrogen escapes, and the oxygen, a very reactive element, forms more oxides in the soil—the oxides that make Mars the red planet. Recall that on Earth the ozone layer protects water vapor from ultraviolet radiation, but Mars never had an oxygen-rich atmosphere, so it never had an ozone layer. Ultraviolet photons from the sun can penetrate deep into the atmosphere and break up molecules. In this way, molecules too massive to leak into space can be lost if they break into lower-mass fragments.

The argon in the Martian atmosphere is evidence that there once was a denser blanket of air. Argon atoms are massive, almost

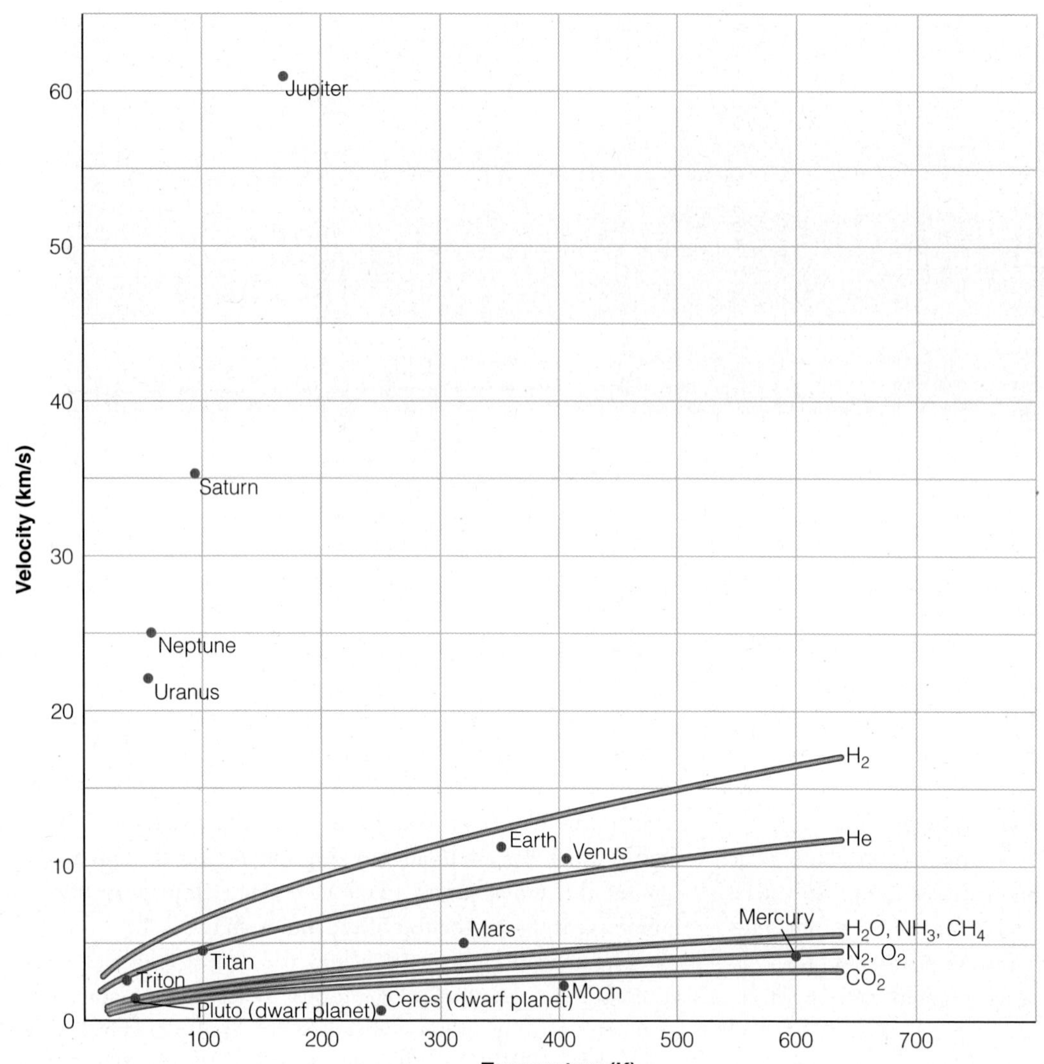


**■ Figure 22-13**

Loss of planetary atmosphere gases. Dots represent the escape velocity and temperature of various solar-system bodies. The lines represent the typical highest velocities of molecules of various masses. The Jovian planets have high escape velocities and can hold onto even the lowest-mass molecules. Mars can hold only the more massive molecules, and the moon has such a low escape velocity that even the most massive molecules can escape.

as massive as a carbon dioxide molecule, and would not be lost easily. In addition, argon is inert and cannot form compounds in the soil. The 1.6 percent argon in the atmosphere of Mars is evidently left over from an ancient atmosphere that may have been 10 to 100 times denser than the present Martian air.

Finally, you should consider the interaction of the solar wind with the atmosphere of Mars. This is not an important process for Earth because Earth has a magnetic field that deflects the solar wind. In contrast, Mars has no magnetic field, and the solar wind interacts directly with the Martian atmosphere. Detailed calculations show that significant amounts of carbon dioxide could have been carried away by the solar wind over the history of the planet. This process would have been most efficient long ago when the sun was more active and the solar wind was stronger. However, you should also keep in mind that Mars probably had a magnetic field when it was younger and still retained significant internal heat. A magnetic field would have protected its atmosphere from the solar wind.

The polar caps contain large amounts of carbon dioxide ice; and, as spring comes to a hemisphere, that ice begins to vaporize and returns to the atmosphere. Meanwhile, at the other pole, carbon dioxide is freezing out and adding to the polar cap there. Dramatic evidence of this cycle appeared when the camera aboard the Mars Odyssey probe sent back images of dark markings on the south polar cap. Evidently as spring comes to the polar cap and the sun begins to peek above the horizon, sunlight penetrates the meter-thick ice and vaporizes carbon dioxide, which bursts out in geysers a few tens of meters high carrying dust and sand. Local winds push the debris downwind to form the fan-shaped dark markings (■ Figure 22-14). These dark markings appear each spring but last only a few months as frozen carbon dioxide returns to the atmosphere.

Although planetary scientists remain uncertain as to how much of an atmosphere Mars has had in its past and how much it has lost, it is a good example for your study of comparative planetology. When you look at Mars, you see what can happen

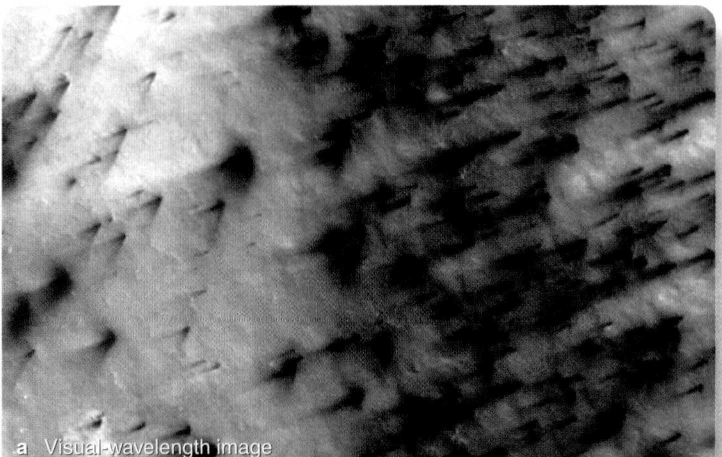

■ **Figure 22-14**

(a) Each spring, spots and fans appear on the ice of the south polar cap on Mars. (b) Studies show the ice is frozen carbon dioxide in a nearly clear layer about a meter thick. High-pressure carbon dioxide gas vaporized by spring sunlight bursts out of the ice in geysers. The gas carries sand and dust hundreds of meters into the air. (NASA; Arizona State University/Ron Miller)

to the atmosphere of a medium-size world. Like its atmosphere, the geology of Mars is probably typical of medium-size worlds.

## The Geology of Mars

If you ever decide to visit another world, Mars may be your best choice. Mars is much more friendly than the moon, Mercury, or Venus. The nights on Mars are deadly cold, but a hot summer day would be comfortable **(Celestial Profile 6).** Mars has weather, complex geology, and signs that water once flowed over its surface. You might even hope to find traces of ancient life hidden in the rocks.

Spacecraft have been visiting Mars for almost 40 years, but the pace has picked up recently. A small armada of spacecraft has gone into orbit around Mars to photograph and analyze its surface, and six spacecraft have landed. Two Viking landers touched down in 1976, and three rovers have landed in recent years. Rovers have an advantage because they are wheeled robots that can be controlled from Earth and directed to travel from feature to feature and make detailed measurements. Pathfinder and its rover, Sojourner, landed in 1997. Rovers Spirit and Opportunity landed in 2004, carrying sophisticated instruments to explore the rocky surface. The Phoenix probe landed in the north polar region of Mars in 2008.

Photographs made by rovers and landers on the surface of Mars, such as Figure 22-12, show reddish deserts of broken rock. These appear to be rocky plains fractured by meteorite impacts, but they don't look much like the surface of Earth's moon. The atmosphere of Mars, thin though it is, protects the surface from the blast of micrometeorites that grinds moon rocks to dust. Also, Martian dust storms may sweep fine dust away from some areas, leaving larger rocks exposed.

Spacecraft orbiting Mars have imaged the surface and measured elevations to reveal that all of Mars is divided into two parts. The southern highlands are heavily cratered, and the number of craters there shows that they must be old. In contrast, the northern lowlands are smooth (■ Figure 22-15) and so remarkably free of craters that they must have been resurfaced no more than a billion years ago. Some astronomers have suggested that volcanic floods filled the northern lowlands and buried the craters there. Growing evidence, however, suggests that the northern

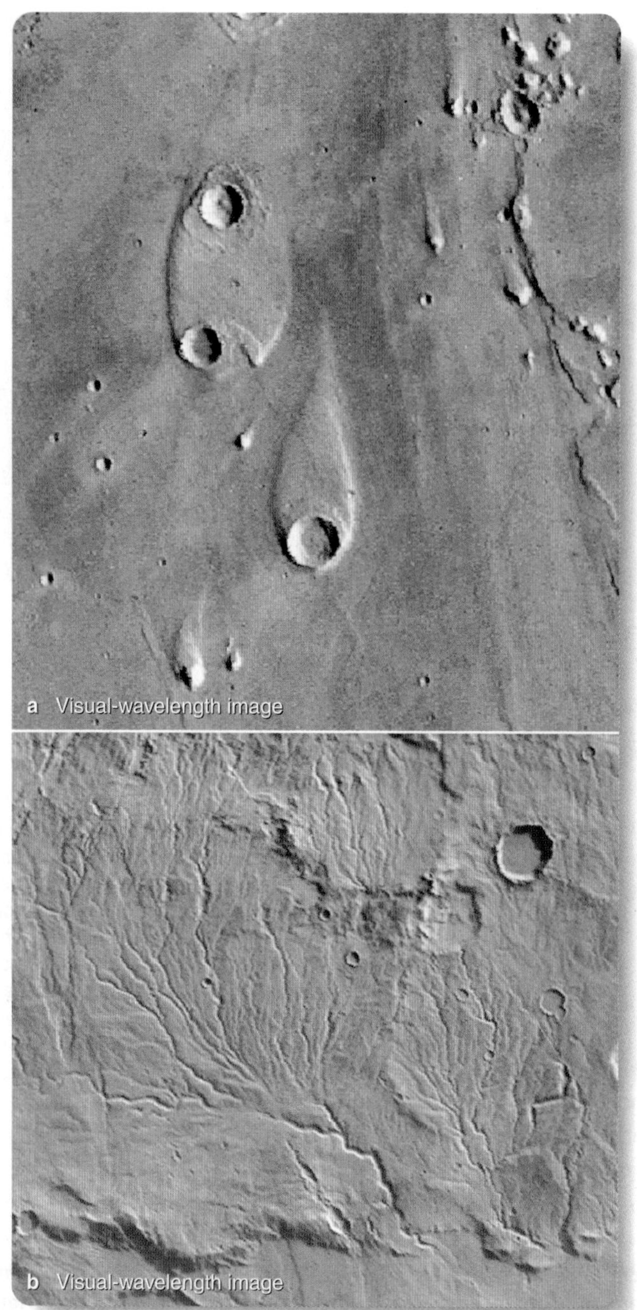


**■ Figure 22-17**

These visual-wavelength images made by the Viking orbiters show some of the features that suggest liquid water on Mars. (a) Outflow channels are broad and shallow and deflect around obstructions such as craters. They appear to have been produced by sudden floods. (a) Valley networks resemble drainage patterns and suggest water flowing over long periods. Crater counts show that both formations are old, but valley networks are older than outflow channels. (NASA © Calvin J. Hamilton)

Many flow features lead into the northern lowlands, and the smooth terrain there has been interpreted as ancient ocean floor. Features along the edges of the lowlands have been compared to shorelines, and many planetary scientists conclude that the northern lowlands were filled by an ocean when Mars was younger. Large, generally circular depressions such as Hellas and Argyre appear to be impact basins that also may once have been flooded by water.

Mars Orbiter photographed the eroded remains of a river delta in an unnamed crater in the old highlands (■ Figure 22-18). Details show that the river flowed for long periods of time, shifting its channel to form meanders and braided channels as rivers on Earth do. The shape of the delta suggests it formed when the river flowed into deeper water and dropped its sediment, much as the Mississippi drops its sediment and builds its delta in the Gulf of Mexico.

Did Mars once have that much water? Deuterium is 5.5 times more abundant than normal (light) hydrogen in the Martian atmosphere, and that suggests that Mars once had about 20 times more water than it has now. Presumably, much of the water was broken up and the normal hydrogen mostly lost to space.

The remaining water on Mars could survive if it were frozen in the crust. High-resolution images and measurements made

**■ Figure 22-18**

This distributary fan was formed where an ancient stream flowed into a flooded crater. Sediment in the moving water was deposited in the still water to form a lobed delta that later became sedimentary rock. Detailed analysis reveals that the stream changed course time after time, and this shows that the stream flowed for an extended period and was not just a short-term flood. Also, the shape of the fan is evidence that a lake persisted in the crater while the fan developed. (NASA/JPL/Malin/Space Science Systems)

from orbit reveal features that suggest subsurface ice. Gullies leading downhill appear to have been eroded recently, judging from their lack of craters; these may have been formed by water seeping from below the surface. Some regions of collapsed terrain appear to be places where subsurface water has drained away. Photos taken over a period of years reveal the appearance of recent landslides that may have been caused by water gushing from crater walls and carrying debris downhill before vaporizing completely. Instruments aboard the Mars Odyssey spacecraft detected water frozen in the soil over large areas of the planet. At latitudes farther than 60 degrees from the equator, water ice may make up more than 50 percent of the surface soil.

If you added a polar bear, changed the colors, and hid the craters in ■ Figure 22-19a, it would look like the broken pack ice on Earth's Arctic Ocean. Mars Express photographed these dust-covered formations near the Martian equator, and the shallow depth of the craters suggests that the ice is still there, just below the surface.

Much of the ice on Mars may be hidden below the polar caps. Radar aboard the Mars Express orbiter was able to penetrate 3.7 km (2.3 mi) below the surface and map ice deposits hidden below the south polar region (Figure 22-19b). There is enough water there, at least 90 percent pure, to cover the entire planet to a depth of 11 meters. The Phoenix probe, which landed in 2008 near Mars's north pole, found water ice not only mixed with the soil as permafrost, but also as small chunks of pure ice, indicating that there once was standing water that had frozen in place.

Rovers Spirit and Opportunity were targeted to land in areas suspected of having had water on their surfaces. Images made from orbit showed flow features at the Spirit landing site, and hematite, a mineral that forms in water, was detected from orbit at the Opportunity landing site. Both rovers reported exciting discoveries, including evidence of past water. Using its analytic instruments, Opportunity found small spherical concretions of hematite (dubbed blueberries) that must have formed in abundant water. Later, Spirit found similar concretions in its area. In other rocks, Opportunity found layers of sediments with ripple marks and crossed layers showing they were deposited in moving water (■ Figure 22-20). Chemical analysis of the rocks at the Opportunity site showed the presence of sulfates much like Epsom salts plus bromides and chlorides. On Earth, these compounds are left behind when bodies of water dry up. Halfway

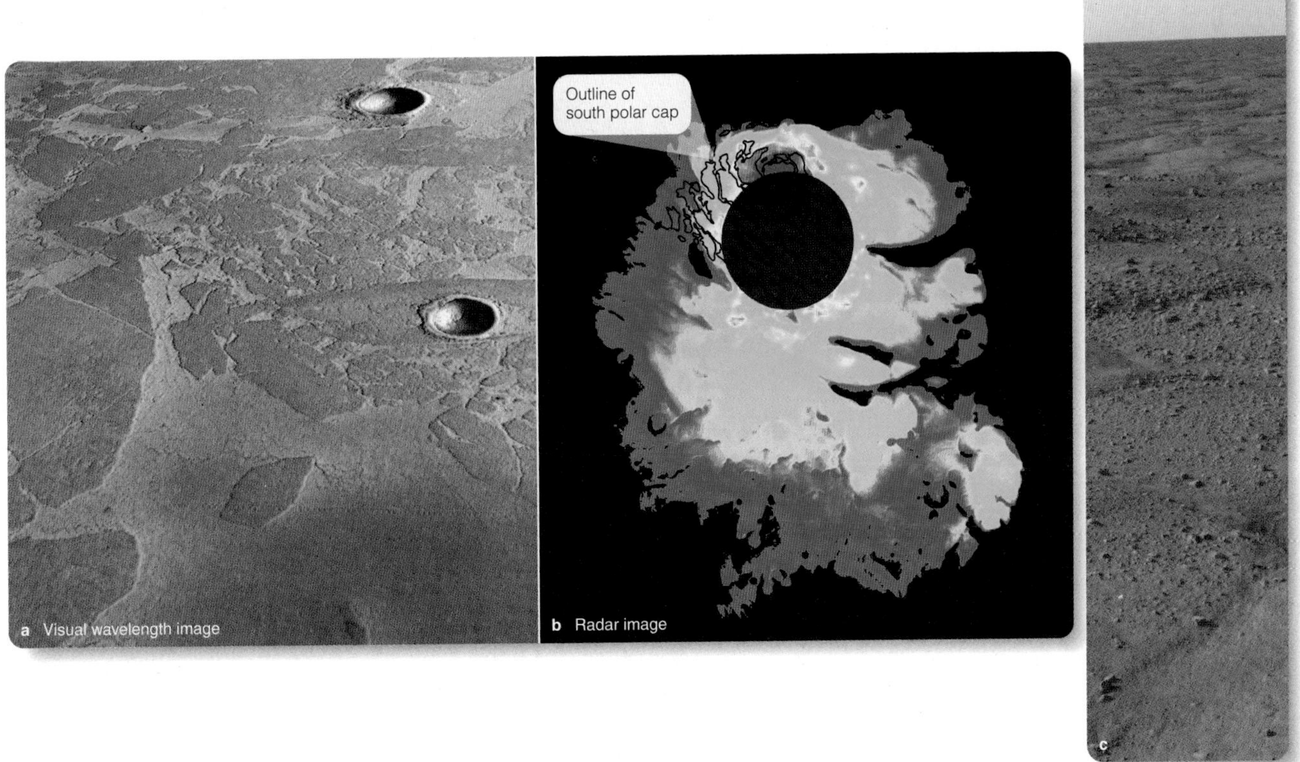


**■ Figure 22-19**

(a) Like broken pack ice, these formations near the equator of Mars suggest floating ice that broke up and drifted apart. Scientists propose that the ice was covered by a protective layer of dust and volcanic ash and may still be present. (ESA/DLR/F. U. Berlin/G. Neukum) (b) Radar aboard the Mars Express satellite probed beneath the surface to image water ice below the south polar cap. The black circle is the area that could not be studied from the satellite's orbit. (NASA/JPL/ESA/Univ. of Rome/MOLA Science Team/USGS) (c) This view from the Phoenix lander shows the landscape of Mars's north polar plain, including polygonal cracks understood to result from seasonal expansion and contraction of ice under the surface. (NASA/JPL-Caltech/University of Arizona.)

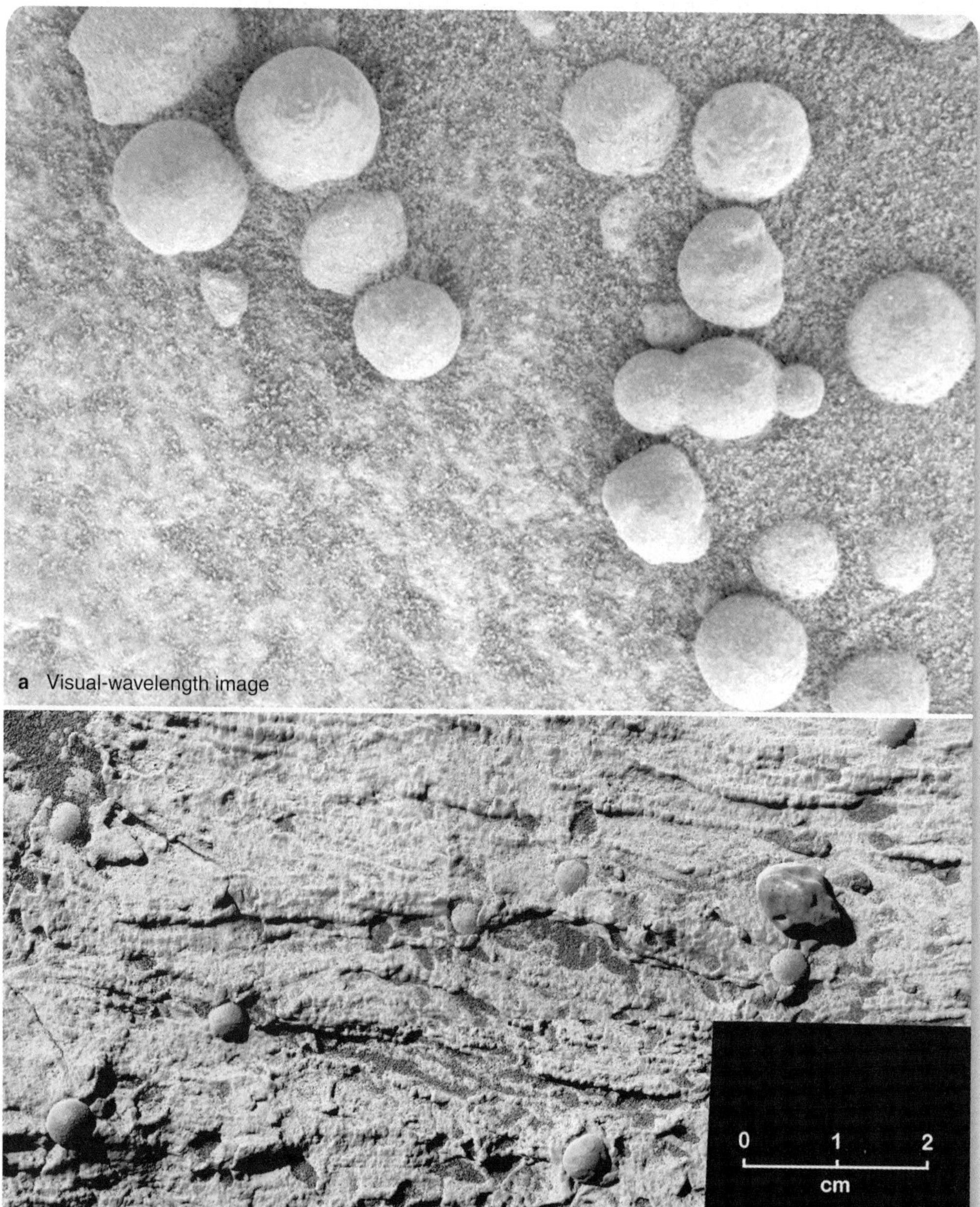


■ **Figure 22-20**

(a) Rover Opportunity photographed these hematite concretions ("blueberries") in a rock near its landing site. The spheres appear to have grown as minerals collected around small crystals in the presence of water. Similar concretions are found on Earth. (b) The layers in this rock were deposited as sand and silt in rapidly flowing water. From the way the layers curve and cross each other, geologists can estimate that the water was at least ten centimeters deep. A few "blueberries" and one small pebble are also visible in this image. (NASA.JPL/Cornell/USGS)

around Mars, Spirit found the mineral goethite, which only forms in the presence of water.

One amazing bit of evidence of water on Mars is the analysis of rock samples from the planet. Of course, no astronaut has ever visited Mars and brought back a rock, but over the history of the solar system occasional impacts by asteroids have blasted giant craters on Mars and ejected bits of rock into space. A few of those bits of rock have fallen to Earth as meteorites, and over 30 have been found and identified (■ Figure 22-21). These meteorites include basalts, as you might expect from a planet so heavily covered by lava flows. They also contain small traces of water and minerals that are deposited by water. Chemical analysis shows that the magma from which the rocks solidified must have contained up to 1.8 percent water. If all of the lava flows on Mars contained that much water and it was all outgassed, it could create a planetwide ocean 20 m deep. That isn't quite enough to explain all the flood features on Mars, but it does show that the planet once had abundant water on its surface.

Mars has water, but it is hidden. When humans reach Mars, they will not need to dig far to find water in the form of ice. They can use solar power to break the water into hydrogen and oxygen. Hydrogen is fuel, and oxygen is the breath of life, so the water on Mars may prove to be buried treasure. Even more exciting is the realization that Mars once had bodies of liquid water

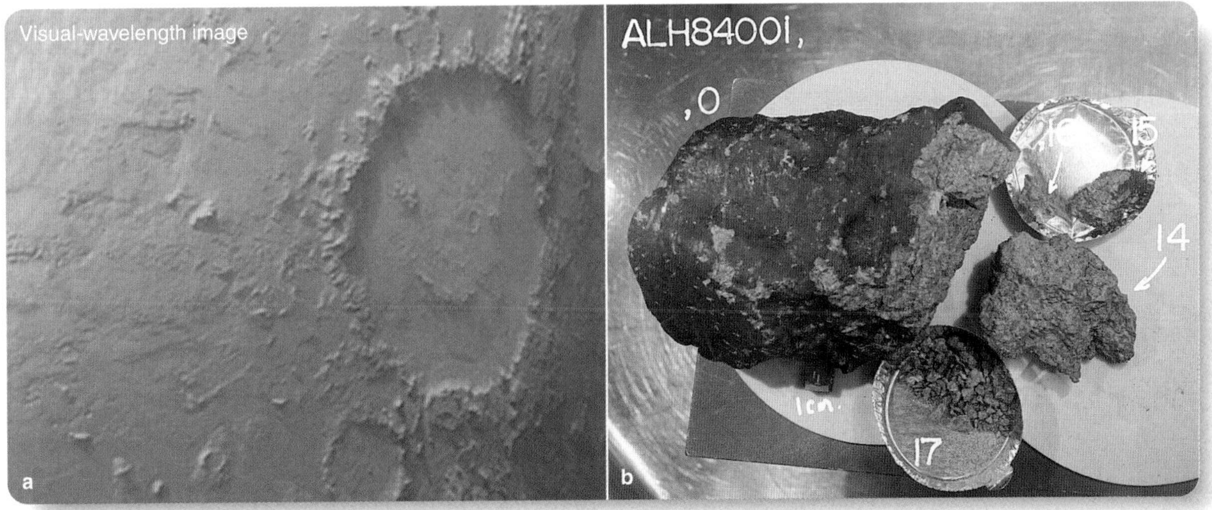


**Figure 22-21**

(a) The impact crater Galle, known for obvious reasons as the Happy Face Crater, is 210 km (130 mi) in diameter. Such large impacts may occasionally eject fragments of the crust into space. A few fragments from Mars have fallen to Earth as meteorites. (Malin Space Science Systems/NASA) (b) Meteorite ALH84001, found in Antarctica, has been identified as an ancient rock from Mars. Minerals found in the meteorite were deposited in water and thus suggest that the Martian crust was once richer in water than it is now. (NASA)

on its surface. It is a desert world now, but someday an astronaut may scramble down an ancient Martian streambed, turn over a rock, and find a fossil. It is even possible that life started on Mars and has managed to persist as the planet gradually became less hospitable. For example, life may have hung on by retreating to limited warm and wet oases underground. You will learn in Chapter 26 about possible evidence of ancient life in Mars meteorite ALH84001 and, even more intriguing, indications in spectra of Mars's atmosphere of current methane production that could be biological.

## A History of Mars

The history of Mars is like a play where all the exciting stuff happens in the first act. After the first two billion years on Mars, most of the activity was over, and it has gone downhill ever since.

Planetary scientists have good evidence that Mars differentiated when it formed and had a hot, molten core. Some of the evidence comes from exquisitely sensitive Doppler-shift measurements of radio signals coming from spacecraft orbiting Mars. Measuring the shifts allowed scientists to map the gravitational field and study the shape of Mars in such detail that they could detect tides on Mars caused by the sun's gravity. Those tides are less than a centimeter high; but, by comparing them with models of the interior of Mars, the scientists can show that Mars has a very dense core, a less dense mantle, and a low-density crust.

Observations made from orbit show that Mars has no overall magnetic field, but it does have traces of magnetism frozen into some sections of old crust. This shows that soon after Mars formed, it had a hot metallic core in which the dynamo effect generated a magnetic field. Because Mars is small, it lost its heat rapidly, and most of its core gradually froze solid. That may be why the dynamo finally shut down. Today Mars probably has a large solid core surrounded by a thin shell of liquid core material in which the dynamo effect is unable to generate a magnetic field.

No one is sure what produced the dramatic difference between the southern highlands and northern lowlands. Powerful convection in the planet's mantle may have pushed crust together to form the southern highlands. The suggestion that a catastrophic impact produced the northern lowlands does not seem to fit with the structure of the highlands, but it is not impossible.

Planetary scientists divide the history of Mars into three periods. The first, the **Noachian period,** extended from the formation of the crust about 4.3 billion years ago until roughly 3.7 billion years ago. During this time the crust was battered by the heavy bombardment as the last of the debris in the young solar system was swept up. The old southern hemisphere survives from this period. The largest impacts blasted out the great basins like Hellas and Argyre very late in the cratering, and there is no trace of magnetic field in those basins. Evidently the dynamo had shut down by then.

The Noachian period included flooding by great lava flows that smoothed some regions. Volcanism in the Tharsis and Elysium regions was very active, and the Tharsis rise grew into a

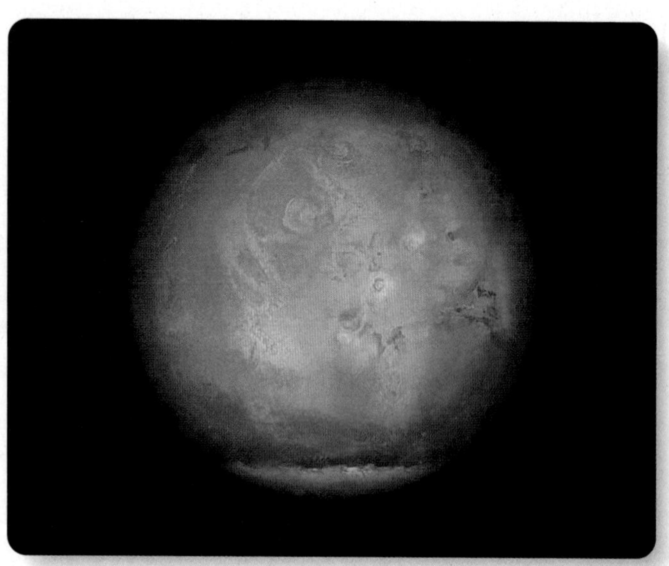

**■ Figure 22-22**

The Tharsis rise is a bulge on the side of Mars consisting of stacked lava flows extending up to 10 km (6 mi) high and over a third the diameter of Mars. It dominates this image of Mars, with a few clouds clinging to the four largest volcanoes. This image was recorded in late winter for the southern hemisphere. Notice the size of the southern polar cap. (NASA/JPL/Malin Space Science Systems)

huge bulge on the side of Mars (■ Figure 22-22). Some of the oldest lava flows on Mars are in the Tharsis rise, but it also contains some of the most recent lava flows. It has evidently been a major volcanic area for most of the planet's history.

The valley networks found in the southern highlands were formed during the Noachian period when water fell as rain or snow and drained down slopes. To keep the water liquid required a higher temperature and higher atmospheric pressure than is present on Mars today. Violent volcanism could have vented gases, including more water vapor that kept the air pressure high. This may have produced episodes in which water flowed over the surface and collected in the northern lowlands and in the deep basins to form oceans and lakes, but it's not known how long those bodies of water survived. A planetwide magnetic field may have protected the atmosphere from the solar wind.

Because Mars is small, it lost its internal heat quickly, and atmospheric gases escaped into space. The **Hesperian period** extended from roughly 3.7 billion years ago to about 3 billion years ago. During this time massive lava flows covered some sections of the surface. Most of the outflow channels date from this period, which suggests that the loss of atmosphere drove Mars to become a deadly cold desert world with its water frozen in the crust. When volcanic heat or large impacts melted subsurface ice, the water could have produced violent floods and shaped the outflow channels.

The history of Mars may hinge on climate variations. Models calculated by planetary scientists suggest that Mars may have once rotated at a much steeper angle to its orbit, as much as 45°. This could have resulted in a generally warmer climate and kept more of the carbon dioxide from freezing out at the poles. The rise of the Tharsis bulge could have tipped the axis to its present 25° and cooled the climate. Some calculations suggest that Mars goes through cycles similar to Earth's Milankovitch cycles (see Chapter 20) as its rotational inclination fluctuates, and this may cause short-term variations in climate, much like the ice ages on Earth.

The third period in the history of Mars, the **Amazonian period,** extended from about 3 billion years ago to the present and was mostly uneventful. The planet has lost much of its internal heat, and the core no longer generates a global magnetic field. The crust of Mars is too thick to be active with plate tectonics, and consequently there are no folded mountain ranges on Mars resembling the ones on Earth. The huge size of the Martian volcanoes clearly shows that crustal plates have not moved on Mars. Volcanism may still occur occasionally on Mars, but the crust has grown too thick for much geological activity beyond slow erosion by wind-borne dust and the occasional meteorite impact.

Planetary scientists cannot tell the story of Mars in great detail, but it is clear that the size of Mars has influenced both its atmosphere and its geology. Neither small nor large, Mars is a medium-sized world, with characteristics intermediate between small Mercury and Earth's moon on one hand, and large Venus and Earth on the other.

## SCIENTIFIC ARGUMENT

***Why doesn't Mars have coronae like those on Venus?***

This argument is a good opportunity to apply the principles of comparative planetology. The coronae on Venus are caused by rising currents of molten magma in the mantle pushing upward under the crust and then withdrawing to leave the circular scars called coronae. Earth, Venus, and Mars have had significant amounts of internal heat, and there is plenty of evidence that they have had rising convection currents of magma under their crusts. Of course, you wouldn't expect to see coronae on Earth; its surface is rapidly modified by erosion and plate tectonics. Furthermore, the mantle convection on Earth seems to produce plate tectonics rather than coronae. Mars, however, is a smaller world and must have cooled faster. There is no evidence of plate tectonics on Mars, and giant volcanoes suggest rising plumes of magma erupting up through the crust at the same point over and over. Perhaps there are no coronae on Mars because the crust of Mars rapidly grew too thick to deform easily over a rising plume. On the other hand, perhaps you could think of the entire Tharsis bulge as a single giant corona.

Planetary scientists haven't explored enough planets yet to see all the fascinating combinations nature has in store. But it does seem likely that the geology of Mars is typical of medium-size worlds. Of course, Mars is not medium in terms of its location. Of the Terrestrial planets, Mars is the farthest from the sun. Build a scientific argument to analyze that factor. **How has the location of Mars affected the evolution of its atmosphere?**

## 22-3 The Moons of Mars

IF YOU COULD CAMP OVERNIGHT on Mars, you might notice its two small moons, Phobos and Deimos. Phobos, shaped like a flattened loaf of bread measuring 20 km × 23 km × 28 km in size, would appear less than half as large in angular diameter as Earth's full moon. Deimos, only 12 km in diameter and three times farther from Mars, would look only $\frac{1}{15}$ the diameter of Earth's moon.

Both moons are tidally locked to Mars, keeping the same side facing the planet as they orbit. Also, both moons revolve around Mars in the same direction that Mars rotates, but Phobos follows such a small orbit that it revolves faster than Mars rotates. If you camped overnight on Mars, you would see Phobos rise in the west, drift eastward across the sky, and set in the east 6 hours later.

### Origin and Evolution

Deimos and Phobos are typical of the small, rocky moons in our solar system (■ Figure 22-23). Their albedos are only about 0.06, making them look as dark as coal. They have low densities, about 2 g/cm$^3$.

Many of the properties of these moons suggest that they are captured asteroids. In the outer parts of the asteroid belt, almost all asteroids are dark, low-density objects like Phobos and Deimos. Massive Jupiter, orbiting just outside the asteroid belt, can scatter such bodies throughout the solar system, so you should not be surprised if a number of them have encountered Mars, the closest Terrestrial planet to the asteroid belt.

However, capturing a passing asteroid into a closed orbit is not so easy that it can be expected to happen often. An asteroid approaches a planet along a hyperbolic (open) orbit and, if it is unimpeded, swings around the planet and disappears back into space. To change the hyperbolic orbit into a closed orbit, the asteroid must somehow be slowed down as it passes. Tidal forces might do this, but in the case of Mars they would be rather weak. Interactions with other moons or grazing collisions with a thick atmosphere might also slow the asteroid enough so it could be captured.

Both Phobos and Deimos have been photographed by nearby spacecraft, and those photos show that the satellites are heavily cratered. Such cratering could have occurred either while the moons were still in the asteroid belt or while they were in orbit around Mars. In any case, the heavy battering has broken the satellites into irregular chunks of rock, and they cannot pull themselves into smooth spheres because their gravity is too weak to overcome the structural strength of the rock. You will

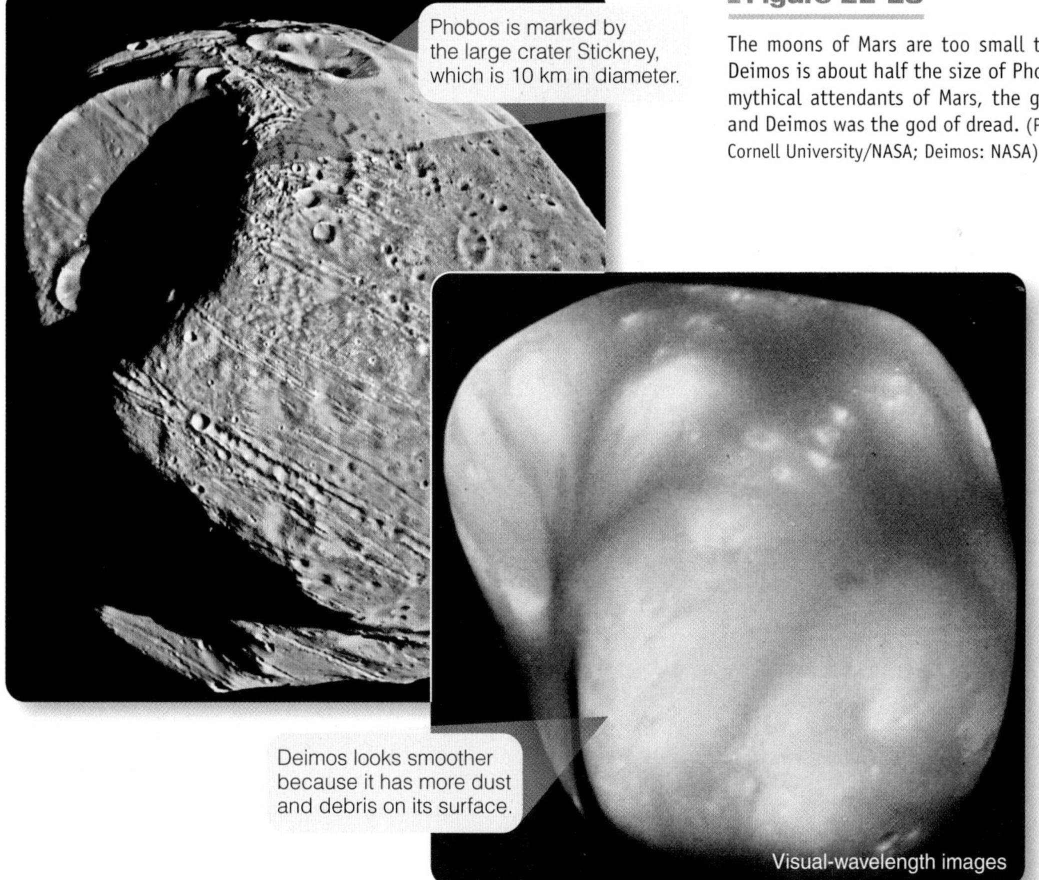
Phobos is marked by the large crater Stickney, which is 10 km in diameter.
Deimos looks smoother because it has more dust and debris on its surface.
Visual-wavelength images

**■ Figure 22-23**

The moons of Mars are too small to pull themselves into spherical shape. Deimos is about half the size of Phobos. The two moons were named for the mythical attendants of Mars, the god of war. Phobos was the god of fear, and Deimos was the god of dread. (Phobos: Damon Simonelli and Joseph Ververka, Cornell University/NASA; Deimos: NASA)

discover in the next chapters that low-mass moons are typically irregular in shape, whereas more massive moons are more spherical.

Images of Phobos reveal a unique set of narrow, parallel grooves (see Figure 22-23). Averaging 150 m wide and 25 m deep, the grooves run from Stickney, the largest crater, to an oddly featureless region on the opposite side of the satellite. One theory suggests that the grooves are deep fractures produced by the impact that formed Stickney. The featureless region opposite Stickney may be similar to the jumbled terrains found on Earth's moon and on Mercury. All these regions were produced by the focusing of seismic waves from a major impact on the far side of the body. High-resolution photographs show that the grooves are lines of pits, suggesting that the pulverized rock material on the surface has drained into the fractures or that gas, liberated by the heat of impact, escaped through the fractures and blew away the dusty soil.

Observations made with the Mars Global Surveyor's infrared spectrometer show that Phobos's surface cools quickly from −4°C to −112°C (from 25°F to −170°F) as it passes from sunlight into the shadow of Mars. Solid rock would retain heat and cool more slowly. To cool as quickly as it does, the dust must be at least a meter deep and very fine. In most photos made by the spacecraft camera, the dust blankets the terrain, but some photos show boulders a few meters in diameter that are thought to be ejecta from impacts.

Deimos looks even smoother than Phobos because of an even thicker layer of dust on its surface (Figure 22-23). This material partially fills craters and covers minor surface irregularities. It seems certain that Deimos experienced collisions in its past, so fractures may be hidden below the debris.

The debris on the surfaces of the moons raises an interesting question. How can the weak gravity of small bodies hold on to fragments from meteorite impacts? Escape velocity on Phobos is only 12 m/s. An athletic astronaut could almost jump into space. Certainly, most fragments from impacts should escape, but some do fall back and accumulate on the surface.

Deimos, smaller than Phobos, has a smaller escape velocity, but it has more debris on its surface because it is farther from Mars. Phobos is close enough to Mars that most ejecta from impacts on Phobos will be drawn into Mars. Deimos, being farther from Mars, is able to keep a larger fraction of its ejecta. Phobos is so close to Mars that tides are making its orbit shrink, and it will fall into Mars or be ripped apart by tidal forces within about 100 million years.

Deimos and Phobos illustrate three principles of comparative planetology that you will find helpful as you explore farther from the sun. First, some satellites are probably captured asteroids. Second, small satellites tend to be irregular in shape and heavily cratered. And third, tidal forces can affect small moons and gradually change their orbits. You will find even stronger tidal effects in Jupiter's satellite system in the next chapter.

### SCIENTIFIC ARGUMENT

***Why would you be surprised if you found volcanism on Phobos or Deimos?***

This is another obvious argument, isn't it? But remember, the purpose of a scientific argument is to test your own understanding, so it is a good way to review. In discussing Earth's moon, Mercury, Venus, Earth, and Mars, you have seen illustrations of the principle that the larger a world is, the more slowly it loses its internal heat. It is the flow of that heat from the interior through the surface into space that drives geological activity such as volcanism and plate motion. A small world, like Earth's moon, cools quickly and remains geologically active for a shorter time than a larger world like Earth. Phobos and Deimos are not just small, they are tiny. However they formed, any interior heat would have leaked away very quickly; with no energy flowing outward, there can be no volcanism.

Some futurists suggest that the first human missions to Mars will not land on the surface of the planet but will build a colony on Phobos or Deimos. These plans are based on speculation that there may be water deep inside the moons that colonists could use. Build an argument based on what you know about water on Mars. **What would happen to water released in the sunlight on the surface of such small worlds?**

### What Are We? Earth-Folk

Space travel isn't easy. We humans made it to the moon, but it took everything we had in the late 1960s. Going back to the moon will be easier next time because the technology will be better, but it will still be expensive and will require people with heroic talent to design, build, and fly the spaceships. Going beyond the moon will be even more difficult.

Going to Mercury or Venus doesn't seem worth the effort. Mercury is barren and dangerous, and the heat and air pressure on Venus may prevent any astronaut from ever visiting its surface. In the next two chapters, you will discover that the Jovian planets and their moons also are not places humans are likely to visit soon. The stars are so far away they may be forever beyond the reach of human spaceships. But Earth has a neighbor.

Astronomically Mars is just up the street, and it isn't such a bad place. You would need a good spacesuit and a pressurized colony to live there, but it isn't impossible. Solar energy and water are abundant. It seems inevitable not only that humans will walk on Mars but that they will someday live there. We Earth-folk have an exciting future. Eventually we will be the Martians.

# Study and Review

## Summary


- Venus is Earth's twin in size but is slightly closer to the sun.
- Venus rotates so slowly that solar heat at the **subsolar point (p. 466)** produces strong atmospheric circulation that circles the planet in only four days.
- The atmosphere of Venus is 95 times thicker than Earth's and composed almost entirely of carbon dioxide.
- Venus is heated by a runaway greenhouse effect. The temperature at the surface of Venus is about 470°C (880°F), hotter than Mercury even though it is farther from the sun.
- The surface of Venus is so hot that compounds have cooked out of the crust to form traces of sulfuric, hydrochloric, and hydrofluoric acids in the atmosphere. The very high clouds on Venus are composed of small droplets of sulfuric acid and sulfur crystals.
- Although it is perpetually hidden below thick clouds, the surface of Venus can be studied by radar mapping, which reveals higher uplands and low rolling plains. Volcanoes, lava flows and channels, and impact craters are detectable. Volcanism is apparently common on Venus, and much of the surface is solidified lava flows. Volcanoes are probably still active on Venus.
- Radar maps can measure altitude, roughness, and in some cases, composition of the surface.
- Landers have analyzed the surface rock and found it to be similar to basalts on Earth.
- **Composite volcanoes (p. 472)** on Earth have steep slopes and are associated with subduction zones. **Shield volcanoes (p. 472)** have shallow slopes and are associated with hot spots. The volcanoes on Venus are shield volcanoes.
- Plate motion across a hot spot produces chains of volcanic peaks. The large size of the volcanic peaks on Venus, and the fact that they are shield volcanoes, shows that the crust is not made of moving plates.
- **Coronae (p. 471)** form where rising currents of magma push the crust up and then withdraw, forming circular faults with associated volcanoes and lava flows.
- Because Venus formed closer to the sun than Earth, it was initially warmer, and whatever oceans it may originally have had were not able to persist and remove carbon dioxide from the atmosphere. The accumulating carbon dioxide produced an intense greenhouse effect, made the planet very hot, and evaporated the remaining water in a runaway greenhouse effect.
- The surface of Venus appears to be about half a billion years old. Planetary scientists suspect that the entire planet was resurfaced by an outpouring of lava.
- Venus has no detectable magnetic field, so its core does not support the dynamo effect, which is a puzzle.
- Venus rotates retrograde (backward). This may have been caused by an off-center impact by a very large planetesimal as the planet was forming or by solar tides raised in its thick atmosphere.
- Mars is smaller than Earth but larger than the moon. It has lost the lower-mass atoms from its atmosphere because of its low escape velocity.
- The atmosphere on Mars is very low-density, consisting mostly of carbon dioxide, and the pressure at the surface is too low to prevent water from boiling away.
- Mars has no magnetic field to protect it from the solar wind, and some of its atmosphere probably has been blasted away over its history by the pressure of the solar wind.
- Although 19th-century astronomers thought they saw networks of canals on Mars, images from spacecraft show that Mars is a dry, desert world on which liquid water does not currently exist.
- The southern hemisphere of Mars is old and heavily cratered, but the northern lowlands are smooth and mostly free of craters.
- Images from spacecraft orbiting Mars reveal **outflow channels (p. 483)** that appear to have been cut by massive floods and **valley networks (p. 483)** that resemble dry riverbeds. Crater counts show that these features are in very old terrain.
- The smooth lowlands of Mars's northern hemisphere may have once contained a liquid-water ocean. Some outflow channels lead into the lowlands, and features resembling shorelines have been found.
- Evidence suggests that whatever water Mars retains is now frozen in the crust as permafrost and as large deposits under the polar caps. Where it seeps out, it can cut gullies and form similar flow features, but rivers, lakes, and oceans cannot now exist on Mars because of the low atmospheric pressure.
- There is evidence that Mars once had a molten core that generated a magnetic field, but it has no detectable magnetic field now. Planetary scientists hypothesize that most of the core has solidified, and the molten outer layer of the core is too small to generate a magnetic field.
- The **Noachian period (p. 487)** extended from the formation of the crust to the end of heavy cratering about 3.7 billion years ago. The valley networks formed during this period, which suggests that the atmosphere was denser then, and water fell as rain or snow.
- The **Hesperian period (p. 488)** began as cratering declined and massive lava flows resurfaced some regions. The climate was colder and the atmosphere thinner, with water frozen in the crust. Massive floods and outflow channels seem to have been produced by sudden melting of subsurface ice.
- The **Amazonian period (p. 488)** extended from about 3 billion years ago to the present and is marked by continued low-rate cratering and erosion by wind, and by small amounts of water seeping from subsurface ice.
- Volcanism has been important throughout the history of Mars, and the Tharsis rise is a huge volcanic uplift. Volcanoes were active in the Noachian period and probably still erupt on Mars.
- Olympus Mons is a very large volcano, but it has not sunk into the crust, and that shows that the crust of Mars is now quite thick.
- The development of the Tharsis rise may have altered the rotation of Mars and changed its climate.
- Mars has captured two asteroids into orbit as moons. Phobos and Deimos are small, irregularly shaped, and cratered. Both are tidally locked to Mars.


## Review Questions

1. Why might you expect Venus and Earth conditions to be similar?
2. What evidence can you cite that Venus and Mars once had more water than at present? Where did that water come from? Where did it go?
3. What features would you look for in high-resolution radar maps of Venus to search for plate tectonics?

4. What evidence shows that Venus has been resurfaced within the last billion years?
5. Why doesn't Mars have mountain ranges like those on Earth? Why doesn't Earth have large volcanoes like those on Mars?
6. What were the canals on Mars? How do they differ from the outflow channels and valley networks on Mars?
7. Propose an explanation for the nearly pure carbon dioxide atmospheres of Venus and Mars. Why is Earth's atmosphere different?
8. What evidence can you cite that the climate on Mars has changed?
9. How can planetary scientists estimate the ages of the outflow channels and valley networks on Mars?
10. Why are Phobos and Deimos nonspherical? Why is Earth's moon much more spherical?
11. **How Do We Know?** How are a weather radar map and an image of a mountain on Venus related?

## Discussion Questions

1. From what you know of Earth, Venus, and Mars, do you expect the volcanoes on Venus and Mars to be active or extinct? Why?
2. If you had a time machine, plus superpowers sufficient to modify or move entire planets, what would you change about Mars, as it was forming, to make its surface environment remain more Earth-like to the present day? How about Venus?
3. Can you make a hypothesis about a single event that might explain both Venus's slow rotation and its geologically recent resurfacing?
4. If humans someday colonize Mars, the biggest problem may be finding water and oxygen. With plenty of solar energy beating down through the thin atmosphere, how might colonizers extract water and oxygen from the Martian environment?

## Problems

1. How long would it take radio signals to travel from Earth to Venus and back if Venus were at its nearest point to Earth? At its farthest point from Earth? (*Hint:* The speed of light is $3.00 \times 10^8$ meters per second.)
2. The Pioneer Venus orbiter circled Venus with a period of 24 hours. What was its average distance above the *surface* of Venus? (*Hint:* See Chapter 5.)
3. Calculate the velocity of Venus in its orbit around the sun. (*Hint:* See Chapter 5.)
4. What is the maximum angular diameter of Venus as seen from Earth? (*Hint:* Use the small-angle formula, Chapter 3.)
5. If the Magellan spacecraft transmitted radio signals down through the clouds on Venus and heard an echo from a certain spot 0.000133 second before the main echo, how high is the spot above the average surface of Venus? (*Hint:* The speed of light is $3.00 \times 10^8$ meters per second.)
6. The smallest feature visible through an Earth-based telescope has an angular diameter of about 1 arc second. If a canal on Mars was just visible when Mars was at its closest to Earth, how wide was the canal? (*Hint:* Use the small-angle formula, Chapter 3.)
7. What is the maximum angular diameter of Phobos as seen from Earth? What surface features should you expect to see from Earth? From the surface of Mars? (*Hints:* Use the small-angle formula, Chapter 3, and use information in the Appendix tables to derive the distances between the objects in question.)
8. What is the maximum angular diameter of Deimos as seen from the surface of Mars? (*Hints:* Use the small-angle formula, Chapter 3, and use information in the Appendix tables to derive the distances between the objects in question.)
9. Deimos is about 12 km in diameter and has a density of 2 g/cm$^3$. What is its mass? (*Hint:* The volume of a sphere is $\frac{4}{3}\pi r^3$.)

## Learning to Look

1. Volcano Sif Mons on Venus is shown in this radar image. What kind of volcano is it, and why is it orange in this image? What color would the rock be if you could see it with your own eyes?

NASA

2. Olympus Mons on Mars is a very large volcano. In this image you can see multiple caldera at the top. What do those caldera and the immense size of the volcano indicate about the geology of Mars?

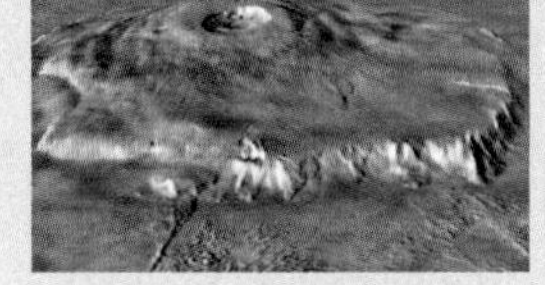

NASA

# 23 Comparative Planetology of Jupiter and Saturn

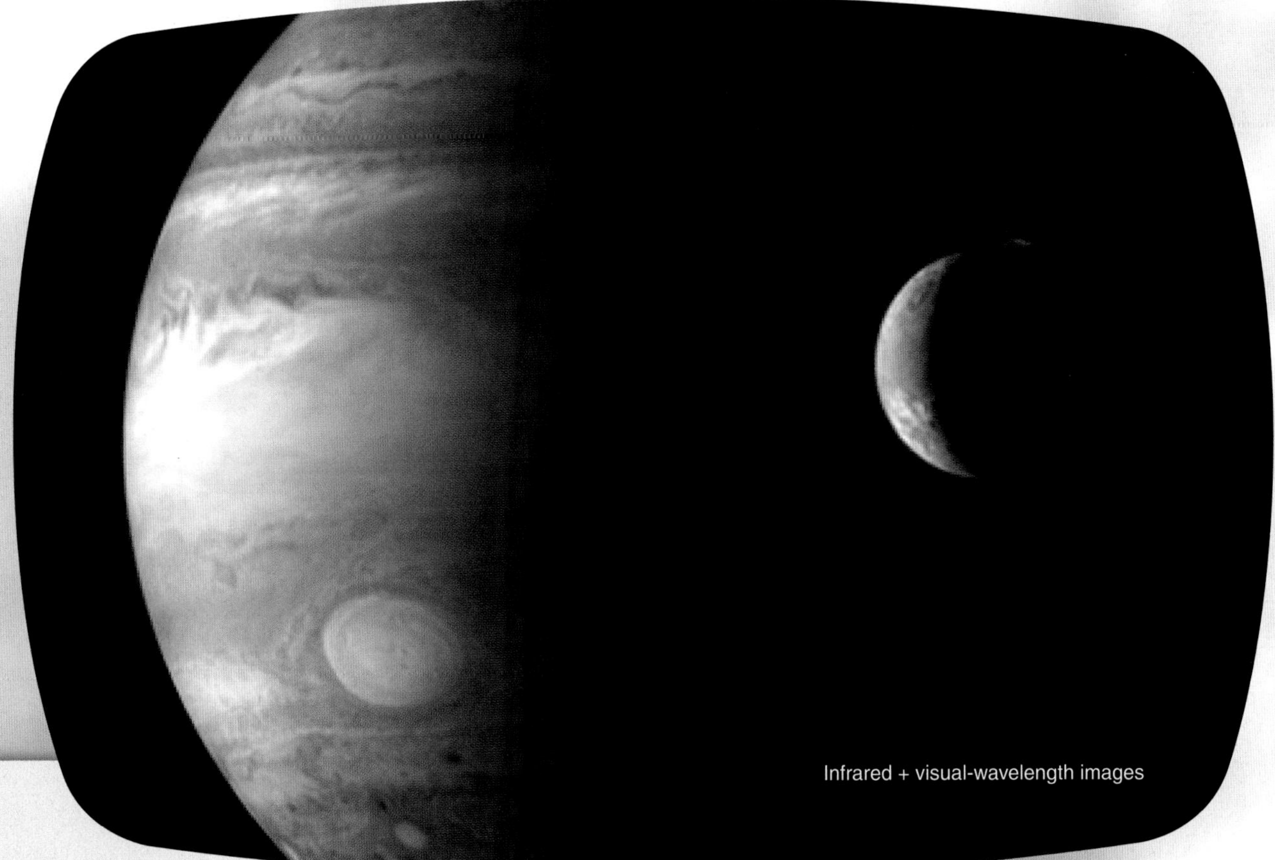


*A montage of images of Jupiter and its volcanic moon Io taken in 2007 by the New Horizons spacecraft during its Jupiter flyby on the way to Pluto and the Kuiper belt. The prominent bluish-white oval in the near infrared image of Jupiter is the Great Red Spot. The Io image combines visual and near infrared data, showing a major eruption in progress on Io's night side at the northern volcano Tvashtar. Incandescent lava glows red beneath a 330-kilometer-high volcanic plume, whose uppermost portions are illuminated by sunlight. The plume appears blue due to scattering of sunlight by small particles in the plume.* (NASA/Johns Hopkins University Applied Physics Laboratory/Southwest Research Institute)

## Guidepost

As you begin this chapter, you leave behind the psychological security of planetary surfaces. You can imagine standing on the moon, on Mars, or even on Venus, but Jupiter and Saturn have no surfaces. Here you face a new challenge—to use comparative planetology to study worlds so unearthly you cannot imagine really being there. On the other hand, Jupiter and Saturn also have extensive systems of moons and rings. Someday humans may walk on some of the moons and watch erupting volcanoes or stroll through methane rain storms, and then journey to the rings and float among the ring particles. As you study these worlds you will find answers to four essential questions:

**How do the outer planets compare with the inner planets?**

**How did Jupiter and Saturn form and evolve?**

**How is Saturn different from Jupiter?**

**How did Jupiter's and Saturn's systems of moons and rings form and evolve?**

After learning about the two largest Jovian planets, in the next chapter you will continue your trip away from the sun and visit their two smaller, and in some ways even stranger, siblings, Uranus and Neptune. It will be interesting, but there is no place like home.

*There is something fascinating about science. One gets such wholesale returns of conjecture out of such a trifling investment of fact.*

— MARK TWAIN, *LIFE ON THE MISSISSIPPI*

WHEN MARK TWAIN WROTE THE SENTENCES that open this chapter, he was poking gentle fun at science, but he was right. The exciting thing about science isn't the so-called facts, the observations in which scientists have greatest confidence. Rather, the excitement lies in the understanding that scientists get by rubbing a few facts together. Science can take you to strange new worlds such as Jupiter and Saturn, and you can get to know them by combining the available observations with known principles of comparative planetology.

## 23-1 A Travel Guide to the Outer Planets

IF YOU TRAVEL MUCH, you know that some cities make you feel at home, and some do not. In this and the next chapter, you will visit worlds that are truly un-Earthly. This travel guide will warn you what to expect.

### The Outer Planets

The outermost planets in our solar system are Jupiter, Saturn, Uranus, and Neptune—the Jovian planets, meaning they resemble Jupiter. ■ Figure 23-1 compares the four outer worlds, and one striking feature, of course, is their sizes. Figure 23-1 shows Earth in scale to the Jovian planets, and it seems tiny in comparison. You can also see that the four Jovian planets can be divided into two pairs, with Jupiter and Saturn being large and nearly the same size, whereas Uranus and Neptune are smaller but very similar in size to each other.

The other feature you will notice immediately when you look at Figure 23-1 is Saturn's rings. They are bright and beautiful and composed of billions of ice particles, each particle following its own orbit around the planet. Astronomers have discovered that Jupiter, Uranus, and Neptune also have rings, but they are not easily detected from Earth and are not visible in this figure. As you visit these worlds in this chapter and the next, you will be able to compare and contrast four giant planets, four moon systems, and four different sets of planetary rings.

■ **Figure 23-1**

The principal worlds of the outer solar system are the four massive but low-density Jovian planets, each much larger than Earth. (NASA/JPL/Space Science Institute/University of Arizona)

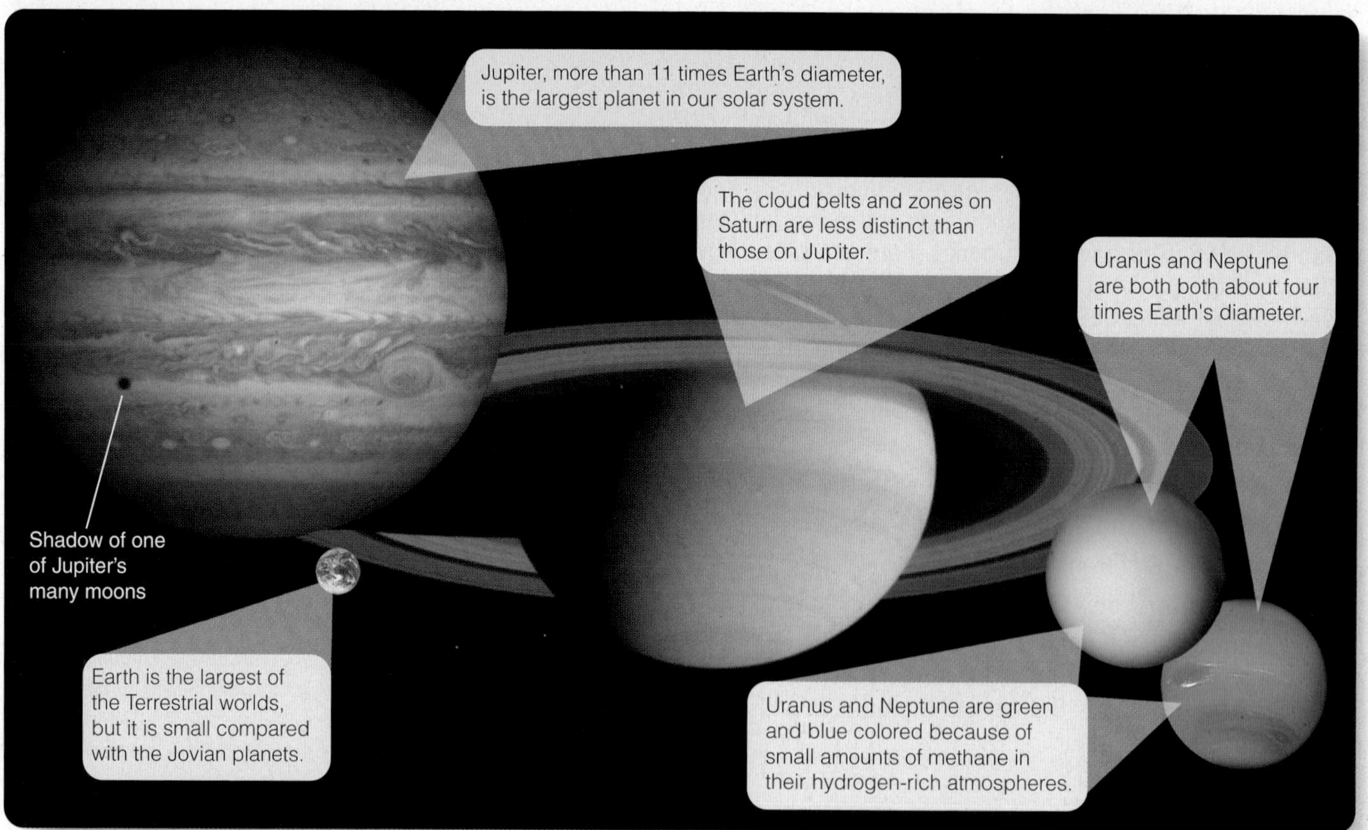

Jupiter, more than 11 times Earth's diameter, is the largest planet in our solar system.
The cloud belts and zones on Saturn are less distinct than those on Jupiter.
Uranus and Neptune are both both about four times Earth's diameter.
Shadow of one of Jupiter's many moons
Earth is the largest of the Terrestrial worlds, but it is small compared with the Jovian planets.
Uranus and Neptune are green and blue colored because of small amounts of methane in their hydrogen-rich atmospheres.

## Atmospheres and Interiors

All the Jovian worlds have hydrogen-rich atmospheres filled with clouds. On Jupiter and Saturn, you can see that the clouds form dark belts and light zones that circle the planets like the stripes on a child's ball. This form of atmospheric circulation is called **belt–zone circulation.** You will find traces of belts and zones on Uranus and Neptune, but they are not very distinct. All the Jovian worlds also have giant circulating storms, the primary example of which is Jupiter's Great Red Spot, which is more than twice the size of Earth. These Jovian storms are comparable to Terrestrial hurricanes, but they can last for centuries. The Great Red Spot storm has been going strong at least since the first telescope observations of Jupiter 400 years ago.

The gaseous atmospheres of the Jovian planets are not very deep. Jupiter's atmosphere makes up only about one percent of its radius. Below that, Jupiter and Saturn are composed of liquid hydrogen, so an older term for these planets, the *gas giants,* reflects a **Common Misconception.** In fact they are made mostly of liquid rather than gas and could more correctly be called the *liquid giants.* Only near their centers could these worlds contain dense material with the composition of rock and metal, but the sizes of these cores are not well known.

Uranus and Neptune are sometimes called the ice giants because they contain a great deal of water, much of which is probably in a solid form. Like Jupiter and Saturn, Uranus and Neptune contain denser material in their cores.

On your visits to the Jovian planets, notice that they are low-density worlds that are rich in hydrogen. Jupiter and Saturn are mostly liquid hydrogen, and even Uranus and Neptune contain a much larger proportion of hydrogen than does Earth. Recall from Chapter 19 that these worlds are hydrogen-rich and low density because they formed in the outer solar nebula where water vapor could freeze to form tremendous amounts of tiny ice particles. These hydrogen-rich ice particles accreted to begin forming the planets; and, once the growing planets became massive enough, they could draw in more hydrogen gas directly by gravitational collapse.

## Satellite Systems

All of the Jovian worlds have large satellite systems. Around each Jovian planet, the moons can be classified into two groups: (1) the **regular satellites,** which tend to be large and orbit in the prograde direction, relatively close to their parent planet, with low inclinations to the planet's equator; versus (2) the **irregular satellites,** which tend to be smaller than the regular satellites, sometimes have retrograde and/or highly inclined orbits, and are generally far from their parent planet. Astronomers have evidence that the regular satellites formed approximately where they are now as the planets formed but that the irregular satellites are mostly, if not all, captured objects.

As you focus on the moons of the Jovian worlds, look for evidence of two processes. The orbits of some moons may have been modified by interactions with other moons, so that they now revolve around their planet in mutual resonances. The same process may allow moons to affect the orbital motions of particles in planetary rings.

The second process allows tides to heat the interiors of some moons and produce geological activity on their surfaces, including volcanoes and lava flows. You have learned that the entire solar system received a heavy bombardment after the planets formed, and heavily cratered surfaces are old, so when you see a section of a moon's surface, or an entire moon, that has few craters, you know that moon must have been geologically active since the end of the heavy bombardment.

### SCIENTIFIC ARGUMENT

***Why do you expect the outer planets to be low-density worlds?***
To build this scientific argument, you need to think about how the planets formed from the solar nebula. In Chapter 19, you discovered that the inner planets could not incorporate ice when they formed because it was too hot near the sun; but, in the outer solar nebula, the growing planets could accumulate lots of ice. Eventually they grew massive enough to grow by gravitational collapse, and that pulled in hydrogen and helium gas. That makes the outer planets low-density worlds.

The outer planets may be unearthly, but they are understandable. For example, extend your argument. **Why do you expect the outer planets to have rings and moons?**

# 23-2 Jupiter

JUPITER IS THE MOST MASSIVE of the Jovian planets, containing more material than all of the other planets combined. This high mass accentuates some processes that are less obvious or nearly absent on the other Jovian worlds. Just as you used Earth as the basis of comparison for your study of the Terrestrial planets, you can examine Jupiter in detail and use it as a standard in your comparative study of the other Jovian planets.

## Surveying Jupiter

Jupiter is extreme because it is big, massive, mostly liquid hydrogen, and very hot inside. The preceding facts are common knowledge among astronomers, but you should demand an explanation of how they know these facts. Often the most interesting thing about a fact isn't the fact itself but how it is known.

At its closest point to Earth, Jupiter is about eight times farther away than Mars, but even a small telescope will reveal that the disk of Jupiter appears more than twice as big as the disk of Mars. If you use the small-angle formula (see Chapter 3), you can compute the diameter of Jupiter—$1.4 \times 10^5$ km, which is about 11 times Earth's diameter (**Celestial Profile 7**).

You can see that Jupiter is massive by watching its moons race around it at high speed. Io is the innermost of the four Galilean moons, and its orbit is just a bit larger than the orbit of our moon around Earth. Io streaks around its orbit in less than two days, whereas Earth's moon takes a month. Jupiter has to be a very massive world to hold on to such a rapidly moving moon (■ Figure 23-2). In fact, you can use the radius of Io's orbit and its orbital period in Newton's version of Kepler's third law (see Chapter 5) to calculate the mass of Jupiter, which is $1.9 \times 10^{27}$ kg, 318 times Earth's mass.

Learning the size and mass of Jupiter is relatively easy, but you might wonder how astronomers know that it is made mostly of hydrogen. The first step is to divide mass by volume to find Jupiter's average density, 1.34 g/cm$^3$. Of course, it is denser at the center and less dense near the surface, but this average density reveals that it can't contain much rock. Rock has a density of 2.5 to 4 g/cm$^3$, so Jupiter must contain material mostly of lower density, such as hydrogen.

Spectra recorded from Earth and from spacecraft visiting Jupiter show that the composition of Jupiter is much like that of the sun—it is mostly hydrogen and helium. This fact was confirmed in 1995 when a probe from the Galileo spacecraft parachuted into the atmosphere and radioed its results back to Earth. Jupiter is mostly hydrogen and helium, with traces of heavier atoms that form molecules such as methane ($CH_4$), ammonia ($NH_3$), and water (■ Table 23-1).

Just as astronomers can build mathematical models of the interiors of stars, they can use the equations that describe gravity, energy, and the compressibility of matter to build mathematical models of the interior of Jupiter. These models reveal that the interior of the planet is mostly liquid hydrogen containing small amounts of heavier elements. The pressure and temperature are higher than the **critical point** for hydrogen, and that means there is no difference between gaseous hydrogen and liquid hydrogen. If you parachuted into Jupiter, you would fall through the gaseous atmosphere and notice the density of the surrounding fluid gradually increasing until you were in a liquid, but you would never splash into a liquid surface.

Roughly a quarter of the way to the center, the pressure is high enough to force the hydrogen to change into **liquid metallic hydrogen,** which is a very good electrical conductor. Because liquid metallic hydrogen has been very difficult to create and study in the laboratory so far, its properties are poorly understood. That is the reason why the models are uncertain about the depth of the transition from normal to metallic liquid hydrogen.

The models are also uncertain about the presence of a heavy element core in Jupiter. The planet contains about 30 Earth masses of elements heavier than helium, but much of that may be suspended in the convectively stirred liquid hydrogen. Measurements by orbiting spacecraft indicate that no more than 10 Earth masses are included in a heavy element core. Some astronomy books refer to this as a rocky core, but, if it exists, it

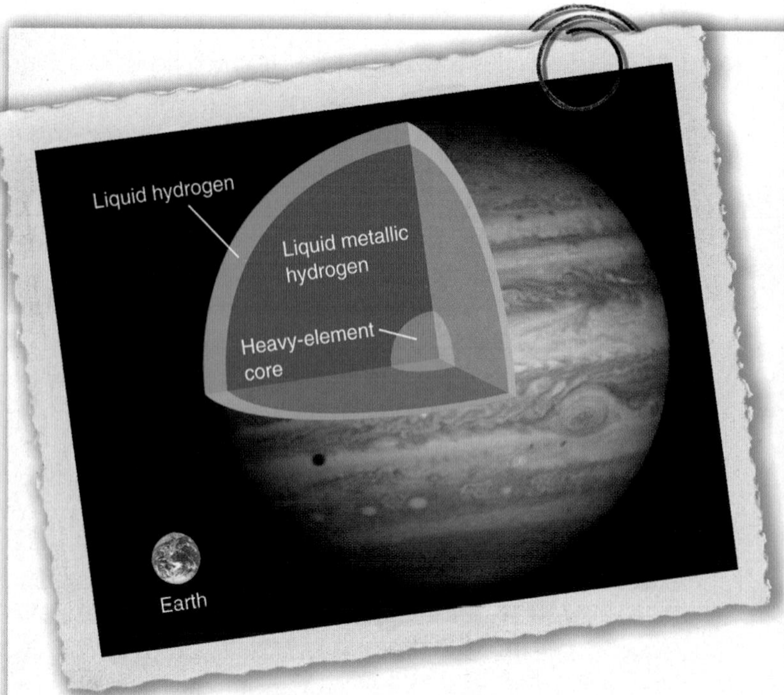


*Jupiter is mostly a liquid planet. It may have a small core of heavy elements not much bigger than Earth.* (NASA/JPL/University of Arizona)

# *Celestial Profile 7: Jupiter*

## *Motion:*

| | |
|---|---|
| Average distance from the sun | 5.20 AU ($7.78 \times 10^8$ km) |
| Eccentricity of orbit | 0.048 |
| Inclination of orbit to ecliptic | 1.3° |
| Average orbital velocity | 13.1 km/s |
| Orbital period | 11.87 y |
| Period of rotation | 9.92 h |
| Inclination of equator to orbit | 3.1° |

## *Characteristics:*

| | |
|---|---|
| Equatorial diameter | $1.43 \times 10^5$ km (11.2 $D_\oplus$) |
| Mass | $1.90 \times 10^{27}$ kg (318 $M_\oplus$) |
| Average density | 1.34 g/cm$^3$ |
| Gravity at base of clouds | 2.54 Earth gravities |
| Escape velocity | 61 km/s (5.4 $V_\oplus$) |
| Temperature at cloud tops | −130°C (−200°F) |
| Albedo | 0.51 |
| Oblateness | 0.064 |

## *Personality Point:*

Jupiter is named for the Roman king of the gods, and it is the largest planet in our solar system. It can be very bright in the night sky, and its cloud belts and four largest moons can be seen through even a small telescope. Its moons are visible even with a good pair of binoculars mounted on a tripod or braced against a wall.

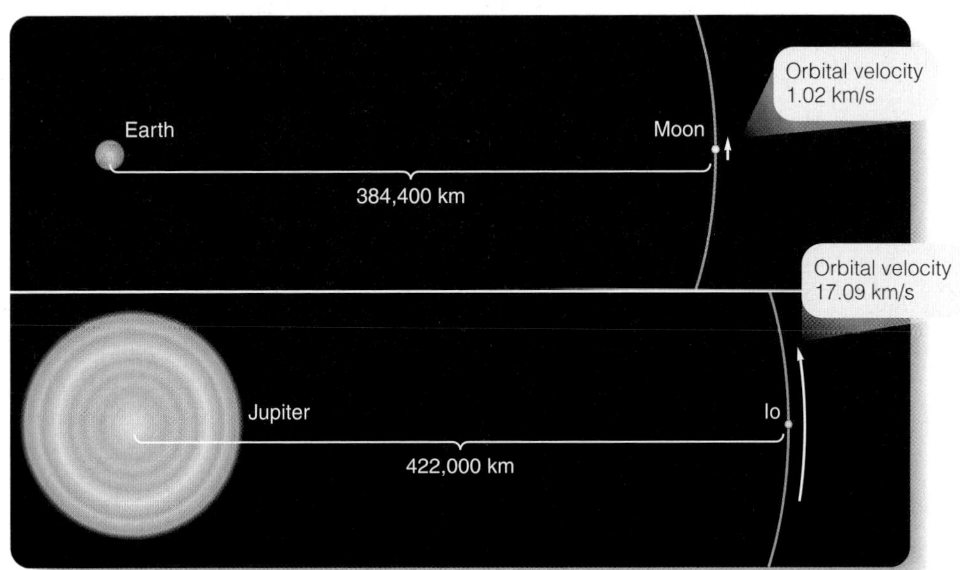


■ **Figure 23-2**

It is obvious that Jupiter is a very massive planet when you compare the motion of Jupiter's moon Io with Earth's moon. Although Io is 10 percent farther from Jupiter, it travels 17 times faster in its orbit than does Earth's moon around Earth. Clearly, Jupiter's gravitational field is much stronger than Earth's, and that means Jupiter must be very massive.

■ **Table 23-1 | Composition of Jupiter and Saturn (by Number of Molecules)**

| Molecule | Jupiter (%) | Saturn (%) |
|---|---|---|
| $H_2$ | 86 | 93 |
| He | 13 | 5 |
| $H_2O$ | 0.1 | 0.1 |
| $CH_4$ | 0.1 | 0.2 |
| $NH_3$ | 0.02 | 0.01 |

cannot be anything like the rock you know on Earth. The center of Jupiter is five or six times hotter than the surface of the sun and is prevented from exploding into vapor only by the tremendous pressure. If there is a core, it is "rocky" only in the sense that it contains heavy elements.

How do astronomers know Jupiter is hot inside? Infrared observations show that Jupiter is glowing strongly in the infrared, radiating 1.7 times more energy than it receives from the sun. That observation, combined with models of its interior, provides an estimate of its internal temperature. You will learn later in this chapter that Saturn resembles Jupiter in this respect.

You can tell that Jupiter is mostly a liquid just by looking at it. If you measure a photograph of Jupiter, you will discover that it is slightly flattened; it is a bit over 6 percent larger in diameter through its equator than through its poles. This is referred to as Jupiter's **oblateness.** The amount of flattening depends on the speed of rotation and on the rigidity of the planet. Jupiter's flattened shape shows that the planet cannot be as rigid as a Terrestrial planet and must have a liquid interior.

Basic observations and the known laws of physics can tell you a great deal about Jupiter's interior. Its vast magnetic field can tell you even more.

## Jupiter's Magnetic Field

As early as the 1950s, astronomers detected radio noise coming from Jupiter and recognized it as synchrotron radiation. That form of radio energy is produced by fast electrons spiraling in a magnetic field, so it was obvious that Jupiter had a magnetic field.

In 1973 and 1974, two Pioneer spacecraft flew past Jupiter, followed in 1979 by two Voyager spacecraft. Those probes found that Jupiter has a magnetic field about 14 times stronger than the Earth's field. Evidently the field is produced by the dynamo effect operating in the highly conductive liquid metallic hydrogen as it is circulated by convection and spun by the rapid rotation of the planet. This powerful magnetic field dominates a huge magnetosphere around the planet. Compare the size of Jupiter's field with that of Earth in ■ Figure 23-3.

Jupiter's magnetic field deflects the solar wind and traps high-energy particles in radiation belts much more intense than Earth's. The radiation is more intense because Jupiter's magnetic field is stronger and can trap and hold more particles, and higher-energy particles, than Earth's field can. The spacecraft passing through the radiation belts received radiation doses equivalent to a billion chest X-rays—at least 100 times the lethal dose for a human. Some of the electronics on the spacecraft were damaged by the radiation.

You will recall that Earth's magnetosphere interacts with the solar wind to produce auroras, and the same process occurs on Jupiter. Charged particles in the magnetosphere leak downward along the magnetic field, and, where they enter the atmosphere,

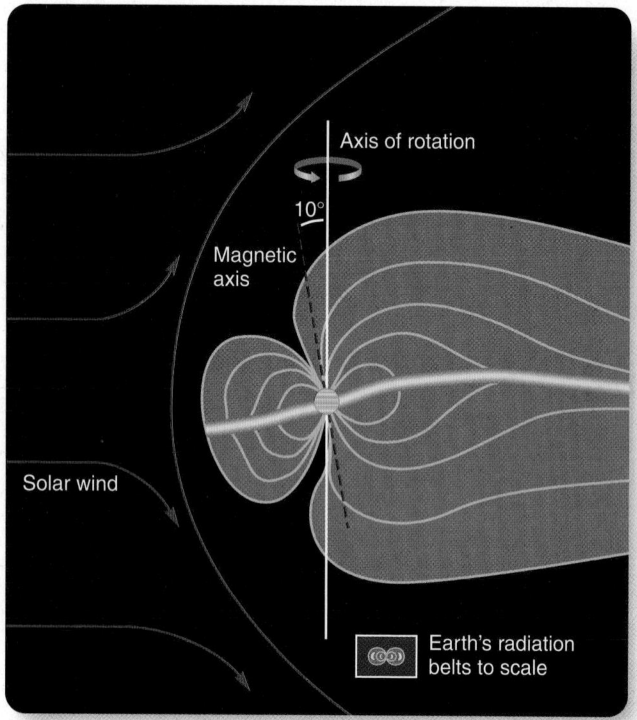


**■ Figure 23-3**

Jupiter's magnetic field is large and powerful. It traps particles from the solar wind to form powerful radiation belts. The rapid rotation of the planet forces the slightly inclined magnetic field to wobble up and down as the planet rotates. Earth's magnetosphere and radiation belts are shown to scale.

they produce auroras 1000 times more luminous than those on Earth. The auroras on Jupiter, like those on Earth, occur in rings around the magnetic poles (■ Figure 23-4).

The four Galilean moons of Jupiter orbit inside the magnetosphere, and some of the heavier ions in the radiation belts come from the innermost moon, Io. As you will see later in this chapter, Io has active volcanoes that spew gas and ash. Because Io orbits with a period of 1.8 days, and Jupiter's magnetic field rotates in only 10 hours, the wobbling magnetic field rushes past Io at high speed, sweeping up stray particles, accelerating them to high energy, and spreading them around Io's orbit in a doughnut of ionized gas called the **Io plasma torus.**

Jupiter's magnetic field interacts with Io to produce a powerful electric current (about a million amperes) that flows through a curving path called the **Io flux tube** from Jupiter out to Io and back to Jupiter. Small spots of bright auroras lie at the two points where the Io flux tube enters Jupiter's atmosphere (Figure 23-4).

Fluctuations in the auroras reveal that the solar wind buffets Jupiter's magnetosphere, but some of the fluctuations seem to be caused by changes in the magnetic dynamo deep inside the planet. In this way, studies of the auroras on Jupiter may help astronomers learn more about its liquid depths.

Jupiter's powerful magnetic field is invisible to your eyes, but the swirling cloud belts are beautifully visible in their complexity.

## Jupiter's Atmosphere

As you learned earlier, Jupiter is a liquid world that has no surface. The gaseous atmosphere blends gradually with the liquid hydrogen interior. Below the clouds of Jupiter lies the largest ocean in the solar system—an ocean that has no surface and no waves.

When you look at Jupiter, all you see are clouds. You can detect a nearly transparent hydrogen and helium atmosphere above the cloud tops by noticing that Jupiter has limb darkening, just as the sun does (see Chapter 8). When you look near the limb of Jupiter (the edge of its disk), the clouds are much dimmer (look back at Figure 23-1) because it is nearly sunset or sunrise along the limb. If you were on Jupiter at that location, you would see the sun just above the horizon, and sunlight arriving there would be dimmed by passing through the planet's atmosphere. In addition, sunlight reflected from the clouds must travel back out through the atmosphere at a steep angle to reach Earth, dimming the light further. Jupiter is brighter near the center of the disk because the sunlight shines nearly straight down on the clouds.

Study **Jupiter's Atmosphere** on pages 500–501 and notice four important ideas:

1. The atmosphere is hydrogen-rich, and the clouds are confined to a shallow layer.
2. The cloud layers lie at certain temperatures within the atmosphere where ammonia ($NH_3$), ammonium hydrosulfide ($NH_4SH$), and water ($H_2O$) can condense to form ice particles.
3. The belt–zone circulation is driven by high- and low-pressure areas related to those on Earth.
4. Finally, the major spots on Jupiter, although they are only circulating storms, can remain stable for decades or even centuries.

Circulation in Jupiter's atmosphere is not totally understood. Observations made by the Cassini spacecraft as it raced past Jupiter on its way to Saturn revealed that the dark belts, which were thought to be entirely regions of sinking gas, contain small rising storm systems too small to have been seen in images by previous probes. Evidently, the general circulation usually attributed to the belts and zones is much more complex when it is observed in more detail. Further understanding of the small-scale motions in Jupiter's atmosphere may have to await future planetary probes.

The highly complex spacecraft that have visited Jupiter are examples of how technology can give scientists the raw data they need to form their understanding of nature. Science is about understanding nature, and Jupiter is an entirely new kind of planet in your study. In fact, Jupiter has another feature that you did not find anywhere among the Terrestrial planets. Jupiter has a ring.

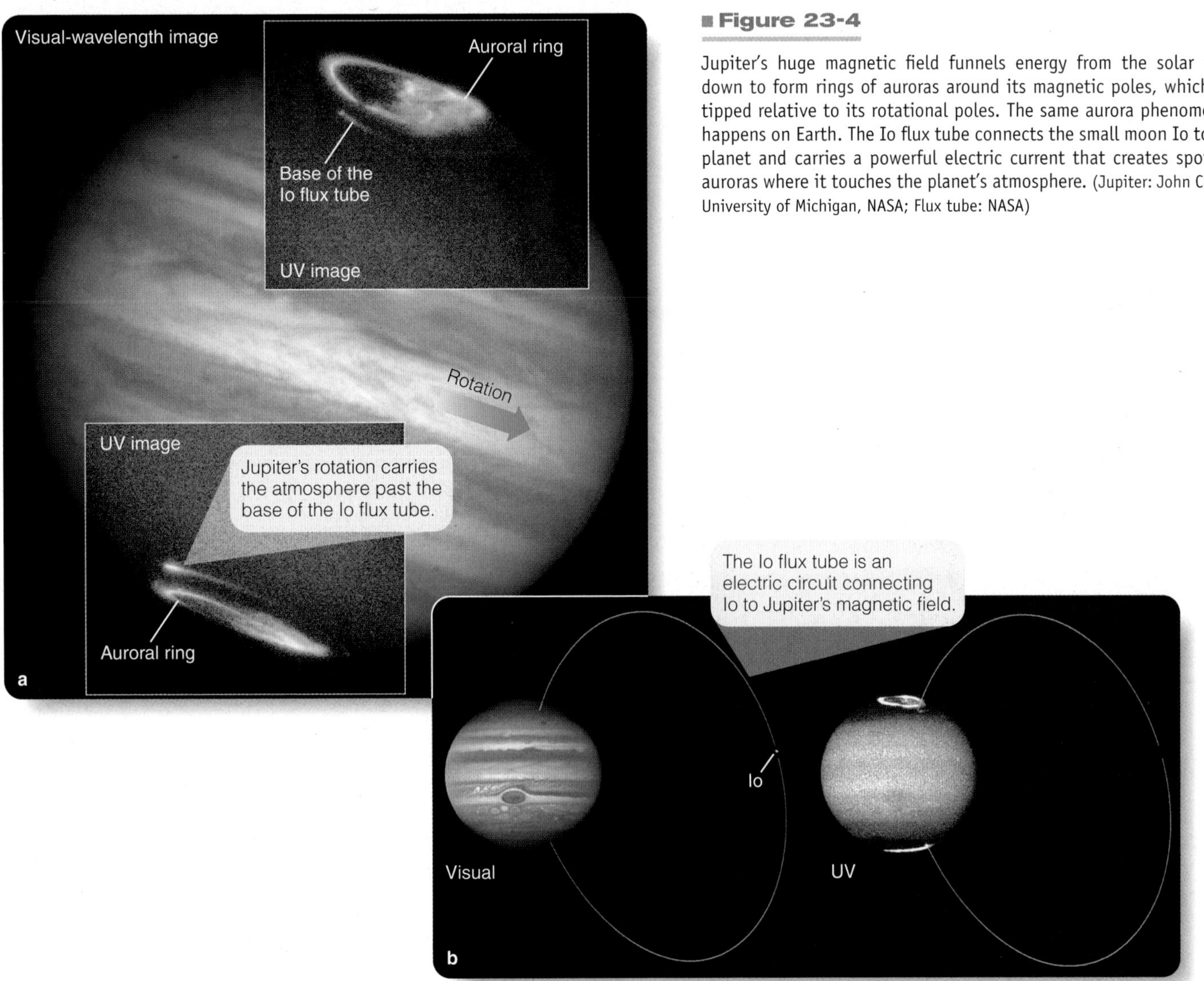


■ **Figure 23-4**

Jupiter's huge magnetic field funnels energy from the solar wind down to form rings of auroras around its magnetic poles, which are tipped relative to its rotational poles. The same aurora phenomenon happens on Earth. The Io flux tube connects the small moon Io to the planet and carries a powerful electric current that creates spots of auroras where it touches the planet's atmosphere. (Jupiter: John Clarke, University of Michigan, NASA; Flux tube: NASA)

## Jupiter's Ring

Astronomers have known for centuries that Saturn has rings, but it was not until 1979 when the Voyager 1 spacecraft sent back photos that Jupiter's ring was discovered. The discovery was confirmed later by difficult ground-based measurements. Less than one percent as bright as Saturn's rings, the ghostly ring around Jupiter is a puzzle. What is it made of? Why is it there? A few simple observations will help you solve some of these puzzles.

Saturn's rings are made of bright ice chunks, but the particles in Jupiter's ring are very dark and reddish. This is evidence that its ring is rocky rather than icy. You can also conclude that the ring particles are mostly microscopic. Photos show that Jupiter's ring is very bright when illuminated from behind (■ Figure 23-5). In other words, it is scattering light forward. Efficient **forward scattering** occurs when particles have diameters roughly the same as the wavelength of light, a few millionths of a meter. Large particles do not scatter light forward, so a ring filled with basketball-size particles would look dark when illuminated from behind. The forward scattering tells you that Jupiter's ring is made mostly of particles about the size of those in smoke.

Larger particles are not entirely ruled out. A sparse component of rocky objects ranging from pieces of gravel to boulders is possible, but objects larger than 1 km would have been detected in spacecraft photos. The vast majority of the ring particles are microscopic dust.

The size of the ring particles is a clue to their origin, and so is their location. They orbit inside the **Roche limit,** the distance from a planet within which a moon cannot hold itself together by its own gravity. If a moon orbits relatively far from its planet, then the moon's gravity will be much greater than the tidal forces caused by the planet, and the moon will be able to hold itself together. If, however, a planet's moon comes inside the Roche limit, the tidal forces can overcome its gravity and pull the moon apart. The International Space Station can orbit inside Earth's Roche limit because it is held together by bolts and welds, and a single large rock can survive inside the Roche limit if it is strong

# Jupiter's Atmosphere

**1** Humans will probably never visit Jupiter's atmosphere. Its cloud layers are deathly cold, and the deeper layers that are warmer have a crushingly high pressure. There is no free oxygen to breathe; the gases are roughly three-quarters hydrogen and a quarter helium, plus small amounts of water vapor, methane, ammonia, and similar molecules. Traces of sulfur and molecules containing sulfur probably make it smell bad. Of course, Jupiter has no surface, so there isn't even a place to stand.

Belts are dark bands of clouds.

Zones are bright bands of clouds.

Shadow of Europa

Jupiter's moon Europa

NASA

NASA/JPL/Univ. of AZ

**1a** The only spacecraft to enter Jupiter's atmosphere was the Galileo probe. Released from the Galileo spacecraft, the probe entered Jupiter's atmosphere in 1995. It parachuted through the upper atmosphere of clear hydrogen, released its heat shield, and then fell through Jupiter's stormy atmosphere until it was crushed by the increasing pressure.

Lightning bolts are common in Jupiter's turbulent clouds.
Artist Concept
Jupiter's atmosphere is a very thin layer of turbulent gas above the liquid interior. It makes up only about 1 percent of the radius of the planet.
NASA/JPL, Univ. of AZ


Hughes Aircraft Co

The Great Red Spot at right is a giant circulating storm in one of the southern zones. It has lasted at least 300 years since astronomers first noticed it after the invention of the telescope. Smaller spots are also circulating storms.

2 The visible clouds on Jupiter are composed of ammonia crystals, but models predict that deeper layers of clouds contain ammonia hydrosulfide crystals, and deeper still lies a cloud layer of water droplets. These compounds are normally white, so planetary scientists think the colors arise from small amounts of other molecules formed in reactions powered by lightning or sunlight.

If you could put thermometers in Jupiter's atmosphere at different levels, you would discover that the temperature rises below the uppermost clouds.

Far below the clouds, the temperature and pressure climb so high the gaseous atmosphere merges gradually with the liquid hydrogen interior and there is no surface.

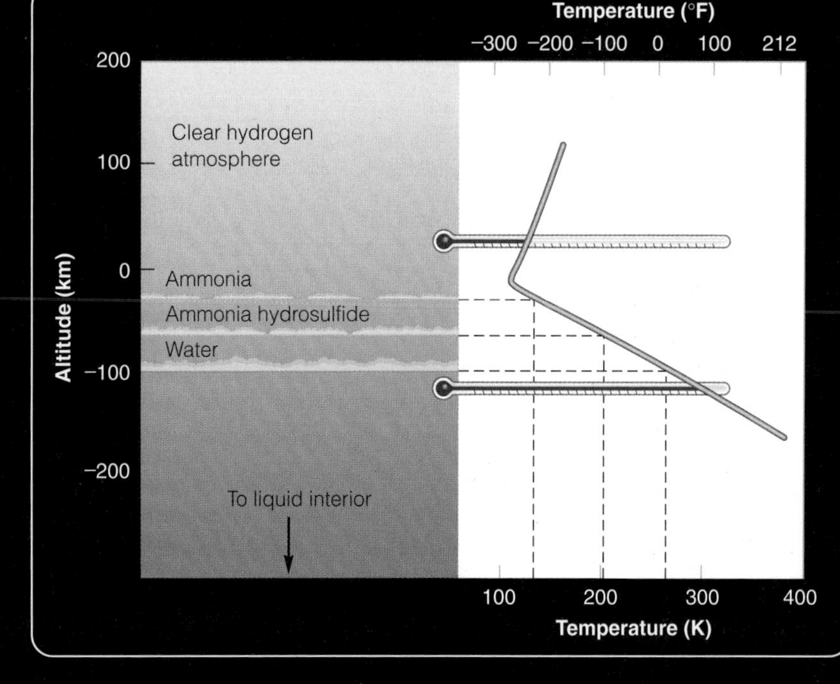


3 On Earth, the temperature difference between the poles and equator drives a wave-shaped high-speed wind that organizes the high- and low-pressure areas into cyclonic circulations familiar from weather maps.

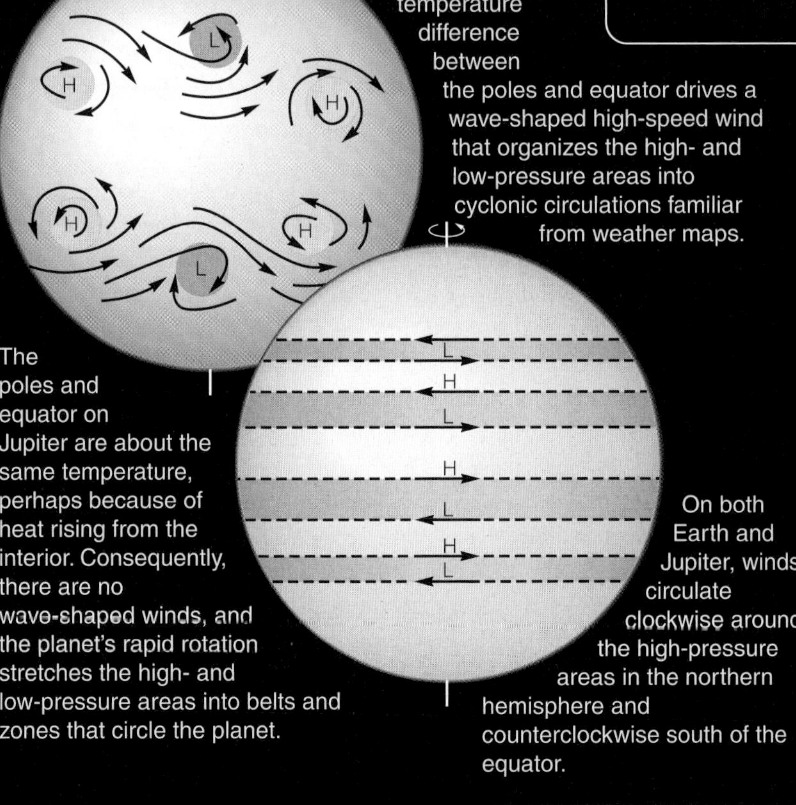


The poles and equator on Jupiter are about the same temperature, perhaps because of heat rising from the interior. Consequently, there are no wave-shaped winds, and the planet's rapid rotation stretches the high- and low-pressure areas into belts and zones that circle the planet.

On both Earth and Jupiter, winds circulate clockwise around the high-pressure areas in the northern hemisphere and counterclockwise south of the equator.

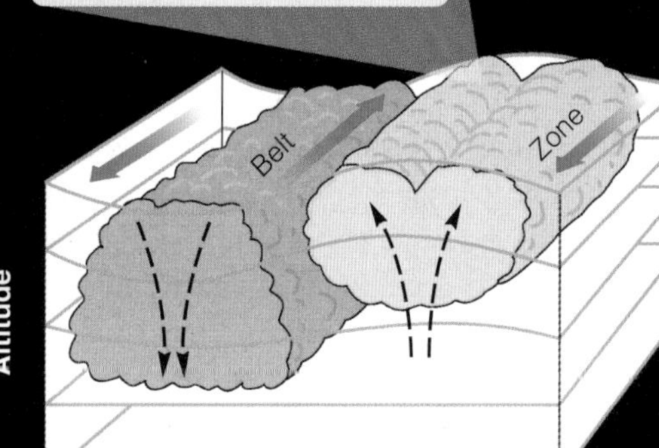

Zones are brighter than belts because rising gas forms clouds high in the atmosphere, where sunlight is strong.
Belt
Zone
Altitude
North
Equator

4 Three circulating storms visible as white ovals since the 1930s merged in 1998 to form a single white oval. In 2006, the storm intensified and turned red like the Great Red Spot. The reason for the red color is unknown, but it may show that the storm is bringing material up from lower in the atmosphere.

Storms in Jupiter's atmosphere may be stable for decades or centuries, but astronomers had never before witnessed the appearance of a new red spot. It may eventually vanish or develop further. Even the Great Red Spot may someday vanish.

Great Red Spot

Red Jr.

Enhanced visible + infrared

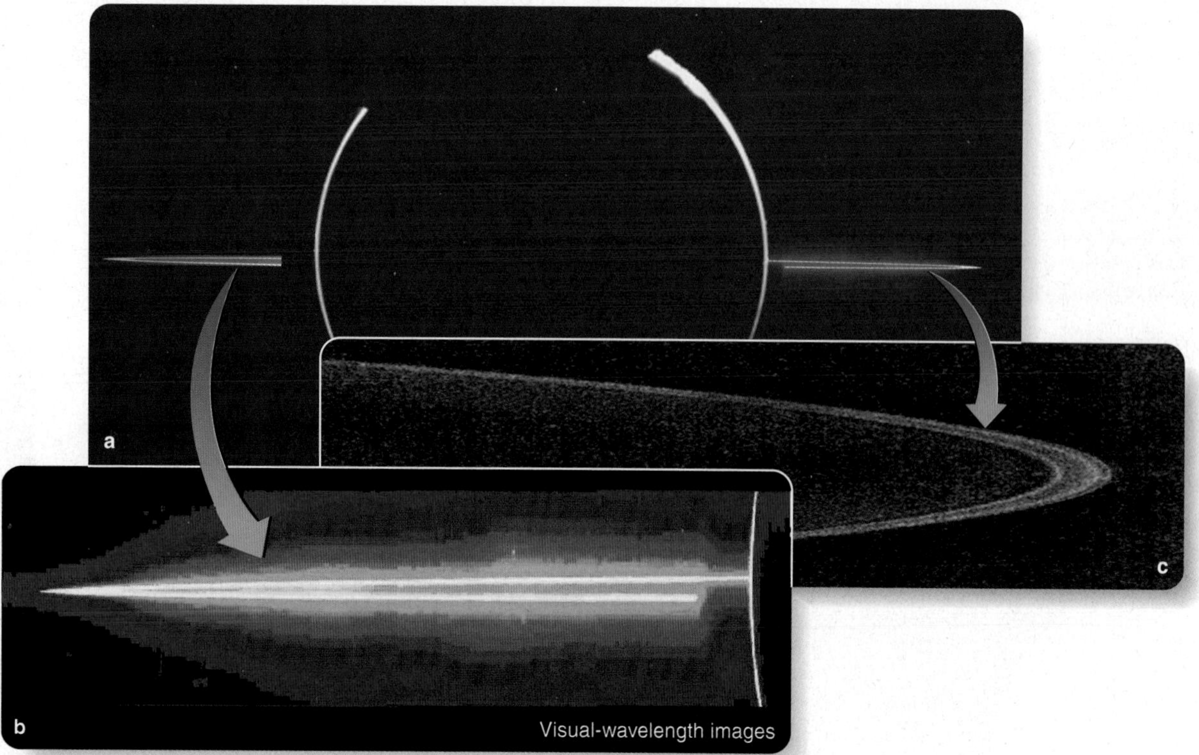


**Figure 23-5**

(a) The main ring of Jupiter, illuminated from behind, glows brightly in this visual-wavelength image made by the Galileo spacecraft while it was within Jupiter's shadow. (b) Digital enhancement and false color reveal the halo of ring particles that extends above and below the main ring. The halo is just visible in panel a. (c) Structure in the ring is probably caused by the gravitational influence of Jupiter's inner moons. (a. and b.: NASA; c. NASA/Johns Hopkins University Applied Physics Lab/Southwest Research Institute)

enough not to break. However, a moon composed of separate rocks and particles held together by their mutual gravity could not survive inside a planet's Roche limit. Tidal forces would destroy such a moon. If a planet and its moon have the same average densities, the Roche limit is at 2.44 times the planet's radius. Jupiter's main ring has an outer radius of 130,000 km (1.8 Jupiter radii) and lies inside Jupiter's Roche limit. The rings of Saturn, Uranus, and Neptune also lie within those planets' respective Roche limits.

Now you can understand the dust in Jupiter's ring. If a dust speck gets knocked loose from a larger rock orbiting inside the Roche limit, the rock's gravity cannot hold the dust speck. And the billions of dust specks in the ring can't pull themselves together to make a larger body—a moon—because of the tidal forces inside the Roche limit.

You can also be sure that the ring particles are not old. The pressure of sunlight and Jupiter's powerful magnetic field alter the orbits of the particles, and they gradually spiral into the planet. Images show faint ring material extending down toward Jupiter's cloud tops, and this is evidently dust specks spiraling inward. Dust is also lost from the ring as electromagnetic effects force it out of the plane of the ring to form a low-density halo above and below the ring (Figure 23-5b). Yet another reason the ring particles can't be old is that the intense radiation around Jupiter can grind dust specks down to nothing in a century or so. For all these reasons, the rings seen today can't be made up of material that has been in the form of small particles for the entire time since the formation of Jupiter.

Obviously, the rings of Jupiter must be continuously resupplied with new material. Dust particles can be chipped off rocks ranging in size from gravel to boulders within the ring, and small moons that orbit near the outer edge of the rings lose particles as they are hit by meteorite impacts. Observations made by the Galileo spacecraft show that the main ring is densest at its outer edge, where the small moon Adrastea orbits, and that another small moon, Metis, orbits inside the ring. Clearly these moons must be structurally strong to withstand Jupiter's tidal forces. Images from the Voyager and Galileo probes also reveal much fainter rings, called the **gossamer rings,** extending twice as far from the planet as the main ring. These gossamer rings are most dense at the orbits of two small moons, Amalthea and Thebe, more evidence that ring particles are being blasted into space by impacts on the moons.

Besides supplying the rings with particles, the moons help confine the ring particles and keep them from spreading outward. You will find that this is an important process in planetary rings when you study the rings of Saturn later in this chapter.

Your exploration of Jupiter reveals that it is much more than just a big planet. It is the gravitational and magnetic center of an entire community of objects. Occasionally the community suffers an intruder.

## Giant Impacts in the Outer Solar System

Comets are very common in the solar system, and Jupiter probably is hit by comets more often than most planets due to its strong gravity. No one had ever seen it happen until 1994, when fragments from a comet disrupted by Jupiter's tidal forces a few years earlier looped back and smashed into the planet. Those impacts are covered more fully, including images, in Chapter 25 within the context of a discussion of the effects on Earth of comet and asteroid impacts.

The comet collision with Jupiter was an astonishing spectacle, but what can it tell you about Jupiter? In fact, the separate impacts were revealing in two ways. First, astronomers used the impacts as probes of Jupiter's atmosphere. By making assumptions about the nature of Jupiter's atmosphere and by using supercomputers, astronomers created models of a high-velocity projectile penetrating into Jupiter's upper atmosphere. By comparing the observed impacts with the models, astronomers were able to fine-tune the models to better represent Jupiter's atmosphere. This method of comparing models with reality is a critical part of science.

Second, the spectacle is a reminder that asteroids and comets, the leftover construction debris from the formation of the solar system, are expected to hit all the planets and their moons occasionally. In 2009, Australian amateur astronomer Anthony Wesley discovered a dark spot that appeared suddenly on Jupiter, and alerted the world scientific community. The spot is apparently the scar of a new impact by an asteroid or comet. Astronomers estimate that Jupiter probably is hit by an object larger than 1 km several times per century. With no solid surface, Jupiter itself does not display its history of impacts, but its moons do. The fact that Jupiter's moons are observed to have radically different crater counts is a clue that their geologic histories have differed.

## The History of Jupiter

Your goal in studying any planet is to be able to tell its story—to describe how it got to be the way it is. While you can understand part of the story of Jupiter, there is still much to learn.

If the solar nebula theory for the origin of the solar system is correct, then Jupiter formed from the colder gases of the outer solar nebula, where ices of water and other molecules were able to condense. Thus, Jupiter grew rapidly and became massive enough to capture hydrogen and helium gas from the solar nebula and form a deep liquid hydrogen envelope. Models are uncertain as to whether a heavy element core survives; it may have been mixed in with the convecting liquid hydrogen envelope, and astronomers estimate that the mass of Jupiter's heavy element core is between 0 and 10 Earth masses.

In the interior of Jupiter, hydrogen exists as liquid metallic hydrogen, a very good electrical conductor. The planet's rapid rotation, coupled with the outward flow of heat from its hot interior, drives a dynamo effect that produces a powerful magnetic field. That vast magnetic field traps high-energy particles from the solar wind to form intense radiation belts and auroras.

The rapid rotation and large size of Jupiter cause belt–zone circulation in its atmosphere. Heat flowing upward from the interior causes rising currents in the bright zones, and cooler gas sinks in the dark belts. As on Earth, winds blow at the margins of these regions, and large spots appear to be cyclonic disturbances. Internal heat has been escaping since Jupiter formed, so you can guess that Jupiter's atmospheric circulation and storms were stronger in the distant past and will diminish in the future.

Although the age of planet building is long past, debris in the form of meteorites and occasional comets continue to hit Jupiter, as it does all the planets. Any debris left over from the formation of Jupiter would have been blown away by sunlight and the solar wind or destroyed by other processes long ago, so the dust trapped in Jupiter's thin ring must be young. It probably comes from meteorites hitting and eroding the innermost moons.

Your study of Jupiter has been challenging because it is so unlike Earth. Most of the features and processes you found on the Terrestrial planets are missing on Jupiter, but, as the prototype of the Jovian worlds, it earns its place as the ruler of the solar system.

### SCIENTIFIC ARGUMENT

***How do astronomers know Jupiter is hot inside?***

A scientific argument is a way to test ideas, and sometimes it is helpful to test even the most basic ideas. You know that something is hot if you touch it and it burns your fingers, but you can't touch Jupiter. You also know something is hot if it is glowing bright red—it is red hot. But Jupiter is not glowing red hot. You can tell that something is hot if you can feel heat when you hold your hand near it. That is, you can detect infrared radiation with your skin. In the case of Jupiter, you would need greater sensitivity than the back of your hand, but infrared telescopes reveal that Jupiter is a source of infrared radiation; it is glowing in the infrared. Sunlight would warm Jupiter a little bit, but it is emitting 70 percent more infrared than it should. That means it must be hot inside. From models of the interior, astronomers conclude that the center must be five or six times hotter than the surface of the sun to make the surface of the planet glow as much as it does in the infrared.

Astronomical understanding is usually based on simple observations, so build an argument to answer the following simple question. **How do astronomers know that Jupiter has a low density?**

## 23-3 Jupiter's Family of Moons

How many moons does Jupiter have? Astronomers are finding many small moons, and the count is now over 60. (You will have to check the Internet to get the latest figure because more moons are discovered every year.) Most of these moons are small and

■ **Figure 23-6**

The Gallilean satellites of Jupiter from left to right are Io, Europa, Ganymede, and Callisto. The circle shows the size of Earth's moon. (NASA)

rocky, and many are probably captured asteroids. Four of the moons, those discovered by Galileo and now called the **Galilean satellites,** are large and have interesting geologies (■ Figure 23-6).

Your study of the moons of Jupiter will illustrate three important principles in comparative planetology. First, a body's composition depends on the temperature of the material from which it formed. This is illustrated by the prevalence of ice as a building material in the outer solar system, where sunlight is weak. You are already familiar with the second principle: that cratering can reveal the age of a surface. Also, as you have seen in your study of the Terrestrial planets, internal heat has a powerful influence over the geology of these larger moons.

## Callisto: An Ancient Surface

The outermost of Jupiter's four large moons, Callisto, is half again as large in diameter as Earth's moon. Like all of Jupiter's larger satellites, Callisto is tidally locked to its planet, keeping the same side forever facing Jupiter. From its gravitational influence on passing spacecraft, astronomers can calculate Callisto's mass, and dividing that mass by its volume shows that its density is 1.8 g/cm$^3$. Ice has a density of about 1 and rock 2.5 to 4 g/cm$^3$, so Callisto must be a mixture of rock and ice.

Images from the Voyager and Galileo spacecraft show that the surface of Callisto is dark, dirty ice heavily pocked with craters (■ Figure 23-7). Old, icy surfaces in the solar system become dark because of dust added by meteorites and because meteorite impacts vaporize water, leaving any dust and rock in the ice behind to form a dirty crust. You may have seen the same thing happen to a city snowbank. As the snow evaporates over a few days, the crud in the snow is left behind to form a dirty rind. Break through that dirty surface, and the snow is much cleaner underneath.

Spectra of Callisto's surface show that in most places it is a 50:50 mix of ice and rock, but some areas are ice free. Nevertheless, the slumped shapes of craters suggests that the outer 10 km of this moon is mostly frozen water; ice isn't very strong, so big piles of it tend to slump under their own weight. The disagreement between the spectra and the shapes of craters can be understood when you recall that the spectra contain information about only the outer 1 mm of the surface, which can be quite dirty, while the shapes of craters tell you about the outermost 10 km, which appear to be rich in ice.

Delicate measurements of the shape of Callisto's gravitational field were made by the Galileo spacecraft as it flew by. Those measurements show that Callisto has never fully differentiated to form a dense core and a lower-density mantle. Its interior is a mixture of rock and ice rather than having distinct layers of different composition. This is consistent with the observation that Callisto has only a weak magnetic field of its own. A strong magnetic field could be generated by the dynamo effect in a liquid convecting core, and Callisto has no core. It does, however,

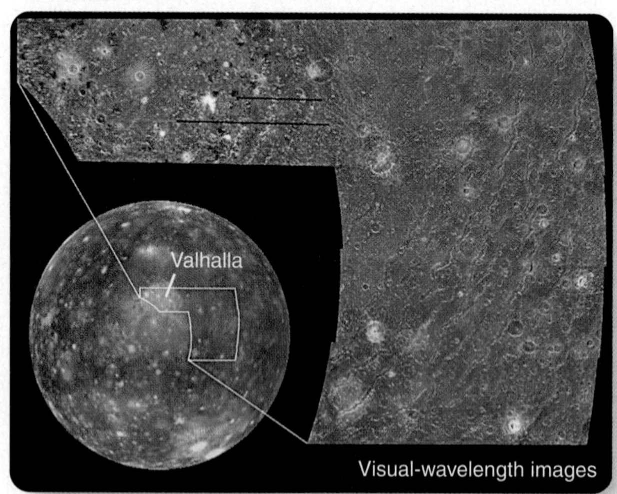


■ **Figure 23-7**

The dark surface of Callisto is dirty ice marked by craters in these images. The youngest craters look bright because they have dug down to cleaner ice. Valhalla is the 4000 km (2500 mi) diameter scar of a giant impact feature, one of the largest in the solar system. Valhalla is so large and old that the icy crust has flowed back to partially heal itself, and the outer rings of Valhalla are shallow troughs marking fractures in the crust. (NASA)

interact with Jupiter's magnetic field in a way that suggests it has a layer of liquid water roughly 10 km thick about 100 km below its icy surface. Slow radioactive decay in Callisto's interior may produce enough heat to keep this layer of water from freezing.

## Ganymede: An Obscure Past

The next Galilean moon inward is Ganymede, larger than Mercury, and over three-quarters the diameter of Mars. In fact, Ganymede is the largest moon in the solar system. Its density is 1.9 g/cm$^3$, and its influence on the Galileo spacecraft reveals that it is differentiated into a rocky core, an ice-rich mantle, and a crust of ice 500 km thick. It may even have a small inner iron core. Ganymede is large enough for radioactive decay to have melted its interior when it formed, allowing rock and iron to sink to its center.

Ganymede's surface hints at an active past. Although a third of the surface is old, dark, and cratered like Callisto's, the rest is marked by bright parallel grooves. Because this bright **grooved terrain** (■ Figure 23-8a) contains fewer craters, it must be younger.

Observations show that the bright terrain was produced when the icy crust broke and water flooded up from below and froze. As the surface broke over and over, sets of parallel groves were formed. Some low-lying regions are smooth and appear to have been flooded by water. Spectra reveal concentrations of salts such as those that would be left behind by the evaporation of mineral-rich water. Also, some features in or near the bright terrain appear to be caldera formed when subsurface water drained away and the surface collapsed (Figure 23-8b).

The Galileo spacecraft found that Ganymede has a magnetic field about 10 percent as strong as Earth's. It even has its own magnetosphere inside the larger magnetosphere of Jupiter. Mathematical models calculated by planetary scientists do not predict that a magnetic field this strong should arise from the dynamo effect in a liquid water mantle layer with the size and location of the one in Ganymede, and there does not appear to be enough heat in Ganymede for it to have a molten metallic core. Thus, the cause of Ganymede's unique magnetic field remains a puzzle. One hypothesis is that the magnetic field is left over and frozen into the rock from a time when Ganymede was hotter and more active.

Ganymede's magnetic field fluctuates with the 10-hour period of Jupiter's rotation. The rotation of the planet sweeps its tilted magnetic field past the moon, and the two fields interact. That interaction reveals that the moon has a layer of liquid water about 170 km (110 mi) below its surface. The data indicate that the water layer is about 5 km (3 mi) thick. It is possible that the water layer was thicker and closer to the surface long ago when the interior of the moon was warmer. That might explain the flooding that appears to have formed the bright grooved terrain.

Ganymede orbits rather close to massive Jupiter, and that exposes it to two unusual processes that many worlds never experience. **Tidal heating,** the frictional heating of a body by changing tides (■ Figure 23-9a), could have heated Ganymede's interior and added to the heat generated by radioactive decay. In its current nearly circular orbit, this moon experiences little or no tidal heating. But, at some point in the past, interactions with the

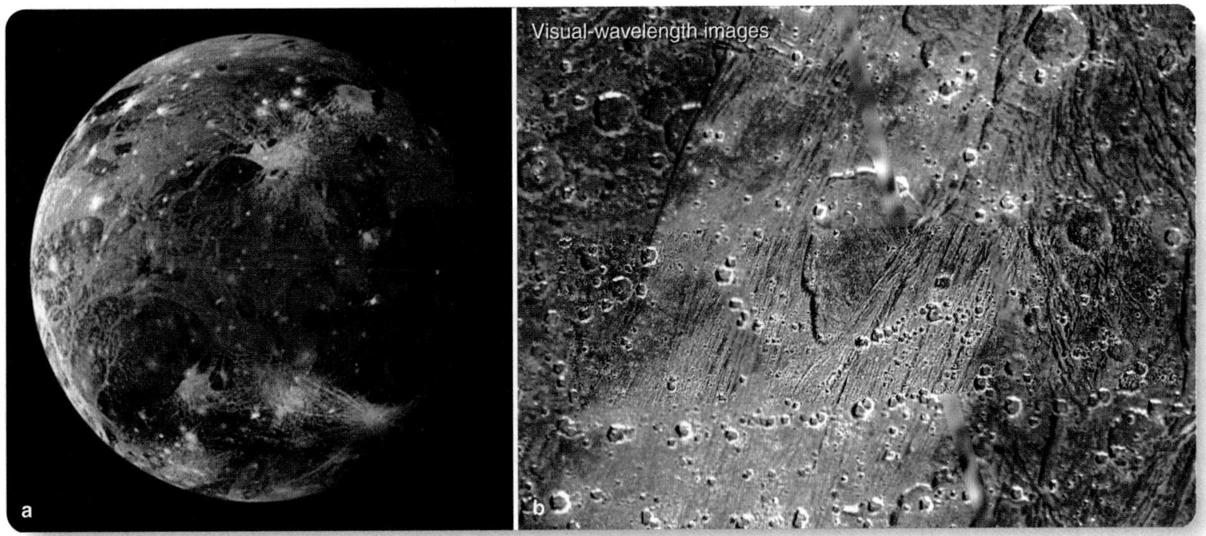


■ **Figure 23-8**

(a) This color-enhanced image of Ganymede shows the frosty poles at top and bottom, the old dark terrain, and the brighter grooved terrain. (b) A band of bright terrain runs from lower left to upper right, and a collapsed area, a possible caldera, lies at the center in this visual-wavelength image. Calderas form where subsurface liquid has drained away, and the bright areas do contain other features probably due to flooding by water. (NASA)

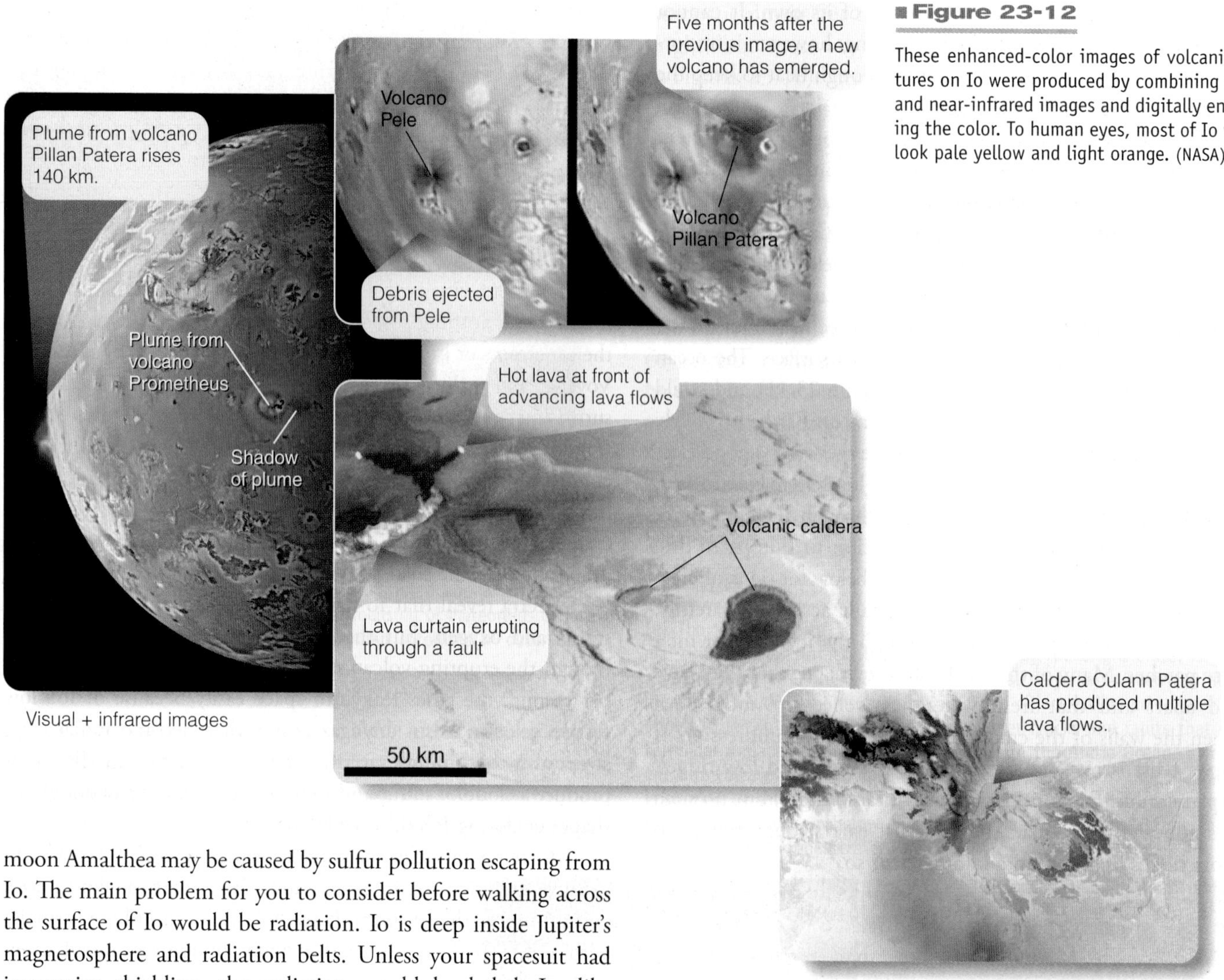

**Figure 23-12**

These enhanced-color images of volcanic features on Io were produced by combining visual and near-infrared images and digitally enhancing the color. To human eyes, most of Io would look pale yellow and light orange. (NASA)



moon Amalthea may be caused by sulfur pollution escaping from Io. The main problem for you to consider before walking across the surface of Io would be radiation. Io is deep inside Jupiter's magnetosphere and radiation belts. Unless your spacesuit had impressive shielding, the radiation would be lethal. Io, like Venus, may be a place that humans will not visit easily.

You can use basic observations to deduce the nature of Io's interior. From its density, 3.5 g/cm$^3$, you can conclude that it is rocky. Spectra reveal no trace of water at all, so there is no ice on Io. In fact, it is the driest world in our solar system. The oblateness of Io caused by its rotation and by the slight distortion produced by Jupiter's gravity gives astronomers more clues to its interior. Model calculations suggest it contains a modest core of iron or iron mixed with sulfur, a deep rocky mantle that is partially molten, and a thin, rocky crust.

The colors of Io have been compared to those of a badly made pizza. The reds, oranges, and browns of Io are caused by sulfur and sulfur compounds, and an early hypothesis proposed that the crust is mostly sulfur. New evidence says otherwise. Infrared measurements show that volcanoes on Io erupt lava with a temperature over 1500°C (2700°F), about 300°C hotter than lavas on Earth. Sulfur on Io would boil at only 550°C, so the volcanoes must be erupting molten rock and not just liquid sulfur. Also, a few isolated mountains exist that are as high as 18 km, twice the height of Mount Everest. Sulfur is not strong enough to support such high mountains. These are all indications that the crust of Io is probably silicate rock.

Volcanism is continuous on Io. Plumes come and go over periods of months, but some volcanic vents, such as Pele, have been active since the Voyager spacecraft first visited Io in 1979 (Figure 23-12). Earth's explosive volcanoes eject lava and ash because of water dissolved in the lava. As rising lava reaches Earth's surface, the sudden decrease in pressure allows the water to come out of solution in the lava, like popping the cork on a bottle of champagne: The water flashes into vapor and blasts material out of the volcano, the process that was responsible for the Mount St. Helens explosion in 1980. But Io is dry. Instead, its volcanoes appear to be powered by sulfur dioxide dissolved in the magma. When the pressure on the magma is released, the sulfur dioxide boils out of solution and blasts gas and ash high above the surface in plumes up to 500 km high. Ash falling back to the surface produces debris layers around the volcanoes, such as that around Pele in Figure 23-12. Whitish areas on the surface are frosts of sulfur dioxide.

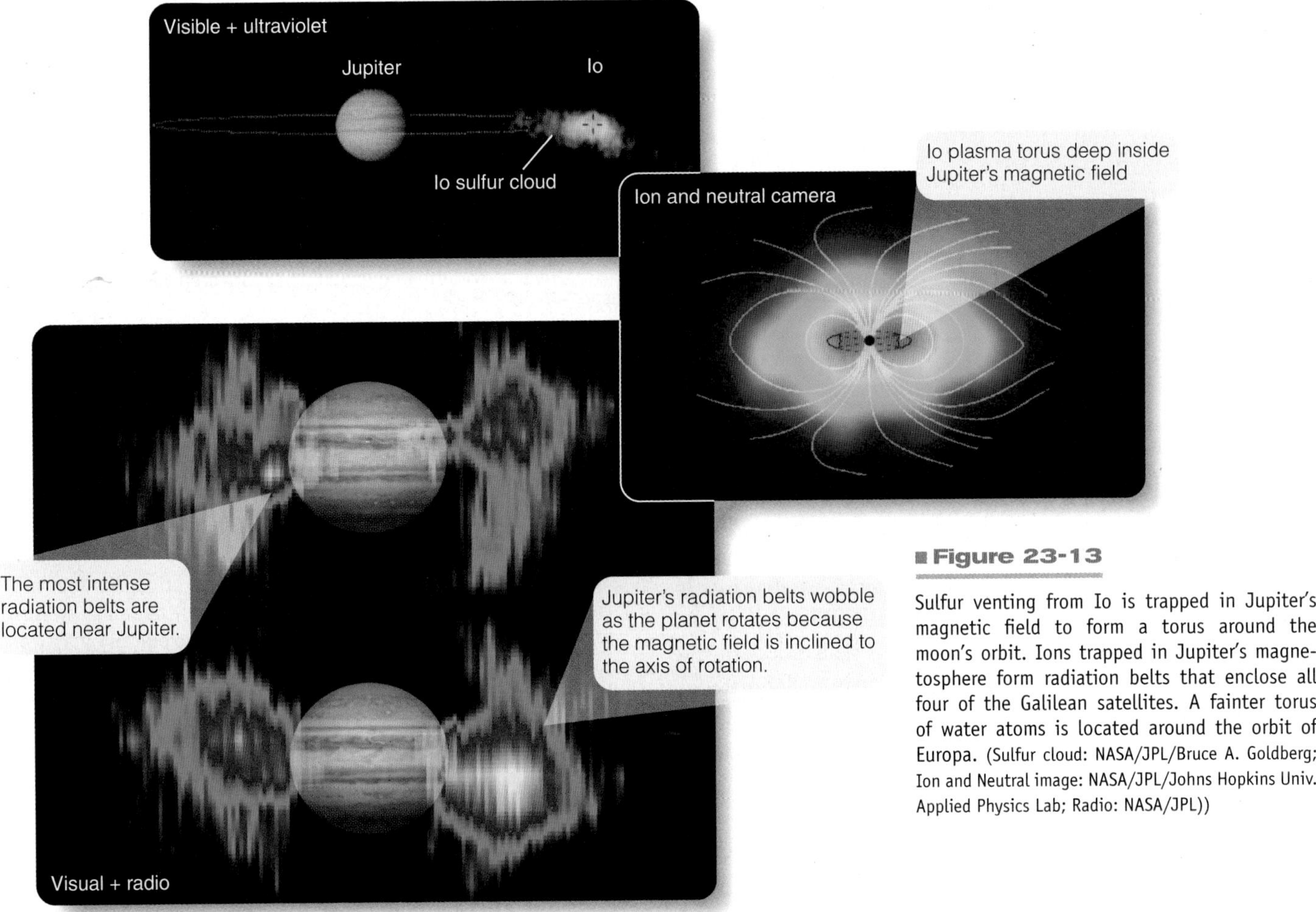
Visible + ultraviolet
Jupiter
Io
Io sulfur cloud
Io plasma torus deep inside Jupiter's magnetic field
Ion and neutral camera
The most intense radiation belts are located near Jupiter.
Jupiter's radiation belts wobble as the planet rotates because the magnetic field is inclined to the axis of rotation.
Visual + radio

**■ Figure 23-13**

Sulfur venting from Io is trapped in Jupiter's magnetic field to form a torus around the moon's orbit. Ions trapped in Jupiter's magnetosphere form radiation belts that enclose all four of the Galilean satellites. A fainter torus of water atoms is located around the orbit of Europa. (Sulfur cloud: NASA/JPL/Bruce A. Goldberg; Ion and Neutral image: NASA/JPL/Johns Hopkins Univ. Applied Physics Lab; Radio: NASA/JPL))

Great lava flows can be detected carrying molten material downhill, burying the surface under layer after layer. Sometimes lava bursts upward through faults to form long lava curtains, a form of eruption seen in Hawaii. Both of these processes are shown in Figure 23-12.

What powers Io? It has abundant internal heat, but it is only 5 percent bigger than Earth's moon, which is cold and dead. Io is too small to have retained heat from its formation or to remain hot from radioactive decay. In fact, the energy blasting out of its volcanoes adds up to about three times more energy than it could make by radioactive decay in its interior.

The answer is that Io is heated by a stronger version of the kind of tidal heating that has affected Ganymede and Europa. Because Io is so close to Jupiter, the tides it experiences are powerful and should have forced Io's orbit to become circular long ago. Io, however, is strongly influenced by its neighboring moons. Io, Europa, and Ganymede are locked in an orbital resonance; in the time it takes Ganymede to orbit once, Europa orbits twice and Io four times. This gravitational interaction keeps the orbits, especially Io's, slightly eccentric; and Io, also being closest to Jupiter, suffers dramatic tides, with its surface rising and falling by about 100 m. For comparison, tides on Earth move the solid ground by only a few centimeters. The resulting friction in Io is enough to melt the interior and drive volcanism. In fact, there is enough energy flowing outward to continually recycle Io's crust. Deep layers melt, are spewed out through the volcanoes to cover the surface, and are later covered themselves until they are buried so deeply that they are again melted.

The four Galilean moons show a clear sequence of more and more tidal heating the nearer they are to Jupiter. The more distant moons have geologies dominated by impacts, while the closer moons are dominated by heat flow from inside and have few craters. What a difference a few hundred thousand miles makes!

## The History of the Galilean Moons

Each time you have finished studying a world, you have tried to summarize its history. Now you have studied a system of four small worlds. Can you tell their story? To do that you need to draw on what you have learned about the moons and on what you have learned about Jupiter and the origin of the solar system (Chapter 19).

The minor, irregular moons of Jupiter are probably captured asteroids, but the regular Galilean satellites seem to be primordial. That is, they formed with Jupiter. Also, they seem to be

■ Table 23-2 | The Galilean Satellites*

| Name | Radius (km) | Density (g/cm³) | Orbital Period (days) |
|---|---|---|---|
| Io | 1821 | 3.528 | 1.769 |
| Europa | 1561 | 3.014 | 3.551 |
| Ganymede | 2631 | 1.942 | 7.155 |
| Callisto | 2410 | 1.8344 | 16.689 |

*For comparison, the radius of Earth's moon is 1738 km, and its density is 3.36 g/cm³

interrelated in that their densities are related to their distance from Jupiter (■ Table 23-2).

From all the evidence, astronomers propose that the four moons formed in a disk-shaped nebula around Jupiter—a mini-solar nebula—in much the same way the planets formed from the solar nebula around the sun. As Jupiter grew massive, it would have formed a hot, dense disk of matter around its equator. The moons could have condensed inside that disk with the innermost moons, Io and Europa, forming from rocky material and the outer moons, Ganymede and Callisto, incorporating more ice. This hypothesis follows the same condensation sequence that led to rocky planets forming near the sun and ice-rich worlds forming farther away.

There are objections to this hypothesis. The disk around Jupiter would have been dense and hot, and moons would have formed rapidly, perhaps in only 1000 years. If the moons formed quickly, the heat of formation released as material fell into the moons would not have leaked away quickly, and they would have grown so hot they would have lost their water. Ganymede and Callisto are rich in water. Furthermore, Callisto has never been hot enough to differentiate and form a core. Also, mathematical models show that moons orbiting in the dense disk would have swept up debris and lost orbital momentum; they would have spiraled into Jupiter within a century.

A newer hypothesis proposes that Jupiter's early disk was indeed dense and hot and may have created moons, but those moons spiraled into the planet and were lost. Only later, as the disk grew thinner and cooler, did the present Galilean moons begin to form. Additional material may have dribbled slowly into the disk, and the moons could have formed slowly enough to retain their water and avoid spiraling into Jupiter. In this scenario, many large moons may have accreted around Jupiter. The Galilean satellites you observe now would thus be only the last batch, formed when the disk of construction material around Jupiter had become thin enough that they did not fall into Jupiter and disappear. The same migration and destruction process may apply, on a larger scale, to planets forming in extrasolar planetary systems (see Chapter 19).

You can combine this hypothesis with what you know about tidal heating to understand the interiors of the moons. The moons formed slowly, over perhaps 100,000 years, and were not heated severely by in-falling material. The inner moons, however, were cooked by tidal heating—possibly enhanced when an orbital resonance developed between Ganymede, Europa, and Io. The innermost moon, Io, was heated so much it lost all of its water, and Europa retained only a small amount. Ganymede was heated enough to differentiate but retained much of its water. Callisto, orbiting far from Jupiter and avoiding orbital resonances, was never heated enough to differentiate. The Galilean moons as they appear today seem to be the result of a combination of slow formation and tidal heating.

The Galilean satellite system is full of clues to the history of the solar system. Understanding that history prepares you to explore farther from the sun.

### SCIENTIFIC ARGUMENT

***What produces Io's internal heat?***

Scientific arguments commonly draw on basic principles that are well understood. In this case, you understand that small worlds lose their internal heat quickly and become geologically inactive. Io is only slightly larger than Earth's moon, which is cold and dead, but Io is full of energy flowing outward.

Clearly, Io must have a powerful source of heat inside, and that heat source is tides. Io's orbit is slightly eccentric (noncircular), so it is sometimes closer to Jupiter and sometimes farther away. This means that Jupiter's powerful gravity sometimes squeezes Io more than at other times, and the flexing of the little moon's interior produces heat through friction. Such tides would rapidly force Io's orbit to become circular, and then tidal heating would end and the planet would become inactive—except that the gravitational tugs of the other moons keep Io's orbit eccentric. Thus, it is the influence of its companions that keeps Io in such an active state.

Io has almost no impact craters, but Callisto has many. Build a new scientific argument drawing on a different principle. **What does the difference in crater distributions on the Galilean satellites tell you about their history?**

## 23-4 Saturn

SATURN HAS PLAYED SECOND FIDDLE to its own rings since Galileo first saw it in 1610. He didn't recognize the rings for what they are, but today they are instantly recognizable as one of the wonders of the solar system. Nevertheless, Saturn itself, not quite 10 times Earth's diameter (**Celestial Profile 8**), is a fascinating planet with a few mysteries of its own. Your exploration of Saturn and its rings can make use of the principles you have learned from Jupiter.

### Surveying Saturn

The basic characteristics of Saturn reveal its composition and interior. Only about a third of the mass of Jupiter and 16 percent smaller in radius, Saturn has an average density of 0.69 g/cm³. It is less dense than water—it would float! Spectra show that its atmosphere is rich in hydrogen and helium (see Table 23-1), and models predict that it is mostly liquid hydrogen with a core of heavy elements.

Infrared observations show that Saturn is radiating 1.8 times as much energy as it receives from the sun, showing that heat is flowing out of its interior. It must be very hot inside Saturn, as is also true for Jupiter. In fact, Saturn's interior is too hot. It should have lost more heat since it formed. Astronomers have calculated models indicating that helium in the liquid hydrogen interior is condensing into droplets and falling inward. The falling droplets, releasing energy as they pick up speed, heat the planet. This heating is similar to the heating produced when a star contracts and may also occur to some extent in the atmospheres of Jupiter, Uranus, and Neptune.

Just as for Jupiter, you can learn more about Saturn's interior from its magnetic field. Spacecraft have found that Saturn's magnetic field is about 20 times weaker than Jupiter's. It also has correspondingly weaker radiation belts. Models comparing Saturn with Jupiter predict that the lower pressure inside Saturn produces a smaller mass of liquid metallic hydrogen. Heat flowing outward causes convection in this conducting layer, and the rapid rotation drives a dynamo effect that produces the magnetic field. Unlike most magnetic fields, Saturn's is not inclined to its axis of rotation, something you can see in ultraviolet images that show rings of auroras around Saturn's poles (■ Figure 23-14). This perfect alignment between Saturn's magnetic axis and the axis of rotation is peculiar, isn't seen in any other planet, and isn't understood.

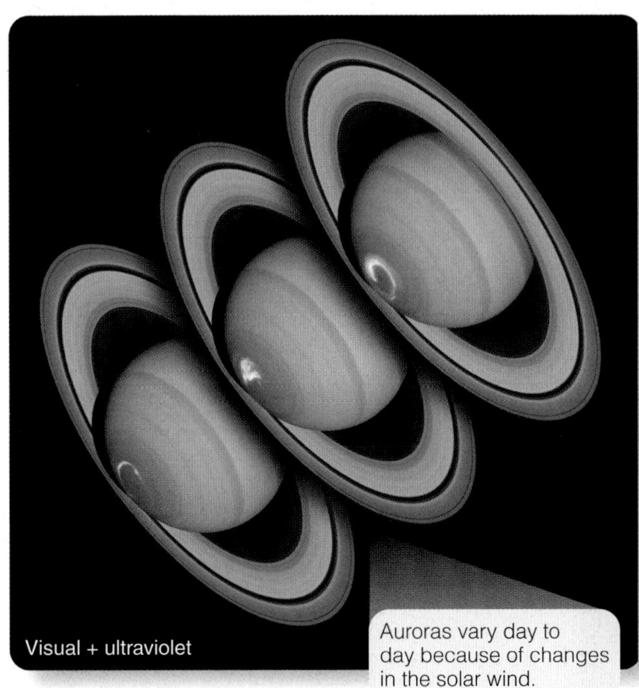

Visual + ultraviolet
Auroras vary day to day because of changes in the solar wind.

**■ Figure 23-14**

Auroras on Saturn occur in rings around the planet's magnetic poles. Because the magnetic field is not inclined very much to the axis of rotation, the rings occur nearly at the planet's geometrical poles. (Compare with Figure 23-4.) (NASA, ESA, J. Clarke, Boston Univ. and X. Levay, STScI)

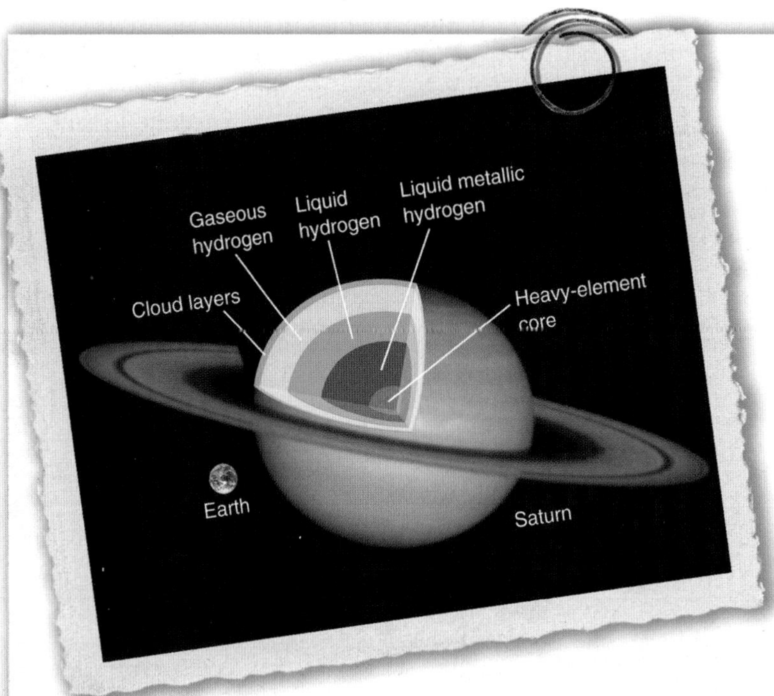


*Saturn's atmosphere blends gradually into its liquid interior. The size of its core is uncertain.* (NASA/STScI)

# *Celestial Profile 8: Saturn*

## *Motion:*

| | |
|---|---|
| Average distance from the sun | 9.54 AU ($1.43 \times 10^9$ km) |
| Eccentricity of orbit | 0.056 |
| Inclination of orbit to ecliptic | 2.5° |
| Average orbital velocity | 9.64 km/s |
| Orbital period | 29.46 y |
| Period of rotation | 10.66 h |
| Inclination of equator to orbit | 26.4° |

## *Characteristics:*

| | |
|---|---|
| Equatorial diameter | $1.21 \times 10^5$ km (9.42 $D_\oplus$) |
| Mass | $5.69 \times 10^{26}$ kg (95.1 $M_\oplus$) |
| Average density | 0.69 g/cm$^3$ |
| Gravity at base of clouds | 1.16 Earth gravities |
| Escape velocity | 35.6 km/s (3.2 $V_\oplus$) |
| Temperature at cloud tops | −180°C (−290°F) |
| Albedo | 0.61 |
| Oblateness | 0.102 |

## *Personality Point:*

The Greek god Cronus was forced to flee when his son Zeus took power. Cronus fled to Italy, where the Romans called him Saturn, protector of the sowing of seed. He was celebrated in a weeklong wild party called the Saturnalia at the time of the winter solstice. Early Christians took over the holiday to celebrate Christmas.

Like Jupiter, Saturn's atmosphere is rich in hydrogen and displays belt–zone circulation, which appears to arise in the same way as the circulation patterns on Jupiter. The light-colored zones are higher clouds formed by rising gas, and the darker belts are lower clouds formed by sinking gas.

Notice, however, that the zones and belt clouds are not very distinct on Saturn compared with Jupiter (■ Figure 23-15a). Measurements from the Voyager and Cassini spacecraft indicate that Saturn's atmosphere is much colder than Jupiter's—something you would expect because Saturn is twice as far from the sun and receives only one-fourth as much solar energy per square meter. The clouds on Saturn form at about the same temperature as the clouds on Jupiter, but those temperature levels are deeper in Saturn's cold atmosphere. Compare the cloud layers in Figure 23-15b with those shown in the diagram on page 501. Because they are deeper in the atmosphere, the cloud layers look dimmer from your viewpoint outside, and a high layer of haze formed by methane crystals makes the cloud layers even more indistinct. The atmospheres of Jupiter and Saturn are actually very similar once you account for the fact that Saturn is colder.

One dramatic difference between Jupiter and Saturn concerns the winds. On Jupiter, winds form the boundaries for each of the belts and zones, but on Saturn the pattern is not the same. Saturn has fewer such winds, but they are much stronger. The eastward wind at the equator of Saturn, for example, blows at 500 m/s (1100 mph), roughly five times faster than the eastward wind at Jupiter's equator. The reason for this difference is not clear.

## Saturn's Rings

Looking at the beauty and complexity of Saturn's rings, an astronomer once said, "The rings are made of beautiful physics." You could add that the physics is actually rather simple, but the result is one of the most amazing sights in our solar system.

In 1609, Galileo became the first to see the rings of Saturn; but perhaps because of the poor optics in his telescopes, he did not recognize the rings as a disk. He drew Saturn as three objects—a central body and two smaller ones on either side. In 1659, Christian Huygens (pronounced approximately, *Howk-gins*) realized that the rings form a disk surrounding but not touching the planet.

Understanding Saturn's rings has required human ingenuity continuing to the present day. In 1859, James Clerk Maxwell (for whom the large mountain on Venus is named)

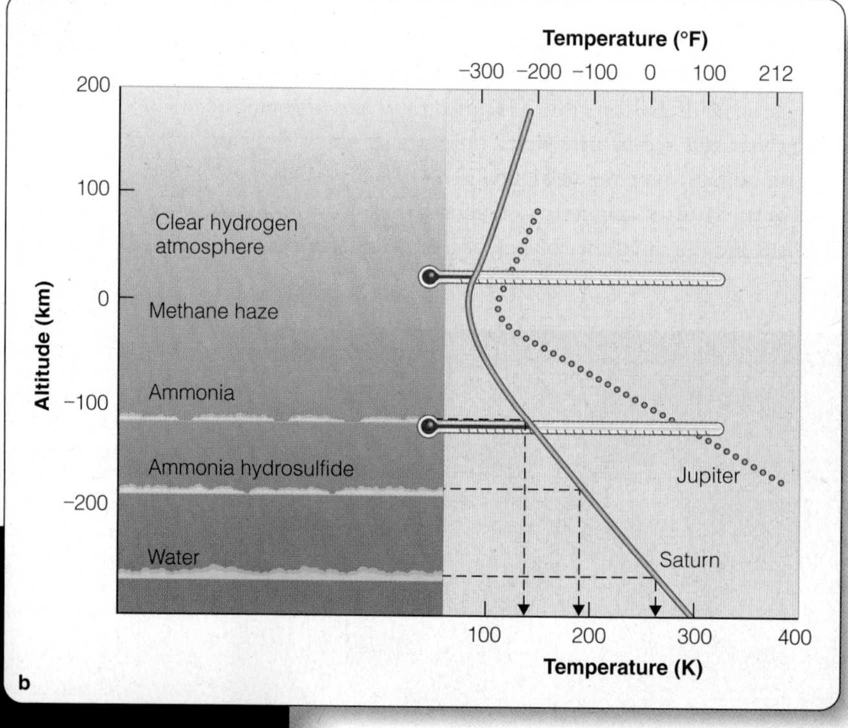


■ **Figure 23-15**

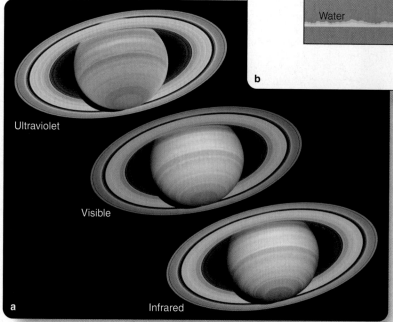

(a) Saturn's belt–zone circulation is not very distinct at visible wavelengths. These images were recorded when Saturn's southern hemisphere was tipped toward Earth. (NASA and E. Karkoschka) (b) Because Saturn is colder than Jupiter, the clouds form deeper in the hazy atmosphere. Notice that the three cloud layers on Saturn form at about the same temperatures as do the three cloud layers on Jupiter.

proved mathematically that solid rings would be unstable. Saturn's rings, he concluded, had to be made of separated particles. In 1867, Daniel Kirkwood demonstrated that gaps in the rings were caused by resonances with some of Saturn's moons. Spectra of the rings eventually showed that the particles were mostly water ice.

Study **The Rings of Saturn** on pages 514–515 and notice three points and a new term:

1. The rings are made up of billions of ice particles, each in its own orbit around the planet. But, just as for Jupiter's rings, the particles observed now in Saturn's rings can't have been there since the planet formed. The rings must be replenished now and then by impacts on Saturn's icy moons or by the disruption of a small moon that moves too close to the planet.

2. The gravitational effects of small moons called *shepherd satellites* can confine some rings in narrow strands or keep the edges of rings sharp. Moons can also produce waves in the rings that are visible as tightly wound ringlets.

3. The ring particles are confined to a thin layer in Saturn's equatorial plane by the gravity of small moons. The small moons in turn are controlled by gravitational interactions with larger, more distant moons. The rings of Saturn, and the rings of the other Jovian worlds, are created from and controlled by the planet's moons. Without the moons, there would be no rings.

Modern astronomers find simple gravitational interactions producing even more complex processes in the rings. Where particles orbit in resonance with a moon, the moon's gravity triggers spiral density waves in much the same way that spiral arms are produced in galaxies. The spiral density waves spread outward through the rings. If the moon follows an orbit that is inclined to the ring plane, the moon's gravity causes a different kind of wave—spiral bending waves—ripples extending above and below the ring plane, which spread inward. Both of these kinds of processes are shown in the inset ring image on page 515.

Many other processes occur in the rings. Specks of dust become electrically charged by sunlight, and Saturn's magnetic field lifts them out of the ring plane. Small moonlets embedded in the rings produce gaps, waves, and scallops in the rings. The Cassini spacecraft has recorded dramatic images (■ Figure 23-16) of the Saturn ring system, including two faint outer rings (E and G) that are rarely detectable from Earth, and these also appear to be related to moons.

The word "particle" in colloquial language connotes tiny specks, but in the context of Saturn's rings, astronomers use that term to refer to any object from snowlike powder grains up to building-sized icy mini-moons (look again at page 514). The larger objects are understood to be aggregates of the smaller ones. The subtle colors of

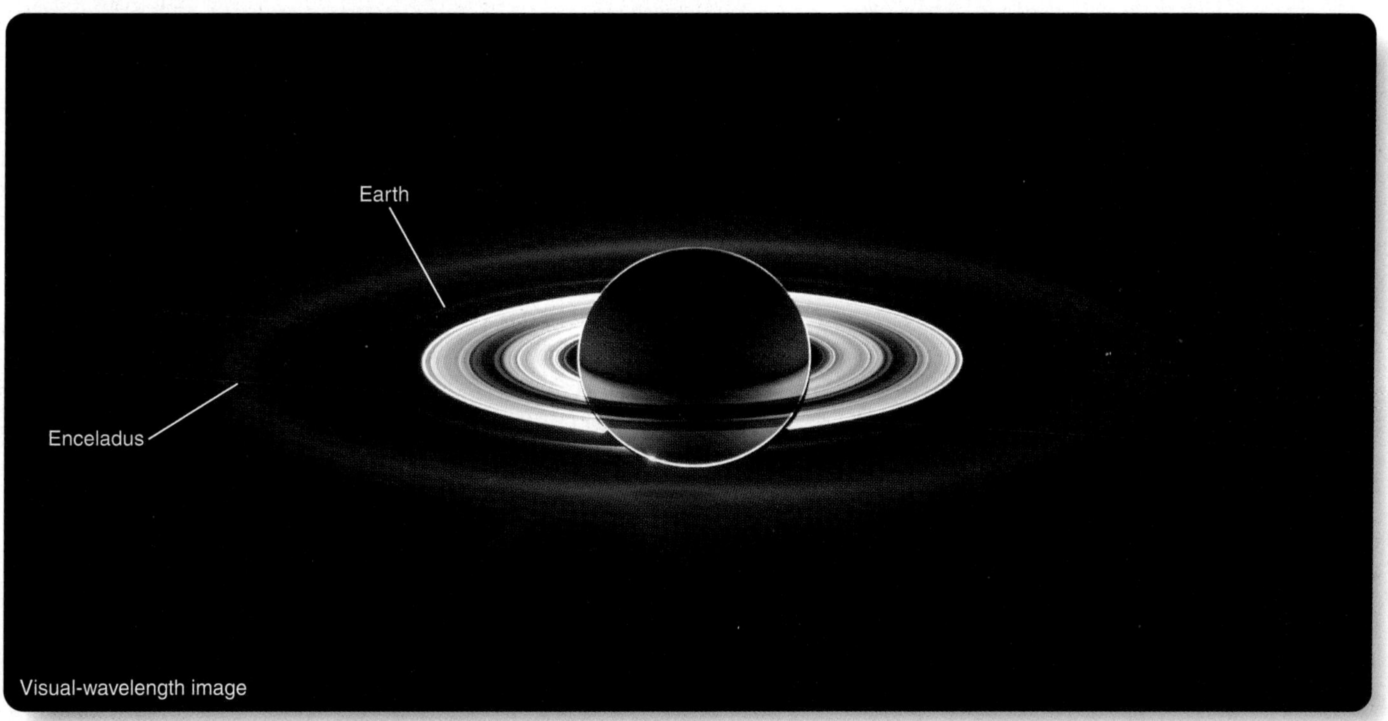


**■ Figure 23-16**

The Cassini spacecraft recorded this image as it passed through Saturn's shadow. Earth is visible as a faint blue dot just inside the G ring, and jets of ice particles vented from the moon Enceladus are visible at the left extreme of the larger E ring. Two faint rings associated with small moons were discovered in this image. (NASA/JPL/Space Science Institute)

# The Ice Rings of Saturn

**1** The brilliant rings of Saturn are made up of billions of ice particles ranging from microscopic specks to chunks bigger than a house. Each particle orbits Saturn in its own circular orbit. Much of what astronomers know about the rings was learned when the Voyager 1 spacecraft flew past Saturn in 1980, followed by the Voyager 2 spacecraft in 1981. The Cassini Spacecraft reached orbit around Saturn in 2004. From Earth, astronomers see three rings labeled A, B, and C. Voyager and Cassini images reveal over a thousand ringlets within the rings.

Saturn's rings can't be leftover material from the formation of Saturn. The rings are made of ice particles, and the planet would have been so hot when it formed that it would have vaporized and driven away any icy material. Rather, the rings must be debris from collisions between passing comets, or other objects, and Saturn's icy moons. Such impacts should occur every 100 million years or so, and they would scatter ice throughout Saturn's system of moons. The ice would quickly settle into the equatorial plane, and some would become trapped in rings. Although the ice may waste away due to meteorite impacts and damage from radiation in Saturn's magnetosphere, new impacts could replenish the rings with fresh ice. The bright, beautiful rings you see today may be only a temporary enhancement caused by an impact that occurred since the extinction of the dinosaurs.

Encke Gap

Cassini Division

A ring

B ring

C ring

As in the case of Jupiter's ring, Saturn's rings lie inside the planet's Roché limit where the ring particles cannot pull themselves together to form a moon.

Because it is so dark, the C ring was once called the crepe ring.

Earth to scale

Visual-wavelength image

The C ring contains boulder-size chunks of ice, whereas most particles in the A and B rings are more like golf balls, down to dust-size ice crystals. Further, C ring particles are less than half as bright as particles in the A and B rings. Cassini observations show that the C ring particles contain less ice and more minerals.

NASA

**1a** An astronaut could swim through the rings. Although the particles orbit Saturn at high velocity, all particles at the same distance from the planet orbit at about the same speed, so they collide gently at low velocities. If you could visit the rings, you could push your way from one icy particle to the next. This artwork is based on a model of particle sizes in the A ring.

**2** Because of collisions among ring particles, planetary rings should spread outward. The sharp outer edge of the A ring and the narrow F ring are confined by **shepherd satellites** that gravitationally usher straying particles back into the rings.

Some gaps in the rings, such as the Cassini Division, are caused by resonances with moons. A particle in the Cassini Division orbits Saturn twice for each orbit of the moon Mimas. On every other orbit, the particle feels a gravitational tug from Mimas. These tugs always occur at the same places in the orbit and force the orbit to become slightly elliptical. Such an orbit crosses the orbits of other particles, which results in collisions, and that removes the particle from the gap.

This image was recorded by the Cassini spacecraft looking up at the rings as they were illuminated by sunlight from below. Saturn's shadow falls across the upper side of the rings.

Visual-wavelength image

Pandora

Prometheus

F ring close up

The F ring is clumpy and sometimes appears braided because of two shepherd satellites.

F ring

Encke Gap

The Encke Gap is not empty. Note the ripples at the inner edge. A small moon orbits inside the gap.

Visual-wavelength images

Waves in the A ring

Saturn does not have enough moons to produce all of its ringlets by resonances. Many are produced by tightly wound waves, much like the spiral arms found in disk galaxies.

Cassini Division

A ring

Encke Gap

This combination of UV images has been given false color to show the ratio of mineral material to pure ice. Blue regions such as the A ring are the purest ice, and red regions such as the Cassini Division are the dirtiest ice. How the particles become sorted by composition is unknown.

Ultraviolet image

NASA/JPL/Space Science Institute

**3** How do moons happen to be at just the right places to confine the rings? That puts the cosmic cart before the horse. The ring particles get caught in the most stable orbits among Saturn's innermost moons. The rings push against the inner moons, but those moons are locked in place by resonances with larger, outer moons. Without the moons, the rings would spread and dissipate.

Saturn's rings are a very thin layer of particles and nearly vanish when the rings turn edge-on to Earth. Although ripples in the rings caused by waves may be hundreds of meters high, the sheet of particles may be only about ten meters thick.

# How Do We Know? | 23-1

## Who Pays for Science?

**Why shouldn't you plan for a career as an industrial paleontologist?** Searching out scientific knowledge can be expensive, and that raises the question of funding. Some science has direct applications, and industry supports such research. For example, pharmaceutical companies have large budgets for scientific research leading to the creation of new drugs. But some basic science is of no immediate practical value. Who pays the bill?

A paleontologist is a scientist who studies ancient life forms by examining fossils of plant and animal remains, and such research does not have commercial applications. Except for the rare Hollywood producer about to release a dinosaur movie, corporations can't make a profit from the discovery of a new dinosaur. The practical-minded stockholders of a company will not approve major investments in such research. Consequently, digging up dinosaurs, like astronomy, is poorly funded by industry.

It falls to government institutions and private foundations to pay the bill for this kind of research. The Keck Foundation has built two giant telescopes with no expectation of financial return, and the National Science Foundation has funded thousands of astronomy research projects for the benefit of society.

The discovery of a new dinosaur or a new galaxy is of no great financial value, but such scientific knowledge is not worthless. Its value lies in what it tells us about the world we live in. Such scientific research enriches our lives by helping us understand what we are. Ultimately, funding basic scientific research is a public responsibility that society must balance against other needs. There isn't anyone else to pick up the tab.

*Sending the Cassini spacecraft to Saturn costs each U.S. citizen 56¢ per year over the life of the project.*

the rings arise from contamination in the ice, and some areas have unusual compositions. The Cassini Division, for instance, contains particles that are richer in rock than most of the ring. No one knows how these differences in composition arise, but they must be related to the way the rings are formed and replenished.

Like a beautiful flower, the rings of Saturn are controlled by many different natural processes. Observations from spacecraft such as Voyager and Cassini will continue to reveal even more about the rings. Such missions are expensive, of course, but they are helping us understand what we are (**How Do We Know? 23-1**).

## The History of Saturn

The farther you journey from the sun, the more difficult it is to understand the history of the planets. Any fully successful history of Saturn should explain its low density, its peculiar magnetic field, and its beautiful rings. Planetary scientists can't tell a complete story yet, but you can understand a few of the principles that affected the formation of Saturn and its rings.

Most of Saturn's story parallels that of Jupiter. Saturn formed in the outer solar nebula, where ice particles were stable. It grew rapidly, becoming massive enough to capture hydrogen and helium gas directly from the nebula. The heavier elements probably form a denser core, and the hydrogen forms a liquid mantle containing liquid metallic hydrogen. The outward flow of heat from the core drives convection currents in this mantle that, coupled with the rapid rotation of the planet, produces its magnetic field. Because Saturn is smaller than Jupiter, it has less liquid metallic hydrogen, and its magnetic field is weaker.

The rings of Saturn definitely are not primordial, meaning, the material in them now has not been in its current form since the formation of the planet. Saturn, like Jupiter, would have been very hot when it formed, and that heat would have vaporized and driven off any nearby small icy particles of leftover material. Also, such a hot Saturn would have had a very distended atmosphere, which would have slowed ring particles by friction and caused the particles to fall into the planet. Finally, the processes that tend to destroy Jupiter's ring particles also apply to Saturn's rings.

Planetary rings do not seem to be stable over 4.6 billion years, so the ring material must have been produced more recently. Saturn's beautiful rings may have been created within just the last 100 million years, an astronomically short time. One suggestion is that a small moon or an icy planetesimal came within Saturn's Roche limit, and tides pulled it apart. At least some of the resulting debris would have settled into the ring plane. Another possibility is that a comet struck one of Saturn's moons. Because both comets and moons in the outer solar system are icy, such a collision would produce icy debris. Bright planetary rings such as Saturn's may be temporary phenomena, forming when violent events produce fresh ice debris and then wasting away as the ice is gradually lost.

## SCIENTIFIC ARGUMENT

***Why do the belts and zones on Saturn look so faint?***

One of the most powerful tools of critical thought is simple comparing and contrasting. You can make that the theme of this scientific argument by comparing and contrasting Saturn with Jupiter. In the atmosphere of Jupiter, the dark belts form in regions where gas sinks, and zones form where gas rises. The rising gas cools and condenses to form icy crystals of ammonia, which are visible as bright clouds. Clouds of ammonia hydrosulfide and water form deeper, below the ammonia clouds, and are not as visible. Saturn is twice as far from the sun as Jupiter, so sunlight is four times dimmer. The atmosphere is colder, and gas currents do not have to rise as far to reach cold levels and form clouds. That means the clouds are deeper in Saturn's atmosphere than in Jupiter's atmosphere. Because the clouds are deeper, they are not as brightly illuminated by sunlight and look dimmer. Also, a layer of methane-ice-crystal haze high above the ammonia clouds makes the clouds even less distinct.

Now build a new argument comparing the ring systems. **How is Saturn's ring system similar to, and different from, Jupiter's ring system?**

## 23-5 Saturn's Moons

SATURN HAS OVER 30 KNOWN SATELLITES—far too many to examine individually—but these moons share characteristics common to icy worlds. Most of them are small and dead, but one is big enough to have an atmosphere and perhaps even oceans or lakes—but not of water.

### Titan

Saturn's largest satellite is a giant ice moon with a thick atmosphere and a mysterious surface. From Earth it is only a dot of light, with no visible detail. Nevertheless, a few basic observations can tell you a great deal about this strange world.

Titan's mass can be estimated from its influence on both passing spacecraft and on other moons. Its mass divided by its volume reveals that its density is 1.9 g/cm$^3$. Its uncompressed density (Chapter 19) is only 1.2 g/cm$^3$. Although it must have a rocky core, it must also contain a large amount of ice.

Titan is a bit larger than the planet Mercury and almost as large as Jupiter's moon Ganymede. Unlike those worlds, Titan has a thick atmosphere. Its escape velocity is low, but it is so far from the sun that it is very cold, and most gas atoms don't move fast enough to escape. (Look back to Figure 22-13.) Methane was detected spectroscopically in 1944, and various hydrocarbons were found beginning in about 1970 (■ Table 23-3), but most of Titan's air is nitrogen with only 1.6 percent methane.

When the Voyager 1 and Voyager 2 spacecraft flew past Saturn in the early 1980s, their cameras could not penetrate Titan's hazy atmosphere (■ Figure 23-17). Measurements showed that the surface temperature is about 94 K (−290°F), and the surface atmospheric pressure is 50 percent greater than on Earth. Model calculations show that in the conditions on Titan, methane could condense from the atmosphere and fall as rain. Some scientists hypothesized that Titan is covered by methane lakes or seas.

**■ Table 23-3 | Some Organic Compounds Detected on Titan**

| | |
|---|---|
| $C_2H_6$ | Ethane |
| $C_2H_2$ | Acetylene |
| $C_2H_4$ | Ethylene |
| $C_3H_4$ | Methylacetylene |
| $C_3H_8$ | Propane |
| $C_4H_2$ | Diacetylene |
| HCN | Hydrogen cyanide |
| $HC_3N$ | Cyanocetylene |
| $C_2N_2$ | Cyanogen |

Sunlight converts methane ($CH_4$) into the gas ethane ($C_2H_6$) plus a collection of other organic molecules.* Some of these molecules produce the smoglike haze; and, as the smog particles gradually settle, they are predicted to deposit smelly, organic goo on the surface. This goo is important because similar organic molecules may have been the precursors of life on Earth. (You will examine this idea further in Chapter 26.)

Before you try to imagine floundering through this strange landscape, you can consider the recent observations made by the Cassini spacecraft, which began exploring Saturn and its moons in 2004.

Infrared cameras and radar instruments on Cassini have been able to see through the hazy atmosphere. The surface is not a featureless ocean of methane, nor an icy plane covered with goo. Rather, the surface consists of icy, irregular highlands and smoother dark areas. There are only a few craters, which suggests geological activity is erasing craters almost as quickly as they are formed.

The Cassini spacecraft released a probe named Huygens, which parachuted down through the atmosphere of Titan and eventually landed on the surface. Huygens radioed back images of the surface as it descended under its parachute, and those images show dark drainage networks that lead into dark smooth areas (Figure 23-17). Those dark regions, which look a lot like bodies of liquid, are actually dry or mostly dry. Precipitation may have washed the black goo off the highlands into the stream channels and lowlands, so that they look smooth and dark even

*Organic molecules are common in living things on Earth but do not have to be derived from living things. In other words, biological substances are organic, but organic substances are not necessarily biological. One chemist defined an organic molecule as "any molecule with a carbon backbone."

**Figure 23-17**

As the Huygens probe descended by parachute through Titan's smoggy atmosphere, it photographed the surface from an altitude of 8 km (5 mi). Although no liquid was present, dark drainage channels lead into the lowlands. Radar images reveal lakes of liquid methane and ethane around the poles. Once the Huygens probe landed on the surface, it radioed back photos showing a level plain and chunks of ice smoothed by a moving liquid.
(ESA/NASA/JPL/USGS/University of Arizona)

Cassini radar image of methane lakes that look dark because they do not reflect radio waves.
d
b Visual-wavelength image
Drainage channels probably were cut by flowing liquid methane.
a
At visual wavelengths, Titan's dense atmosphere hides its surface.
Possible methane fog
Icy grapefruit-size "rocks" on Titan are bathed in orange light from its hazy atmosphere.
c

though the liquid had temporarily evaporated. If you visit Titan, you may get caught in a shower of methane rain, but it may not rain often at your landing site.

When the Huygens probe landed on Titan's frigid surface, it radioed back measurements and images. The surface is mostly frozen water ice with some methane mixed in. The sunlight is orange because it has filtered down through the orange haze. Rocks littering the ground are actually steel-hard chunks of supercold water ice smoothed by erosion. Some rest in small depressions, suggesting that a liquid has flowed around them. You can see these depressions around the rocks in Figure 23-17c.

Radar observations made as the Cassini probe orbited past Titan have revealed lakes of liquid methane in its polar regions, confirming the earlier hypotheses. Some of those lakes are as large as Lake Superior. Evaporation from the lakes can maintain the 1.6 percent methane gas in the atmosphere, but sunlight eventually destroys methane, so Titan must have a large supply of methane ice. Ice volcanoes on Titan may occasionally vent methane into the atmosphere.

By the way, before you go to Titan, check your spacesuit for leaks. Nitrogen is not a very reactive gas, but methane is used as cooking gas on Earth and is highly flammable. Of course, there is no free oxygen on Titan, so you are safe so long as your spacesuit does not leak oxygen.

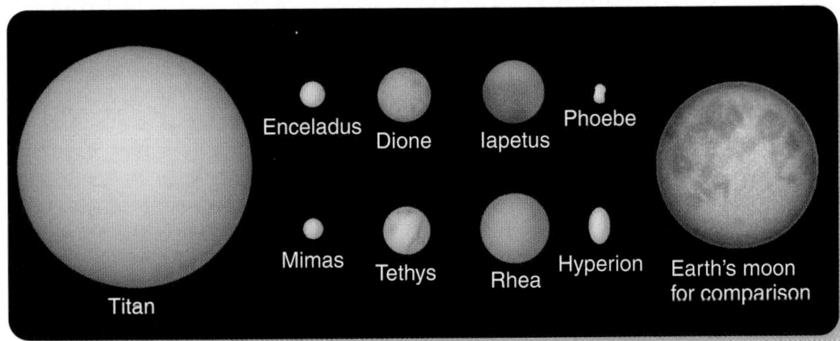


■ **Figure 23-18**

A few of Saturn's moons compared with Earth's moon at the right. In general, larger moons are round and more likely to show signs of geological activity. Small moons such as Phoebe and Hyperion are cratered and do not have enough gravity to overcome the strength of their own material and squeeze themselves into a spherical shape.

## The Smaller Moons

In addition to Titan, Saturn has a large family of smaller moons. They are mixtures of rock and ice and are heavily cratered. Some of the smallest are probably captured objects and are geologically dead, but some of the larger moons show traces of geological activity. You can compare the sizes of a few of these moons in ■ Figure 23-18.

The moon Phoebe orbits on the outer fringes of Saturn's satellite family, and it moves in the retrograde direction—it orbits backward. It is quite small, only about 210 km (130 mi) in diameter, but it is nevertheless the largest of Saturn's irregular satellites. Phoebe's surface is dark, with an albedo of only 6 percent, and heavily cratered (■ Figure 23-19). Traces of ice are detected where impacts have excavated deeper layers or where landslides have exposed fresh material. The density of Phoebe is 1.6 g/cm$^3$, which is high enough to show that it contains a significant amount of rock. It seems unlikely that Phoebe came from the asteroid belt, where ice is relatively rare. It is more likely to be a captured Kuiper belt object that originated in an orbit beyond Neptune.

Larger regular moons such as Tethys, which has a diameter of over 1000 km (620 mi), are icy and cratered, but they show some signs of geological activity. Some smooth areas on Tethys appear to have been resurfaced by flowing water "lava," and long cracks and grooves may have formed when geological activity strained the icy crust (Figure 23-19).

■ **Figure 23-19**

Phoebe and Tethys have ancient cratered surfaces, but they differ in interesting ways. Phoebe is only $\frac{1}{5}$ the diameter of Tethys and shows no sign of internal heat causing surface activity. Tethys has smooth areas and long cracks on its surface, showing that it has been active. Phoebe is probably a captured Kuiper belt object, but Tethys likely formed with Saturn. (NASA/JPL/Space Science Institute)

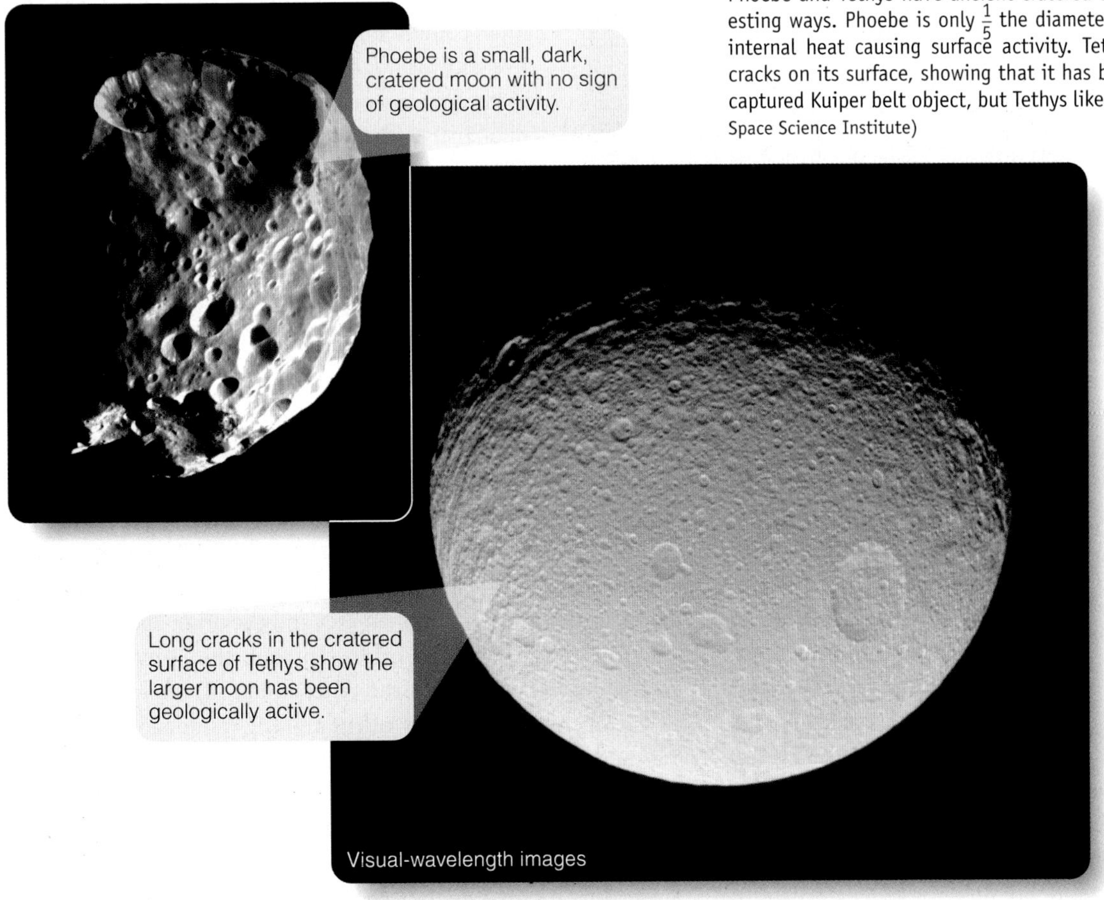

Phoebe is a small, dark, cratered moon with no sign of geological activity.
Long cracks in the cratered surface of Tethys show the larger moon has been geologically active.
Visual-wavelength images

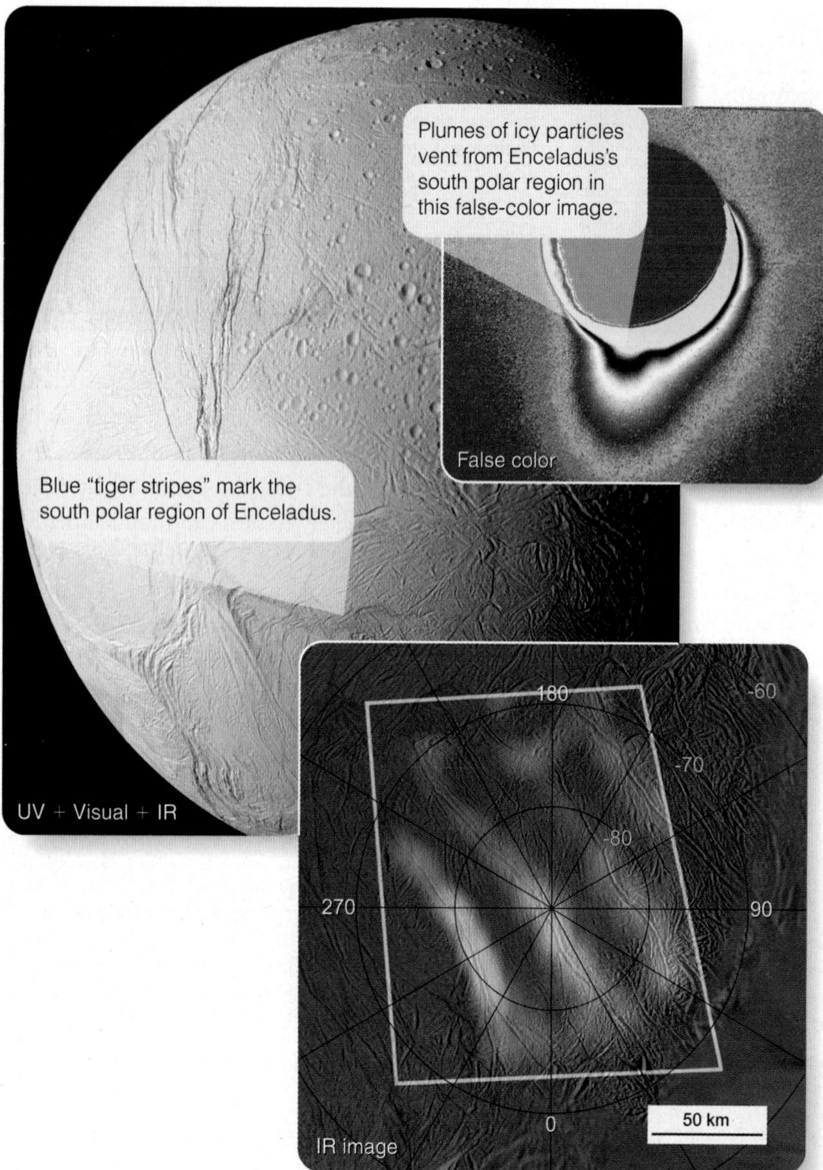
Plumes of icy particles vent from Enceladus's south polar region in this false-color image.
False color
Blue "tiger stripes" mark the south polar region of Enceladus.
UV + Visual + IR
180
-60
-70
-80
270
90
0
50 km
IR image

**■ Figure 23-20**

The bright, clean icy surface of Enceladus does not look old. Some areas have few craters, and the numerous cracks and lanes of grooved terrain resemble the surface of Jupiter's moon Ganymede. Enceladus is venting water, ice, and organic molecules from geysers near its south pole. A thermal infrared image reveals internal heat leaking to space from the "tiger stripe" cracks where the geysers are located. (NASA/JPL/Space Science Institute)

With a diameter of 520 km (320 mi), the small moon Enceladus isn't much larger than Phoebe, but Enceladus shows dramatic signs of geological activity (■ Figure 23-20). For one thing, Enceladus has an albedo of 0.9. That is, it reflects 90 percent of the sunlight that hits it, and that makes it the most reflective object in the solar system. You know that old icy surfaces become dark, so the surface of Enceladus must be quite young. Look closely at the surface and you will see that some regions have few craters, and that grooves and cracks are common. Observations made by the Cassini spacecraft show that Enceladus has a tenuous atmosphere of water vapor and nitrogen. It is too small to keep such an atmosphere, so it must be releasing gas continuously. Cassini detected a large cloud of water vapor over the moon's south pole where water vents through cracks and produces ice-crystal jets extending hundreds of kilometers above the surface. As these ice crystals escape into space, they replenish Saturn's E ring, which is densest at the position of Enceladus. Infrared images made by Cassini show significant amounts of heat escaping to space through the same cracks from which the water is venting (Figure 23-20).

The possibility of liquid water below the icy crust of Enceladus has excited those scientists searching for life on other worlds. You will read more about this possibility in Chapter 26. Nevertheless, it will be a long time before explorers can drill through the crust and analyze the water below for signs of living things.

Of course, you are wondering how a little moon like Enceladus can have heat flowing up from its interior. With a density of 1.6 g/cm$^3$, Enceladus must contain a significant rocky core, but radioactive decay is not enough to keep it active. A clue lies in the moon's orbit. Enceladus orbits Saturn in a resonance with the larger moon Dione. Each time Dione orbits Saturn once, Enceladus orbits twice. That means Dione's gravitational tugs on Enceladus always occur in the same places and make the orbit of the little moon slightly eccentric. As Enceladus follows that eccentric orbit around Saturn, tides flex it, and tidal heating warms the interior. You saw how resonances keep some of Jupiter's moons active, so you can add Enceladus to the list.

More complicated gravitational interactions between moons are dramatically illustrated by the two small moons that shepherd the F ring (page 515). An even more peculiar pair of moons is known as the coorbital satellites. These two irregularly shaped moonlets have orbits separated by only 100 km. Because one moon is about 200 km in diameter and the other about 100 km, they cannot pass in their orbits. Instead, the innermost moon gradually catches up with the outer moon. As the moons draw closer together, the gravity of the trailing moon slows the leading moon and makes it fall into a lower orbit. Simultaneously, the gravity of the leading moon pulls the trailing moon forward, and it rises into a higher orbit (■ Figure 23-21). The higher orbit has a longer period, so the trailing moon begins to fall behind the leading moon, which is now in a smaller, faster orbit. In this

**Coorbital Moons**

The inner moon orbits faster and overtakes the outer moon.
(Not to scale)

The gravitational interaction pulls the outer moon backward and the inner moon forward.

As they approach, the inner moon moves to a higher orbit, and the outer moon sinks to a lower orbit.


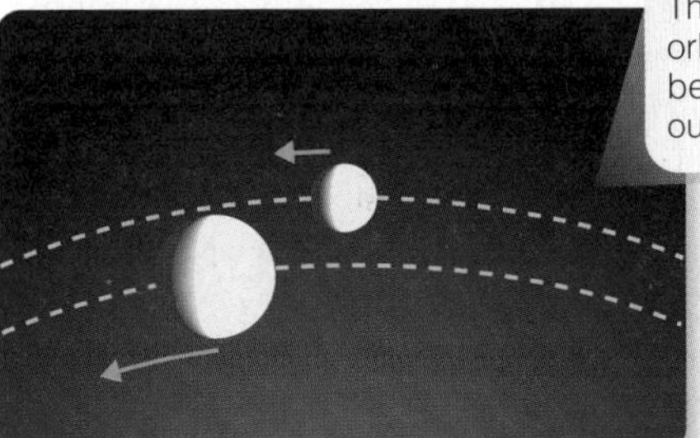

The moons have changed orbits, and the inner moon begins gaining on the outer moon.


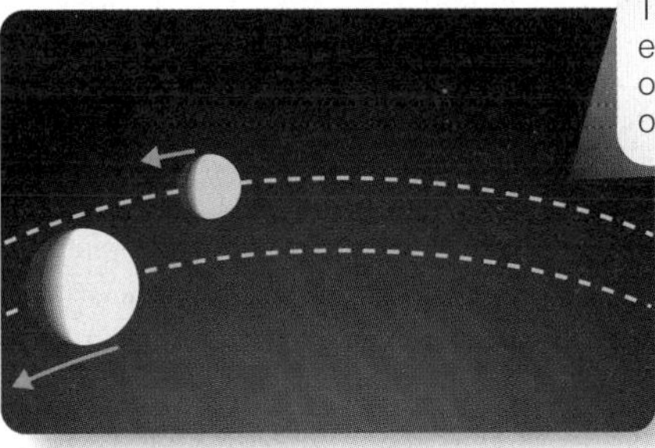

The inner moon will eventually overtake the outer moon from behind once again.


■ **Figure 23-21**

Saturn's coorbital moons follow nearly identical orbits. The moon in the lower orbit travels faster and always overtakes the other moon from behind. The moons interact and change orbits over and over again.

elegant dance, these moons exchange orbits and draw apart only to meet again and again. It seems very likely that these two moons are fragments of a larger moon destroyed by a major impact.

In addition to waltzing coorbital moons, the Voyager spacecraft discovered small moonlets trapped at the L4 and L5 Lagrangian points (see Figure 13-5) in the orbits of Dione and Tethys. These points of stability lie 60° ahead of and 60° behind the two moons, and small moonlets can become trapped in these regions. (You will see in Chapter 25 that some asteroids are trapped in the Lagrangian points of Jupiter's and Neptune's orbits around the sun.) This gravitational curiosity is not unique to the Saturn system, and it indicates that interactions between moons can dramatically alter their orbits.

Saturn has too many moons to discuss in detail here, but you should meet at least one more, Iapetus (pronounced *Ee-yap-eh-tus*). It is literally an odd ball: Iapetus is an asymmetric moon. Its trailing side, the side that always faces backward as it orbits Saturn, is old, cratered, icy, and about as bright as dirty snow. Its leading side, the side that always faces forward in its orbit, is also old and cratered, but it is much darker than you would expect. It has an albedo of only 4 percent—about as dark as fresh asphalt on a highway (■ Figure 23-22). The origin of this dark material is unknown, but theorists suspect that the little moon has swept

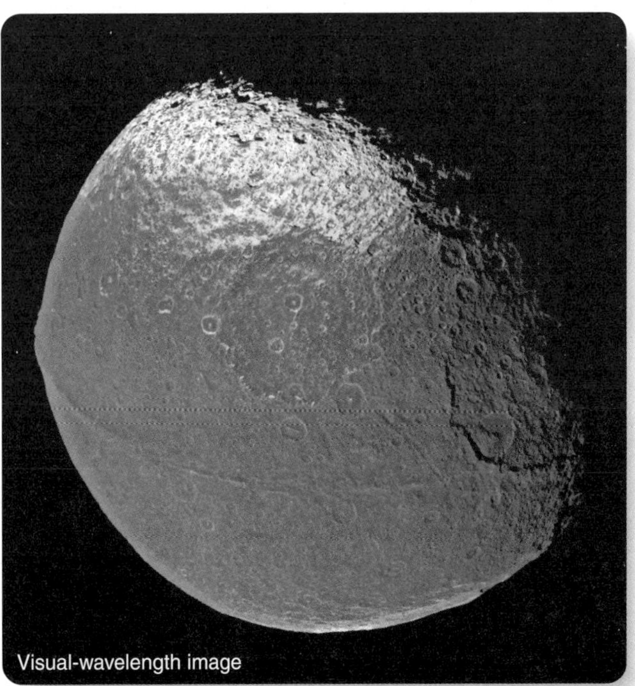


■ **Figure 23-22**

Like the windshield of a speeding car, the leading side of Saturn's moon Iapetus seems to have accumulated a coating of dark material. The poles and trailing side of the moon are much cleaner ice. The equatorial ridge is 20 km (12 mi) wide and up to 13 km (8 mi) high. It stretches roughly 1,300 km (800 mi) along the moon's equator. (NASA/JPL/Space Science Institute)

up dark, silicon- and carbon-rich material on its leading side. The source of the material covering the leading side of Iapetus could be meteorites striking and eroding the carbonaceous surface of the outermost moon, Phoebe, and tossing the resulting dust into space to be scooped up eventually by Iapetus.

Another odd feature on Iapetus shows up in Cassini images—an equatorial ridge that stands as high as 13 km (8 mi) in some places. You can see the ridge clearly in Figure 23-21. The origin of this ridge is unknown, but it is not a minor feature. At that height, it is over 50 percent higher than Mount Everest, and it extends for a long distance across the surface. That is a big pile of rock and ice. The ridge sits atop an equatorial bulge, and both ridge and bulge may have formed when Iapetus was young, spun rapidly, and was still mostly molten.

Saturn's moons illustrate a number of principles of comparative planetology. Small moons are irregular in shape, and old surfaces are dark and cratered. Resonances can trigger tidal heating, and that can in turn resurface moons and outgas atmospheres. Small moons can't keep atmospheres, but big, cold moons can. You are an expert in all of this, so you are ready to wonder where the moons came from.

## The Origin of Saturn's Moons

Jupiter's four Galilean satellites seem clearly related to one another, and you can safely conclude that they formed with Jupiter. No such simple relationships link Saturn's satellites. That seems to indicate that, unlike Jupiter, Saturn was not enough of a heat source during that system's formation to cause the densities of its regular moons to follow the condensation sequence. Planetary scientists also suspect that comet impacts have so badly fractured the regular moons that they no longer show much evidence of their common origin. Understanding the origin of Saturn's moons is also difficult because the moons interact gravitationally so that the orbits they now occupy may differ significantly from their earlier orbits.

The complex orbital relationships of Saturn's moons and their evidently intense cratering suggest that the moons have interacted and may have collided with each other, with comets, and with large planetesimals in the past. Nevertheless, as with Jupiter's moons, astronomers hypothesize that most or all of Saturn's regular moons formed with the planet and that the irregular moons are captured. As you continue your exploration of the outer solar system, you can be alert for the presence of more such small, icy worlds.

### SCIENTIFIC ARGUMENT

***What features on Enceladus suggest that it has been active?***
This argument is based on comparative planetology applied, in this case, to moons. The smaller moons of Saturn are icy worlds battered by impact craters, and you can suspect that they are cold and old. Small worlds lose their heat quickly; and, with no internal heat, there is no geological activity to erase impact craters. A small, icy world covered with craters is exactly what you would expect in the outer solar system. Enceladus, however, is peculiar. Although it is small and icy, its surface is highly reflective, and some areas contain fewer craters than you would expect. In fact, some regions seem almost free of craters. Grooves and faults mark some regions of the little moon and suggest motion in the crust. These features should have been destroyed long ago by impact cratering, so you must suppose that the moon has been geologically active at some time since the end of the heavy bombardment at the conclusion of planet building. The water vents discovered at the south pole of Enceladus show the moon is still active.

If you look at Titan, Saturn's largest moon, you see quite a different world. Build an argument based on a different principle of comparative planetology. **How can such a small world as Titan keep a thick atmosphere?**

### What Are We? Basic Scientists

People often describe science that has no known practical value as basic science or basic research. The exploration of distant worlds would be called basic science, and it is easy to argue that basic science is not worth the effort and expense because it has no known practical use. Of course, the problem is that no one has any way of knowing what knowledge will be of use until that knowledge is acquired.

In the middle of the 19th century, Queen Victoria asked physicist Michael Faraday what good his experiments with electricity and magnetism were. He answered, "Madam, what good is a baby?" Of course, Faraday's experiments were the beginning of the electronic age. Many of the practical uses of scientific knowledge that fill your world—digital electronics, synthetic materials, modern vaccines—began as basic research. Basic scientific research provides the raw materials that technology and engineering use to solve problems, so to protect its future, the human race must continue its struggle to understand how nature works.

Basic scientific research has yet one more important use that is so valuable it seems an insult to refer to it as merely practical. Science is the study of nature, and as you learn more about how nature works, you learn more about what your existence in this universe means. The seemingly impractical knowledge gained from space probes visiting other worlds tells you about your own planet and your own role in the scheme of nature. Science tells us where we are and what we are, and that knowledge is beyond value.

# Study and Review

## Summary


- The outer planets, Jupiter, Saturn, Uranus, and Neptune, are much larger than Earth and lower in density. These Jovian planets are rich in hydrogen. All four of the Jovian planets have rings, large satellite systems, and shallow atmospheres above liquid hydrogen mantles.
- Jupiter and Saturn show strong **belt–zone circulation (p. 495)**, but this is harder to see on Uranus and Neptune. Belts are low-pressure areas where gas is sinking, and zones are high-pressure areas where gas is rising.
- Moons in the Jovian satellite systems interact gravitationally, and some moons have been heated by tides to produce geological activity. Most are old and cratered. **Regular satellites (p. 495)** are generally larger, are close to the parent planet, and have low orbital inclinations; **irregular satellites (p. 495)** are generally small, are far from the parent planet, and have high orbital inclinations.
- Simple observations made from Earth show that Jupiter is 11 times Earth's diameter and 318 times Earth's mass. From that you can calculate its density, which is much lower than Earth's. It is rich in hydrogen and helium and cannot contain more than a small core of heavy elements.
- Not far below Jupiter's clouds, the temperature and pressure exceed the **critical point (p. 496)**, which means there is no difference between gaseous and liquid hydrogen. Consequently, the liquid hydrogen layer has no surface; the transition from gaseous hydrogen to liquid hydrogen is gradual.
- Jupiter's atmospheric composition is much like that of the sun—mostly hydrogen and helium with smaller amounts of heavier elements.
- Infrared observations show that Jupiter radiates more heat than it receives from the sun, so its interior must be five or six times hotter than the sun's surface and is prevented from flashing into vapor by the high pressure.
- The pressure inside Jupiter converts much of its hydrogen into **liquid metallic hydrogen (p. 496)**, and the dynamo effect generates a powerful magnetic field that produces rings of auroras around the planet's magnetic poles, interacts with the small moon Io, and traps high-energy particles to form intense radiation belts.
- The **oblateness (p. 497)** of Jupiter arises because its hydrogen envelope is highly fluid and the planet rotates rapidly.
- Ionized atoms from Jupiter's inner moon Io are swept up by Jupiter's rapidly spinning magnetic field to form the **Io plasma torus (p. 498)**, which encloses the orbit of the moon. Powerful electrical currents flow through the **Io flux tube (p. 498)** and produce spots of aurora where it enters Jupiter's atmosphere.
- Jupiter's shallow atmosphere is rich in hydrogen, with three layers of clouds at the temperatures where ammonia, ammonium hydrosulfide, and water condense. High- and low-pressure areas form belt–zone circulation.
- Spots on Jupiter, such as the Great Red Spot, are long-lasting, circulating storm systems.
- **Forward scattering (p. 499)** shows that Jupiter's ring is composed of tiny dust specks orbiting inside Jupiter's **Roche limit (p. 499)**. The dust particles cannot have survived since the formation of the planet; rather, they are being produced by meteorite impacts on some of Jupiter's inner moons.
- A small moon can orbit inside a planet's Roche limit and survive if it is a solid piece of rock strong enough to endure the tidal forces trying to pull it apart.
- The dimmer **gossamer rings (p. 502)** lie near the orbits of two moons, adding evidence that the rings are sustained by particles from moons.
- Impacts on moons and planets must be common in the history of the solar system. Jupiter was hit by the fragmented head of a comet in 1994. Such impacts had not been seen before.
- At least some of Jupiter's small moons are captured asteroids.
- The four **Galilean satellites (p. 504)** appear to have formed with Jupiter. They have higher densities closer to Jupiter and lower farther from Jupiter, similar to the dependence of planet densities on distance from the sun caused by the condensation sequence.
- Callisto, the outermost Galilean satellite, is composed of ice and rock and has an old and cratered surface. Unlike the three inner moons, Callisto is not caught in an orbital resonance and has never been active.
- Ganymede, Europa, and the innermost moon, Io, are locked in an orbital resonance, and that causes **tidal heating (p. 505)**. Ganymede's surface is old and cratered in some areas, but bright **grooved terrain (p. 505)** must have been produced by a past episode of geological activity.
- The inward focusing of meteorites should expose moons near massive planets to more cratering impacts, so it is surprising that Europa and Io have almost no craters, but this is explained by geological activity caused by tidal heating.
- Europa is mostly rock with a thin icy crust that contains only a few scars of past craters. Cracks and lines show that the crust has broken repeatedly, and a subsurface ocean probably vents through the crust and deposits ice to cover craters as fast as they form. The subsurface ocean might be a place to look for life.
- Io is strongly heated by tides and has no water at all. Over 150 volcanoes erupt molten rock and throw ash high above the surface. No impact craters are visible because they are destroyed or buried as fast as they form. Sulfur compounds color the surface yellow and orange and vent into space to be caught in Jupiter's magnetic field.
- Saturn must have formed much as Jupiter did, but it is slightly smaller than Jupiter and less dense than water. It has a hot interior but contains less liquid metallic hydrogen, so its magnetic field is weaker than Jupiter's and is, for some reason, closely aligned with its axis of rotation.
- Saturn is twice as far from the sun as Jupiter and is much colder. The three cloud layers seen on Jupiter form deeper in Saturn's hazy atmosphere and are not as clearly visible.
- Saturn's rings are composed of ice particles and cannot have lasted since the formation of the planet. The rings must receive occasional additions of ice particles when a small moon wanders inside Saturn's Roche limit and is pulled apart by tides or when comets hit the planet's icy moons.
- Icy particles can become trapped in stable places among the orbits of the innermost small moons, those within the Roche limit. Resonances with outer moons can produce gaps in the rings and generate waves that move like ripples through the rings. Small **shepherd satellites (p. 515)** can confine sections of the ring to produce sharp edges, ripples, or narrow ringlets.
- Without moons to confine them, the rings would have spread outward and dissipated long ago.
- Some of Saturn's smaller moons, such as Phoebe, are probably captured asteroids or Kuiper belt objects. All of the moons are mixtures of rock and ice.
- Titan, the largest moon, is so big it can retain a dense atmosphere of nitrogen with a small amount of methane. The methane condenses from

the atmosphere, falls as rain, and drains over the surface, washing dark, organic material into drainage channels and on into the lowlands.

- The methane in Titan's atmosphere can be destroyed by sunlight, so it must be continuously replenished. It probably vents from the icy crust and may form methane volcanoes. Lakes of liquid methane have been found in the moon's polar regions.
- Some of the larger moons, such as Tethys, have old, dark, cratered surfaces with cracks and smoothed areas that suggest past activity. Enceladus, a rather small moon, is the most reflective object in the solar system and has large smooth areas, so it must have been very active. It orbits in a resonance with the moon Dione, and that may cause tidal heating inside Enceladus.
- Water vapor has been found above the south pole of Enceladus where water vents into space and forms ice crystals, which resupply Saturn's E ring.
- The moon Iapetus has a bright icy surface on its trailing side, but its leading side is coated with very dark material, possibly debris from Phoebe. In addition, Iapetus has a long equatorial ridge higher than Mount Everest on Earth. The ridge and the moon's equatorial bulge may have formed when the moon was young, molten, and spinning rapidly.
- The origin and evolution of Saturn's moons is not as clear as for Jupiter's Galilean moons. Orbital interactions and impacts have been important to these moons.

## Review Questions

1. Why is Jupiter more oblate than Earth? Do you expect all Jovian planets to be oblate? Why or why not?
2. How do the interiors of Jupiter and Saturn differ? How does this affect their magnetic fields?
3. What is the difference between a belt and a zone?
4. How can you be certain that Jupiter's ring does not date from the formation of the planet?
5. If Jupiter had a satellite the size of our own moon orbiting outside the orbit of Callisto, what would you predict for its density and surface features?
6. Why are there no craters on Io and few on Europa? Why should you expect Io to suffer more impacts per square kilometer than Callisto?
7. Why are the belts and zones on Saturn less distinct than those on Jupiter?
8. If Saturn had no moons, what do you suppose its rings would look like?
9. Where did the particles in Saturn's rings come from?
10. How can Titan keep an atmosphere when it is smaller than airless Ganymede?
11. If you piloted a spacecraft to visit Saturn's moons and wanted to land on a geologically old surface, what features would you look for? What features would you avoid?
12. Why does the leading side of some satellites differ from the trailing side?
13. **How Do We Know?** Why would you expect research in archaeology to be less well funded than research in chemistry?

## Discussion Questions

1. Some astronomers argue that Jupiter and Saturn are unusual, while other astronomers argue that all solar systems should contain one or two such giant planets. What do you think? Support your argument with evidence.
2. Why don't the Terrestrial planets have rings?

## Problems

1. What is the maximum angular diameter of Jupiter as seen from Earth? What is the minimum? Repeat this calculation for Saturn and Titan. (*Hint:* Use the small-angle formula, Chapter 3.)
2. The highest-speed winds on Jupiter are in the equatorial jet stream, which has a velocity of 150 m/s. How long does it take for these winds to circle Jupiter?
3. What are the orbital velocity and period of a ring particle at the outer edge of Jupiter's ring? At the outer edge of Saturn's A ring? (*Hints:* The radius of the edge of the A ring is 136,500 km. Also, see Chapter 5.)
4. What is the angular diameter of Jupiter as seen from the surface of Callisto? (*Hint:* Use the small-angle formula, Chapter 3.)
5. What is the escape velocity from the surface of Ganymede if its mass is $1.5 \times 10^{23}$ kg and its radius is $2.6 \times 10^{3}$ km? (*Hint:* See Chapter 5.)
6. If you were to record the spectrum of Saturn and its rings, you would find light from one edge of the rings redshifted and light from the other edge blueshifted. If you observed at a wavelength of 500 nm, what difference in wavelength should you expect between the two edges of the rings? (*Hints:* See Problem 3 and Chapter 7.)
7. What is the difference in orbital velocity between particles at the outer edge of Saturn's B ring and particles at the inner edge of the B ring? (*Hint:* The outer edge of the B ring has a radius of 117,500 km, and the inner edge has a radius of 92,000 km.)
8. What is the difference in orbital velocity between the two coorbital satellites if the semimajor axes of their orbits are 151,400 km and 151,500 km? The mass of Saturn is $5.7 \times 10^{26}$ kg. (*Hint:* See Chapter 5.)

## Learning to Look

1. This image shows a segment of the surface of Jupiter's moon Callisto. Why is the surface dark? Why are some craters dark and some white? What does this image tell you about the history of Callisto?

NASA/JPL

2. The Cassini spacecraft recorded this image of Saturn's A ring and the Encke Gap. What do you see in this photo that tells you about processes that confined and shape planetary rings?

NASA/JPL/Space Science Institute

# 24 Uranus, Neptune, and the Dwarf Planets

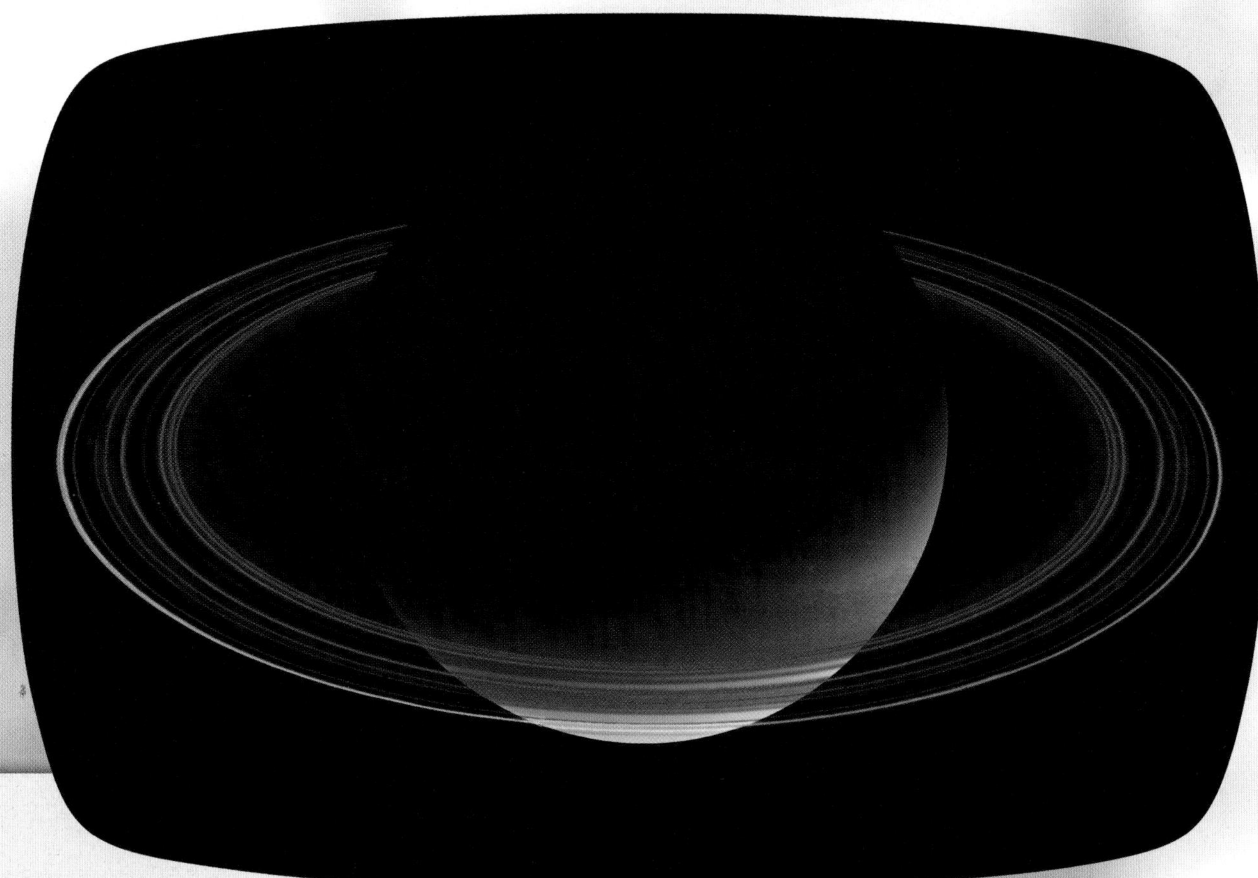

*Uranus is a cloudy, Jovian world far from the sun. It is orbited by dark, rocky particles that make up narrow rings much enhanced in this artist's impression.* (Bryan Brandenburg)

## Guidepost

Two planets circle the sun in the twilight beyond Saturn. You will find Uranus and Neptune substantially different from Jupiter and Saturn but still recognizable as Jovian planets. As you explore further you will also discover a family of dwarf planets, including Pluto, which appear to be leftover planet construction material. This chapter will help you answer four essential questions:

- **How are Uranus and Neptune similar to, and different from, Jupiter and Saturn?**
- **How did Uranus and Neptune, along with their rings and moons, form and evolve?**
- **How is Neptune different from Uranus?**
- **How are Pluto and the dwarf planets related to the origin of the solar system?**

As you finish this chapter, you will have visited all of the major worlds in our solar system. But there is more to see. Vast numbers of small rocky and icy bodies orbit among the planets, and the next chapter will introduce you to these fragments from the age of planet building.

*A good many things go around in the dark besides Santa Claus.*

— HERBERT HOOVER

OUT IN THE DARKNESS BEYOND SATURN, out where sunlight is 1000 times fainter than on Earth, there are objects orbiting the sun that Aristotle, Galileo, and Newton never imagined. They knew about Mercury, Venus, Mars, Jupiter, and Saturn, but our solar system also includes worlds that were not discovered until after the invention of the telescope. The stories of those discoveries highlight the process of scientific discovery, and the characteristics of these dimly lit worlds will reveal more of nature's secrets from the birth of the solar system.

## 24-1 Uranus

IN MARCH 1781, Benjamin Franklin was in France raising money, troops, and arms for the American Revolution. George Washington and his colonial army were only six months away from the defeat of Cornwallis at Yorktown and the end of the war. In England, King George III was beginning to show signs of madness. And a German-speaking music teacher in the English resort city of Bath was about to discover the planet Uranus.

### The Discovery of Uranus

William Herschel (■ Figure 24-1) came from a musical family in Hanover, Germany, but emigrated to England as a young man and eventually obtained a prestigious job as the organist at the Octagon Chapel in Bath.

To compose exercises for his students and choral and organ works for the chapel, Herschel studied the mathematical principles of musical harmony from a book by Professor Robert Smith of Cambridge. The mathematics in the book were so interesting that Herschel searched out other works by Smith, including a book on optics. Of course, it is not surprising that an 18th-century book on optics written by a professor in England relied heavily on Isaac Newton's discoveries and described the principles

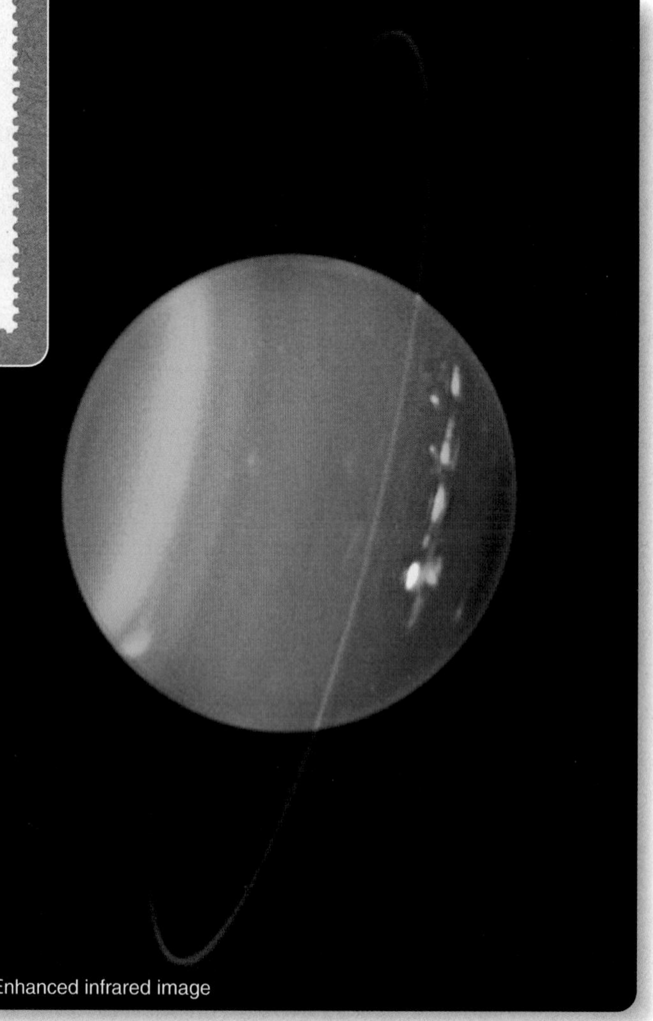


■ **Figure 24-1**

(a) When William Herschel discovered Uranus in 1781, he saw only a tiny green-blue dot. He never knew how interesting that planet is. Even the designer of this stamp marking the passage of Comet Halley in 1986 had to guess at the planet's appearance. (b) An image of Uranus recorded by a near-infrared camera on the Keck telescope shows banding and cloud features. Seasonal changes are observed on Uranus as spring comes to the northern hemisphere. The rings look red in this view because of image processing. (Lawrence Sromovsky, UW-Madison Space Science and Engineering Center)

of Newton's reflecting telescope (see Chapter 6). William Herschel and his brother Alexander began building telescopes, and William went on to study astronomy in his spare time.

Herschel's telescopes were similar to Newton's in that they had metal mirrors, but Herschel's were much larger. Newton's telescope had a mirror about 1 in. in diameter, but Herschel developed ways of making much larger mirrors, and he soon had telescopes as long as 20 ft. One of his favorite telescopes was 7 ft long and had a mirror 6.2 in. in diameter. Using this telescope, he began the research project that led to the discovery of Uranus.

Herschel did not set out to search for a planet; he was trying to detect stellar parallax produced by Earth's motion around the sun (see Figure 9-3). No one had yet detected this effect, although by the 1700s all astronomers accepted the idea that Earth moved. Galileo had pointed out that parallax might be detected if a nearby star and a very distant star lay so nearly along the same line of sight that they looked like a very close double star through a telescope. In such a case, Earth's orbital motion would produce a parallactic shift in the position of the nearby star with respect to the more distant star (■ Figure 24-2). Herschel began to examine all stars brighter than eighth magnitude to search for double stars that might show parallax. That project alone took over two years. In the process, of course, he found many double stars that were binary stars that are actually close together in space and orbiting each other (see Chapter 9).

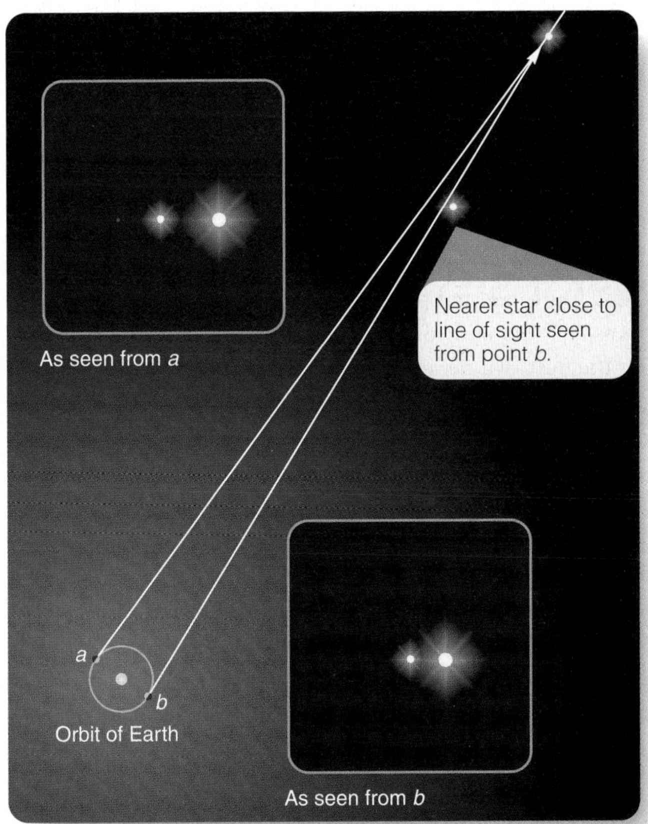


**■ Figure 24-2**

A double star consisting of a nearby star and a more distant star could be used to detect stellar parallax. Seen from point *b* in Earth's orbit, the two stars appear closer together than from point *a*. The effect is actually much too small to detect by too small to detect by the unaided eye.

On the night of March 13, 1781, Herschel set up the 7-ft-long telescope in his back garden and continued his work. He later wrote, "In examining the small stars in the neighborhood of H Geminorum, I perceived one that appeared visibly larger than the rest." As seen from Earth, Uranus is never larger in angular diameter than 3.7 arc seconds, so Herschel's detection of the disk illustrates the quality of his telescope and his eye. At first he suspected that the object was a comet, but other astronomers quickly realized that it was a planet orbiting the sun beyond Saturn.

The discovery of Uranus made Herschel world famous. Since antiquity, astronomers had known of five planets—Mercury, Venus, Mars, Jupiter, and Saturn—and they had supposed that the list was complete. Herschel's discovery extended the classical universe by adding a new planet. The English public accepted Herschel as their astronomer-hero, and, having named the new planet *Georgium Sidus* (George's Star) after King George III, Herschel received a royal pension. The former music teacher was welcomed into court society, where he eventually met and married a wealthy widow and took his place as one of the great English astronomers. His new financial position allowed him to build large telescopes on his estate and, with his sister Caroline, also a talented astronomer, he attempted to map the extent of the universe. You read about their "star gauging" research in Chapter 15.

Continental astronomers were less than thrilled that an Englishman had made such a great discovery, and even some professional English astronomers thought Herschel a mere amateur. They called his discovery a lucky accident. Herschel defended himself by making three points. First, he had built some of the finest-quality telescopes then in existence. Second, he had been conducting a systematic research project and would have found Uranus eventually because he was inspecting all of the brighter stars visible with his telescope. And third, he had great experience seeing fine detail with his telescopes. As a musician, he knew the value of practice and applied it to the business of astronomical observing. In fact, records show that other astronomers had seen Uranus at least 17 times before Herschel, but each time they failed to notice that it was not a star. They plotted Uranus on their charts as if it were just another faint star.

This illustrates one of the ways in which scientific discoveries are made. Often, discoveries seem accidental, but on closer examination you find that the scientist has earned the right to the discovery through many years of study and preparation (**How Do We Know? 24-1**). To quote a common saying, "Luck is what happens to people who work hard."

Continental astronomers, especially the French, insisted that the new planet not be named after an English king. They, along with many other non-English astronomers, stubbornly called the planet Herschel. Years later, German astronomer Johann Bode suggested the name Uranus, one of the most senior of the Greek gods.

# How Do We Know? 24-1

## Scientific Discoveries

**Why didn't Galileo expect to discover Jupiter's moons?** In 1928, Alexander Fleming noticed that bacteria in a culture dish were avoiding a spot of mold. He went on to discover penicillin. In 1895, Conrad Roentgen noticed a fluorescent screen glowing in his laboratory when he experimented with other equipment. He discovered X-rays. In 1896, Henri Becquerel stored a uranium mineral on a photographic plate safely wrapped in black paper. The plate was later found to be fogged, and Becquerel discovered natural radioactivity. Like many discoveries in science, these seem to be accidental; but, as you have seen in this chapter, "accidental" doesn't quite describe what happened.

The most important discoveries in science are those that totally change the way people think about nature, and it is very unlikely that anyone would predict such discoveries. For the most part, scientists work within a paradigm (look back at **How Do We Know? 4-1**), a set of models, hypotheses, theories, and expectations about nature, and it is very difficult to imagine natural events that lie beyond that paradigm. Ptolemy, for example, could not have imagined galaxies because they were not part of his geocentric paradigm. That means that the most important discoveries in science are almost always unexpected.

An unexpected discovery, however, is not the same as an accidental discovery. Fleming discovered penicillin in his culture dish not because he was the first to see it, but because he had studied bacterial growth for many years; so, when he saw what many others must have seen before, he recognized it as important. Roentgen realized that the glowing screen in his lab was important, and Becquerel didn't discard that fogged photographic plate. Long years of experience prepared them to recognize the significance of what they saw.

A historical study has shown that each time astronomers build a telescope that greatly surpasses existing telescopes in capability, their most important discoveries are unexpected. Herschel didn't expect to discover Uranus with his 7-foot telescope, and modern astronomers didn't expect to discover evidence of dark energy with the Hubble Space Telescope.

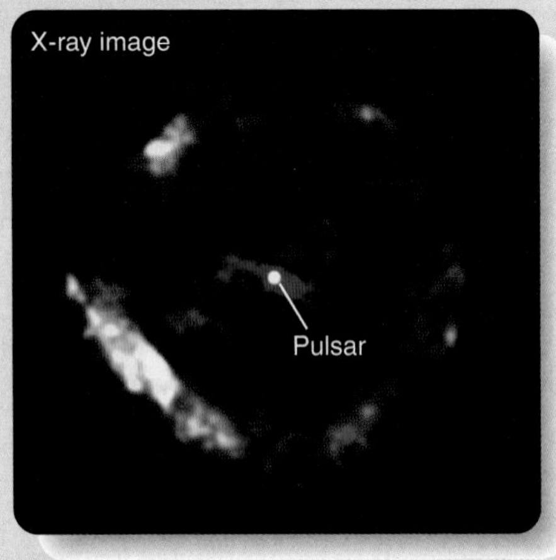


*The discovery of pulsars, spinning neutron stars, was totally unexpected, as important scientific discoveries often are.* (NASA/McGill/V. Kaspi et al.)

Scientists pursuing basic research are rarely able to explain the potential value of their work, but that doesn't make their discoveries accidental. They earn their right to those lucky accidents.

Over the half-century following the discovery of Uranus, astronomers noted that Newton's laws did not exactly predict the observed position of the planet. Tiny variations in the orbital motion of Uranus eventually led to the discovery of Neptune, a controversial story you will read later in this chapter.

## The Motion of Uranus

Uranus orbits nearly 20 AU from the sun and takes 84 years to go around once (**Celestial Profile 9**). The ancients thought of Saturn as the slowest of the planets, but Saturn orbits in a bit over 29 years. Uranus, being farther from the sun, moves even slower than Saturn and has a longer orbital peri od.

The rotation of Uranus is peculiar. Earth rotates approximately upright in its orbit. That is, its axis of rotation is inclined only 23.5° from the perpendicular to its orbit. The other planets have similarly moderate axial inclinations. Uranus, in contrast, rotates on an axis that is inclined 97.9° from the perpendicular to its orbit. It rotates on its side (■ Figure 24-3).

Because of its odd axial tilt, seasons on Uranus are extreme. The first good photographs of Uranus were taken in 1986, when the Voyager 2 spacecraft flew past. At that time, Uranus was in the segment of its orbit in which its south pole faced the sun. Consequently, its southern hemisphere was bathed in continuous sunlight, and a creature living on Uranus (an unlikely possibility, as you will discover later) would have seen the sun near the planet's south celestial pole. The sun was at southern solstice on Uranus in 1986, and you can see this at lower left in Figure 24-3. Over the next two decades, Uranus moved about a quarter of the way around its orbit, and, with the sun shining down from above the planet's equator, a citizen of Uranus would see the sun rise and set with the rotation of the planet. The sun reached equinox on Uranus in December 2007, and you can see that geometry by looking at the lower right in Figure 24-3. As Uranus continues along its orbit, the sun approaches the planet's north celestial pole, and the southern hemisphere of the planet experiences a lightless winter lasting 21 Earth years. In other words, the ecliptic on Uranus passes very near the planet's celestial poles, and the result is strong seasonal variation.

## The Atmosphere of Uranus

Like Jupiter and Saturn, Uranus has no surface. The gases of its atmosphere—mostly hydrogen, 15 percent helium, and a few percent methane, ammonia, and water vapor—blend gradually into a fluid interior.

Seen through Earth-based telescopes, Uranus is a small, featureless greenish-blue disk. The green-blue color arises because the atmosphere contains methane, a good absorber of longer-wavelength photons. As sunlight penetrates into the atmosphere and is scattered back out, the longer-wavelength (red) photons are more likely to be absorbed. That means that the sunlight reflecting off Uranus and then entering your eye is richer in blue photons, giving the planet a blue color.

As Voyager 2 drew closer to the planet in late 1985, astronomers studied the images radioed back to Earth. Uranus was a pale green-blue ball with no obvious clouds, and only when the images were carefully computer enhanced was any banded structure detected (■ Figure 24-4). A few very high clouds of methane ice particles were detected, and their motions allowed astronomers to make the first good measurement of the planet's rotation period.

You can understand the nearly featureless appearance of the atmosphere by studying the temperature profile of Uranus shown in ■ Figure 24-5. The atmosphere of Uranus is much colder than that of Saturn or Jupiter. Consequently, the three cloud layers of ammonia, ammonia hydrosulfide, and water that form the belts and zones in the atmospheres of Jupiter and Saturn lie very deep in the atmosphere of Uranus. These cloud layers, if they exist at all in Uranus, are not visible because of the thick atmosphere of hydrogen through which an observer has to look. The clouds that are visible on Uranus are clouds of methane ice crystals, which form at such a low temperature that they occur high in the atmosphere of Uranus. Figure 24-5 shows that there can be no methane clouds on Jupiter because that planet is too warm. The coldest part of Saturn's atmosphere is just cold enough to form a thin methane haze high above its more visible cloud layers. (See Figure 23-15b.)

The clouds and atmospheric banding that are faintly visible on Uranus appear to be the result of belt–zone circulation, which is a bit surprising. Uranus rotates on its side, so solar energy strikes its surface with geometry quite different than that for Jupiter and Saturn. Evidently belt–zone circulation is dominated by the rotation of the planet and not by the direction of sunlight.

The Voyager 2 images from 1986 made some astronomers expect that Uranus was always a nearly featureless planet, but later observations have revealed that Uranus has seasons. Since 1986, Uranus has moved along its orbit, and spring has come to its northern hemisphere. The Hubble Space Telescope and giant Earth-based telescopes have detected changing clouds on Uranus, including a dark cloud that may be a vortex resembling the spots on Jupiter (■ Figure 24-6). The clouds appear to be part of a seasonal cycle on Uranus, but its year lasts 84 Earth years, so you will have to be patient to see summer come to its northern hemisphere.

## The Interior of Uranus

Astronomers cannot describe the interiors of Uranus and Neptune as accurately as they can the interiors of Jupiter and Saturn. Observational data are sparse, and the materials inside these planets are not as easy to model as simple liquid hydrogen.

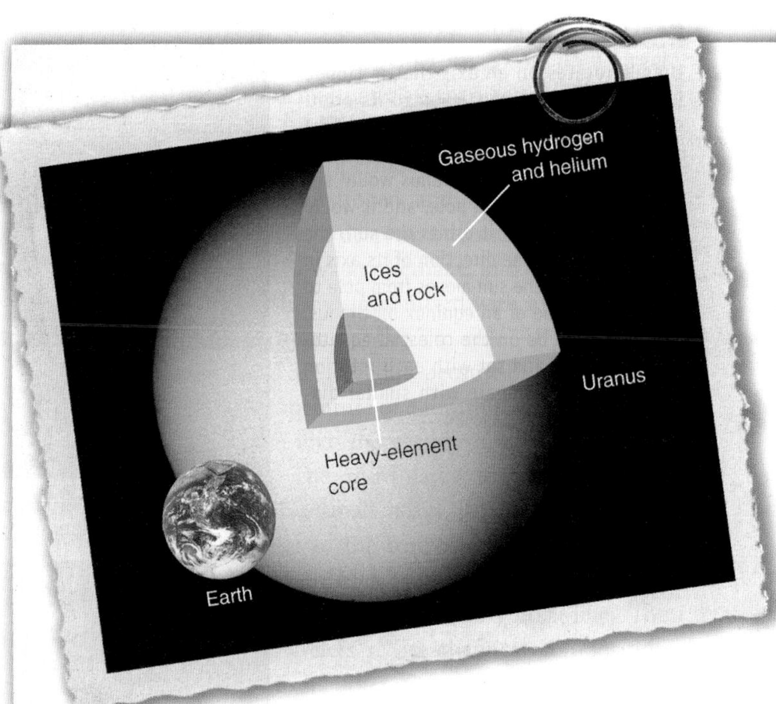


*Uranus rotates on its side. When Voyager 2 flew past in 1986, the planet's south pole was pointed almost directly at the sun.* (NASA)

# *Celestial Profile 9: Uranus*

## *Motion:*

| | |
|---|---|
| Average distance from the sun | 19.2 AU ($2.87 \times 10^9$ km) |
| Eccentricity of orbit | 0.046 |
| Inclination of orbit to ecliptic | 0.8° |
| Average orbital velocity | 6.8 km/s |
| Orbital period | 84.0 y |
| Period of rotation | 17.23 h |
| Inclination of equator to orbit | 97.9° (retrograde rotation) |

## *Characteristics:*

| | |
|---|---|
| Equatorial diameter | $5.11 \times 10^4$ km (4.01 $D_\oplus$) |
| Mass | $8.69 \times 10^{25}$ kg (14.5 $M_\oplus$) |
| Average density | 1.29 g/cm$^3$ |
| Gravity | 0.9 Earth gravity |
| Escape velocity | 22 km/s (2.0 $V_\oplus$) |
| Temperature above cloud tops | −220°C (−365°F) |
| Albedo | 0.35 |
| Oblateness | 0.023 |

## *Personality Point:*

Most creation stories begin with a separation of opposites, and Greek mythology is no different. Uranus (the sky) separated from Gaia (Earth) who was born from the void, Chaos. They gave birth to the giant Cyclops, Cronos (Saturn, father of Zeus), and his fellow Titans. Uranus is sometimes called the starry sky, but the sun (Helios), moon (Selene), and the stars were born later, so Uranus, one of the most ancient gods, began as the empty, dark sky.

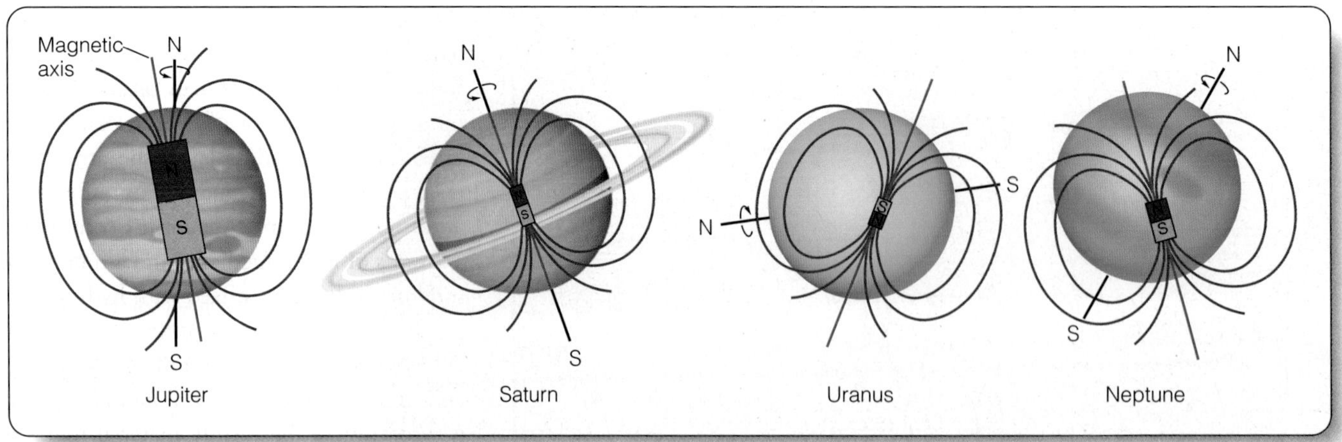


■ **Figure 24-7**

The magnetic fields of Uranus and Neptune are peculiar. While the magnetic axis of Jupiter is tipped only 10° from its axis of rotation, and the magnetic axis of Saturn is not tipped at all, the magnetic axes of Uranus and Neptune are tipped at large angles. Furthermore, the magnetic fields of Uranus and Neptune are offset from the center of the planet. This suggests that the dynamo effect operates differently in Uranus and Neptune than in Jupiter, Saturn, and Earth.

solar wind and traps some charged particles to create weak radiation belts in the planet's magnetosphere. High-speed electrons spiraling along the magnetic field produce synchrotron radio emission just as around Jupiter, and the Voyager 2 spacecraft recorded this radiation fluctuating with a period of 17.2 hours—the period of rotation of the magnetic field and, presumably, the planetary interior.

The magnetic field and the high inclination of the planet produce some peculiar effects. As is the case for all planets with magnetic fields, the solar wind deforms the magnetosphere and draws it out into a long tail extending away from the planet in the direction opposite the sun. The rapid rotation of Uranus and its high inclination give the magnetosphere and its long extension a corkscrew shape. At the time Voyager 2 flew past in 1986, the south pole of Uranus was pointed nearly at the sun, and once during each rotation the solar wind poured down into the south magnetic pole. The resulting interaction produced strong auroras that Voyager 2 detected in the ultraviolet at both magnetic poles (■ Figure 24-8). In the years since, Uranus has moved farther

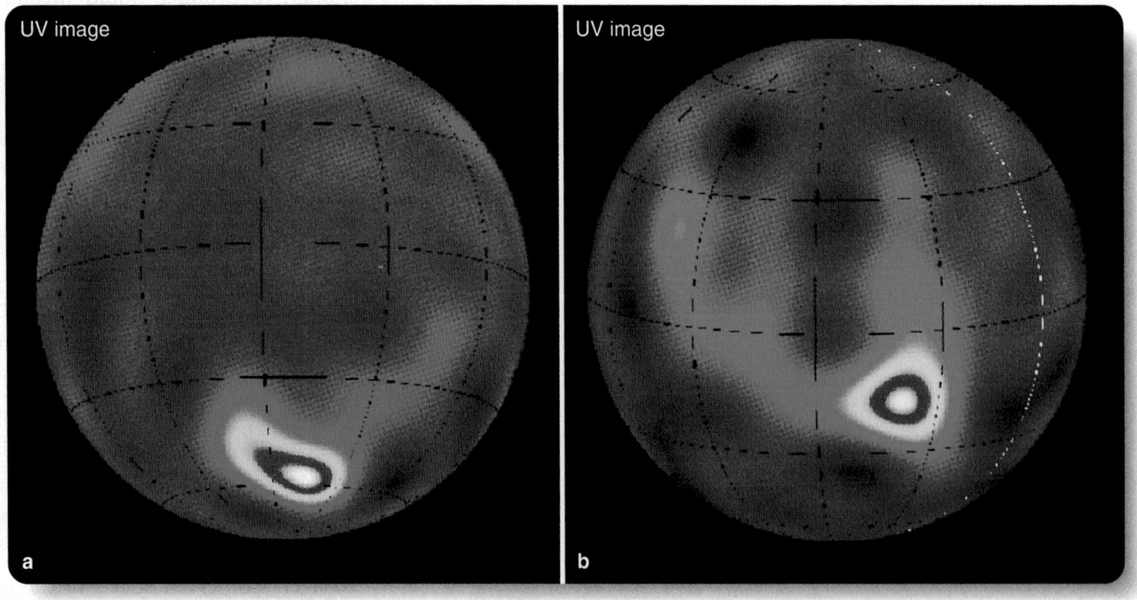


■ **Figure 24-8**

Auroras on Uranus were detected in the ultraviolet by the Voyager 2 spacecraft when it flew by in 1986. These maps of opposite sides of Uranus show the location of auroras near the magnetic poles. Recall that the magnetic field is highly inclined and offset from the planet's center, so the magnetic poles do not lie near the poles defined by rotation. The white dashed line marks zero longitude. (Courtesy Floyd Herbert, LPL)

around its orbit, so the geometry of its interaction with the solar wind is now different. Unfortunately, no spacecraft orbit Uranus at present, and there is no way to observe effects of these changes in detail.

Like the magnetic field, the temperature of Uranus can reveal something about its interior. Jupiter and Saturn are warmer than you would expect, given the amount of energy they receive from the sun, and this means that heat is leaking out from their hot interiors. Uranus, in contrast, is about the temperature you would expect for a world at its distance from the sun. Apparently, it has lost much of its interior heat. Yet, it must have some internal heat to cause convection in the fluid mantle and drive the dynamo effect. The temperature in its core is estimated from models to be about 8000 K. The decay of natural radioactive elements would generate heat, but some astronomers have suggested that the slow settling of heavier elements through the fluid mantle could also release energy to warm the interior.

Laboratory studies of methane show that it can break down under the temperature and pressure found inside Uranus and form various compounds plus pure carbon in the form of diamonds. If this happens in Uranus, the diamond crystals would fall inward, warming the interior through friction. Determining for sure whether a planetwide rain of diamonds actually exists inside Uranus is probably forever beyond human reach.

For a Jovian world, Uranus seems small and mostly featureless. But now you are ready to visit one of its best attractions—its rings.

## The Rings of Uranus

Both Uranus and Neptune have rings that are more like those of Jupiter than those of Saturn. They are dark, faint, not easily visible from Earth, and confined by shepherd satellites.

Study **The Rings of Uranus and Neptune** on pages 534–535 and notice three important points about the rings of Uranus and one new term:

1. The rings of Uranus were discovered during an *occultation* when Uranus crossed in front of a star.

2. The rings are made up of a thin layer of very dark boulders. They are confined by small moons, and except for the outermost rings, they contain little dust.

3. Like the rings of Jupiter and Saturn, the rings around Uranus and Neptune cannot survive for long periods. All the Jovian rings need to be resupplied with material from impacts on moons.

When you read about Neptune's rings later in this chapter, you will return to this artwork and see how closely the two ring systems compare.

Images made with the Hubble space telescope in 2003 and 2005 have revealed two larger, fainter, dustier rings lying outside the previously known ring system (■ Figure 24-9). The larger of

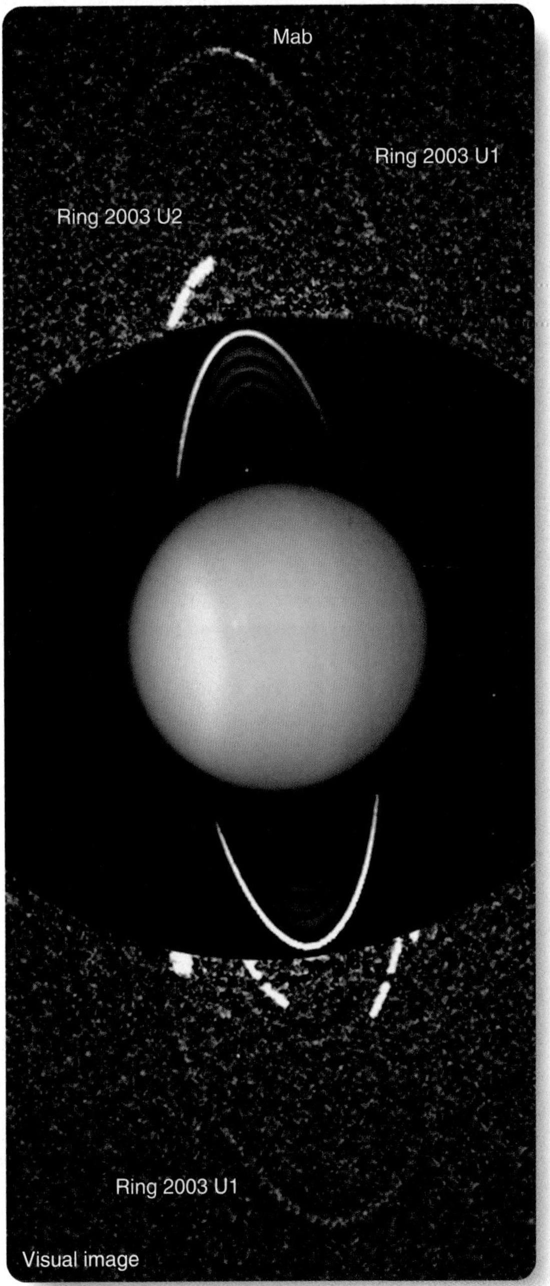


■ **Figure 24-9**

Two newly discovered rings orbit Uranus far outside the previously known rings. The outermost ring follows the orbit of the small moon Mab, only 12 km (7 mi) in diameter. The short bright arcs in this photo were caused by moons moving along their orbits during the long time exposure. (NASA, ESA, M. Showalter, SETI Institute)

these rings coincides with the orbit of the small moon Mab and is probably replenished by particles blasted off of the moon by meteorite impacts. The smaller of the rings is confined between the orbits of the moons Portia and Rosalind.

As you read about planetary rings, notice their close relationship with moons. Because of collisions among ring particles, planetary rings tend to spread outward, almost like an expanding

# The Rings of Uranus and Neptune

**1** The rings of Uranus were discovered in 1977, when Uranus crossed in front of a star. During this **occultation**, astronomers saw the star dim a number of times before and again after the planet crossed over the star. The dips in brightness were caused by rings circling Uranus.

More rings were discovered by Voyager 2. The rings are identified in different ways depending on when and how they were discovered.

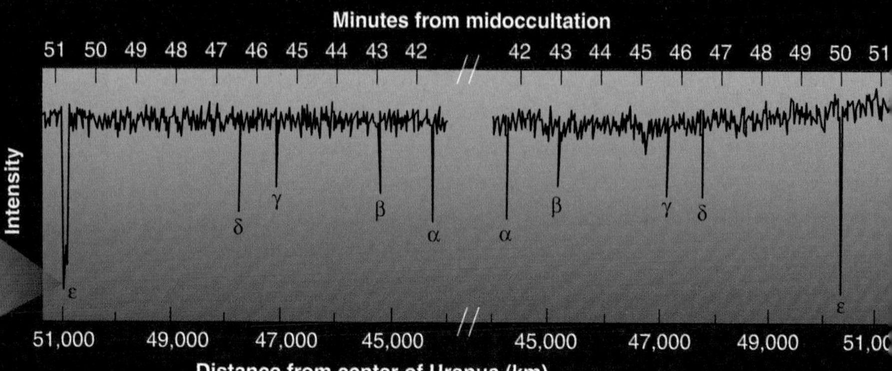


Notice the eccentricity of the ε ring. It lies at different distances on opposite sides of the planet.

**2** The albedo of the ring particles is only about 0.015, darker than lumps of coal. If the ring particles are made of methane-rich ices, radiation from the planet's radiation belts could break the methane down to release carbon and darken the ices. The same process may darken the icy surface of Uranian moons.

The narrowness of the rings suggests they are shepherded by small moons. Voyager 2 found Ophelia and Cordelia shepherding the ε ring. Other small moons must be shepherding the other narrow rings. Such moons must be structurally strong to hold themselves together inside the planet's Roche limit.

ε
λ
η
γ
δ
β
α
6
5
4
1986 U2R
Ophelia
Cordelia

The eccentricity of the ε ring is apparently caused by the eccentric orbits of Ophelia and Cordelia.

**2a** When the Voyager 2 spacecraft looked back at the rings illuminated from behind by the sun, the rings were not bright. That is, the rings are not bright in forward-scattered light. That means they must not contain small dust particles. The nine main rings contain particles no smaller than meter-sized boulders.

Uranus

**3** Ring particles don't last forever as they collide with each other and are exposed to radiation. The rings of Uranus may be resupplied with fresh particles occasionally as impacts on icy moons scatter icy debris.

Collisions among the large particles in the ring produce small dust grains. Friction with Uranus's tenuous upper atmosphere plus sunlight pressure act to slow the dust grains and make them fall into the planet. The Uranian rings actually contain very little dust.

The rings of Neptune are bright in forward-scattered light, as in the image above, and that indicates that the rings contain significant amounts of dust. The ring particles are as dark as those that circle Uranus, so they probably also contain methane-rich ice darkened by radiation.

**4** The brightness of Neptune is hidden behind the black bar in this Voyager 2 image. Two narrow rings are visible, and a wider, fainter ring lies closer to the planet. More ring material is visible between the two narrow rings.

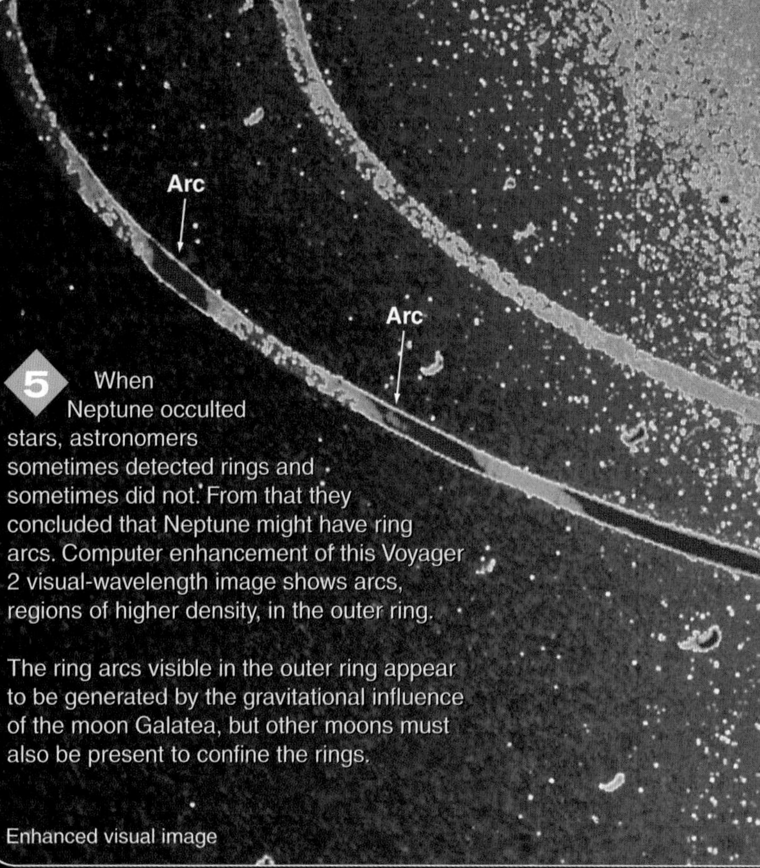


**5** When Neptune occulted stars, astronomers sometimes detected rings and sometimes did not. From that they concluded that Neptune might have ring arcs. Computer enhancement of this Voyager 2 visual-wavelength image shows arcs, regions of higher density, in the outer ring.

The ring arcs visible in the outer ring appear to be generated by the gravitational influence of the moon Galatea, but other moons must also be present to confine the rings.

**4a** Neptune's rings lie in the plane of the planet's equator and inside the Roche limit. The narrowness of the rings suggests that shepherd moons must confine them, and a few such moons have been found among the rings. There must be more undiscovered small moons to confine the rings completely.

Naiad

Galatea

Thalassa

Adams

LeVerrier

Despina

Galle

**4b** Neptune's rings have been given names associated with the planet's history. English astronomer Adams and French astronomer LeVerrier predicted the existence of Neptune from the motion of Uranus. The German astronomer Galle discovered the planet in 1846 based on LeVerrier's prediction.

**4c** Like the rings of the other Jovian planets, the ring particles that orbit Neptune cannot have survived since the formation of the planet. Occasional impacts on Neptune's moons must scatter debris and resupply the rings with fresh particles.

Neptune

gas. If a planet had no moons, its rings would spread out into a more and more tenuous sheet until they were gone. The spreading rings can be anchored by small shepherd moons, which interact gravitationally with wandering ring particles, absorb orbital energy, and cause the particles to remain within the rings. As they gain orbital energy, these small moons move slowly outward, but they can be anchored in turn by orbital resonances with larger, more distant moons that are so massive they do not get pushed outward significantly. In this way, a system of moons can confine and preserve a system of planetary rings.

## The Moons of Uranus

Uranus has five large regular moons that were discovered from Earth-based observations (■ Figure 24-10). Those five moons, from the outermost inward, are Oberon, Titania, Umbriel, Ariel, and Miranda. The names Umbriel and Ariel are names from Alexander Pope's *The Rape of the Lock,* and the rest are from Shakespeare's *A Midsummer Night's Dream* and *The Tempest* (an Ariel also appears in *The Tempest*). Spectra show that the moons contain frozen water, although their surfaces are dark. Planetary scientists assumed they were made of dirty ices, but little more was known of the moons before Voyager 2 flew through the system.

In addition to imaging the known moons, the Voyager 2 cameras discovered ten more moons too small to have been seen from Earth. Since then, the construction of new-generation telescopes and the development of new imaging techniques (see Chapter 6) have allowed astronomers to find even more small moons orbiting Uranus. Roughly 30 are currently known, but there are almost certainly more to be found.

The smaller moons are all as dark as coal. They are icy worlds with surfaces that have been darkened by impacts vaporizing ice and concentrating embedded dirt. In addition, they orbit inside the radiation belts, and that radiation can convert methane trapped in their ices into dark carbon deposits to further darken their surfaces.

More is known about the five larger moons. They are all tidally locked to Uranus, and that means their south poles were pointed toward the sun in 1986. Voyager could not photograph their northern hemispheres, so the analysis of their geology must depend on images of only half their surfaces. The densities of the moons suggest that they contain relatively large rock cores surrounded by icy mantles, as shown in Figure 24-10.

Oberon, the outermost of the large moons, has a cratered surface, but there is visible evidence that it was once geologically active (■ Figure 24-11). A large fault crosses the sunlit hemisphere, and dark material, perhaps dirty water "lava," appears to have flooded the floors of some craters.

Titania is the largest of the five moons and has a heavily cratered surface, but it has no very large craters (Figure 24-11). This suggests that after the end of the heavy bombardment, the young Titania underwent an active phase in which its surface was flooded with water that covered early craters with fresh ice. Since then, the craters that have formed are not as large as the largest of those that were erased. The network of faults that crosses Titania's surface is another sign of past activity.

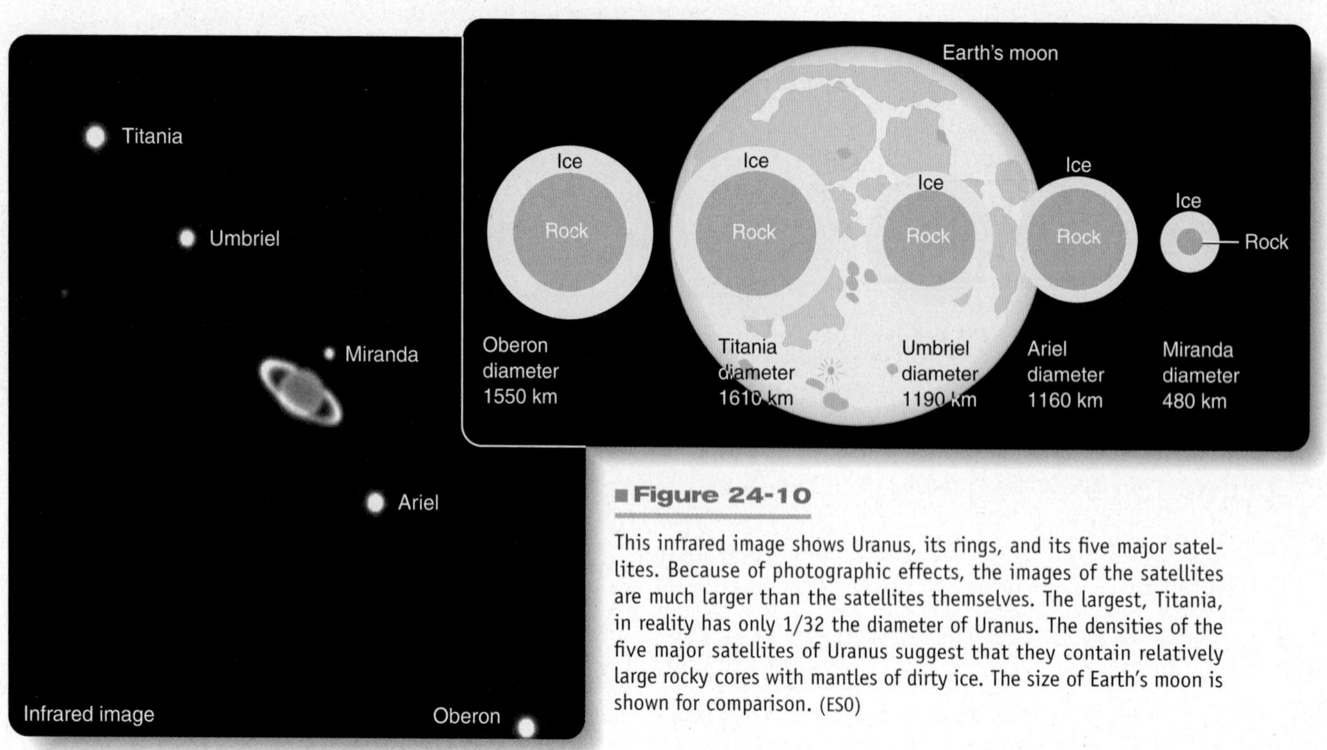


**■ Figure 24-10**

This infrared image shows Uranus, its rings, and its five major satellites. Because of photographic effects, the images of the satellites are much larger than the satellites themselves. The largest, Titania, in reality has only 1/32 the diameter of Uranus. The densities of the five major satellites of Uranus suggest that they contain relatively large rocky cores with mantles of dirty ice. The size of Earth's moon is shown for comparison. (ESO)

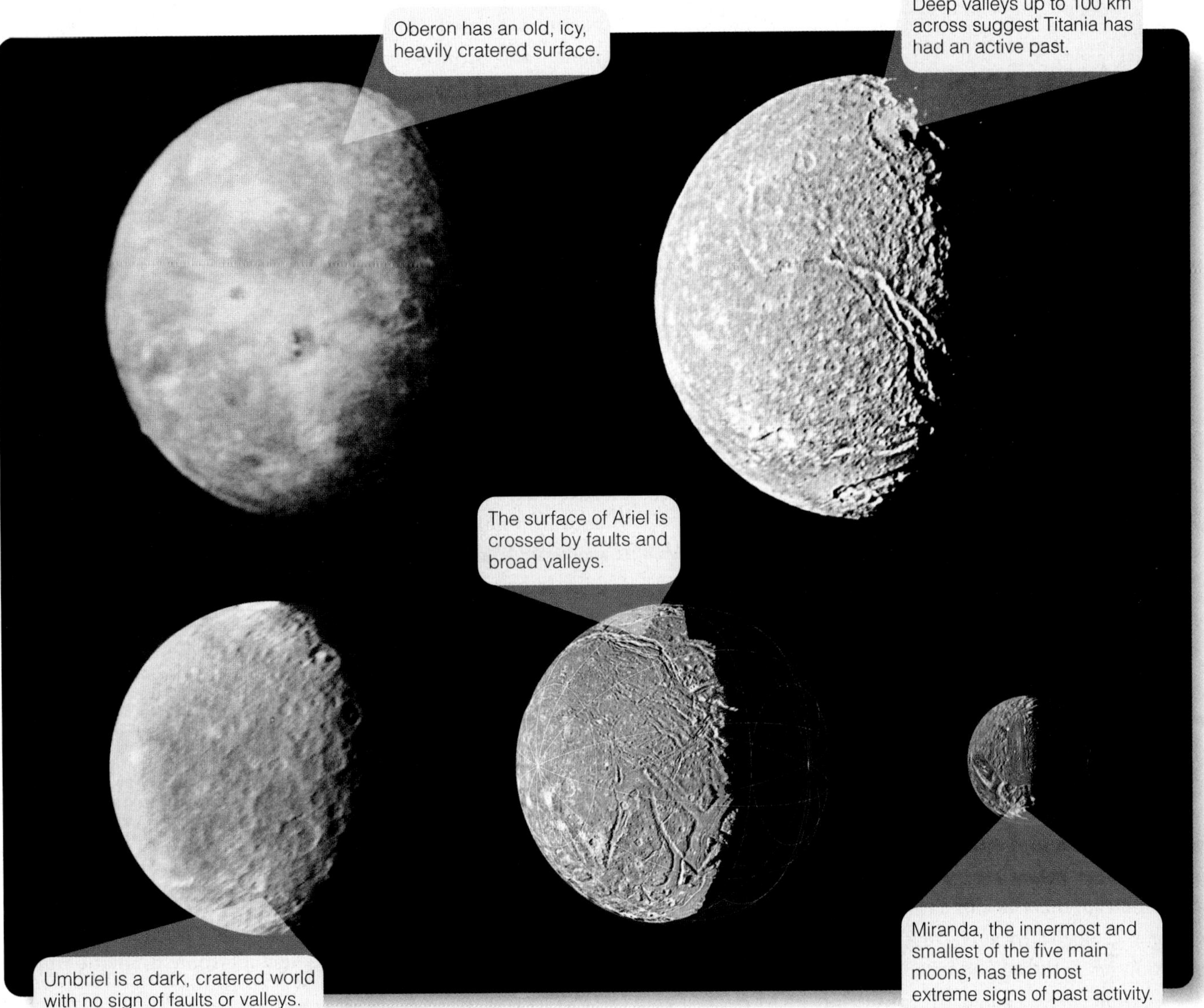


**Figure 24-11**

The five largest moons of Uranus, shown here in correct relative size scale, range from the largest, Titania, 46 percent the diameter of Earth's moon, down to Miranda, only 14 percent the diameter of Earth's moon. For a better view of Miranda, see the next figure. (Ariel and Miranda: U.S. Geological Survey, Flagstaff, Arizona; Other images: NASA)

Umbriel, the next moon inward, is a dark, cratered world with no sign of faults or surface activity (Figure 24-11). It is the darkest of the moons, with an albedo of only 0.16 compared with 0.25 to 0.45 for the other moons. Its crust is a mixture of rock and ice. A bright crater floor in one region suggests that clean ice may lie at shallow depths in some regions.

Ariel has the brightest surface of the five major Uranian satellites and shows clear signs of geological activity. It is crossed by faults over 10 km deep, and some regions appear to have been smoothed by resurfacing, as you can see in Figure 24-11. Crater counts show that the smoothed regions are in fact younger than the other regions. Ariel may have been subject to tidal heating caused by an orbital resonance with Miranda and Umbriel.

Miranda is a mysterious moon. As you can see in Figure 24-11, it is the smallest of the five large moons, but it appears to have been the most active. In fact, its active past appears to have been quite unusual. Miranda is marked by oval patterns of grooves known as **ovoids** (■ Figure 24-12). These features were originally hypothesized to have been produced when mutual gravitation pulled together fragments of an earlier moon shattered by a major impact. More recent studies of the ovoids show that they are associated with faults, ice-lava flows, and rotated blocks of crust, suggesting that a major impact was not involved. Rather, the ovoids may have been created by internal heat driving large but very slow convection currents in Miranda's icy mantle.

Certainly, Miranda has had a violent past. Near the equator, a huge cliff rises 20 km (12 mi). If you stood in your spacesuit at

**■ Figure 24-12**

Miranda, the smallest of the five major moons of Uranus, is only 480 km (290 mi) in diameter, but its surface shows signs of activity. This photomosaic of Voyager 2 images reveals that it is marked by great oval systems of grooves. The smallest features detected are about 1.5 km (1 mi) in diameter. (U.S. Geological Survey, Flagstaff, Arizona)

the top of the cliff and dropped a rock over the edge, it would fall for 10 minutes before hitting the bottom. Nevertheless, crater counts indicate that the cliff and the ovoids are old. Miranda is no longer geologically active, but you can read hints of its active past on its disturbed surface.

If the ovoids were caused by convection in Miranda's icy mantle, then heat rising from its interior must have been the dominant factor. Miranda is so small that its heat of formation must have been lost quickly, and there is no reason to expect it to have been strongly heated by radioactive decay. Tidal heating could have occurred if Miranda's orbit was once made slightly eccentric by a resonance with other moons. Your knowledge of tidal heating in other moons can help you understand Miranda.

## A History of Uranus

The challenge of comparative astronomy is to tell the story of a world, and Uranus may present the biggest challenge of all the objects in our solar system. Not only is it so distant that it is difficult to study, but it is also peculiar in many ways.

Uranus formed from the solar nebula, as did the other Jovian planets, but calculations show that Uranus and Neptune could not have grown to their present size in the slow-moving orbits they now occupy so far from the sun. Computer models of planet formation in the solar nebula suggest that Uranus and Neptune formed closer to the sun, in the neighborhood of Jupiter and Saturn. Gravitational interactions among the Jovian planets could have gradually moved Uranus and Neptune outward to their present locations. One such model has Neptune forming closer to the sun than Uranus and the two planets later switching places as they moved out. As you will learn later in this chapter, the migration of giant planets could have triggered the late heavy bombardment episode. These are interesting hypotheses, and they illustrate how uncertain the histories of the outer planets really are.

The highly inclined axis of Uranus may have originated late in its formation when it was struck by a planetesimal that would have to have been as large as Earth. That impact also may have disturbed the interior of the world and caused it to lose much of its heat. Uranus now radiates less than 10 percent more heat than it receives from the sun. There is now just enough heat flowing outward to drive circulation in its slushy mantle and generate its magnetic field. A more recent study shows that tidal interactions with Saturn could have altered Uranus's axis of rotation as it was pushed outward. In this case, an older catastrophic hypothesis is being challenged by a newer evolutionary hypothesis (**How Do We Know? 19-1**).

Impacts have been important in the history of the moons and rings of Uranus. All of the moons are cratered, and some show signs of large impacts. Meteorites and the nuclei of comets striking the moons may create debris that becomes trapped among the orbits of the smaller moons to produce the narrow rings. The ring particles observed now could not have lasted since the formation of the planet, so they must be replenished with fresh material now and then. When the final history of Uranus is written, it will surely include some dramatic impact events.

### SCIENTIFIC ARGUMENT

***How do astronomers know what the interior of Uranus is like?***
This argument must combine observation with theory. Obviously, you can't see inside the planet, so planetary scientists are limited to a few basic observations that they must connect with models in a chain of inference to describe the interior. First, the size of Uranus can be found from its angular diameter and distance, and you can find its mass by observing the orbital radii and orbital periods of its moons. Its mass divided by its volume equals its density, which implies that it must contain a certain proportion of dense material such as ice and rock, more than there is inside Saturn. Add to that the chemical composition obtained from spectra, and you have the data needed to build a mathematical model of the interior. Such models predict a core of heavy elements and a mantle of ices mixed with heavier material of rocky composition.

Going from observable properties to unobservable properties is the heart of astronomical research. Now use this approach to build a new argument: **What observational properties of the rings of Uranus show that small moons must orbit among the rings?**

## 24-2 Neptune

URANUS AND NEPTUNE are often discussed together. They are about the same size and density and are very similar planets in some ways. They do differ, however, in certain respects. Unlike Uranus, Neptune has a significant amount of heat flowing out from its interior. Also, Neptune has especially complicated ring and satellite systems. Even the discovery of Neptune differed from the discovery of Uranus.

### The Discovery of Neptune

The discovery of Neptune triggered one of the greatest controversies in the history of science. For over a century people told the story and took sides, but only in the last few years has the real story become known. You can follow the story and decide which side you favor.

In 1843, the young English astronomer John Couch Adams (1819–1892) completed his degree in astronomy and immediately began the analysis of one of the great problems of 19th-century astronomy. William Herschel had discovered Uranus in 1781, but earlier astronomers had seen the planet as early as 1690 and had mistakenly plotted it on charts as a star. When 19th-century astronomers tried to combine all those data, they didn't quite fit together. No planet obeying Newton's laws of motion and controlled only by the gravity of the sun and the other known planets could follow such an orbit.

Some astronomers suggested that the gravitational attraction of an undiscovered planet was causing the discrepancies. Adams began with the observed variations from the predicted positions of Uranus, never more than 2 arc minutes, and by October 1845 he had, through a laborious and difficult calculation, computed the orbit of the undiscovered planet. He sent his prediction to the Astronomer Royal, Sir George Airy, who passed it on to an observer who began a painstaking search of the area star by star.

Meanwhile, the French astronomer Urbain Jean Leverrier (1811–1877) made the same calculations and sent his predicted position of the planet to Johann Galle at the Berlin Observatory. Galle received Leverrier's prediction on the afternoon of September 23, 1846, and, after searching for 30 minutes that evening, found Neptune. It was only 1° from the position predicted by Leverrier.

The discovery of a new planet caused a sensation, but English astronomers didn't like a Frenchman getting all the credit. After all, they said, the planet was found only 2° from the position predicted by the orbit computed by their own astronomer, Adams. When the English announced Adams's work, the French suspected that he had plagiarized the calculations, and the controversy was bitter, as controversies between England and France often are. For over a century, historians of science have repeated the story of the young English astronomer who missed his chance because the astronomers in charge of the search were careless and slow.

Original papers related to Adam's calculations were lost for decades, but when they were found in 1998, they painted a different picture. Adams did the calculations correctly, but he computed only the orbit for the new planet. He didn't actually calculate its position in the sky along the orbit; that was left to the English astronomers conducting the search. Once the new planet was discovered, the English astronomers, out of national pride, pressed Adam's case further than they should have.

For 150 years, astronomers took sides or gave both astronomers equal credit. In fact, it seems clear that Leverrier deserves credit for making an accurate, useful prediction of a position on the sky and then pressing it aggressively on an astronomer who had the right skills to make the search. You should still give both astronomers credit for solving a difficult problem that was one of the great challenges of the age. A modern analysis shows that both Leverrier and Adams made unwarranted assumptions about the undiscovered planet's distance from the sun. By good fortune their assumptions made no difference in the 19th century, and the planet was close to the positions they predicted. Do you think they deserve less credit for that reason? They tried, and the other astronomers of the world didn't.

Leverrier and Adams could have been beaten to the discovery had Galileo paid a little less attention to Jupiter and a little more attention to what he saw in the background. Modern studies of Galileo's notebooks show that he saw Neptune on December 24, 1612, and again on January 28, 1613, but he plotted it as a star in the background of drawings of Jupiter. It is interesting to speculate about the response of the Inquisition had Galileo proposed that a planet existed beyond Saturn. Unfortunately for history, but perhaps fortunately for Galileo, he did not recognize Neptune as a planet, and its discovery had to wait another 234 years.

Historians of science have recounted the discovery of Neptune as a triumph for Newtonian physics: The three laws of motion and the law of gravity had proved sufficient to predict the position and orbit of an unseen planet. Thus, the discovery of Neptune was fundamentally different from the discovery of Uranus. Uranus was discovered "accidentally" in the course of Herschel's attempt to systematically observe the entire sky, whereas the existence of Neptune was predicted by Leverrier and Adams using basic laws.

### The Atmosphere and Interior of Neptune

Little was known about Neptune before the Voyager 2 spacecraft swept past it in 1989. Seen from Earth, Neptune is a tiny, blue-green dot never more than 2.3 arc seconds in diameter. Before Voyager 2, astronomers knew Neptune is almost four times the diameter of Earth, or about 4 percent smaller in diameter than Uranus (**Celestial Profile 10**), with a mass about 17 times that of Earth, 20 percent more than Uranus. Spectra revealed that, as in the case of Uranus, its blue-green color was caused by methane in its hydrogen-rich atmosphere absorbing red light. Neptune's density showed that it is a Jovian planet rich in hydrogen, but almost no detail was visible from Earth, so even its period of rotation was uncertain.

Voyager 2 passed only 4900 km (3050 mi) above Neptune's cloud tops, closer than any spacecraft had ever come to one of the Jovian planets. The images it captured revealed that Neptune is marked by dramatic belt–zone circulation parallel to the planet's equator. Voyager 2 also saw at least four cyclonic disturbances. The largest, dubbed the Great Dark Spot, looked similar to the Great Red Spot on Jupiter (■ Figure 24-13). Neptune's Great Dark Spot was located in the southern hemisphere and rotated around its center counterclockwise, with a period of about 16 days. Like the Great Red Spot, it appeared to be caused by gas rising from its planet's interior. Unexpectedly, when the Hubble Space Telescope began imaging Neptune in 1994, the Great Dark Spot was gone, and new cloud features were seen appearing and disappearing in Neptune's atmosphere (Figure 24-13). Evidently, the cyclonic disturbances on Neptune are not nearly as long lived as Jupiter's Great Red Spot, which has persisted for at least 400 years.

The Voyager 2 images reveal other cloud features standing out against the deep blue of the methane-rich atmosphere. The white clouds are made of crystals of frozen methane and range up to 50 km above the deeper layers, just where the temperature in Neptune's atmosphere is low enough for rising methane to freeze into crystals. (See Figure 24-5.) Presumably these features are related to rising convection currents that produce clouds high in Neptune's atmosphere where they catch sunlight and appear bright. Special filters can reveal these cloud belts in visual-wavelength images and at infrared wavelengths (■ Figure 24-14). Some observations made by the Hubble Space Telescope suggest that atmospheric activity on Neptune may be related to flares and other eruptions on the sun, but more data are needed to explore this connection.

As on the other Jovian worlds, winds circle Neptune parallel to its equator, but Neptune's winds blow at very high speeds and tend to blow backward—against the rotation of the planet. Why Neptune should have such high-speed retrograde winds is not understood, and it is part of the larger problem of understanding belt–zone circulation.

Now that you have seen belts and zones on all four of the Jovian planets (assuming that the very faint clouds observed on Uranus are in fact traces of belt–zone circulation), you can ask what drives this circulation. Because belts and zones remain parallel to a planet's equator even when the planet rotates at a high inclination, as in the case of Uranus, it seems reasonable to believe that the atmospheric circulation is dominated by the rotation of the planet, and perhaps also by heat flow and circulation currents in the liquid interior, but not by solar heating.

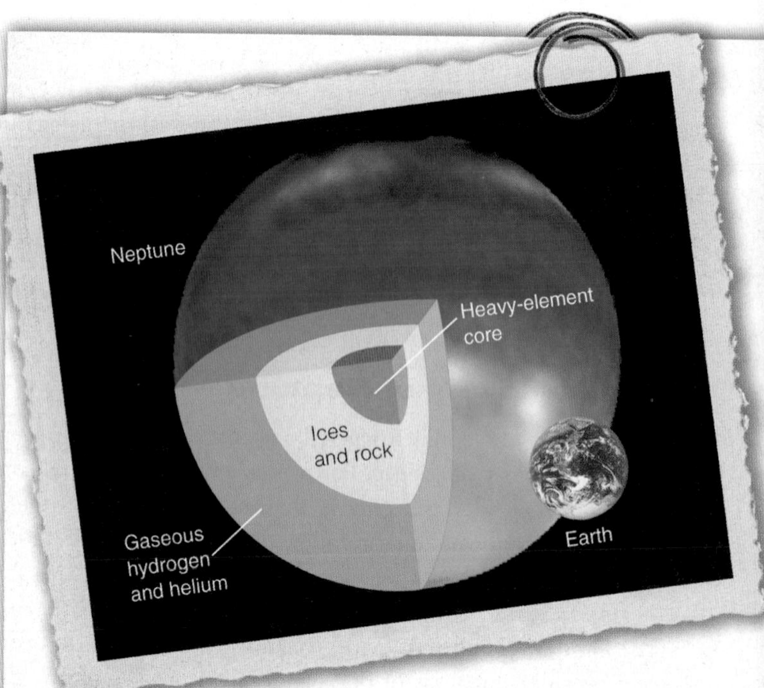


*Neptune was tipped slightly away from the sun when the Hubble Space Telescope recorded this image. The interior is much like Uranus's, but there is more heat.* (NASA)

## *Celestial Profile 10: Neptune*

### *Motion:*

| | |
|---|---|
| Average distance from the sun | 30.1 AU ($4.50 \times 10^9$ km) |
| Eccentricity of orbit | 0.010 |
| Inclination of orbit to ecliptic | 1.8° |
| Average orbital velocity | 5.4 km/s |
| Orbital period | 164.8 y |
| Period of rotation | 16.05 h |
| Inclination of equator to orbit | 28.8° |

### *Characteristics:*

| | |
|---|---|
| Equatorial diameter | $4.95 \times 10^5$ km (3.93 $D_{\oplus}$) |
| Mass | $1.03 \times 10^{26}$ kg (17.2 $M_{\oplus}$) |
| Average density | 1.66 g/cm³ |
| Gravity | 1.2 Earth gravities |
| Escape velocity | 25 km/s (2.2 $V_{\oplus}$) |
| Temperature at cloud tops | −215°C (−355°F) |
| Albedo | 0.35 |
| Oblateness | 0.017 |

### *Personality Point:*

Because the planet Neptune looked so blue, astronomers named it after the Roman god of the sea, Neptune (Poseidon to the Greeks). His wife was Amphitrite, granddaughter of Ocean, one of the Titans. Neptune controlled the storms and waves and was a powerful god, not to be trifled with. His three-pronged trident became the symbol for the planet Neptune.

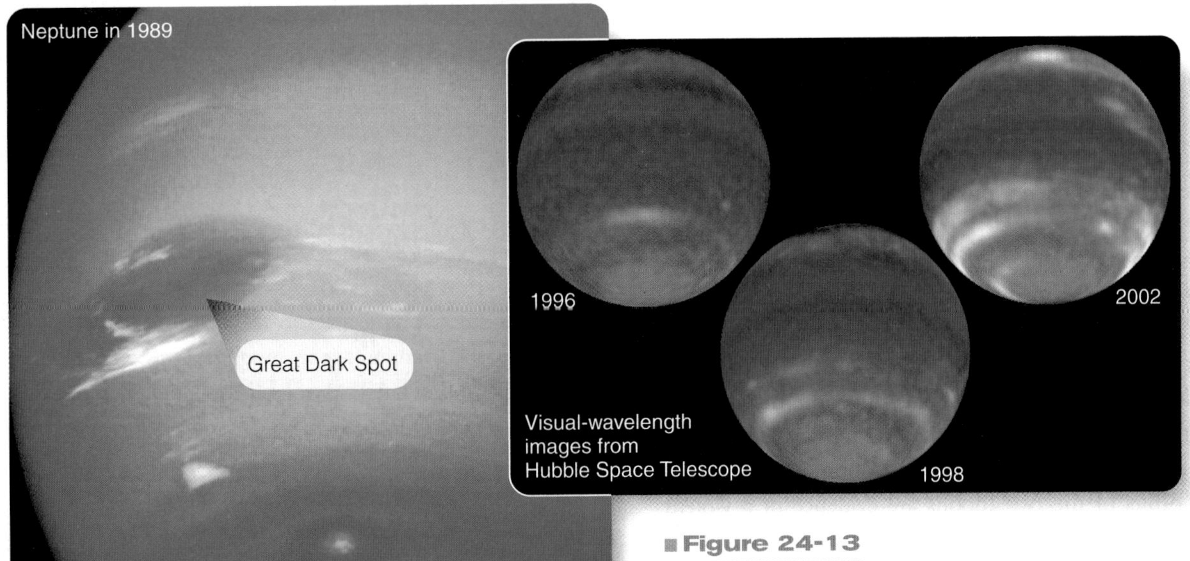


**Figure 24-13**

Because Neptune's equator is inclined almost 29 degrees in its orbit, it experiences seasons. Because of the planet's orbital period, each season lasts about 40 years. Since Voyager 2 visited in 1989, spring has come to the southern hemisphere, and the weather has clearly changed, which is surprising because sunlight on Neptune is 900 times dimmer than on Earth. (NASA, L. Sromovsky, and P. Fry, University of Wisconsin-Madison.)

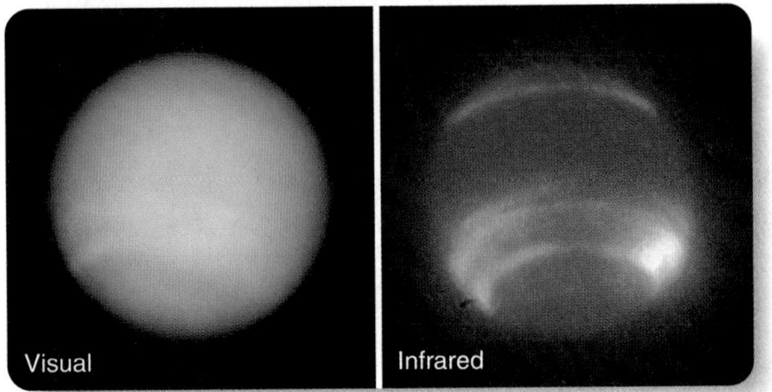


**Figure 24-14**

(a) This visual-wavelength image of Neptune was made through filters that make belts of methane stand out. (NASA and Erich Karkoschka) (b) The infrared image at right was recorded using one of the Keck 10-meter telescopes and adaptive optics. The location of methane cloud belts around the planet are evident. (U. C. Berkeley/W. M. Keck Observatory)

The same observations that helped define the interior of Uranus can be applied to Neptune. Models suggest that the interior contains a small heavy-element core, surrounded by a deep mantle of slushy or solid water mixed with heavier material having a chemical composition resembling rock. Neptune's magnetic field is a bit less than half as strong as Earth's and is tipped 47° from the axis of rotation. It is also offset 55 percent of the way to the surface (Figure 24-7). As in the case of Uranus, Neptune's field is probably generated by the dynamo effect acting in the conducting fluid mantle rather than in the planet's core.

Neptune has more internal heat than Uranus, and part of that heat may be generated by radioactive decay in the minerals in its interior. Some of the energy also may be released by denser material falling inward, including, as in the case of Uranus, diamond crystals formed by the disruption of methane.

Neptune is a tantalizing world just big enough to be imaged by the Hubble Space Telescope but far enough away to make it difficult to study. The data from Voyager 2 revealed one detail that had been nearly undetectable from Earth—rings.

## The Rings of Neptune

Astronomers on Earth saw hints of rings earlier when Neptune occulted stars, but the rings were not firmly recognized until Voyager 2 flew past the planet in 1989.

Look again at **The Rings of Uranus and Neptune** on pages 534–535 and compare the rings of Neptune with those of Uranus. Notice two additional points:

4. Neptune's rings, named after the astronomers involved in the discovery of the planet, are similar to those of Uranus but contain more small dust particles that forward-scatter light.
5. Also notice a new way that rings can interact with moons: One of Neptune's moons produces short arcs in the outermost ring, and a similar arc has been found in Saturn's rings.

Like the rings of Uranus, Saturn, and Jupiter, the particles observed now in Neptune's rings can't be primordial. That is, they can't have lasted in their present form since the formation of Neptune. Evidently, impacts on moons occasionally scatter debris through the satellite system, and some of it falls into the places where the orbits of ring particles are most stable among the orbits of the moons.

## 24-3 The Dwarf Planets

IN 1930, A WORLD WAS FOUND orbiting beyond Neptune. Although a solid object smaller than Earth's moon rather than a low-density Jovian planet, the public welcomed it as the ninth planet, and it was named Pluto. At the end of the 20th century, with much improved telescopes, astronomers found more such small worlds, and it became clear that Pluto was just one of a large family of similar objects.

In 2006, the International Astronomical Union voted to move Pluto out of the family of planets and make it part of a larger family of small worlds labeled "dwarf planets." To understand this controversial subject, you can start with Pluto and its discovery.

### The Discovery of Pluto

Percival Lowell (1855–1916) was fascinated with the idea that an intelligent race built the canals he thought he could see on Mars (see Chapter 22). Lowell founded Lowell Observatory in Flagstaff, Arizona, primarily for the study of Mars. Later, some say motivated to improve the reputation of his observatory, he began to search for a planet beyond Neptune.

Lowell used the same method that Adams and Leverrier had used to predict the position of Neptune. Working from what were understood to be irregularities in the motion of Neptune, Lowell predicted the location of an undiscovered planet beyond Neptune. He concluded it would contain about 7 Earth masses and would look like a 13th-magnitude object in eastern Taurus. Lowell searched for the planet photographically until his death in 1916.

In the late 1920s, 22-year-old amateur astronomer Clyde Tombaugh began using a homemade 9-in. telescope to sketch Jupiter and Mars from his family's wheat farm in western Kansas. He sent his drawings to Lowell Observatory, and the observatory director Vesto Slipher (see Chapter 16) hired him without an interview. The young Tombaugh bought a one-way train ticket for Flagstaff not knowing what his new job would be like.

Slipher set Tombaugh to work photographing the sky along the ecliptic around the predicted position of the planet. The search technique was a classic method in astronomy. Tombaugh obtained pairs of 14 × 17-inch glass plates exposed two or three days apart. To search a pair of plates, he mounted them in a blink comparator, a machine that allowed him to look through a microscope at a small spot on one plate and then, at the flip of a lever, see the same spot on the other plate. As he blinked back and forth, the star images did not move, but a planet would have moved along its orbit during the two or three days that elapsed before the second plate was exposed. So Tombaugh searched the giant plates, star image by star image, looking for an image that moved. A single pair of plates could contain 400,000 star images. He searched pair after pair and found nothing.

The observatory director turned to other projects, and Tombaugh, working alone, expanded his search to cover the entire ecliptic. For almost a year, Tombaugh exposed plates by night and blinked plates by day. Then, on February 18, 1930, nearly a year after he had left Kansas, a quarter of the way through a pair of plates, he found a 15th magnitude image that moved (■ Figure 24-17). He later remembered, " 'Oh,' I thought, 'I had better look at my watch; this could be a historic moment. It was within about 2 minutes of 4 PM [MST].' " The discovery was announced on March 13, the 149th anniversary of the discovery of Uranus and the 75th anniversary of the birth of Percival Lowell. The object was named Pluto after the god of the underworld and, also in a way, after Lowell, because the first two letters in Pluto are the initials of Percival Lowell.

The discovery of Pluto seemed a triumph of discovery by prediction, but Tombaugh sensed something was wrong from the first moment he saw the image. It was moving in the right direction by the right amount, but it was 2.5 magnitudes too faint. Clearly, Pluto was not the 7-Earth-mass planet that Lowell had predicted. The faint image implied that Pluto was a small world with a mass too low to noticeably alter the motion of Neptune.

Later analysis has shown that the variations in the motion of Neptune, which Lowell used to predict the location of Pluto, were random uncertainties of observation and could not have led to a trustworthy prediction. The discovery of the new planet only 6° from Lowell's predicted position was apparently an accident.

### Pluto as a World

Pluto is very difficult to observe from Earth. Only a bit larger than 0.1 arc second in diameter, it is only 65 percent the diameter of Earth's moon and shows little surface detail even when observed with the Hubble Space Telescope, although as you will learn,

■ **Figure 24-17**

Pluto is small and far away, so its image is indistinguishable from that of a star on most photographs. Clyde Tombaugh discovered the planet in 1930 by looking for an object that moved relative to the stars on a pair of photographs taken a few days apart. (Lowell Observatory photographs)

astronomers have used a clever trick to make low-resolution maps of Pluto and its moon Charon. If all goes well the New Horizons probe, due to arrive in 2015, will send the first close-up images of Pluto and its moons.

Most planetary orbits in our solar system are nearly circular, but Pluto's is significantly eccentric. In fact, from 1979 to 1999, Pluto was closer to the sun than Neptune. The two worlds will never collide, however, because Pluto's orbit is inclined 17° to the ecliptic and because, as you will learn later in this chapter, Pluto and Neptune orbit in resonance with each other so they never come close together.

If you land on the surface of Pluto, your spacesuit will have to work hard to keep you warm. Orbiting so far from the sun, Pluto is cold enough to freeze most compounds that you think of as gases, and spectroscopic observations have found evidence of solid nitrogen ice on its surface with traces of frozen methane and carbon monoxide. The maximum daytime temperature of about 55 K (−360°F) is enough to vaporize some of the nitrogen and carbon monoxide and a little of the methane to form a thin atmosphere around Pluto. This atmosphere was detected in 1988 when Pluto occulted a distant star, and the starlight was observed to fade gradually rather than winking out suddenly.

Pluto's largest moon was discovered on photographs in 1978. It is very faint and about half the diameter of Pluto (■ Figure 24-18). The moon was named Charon after the mythological ferryman who transports souls across the river Styx into the underworld. Two smaller moons, named Nix and Hydra, were found in 2005 and confirmed in 2006 by astronomers using the Hubble Space Telescope.

The discovery of Pluto's moons is important for a number of reasons. Charon, the largest and easiest to track, orbits Pluto in a nearly circular orbit in the plane of Pluto's equator. Observations show that the moon and Pluto are tidally locked to each other and that Pluto's axis of rotation is highly inclined to its orbit around the sun (■ Figure 24-19). Furthermore, tracking the orbital motion of Charon allowed calculation of the mass of Pluto. Charon orbits 19,640 km from Pluto with an orbital period of 6.387 days. Kepler's third law reveals that the mass of the system is $6.5 \times 10^{-9}$ solar masses or only about 0.002 Earth mass. Most of that mass is Pluto, which seems to be about 12 times more massive than Charon.

You know that the mass of a body is important in astronomy because mass divided by volume is density. The density of Pluto is about 2 g/cm$^3$, and the density of Charon is just a bit less. Those densities indicate that Pluto and Charon must contain about 35 percent ice and 65 percent rock. Spectra of Charon show that the small moon has a surface that is mostly water ice with not much evidence of other volatiles that are detected in spectra of Pluto. Perhaps Charon has lost its more volatile compounds because of its lower escape velocity. Water ice at Charon's surface temperature is no more volatile than is a piece of rock on Earth, so a water ice surface could last a long time.

Another reason Pluto's moon Charon has proved important is that, as Charon and Pluto orbit the sun, occasionally their mutual orbit is seen edge-on from Earth. During those times, astronomers can watch Pluto and Charon eclipse each other; and, by carefully measuring the combined light from the two objects, they can produce crude maps (■ Figure 24-20). Those

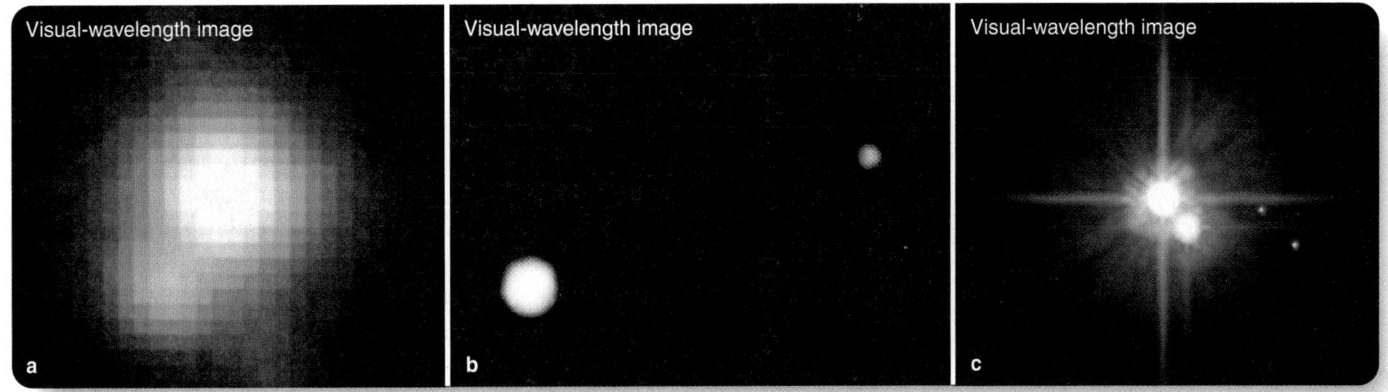


**■ Figure 24-18**

(a) A high-quality, ground-based photo shows Pluto and its moon, Charon, badly blurred by seeing. (NASA) (b) The Hubble Space Telescope image clearly separates the planet and its moon and allows more accurate measurements of the position of the moon. (R. Albrecht, ESA/ESO Space Telescope European Coordinating Facility, NASA) (c) A long-exposure photograph made in 2006 with the Hubble Space Telescope confirmed discovery of two more moons of Pluto that were named Nix and Hydra. (NASA, ESA, H.Weaver/JHU/APL, A. Stern/SwRI)

observations reveal that Pluto has a surprising amount of albedo variation on its surface, but nobody knows what the dark and light features might be. Understanding that will have to wait until the New Horizons probe flies by in 2015.

Both Pluto and Charon go through dramatic seasons much like those on Uranus as they circle the sun with their highly inclined rotation axes. This should cause large changes in Pluto's atmosphere as the planet grows warmer when it is closest to the sun, as it was in the late 1980s, and then freezes as it draws away.

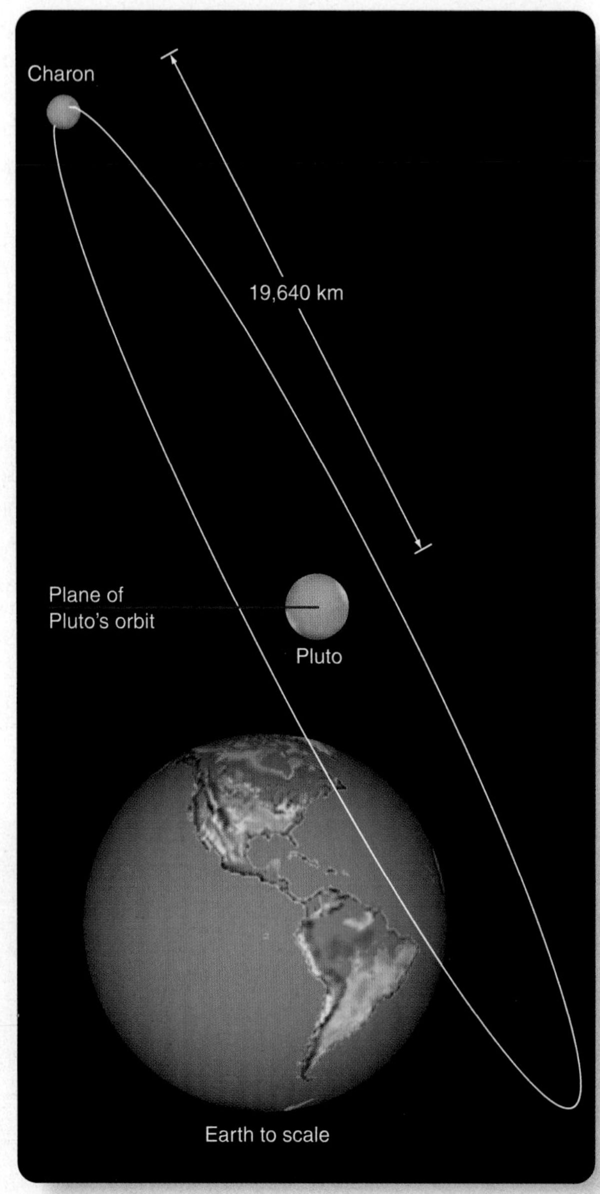


**■ Figure 24-19**

The nearly circular orbit of Charon is only a few times bigger than Earth. It is shown here nearly edge-on, and consequently it looks elliptical in this diagram. Charon's orbit and the equator of Pluto are inclined 119.6° to the plane of Pluto's orbit around the sun.

## The Family of Dwarf Planets

Perhaps the most interesting thing about Pluto is that it is not alone. In Chapter 19, you read about the Kuiper belt objects that orbit beyond Neptune. There are many thousands of such objects, and some of them are quite large. At least one is larger than Pluto.

The object known as Eris, discovered in 2003, is 4 percent larger in diameter than Pluto and 27 percent more massive. It orbits about 1.7 times farther away from the sun than does Pluto. Eris's orbit is more eccentric and more highly inclined than Pluto's, but it seems to be a similar object. The discovery of Eris led the International Astronomical Union to recognize a new class of solar system objects called **dwarf planets**—objects that orbit the sun and are large enough to assume a spherical shape but not large enough to have sufficient gravitational influence to absorb or otherwise clear away remaining objects in the same region. In contrast, the major planets, including Earth, were able to do that type of regional clearing in the process of growing from protoplanets into planets. Three solar system objects are clearly dwarf planets—Pluto and Eris in the Kuiper belt, plus Ceres, the 900-km-diameter asteroid that orbits within the asteroid belt between Mars and Jupiter and is nearly twice as big as the next largest asteroids, Pallas and Vesta.

Roughly ten other Kuiper belt objects are known to have sizes approaching that of Pluto and Eris, and there are probably others yet to be discovered. Two large objects named Sedna and Orcus are each about 63 percent the size of Pluto. Another object, called Quaoar (pronounced *kwah-o-wahr*), is 50 percent the diameter of Pluto. The ten largest Kuiper belt objects are considered candidate dwarf planets, pending determination of their shape. One strange object that may complicate things is Haumea, with such a rapid spin—once every 4 hours—that it is shaped like a flattened cigar. One of its dimensions is nearly as large as Pluto's diameter. Haumea is not spherically shaped, but if it were not spinning so rapidly, it might be a sphere and meet the definition of a dwarf planet. Should the effect of Haumea's rotation be counted in considering its status, or not?

Some astronomers argue that Charon, Pluto's big spherical moon, should be a member of the dwarf planets even though it orbits another world and not the sun. Other astronomers are upset that Ceres, a rocky asteroid, is included. The classification may seem arbitrary until you begin thinking about how the dwarf planets formed. Some astronomers refer to them as oligarchs. The term *oligarch* is usually applied to business or political leaders who are the biggest, meanest dudes in town. They are not alone, but they are the bosses. The dwarf planets appear to have been bodies in the solar nebula that grew more rapidly than their neighbors and became dominant, but never got big enough to take over completely and sweep up all objects orbiting nearby. Planets such as Earth and Jupiter cleared their orbital lanes

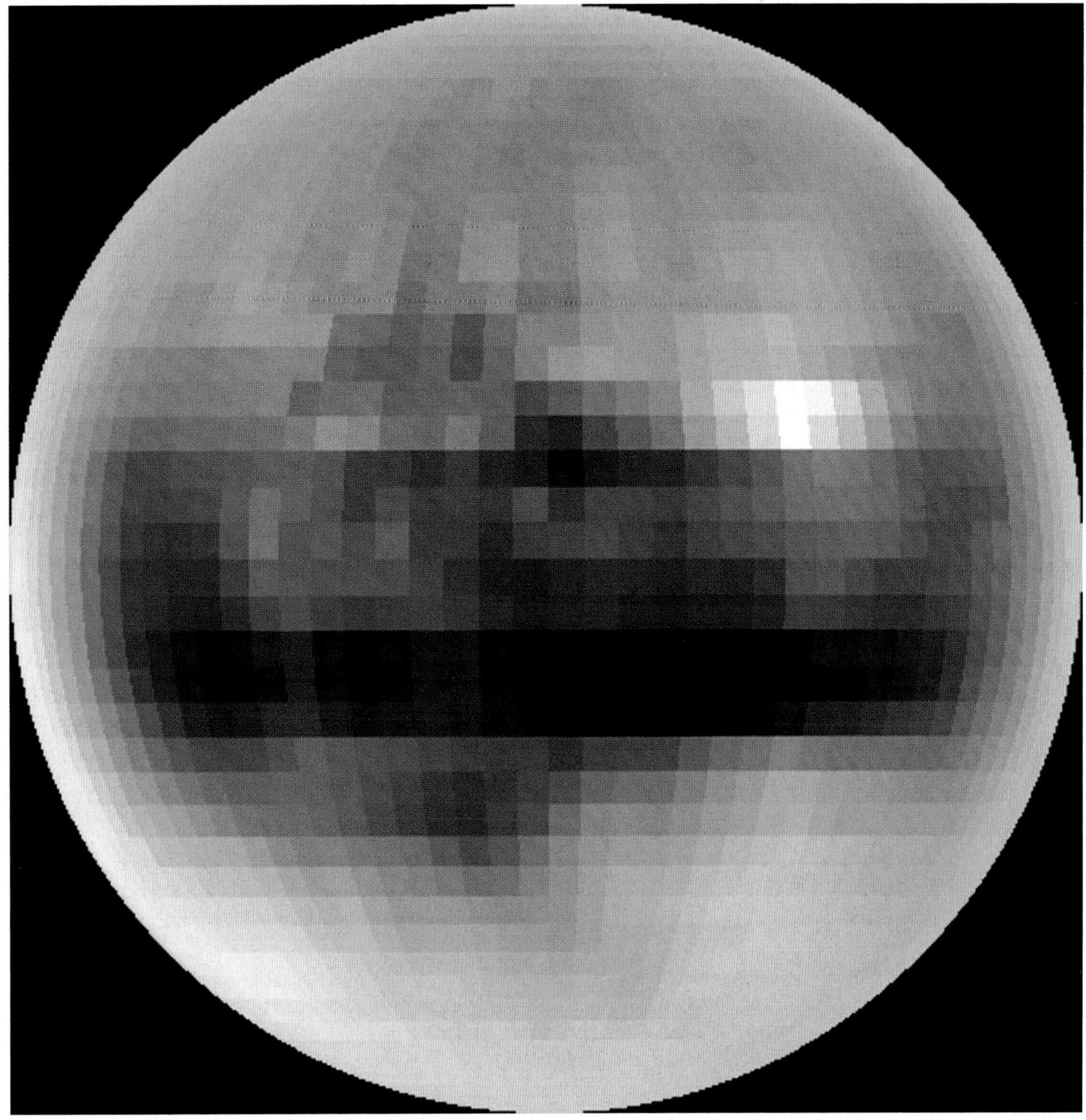

**Figure 24-20**

Observing Pluto's brightness variations as its moon Charon occasionally moves across the planet's disk allowed astronomers to make this low-resolution map of Pluto. These are approximately true colors. (E. Young/SwRI)

around the sun, but the dwarf planets never got quite big enough to do that.

## Pluto and the Plutinos

No, this section is not about a 1950s rock-and-roll band. It is about the history of the dwarf planets, and it will take you back billions of years to watch the outer planets form.

You should begin by eliminating an old idea. As soon as Pluto was discovered, astronomers realized that it was much smaller than its Jovian neighbors, and some suggested that it was a moon of Neptune that had escaped. The orbits of Neptune and Pluto actually cross, although they will never collide. That escaped-moon hypothesis for Pluto has been totally abandoned. It is hard to understand how Pluto could have escaped from Neptune and reached its present orbit. And how did it get moons, if it was once a moon itself? Modern astronomers have a much better hypothesis, and it recognizes that Pluto is just one of the dwarf planets that is related to many other objects in the Kuiper belt.

Dozens, perhaps hundreds, of Kuiper belt objects are, like Pluto, caught in a 3:2 resonance with Neptune. That is, they orbit the sun twice while Neptune orbits three times. You learned about orbital resonances when you studied Jupiter's Galilean moons. A 3:2 resonance with Neptune makes the orbiting bodies immune to any disturbing gravitational influence from Neptune so their orbits are more stable. Because

Pluto is also caught in the same 3:2 resonance, these Kuiper belt objects have been named **plutinos.**

How did the plutinos get caught in resonances with Neptune? Computer models of the formation of the planets suggest that Uranus and Neptune may have formed closer to the sun, where the solar nebula was denser. Sometime later, gravitational interactions with Jupiter and Saturn gradually shifted the two ice giants outward, and, as Neptune migrated outward, its orbital resonances could have swept up small bodies like nets pushed in front of a fishing boat. Other Kuiper belt objects are caught in other stabilizing resonances, and they were apparently swept up in the same way. The plutinos in 3:2 resonance and other Kuiper belt objects in other resonances with Neptune are evidence that Neptune really did migrate outward. This planet migration could have scattered small bodies throughout the solar system and caused the apparent late heavy bombardment event during which Earth, the moon, and other solar system bodies suffered a brief but devastating increase in cratering rate.

Some astronomers are angry that the International Astronomical Union "demoted" Pluto, but it is actually a more interesting world once you realize that it is the best studied of the dwarf planets. In the inner solar system, only the asteroid Ceres was able to grow fast enough to become a dwarf planet, but in the outer solar system huge numbers of icy bodies formed, ranging from pebbles to the oligarchs now recognized as dwarf planets. As they are understood better, the dwarf planets will reveal more secrets from the age of planet building.

## SCIENTIFIC ARGUMENT

***Why is Earth a planet and not a dwarf planet?***

This question calls for an argument based on a scientific classification. If it is to be useful, a classification must be based on real characteristics, so you need to think about the definition of the dwarf plants. According to the International Astronomical Union, a dwarf planet must orbit the sun, not be a satellite of a planet, and be spherical. Earth meets these characteristics, but there is one more requirement. By definition, a dwarf planet must not have been large enough to clear out most of the smaller objects near its orbit. As Earth grew in the solar nebula, it absorbed the small bodies that orbited the sun in similar orbits. In other words, Earth cleared its traffic lane around the sun. Pluto never became massive enough to clear its lane (the Kuiper belt), so Pluto is defined as a dwarf planet. Because Earth grew large enough to sweep its traffic lane clear, Earth is defined as a planet.

The dwarf planets are interesting for what they can reveal about the origin of the solar system. **How do the orbital resonances between the plutinos and Neptune confirm a hypothesis for the formation of the Jovian planets?**

## What Are We? Trapped

No person has ever been farther from Earth than the moon. We humans have sent robotic spacecraft to explore the worlds in our solar system beyond Earth's moon, but no human has ever set foot on them. We are trapped on Earth.

We lack the technology to leave Earth easily. Getting away from Earth's gravitational field calls for very large rockets. America built huge rockets in the 1960s and early 1970s to send astronauts to the moon, but such rockets no longer exist. The best technology today can carry astronauts just a few hundred kilometers above Earth's surface to orbit above the atmosphere. We can probably reach Mars with a few decades of effort, but going beyond may take more resources than Earth can provide.

There is another reason we Earthlings are trapped on our planet. We have evolved to fit the environment on Earth. None of the planets or moons you explored beyond Earth would welcome you. Radiation belts, extreme heat or cold, and lack of air are obvious problems, but Earthlings have evolved to live with one Earth gravity. Astronauts in orbit for just a few weeks suffer biomedical problems because their muscles and bones no longer feel Earth's gravity. Could humans live for years in the weak gravity on Mars? We may be trapped on Earth not because we lack big rockets but because we need Earth's environment.

It seems likely that we need Earth more than it needs us. The human race is changing the world we live on at a terrific pace, and some of those changes are making Earth less hospitable. All of your exploring of un-Earthly worlds serves to remind you of the nurturing beauty of our home planet. It is probably the only one we will ever have.

# Study and Review

## Summary

- Uranus was discovered by William Herschel in 1781.
- Although Uranus is a Jovian planet, it is significantly smaller and less massive than Jupiter. Uranus is an ice giant planet about four times the diameter of Earth. Uranus rotates "on its side," with its pole nearly in the plane of its orbit, causing it to have extreme seasons.
- The atmosphere of Uranus is mostly hydrogen and helium with some methane, which absorbs longer-wavelength photons and gives the planet a greenish-blue color.
- The atmosphere of Uranus is so cold that only methane clouds are visible, and some traces of belt–zone circulation can be seen. As spring has come to the northern hemisphere, more cloud features have developed, suggesting that Uranus follows a seasonal cycle and is not always as bland as it was when Voyager 2 flew past in 1986.
- Model calculations indicate that Uranus has a small core of dense matter and a deep slushy mantle of ice, water, and rock. Convection in the mantle may produce its highly inclined offset magnetic field.
- Uranus emits about the predicted amount of heat for a planet at its distance from the sun, which suggests that it is not extremely hot inside.
- The rings of Uranus were discovered during an **occultation (p. 534)** as the planet crossed in front of a star.
- The rings of Uranus are composed mostly of boulder-size objects darkened by radiation and trapped among the orbits of small moons. The ring material is probably produced by impacts on icy moons.
- The five large moons of Uranus appear to be icy and old, although some show signs of geological activity. Miranda appears to have suffered from dramatic activity, as shown by **ovoids (p. 537)** on its surface. Tidal heating is a likely source of the energy that drove this activity.
- Uranus appears to have formed slowly and never became massive enough to trap large amounts of hydrogen and helium from the solar nebula. Uranus and Neptune are thought to have formed closer to the sun and were moved outward by gravitational interactions with Jupiter and Saturn.
- An impact by a large planetesimal while Uranus was forming may have caused the planet to rotate around a highly inclined axis, but it is also possible that tidal interactions with Saturn altered the rotation of Uranus as it was moved outward.
- Neptune was discovered in 1846 based on a position computed from irregularities it was causing in the orbital motion of Uranus.
- Neptune is an ice giant quite similar in size to Uranus. It has a core of denser material; a mantle of ice, water, and rock; and an atmosphere of hydrogen and helium with traces of methane, which gives the planet a blue color.
- Methane clouds come and go in Neptune's cold atmosphere and appear to follow a belt–zone circulation.
- Circulation in the liquid mantle gives rise to Neptune's magnetic field.
- Neptune has more internal heat than Uranus, and the sinking of dense material, including diamonds, may be adding to the internal heat.
- The rings of Neptune probably formed when impacts on moons scattered icy debris into stable places among the orbits of the small moons. Forward scattering shows that the rings contain more small dust particles than those of Uranus. Arcs in the rings appear to be caused by the gravitational influence of a small moon or moons.
- Triton, Neptune's largest moon, follows a retrograde orbit, and Nereid follows a long eccentric orbit. These two moons' odd orbits may have been caused by the gravitational influence of a massive planetesimal or by the capture of Triton.
- Triton has an icy surface and a thin atmosphere of nitrogen. The lack of many craters, flooded areas, and the presence of cracks and faults suggest that the moon may still be active. Sunlight and heat from the interior appear to trigger nitrogen geysers in the crust. Tidal heating during a close approach to Neptune when it was captured into orbit is another possible source of heat.
- Neptune formed slowly, as did Uranus, and never accumulated a deep atmosphere of hydrogen and helium before the solar nebula was blown away.
- Pluto was discovered in 1930 during a search for a large planet orbiting beyond Neptune and disturbing Neptune's orbital motion. That discovery was in a sense an accident, because Pluto is much too small to affect Neptune, and more observations of Neptune indicate its orbital motion is actually undisturbed.
- No spacecraft has yet visited Pluto, but the New Horizons probe is on its way for a flyby in 2015.
- Spectroscopic observations indicate Pluto has a frigid crust of solid nitrogen ice with traces of frozen methane and carbon monoxide. Its thin atmosphere is mostly nitrogen.
- Pluto's moon Charon orbits in an orbit highly inclined to their orbit around the sun. Charon and Pluto are tidally locked to face each other. That means Pluto rotates on its side much like Uranus.
- Careful measurements of the brightness of Charon and Pluto on occasions when they move in front of each other have allowed astronomers to construct low-resolution maps of both objects.
- The density of Pluto shows that it and Charon must contain about two-thirds rock and one-third ices.
- The **dwarf planets (p. 546)** are small bodies that orbit the sun and do not orbit a planet; they are spherical, like the major planets, but unlike the major planets they are not large enough to have cleared their orbital lanes of other objects.
- Pluto, another Kuiper belt object named Eris, and the asteroid Ceres are classified as dwarf planets. About 10 other Kuiper belt objects are candidate dwarf planets pending the determination of their shapes. At present more than a thousand Kuiper belt objects are known, most of which are too small to be dwarf planets.
- The dwarf planets grew large in the solar nebula but never became large enough capture or eject other objects orbiting nearby.
- Pluto and similar Kuiper belt objects called **plutinos (p. 548)** are caught in a stabilizing 3:2 resonance with Neptune. Many other Kuiper belt objects orbit in other resonances with Neptune. These objects are evidence that the Jovian planets interacted soon after formation, pushed Neptune outward in the solar system, and swept up remnant planetesimals in these resonances.

## Review Questions

1. Describe the location of the equinoxes and solstices in the Uranian sky. What are seasons like on Uranus?
2. Why is belt–zone circulation difficult to detect on Uranus?
3. Discuss the origin of the rings of Uranus and Neptune. Cite evidence to support hypotheses.
4. How do the magnetic fields of Uranus and Neptune suggest that the mantles inside those planets are fluid?

*When they shall cry "PEACE, PEACE" then cometh sudden destruction!*

*COMET'S CHAOS?—What Terrible events will the Comet bring?*

— FROM A PAMPHLET PREDICTING THE END OF THE WORLD BECAUSE OF THE APPEARANCE OF COMET KOHOUTEK IN 1973

One afternoon in 1954, while Mrs. E. Hulitt Hodges of Sylacauga, Alabama, lay napping on her living room couch, an explosion and a sharp pain jolted her awake. Analysis of the brick-sized rock that smashed through the ceiling and bruised her left leg showed that it was a meteorite. Mrs. Hodges is the only person known to have been injured by a meteorite. Coincidentally, she lived right across the street from the Comet Drive-In Theater.

Meteorites arrive from space all over the Earth every day, although not as spectacularly as the one that struck Mrs. Hodges. You will learn in this chapter that meteorites are fragments of asteroids, and that asteroids, as well as their icy cousins the comets, carry precious clues about conditions in the solar nebula from which the sun and planets formed. Because you cannot easily visit comets and asteroids, you can begin by learning about the pieces of those bodies that come to you.

## 25-1 Meteoroids, Meteors, and Meteorites

You learned some things about meteorites in Chapter 19 when you studied evidence for the age of the solar system. There you saw that the solar system includes small particles called *meteoroids.* Some of them collide with Earth's atmosphere at speeds of 10 to 70 km/s. Friction with the air heats the meteoroids enough so that they glow, and you see them vaporize as streaks across the night sky. Those streaks are called *meteors* ("shooting stars"). If a meteoroid is big enough and holds together well enough, it can survive its plunge through the atmosphere and reach Earth's surface. Once the object strikes Earth's surface, it is called a *meteorite* ("-ite" being the Greek root for "rock"). As you will learn later in this chapter, the largest of those objects can blast out craters on Earth's surface, but such big impacts are extremely rare. The great majority of meteorites are too small to form craters.

What can meteorites and meteors tell you about the origin of the solar system? To answer that question, you can consider their compositions and orbits.

### Composition of Meteorites and Meteors

One of the best places to look for meteorites turns out to be certain parts of Antarctica—not because more meteorites fall there but because they are easy to recognize. No Earth rocks are on top of the Antarctic ice cap; the nearest native rocks are buried under the ice. Any rock you find there must have fallen from space. (For similar reasons, another good place to find meteorites is the Sahara desert, where deep layers of sand keep Earth rocks completely out of sight.) The slow flow of the Antarctic ice cap from the center of the continent toward the ocean concentrates meteorites in areas of especially good hunting where the moving ice runs into mountain barriers, slows down, and evaporates. Teams of scientists travel to Antarctica and ride snowmobiles in systematic sweeps across the ice each South Pole summer to recover meteorites (■ Figure 25-1). A four- to eight-person team can find a thousand meteorites during a single two-month field season. After 25 years of work, more than half of the 40,000 meteorites in human hands are from Antarctica.

Meteorites that are seen to fall are called **falls;** a fall is known to have occurred at a given time and place, and thus the meteorite is well documented. A meteorite that is discovered on or in the ground, but was not seen to fall, is called a **find.** Such a meteorite could have fallen thousands of years ago. The distinction between *falls* and *finds* will be important as you analyze the different kinds of meteorites.

Meteorites can be divided into three broad composition categories. **Iron meteorites** (■ Figure 25-2a) are solid chunks of iron and nickel. **Stony meteorites** (Figure 25-2b) are silicate masses that resemble Earth rocks. **Stony-iron meteorites** (Figure 25-2c) are mixtures of iron and stone. **Carbonaceous chondrites** (pronounced *kon-drite;* Figure 25-2d) are a special type of stony meteorite.

Iron meteorites are easy to recognize because they are heavy, dense lumps of metal—a magnet will stick to them. That explains an important statistic. Iron meteorites make up 66 percent of finds (■ Table 25-1) but only 6 percent of falls. Why? Because an iron meteorite doesn't look like an ordinary rock. If you trip over one on a hike, you are more likely to recognize it as something odd, carry it home, and show it to the local museum. Also, some stony meteorites deteriorate rapidly when exposed to weather; irons are made of stronger material and generally survive longer. That means there is a **selection effect** that makes it more likely that iron meteorites will be found (**How Do We Know? 25-1**). The fact that only 6 percent of falls are irons shows that iron meteoroids, although easier to find on Earth, are relatively rare in space.

When iron meteorites are sliced open, polished, and etched with acid, they reveal regular bands called **Widmanstätten patterns** (pronounced *Veed-mahn-state-en*) (■ Figure 25-3). Those patterns are caused by certain alloys of iron and nickel that formed crystals as the molten metal cooled and solidified long ago. The size and shape of the bands indicate that the molten metal cooled very slowly, no faster than 20 degrees Kelvin per million years.

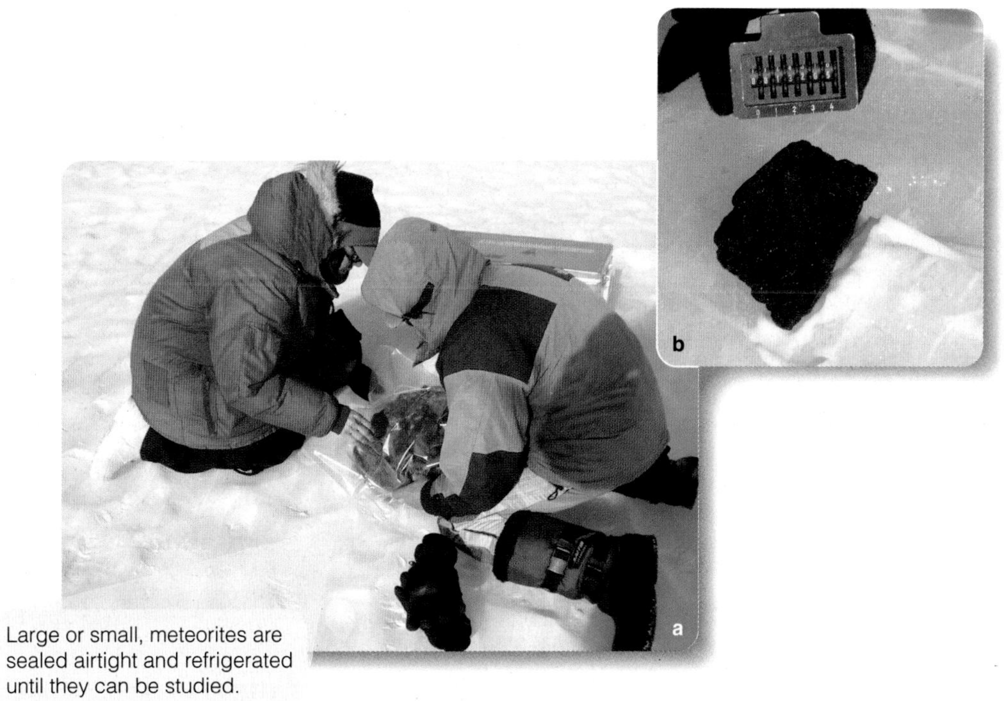

Large or small, meteorites are sealed airtight and refrigerated until they can be studied.

**Figure 25-1**

Braving bitter cold and high winds, teams of scientists riding snowmobiles search for meteorites that fell long ago in Antarctica and are exposed as the ice evaporates. When a meteorite is found, it is photographed where it lies, assigned a number, and placed in an airtight bag. Thousands of meteorites have been collected in this way, including rare meteorites from the moon and Mars. (Courtesy Monika Kress)

The Widmanstätten pattern tells you that the metal in iron meteorites was once molten and must have been well insulated to cool so slowly. Such slow cooling indicates a location inside bodies at least 30 km in diameter. (In comparison, a small lump of molten metal exposed in space would cool in just a few hours.) On the other hand, the iron meteorites do not show effects of the very high pressures that would exist deep inside a planet. Evidently, iron meteorite material formed in the cooling interiors of planetesimal-sized objects, not planets. You will find this is one important clue to the origin of meteorites.

A small fraction of falls are meteorites made of mixed iron and stone (Figure 25-2b). These stony-iron meteorites appear to have solidified from both molten iron and rock—the kind of environment you might expect to find deep inside a planetesimal at the boundary between a liquid metal core and a rocky mantle.

In contrast to irons and stony-irons, stony meteorites (Figure 25-2c) are common among falls (Table 25-1). Although there are many different types of stony meteorites, you can classify them into two main categories depending on their physical properties and chemical content, **chondrites** and **achondrites.**

Chondrites look like dark gray, granular rocks (Figure 25-2d). The classification of meteorites has become quite complicated, and there are many types of chondrites, but in general they contain some volatiles including water and organic (carbon) compounds. A few chondrites appear to have formed in the presence of liquid water.

Most types of chondrites also contain **chondrules,** small round bits of glassy rock only a few millimeters across. To be glassy rather than crystalline, the chondrules must have cooled from a molten state quickly, within a few hours. One hypothesis is that chondrules are bits of matter from the inner part of the solar nebula, near the sun, that were blown outward by winds or jets to cooler parts of the nebula where they condensed and were later incorporated into larger rocks. Another hypothesis is that the chondrules were once solid bits of matter that were melted by shock waves spreading through the solar nebula and then resolidified. The presence of chondrule particles inside chondrite meteorites indicates that those rocks have never been melted since they formed because melting would have destroyed the chondrules.

Among the chondrites, the carbonaceous chondrites are rare but quite important. These dark gray, rocky meteorites are

Iron meteorites are very heavy for their size and have a dark, irregular surface.

A stony-iron meteorite cut and polished reveals a mixture of iron and rock.

Stony meteorites tend to have a fusion crust caused by melting in Earth's atmosphere.

Chondrules are small, glassy spheres found in chondrites.

a

b

c

d

**■ Figure 25-2**

The three main types of meteorites—(a) irons, (b) stony-irons, and (c) stones—have distinctive characteristics. (d) Carbonaceous chondrite meteorites are a rare type of stony meteorite that is rich in carbon, making the rock very dark. (Lab photos courtesy of Russell Kempton, New England Meteoritical)

**■ Table 25-1 | Proportions of Meteorites**

| Type | Falls (%) | Finds (%) |
|---|---|---|
| Iron | 6 | 66 |
| Stony-iron | 2 | 8 |
| Stony | 92 | 26 |

especially rich in water, other volatiles, and organic compounds. Those substances all would have been lost if the meteoroid had been heated even to room temperature.

One of the most important meteorites ever studied was a carbonaceous chondrite seen falling in 1969 near the little Mexican village of Allende (pronounced *ah-yen-day*). About 2 tons of fragments were recovered. Studies of the Allende meteorite disclosed that it contained chondrules, water, complex organic compounds including amino acids, and a number of small, irregular inclusions

**■ Figure 25-3**

Sliced, polished, and etched with acid, iron meteorites show what is called a Widmanstätten pattern of large crystals, indicating that this material cooled very slowly from a molten state and must have been inside a fairly large object. (Photo courtesy of Russell Kempton, New England Meteoritical Services)

## How Do We Know? 25-1

### Selection Effects

***How is a red insect like a red car?*** Scientists must plan ahead and design their research projects with great care. Biologists studying insects in the rain forest, for example, must choose which ones to catch. They can't catch every insect they see, so they might decide to catch and study any insect that is red. If they are not careful, a selection effect could bias their data and lead them to incorrect conclusions without their ever knowing it.

For example, suppose you needed to measure the speed of cars on a highway. There are too many cars to measure every one, so you might reduce the workload and measure only red cars. It is quite possible that this selection criterion will mislead you because people who buy red cars may be more likely to be younger and drive faster. Should you instead measure only brown cars? No, because older, more sedate people might tend to buy brown cars. Only by very carefully designing your experiment can you be certain that the cars you measure are traveling at typical speeds.

Astronomers understand that what you see through a telescope depends on what you notice, and that is powerfully influenced by what are called **selection effects**. The biologists in the rain forest, for example, should not catch and study only red insects. Often, the most brightly colored insects are poisonous or at least taste bad to predators. Catching only red insects could produce a result highly biased by a selection effect.

*Things that are bright and beautiful, such as spiral galaxies, may attract a disproportionate amount of attention. Scientists must be aware of such selection effects.* (Hubble Heritage Team/STScI/AURA/NASA)

rich in calcium, aluminum, and titanium (■ Figure 25-4). Now called **CAIs,** for calcium–aluminum-rich inclusions, these bits of matter are highly refractory; that is, they vaporize or condense only at very high temperatures.

If you could scoop out a portion of the sun's photosphere and cool it, the first particles to solidify would have the chemical composition of CAIs. As the temperature fell, other materials would become solid in accord with the condensation sequence described in Chapter 19. When the material finally reached room temperature, you would find that almost all of the hydrogen, helium, and some other gases such as argon and neon had escaped and that the remaining lump would have almost exactly the same overall chemical composition as the Allende meteorite. You can understand this is evidence that the Allende meteorite is a very old sample of the solar nebula, confirmed by the fact that the CAIs have radioactive ages equal to the oldest of any other solar system material.

Another large load of carbonaceous chondrite material arrived on Earth in the year 2000 at Tagish Lake in the Canadian Arctic. Analysis of that meteorite produced a surprise: It has noticeably less complex organics than Allende. Scientists are not sure whether this means that the Tagish organics formed so early in the solar system's history that chemical reactions had not yet advanced to the stage of making Allende's complex compounds, or whether the Tagish material was once heated just enough to break down big molecules into smaller ones.

The condensing solar nebula should have incorporated volatiles and organics into solid particles as they formed. If that material had later been heated it would have lost the volatiles, and many of the organic compounds would have been destroyed. The chondrites show properties ranging from

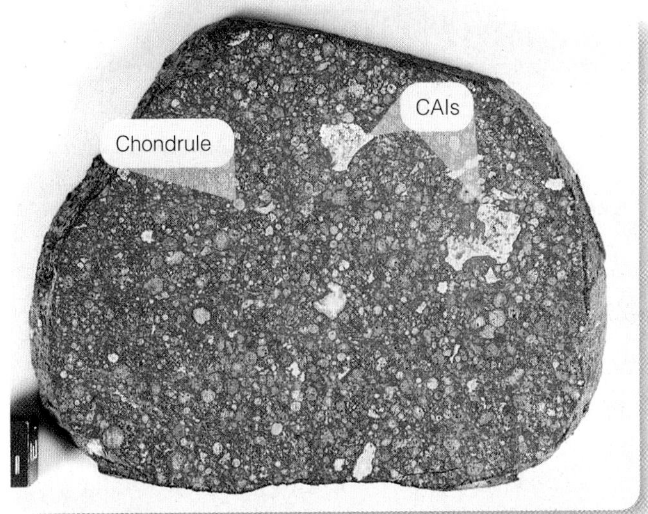


**■ Figure 25-4**

A sliced portion of the Allende carbonaceous chondrite meteorite, showing irregularly shaped white inclusions called CAIs that were probably the first solid material to condense as the solar system formed. (NASA)

carbonaceous chondrites, most of which have avoided being heated or modified, to other chondrites in which the material was slightly heated and somewhat altered from the form in which it first solidified. Chondrites in general offer us the best direct information about conditions and processes occurring in the earliest days of the solar nebula when planetesimals and planets were forming.

Stony meteorites called achondrites contain no chondrules and also lack volatiles. These rocks appear to have been subjected to intense heat that melted the chondrules and completely drove off the volatiles, leaving behind rock with composition similar to Earth's basalts.

The different types of meteorites evidently had a wide variety of histories. Some achondrites seem like pieces of lava flows, whereas stony-iron and iron meteorites apparently were once deep inside the molten interiors of differentiated objects. The differences between various chondrites are thought to result from: (1) meteoritic material solidifying at different distances from the sun with different chemical compositions due to the temperature-dependent condensation sequence, (2) each location in the nebula having some material transported and mixed in from other locations, and (3) processes that altered meteoritic material after condensation. Meteorites provide evidence that the early history of the solar system was complex.

## Orbits of Meteors and Meteorites

Meteoroids are much too small to be visible through even the largest telescope. They are visible only when they fall into Earth's atmosphere and are heated by friction with the air. A typical meteoroid has roughly the mass of a paper clip and vaporizes at an altitude of about 80 km (50 mi) above Earth's surface. The meteor trail points back along the path of the meteoroid, so if you study the direction and speed of meteors, you can get clues to their orbits in the solar system before they encountered Earth.

One way to backtrack meteor trails is to observe **meteor showers.** On any clear night, you can see 3 to 15 meteors per hour, but on some nights you can see a shower of hundreds of meteors per hour that are obviously related to each other. To confirm this, try observing a meteor shower. Pick a shower from ■ Table 25-2 and on the appropriate night stretch out in a lawn chair and watch a large area of the sky. When you see a meteor, sketch its path on the appropriate sky chart from the back of this book. In just an hour or so you will discover that all or almost all of the meteors you see seem to come from a single area of the sky, called the **radiant** of the shower (■ Figure 25-5a). Meteor showers are named after the constellation from which they seem to radiate; for example, the Perseid shower seen in mid-August radiates from the constellation Perseus.

Observing a meteor shower is a natural fireworks show, but it is even more exciting when you understand what a meteor shower tells you. The fact that the meteors in a shower appear to come from a single point in the sky means that the meteoroids were traveling through space along parallel paths. When they encounter Earth and are vaporized in the upper atmosphere, you see their fiery tracks in perspective, so they appear to come from a single radiant point, just as railroad tracks seem to come from a single point on the horizon (Figure 25-5b).

■ **Table 25-2 | Meteor Showers**

| Shower | Dates | Hourly Rate | Radiant* R. A. | Radiant* Dec. | Associated Comet |
|---|---|---|---|---|---|
| Quadrantids | Jan. 2–4 | 30 | 15h 24m | 50° | |
| Lyrids | April 20–22 | 8 | 18h 4m | 33° | 1861 I |
| η Aquarids | May 2–7 | 10 | 22h 24m | 0° | Halley? |
| δ Aquarids | July 26–31 | 15 | 22h 36m | −10° | |
| Perseids | Aug. 10–14 | 40 | 3h 4m | 58° | Swift-Tuttle |
| Orionids | Oct. 18–23 | 15 | 6h 20m | 15° | Halley? |
| Taurids | Nov. 1–7 | 8 | 3h 40m | 17° | Encke |
| Leonids | Nov. 14–19 | 6 | 10h 12m | 22° | 1866 I Temp |
| Geminids | Dec. 10–13 | 50 | 7h 28m | 32° | |

*R. A. and Dec. give the celestial coordinates (right ascension and declination) of the radiant for each shower.

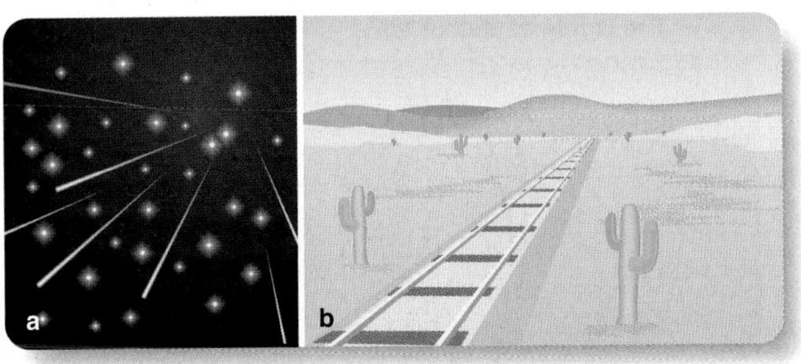

■ **Figure 25-5**

(a) Meteors in a meteor shower enter Earth's atmosphere along parallel paths, but perspective makes them appear to diverge from a single point in the sky. (b) Similarly, parallel railroad tracks appear to diverge from a point on the horizon.

Studies of meteor shower radiants reveal that those meteoroids are orbiting the sun along the paths of comets. As you learned in Chapter 19, the vaporizing head of a comet releases bits of rock that become spread along its entire orbit (■ Figure 25-6). When Earth passes through this stream of material, you see a meteor shower. In some cases the comet has wasted away and is no longer visible, but in other cases the comet is still prominent, although located somewhere else along its orbit. For example, each May, Earth comes near the orbit of Comet Halley, and you can see the Eta Aquarid shower. Each October, Earth passes near the other side of the orbit of Comet Halley, and the Orionid shower appears.

Even when there is no shower, you still can occasionally see meteors, which are called **sporadic meteors** because they are not part of specific showers. Many of these are produced by stray bits of matter that were released long ago by comets. Such comet debris gradually get spread throughout the inner solar system, away from their original tracks, and bits collide with Earth even when there is no shower.

Another way to backtrack meteor trails is to photograph the same meteor from two locations on Earth a few miles apart.

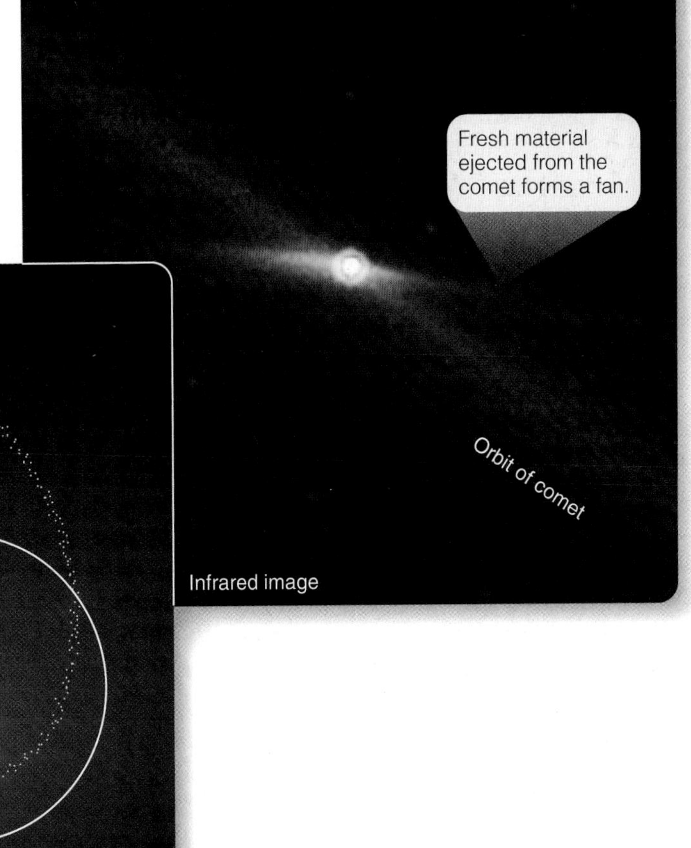


■ **Figure 25-6**

(a) Meteors in a shower are debris left behind as a comet's icy nucleus vaporizes. The rocky and metallic bits of matter spread along the comet's orbit. If Earth passes through such material, you can see a meteor shower. (b) In this infrared image of Comet Encke, the millimeter-sized bits of rock along its orbit glow as they are warmed by sunlight. The Taurid meteor shower occurs every October when Earth crosses Encke's orbit. (NASA/JPL-Caltech/M. Kelley, Univ. of Minnesota)

Then, you can use triangulation to find the altitude, speed, and direction of the meteor as it moves through the atmosphere, and work backward to calculate its orbit before it entered Earth's atmosphere. These studies confirm that meteors belonging to showers, as well as some sporadic meteors, have orbits that are similar to the orbits of comets. In contrast, a few sporadic meteors, including *all* observed meteorite falls, have orbits that lead back to the asteroid belt between Mars and Jupiter. From this you can conclude that meteors have a dual source: Many come from comets, but a few come from the asteroid belt. Meteors that are big and durable enough to become meteorites on the ground appear always to come from the asteroid belt.

It is a **Common Misconception** that a bright meteor disappearing behind a distant hill or line of trees probably landed just a mile or two away. This has triggered hilarious "wild goose chases" as police, fire companies, and TV crews try to find the impact site. Almost every meteor you see vaporizes high above Earth's surface. Only rarely does a meteor become a meteorite by reaching the ground, and it can land as much as 100 miles from where you are standing when you see it.

## Origin of Meteoroids and Meteorites

Evidence you have already encountered suggests that many meteorites are fragments of parent bodies that were large enough to grow hot from radioactive decay or other processes. They then melted and differentiated to form iron-nickel cores and rocky mantles. The molten iron cores would have been well insulated by the thick rocky mantles, so that the iron would have cooled slowly enough to produce big crystals that result in Widmanstätten patterns. Some stony meteorites that have been strongly heated appear instead to have come from the mantles or surfaces of such bodies. Stony-iron meteorites apparently come from boundaries between stony mantles and iron cores. Collisions could break up such differentiated bodies and produce different kinds of meteorites (■ Figure 25-7). In contrast, many chondrites are probably fragments of smaller bodies that never melted, and carbonaceous chondrites may be from unaltered bodies that formed especially far from the sun.

These hypotheses trace the origin of meteorites to planetesimal-like parent bodies, but they leave you with a puzzle. The small meteoroids now flying through the solar system cannot have existed in their present form since the formation of the solar system because they would have been swept up by the planets in a billion years or less. They could not have survived traveling in their current orbits for the full 4.6 billion years that the solar system has existed. Nevertheless, when the orbits of meteorite falls are determined, those orbits lead back into the asteroid belt. Thus, all the evidence together indicates that the meteorites now in museums around the world were produced by asteroid collisions within the last billion years. Astronomers conclude that nearly all meteors are material from comets, but meteorites are pieces of shattered asteroids.

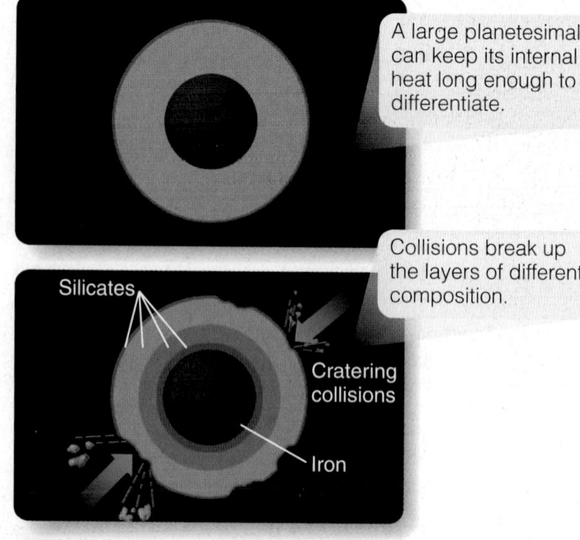
The Origin of Meteorites
A large planetesimal can keep its internal heat long enough to differentiate.
Collisions break up the layers of different composition.
Silicates
Cratering collisions
Iron

Meteorites from deeper in the planetesimal were heated to higher temperatures.

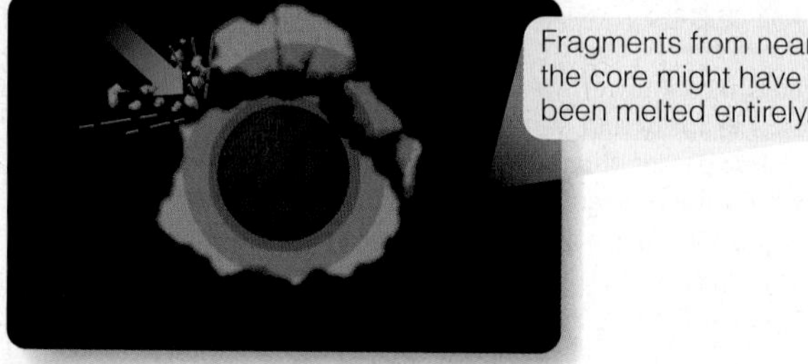
Fragments from near the core might have been melted entirely.

Fragments of the iron core would fall to Earth as iron meteorites.

■ **Figure 25-7**

Some of the planetesimals that formed early in the solar system's history may have differentiated—that is, melted and separated into layers of different density and composition, as did the Terrestrial planets. The fragmentation of such a planetesimal could produce different types of meteorites. (Adapted from a diagram by Clark Chapman)

## SCIENTIFIC ARGUMENT

***How can you say that meteors come from comets, but meteorites come from asteroids?***

To begin with, remember the distinction between meteors and meteorites. A meteor is the streak of light seen in the sky when a particle from space is heated by friction with Earth's atmosphere. A meteorite is a piece of space material that actually reaches the ground.

The distinction between comet and asteroid sources must take into account two very strong effects that prevent you from finding meteorites that originated in comets. First, evidence that you will learn more about later in this chapter indicates that cometary material is physically weak, so comet particles vaporize in Earth's atmosphere easily. Very few ever reach the ground. Second, even if a comet particle reached the ground, it would be so fragile that it would weather away rapidly, and you would be unlikely to find it before it disappeared. Asteroidal particles, however, are made from rock and metal and are stronger. They are more likely to survive their plunge through the atmosphere and, afterward, more likely to survive erosion on the ground. Every known meteorite is from the asteroids—not a single meteorite is known to be cometary. In contrast, meteor tracks show that most meteors you see come from comets, and very few are from the asteroid belt.

Now build a new argument. **What evidence suggests that meteorites were once part of larger bodies broken up by impacts?**

# 25-2 Asteroids

ASTEROIDS ARE DISTANT OBJECTS too small to study in detail with Earth-based telescopes. Astronomers nevertheless have learned a surprising amount about those little worlds, using spacecraft and space telescopes.

## Properties of Asteroids

Evidence from meteorites shows that the asteroids are the last remains of the rocky planetesimals that built the Terrestrial planets 4.6 billion years ago.

Study **Observations of Asteroids** on pages 560–561 and notice four important points:

1. Most asteroids are irregular in shape and battered by impact cratering. Many asteroids seem to be rubble piles of broken fragments.
2. Some asteroids are double objects or have small moons in orbit around them. This is further evidence that asteroids have suffered collisions.
3. A few asteroids show signs of geological activity that probably happened on their surfaces when those asteroids were young.
4. Asteroids can be classified by their albedos, colors, and spectra to reveal clues to their compositions. This also allows them to be compared to meteorites in labs on Earth.

## The Asteroid Belt

The first asteroid was discovered on January 1, 1801 (the first night of the 19th century) by the Sicilian monk Giuseppe Piazzi. It was later named Ceres after the Roman goddess of the harvest (and source of our word *cereal*).

Astronomers were excited by Piazzi's discovery because there seemed to be a pattern to the location of planet orbits, except for a wide gap between Mars and Jupiter where the pattern implied a planet "ought" to exist at an average distance from the sun of 2.8 AU. Ceres fit the pattern: Its average distance from the sun is 2.77 AU. But Ceres is much smaller than the planets, and three even smaller objects—Pallas, Juno, and Vesta—were discovered within a few years, all orbiting between Mars and Jupiter, so astronomers decided that Ceres and the other asteroids should not be considered true planets. As you learned in Chapter 24, Ceres has now been re-classified as a dwarf planet because it has enough gravitational strength to squeeze itself into a spherical shape but not enough to have absorbed or cleared away the rest of the asteroids.

Today over 100,000 asteroids have well-charted orbits. Only three are larger than 400 km in diameter (■ Figure 25-8), and most are much smaller. Astronomers are sure that all the large

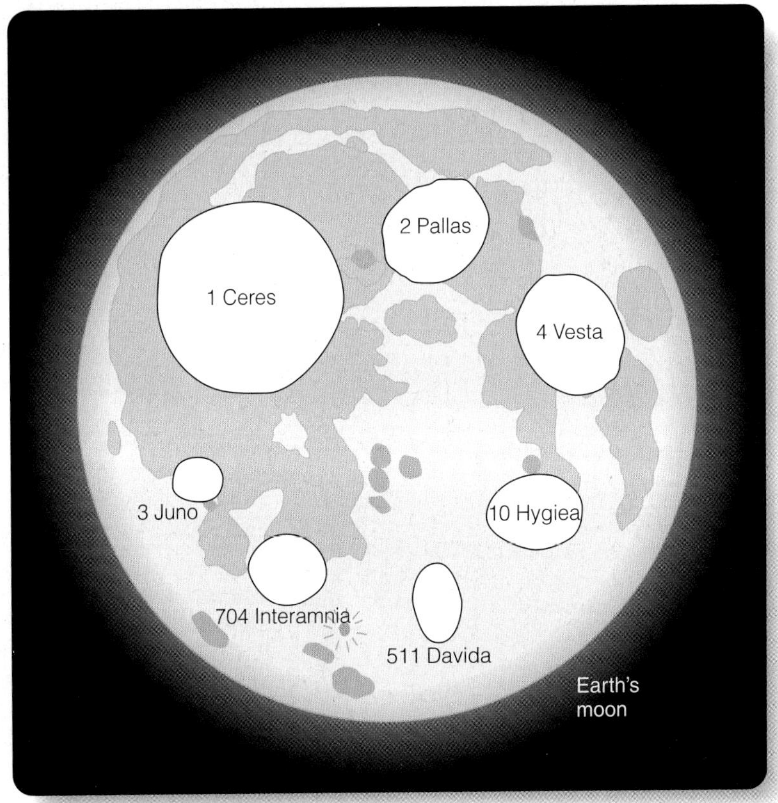


■ **Figure 25-8**

The relative size and approximate shape of the larger asteroids are shown here compared with the size of Earth's moon. Smaller asteroids can be highly irregular in shape.

# Observations of Asteroids

**1** Seen from Earth, asteroids look like faint points of light moving in front of distant stars. Not many years ago they were known mostly for drifting slowly through the field of view and spoiling long time exposures. Some astronomers referred to them as "the vermin of the sky." Spacecraft have now visited asteroids, and the images radioed back to Earth show that the asteroids are mostly small, gray, irregular worlds heavily cratered by impacts.

The Near Earth Asteroid Rendezvous (NEAR) spacecraft visited the asteroid Eros in 2000 and found it to be heavily cratered by collisions and covered by a layer of crushed rock ranging from dust to large boulders. The NEAR spacecraft eventually landed on Eros.

Visual-wavelength image

Eros appears to be a solid fragment of rock.

10 km

NASA

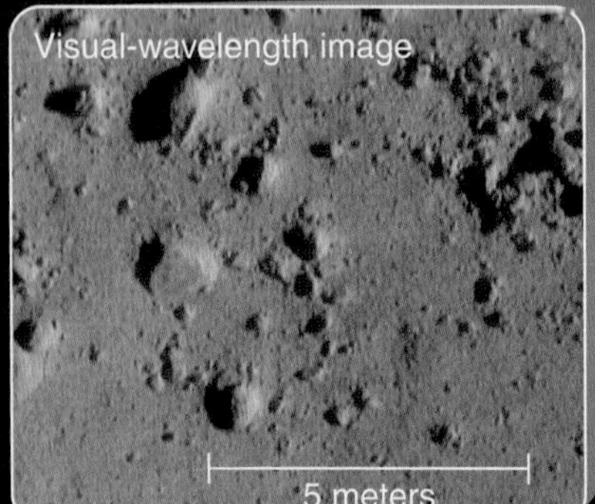


Most asteroids are too small for their gravity to pull them into a spherical shape. Impacts break them into irregularly shaped fragments.

Visual

NASA

50 km

The surface of Mathilde is very dark rock.

**1a** The mass of an asteroid can be found from its gravitational influence on passing spacecraft. Its volume can be measured using images made from a range of perspectives. The density is mass divided by volume. Mathilde, at left, has such a low density that it cannot be solid rock. Like many asteroids, Mathilde may be a rubble pile of broken fragments with large empty spaces between fragments.

If you walked across the surface of an irregularly shaped asteroid such as Eros, you would find gravity very weak; and in many places, it would not be perpendicular to the surface.

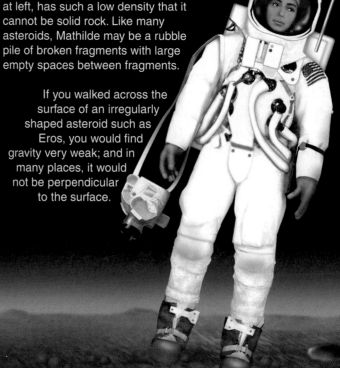

Enhanced visual image

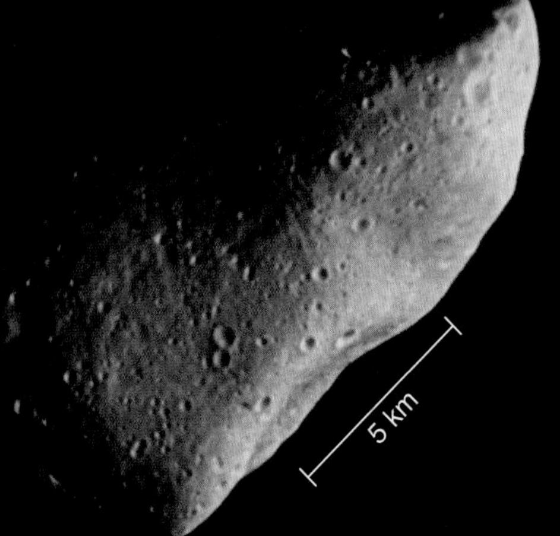

Like most asteroids, Gaspra would look gray to your eyes; but, in this enhanced image at left, color differences probably indicate difference in mineralogy.

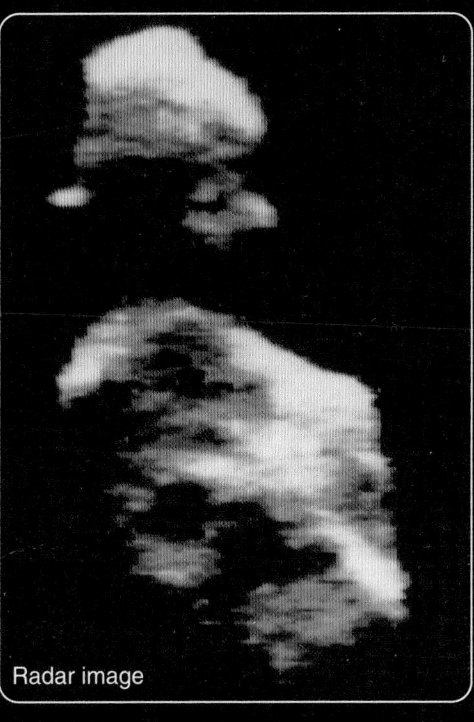


**2** Asteroids that pass near Earth can be imaged by radar. The asteroid Toutatis is revealed to be a double object—two objects orbiting close to each other or actually in contact.

Ida

NASA

Dactyl

30 km

Enhanced visual + infrared

Double asteroids are more common than was once thought, reflecting a history of collisions and fragmentation. The asteroid Ida is orbited by a moon Dactyl only about 1.5 km in diameter.

Occasional collisions among the asteroids release fragments, and Jupiter's gravity scatters them into the inner solar system as a continuous supply of meteorites.

**3** The large asteroid Vesta, as shown at right, provides evidence that some asteroids once had geological activity. No spacecraft has visited it, but its spectrum resembles that of solidified lava. Images made by the Hubble Space Telescope allow the creation of a model of its shape. It has a huge crater at its south pole. A family of small asteroids is evidently composed of fragments from Vesta, and a certain class of meteorites, spectroscopically identical to Vesta, are believed to be fragments from the asteroid. The meteorites appear to be solidified basalt.

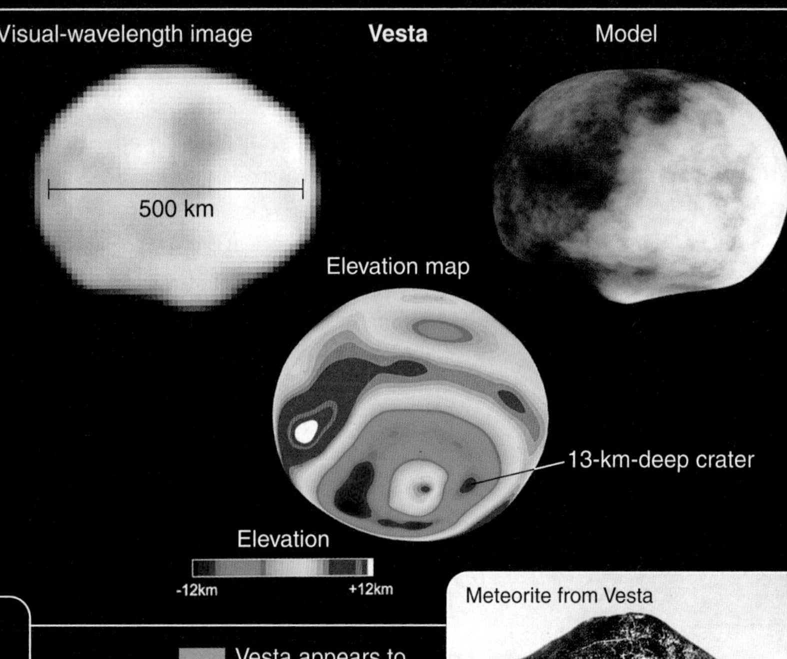

Courtesy of Russell Kemton, New England Meteoritical

**3a** Vesta appears to have had internal heat at some point in its history, perhaps due to the decay of radioactive minerals. Lava flows have covered at least some of its surface.

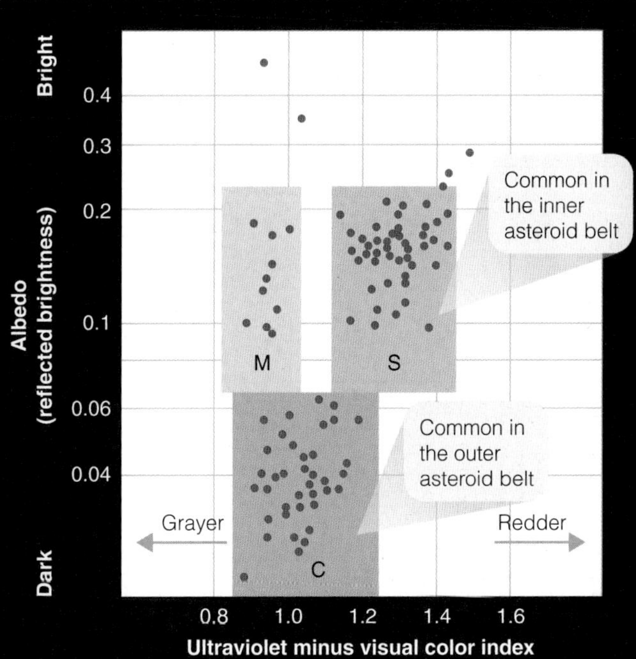


**4** Although asteroids would look gray to your eyes, they can be classified according to their albedos (reflected brightness) and spectroscopic colors. As shown at left, S-types are brighter and tend to be reddish. They are the most common kind of asteroid and appear to be the source of the most common chondrites.

M-type asteroids are not too dark but are also not very red. They may be mostly iron-nickel alloys.

C-type asteroids are as dark as lumps of sooty coal and appear to be carbonaceous.

asteroids in the asteroid belt have been discovered but are also sure that many small asteroids remain undiscovered.

Movies and TV have created a **Common Misconception** that flying through an asteroid belt is a hair-raising plunge requiring constant dodging left and right. The asteroid belt between Mars and Jupiter is actually mostly empty space. In fact, if you were standing on an asteroid, it would be many months or years between sightings of other asteroids.

If you discover an asteroid you are allowed to choose a name for it, and asteroids have been named for spouses, lovers, dogs, politicians, and others.* Once an orbit has been calculated, the asteroid is assigned a number listing its order in the catalog known as the *Ephemerides of Minor Planets.* Thus, Ceres is officially known as 1 Ceres, Pallas as 2 Pallas, and so on.

The distribution of asteroids in the belt is strongly affected by Jupiter's gravitation. Certain orbits in the belt that are almost free of asteroids are called **Kirkwood gaps** after their discoverer, Daniel Kirkwood (■ Figure 25-9). These missing orbits are at certain distances from the sun where an asteroid would find itself in a resonance with Jupiter. For example, an asteroid with an average distance from the sun of 3.28 AU will go exactly twice around the sun in the time it takes Jupiter to go once. Such an asteroid would pass Jupiter at the same place in space every second orbit and be tugged outward. The cumulative perturbations would rapidly change the asteroid's orbit until it was no longer in resonance with Jupiter. Thus, Jupiter effectively eliminates objects from the orbit resonance. The example given represents a 2:1 resonance, but gaps occur in the asteroid belt at many other resonances, including 3:1, 5:2, and 7:3. You will recognize that Kirkwood gaps in the asteroid belt are produced in the same way as some of the gaps in Saturn's rings (see Chapter 23) that were also discovered by Kirkwood.

Computer models show that the motion of asteroids in Kirkwood gaps is described by a theory in mathematics that deals with chaotic behavior. As an example, consider how the smooth motion of water sliding over the edge of a waterfall decays rapidly into a chaotic jumble. The same theory of chaos that describes the motion of the water shows how the slowly changing orbit of an asteroid within one of the Kirkwood gaps can suddenly (astronomically speaking) become a long, eccentric orbit that carries the asteroid into the inner solar system.

## Asteroids Outside the Main Belt

You don't have to go all the way to the asteroid belt if you want to visit an asteroid; some follow orbits that cross the orbits of the Terrestrial planets and come near Earth. Others wander far away, among the Jovian worlds. There are also asteroids that share orbits with the planets (■ Figure 25-10).

**Apollo-Amor objects** are asteroids with orbits that carry them into the inner solar system. Amor objects follow orbits that cross the orbit of Mars but don't reach the orbit of Earth, whereas Apollo objects have Earth-crossing orbits. About 3000 Apollo and Amor objects have been found so far. The influences of Jupiter and other planets act to continually change their orbits. Astronomers calculate that about one-third of Apollo-Amors will be thrown into the sun, a few will be ejected from the solar system, and, as you will discover later in this chapter, some are doomed to collide with a planet—perhaps ours.

Several research teams are now intent on identifying **Near-Earth Objects (NEOs),** including Apollo-Amor objects. For example, LONEOS (Lowell Observatory Near-Earth Object Search) is searching the entire sky visible from Lowell Observatory in northern Arizona once a month. The LINEAR (Lincoln Near-Earth Asteroid Research) telescope in New Mexico (■ Figure 25-11) and the NEAT (Near-Earth Asteroids Tracking) facilities in California and Hawaii also have been successful in finding NEOs, as well as new main-belt asteroids and Kuiper belt objects. The combined searches are expected to be able to locate at least 90 percent of the NEOs larger than 1 km in diameter by 2011.

It is easy to hypothesize that the Apollo-Amor objects are rocky asteroids that have been sent into their unusual orbits by collisions in the main asteroid belt or by planetary perturbations,

*Some sample asteroid names: Chicago, Vaticana, Noel, Ohio, Tea, Gaby, Fidelio, Hagar, Geisha, Tata, Mimi, Dulu, Tito, Zulu, Zappafrank, and Garcia (after the late musicians Frank Zappa and Jerry Garcia, respectively).

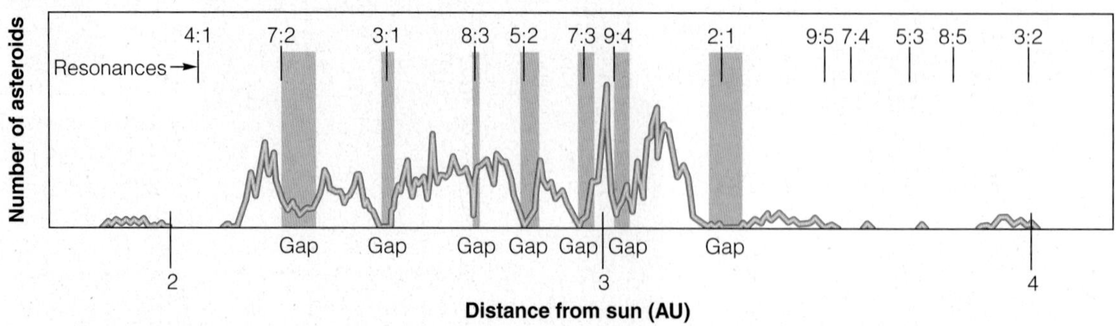


■ **Figure 25-9**

The red curve in this plot shows the number of asteroids at different distances from the sun. Purple bars mark Kirkwood gaps, where there are few asteroids. Note that these gaps match resonances with the orbital motion of Jupiter.

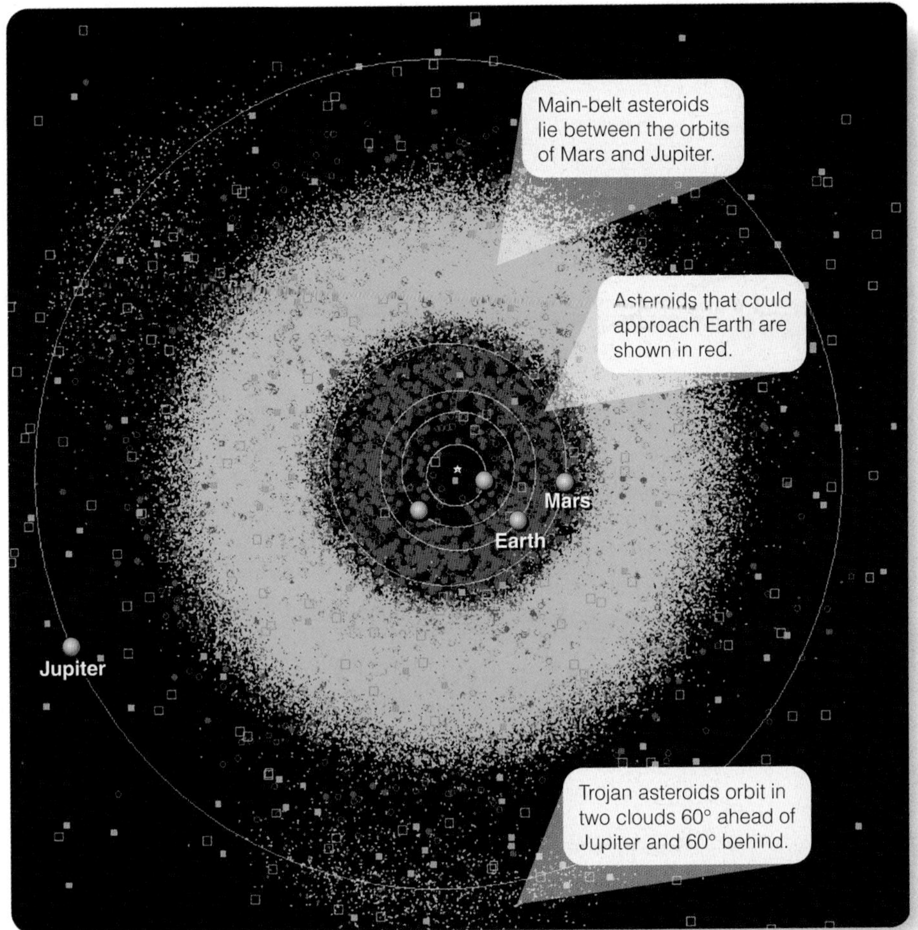


**Figure 25-10**

This diagram plots the position of known asteroids between the sun and the orbit of Jupiter on a specific day. Most asteroids are in the main belt. Squares, filled or empty, show the location of known comets. Although asteroids and comets are small bodies and lie far apart, there are a great many of them in the inner solar system. (Minor Planet Center)

for example Jupiter's Kirkwood gap-clearing effect that you learned about in the previous section. There is evidence that a few of these objects instead may be comets that became trapped in short orbits that kept them in the inner solar system so they have exhausted their volatiles. You can see from this that the distinction between comets and asteroids is not sharply defined.

Jupiter ushers two groups of nonbelt asteroids around its own orbit. These objects have become trapped in the Lagrangian points along Jupiter's orbit that are 60° ahead of and 60° behind the planet. Lagrangian points are regions like cosmic sinkholes where gravitational effects of the two larger bodies, in this case the sun and Jupiter, combine to trap small bodies (Figure 25-10). (For an example of Lagrangian points in a stellar context, see the positions labeled $L_4$ and $L_5$ in Figure 13-5.) The Jupiter Lagrangian-point objects are called **Trojan asteroids** because individual asteroids have been named after the heroes of the Trojan War (588 Achilles, 624 Hektor, 659 Nestor, and 1143 Odysseus, for example). Almost 2000 Trojan asteroids are known, but only the brightest have been given names. Some astronomers speculate that there may be as many Trojan asteroids as asteroids in the main belt. Astronomers have also found a few objects in the Lagrangian points of the orbits of Mars and Neptune. Other planets, including Earth, may have undiscovered Lagrangian point asteroids trapped in their orbits.

There are other nonbelt asteroids beyond the main belt. The object Chiron, found in 1977, is about 170 km (110 mi) in diameter. Its orbit carries it from the orbit of Uranus to just inside the orbit of Saturn. Although it was first classified as an asteroid, Chiron surprised astronomers ten years after its discovery by suddenly brightening as it released jets of vapor and dust. Old photographs were found showing that Chiron had done this before. Astronomers now suspect Chiron has a rocky crust covering deposits of ices such as solid nitrogen, methane, and carbon monoxide. You will learn in the next section that this more resembles the characteristics of comets than asteroids. Objects like Chiron with orbits between, or crossing, orbits of the Jovian planets are called **centaurs.** The characteristics of centaurs show you yet again that the distinction between asteroids and comets is not clear cut.

As technology allows astronomers to detect smaller and more distant objects, they are learning that our solar system contains large numbers of these small bodies. The challenge is to explain their origin.

## Origin and History of the Asteroids

An old hypothesis proposed that asteroids are the remains of a planet that exploded. Planet-shattering death rays may make for exciting science-fiction movies, but in reality planets do not explode. The gravitational field of a planet holds the mass together so tightly that completely disrupting the planet would take tremendous energy. In addition, the present-day total mass of the asteroids is only about one-twentieth the mass of Earth's moon, hardly enough to be the remains of a planet.

Astronomers have evidence that the asteroids are the remains of material lying 2 to 4 AU from the sun that was unable to form a planet because of the gravitational influence of Jupiter, the next planet outward. Over the 4.6-billion-year history of the solar system, most of the objects originally in the asteroid belt have

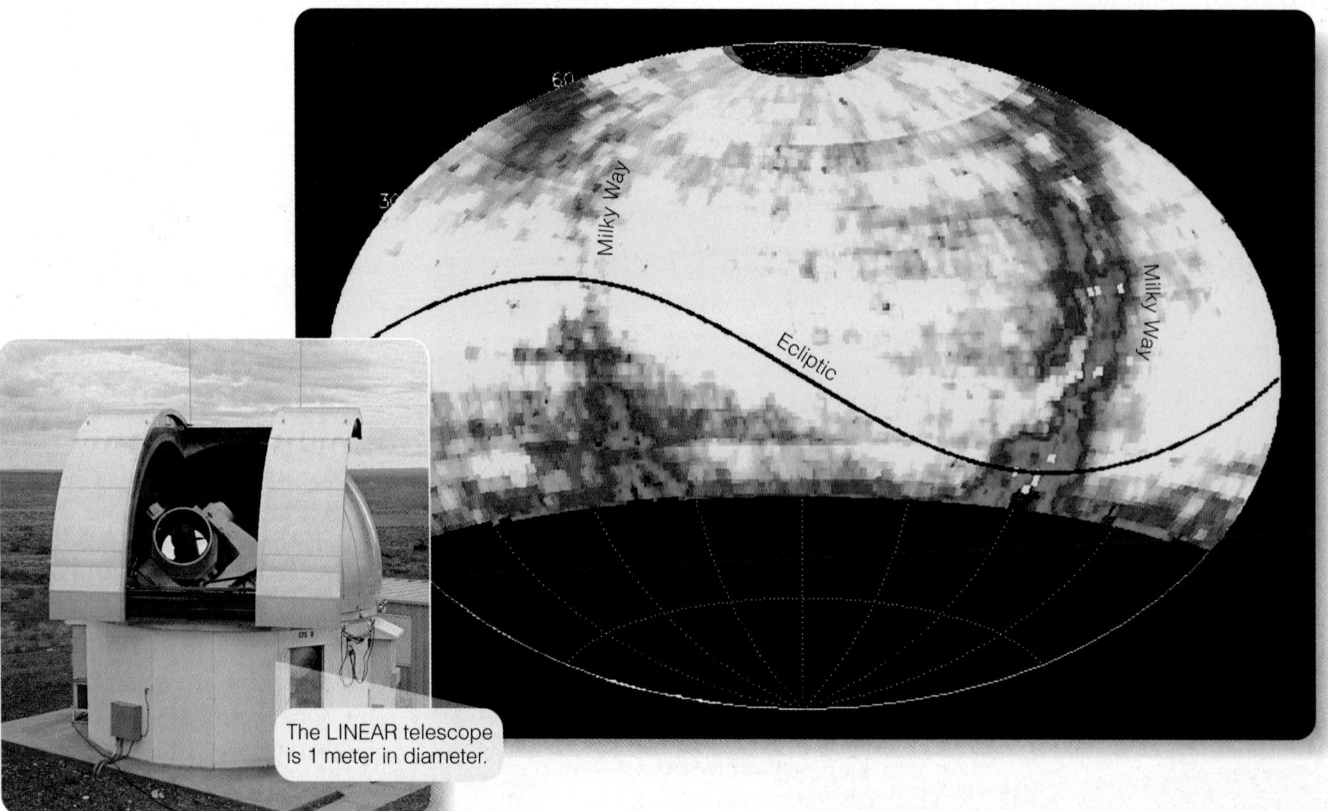


**Figure 25-11**

(a) The LINEAR telescope searches for asteroids every clear night when the moon is not bright. A diagram showing color codes for the thoroughness of the LINEAR search over the entire sky for one year. The green (poorly searched) region is the plane of the Milky Way, where asteroids are difficult to discover against a dense starry background. (MIT/Lincoln Labs)

collided, fragmented, and been covered with craters. Some asteroids were perturbed by the gravity of Jupiter and other planets into orbits that collided with planets or the sun, caused them to be captured as planetary satellites, or ejected them from the solar system. The present-day asteroids are understood to be a very minor remnant of the original mass in that zone.

Collisions among asteroids must have been occurring since the formation of the solar system (look again at Figure 25-7 and page 560). Astronomers have found evidence of catastrophic impacts powerful enough to shatter an asteroid. Early in the 20th century, astronomer Kiyotsugu Hirayama discovered that some groups of asteroids share similar orbits. Each group is distinct from other groups, but asteroids within a given group have the same average distance from the sun, the same eccentricity, and the same inclination. Up to 20 of these **Hirayama families** are known. Modern observations show that the asteroids in a family typically share similar spectroscopic characteristics. Apparently, each family was produced by a catastrophic collision that broke a single asteroid into fragments that continue traveling along similar orbits around the sun. Evidence shows that one family was produced only 5.8 million years ago in a collision between asteroids estimated to have been 3 km and 16 km in diameter, traveling at relative speeds of about 5 km/s (11,000 mph), a typical speed for asteroid collisions. It seems that the fragmentation of asteroids is a continuing process.

In 1983, the Infrared Astronomy Satellite detected the infrared glow of sun-warmed dust scattered in bands throughout the asteroid belt. These dust bands appear to be the products of past collisions. The dust will eventually be destroyed, but because collisions occur constantly in the asteroid belt, new dust bands will presumably be produced as the present bands dissipate. The interplanetary dust in our solar system is analogous to dust in extrasolar planetary debris disks, produced astronomically recently by collisions of remnant plantesimals (see Chapter 19).

Even though most of the planetesimals originally in the main belt have been lost or destroyed, the objects left behind carry clues to their origin in their albedos and spectroscopic colors (p. 561). **C-type asteroids** have albedos less than 0.06 and would look very dark to your eyes. They are probably made of carbon-rich material similar to that in carbonaceous chondrite meteorites. C-type asteroids are more common in the outer asteroid belt. It is cooler there, and the condensation sequence (see Chapter 19) predicts that carbonaceous material would form more easily in the outer belt than in the inner belt.

**S-type asteroids** have albedos of 0.1 to 0.2, so they would look brighter as well as redder than C-types; S-types may be composed of rocky material. **M-type** asteroids are also bright but not as red as the S-types; they seem to be metal-rich and may be fragments from iron cores of differentiated asteroids. A few other types of asteroids are known, and a number of individual asteroids have been found that are unique, but these three classes include most of the known asteroids.

Although S-type asteroids are very common in the inner asteroid belt, their colors and albedos are different from those of chondrites, the most common kind of meteorite. This represented a puzzle to astronomers; shouldn't the common type of asteroid nearest Earth be the source of the most common type of meteorite that hit Earth? New evidence from the analysis of moon rocks and from observations of Eros, an S-type asteroid, shows that bombardment by micrometeorites and solar wind particles can redden and darken rocky materials until they have the colors and albedos of S-type asteroids. It therefore seems likely that chondrite meteorites are in fact fragments of S-type asteroids.

In March 2009 a small asteroid (2–3 meters in diameter, about the size of a small truck) was spotted by the NEO detection network on a collision course with Earth. Astronomers were able to observe it in space before impact and discovered that its colors and albedo matched the fairly rare F-type asteroids that are mostly in the outer belt. The asteroid entered Earth's atmosphere over the desert of northern Sudan and was witnessed exploding. Scientists Peter Jenniskens from the SETI Institute and Muawia Shaddad of the University of Khartoum organized a team of local faculty and students to search for pieces of the object. They ultimately found about 4 kilograms (9 pounds) of fragments corresponding to the rate ureilite meteorite type. For the first time, planetary scientists were able to make a definite connection between an asteroid observed in space and meteorites with properties measured in an Earth laboratory.

As you saw in the case of Vesta, a few asteroids may once have been geologically active, with lava flowing on their surfaces when they were young. Perhaps they incorporated especially large amounts of short-lived radioactive elements such as aluminum-26. Those radioactive elements were probably produced by a supernova explosion that could also have been the trigger for the formation of the sun and planets while seeding the young solar system with its nucleosynthesis products (see Chapter 13 and Figure 11-2).

Not all large asteroids have been active. Ceres, 900 km in diameter, is almost twice as big as Vesta, but it shows no spectroscopic sign of past activity and evidently has an ice-rich mantle. The puzzling differences between those two large asteroids will be investigated by the Dawn spacecraft that is on its way to orbit Vesta in 2011 and Ceres in 2015.

Although there are still mysteries to solve, you can understand the story of the asteroids. They are fragments of planetesimals, some of which differentiated, developed molten metal cores, in a few cases even had lava flows on their surfaces, and then cooled slowly. The largest asteroids astronomers see today may be nearly unbroken examples of original planetesimals, but the smaller asteroids are fragments produced by 4.6 billion years of collisions.

### SCIENTIFIC ARGUMENT

***What is the evidence that asteroids have been fragmented?***
First, your argument might note that the solar nebula theory predicts that planetesimals collided and either stuck together or fragmented. This is suggestive, but it is not evidence. A theory can never be used as evidence to support some other theory or hypothesis. Evidence means observations or the results of experiments, so your argument must cite observations. The spacecraft photographs of asteroids show irregularly shaped little worlds heavily scarred by impact craters. Further evidence indicates some asteroids may be pairs of bodies split apart but still in contact, and images of asteroid Ida reveal a small satellite, Dactyl. Other asteroids with moons have been found. These double asteroids and asteroids with moons probably reveal the results of fragmenting collisions between asteroids. Furthermore, meteorites appear to have come from the asteroid belt astronomically recently, so fragmentation must be a continuing process there.

Now build an argument to combine what you know of meteorites with your experience with asteroids. **What evidence could you cite to show what the first planetesimals were like?**

## 25-3 Comets

Few sights in astronomy are more beautiful than a bright comet hanging in the night sky (■ Figure 25-12). It is a **Common Misconception** that comets whiz rapidly across the sky like meteors. Actually, comets move with the stately grace of great ships at sea, their motion hardly apparent; night by night they shift position slightly against the background stars, and they may remain visible for weeks.

Throughout history, comets have been considered omens of doom. Comets may be beautiful, but are also so strange in appearance that they can create some instinctive alarm. Even recent appearances of bright comets have caused predictions of the end of the world. In 1910, Comet Halley was spectacular, and it was frightening to some people. Comet Kohoutek in 1973, Comet Halley returning in 1986, and Comet Hale–Bopp in 1997 also caused concern among the superstitious.*

Faint comets are common; several dozen are discovered each year. Truly bright comets appear about once per decade. Comet McNaught in 2007 was bright enough to be classed

*Comets are now named after the person or persons who discover them. Comet Halley, in contrast, which has been seen occasionally for at least the last 2,250 years, has no known "discoverer." It is named after Edmund Halley, an English astronomer who was a friend of Isaac Newton and who was the first to realize that certain comets appearing at 76-year intervals were actually the same comet on a repeating orbit. Halley correctly predicted that the comet would be seen again in a certain year.

■ **Figure 25-12**

Comet McNaught swept through the inner solar system in 2007 and was a dramatic sight in the southern sky. Seen here from Australia, the comet was on its way back into deep space after making its closest approach to the sun ten days earlier. Comet McNaught began this passage with a period of about 300,000 years, but gravitational perturbations by the planets changed its orbit shape from elliptical to hyperbolic, so it will never return but instead is leaving the solar system to journey forever in interstellar space. (© John White)

with other great comet appearances such as Comet Halley in 1910. An average person might see five or ten bright comets in a lifetime. While everyone can enjoy the beauty of comets, astronomers study them because they are messengers from the past carrying cargos of information about the origin of our solar system.

## Properties of Comets

As always, you should begin your study of a new kind of object by summarizing its observational properties. What do comets look like, and how do they behave?

Study **Observations of Comets** on pages 568–569 and notice three important properties of comets plus three new terms:

1. Comets have two kinds of tails, shaped by the solar wind and solar radiation. Gas and dust released by a comet's icy nucleus produces a head or *coma* and are then blown outward, away from the sun. The gas produces a *type I,* or *gas, tail,* and the dust produces a *type II,* or *dust, tail.*

2. Comet dust produces not only one of the two types of comet tails but also spreads throughout the solar system. Some of those comet dust particles later encounter the Earth and are seen as meteors.

3. Evidence shows that comet nuclei are fragile and can break into pieces easily.

Astronomers can put these and other observations together to study the structure of comet nuclei.

## The Geology of Comet Nuclei

The nuclei of comets are quite small and cannot be studied in detail using Earth-based telescopes. Nevertheless, when a comet nucleus approaches the sun, it emits material that forms into a coma (head) and tail that can be millions of kilometers in size and is easily observed.

Spectra of comet comae (plural of *coma*) and tails indicate the nuclei must contain ices of water and other volatile compounds such as carbon dioxide, carbon monoxide, methane, ammonia, and so on. These are the kinds of compounds that would have condensed in cold regions of the solar nebula. This convinces astronomers that comets are ancient samples of the gases and dust from which the outer planets formed. As the ices absorb energy from sunlight, they sublime—change from a solid directly into a gas. The gases break down and combine chemically, producing other substances found in comet spectra. For example, vast clouds of hydrogen gas observed around the heads of comets are understood to derive from the breakup of ice molecules.

Five spacecraft flew past the nucleus of Comet Halley when it visited the inner solar system in 1985 and 1986. Other spacecraft flew past the nuclei of Comet Borrelly in 2001, Comet Wild 2 (pronounced *vildt-two*) in 2004, and Comet Tempel 1 in 2005. Images show that all these comet nuclei are 1 to 10 km across, similar in size to many asteroids, and irregular in shape (■ Figure 25-13). In general, these nuclei are darker than a lump of coal, which suggests composition similar to carbon-rich carbonaceous chondrite meteorites described earlier in this chapter.

The mass and density of comet nuclei can be calculated from their gravitational influence on passing spacecraft. Comet nuclei appear to have densities between 0.1 and 0.25 g/cm$^3$, much less than the density of ice. Also, as you will learn later in this chapter, comets subjected to tidal stresses from Jupiter or the sun come apart very easily. Comet nuclei have been described as dirty snowballs or icy mudballs, but that seems to be incorrect; their shapes, low densities, and lack of material strength suggest that comets are not solid objects. The evidence leads astronomers to conclude that most comet nuclei must be fluffy mixtures of ices and dust with significant amounts of empty space. On the other hand, images of the nucleus of Comet Wild 2 revealed cliffs, pinnacles, and other features that show the material has enough strength to stand against the weak gravity of the comet.

Photographs of comet comae (Figure 25-13) often show jets springing from the nucleus into the coma and being swept back by the pressures of sunlight and the solar wind to form the tail. Studies of the motions of these jets as the nucleus rotates reveal that they originate from small active regions that may be similar

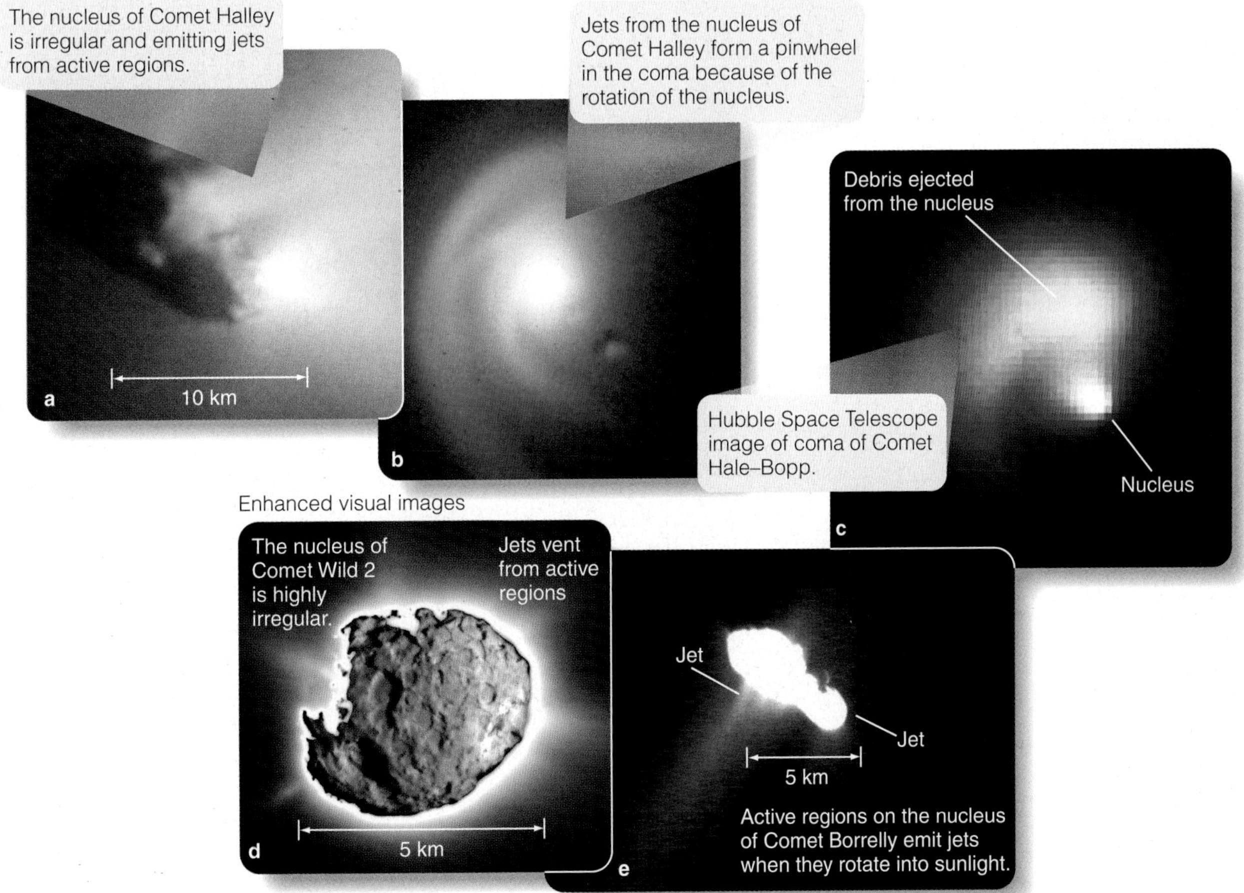
The nucleus of Comet Halley is irregular and emitting jets from active regions.
10 km
a
Jets from the nucleus of Comet Halley form a pinwheel in the coma because of the rotation of the nucleus.
b
Debris ejected from the nucleus
Hubble Space Telescope image of coma of Comet Hale–Bopp.
Nucleus
c
Enhanced visual images
The nucleus of Comet Wild 2 is highly irregular.
Jets vent from active regions
5 km
d
Jet
Jet
5 km
Active regions on the nucleus of Comet Borrelly emit jets when they rotate into sunlight.
e

**■ Figure 25-13**

Visual-wavelength images made by spacecraft and by the Hubble Space Telescope show how the nuclei of comets produce jets of gases from regions where sunlight vaporizes ices. (Halley nucleus: © 1986 Max-Planck Institute; Halley coma: Steven Larson; Comets Borrelly, Hale–Bopp, and Wild 2: NASA)

# Observations of Comets

**1** A **type I** or **gas tail** is produced by ionized gas carried away from the nucleus by the solar wind. The spectrum of a gas tail is an emission spectrum. The atoms are ionized by the ultraviolet light in sunlight. The wisps and kinks in gas tails are produced by the magnetic field embedded in the solar wind.

Spectra of gas tails reveal atoms and ions such as $H_2O$, $CO_2$, CO, H, OH, O, S, C, and so on. These are released by the vaporizing ices or produced by the breakdown of those molecules. Some gases, such as hydrogen cyanide (HCN), must be formed by chemical reactions.



**1a** A **type II** or **dust tail** is produced by dust that was contained in the vaporizing ices of the nucleus. The dust is pushed gently outward by the pressure of sunlight, and it reflects an absorption spectrum, the spectrum of sunlight. The dust is not affected by the magnetic field of the solar wind, so dust tails are more uniform than gas tails. Dust tails are often curved because the dust particles follow their individual orbits around the sun once they leave the nucleus. Because of the forces acting on them, both gas and dust tails extend away from the sun.

**1b** The nucleus of a comet (not visible here) is a small, fragile lump of porous rock containing ices of water, carbon dioxide, ammonia, and so on. Comet nuclei can be 1 to 100 km in diameter.

The **coma** of a comet is the cloud of gas and dust that surrounds the nucleus. It can be over 1,000,000 km in diameter, bigger than the sun.

**1c** Comet Mrkos in 1957 shows how the gas tail can change from night to night due to changes in the magnetic field in the solar wind.

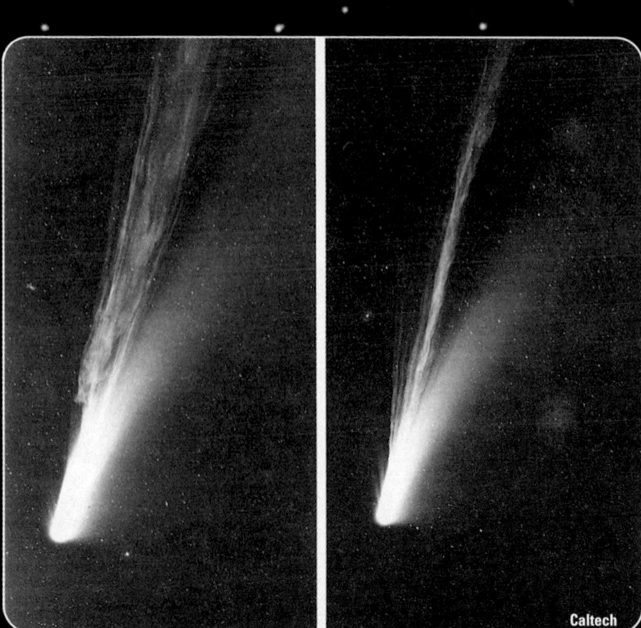

Visual-wavelength images

Visual-wavelength images

**2** As the ices in a comet nucleus vaporize, they release dust particles that not only form the dust tail but also spread throughout the solar system.

The Deep Impact spacecraft released an instrumented probe into the path of Comet Tempel 1. When the comet slammed into the probe at 10.2 km/s as shown at right, huge amounts of gas and dust were released. From the results, scientists conclude that the nucleus of the comet is rich in dust finer than the particles of talcum powder. The nucleus is marked by craters, but it is not solid rock. It is about the density of fresh-fallen snow.

NASA/JPL-Caltech/UMD

Only seconds before impact, craters are visible on the dark surface.

Images of Comet Tempel 1 from the flyby probe 13 seconds after the impact probe hit. Gas and dust are thrown out of the impact crater.

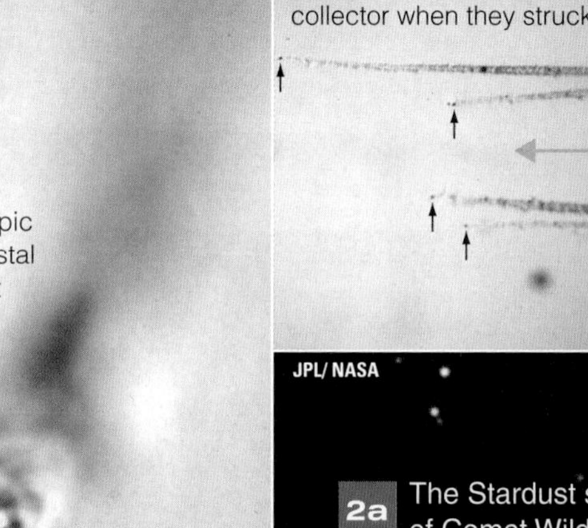


Dust particles (arrows) were embedded in the collector when they struck at high velocity.

JPL/ NASA

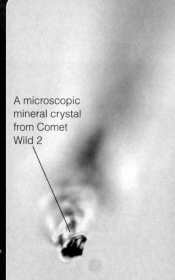

A microscopic mineral crystal from Comet Wild 2

NASA

**2a** The Stardust spacecraft flew past the nucleus of Comet Wild 2 and collected dust particles (as shown above) in an exposed target that was later parachuted back to Earth. The dust particles hit the collector at high velocity and became embedded, but they can be extracted for study.

Some of the collected dust is made of high-temperature minerals that could only have formed near the sun. This suggests that material from the inner solar nebula was mixed outward and became part of the forming comets in the outer solar system. Other minerals found include olivine, a very common mineral on Earth but not one that scientists expected to find in a comet.

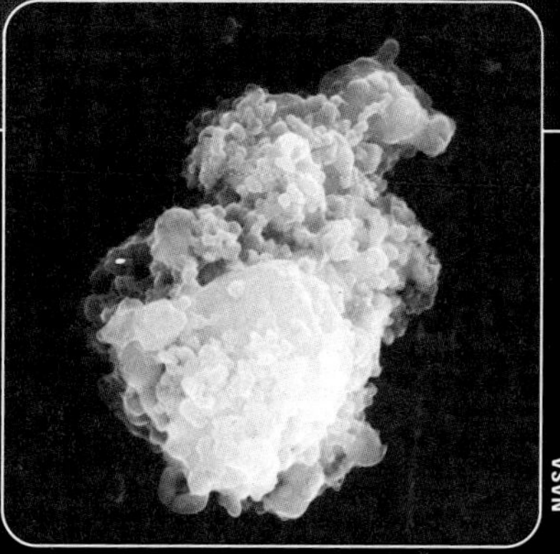

NASA

This dust particle was collected by spacecraft above Earth's atmosphere. It is almost certainly from a comet.

**3** The nuclei of comets are not strong and can break up. In 2006, Comet Schwassmann-Wachmann 3 broke into a number of fragments that themselves fragmented. Fragment B is shown at the right breaking into smaller pieces. The gas and dust released by the breakup made the comet fragments bright in the night sky, and some were visible with binoculars. As its ices vaporize and its dust spreads, the comet may totally disintegrate and leave nothing but a stream of debris along its previous orbit.

Comets most often break up as they pass close to the sun or close to a massive planet like Jupiter. Comet LINEAR broke up in 2000 as it passed by the sun. The comet that hit Jupiter in 1994 was first ripped to pieces by tidal stresses from Jupiter's gravity. Comets can also fragment far from planets, perhaps because of the vaporization of critical structural areas of ice.

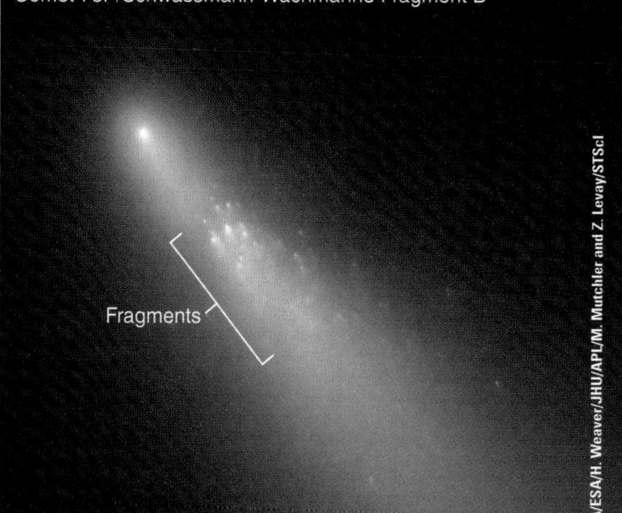


Comet 73P/Schwassmann-Wachmann3 Fragment B

NASA/ESA/H. Weaver/JHU/APL/M. Mutchler and Z. Levay/STScI

**■ Figure 25-14**

The crusts of comets are evidently delicate mixtures of rock, ice, and dust. The dust is ejected along with gases as the ices in a comet vaporize in sunlight, as shown in this artist's conception. (NASA/NSSDC; Tom Herbst, Max-Planck-Institut für Astronomie, Heidelberg; Doug Hamilton, Max-Planck-Institut für Kernphysik, Heidelberg; Hermann Böehnhardt, Universitäts-Sternwarte, München; and Jose Luis Ortiz Moreno, Instituto de Astrofisica)

to volcanic faults or vents (■ Figure 25-14). As the rotation of a comet nucleus carries an active region into sunlight, it begins venting gas and dust, and as it rotates into darkness it shuts down.

The nuclei of comets appear to have a crust of rocky dust left behind as the ices vaporize. Breaks in that crust can expose ices to sunlight, and vents can occur in those regions. It also seems that some comets have large pockets of volatiles buried below the crust. When one of those pockets is exposed and begins to vaporize, the comet can suffer a dramatic outburst, as Comet Holmes did in 2007 (■ Figure 25-15).

Astronomers have devised ways to study comet material more directly. The Stardust spacecraft passed through the tail of Comet Wild 2 in 2004, collected dust particles that had been ejected from the comet's nucleus, and returned the samples in a sealed capsule to Earth in 2006 for analysis. In 2005, the Deep Impact spacecraft released an instrumented impactor probe into the path of comet Tempel 1. As planned, the nucleus of the comet ran into the impactor at almost 10 km/s (22,000 mph). The probe broke through the crust of the nucleus and blasted vapor and dust out into space where the Deep Impact "mother ship," as well as the Spitzer and Hubble space telescopes and observatories on Earth, could analyze it (page 569). Those missions produced the surprising discovery that some comet dust is crystalline and must have formed originally in very warm environments close to the sun, but then was incorporated somehow into comet nuclei in the cold outer solar system.

The Solar and Heliospheric Observatory (SOHO) spacecraft was put into space to observe the sun, but it has also discovered over a thousand comets, called "sun grazers," that come very close to the sun, in some cases 70 times closer to the sun's surface than the planet Mercury (■ Figure 25-16). As many as three comets per week plunge into the sun and are destroyed. Most sun grazers belong to one of four groups, and the comets in each group have very similar orbits. Like the Hirayama families of asteroids, these comet groups appear to be made up of fragments of larger comet nuclei. The original comet may have been ripped apart by the violence of gases superheated near the sun and bursting through the crust, or by solar tidal forces, or both.

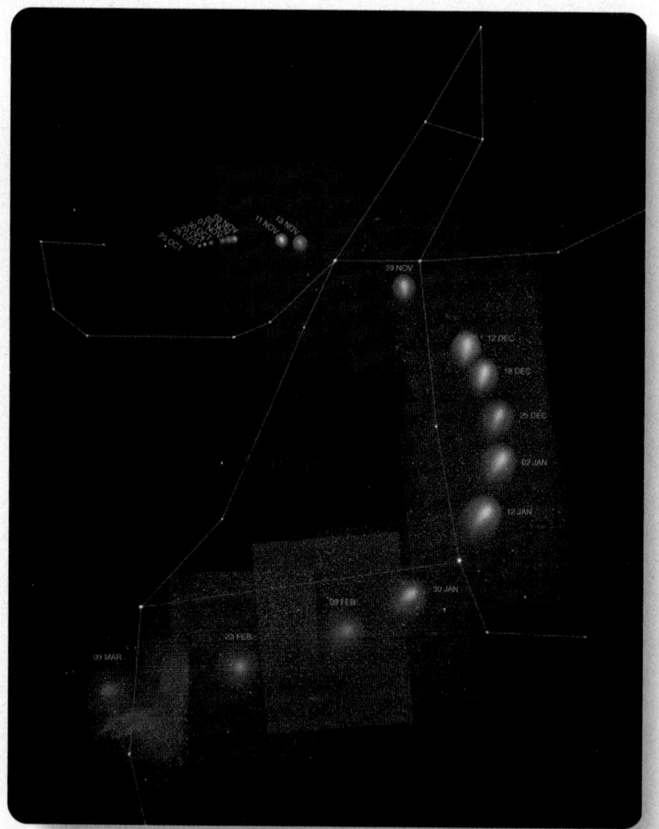


**■ Figure 25-15**

Composite of 19 snapshots of Comet Holmes, showing its changing brightness and position spanning the period from October 2007 until March 2008. During its outburst in late October 2007, the comet brightened by a factor of about 500,000 as a large pocket of volatile material exploded through its crust and spread into space. (John Pane)

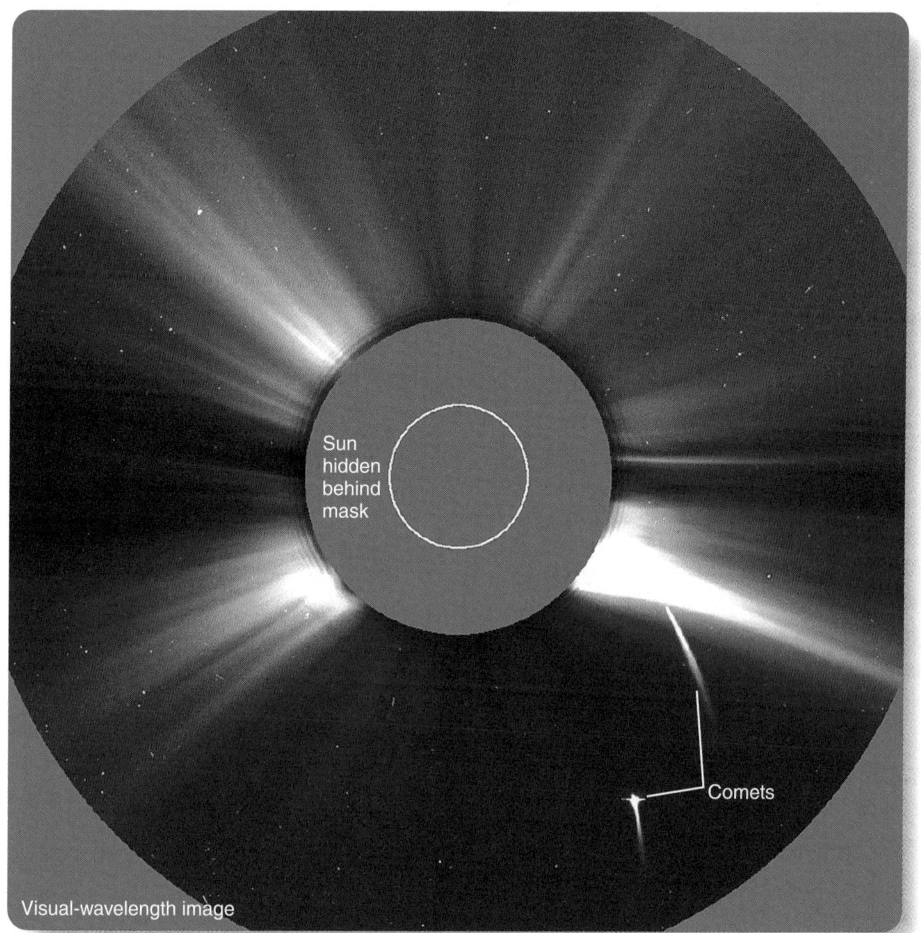


■ **Figure 25-16**

The SOHO observatory can see comets rounding the sun on very tight orbits. Some of these sun-grazing comets, like the two shown here, are destroyed by the heat and are not detected emerging on the other side of the sun. (SOHO/NASA)

Sun-grazing comets can be destroyed quickly by the sun, but even normal comets suffer from the effects of solar heating. Each passage around the sun vaporizes many millions of tons of ices, so the nucleus slowly loses its ices until there is nothing left but dust and rock moving along an orbit around the sun. The eventual fate of a comet is clear, but a more important question is its origin.

## The Origin and History of Comets

Family relationships among the comets can give you clues to their origin. Most comets have long, elliptical orbits with periods greater than 200 years and are known as long-period comets. The long-period comet orbits are randomly inclined to the plane of the solar system, so those comets approach the inner solar system from all directions. Long-period comets revolve around the sun in about equal numbers in prograde orbits (the same direction in which the planets move) and retrograde orbits.

In contrast, about 100 or so of the 600 well-studied comets have orbits with periods less than 200 years. These short-period comets usually follow orbits that lie within 30° of the plane of the solar system, and most revolve around the sun prograde. Comet Halley, with a period of 76 years, is an unusual short-period comet with a retrograde orbit.

A comet cannot survive long in an orbit that brings it into the inner solar system. The heat of the sun vaporizes ices and reduces comets to inactive bodies of rock and dust; such comets can last at most 100 to 1000 orbits around the sun. Astronomers calculate that even before a comet completely vaporizes from solar heating, it can't survive more than about half a million years crossing the orbits of the planets, especially Jupiter, without having its path rerouted into the sun or out of the solar system or colliding with one of the planets. Therefore, comets visible in our skies now can't have survived in their present orbits for 4.6 billion years since the formation of the solar system, and that means there must be a continuous supply of new comets. Where do they come from?

In the 1950s, astronomer Jan Oort proposed that the long-period comets are objects that fall inward from what has become known as the **Oort cloud,** a spherical cloud of icy bodies that extends from about 10,000 to 100,000 AU from the sun (■ Figure 25-17). Astronomers estimate that the cloud contains several trillion ($10^{12}$) icy bodies. Far from the sun, they are very cold, lack comae and tails, and are invisible from Earth. The gravitational influence of occasional passing stars can perturb a few of these objects and cause them to fall into the inner solar system, where the heat of the sun warms their ices and transforms them into comets. The fact that long-period comets are observed to fall inward from all directions is explained by their Oort cloud reservoir being spherically symmetric around the sun and inner solar system.

It is not surprising that stars pass close enough to affect the Oort cloud. For example, data from the Hipparcos satellite show

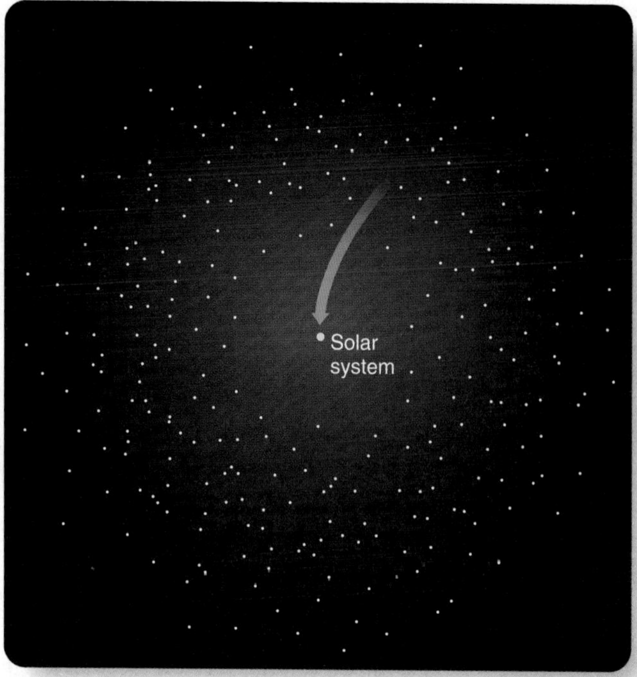


**■ Figure 25-17**

Long-period comets appear to originate in the spherical Oort cloud. Objects that fall into the inner solar system from that cloud arrive from all directions.

that the star Gliese 710 will pass within 1 light-year (about 63,000 AU) of the sun, crossing through the Oort cloud, in about a million years. The result may be a shower of Oort cloud objects perturbed into the inner solar system, where, warmed by the sun, they will become comets.

Saying that comets come from the Oort cloud only pushes the mystery back one step. How did those icy bodies get there? In preceding chapters, you have studied the origin and evolution of our solar system so carefully that the answer may leap out at you. Those Oort cloud comets are some of the icy planetesimals that formed in the outer solar nebula. The bodies in the Oort cloud, however, could not have formed at their present location because the solar nebula would not have been dense enough so far from the sun. And, if they had formed from the solar nebula, they would be distributed in a disk and not in a sphere. Astronomers think the Oort cloud planetesimals formed in the outer solar system near the present orbits of the giant planets. As those planets grew, they swept up many of these planetesimals, but they also would have ejected some out of the solar system. Most of those ejected objects vanished into space, but perhaps 10 percent had their orbits modified by the gravity of stars passing nearby and became part of the Oort cloud. Some of those later became the long-period comets.

Long-period comets originate in the Oort cloud, and some of the short-period comets do also. A few short-period comets, including Comet Halley, appear to have begun as long-period comets from the Oort cloud and had their orbits altered by a close encounter with Jupiter. However, that process can't explain all of the short-period comets: Some follow orbits that could not have been produced by Oort cloud objects interacting with Jupiter or the other planets. There must be another source of icy bodies in our solar system. To find the answer, you need only look beyond the orbit of Neptune and study the icy Kuiper belt objects.

## Comets from the Kuiper Belt

You first met the Kuiper belt in Chapter 19 when you studied the origin of the solar system. In Chapter 24, you learned about the largest Kuiper belt objects as examples of dwarf planets. In this chapter, it is important to study the smaller bodies of the Kuiper belt because they are one of the sources of comets.

In 1951, astronomer Gerard Kuiper proposed that the formation of the solar system should have left behind a belt of small, icy planetesimals beyond the Jovian planets and in the plane of the solar system. Such objects were first discovered in 1992 and are now known as Kuiper belt objects (KBOs).*

The Kuiper belt objects are small, icy bodies (■ Figure 25-18) that orbit in the plane of the solar system extending from the orbit of Neptune out to about 50 AU from the sun. Some objects are known to loop out as far as 1000 AU, but those may have been scattered into those orbits by gravitational interactions with passing stars. The entire Kuiper belt, containing as many as 100,000 objects 100 km or larger in diameter and hundreds of millions of smaller bodies, would be hidden behind the yellow dot representing the solar system in Figure 25-17. Some Kuiper belt objects are as large, or larger than, Pluto, but most are quite small.

Can this belt of ancient, icy worlds generate short-period comets? Because Kuiper belt objects orbit in the same direction as the planets and in the plane of the solar system, it is possible for an object perturbed inward by the influence of the giant planets to move into an orbit resembling those of the short-period comets. Rare collisions and interactions among the KBOs could also add to a continuous supply of small, icy bodies from the Kuiper belt sent into the inner solar system.

Comets vary in brightness and orbit. Nevertheless, there are two basic types of comets in our solar system. Some originate in the Oort cloud far from the sun. Others come from the Kuiper belt just beyond Neptune. They all share one characteristic—they are ancient icy bodies that were born when the solar system was young.

*In 1943 and 1949, astronomer Kenneth Edgeworth published papers that included a paragraph speculating about objects beyond Pluto. Consequently, you may occasionally see the Kuiper belt referred to as the Edgeworth–Kuiper belt, but most astronomers refer to this part of the solar system as the Kuiper belt.

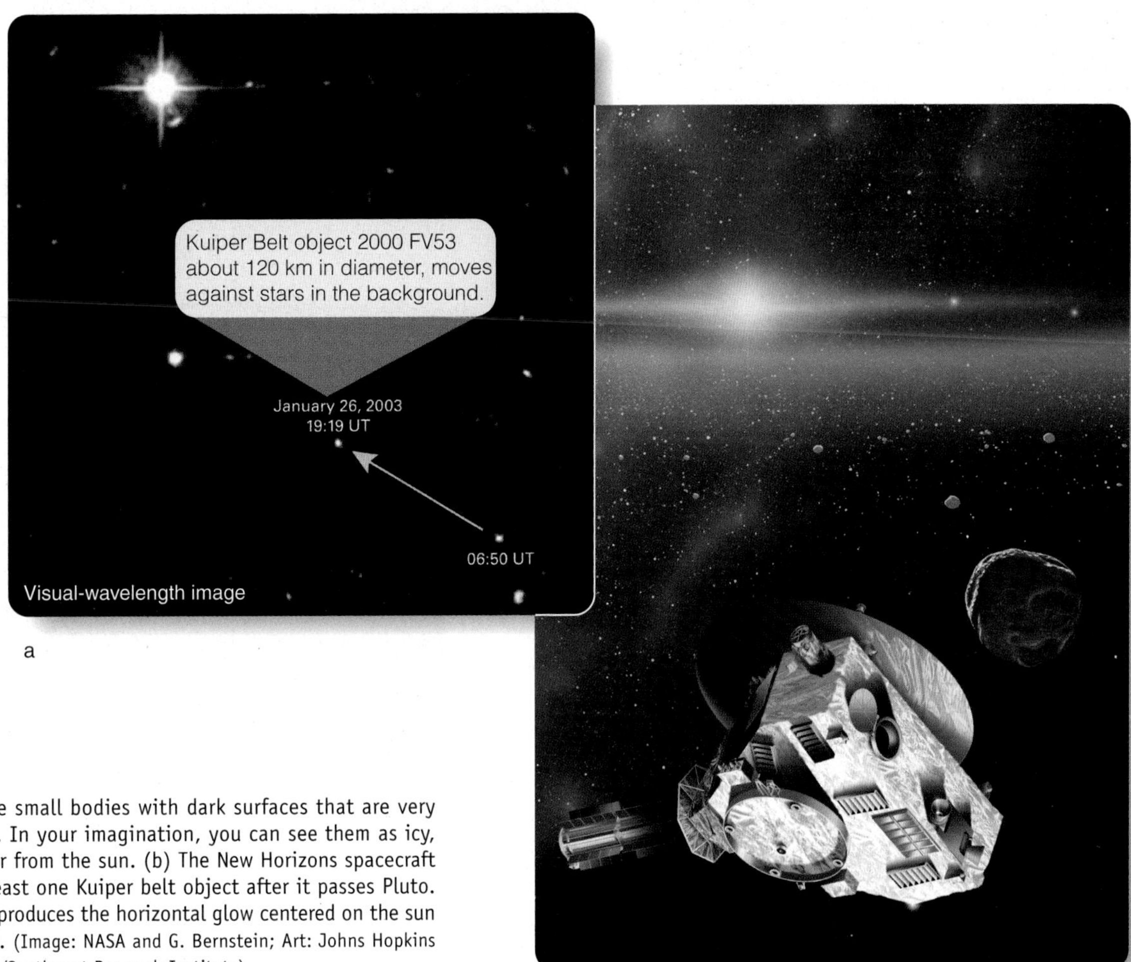
Kuiper Belt object 2000 FV53 about 120 km in diameter, moves against stars in the background.
January 26, 2003
19:19 UT
06:50 UT
Visual-wavelength image
a
b

**Figure 25-18**

(a) Kuiper belt objects are small bodies with dark surfaces that are very hard to detect from Earth. In your imagination, you can see them as icy, cratered worlds orbiting far from the sun. (b) The New Horizons spacecraft is planned to fly past at least one Kuiper belt object after it passes Pluto. Dust, mostly from comets, produces the horizontal glow centered on the sun in this artist's conception. (Image: NASA and G. Bernstein; Art: Johns Hopkins University Applied Physics Lab/Southwest Research Institute)

### SCIENTIFIC ARGUMENT

***How do comets help explain the formation of the planets?***

This argument must refer to the solar nebula hypothesis. The planetesimals that formed in the inner solar nebula were warm and could not incorporate much ice. The asteroids are understood to be the last remains of such rocky bodies. On the other hand, planetesimals in the outer solar system contained large amounts of ices. Many of them were destroyed by being accreted together to make the Jovian planets, but some survived intact. The icy bodies of the Oort cloud and the Kuiper belt may be the solar system's last surviving icy planetesimals. When those icy objects have their orbits perturbed by the gravity of the planets or passing stars, some are redirected into the inner solar system where you see them as comets. The gases released by comets indicate that they are rich in volatile materials such as water and carbon dioxide. These are the ices you would expect to find in the icy planetesimals. Comets also contain dust with rocklike chemical composition, and the planetesimals must have included large amounts of such dust frozen into the ices when they formed. Thus, the nuclei of comets seem to be frozen samples of the original outer solar nebula.

Nearly all of the mass of a comet is in the nucleus, but the light you see comes from the coma and the tail. Build a new argument to discuss observations. **What do spectra of comets tell you about the process that converts dirty ice into a comet?**

## 25-4 Asteroid and Comet Impacts

METEORITE IMPACTS ON HOMES of the type described at the start of this chapter are not common. Most meteors are small particles ranging from a few centimeters down to microscopic dust. Astronomers estimate that Earth gains about 40,000 tons of mass per year from meteorites of all sizes. (That may seem to you like a lot, but it is less than a hundred-thousandth of a trillionth of Earth's total mass.) Statistical calculations indicate that a meteorite large enough to cause some damage, like the one that hit Mrs. Hodges and her house in 1954, strikes a building somewhere in the world about once every 16 months.

Objects a few tens of meters or less in diameter are likely to fragment and explode in Earth's atmosphere without reaching the surface, but the shock waves from their explosions could still cause serious damage on Earth's surface. Declassified data from military satellites show that Earth is hit about once a week by meter-size asteroids. What happens when even larger solar system objects collide with Earth?

**■ Figure 25-19**

(a) The Barringer Meteorite Crater (near Flagstaff, Arizona) is nearly a mile in diameter and was formed about 50,000 years ago by the impact of an iron meteorite estimated to have been roughly 50 meters in diameter. Notice the raised and deformed rock layers all around the crater. The brick museum building visible on the far rim at right provides some idea of scale. (M. A. Seeds) (b) Like all larger-impact features, the Barringer Meteorite Crater has a raised rim and scattered ejecta. (USGS)

## Barringer Crater

Barringer Crater near Flagstaff, Arizona, is 1.2 km ($\frac{3}{4}$ of a mile) in diameter and 200 m (650 ft) deep. It seems quite large when you stand on the edge, and the hike around it, though beautiful, is long and dry (■ Figure 25-19). Barringer Crater was the subject of controversy among geologists for years as to whether it was caused by a volcanic event or a large meteorite impact. Finally, in 1963, Eugene Shoemaker proved in his doctoral thesis that the crater must be the result of an impact because quartz crystals in and around it had been subjected to pressures much higher than can be produced by a volcano.

Further studies showed Barringer Crater was created approximately 50,000 years ago by a meteorite estimated to have been about 50 m (160 ft) in diameter, as large as a good-sized building, that hit at a speed of 11 km/s, releasing as much energy as a large thermonuclear bomb. An object of that size could be called either a large meteorite or a small asteroid. Debris at the site shows that the impactor was composed of iron.

## The Tunguska Event

On a summer morning in 1908, reindeer herders and homesteaders in central Siberia were startled to see a brilliant blue-white fireball brighter than the sun streak across the sky. Still descending, it exploded with a blinding flash and an intense pulse of heat. One eyewitness account states:

> *The whole northern part of the sky appeared to be covered with fire. . . . I felt great heat as if my shirt had caught fire . . . there was a . . . mighty crash. . . . I was thrown on the ground about [7 meters] from the porch. . . . A hot wind, as from a cannon, blew past the huts from the north.*

The blast was heard up to 1000 km away, and the resulting pulse of air pressure circled Earth twice. For a number of nights following the blast, European astronomers, who knew nothing of the explosion, observed a glowing reddish haze high in the atmosphere.

When members of a scientific expedition arrived at the site in 1927, they found that the blast had occurred above the Stony Tunguska River valley and had flattened trees in an irregular pattern extending out 30 km (20 mi) (■ Figure 25-20). The trees were knocked down pointing away from the center of the blast, and limbs and leaves had been stripped away. The trunks of trees at the very center of the area were still standing, although they had lost all their limbs. No crater has been found, so it seems that the explosion, estimated to have equaled 12 megatons (12 million tons) of TNT, occurred at least a few kilometers above the ground.

In the early 1980s, a detailed analysis of all the Tunguska evidence suggested that the impactor's speed and direction resembled the orbits of Apollo objects. In 1993, astronomers produced computer models of objects entering Earth's atmosphere at various speeds and concluded that the fragile icy head of a comet would have exploded much too high in the

atmosphere. On the other hand, a dense iron-rich meteorite would be so strong that it would have survived to reach the ground and would have formed a large crater. Therefore, the most likely candidate for the Tunguska object seems to be a stony asteroid about 30 m in diameter, perhaps one-tenth the mass of the Barringer impactor. The models indicate that an object of this size with moderate material strength would have fragmented and exploded at just about the right height to produce the observed blast. This conclusion is consistent with modern studies of the Tunguska area showing that thousands of tons of powdered material with a composition resembling carbonaceous chondrites are scattered in the soil.

■ **Figure 25-20**

The 1908 Tunguska event in Siberia destroyed an area the size of a large city. Here the area of destruction is superimposed on a map of Washington, D.C., and its surrounding beltway. In the central area, trees were burned; in the outer area, trees were blown down in a pattern away from the path of the impactor.

## Big Impacts

There are some very big craters in the solar system, for example on the moon (page 446), that show what can happen when a full-sized asteroid or comet collides with a planet. Also, Earthlings watched in awe in 1994 as fragments from the nucleus of Comet Shoemaker–Levy 9 (abbreviated SL-9) slammed into Jupiter and produced impacts equaling millions of megatons (that is, trillions of tons) of TNT (■ Figure 25-21). Note that the Shoemaker in Shoemaker–Levy refers to Carolyn and Eugene Shoemaker who co-discovered the comet with David Levy. Eugene Shoemaker is the person whose analysis showed that Barringer Crater in Arizona is an impact crater.

As you know, Jupiter does not have a solid surface, so SL-9 did not leave any permanent craters, but astronomers have found chains of craters on other solar system objects that seem to have been formed by fragmented comets (■ Figure 25-22). Evidently events like the SL-9 collision with Jupiter have occurred many times in the history of the solar system.

What would happen if an object the size of SL-9, or even larger, were to hit Earth? Sixty-five million years ago, at the end of the Cretaceous period, over 75 percent of the species on Earth, including the dinosaurs, became extinct. Scientists have found a thin layer of clay all over the world that was laid down at that time, and it is rich in the element iridium—common in meteorites but rare in Earth's crust. This suggests that an impact occurred that was large enough to have altered Earth's climate and caused the worldwide extinction.

**■ Figure 25-21**

In 1992 Comet Shoemaker–Levy 9 passed within 1.3 planetary radii of Jupiter's center, well within its Roche limit, and tidal forces ripped the nucleus into more than 20 pieces. The fragmented pieces were as large as a few kilometers in diameter and spread into a long string of objects that looped away from Jupiter and then fell back to strike the planet, producing massive impacts over a period of six days in July 1994. Note that the impact flash in panel (c) is larger than Earth. (NASA)

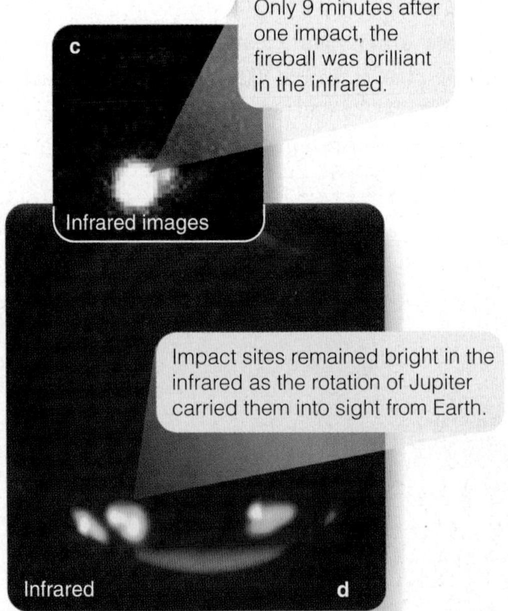


Mathematical models combined with observations create a plausible scenario of a major impact on Earth. Of course, creatures living near the site of the impact would die in the initial shock, but then things would get bad elsewhere. An impact at sea would create tsunamis (tidal waves) many hundreds of meters high that would sweep around the world, devastating regions far inland from coasts. On land or sea, a major impact would eject huge amounts of pulverized rock high above the atmosphere. As this material fell back, Earth's atmosphere would be turned into a glowing oven of red-hot meteorites streaming through the air, and the heat would trigger massive forest fires around the world. Soot from such fires has been found in the layers of clay laid down at the end of the Cretaceous period. Once the firestorms cooled, the remaining dust in the atmosphere would block sunlight and produce deep darkness for a year or more, killing off most plant life. At the same time, if the impact site were at or near limestone deposits, large amounts of carbon dioxide could be released into the atmosphere and produce intense acid rain.

Geologists have located a crater at least 180 km (110 mi) in diameter centered near the village of **Chicxulub** (pronounced *cheek-shoe-lube*) in the northern Yucatán region of Mexico (■ Figure 25-23). Although the crater is now completely covered by sediments, mineral samples show that it contains shocked quartz typical of impact sites and that it is the right age. The impact of an object 10 to 14 km in diameter formed the crater about 65 million years ago, just when the dinosaurs and many other species died out. Most scientists now conclude that this is the scar of the impact that ended the Cretaceous period.

There are a number of major extinctions in the fossil record, and at least some of these were probably caused by large impacts. Large asteroid impacts on Earth happen very rarely from a human perspective, but they happen often relative to geological

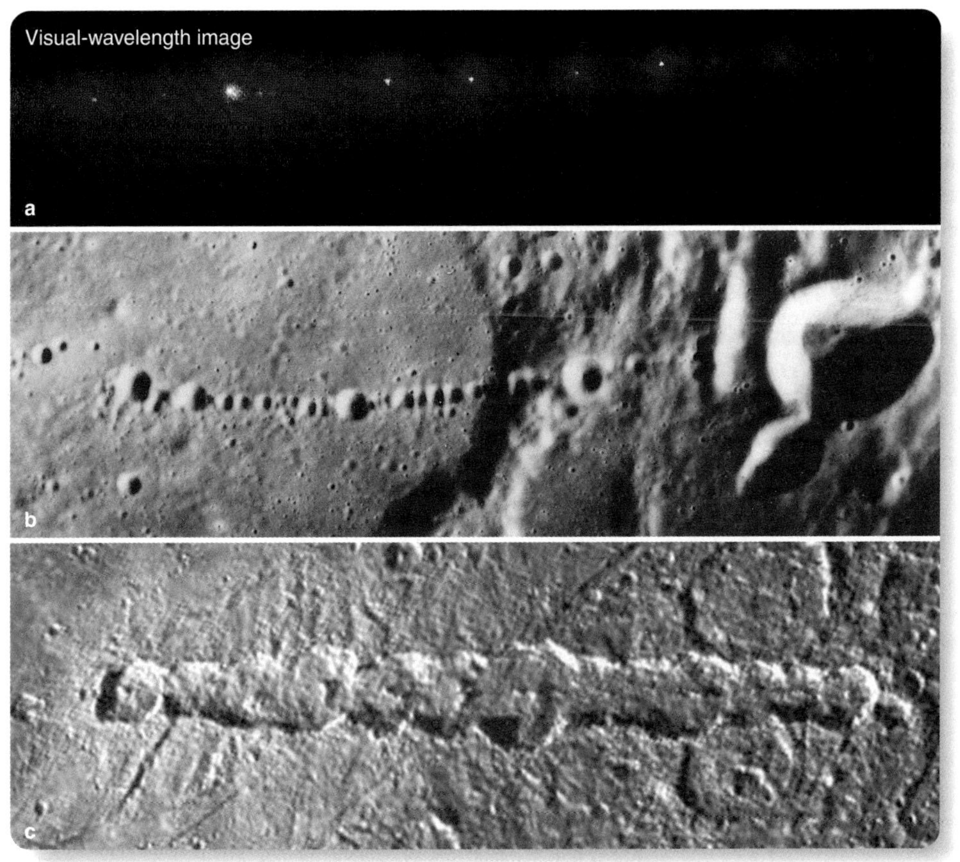


■ **Figure 25-22**

(a) Closeup image of the Comet Shoemaker–Levy 9 fragment train on the way to colliding with Jupiter. (b) A 40-km-long crater chain on Earth's moon, and (c) a 140-km-long crater chain on Jupiter's moon Callisto, probably formed by the impact of fragmented comet nuclei similar to Shoemaker–Levy 9. (NASA)

and astronomical time scales. For example, astronomers estimate an Apollo object hits Earth once every 250,000 years on average. A typical Apollo with a diameter of 1 km would strike with the power of a 100,000-megaton bomb and dig a crater more than 10 km in diameter. The good news is that we are certain that no known Apollo object will hit Earth in the foreseeable future; the bad news is that there are about 1000 of them 1 km in size or larger.

Asteroid 2004 MN4 was initially predicted to have a 2.6 percent chance of striking Earth in 2029. That object is about 400 m ($\frac{1}{4}$ mile) in diameter, large enough to do significant damage over a wide area but not large enough to alter Earth's global climate. Fortunately, further observations and calculations revealed that the object will not hit Earth. There will be no impact by 2004 MN4 in 2029, but there are plenty more asteroids in Earth-crossing orbits to be discovered. For example, a rock designated 2009 DD45, estimated to have been 30 m (100 ft) in diameter, about the size of the Tunguska impactor, passed only 64,000 km (40,000 mi) from Earth in March 2009. That is only twice the distance of humanity's geosynchronous communications and weather satellites, a

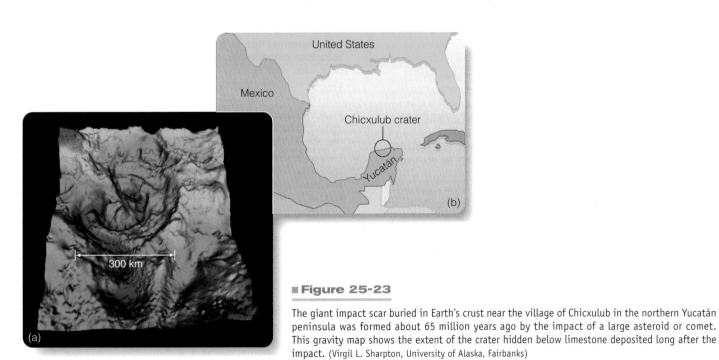

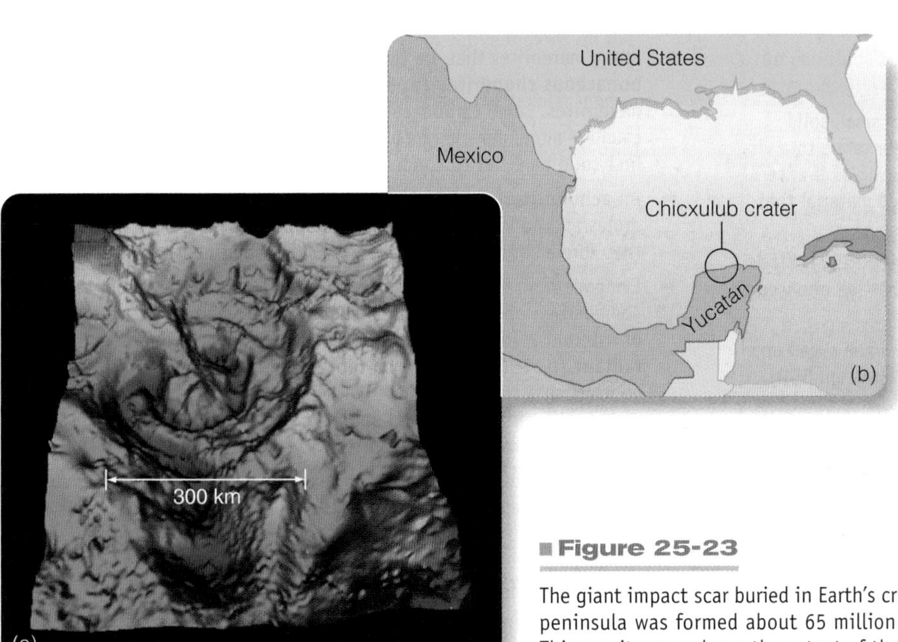


■ **Figure 25-23**

The giant impact scar buried in Earth's crust near the village of Chicxulub in the northern Yucatán peninsula was formed about 65 million years ago by the impact of a large asteroid or comet. This gravity map shows the extent of the crater hidden below limestone deposited long after the impact. (Virgil L. Sharpton, University of Alaska, Fairbanks)

telescopes is about 1 arc second and of Hubble about 0.1 arc second. Ceres's average distance from the sun is 2.8 AU.)

6. What is the orbital period of Ceres? (*Hints:* Use Kepler's third law and refer to the previous problem for Ceres's average distance from the sun.)
7. At what distances from the sun would you expect to find Kirkwood gaps where the orbital period of asteroids is one-half of, and one-third of, the orbital period of Jupiter? Compare your results with Figure 25-9. (*Hint:* Use Kepler's third law.)
8. If the velocity of the solar wind is about 400 km/s and the visible tail of a comet is $1 \times 10^8$ km long, how long does it take a solar wind atom to travel from the nucleus to the end of the visible tail?
9. If you saw Comet Halley when it was 0.7 AU from Earth and it had a visible tail 5° long, how long was the tail in kilometers? Suppose that the tail was not perpendicular to your line of sight. Is your first answer too large or too small? (*Hint:* Use the small-angle formula, Chapter 3.)
10. What is the orbital period of a comet nucleus in the Oort cloud? What is its orbital velocity? (*Hints:* Use Kepler's third law. The circumference of a circular orbit $= 2\pi r$. Refer to the text for typical Oort cloud object distances from the sun.)
11. The mass of an average comet's nucleus is about $10^{12}$ kg. If the Oort cloud contains $2 \times 10^{11}$ comet nuclei, what is the mass of the cloud in Earth masses? Compare that with Jupiter's mass. (*Hint:* See Appendix Table A-10.)

## Learning to Look

1. What do you see in the image to the right that tells you the size of planetesimals when the solar system was forming?

Russell Kempton, New England Meteoritical Services

2. Discuss the surface of the asteroid Mathilde, pictured to the right. What do you see that tells you something about the history of the asteroids?

NASA

3. What do you see in this image of the nucleus of Comet Borrelly that tells you how comets produce their comae and tails?

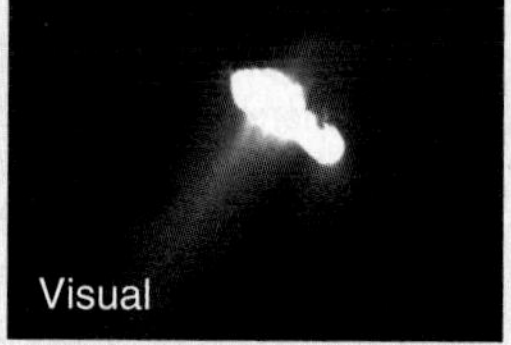


NASA

# 26 Astrobiology: Life on Other Worlds

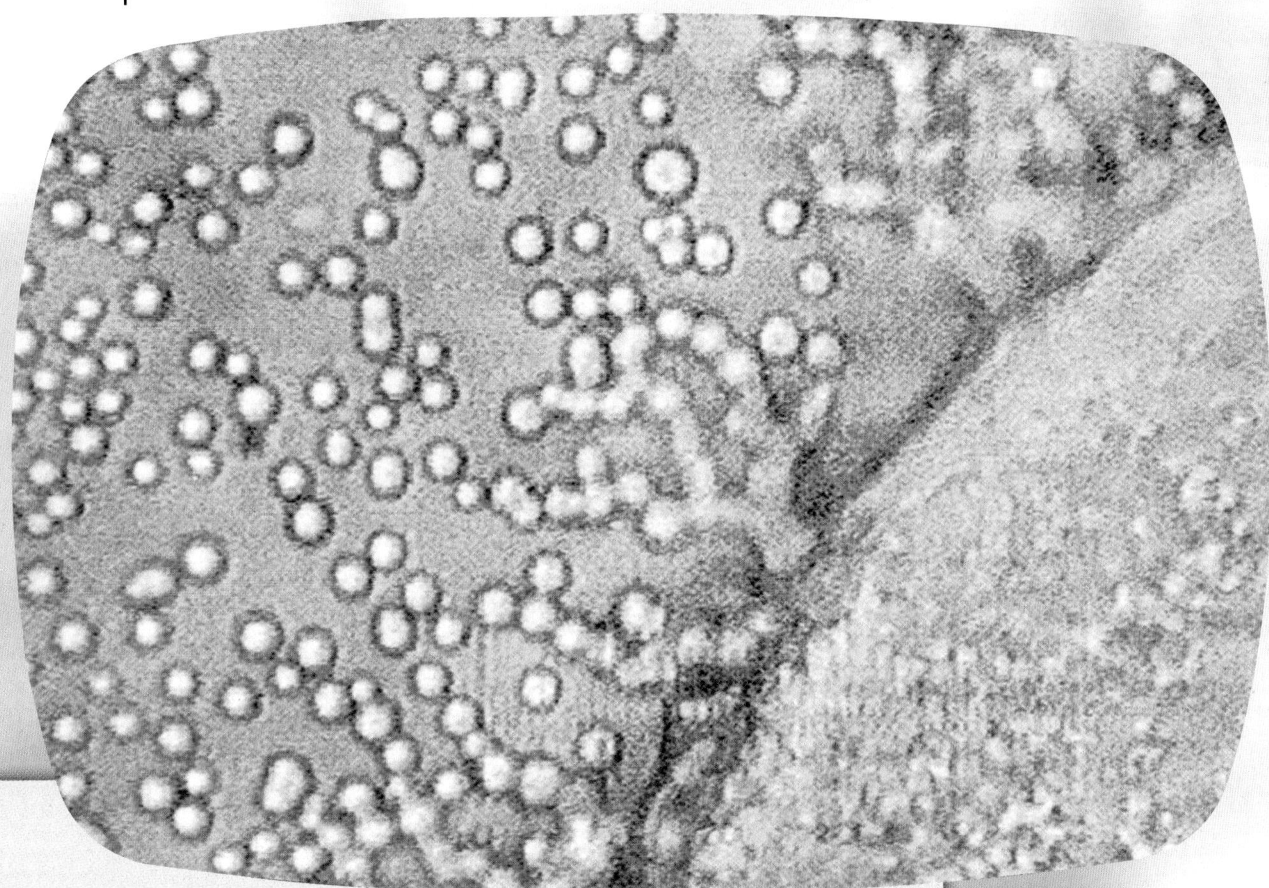

*Single amino acids can be assembled into long proteinlike molecules. When such material cools in water, it often forms microspheres, tiny globules with double-layered boundaries similar to cell membranes. Microspheres may have been an intermediate stage in the evolution of life between complex but nonliving molecules and living cells holding molecules reproducing genetic information.* (Sidney Fox and Randall Grubbs)

## Guidepost

This chapter is either unnecessary or vital. If you believe that astronomy is the study of the physical universe above the clouds, then you are done; the previous 25 chapters completed your study of astronomy. But, if you believe that astronomy is the study not only of the physical universe but also of your role as a living being in the evolution of the universe, then everything you have learned so far from this book has been preparation for this final chapter.

As you read this chapter, you will ask four important, related questions:

- **What is life?**
- **How did life originate on Earth?**
- **Could life begin on other worlds?**
- **Can humans on Earth communicate with civilizations on other worlds?**

You won't get more than the beginnings of answers to those questions here, but often in science asking a question is more important than getting an immediate answer.

You have explored the universe from the phases of the moon to the big bang, from the origin of Earth to the death of the sun. Astronomy is meaningful, not just because it is about the universe but because it is also about you. Now that you know astronomy, you see yourself and your world in a different way. Astronomy has changed you.

*Did I solicit thee from darkness to promote me?*

— ADAM, TO GOD, IN JOHN MILTON'S *PARADISE LOST*

As a living thing, you have been promoted from darkness. The atoms of carbon, oxygen, and other heavy elements that are necessary components of your body did not exist at the beginning of the universe but were built up by successive generations of stars.

The elements from which you are made are common everywhere in the observable universe, so it is possible that life began on other worlds and evolved to intelligence. If so, perhaps those other civilizations will be detected from Earth. Future astronomers may discover distant alien species completely different from any life on Earth.

Your goal in this chapter is to try to understand truly intriguing puzzles—the origin and evolution of life on Earth and what that tells you about whether there is life on other worlds (**How Do We Know? 26-1**). This new hybrid field of study is called **astrobiology.**

## 26-1 The Nature of Life

What is life? Philosophers have struggled with that question for thousands of years, and it is not possible to answer it completely in one chapter or even one book. An attempt at a general definition of what living things do, distinguishing them from nonliving things, might be: Life is a *process* by which an organism extracts energy from the surroundings, maintains itself, and modifies the surroundings to foster its own survival and reproduction.

One very important observation is that all living things on Earth, no matter how apparently different, share certain characteristics in how they perform the process of life.

### The Physical Basis of Life

The physical basis of life on Earth is the element carbon (■ Figure 26-1). Because of the way carbon atoms bond to each other and to other atoms, they can join into long, complex, stable chains that are capable, among many other feats, of storing and transmitting information. A large amount of information is necessary to control the activities and maintain the forms of living things.

It is possible that life on other worlds could use silicon instead of carbon. Silicon is right below carbon in the periodic table (Appendix Table A-16), which means that it shares many of carbon's chemical properties. But life based on silicon rather than carbon seems unlikely because silicon chains are harder to assemble and disassemble than their carbon counterparts and can't be as lengthy. Science fiction has proposed even stranger life forms based on, for example, electromagnetic fields and ionized gas, and none of these possibilities can be ruled out. Those hypothetical life forms make for fascinating speculation, but for now they can't be studied systematically in the way that life on Earth can.

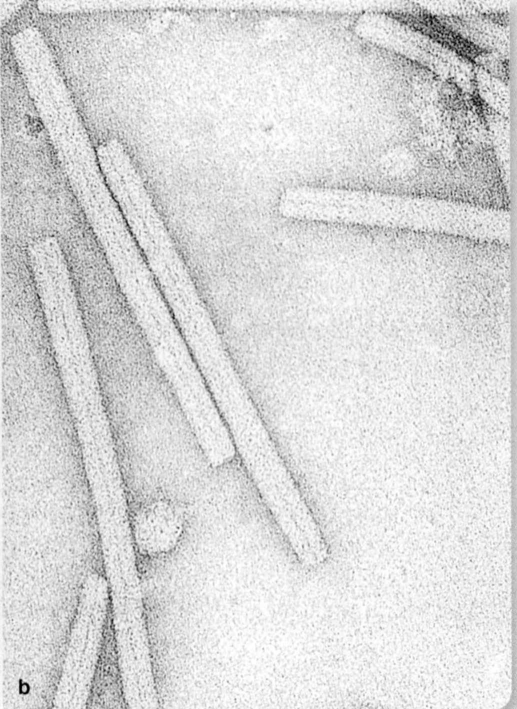

■ **Figure 26-1**

All living things on Earth are based on carbon chemistry. Even the long molecules that carry genetic information, DNA and RNA, have a framework defined by chains of carbon atoms. (a) Ana, a complex mammal, contains about 60 AU of DNA. (Jamie Backman) (b) Each rod-like tobacco mosaic virus contains a single spiral strand of RNA about 0.01 mm long as its genetic material. (L. D. Simon)

# How Do We Know? | 26-1

## The Nature of Scientific Explanation

**Must science and religion be in conflict?** Science is a way of understanding the world around you, and at the heart of that understanding are explanations that science gives for natural phenomena. Whether you call these explanations stories, histories, hypotheses, or theories, they are attempts to describe how nature works based on evidence and intellectual honesty. While you may take these explanations as factual truth, you can understand that they are not the only explanations that describe the universe.

A separate class of explanations involves religion. For example, the Old Testament description of creation does not fit well with scientific observations, but it is a way of understanding the universe nonetheless. Religious explanations are based partly on faith rather than on strict rules of logic and evidence, and it is wrong to demand that they follow the same rules as scientific explanations. In the same way, it is wrong to demand that scientific explanations take into account religious beliefs. The so-called conflict between science and religion arises when people fail to recognize that science and religion are different ways of knowing about the universe.

Scientific explanations are very compelling because science has been very successful at producing technological innovations that have changed the world you live in. From new vaccines, to digital music players, to telescopes that can observe the most distant galaxies, the products of the scientific process are all around you. Scientific explanations have provided tremendous insights into the workings of nature. Many people are attracted to the suggestion, made by evolutionary biologist Stephen Jay Gould and others, that religious explanations and scientific explanations should be considered as "separate magisteria." In other words, religion and science are devoted to different realms of the mystery of existence.

Science and religion offer differing ways of explaining the universe, but the two ways follow separate rules and cannot be judged by each other's standards. The trial of Galileo can be understood as a conflict between these two ways of knowing.

*Galileo's telescope gave him a new way to know about the universe.*

This chapter is concerned with the origin and evolution of life as it is on Earth, based on carbon, not because of lack of imagination but because it is the only form of life about which we know anything.

Even carbon-based life has its mysteries. What makes a lump of carbon-based molecules a living thing? An important part of the answer lies in the transmission of information from one molecule to another.

## Information Storage and Duplication

Most actions performed by living cells are carried out by molecules that are built within the cells. Cells must store recipes for making all those molecules, as well as how and when to use them, and then somehow pass the recipes on to their offspring.

Study **DNA: The Code of Life** on pages 584–585 and notice four important points and seven new terms:

1. The chemical recipes of life are stored in each cell as information on *DNA (deoxyribonucleic acid)* molecules that resemble a ladder with rungs that are composed of chemical bases. The recipe information is expressed by the sequence of the rungs, providing instructions to guide specific chemical reactions within the cell.

2. The instructions stored in DNA are genetic information passed along to offspring. DNA instructions normally are expressed by being copied into a messenger molecule called *RNA (ribonucleic acid).* The RNA molecule travels to a location in the cell where its message causes a sequence of molecular units called *amino acids* to be connected into large molecules called *proteins.* Proteins serve as the cell's basic structural molecule or as *enzymes* that control chemical reactions.

3. The DNA molecule reproduces itself when a cell divides so that each new cell contains a copy of the original information. A sequence of DNA that composes one instruction is called a *gene.* Genes are organized into long coiled chains called *chromosomes.* The genes linked on one chromosome are normally passed on to offspring together.

To produce viable offspring, a cell must be able to make copies of its DNA. Surprisingly, it is important for the continued existence of all life that the copying process includes mistakes.

## DNA: The Code of Life

**1** The key to understanding life is information — the information that guides all of the processes in an organism. In most living things on Earth, that information is stored on a long spiral molecule called **DNA (deoxyribonucleic acid).**

**1a** The DNA molecule looks like a spiral ladder with rails made of phosphates and sugars. The rungs of the ladder are made of four chemical bases arranged in pairs. The bases always pair the same way. That is, base A always pairs with base T, and base G always pairs with base C.

**1b** Information is coded on the DNA molecule by the order in which the base pairs occur. To read that code, molecular biologists have to "sequence the DNA." That is, they must determine the order in which the base pairs occur along the DNA ladder.

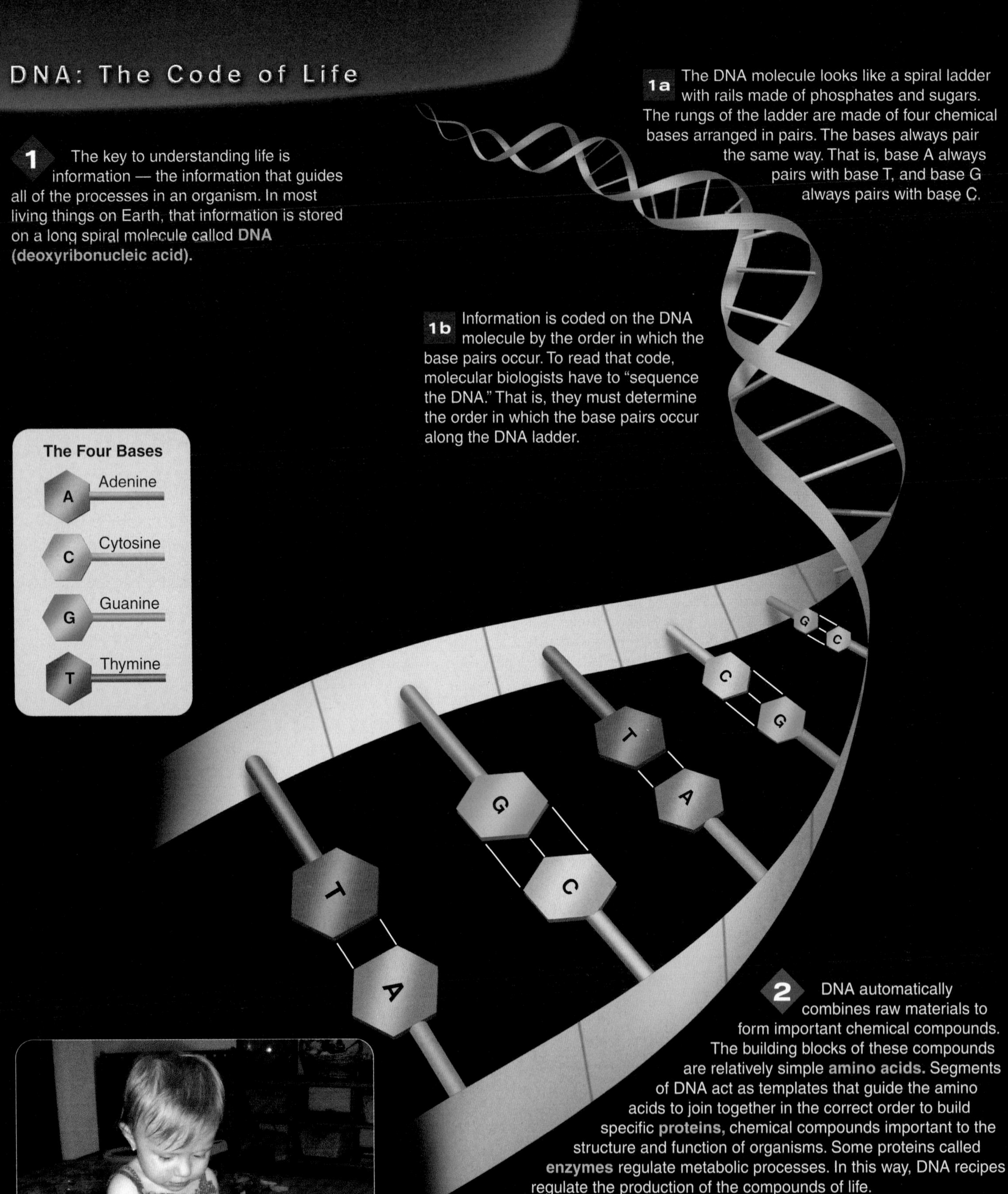


**2** DNA automatically combines raw materials to form important chemical compounds. The building blocks of these compounds are relatively simple **amino acids.** Segments of DNA act as templates that guide the amino acids to join together in the correct order to build specific **proteins,** chemical compounds important to the structure and function of organisms. Some proteins called **enzymes** regulate metabolic processes. In this way, DNA recipes regulate the production of the compounds of life.

Jamie Backman

The traits you inherit from your parents, the chemical processes that animate you, and the structure of your body are all encoded in your DNA. When people say "you have your mother's eyes," they are talking about DNA codes.

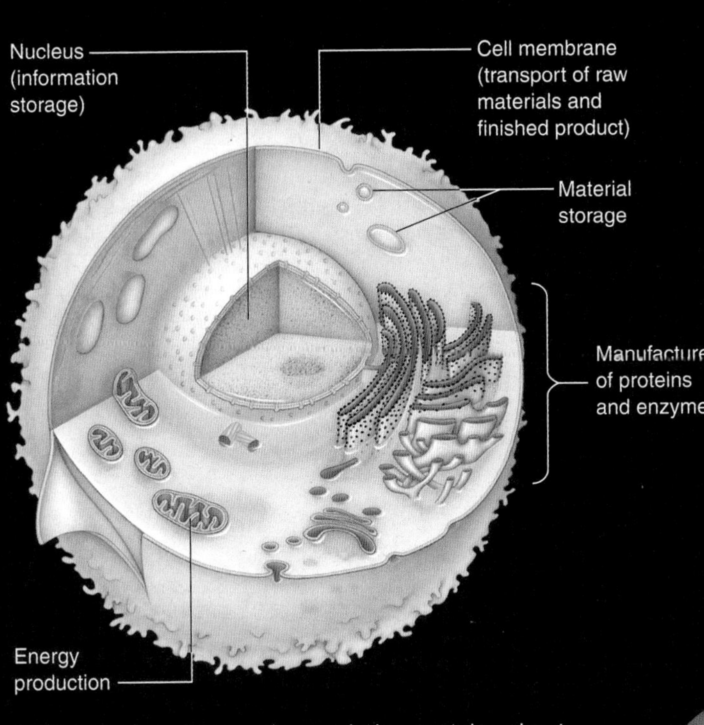


**2a** A cell is a tiny factory that uses the DNA code to manufacture chemicals. Most of the DNA remains safe in the nucleus of a cell, and the code is copied to create a molecule of **RNA (ribonucleic acid).** Like a messenger carrying blueprints, the RNA carries the code out of the nucleus to the work site where the proteins and enzymes are made.

**2b** A single cell from a human being contains about 1.5 meters of DNA containing about 4.5 billion base pairs — enough to record the entire works of Shakespeare 200 times. A typical adult human contains a total of about 600 AU of DNA. Yet the DNA in each cell, only 1.5 meters in length, contains all of the information to create a new human. A clone is a new creature created from the DNA code found in a single cell.

Original DNA

Copy DNA

Copy DNA

**3** DNA, coiled into a tight spiral, makes up the **chromosomes** that are the genetic material in a cell. A **gene** is a segment of a chromosome that controls a certain function. When a cell divides, each of the new cells receives a copy of the chromosomes, as genetic information is handed down to new generations.

**3a** To divide, a cell must duplicate its DNA. The DNA ladder splits, and new bases match to the exposed bases of the ladder to build two copies of the original DNA code. Because the base pairs almost always match correctly, errors in copying are rare. One set of the DNA code goes to each of the two new cells.

**Cell Reproduction by Division**

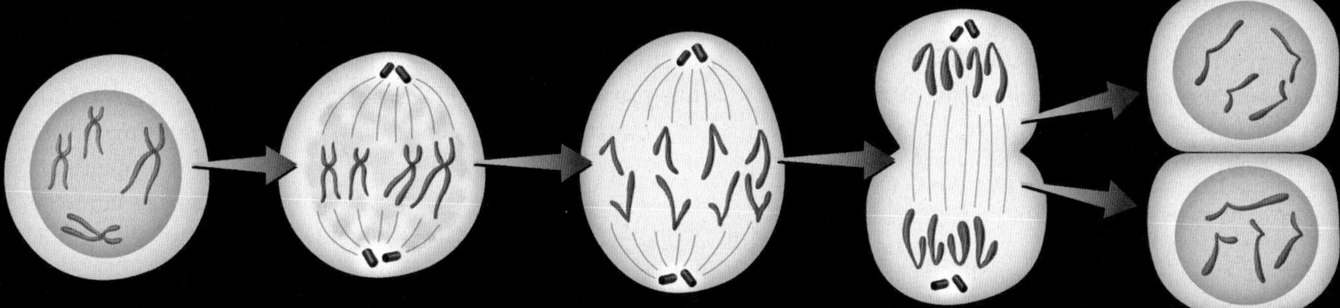

As a cell begins to divide, its DNA duplicates itself.

The duplicated chromosomes move to the middle.

The two sets of chromosomes separate, and . . .

the cell divides to produce . . .

two cells, each containing a full set of the DNA code.

## Modifying the Information

Earth's environment changes continuously. To survive, species must change as their food supply, climate, or home terrain changes. If the information stored in DNA could not change, then life would go extinct quickly. The process by which life adjusts itself to changing environments is called **biological evolution.**

When an organism reproduces, its offspring receive a copy of its DNA. Sometimes external effects such as natural radiation alter the DNA during the parent organism's lifetime, and sometimes mistakes are made in the copying process, so that occasionally the copy is slightly different from the original. Offspring born with random alterations to their DNA are called **mutants.** Most mutations make no difference, but some mutations are fatal, killing the afflicted organisms before they can reproduce. In rare but vitally important cases, a mutation can actually help an organism survive.

These changes produce variation among the members of a species. All of the squirrels in the park may look the same, but they carry a range of genetic variation. Some may have slightly longer tails or faster-growing claws. These variations make almost no difference until the environment changes. For example, if the environment becomes colder, a squirrel with a heavier coat of fur will, on average, survive longer and produce more offspring than its normal contemporaries. Likewise, the offspring that inherit this beneficial variation will also live longer and have more offspring of their own. In contrast, squirrels containing DNA recipes for thin fur coats will gradually decrease in number. These differing rates of survival and reproduction are examples of **natural selection.** Over time, the beneficial variation increases in frequency, and a species can evolve until the entire population shares the trait. In this way, natural selection adapts species to their changing environments by selecting, from the huge array of random variations, those that would most benefit the survival of the species.

It is a **Common Misconception** that evolution is random, but that is not true. The underlying mechanisms creating variation within each species may be random, but natural selection is not random because progressive changes in a species are directed by changes in the environment.

### SCIENTIFIC ARGUMENT

***Why is it important that errors occur in copying DNA?***

Sometimes the most valuable scientific arguments are those that challenge what seems like common sense. It appears obvious that mistakes shouldn't be made in copying DNA, but in fact variation is necessary for long-term survival of a species. For example, the DNA in a starfish contains all the information the starfish needs to grow, develop, survive, and reproduce. The information must be passed on to the starfish's offspring for them to survive. That information must change, however, if the environment changes. A change in the ocean's temperature may kill the specific shellfish that the starfish eat. If none of the starfish are able to digest another kind of food—if all the starfish have exactly the same DNA—they all will die. But if a few starfish are born with the ability to make enzymes that can digest a different kind of shellfish, the species may be able to carry on.

Variations in DNA are caused both by external factors such as natural radiation and by occasional mistakes in the copying process. The survival of life depends on this delicate balance between mostly reliable reproduction and the introduction of small variations in DNA. Now build a new argument. **Why does the DNA copying process need to be mostly reliable?**

## 26-2 Life in the Universe

LIFE AS WE KNOW IT consists of just the single example of life on Earth. It is OK to think of all life on Earth as being just a single type of life, because, as you learned in the previous section, all living things on Earth have the same physical basis: the same chemistry and the same genetic code alphabet. How life began on Earth and then developed and evolved into its present variety is the only solid information you have to work with, when considering what might be possible on other worlds.

Everything currently known about life on Earth indicates that the same natural processes should lead to the origin of life on some fraction of other planets with liquid water. If there is life on other worlds, does it use DNA and RNA to carry the information for life processes, or different molecules playing the same role, or some radically different scheme? There is no way to know unless another example of life is found on another world. If and when that day comes, even if the non-Earthly life is a simple one-celled organism, the discovery will be one of the most important in the history of science. It will complete the journey of human understanding, begun in the Copernican revolution, of progressive realizations that Earth is not unique.

### Origin of Life on Earth

It is obvious that the 4.5 billion chemical bases that make up human DNA did not come together in the right order just by chance. The key to understanding the origin of life lies in picturing the processes of evolution running "backward." The complex interplay of environmental factors with the DNA of generation after generation of organisms drove some life forms to become more sophisticated over time, until they became the unique and specialized creatures alive today. Imagining this process in reverse leads to the idea that life on Earth began with very simple forms.

Biologists hypothesize that the first living things would have been carbon-chain molecules able to copy themselves. Of course, this is a scientific hypothesis for which you can seek evidence. What evidence exists regarding the origin of life on Earth?

The oldest fossils are the remains of sea creatures, and this indicates that life began in the sea. Identifying the oldest fossils is not easy, however. Ancient rocks from western Australia that are at least 3.4 billion years old contain features that biologists identify as **stromatolites,** fossilized remains of colonies of single-celled organisms (■ Figure 26-2). Fossils this old are difficult to recognize because the earliest living things did not contain easily preserved hard parts like bones or shells and because the individual organisms were microscopic. Thus the evidence, though scarce, indicates that simple organisms lived in Earth's oceans less than 1.2 billion years after Earth formed. Stromatolite colonies of microorganisms are more complex than individual cells, so you can imagine there probably were earlier, simpler organisms. Where did those first simplest organisms come from?

An important experiment performed by Stanley Miller and Harold Urey in 1952 sought to re-create the conditions in which life on Earth began. The **Miller experiment** consisted of a sterile, sealed glass container holding water, hydrogen, ammonia, and methane. An electric arc inside the apparatus made sparks to simulate the effects of lightning in Earth's early atmosphere (■ Figure 26-3).

Miller and Urey let the experiment run for a week and then analyzed the material inside. They found that the interaction between the electric arc and the simulated atmosphere had produced many organic molecules from the raw material of the experiment, including such important building blocks of life as amino acids. (Recall that an organic molecule is simply a molecule with a carbon-chain structure and need not be derived from a living thing: "Organic" does not necessarily imply "biological.")

When the experiment was run again using different energy sources such as hot silica to represent molten lava spilling into the ocean, similar molecules were produced. Even a source of ultraviolet radiation representing the small amount of UV in sunlight was sufficient to produce complex organic molecules.

**■ Figure 26-2**

(a) A fossil stromatolite from western Australia that is more than 3 billion years old, constituting some of the oldest evidence of life on Earth. Stromatolites are formed, layer upon layer, by mats of bacteria living in shallow water where they are covered repeatedly by sediments. (Chip Clark, National Museum of Natural History) (b) Artist's conception of a scene on the young Earth, 3 billion years ago, with stromatolite bacterial mats growing near the shores of an ocean. (Mural by Peter Sawyer)

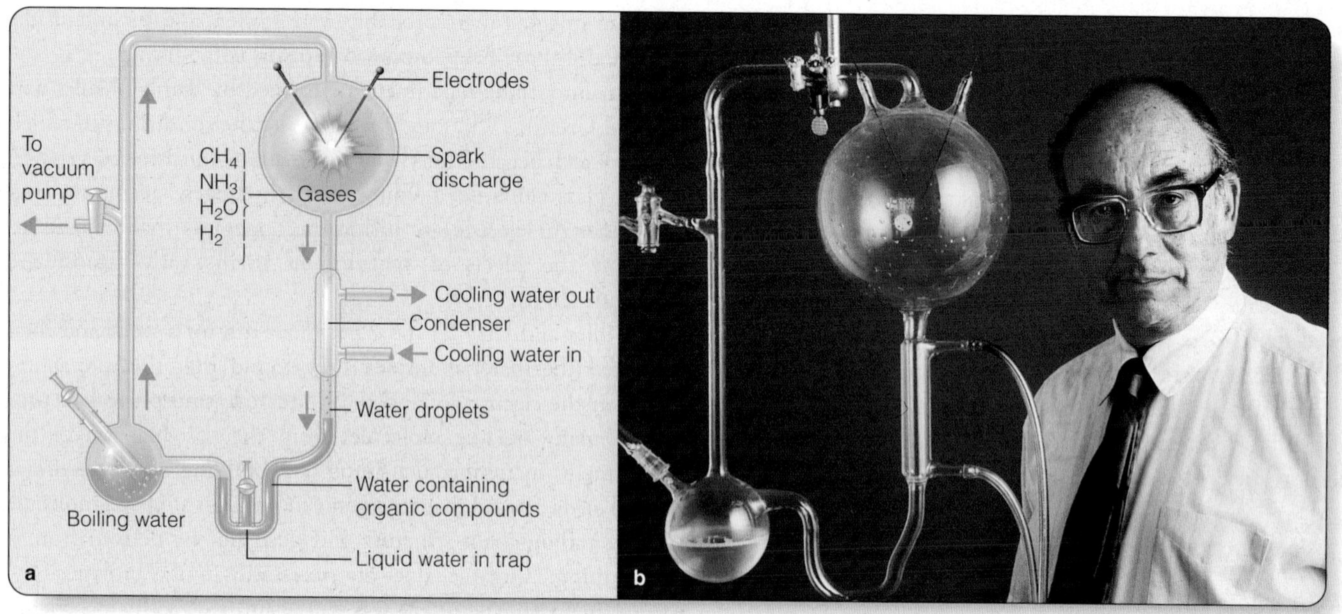


**Figure 26-3**

(a) The Miller experiment circulated gases through water in the presence of an electric arc. This simulation of primitive conditions on Earth produced many complex organic molecules, including amino acids, the building blocks of proteins. (b) Stanley Miller with a Miller apparatus. (Stanley Miller)

Scientists are professionally skeptical about scientific findings (see "How Do We Know?" 19-2), and they have reevaluated the Miller–Urey experiment in light of new information. According to updated models of the formation of the solar system and Earth (see Chapters 19 and 20), Earth's early atmosphere probably consisted mostly of carbon dioxide, nitrogen, and water vapor instead of the mix of hydrogen, ammonia, methane, and water vapor assumed by Miller and Urey. When gases corresponding to the newer understanding of the early Earth atmosphere are processed in a Miller apparatus, lesser, but still significant, amounts of organic molecules are produced.

The Miller experiment is important because it shows that complex organic molecules form naturally in a wide variety of circumstances. Lightning, sunlight, and hot lava are just some of the energy sources that can naturally rearrange common simple molecules into the complex molecules that make life possible. If you could travel back in time, you would expect to find Earth's early oceans filled with a rich mixture of organic compounds called the **primordial soup.**

Many of these organic compounds would have been able to link up to form larger molecules. Amino acids, for example, can link together to form proteins by joining ends and releasing a water molecule (■ Figure 26-4). That reaction, however, does not

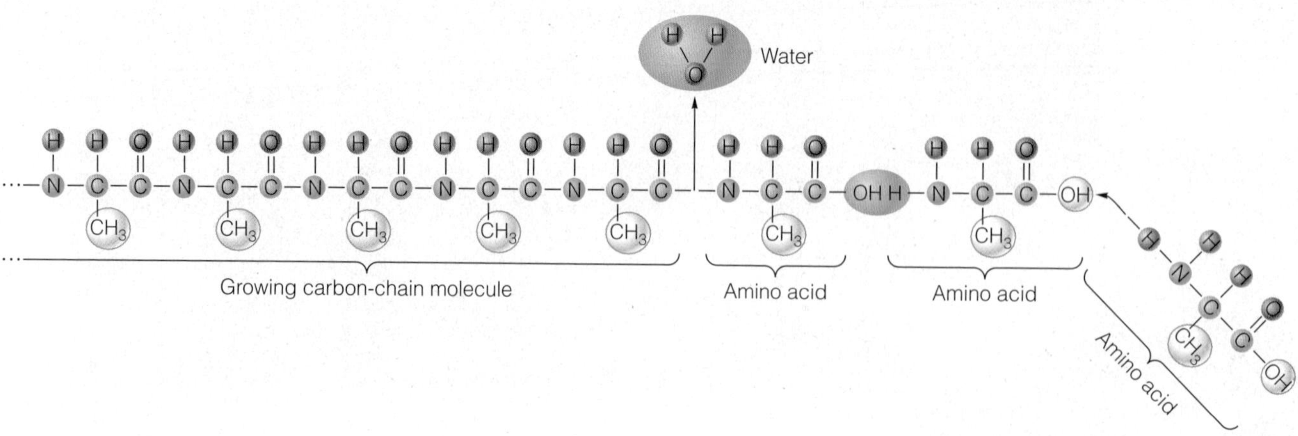


**Figure 26-4**

Amino acids can link together via the release of a water molecule to form long carbon-chain molecules. The amino acid in this hypothetical example is alanine, one of the simplest.

proceed easily in a water solution. Scientists hypothesize that this step may have been more likely to happen on shorelines or in sun-warmed tidal pools where organic molecules from the primordial soup could have been concentrated by water evaporation. The production of large organic molecules may have been aided in such semidry environments by clay crystals acting as templates to hold the organic subunits close together.

These complex organic molecules were still not living things. Even though some proteins may have contained hundreds of amino acids, they did not reproduce but rather linked and broke apart at random. Because some molecules are more stable than others, and some bond together more easily than others, scientists hypothesize that a process of **chemical evolution** eventually concentrated the various smaller molecules into the most stable larger forms. Eventually, according to the hypothesis, somewhere in the oceans, after sufficient time, a molecule formed that could copy itself. At that point, the natural selection and chemical evolution of molecules became the biological evolution of living things.

An alternate theory for the origin of life holds that reproducing molecules may have arrived here from space. Astronomers have found a wide variety of organic molecules in the interstellar medium, and similar compounds have been found inside meteorites (■ Figure 26-5). The Miller experiment showed how easy it is for complex organic molecules to form naturally from simpler compounds, so it is not surprising to find them in space. Although speculation is fun, the hypothesis that life arrived on Earth from space is presently more difficult to test than the hypothesis that Earth's life originated on Earth.

Whether the first reproducing molecules formed here on Earth or in space, the important thing is that they could have formed by natural processes. Scientists know enough about those processes to feel confident about them, even though some of the steps remain unknown.

The details of the origin of the first cells are unknown. The structure of cells may have arisen automatically because of the way molecules interact during chemical evolution. If a dry mixture of amino acids is heated, the acids form long, proteinlike molecules that, when poured into water, collect to form microscopic spheres that behave in ways similar to cells (pictured in the image on the opening page to this chapter). They have a thin membrane surface, they absorb material from their surroundings, they grow in size, and they divide and bud just as cells do. However, they contain no large molecule that copies itself, so they are not alive. The first reproducing molecule to be surrounded by a protective membrane, resulting in the first cell, would have gained an important survival advantage over other reproducing molecules.

**■ Figure 26-5**

A piece of the Murchison meteorite, a carbonaceous chondrite (see Chapter 25) that fell near Murchison, Australia, in 1969. Analysis of the interior of the meteorite revealed the presence of amino acids. Whether the first chemical building blocks of life on Earth originated in space is a matter of debate, but the amino acids found in meteorites illustrate how commonly amino acids and other complex organic molecules occur in the universe, even in the absence of living things. (Chip Clark, National Museum of Natural History)

## Geologic Time and the Evolution of Life

Biologists infer that the first cells must have been simple single-celled organisms similar to modern bacteria. As you learned earlier, these kinds of cells are preserved in stromatolites (Figure 26-2), mineral formations produced by layers of bacteria and shallow ocean sediments. Stromatolite fossils are found in rocks with radioactive ages of 3.4 billion years, and living stromatolites still form in some places today.

Stromatolites and other photosynthetic organisms would have begun adding oxygen, a product of photosynthesis, to Earth's early atmosphere. An oxygen abundance of only 0.1 percent would have created an ozone screen, protecting organisms from the sun's ultraviolet radiation and later allowing life to colonize the land.

Over the course of eons, the natural processes of evolution gave rise to stunningly complex **multicellular** life forms with their own widely differing ways of life. It is a **Common Misconception** to imagine that life is too complex to have evolved from such simple beginnings. It is possible because small variations can accumulate, although that accumulation requires great amounts of time.

There is little evidence of anything more than simple organisms on Earth until about 540 million years ago, almost 3 billion years after the earliest signs of life, at which time fossils indicate life suddenly developing into a wide variety of complex forms such as the trilobites (■ Figure 26-6). This sudden increase in complexity is known as the **Cambrian explosion** and marks the beginning of the Cambrian period.

■ **Figure 26-6**

(a) Trilobites made their first appearance in the Cambrian oceans. The smallest were almost microscopic, and the largest were bigger than dinner plates. This example, about the size of your hand, lived 400 million years ago in an ocean floor that is now a limestone deposit in Pennsylvania. (Franklin & Marshall College/Grundy Observatory/Michael Seeds) (b) In this artist's conception of a Cambrian sea bottom, Anomalocaris (near at right center and looming at upper right), about human-hand-sized, had specialized organs including eyes, coordinated fins, gripping mandibles, and a powerful, toothed maw. Notice Opabinia at center right with its long snout. (Smithsonian and D. W. Miller)

If you represented the entire history of Earth on a scale diagram, the Cambrian explosion would be near the top of the column, as shown at the left of ■ Figure 26-7. The emergence of most animals familiar to you today, including fishes, amphibians, reptiles, birds, and mammals, would be crammed into the topmost part of the chart, above the Cambrian explosion.

If you magnify that portion of the diagram, as shown on the right side of Figure 26-7, you can get a better idea of when these events occurred in the history of life. Humanoid creatures have walked on Earth for about 4 million years. This is a long time by the standard of a human lifetime, but it makes only a narrow red line at the top of the diagram. All of recorded history would be a microscopically thin line at the very top of the column.

To understand just how thin that line is, imagine that the entire 4.6-billion-year history of the Earth has been compressed onto a yearlong video and that you began watching this video on January 1. You would not see any signs of life until March or early April, and the slow evolution of the first simple forms would take the next six or seven months. Suddenly, in mid-November, you would see the trilobites and other complex organisms of the Cambrian explosion.

You would see no life of any kind on land until November 28, but once life appeared it would diversify quickly, and by December 12 you would see dinosaurs walking the continents. By the day after Christmas they would be gone, and mammals and birds would be on the rise.

If you watched closely, you might see the first humanoid forms by late afternoon on New Year's Eve, and by late evening you could see humans making the first stone tools. The Stone Age would last until 11:59 PM, after which the first towns, and then cities, would appear. Suddenly things would begin to happen at lightning speed. Babylon would flourish, the pyramids would rise, and Troy would fall. The Christian era would begin 14 seconds before the New Year. Rome would fall, then the Middle Ages and the Renaissance would flicker past. The American and French revolutions would occur one-and-a-half seconds before the end of the video.

By imagining the history of Earth as a yearlong video, you have gained some perspective on the rise of life. Tremendous amounts of time were needed for the first simple living things to evolve in the oceans. As life became more complex, new forms arose more and more quickly as the hardest problems—how to reproduce, how to take energy efficiently from the environment, how to move around—were "solved" by the process of biological evolution. The easier problems, like what to eat, where to live, and how to raise young, were managed in different ways by different organisms, leading to the diversity that is seen today. Intelligence—that which appears to set humans apart from other animals—may be a unique solution to an evolutionary problem posed to humanity's ancient ancestors. A smart animal is better able to escape predators, outwit its prey, and feed and shelter itself and its offspring, so under certain conditions evolution is likely to naturally select for intelligence.

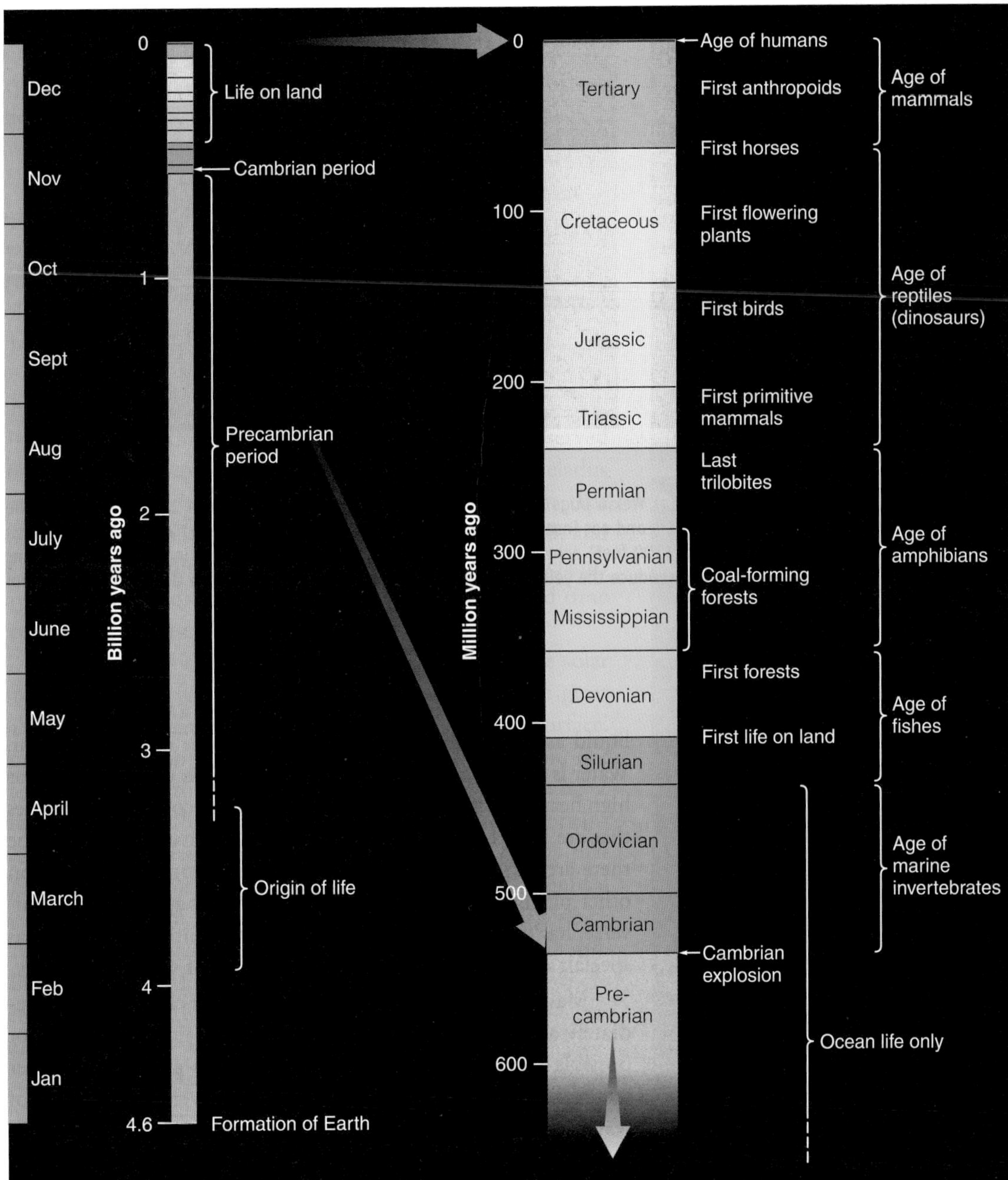


**Figure 26-7**

Complex life has developed on Earth only recently. If the entire history of Earth were represented in a time line (left), you have to examine the end of the line closely to see details such as life leaving the oceans and dinosaurs appearing. The age of humans would still be only a thin line at the top of your diagram. If the history of Earth were a year-long videotape, humans would not appear until the last hours of December 31.

## Extremophiles

Scientists on Earth are finding life in places previously judged inhospitable, such as the bottoms of ice-covered lakes in Antarctica, far underground inside solid rock, and among the cinders at the summits of extinct volcanoes (■ Figure 26-8a). An organism that can survive and even thrive in an extreme environment is called an **extremophile.** Maybe you have friends like that.

Linguists can figure out the vocabulary of the long-vanished Indo-European language by comparing words in modern languages such as English, Spanish, Russian, Greek, and Hindi that evolved from it. In much the same way, biologists can work out the DNA sequences of ancient species by comparing the sequences of their present-day descendants. This type of analysis indicates that the creature that was ancestor to all life on Earth today resembled present-day single-celled organisms known as **archaea,** especially the extremophile types of archaea that are tolerant of high temperatures, called **thermophiles** (Figure 26-8b). Some biologists think this is evidence that life began near volcanic vents on the sea

fossils of ancient Martian microorganisms could possibly be non-biological mineral formations instead.

In 2009 astronomers observing the Martian atmosphere announced detection of faint traces of methane, a substance made abundantly by living things on Earth. Methane would be destroyed in the Martian environment by solar UV radiation and chemical reactions, so the methane that is present must have been produced recently. There are also geological processes that can emit methane, but those processes are connected with volcanism. The map of Martian atmospheric methane concentrations in Figure 26-9c shows methane primarily in locations that are not volcanic provinces. Perhaps organisms living in the Martian soil are making methane right now. This evidence regarding potential life on Mars remains highly controversial. Conclusive evidence of life on Mars may have to wait until a geologist from Earth can scramble down dry Martian streambeds and crack open rocks looking for fossils, or drill into the soil seeking signs of metabolizing microorganisms.

There is no strong evidence for the existence of life in the solar system other than on Earth. Now your search will take you to distant planetary systems.

## Life in Other Planetary Systems

Could life exist in other planetary systems? You already know that there are many different kinds of stars and that many of these stars have planetary systems. As a first step toward answering this question, you can try to identify the kinds of stars that seem most likely to have stable planetary systems where life could evolve.

If a planet is to be a suitable home for living things, it must be in a stable orbit around its sun. That is easy in a planetary system like our own, but planet orbits in binary star systems would be unstable unless the component stars are very close together or very far apart. Astronomers can calculate that, in binary systems with stars separated by intermediate distances of a few AU, the planets should eventually be swallowed up by one of the stars or ejected from the system. Half the stars in the galaxy are members of binary systems, and many of them are unlikely to support life on planets.

Moreover, just because a star is single does not necessarily make it a good candidate for sustaining life. Earth required perhaps as much as 1 billion years to produce the first cells and 4.6 billion years for intelligence to emerge. Massive stars that shine for only a few million years do not meet this criterion. If the history of life on Earth is representative, then stars more massive and luminous than about spectral type F5 last too short a time for complex life to develop. Main-sequence stars of types G and K, and possibly some of the M stars, are the best candidates.

The temperature of a planet is also important, and that depends on the type of star it orbits and its distance from the star. Astronomers have defined a **habitable zone** around a star as a region within which orbiting planets have temperatures permitting the existence of liquid water. The sun's habitable zone extends from near the orbit of Venus to the orbit of Mars, with Earth right in the middle. A low-luminosity star has a small and narrow habitable zone, whereas a high-luminosity star has a large and wide one.

Stable planets inside the habitable zones of long-lived stars are the places where life seems most likely, but, given the tenacity and resilience of Earth's life forms, there might be other, seemingly inhospitable, places in the universe where life exists. You should also note that three of the environments considered as possible havens for life—Europa, Titan, and Enceladus—are in the outer solar system, far outside the sun's conventionally defined habitable zone.

### SCIENTIFIC ARGUMENT

***What evidence indicates that life is possible on other worlds?***

A good scientific argument involves careful analysis of evidence. Fossils on Earth show that life originated in the oceans at least 3.4 billion years ago, and biologists have outlined likely chemical processes that, over long time intervals, could have changed simple organic compounds into reproducing molecules inside membranes, the first simple life forms. Meager fossil evidence indicates that life developed slowly at first. The pace of evolution quickened about half a billion years ago, when life took on complex forms. Later, life emerged onto the land and continued evolving rapidly into diverse forms. Intelligence is a relatively recent development: It is only a few million years old.

If this evolutionary process occurred on Earth, it seems reasonable that it could have occurred on other worlds as well. Earth-like worlds could be plentiful in the universe. Life may begin and eventually evolve to intelligence on any world where conditions are right. Now make a related argument. **What are the conditions you should expect on other worlds that host life?**

## 26-3 Intelligent Life in the Universe

COULD INTELLIGENT LIFE ARISE on other worlds? To try to answer this question, you can estimate the chances of any type of life arising on other worlds, then assess the likelihood of that life developing intelligence. If other civilizations exist, it is possible humans eventually may be able to communicate with them. Nature puts restrictions on the pace of such conversations, but the main problem lies in the unknown life expectancy of civilizations.

### Travel Between the Stars

The distances between stars are almost beyond comprehension. The fastest human device ever launched, the *New Horizons* probe currently on its way to Pluto and the Kuiper belt (see Chapter 24), will take about 90,000 years to travel the distance to the nearest star, 4 light-years. The obvious way to overcome these huge distances is with tremendously fast spaceships, but even the closest stars are many light-years away.

Nothing can exceed the speed of light, and accelerating a spaceship close to the speed of light takes huge amounts of energy. Even if you travel slower than light, your rocket would still require massive amounts of fuel. If you wanted to pilot a spaceship with a mass of 100 tons (about the size of a fancy yacht) to the nearest star, and you traveled at half the speed of light so as to arrive in eight years, the trip would require 400 times as much energy as the entire United States consumes in a year. Don't even think about how much fuel the starship *Enterprise* needs.

These limitations not only make it difficult for humans to leave the solar system, but they would also make it difficult for aliens to visit Earth. Reputable scientists have studied "unidentified flying objects" (UFOs) and have never found any evidence that Earth is being visited or has ever been visited by aliens (**How Do We Know? 26-2**). Humans are unlikely ever to meet aliens face-to-face. However, communication by electromagnetic signals across interstellar distances takes relatively little energy.

## Radio Communication

Nature puts restrictions on travel through space, and it also restricts the possibility of communicating with distant civilizations by radio. One restriction is based on simple physics: Radio signals are electromagnetic waves and travel at the speed of light. Due to the distances between the stars, the speed of radio waves would severely limit humanity's ability to carry on normal conversations with distant civilizations. Decades could elapse between asking a question and getting an answer.

So, rather than try to begin a conversation, one group of astronomers decided in 1974 to broadcast a message of greeting toward the globular cluster M13, 26,000 light years away, using the Arecibo radio telescope (see Figure 6-21b). When the signal arrives 26,000 years in the future, alien astronomers may be able to understand it because the message is **anticoded,** meaning that it is intended to be decoded by beings about whom we know nothing except that they build radio telescopes. The message is a string of 1679 pulses and gaps. Pulses represent 1s, and gaps represent 0s. The string can be arranged in two dimensions in only two possible ways: as 23 rows of 73 or as 73 rows of 23. The second arrangement forms a picture containing information about life on Earth (■ Figure 26-10).

What are the chances that a signal like the Arecibo message would be heard across interstellar distances? Surprisingly, a radio dish the size of the Arecibo telescope, located anywhere in the Milky Way Galaxy, could detect the output from "our" Arecibo. The human race's modest technical capabilities already can put us into cosmic chat rooms.

Although the 1974 Arecibo beacon was the only powerful signal sent purposely from Earth to other star systems, Earth is sending out many other signals more or less accidentally. Short-wave radio signals, including TV and FM, have been leaking into space for the last 60 years or so. Any civilization within 60 light-years could already have detected Earth's civilization. That works both ways: Alien signals, whether intentional messages of friendship or the blather of their equivalent to daytime TV, could be arriving at Earth now. Groups of astronomers from several countries are pointing radio telescopes at the most likely stars and listening for alien civilizations.

Which channels should astronomers monitor? Signals with wavelengths longer than 100 cm would get lost in the background noise of our Milky Way Galaxy, while wavelengths shorter than about 1 cm are mostly absorbed in Earth's atmosphere. Between those wavelengths is a radio window that is open for communication. Even this restricted window contains millions of possible radio-frequency bands and is too wide to monitor easily, but astronomers may have thought of a way to narrow the search. Within this broad radio window lie the 21-cm spectral line of neutral hydrogen and the 18-cm line of OH (■ Figure 26-11). The interval between those lines has low background interference and is named the **water hole** because H plus OH yields water. Any civilizations sophisticated enough to do radio astronomy would know of these lines and might appreciate their significance in the same way as do Earthlings.

A number of searches for extraterrestrial radio signals have been made, and some are now under way. This field of study is known as **SETI, Search for Extra-Terrestrial Intelligence,** and it has generated heated debate among astronomers, philosophers, theologians, and politicians. You might imagine that the discovery of real alien intelligence would cause a huge change in humanity's worldview, akin to Galileo's discovery that the moons of Jupiter do not go around the Earth. Congress funded a NASA SETI search for a short time but ended support in the early 1990s. In fact, the annual cost of a major search is only about as much as a single Air Force helicopter, but much of the reluctance to fund searches stems from issues other than cost. Segments of the population, including some members of Congress, considered the idea of extraterrestrial beings as so outlandish that continued public funding for the search became impossible.

In spite of the controversy, the search continues. The NASA SETI project canceled by Congress was renamed Project Phoenix and completed using private funds. The SETI Institute, founded in 1984, managed Project Phoenix plus several other important searches and is currently building a new radio telescope array in northern California, in collaboration with the University of California, Berkeley, and partly funded by Paul Allen of Microsoft (■ Figure 26-12).

There is even a way for you to help with searches. The Berkeley SETI team (*Note:* they are separate from the SETI Institute), with the support of the Planetary Society, has

## UFOs and Space Aliens

**Has Earth been visited by aliens?** If you conclude that there is likely to be life on other worlds, then you might be tempted to use UFO sightings as evidence to test your hypothesis. Scientists don't do this for two reasons.

First, the reputation of UFO sightings and alien encounters does not inspire confidence that these data are reliable. Most people hear of such events in grocery store tabloids, daytime talk shows, or sensational "specials" on viewer-hungry cable networks. You should take note of the low reputation of the media that report UFOs and space aliens. Most of these reports, like the reports that Elvis is alive and well, are simply made up for the sake of sensation, and you cannot use them as reliable evidence.

Second, the few UFO sightings that are not made up do not survive careful examination. Most are mistakes and unintentional misinterpretations, committed by honest people, of natural events or human-made objects. It is very important to realize that experts have studied these incidents over many decades and found *none* that are convincing to the professional scientific community. In short, despite false claims to the contrary on TV shows, there is no dependable evidence that Earth has ever been visited by aliens.

In a way, that's too bad. A confirmed visit by intelligent creatures from beyond our solar system would answer many questions. It would be exciting, enlightening, and, like any real adventure, a bit scary. Most scientists would love to be part of such a discovery. But, scientists must professionally pay attention to what is supported by evidence rather than what might be thrilling. There is not yet any direct evidence of life on other worlds.

*Flying saucers from space are fun to think about, but there is no evidence that they are real.*

recruited about 4 million owners of personal computers that are connected to the Internet. You can download a screen saver that searches data files from the Arecibo radio telescope for signals whenever you are not using the computer. For information, locate the seti@home project at http://setiathome.ssl.berkeley.edu/.

The search continues, but radio astronomers struggle to hear anything against the worsening babble of radio noise from human civilization. Wider and wider sections of the electromagnetic spectrum are being used for Earthly communication, and this, combined with stray electromagnetic radiation from electronic devices including everything from computers to refrigerators, makes hearing faint radio signals difficult. It would be ironic if humans fail to detect faint radio signals from another world because our own world has become too noisy. One alternate search strategy is to look for rapid flashes of laser light at optical or near-infrared wavelengths. Such extraterrestrial signals, if they exist, would have the advantage of being easily distinguished from natural light sources but the disadvantage of being blocked by interstellar dust. Ultimately, the chance of success for any of the searches depends on the number of inhabited worlds in the galaxy.

### How Many Inhabited Worlds?

Given enough time, the searches will find other worlds with civilizations, assuming that there are at least a few out there. If intelligence is common, scientists should find signals relatively soon—within the next few decades—but if intelligence is rare, it may take much longer.

Simple arithmetic can give you an estimate of the number of technological civilizations in the Milky Way Galaxy with which you might communicate, $N_c$. The formula proposed for discussions about $N_c$ is named the **Drake equation** after the radio astronomer Frank Drake, a pioneer in the search for extraterrestrial intelligence. The version of the Drake equation presented here is modified slightly from its original form:

$$N_c = N_* \cdot f_P \cdot n_{HZ} \cdot f_L \cdot f_I \cdot f_S$$

$N_*$ is the number of stars in our galaxy, and $f_P$ represents the fraction of stars that have planets. If all single stars have planets, $f_P$ is about 0.5. The factor $n_{HZ}$ is the average number of planets in each planetary system suitably located in the habitable zone—meaning, for the sake of the present discussion, the number of planets per planetary system possessing substantial amounts of liquid water. The conventional habitable zone in our system

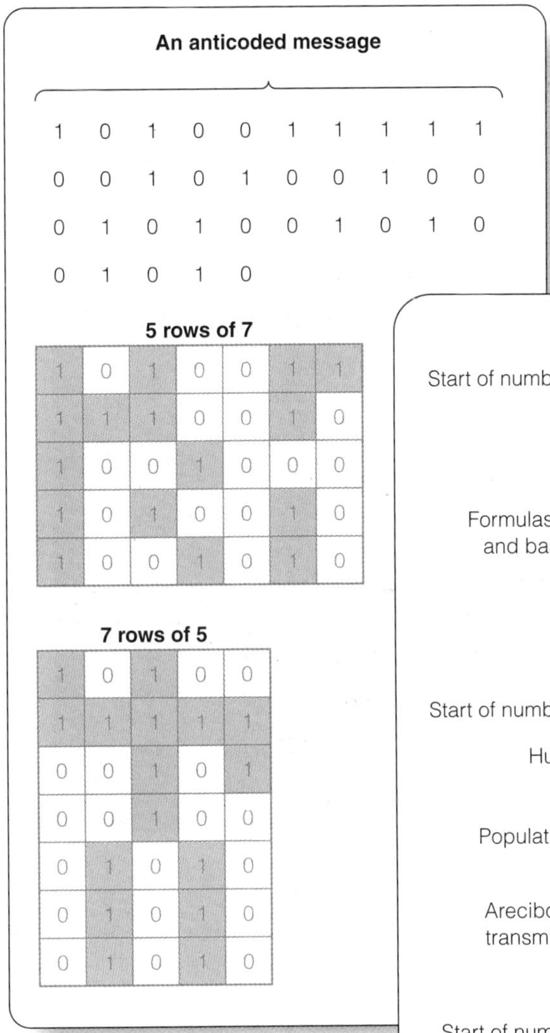

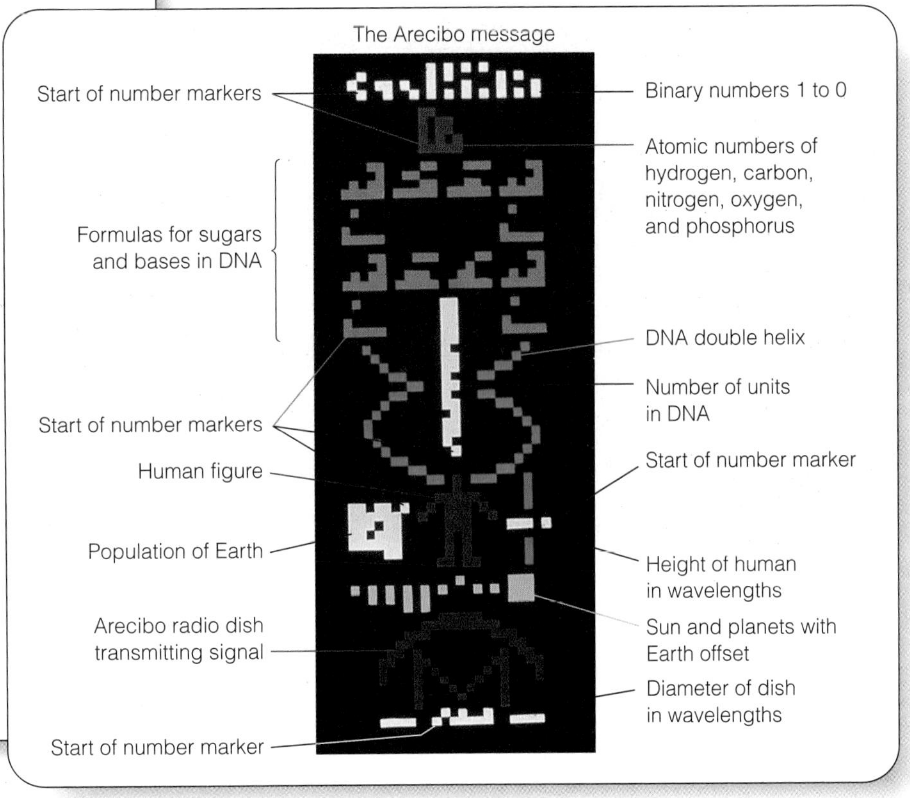

An anticoded message

1 0 1 0 0 1 1 1 1 1
0 0 1 0 1 0 0 1 0 0
0 1 0 1 0 0 1 0 1 0
0 1 0 1 0

5 rows of 7

| 1 | 0 | 1 | 0 | 0 | 1 | 1 |
|---|---|---|---|---|---|---|
| 1 | 1 | 1 | 0 | 0 | 1 | 0 |
| 1 | 0 | 0 | 1 | 0 | 0 | 0 |
| 1 | 0 | 1 | 0 | 0 | 1 | 0 |
| 1 | 0 | 0 | 1 | 0 | 1 | 0 |

7 rows of 5

| 1 | 0 | 1 | 0 | 0 |
|---|---|---|---|---|
| 1 | 1 | 1 | 1 | 1 |
| 0 | 0 | 1 | 0 | 1 |
| 0 | 0 | 1 | 0 | 0 |
| 0 | 1 | 0 | 1 | 0 |
| 0 | 1 | 0 | 1 | 0 |
| 0 | 1 | 0 | 1 | 0 |

The Arecibo message

Start of number markers
Formulas for sugars and bases in DNA
Start of number markers
Human figure
Population of Earth
Arecibo radio dish transmitting signal
Start of number marker
Binary numbers 1 to 0
Atomic numbers of hydrogen, carbon, nitrogen, oxygen, and phosphorus
DNA double helix
Number of units in DNA
Start of number marker
Height of human in wavelengths
Sun and planets with Earth offset
Diameter of dish in wavelengths

**Figure 26-10**

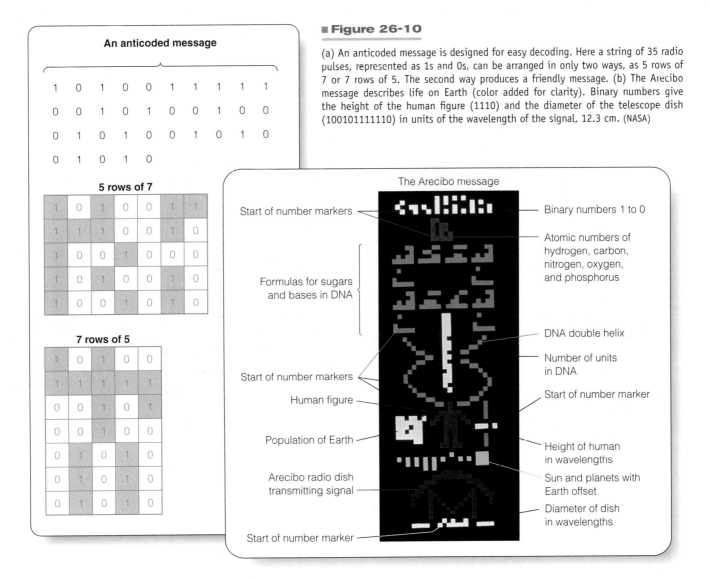

(a) An anticoded message is designed for easy decoding. Here a string of 35 radio pulses, represented as 1s and 0s, can be arranged in only two ways, as 5 rows of 7 or 7 rows of 5. The second way produces a friendly message. (b) The Arecibo message describes life on Earth (color added for clarity). Binary numbers give the height of the human figure (1110) and the diameter of the telescope dish (100101111110) in units of the wavelength of the signal, 12.3 cm. (NASA)

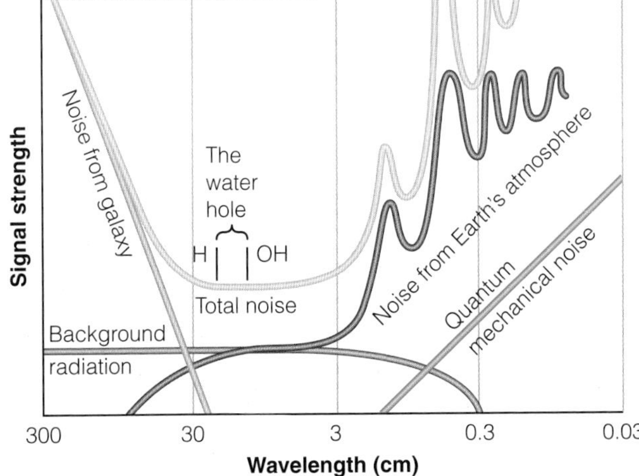


**Figure 26-11**

Radio noise from various astronomical sources and Earth's atmospheric opacity make it difficult to detect distant signals at wavelengths longer than 100 cm or shorter than 1 cm. In this range, wavelengths of radio emission lines from H atoms and from OH molecules mark a small wavelength range named the water hole that may be a likely channel for interstellar communication.

contains Earth's orbit, and, arguably, Venus and Mars as well. Europa and Enceladus in our solar system show that liquid water can exist due to tidal heating outside the conventional habitable zone. Thus, $n_{HZ}$ may be larger than had been originally thought. The factor $f_L$ is the fraction of suitable planets on which life begins, and $f_I$ is the fraction of those planets where life evolves to intelligence.

The six factors on the right-hand side of the Drake equation can be roughly estimated, with decreasing certainty as you proceed from left to right. The final factor is extremely uncertain. That factor $f_S$ is the fraction of a star's life during which an intelligent species is communicative. If a society survives at a technological level for only 100 years, the chances of communicating with it are small. On the other hand, a society that stabilizes and remains technologically capable for a long time is much more

**■ Figure 26-12**

Part of the Allen Telescope Array (ATA), which is now being built in California, is planned to eventually include 350 radio dishes, each 6 meters in diameter, in an arrangement designed to maximize their combined angular resolution. As radio astronomers aim the telescope at galaxies and nebulae of scientific interest, state-of-the-art computer systems will analyze stars in the field of view searching for signals from distant civilizations. (SETI Institute)

likely to be detected. For a star with a life span of 10 billion years, $f_S$ might conceivably range from $10^{-8}$ for extremely short-lived societies to $10^{-4}$ for societies that survive for a million years. ■ Table 26-1 summarizes what many scientists consider a reasonable range of values for $f_S$ and the other factors.

If the optimistic estimates are true, there could be a communicative civilization within a few tens of light-years from Earth. On the other hand, if the pessimistic estimates are true, Earth may be the only planet that is capable of communication within thousands of the nearest galaxies.

**■ Table 26-1 | The Number of Technological Civilizations per Galaxy**

| Estimates | Variables | Pessimistic | Optimistic |
|---|---|---|---|
| $N_*$ | Number of stars per galaxy | $2 \times 10^{11}$ | $2 \times 10^{11}$ |
| $f_P$ | Fraction of stars with planets | 0.1 | 0.5 |
| $n_{HZ}$ | Number of planets per star that lie in habitable zone for longer than 4 billion years | 0.01 | 1 |
| $f_L$ | Fraction of suitable planets on which life begins | 0.01 | 1 |
| $f_I$ | Fraction of planets with life where life evolves to intelligence | 0.01 | 1 |
| $f_S$ | Fraction of star's existence during which a technological society survives | $10^{-8}$ | $10^{-4}$ |
| $N_c$ | Number of communicative civilizations per galaxy | $2 \times 10^{-4}$ | $1 \times 10^{7}$ |

## SCIENTIFIC ARGUMENT

***Why does the number of civilizations that could be detected depend on how long civilizations survive at a technological level?***

This scientific argument depends on the timing of events. If you turned a radio telescope to the sky and scanned millions of frequency bands for many stars, you would be taking a snapshot of the universe at a particular time. Broadcasts from other civilizations must be arriving at that same time if they are to be detected. If most civilizations survive for a long time, there is a much greater chance that you will detect one of them in your snapshot than if civilizations tend to disappear quickly due to, for example, nuclear war or environmental collapse. If most civilizations last only a short time, there may be none capable of transmitting during the cosmically short interval when Earthlings are capable of building radio telescopes to listen for them.

The speed at which astronomers can search for signals is limited because computers must search many frequency intervals, but not all frequencies inside Earth's radio window are subject to intensive search. Build a new argument to explain: **Why is the water hole an especially good frequency band in which to listen?**

## What Are We? Matter and Spirit

There are over 4000 religions around the world, and nearly all hold that humans have a dual nature: We are physical objects made of atoms, but we are also spiritual beings. Science is unable to examine the spiritual side of existence, but it can tell us about our physical nature.

The matter you are made of appeared in the big bang and was cooked into a wide range of elements inside stars. Your atoms may have been inside at least two or three generations of stars. Eventually, your atoms became part of a nebula that contracted to form our sun and the planets of the solar system

Your atoms have been part of Earth for the last 4.6 billion years. They have been recycled many times through dinosaurs, stromatolites, fish, bacteria, grass, birds, worms, and other living things. You are using your atoms now, but when you are done with them, they will go back to Earth and be used again and again.

When the sun swells into a red giant star and dies in a few billion years, Earth's atmosphere and oceans will be driven away, and at least the outer few kilometers of Earth's crust will be vaporized and blown outward to become part of the nebula around the white-dwarf remains of the sun. Your atoms are destined to return to the interstellar medium and will become part of future generations of stars and planets.

The message of astronomy is that humans are not just observers: We are participants in the universe. Among all of the galaxies, stars, planets, planetesimals, and bits of matter, humans are objects that can think, and that means we can understand what we are.

Is the human race the only thinking species? If so, we bear the sole responsibility to understand and admire the universe. The detection of signals from another civilization would demonstrate that we are not alone, and such communication would end the self-centered isolation of humanity and stimulate a reevaluation of the meaning of human existence. We may never realize our full potential as humans until we communicate with non-human intelligent life.

# Study and Review

## Summary

- Consideration of the possibility that life exists elsewhere in the universe, and the study of life and its origin on Earth as it relates to life on other worlds, is called **astrobiology (p. 582)**.
- Life can be defined as a process that extracts energy from the surroundings, maintains an organism, and modifies the surroundings to promote the organism's survival.
- Living things have a physical basis—the arrangement of matter and energy that makes life possible. Life on Earth is based on carbon chemistry occurring in bags of water (cells).
- Living things must have a controlling unit of information that can be passed to successive generations.
- Genetic information for life on Earth is stored in long carbon-chain molecules such as **DNA (deoxyribonucleic acid) (p. 584)**.
- The DNA molecule stores information in the form of chemical bases linked together like the rungs of a ladder. Copied by the **RNA (ribonucleic acid) (p. 585)** molecule, the patterns of bases act as recipes for connecting together **amino acid (p. 584)** subunits to construct **proteins (p. 584)**, including **enzymes (p. 584)** that are respectively the main structural and control components of the life process.

- The unit of heredity is a **gene (p. 585)**, a piece or several pieces of DNA that in most cases specifies the construction of one particular type of protein molecule. Genes are connected together in structures called **chromosomes (p. 585)**, that are essentially single long DNA molecules.
- When a cell divides, the chromosomes split lengthwise and duplicate themselves so that each of the new cells can receive a copy of the genetic information.
- **Biological evolution (p. 586)** is the process by which life adjusts itself to changing environments.
- Errors in duplication or damage to the DNA molecule can produce **mutants (p. 586)**, organisms that contain new DNA information and therefore have new properties. Variation in genetic codes can become widespread among individuals in a species. **Natural selection (p. 586)** determines which of these variations are best suited to survive, and the species evolves to fit its environment.
- Evolution is not random. Genetic variation is random, but natural selection is controlled by the environment.
- The oldest definitely identified fossils on Earth, structures called **stromatolites (p. 587)** that are composed of stacks of bacterial mats and sediment layers, are at least 3.4 billion years old. Those fossils provide evidence that life began in the oceans.
- Fossil evidence indicates that life began on Earth as simple single-celled organisms like bacteria and much later evolved into more complex, **multicellular (p. 589)** creatures.
- The **Miller experiment (p. 587)** shows that the chemical building blocks of life form naturally under a wide range of circumstances.
- Scientists hypothesize that **chemical evolution (p. 589)** occurred before biological evolution. Chemical evolution concentrated simple molecules into a diversity of larger stable organic molecules dissolved in the young Earth's oceans, but those molecules did not reproduce copies of themselves. The hypothetical organic-rich water is sometimes referred to as the **primordial soup (p. 588)**. Biological evolution began when molecules developed the ability to make copies of themselves.
- Life forms did not become large and complex until about 0.5 billion years ago, during what is called the **Cambrian explosion (p. 589)**.
- Life emerged from the oceans only about 0.4 billion years ago, and human intelligence developed over the last 4 million (0.004 billion) years.
- Life as it is known on Earth requires liquid water and thus a specific range of temperatures.
- Organisms that thrive in extreme environments are called **extremophiles (p. 591)**. Genetic evidence indicates that the common ancestor of all Earth life was a **thermophile (p. 591)**, a heat-tolerant version of present-day single-celled organisms called **archaea (p. 591)**.
- No other planet in our solar system appears to harbor life at present. Most are too hot or too cold, although life might have begun on Mars before the planet became too cold and dry. If so, life conceivably could persist today on Mars in limited hospitable environments.
- Liquid water exists, and therefore Earth-like life is at least possible, under the surfaces of Jupiter's moons Europa and Ganymede and Saturn's moon Enceladus. Saturn's moon Titan has abundant organic compounds but does not have liquid water.
- Because the origin of life and its evolution into intelligent creatures took so long on Earth, scientists do not consider middle- and upper-main-sequence stars, which shine for astronomically short time spans, as likely hosts for life-bearing planets.
- Main-sequence G and K stars are thought to be likely candidates to host planets with life. Scientists are not sure whether the M stars are also good candidates.
- The **habitable zone (p. 594)** around a star, within which planets can have liquid water on their surfaces, may be larger than scientists had expected, given the wide variety of living things now found in extreme environments on Earth. Tidally heated moons orbiting large planets could have liquid water at any distance from a star.
- Because of distance, speed, and fuel, travel between the stars seems almost impossible for humans or for aliens who might visit Earth.
- Communication between planetary systems using electromagnetic signals such as radio waves or laser beams may be possible, but a real conversation would be difficult because of long travel times for such signals.
- Broadcasting a radio (or light) beacon of pulses would distinguish the signal from naturally occurring emission and identify the source as a technological civilization. The signal can be **anticoded (p. 595)** in the hope it would be possible for another civilization to decode.
- One good part of the radio spectrum for communication is called the **water hole (p. 595)**, the wavelength range from the 21-cm spectral line of hydrogen to the 18-cm line of OH. Even so, millions of radio wavelengths need to be tested to fully survey the water hole for a given target star.
- Sophisticated searches are now underway to detect radio transmissions from civilizations on other worlds, but such **SETI (Search for Extra-Terrestrial Intelligence) (p. 595)** programs are hampered by limited computer power and radio noise pollution from human civilization.
- The number of civilizations in our galaxy that are at a technological level and able to communicate while humans are listening can be estimated by the **Drake equation (p. 596)**. This number is limited primarily by the lifetimes of their and our civilizations.

## Review Questions

1. If life is based on information, what is that information?
2. How does the DNA molecule produce a copy of itself?
3. What would happen to a life form if the genetic information handed down to offspring was copied extremely inaccurately? How would that endanger the future of the life form?
4. What would happen to a life form if the information handed down to offspring was always the same? How would that endanger the future of the life form?
5. Give an example of natural selection acting on new DNA patterns to select the most advantageous characteristics.
6. What evidence do scientists have that life on Earth began in the sea?
7. Why do scientists generally think that liquid water is necessary for the origin of life?
8. What is the difference between chemical evolution and biological evolution?
9. What is the significance of the Miller experiment?
10. How does intelligence make a creature more likely to survive?
11. Why are upper-main-sequence (high-luminosity) stars unlikely sites for intelligent civilizations?
12. Why is it reasonable to suspect that travel between stars is nearly impossible?
13. How does the stability of technological civilizations affect the probability that Earth can communicate with them?
14. What is the water hole, and why is it a good "place" to search for extraterrestrial civilizations?
15. Why is it difficult to anticode a message? In other words, why is it hard to make a message that potentially can be understood by completely unknown recipients?
16. **How Do We Know?** How do science and religion have complementary explanations of the world?
17. **How Do We Know?** Why are scientists sure Earth has never been visited by aliens?

## Discussion Questions

1. Do you expect that hypothetical alien recipients of the Arecibo message will be able to decode it? Why or why not?
2. How do you think the detection of extraterrestrial intelligence would be received by the public? Would it be likelier to upset, or confirm, human's beliefs about themselves and the world?
3. What do you think it would mean if decades of careful searches for radio signals for extraterrestrial intelligence turn up nothing?

## Problems

1. A single human cell encloses about 1.5 m of DNA, containing 4.5 billion base pairs. What is the spacing between these base pairs in nanometers? That is, how far apart are the rungs on the DNA ladder?
2. If you represent Earth's history by a line 1 m long, how long a segment would represent the 400 million years since life moved onto the land? How long a segment would represent the 4-million-year history of human life?
3. If a human generation, the average time from birth to childbearing, has been 20 years long, how many generations have passed in the last 1 million years?
4. If a star must remain on the main sequence for at least 4 billion years for life to evolve to intelligence, what is the most massive a star can be and still possibly harbor intelligent life on one of its planets? (*Hints:* See Chapter 12 and Appendix Table A-7.)
5. If there are about $1.4 \times 10^{-4}$ stars like the sun per cubic light-year, how many lie within 100 light-years of Earth? (*Hint:* The volume of a sphere is $\frac{4}{3}\pi r^3$.)
6. Mathematician Karl Gauss suggested planting forests and fields in gigantic geometric figures as signals to possible Martians that intelligent life exists on Earth. If Martians had telescopes that could resolve details no smaller than 1 arc second, how large would the smallest element of Gauss's signal have to be for it to be visible at Mars's closest approach to Earth? (*Hints:* See Appendix Table A-10 and use the small-angle formula, Chapter 3.)
7. If you detected radio signals with an average wavelength of 20 cm and suspected that they came from a civilization on a distant Earth-like planet, roughly how much of a change in wavelength should you expect to see because of the orbital motion of the distant planet? (*Hints:* See the equation for the Doppler effect in Chapter 7; the Earth's orbital velocity is 30 km/s.)
8. Calculate the number of communicative civilizations per galaxy using your own estimates of the factors in Table 26-1.

## Learning to Look

1. The star cluster shown in the image to the right contains cool red giants and main-sequence stars from hot blue stars all the way down to red dwarfs. Discuss the likelihood that planets orbiting any of these stars might be home to life. (*Hint:* Estimate the age of the cluster.)

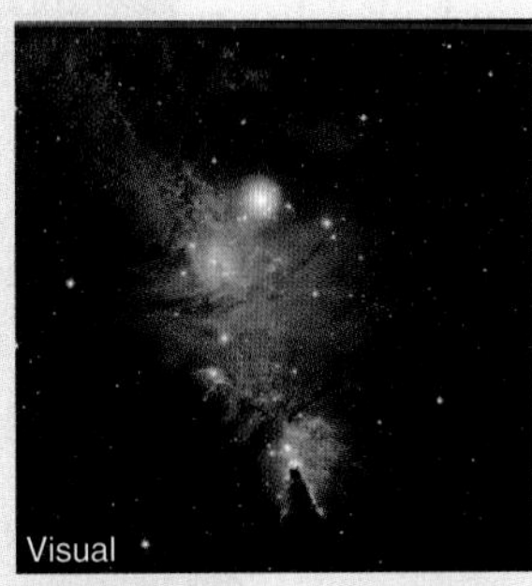

NASA

2. If you could search for life in the galaxy shown in the image to the right, would you look among disk stars or halo stars? Discuss the factors that influence your decision.

ESO

# Appendix A
# Units and Astronomical Data

## Introduction

The metric system is used worldwide as the system of units, not only in science but also in engineering, business, sports, and daily life. Developed in 18th-century France, the metric system has gained acceptance in almost every country in the world because it simplifies computations.

A system of units is based on the three fundamental units for length, mass, and time. Other quantities, such as density and force, are derived from these fundamental units. In the English (or British) system of units (commonly used only in the United States, Tonga, and Southern Yemen, but, ironically, not in Great Britain) the fundamental unit of length is the foot, composed of 12 inches. The metric system is based on the decimal system of numbers, and the fundamental unit of length is the meter, composed of 100 centimeters.

Because the metric system is a decimal system, it is easy to express quantities in larger or smaller units as is convenient. You can give distances in centimeters, meters, kilometers, and so on. The prefixes specify the relation of the unit to the meter. Just as a cent is 1/100 of a dollar, so a centimeter is 1/100 of a meter. A kilometer is 1000 m, and a kilogram is 1000 g. The meanings of the commonly used prefixes are given in ■ Table A-1.

■ **Table A-1 | Metric Prefixes**

| Prefix | Symbol | Factor |
|---|---|---|
| Mega | M | $10^6$ |
| Kilo | k | $10^3$ |
| Centi | c | $10^{-2}$ |
| Milli | m | $10^{-3}$ |
| Micro | μ | $10^{-6}$ |
| Nano | n | $10^{-9}$ |

■ **Table A-2 SI** *(Système International)* **Metric Units**

| Quantity | SI Unit |
|---|---|
| Length | Meter (m) |
| Mass | Kilogram (kg) |
| Time | Second (s) |
| Force | Newton (N) |
| Energy | Joule (J) |

## The SI Units

Any system of units based on the decimal system would be easy to use; but, by international agreement, the preferred set of units, known as the *Système International d'Unités* (SI units) is based on the meter, kilogram, and second. These three fundamental units define the rest of the units, as given in ■ Table A-2.

The SI unit of force is the newton (N), named after Isaac Newton. It is the force needed to accelerate a 1 kg mass by 1 m/s$^2$, or the force roughly equivalent to the weight of an apple at Earth's surface. The SI unit of energy is the joule (J), the energy produced by a force of 1 N acting through a distance of 1 m. A joule is roughly the energy in the impact of an apple falling off a table.

### Exceptions

Units can help you in two ways. They make it possible to make calculations, and they can help you to conceive of certain quantities. For calculations, the metric system is far superior, and it is used for calculations throughout this book.

Americans commonly use the English system of units, so for conceptual purposes this book also expresses quantities in English units. Instead of saying the average person would weigh 133 N on the moon, it might be more helpful to some readers for that weight to be expressed as 30 lb. Consequently, this text commonly gives quantities in metric form followed by the English form in parentheses: The radius of the moon is 1738 km (1080 mi).

In SI units, density should be expressed as kilograms per cubic meter, but no human hand can enclose a cubic meter, so that unit does not help you grasp the significance of a given density. This book refers to density in grams per cubic centimeter. A gram is roughly the mass of a paperclip, and a cubic centimeter is the size of a small sugar cube, so you can easily conceive of a density of 1 g/cm$^3$, roughly the density of water. This is not a bothersome departure from SI units because you will not have to make complex calculations using density.

## Conversions

To convert from one metric unit to another (from meters to kilometers, for example), you have only to look at the prefix. However, converting from metric to English or English to metric is more complicated. The conversion factors are given in ■ Table A-3.

**Table A-3 | Conversion Factors Between British and Metric Units**

| | |
|---|---|
| 1 inch = 2.54 centimeters | 1 meter = 39.36 inches = 3.28 feet |
| 1 foot = 0.3048 meter | 1 kilometer = 0.6214 mile |
| 1 mile = 1.609 kilometers | 1 kilogram = 0.0685 slug |
| 1 slug = 14.59 kilograms | 1 Newton = 0.2248 pound |
| 1 pound = 4.448 Newtons | 1 Joule = 0.7376 foot-pound |
| 1 foot-pound = 1.356 Joules | 1 Joule/s = 0.001341 horsepower |
| 1 horsepower = 745.7 Joules/s | 1 Joule/s = 1 Watt |
| 1 centimeter = 0.394 inch | |

Example: The radius of the moon is 1738 km. What is this in miles? Table A-3 indicates that 1.000 mile equals 1.609 km, so

$$1738 \text{ km} \times \frac{(1.000 \text{ mi})}{(1.609 \text{ km})} = 1080 \text{ mi}$$

## Temperature Scales

In astronomy, as in most other sciences, temperatures are expressed on the Kelvin scale, although the centigrade (or Celsius) scale is also used. The Fahrenheit scale commonly used in the United States is not used in scientific work.

Temperatures on the Kelvin scale are measured from absolute zero, the temperature of an object that contains no extractable heat. In practice, no object can be as cold as absolute zero, although laboratory apparatuses have reached temperatures lower than $10^{-6}$ K. The Kelvin scale is named after the Scottish mathematical physicist William Thomson, Lord Kelvin (1824–1907).

The centigrade scale refers temperatures to the freezing point of water (0°C) and to the boiling point of water (100°C). One degree centigrade is $\frac{1}{100}$ of the temperature difference between the freezing and boiling points of water, thus the prefix *centi.* The centigrade scale is also called the Celsius scale after its inventor, the Swedish astronomer Anders Celsius (1701–1744).

The Fahrenheit scale fixes the freezing point of water at 32°F and the boiling point at 212°F. Named after the German physicist Gabriel Daniel Fahrenheit (1686–1736), who made the first successful mercury thermometer in 1720, the Fahrenheit scale is used routinely only in the United States.

It is easy to convert temperatures from one scale to another using the information given in ■Table A-4.

**Table A-4 Temperature Scales and Conversion Formulas**

| | Kelvin (K) | Centigrade (°C) | Fahrenheit (°F) |
|---|---|---|---|
| Absolute zero | 0 K | −273°C | −459°F |
| Freezing point of water | 273 K | 0°C | 32°F |
| Boiling point of water | 373 K | 100°C | 212°F |

Conversions:

K = °C + 273

°C = $(\frac{5}{9})$(°F − 32)

°F = $(\frac{9}{5})$(°C) + 32

## Powers of 10 Notation

Powers of 10 make writing very large numbers much simpler. For example, the nearest star is about 43,000,000,000,000 km from the sun. Writing this number as $4.3 \times 10^{13}$ km is much easier.

Very small numbers can also be written with powers of 10. For example, the wavelength of visible light is about 0.0000005 m. In powers of 10 this becomes $5 \times 10^{-7}$ m.

The powers of 10 used in this notation appear below. The exponent tells you how to move the decimal point. If the exponent is positive, move the decimal point to the right. If the exponent is negative, move the decimal point to the left. For example, $2.0000 \times 10^3$ equals 2000, and $2 \times 10^{-3}$ equals 0.002.

$$10^5 = 100{,}000$$
$$10^4 = 10{,}000$$
$$10^3 = 1{,}000$$
$$10^2 = 100$$
$$10^1 = 10$$
$$10^0 = 1$$
$$10^{-1} = 0.1$$
$$10^{-2} = 0.01$$
$$10^{-3} = 0.001$$
$$10^{-4} = 0.0001$$

If you use scientific notation in calculations, be sure you correctly enter the numbers into your calculator. Not all calculators accept scientific notation, but those that can have a key labeled EXP, EEX, or perhaps EE that allows you to enter the exponent of ten. To enter a number such as $3 \times 10^8$, press the keys 3 EXP 8. To enter a number with a negative exponent, you must use the change-sign key, usually labeled +/− or CHS. To enter the number $5.2 \times 10^{-3}$, press the keys 5.2 EXP +/− 3. Try a few examples.

To read a number in scientific notation from a calculator you must read the exponent separately. The number $3.1 \times 10^{25}$ may appear in a calculator display as 3.1 25 or on some calculators as 3.1 $10^{25}$. Examine your calculator to determine how such numbers are displayed.

# Astronomy Units and Constants

Astronomy, and science in general, is a way of learning about nature and understanding the universe. To test hypotheses about how nature works, scientists use observations of nature. The tables that follow contain some of the basic observations that support science's best understanding of the astronomical universe. Of course, these data are expressed in the form of numbers, not because science reduces all understanding to mere numbers but because the struggle to understand nature is so demanding that science must use every valid means available. Quantitative thinking—reasoning mathematically—is one of the most powerful techniques ever invented by the human brain. Thus, these tables are not nature reduced to mere numbers but numbers supporting humanity's growing understanding of the natural world around us.

**Table A-5 | Astronomical Constants**

| | |
|---|---|
| Velocity of light ($c$) | $= 3.00 \times 10^8$ m/s |
| Gravitational constant ($G$) | $= 6.67 \times 10^{-11}$ m$^3$/s$^2$kg |
| Mass of H atom | $= 1.67 \times 10^{-27}$ kg |
| Mass of Earth ($M_\oplus$) | $= 5.98 \times 10^{24}$ kg |
| Earth equatorial radius ($R_\oplus$) | $= 6.38 \times 10^3$ km |
| Mass of sun ($M_\odot$) | $= 1.99 \times 10^{30}$ kg |
| Radius of sun ($R_\odot$) | $= 6.96 \times 10^8$ m |
| Solar luminosity ($L_\odot$) | $= 3.83 \times 10^{26}$ J/s |
| Mass of moon | $= 7.35 \times 10^{22}$ kg |
| Radius of moon | $= 1.74 \times 10^3$ km |

**Table A-6 | Units Used in Astronomy**

| | |
|---|---|
| 1 Angstrom (Å) | $= 10^{-8}$ cm |
| | $= 10^{-10}$ m |
| | = 10 nm |
| 1 astronomical unit (AU) | $= 1.50 \times 10^{11}$ m |
| | $= 93.0 \times 10^6$ mi |
| 1 light-year (ly) | $= 6.32 \times 10^4$ AU |
| | $= 9.46 \times 10^{15}$ m |
| | $= 5.88 \times 10^{12}$ mi |
| 1 parsec (pc) | $= 2.06 \times 10^5$ AU |
| | $= 3.09 \times 10^{16}$ m |
| | = 3.26 ly |
| 1 kiloparsec (kpc) | = 1000 pc |
| 1 megaparsec (Mpc) | = 1,000,000 pc |

## Table A-7 | Properties of Main-Sequence Stars

| Spectral Type | Absolute Visual Magnitude ($M_v$) | Luminosity* | Temp. (K) | $\lambda_{max}$ (nm) | Mass* | Radius* | Average Density (g/cm$^3$) |
|---|---|---|---|---|---|---|---|
| O5 | −5.8 | 501,000 | 40,000 | 72.4 | 40 | 17.8 | 0.01 |
| B0 | −4.1 | 20,000 | 28,000 | 100 | 18 | 7.4 | 0.1 |
| B5 | −1.1 | 790 | 15,000 | 190 | 6.4 | 3.8 | 0.2 |
| A0 | +0.7 | 79 | 9900 | 290 | 3.2 | 2.5 | 0.3 |
| A5 | +2.0 | 20 | 8500 | 340 | 2.1 | 1.7 | 0.6 |
| F0 | +2.6 | 6.3 | 7400 | 390 | 1.7 | 1.4 | 1.0 |
| F5 | +3.4 | 2.5 | 6600 | 440 | 1.3 | 1.2 | 1.1 |
| G0 | +4.4 | 1.3 | 6000 | 480 | 1.1 | 1.0 | 1.4 |
| G5 | +5.1 | 0.8 | 5500 | 520 | 0.9 | 0.9 | 1.6 |
| K0 | +5.9 | 0.4 | 4900 | 590 | 0.8 | 0.8 | 1.8 |
| K5 | +7.3 | 0.2 | 4100 | 700 | 0.7 | 0.7 | 2.4 |
| M0 | +9.0 | 0.1 | 3500 | 830 | 0.5 | 0.6 | 2.5 |
| M5 | +11.8 | 0.01 | 2800 | 1000 | 0.2 | 0.3 | 10.0 |
| M8 | +16 | 0.001 | 2400 | 1200 | 0.1 | 0.1 | 63 |

*Luminosity, mass, and radius are given in terms of the sun's luminosity, mass, and radius.

## Table A-8 | The 15 Brightest Stars

| Star | Name | Apparent Visual Magnitude ($m_v$) | Spectral Type | Absolute Visual Magnitude ($M_v$) | Distance (ly) |
|---|---|---|---|---|---|
| α CMa A | Sirius | −1.47 | A1 | 1.4 | 8.7 |
| α Car | Canopus | −0.72 | F0 | −3.1 | 98 |
| α Cen | Rigil Kentaurus | −0.01 | G2 | 4.4 | 4.3 |
| α Boo | Arcturus | −0.06 | K2 | −0.3 | 36 |
| α Lyr | Vega | 0.04 | A0 | 0.5 | 26.5 |
| α Aur | Capella | 0.05 | G8 | −0.6 | 45 |
| β Ori A | Rigel | 0.14 | B8 | −7.1 | 900 |
| α CMi A | Procyon | 0.37 | F5 | 2.7 | 11.3 |
| α Ori | Betelgeuse | 0.41 | M2 | −5.6 | 520 |
| α Eri | Achernar | 0.51 | B3 | −2.3 | 118 |
| β Cen AB | Hadar | 0.63 | B1 | −5.2 | 490 |
| α Aql | Altair | 0.77 | A7 | 2.2 | 16.5 |
| α Tau A | Aldebaran | 0.86 | K5 | −0.7 | 68 |
| α Cru | Acrux | 0.90 | B2 | 3.5 | 260 |
| α Vir | Spica | 0.91 | B1 | −3.3 | 220 |

## Table A-9 | The 15 Nearest Stars

| Name | Absolute Magnitude ($M_v$) | Distance (ly) | Spectral Type | Apparent Visual Magnitude ($m_v$) |
|---|---|---|---|---|
| Sun | 4.83 | | G2 | −26.8 |
| Proxima Cen | 15.45 | 4.28 | M5 | 11.05 |
| α Cen A | 4.38 | 4.3 | G2 | 0.1 |
| α Cen B | 5.76 | 4.3 | K5 | 1.5 |
| Barnard's Star | 13.21 | 5.9 | M5 | 9.5 |
| Wolf 359 | 16.80 | 7.6 | M6 | 13.5 |
| Lalande 21185 | 10.42 | 8.1 | M2 | 7.5 |
| Sirius A | 1.41 | 8.6 | A1 | −1.5 |
| Sirius B | 11.54 | 8.6 | white dwarf | 7.2 |
| Luyten 726-8A | 15.27 | 8.9 | M5 | 12.5 |
| Luyten 726-8B (UV Cet) | 15.8 | 8.9 | M6 | 13.0 |
| Ross 154 | 13.3 | 9.4 | M5 | 10.6 |
| Ross 248 | 14.8 | 10.3 | M6 | 12.2 |
| ϵ Eri | 6.13 | 10.7 | K2 | 3.7 |
| Luyten 789-6 | 14.6 | 10.8 | M7 | 12.2 |

## Table A-10 | Properties of the Planets

**Physical Properties (Earth = ⊕)**

| Planet | Equatorial Radius | | Mass (⊕ = 1) | Average Density (g/cm$^3$) | Surface Gravity (⊕ = 1) | Escape Velocity (km/s) | Sidereal Period of Rotation | Inclination of Equator to Orbit |
|---|---|---|---|---|---|---|---|---|
| | (km) | (⊕ = 1) | | | | | | |
| Mercury | 2439 | 0.38 | 0.056 | 5.44 | 0.38 | 4.3 | 58.65d | 0° |
| Venus | 6052 | 0.95 | 0.815 | 5.24 | 0.90 | 10.3 | 243.02d | 177° |
| Earth | 6378 | 1.00 | 1.000 | 5.50 | 1.00 | 11.2 | 23.93h | 23°.5 |
| Mars | 3396 | 0.53 | 0.108 | 3.94 | 0.38 | 5.0 | 24.62h | 25°.3 |
| Jupiter | 71,494 | 11.20 | 317.8 | 1.34 | 2.54 | 61 | 9.92h | 3°.1 |
| Saturn | 60,330 | 9.42 | 95.2 | 0.69 | 1.16 | 35.6 | 10.23h | 26°.4 |
| Uranus | 25,559 | 4.01 | 14.5 | 1.19 | 0.92 | 22 | 17.23h | 97°.9 |
| Neptune | 24,750 | 3.93 | 17.2 | 1.66 | 1.19 | 25 | 16.05h | 28°.8 |

**Orbital Properties**

| Planet | Semimajor Axis ($a$) | | Orbital Period ($P$) | | Average Orbital Velocity (km/s) | Orbital Eccentricity | Inclination to Ecliptic |
|---|---|---|---|---|---|---|---|
| | (AU) | (10$^6$ km) | (y) | (days) | | | |
| Mercury | 0.39 | 57.9 | 0.24 | 87.97 | 47.89 | 0.206 | 7°.0 |
| Venus | 0.72 | 108.2 | 0.62 | 224.68 | 35.03 | 0.007 | 3°.4 |
| Earth | 1.00 | 149.6 | 1.00 | 365.26 | 29.79 | 0.017 | 0° * |
| Mars | 1.52 | 227.9 | 1.88 | 686.95 | 24.13 | 0.093 | 1°.9 |
| Jupiter | 5.20 | 778.3 | 11.87 | 4334.3 | 13.06 | 0.048 | 1°.3 |
| Saturn | 9.54 | 1427.0 | 29.46 | 10,760 | 9.64 | 0.056 | 2°.5 |
| Uranus | 19.18 | 2869.0 | 84.01 | 30,685 | 6.81 | 0.046 | 0°.8 |
| Neptune | 30.06 | 4497.1 | 164.79 | 60,189 | 5.43 | 0.010 | 1°.8 |

* By definition.

## Table A-11 | Principal Satellites of the Solar System

| Planet | Satellite | Radius (km) | Distance from Planet ($10^3$ km) | Orbital Period (days) | Orbital Eccentricity | Orbital Inclination |
|---|---|---|---|---|---|---|
| Earth | Moon | 1738 | 384.4 | 27.32 | 0.055 | 5°.1 |
| Mars | Phobos | 14 × 12 × 10 | 9.4 | 0.32 | 0.018 | 1°.0 |
| | Deimos | 8 × 6 × 5 | 23.5 | 1.26 | 0.002 | 2°.8 |
| Jupiter | Amalthea | 135 × 100 × 78 | 182 | 0.50 | 0.003 | 0°.4 |
| | Io | 1820 | 422 | 1.77 | 0.000 | 0°.3 |
| | Europa | 1565 | 671 | 3.55 | 0.000 | 0°.5 |
| | Ganymede | 2640 | 1071 | 7.16 | 0.002 | 0°.2 |
| | Callisto | 2420 | 1884 | 16.69 | 0.008 | 0°.2 |
| | Himalia | ~ 85 | 11,470 | 250.6 | 0.158 | 27°.6 |
| Saturn | Janus | 110 × 80 × 100 | 151.5 | 0.70 | 0.007 | 0°.1 |
| | Mimas | 196 | 185.5 | 0.94 | 0.020 | 1°.5 |
| | Enceladus | 250 | 238.0 | 1.37 | 0.004 | 0°.0 |
| | Tethys | 530 | 294.7 | 1.89 | 0.000 | 1°.1 |
| | Dione | 560 | 377 | 2.74 | 0.002 | 0°.0 |
| | Rhea | 765 | 527 | 4.52 | 0.001 | 0°.4 |
| | Titan | 2575 | 1222 | 15.94 | 0.029 | 0°.3 |
| | Hyperion | 205 × 130 × 110 | 1484 | 21.28 | 0.104 | ~0°.5 |
| | Iapetus | 720 | 3562 | 79.33 | 0.028 | 14°.7 |
| | Phoebe | 110 | 12,930 | 550.4 | 0.163 | 150° |
| Uranus | Miranda | 242 | 129.9 | 1.41 | 0.017 | 3°.4 |
| | Ariel | 580 | 190.9 | 2.52 | 0.003 | 0° |
| | Umbriel | 595 | 266.0 | 4.14 | 0.003 | 0° |
| | Titania | 805 | 436.3 | 8.71 | 0.002 | 0° |
| | Oberon | 775 | 583.4 | 13.46 | 0.001 | 0° |
| Neptune | Proteus | 205 | 117.6 | 1.12 | ~0 | ~0° |
| | Triton | 1352 | 354.59 | 5.88 | 0.00 | 160° |
| | Nereid | 170 | 5588.6 | 360.12 | 0.76 | 27.7° |

## Table A-12 | Meteor Showers

| Shower | Dates | Hourly Rate | Radiant of shower R.A. | Radiant of shower Dec. | Associated Comet |
|---|---|---|---|---|---|
| Quadrantids | Jan. 2–4 | 30 | $15^h24^m$ | 50° | |
| Lyrids | April 20–22 | 8 | $18^h4^m$ | 33° | 1861 I |
| $\eta$ Aquarids | May 2–7 | 10 | $22^h24^m$ | 0° | Halley? |
| $\delta$ Aquarids | July 26–31 | 15 | $22^h36^m$ | −10° | |
| Perseids | Aug. 10–14 | 40 | $3^h4^m$ | 58° | Swift-Tuttle |
| Orionids | Oct. 18–23 | 15 | $6^h20^m$ | 15° | Halley? |
| Taurids | Nov. 1–7 | 8 | $3^h40^m$ | 17° | Encke |
| Leonids | Nov. 14–19 | 6 | $10^h12^m$ | 22° | 1866 I Temp |
| Geminids | Dec. 10–13 | 50 | $7^h28^m$ | 32° | |

### Table A-13 | Greatest Elongations of Mercury*

| Evening Sky | Morning Sky |
|---|---|
| Dec. 18, 2009 | Jan. 27, 2010 |
| Apr. 8, 2010** | May 26, 2010 |
| Aug. 7, 2010 | Sept. 19, 2010** |
| Dec. 1, 2010 | Jan. 9, 2011 |
| March 23, 2011** | May 7, 2011 |
| July 20, 2011 | Sept. 3, 2011** |
| Nov. 14, 2011 | Dec. 23, 2011 |
| March 5, 2012** | April 18, 2012 |
| July 1, 2012 | Aug. 16, 2012 |
| Oct. 26, 2012 | Dec. 4, 2012 |
| Feb. 16, 2013 | March 31, 2013 |
| June 12, 2013 | July 30, 2013 |
| Oct. 9, 2013 | Nov. 18, 2013 |

*Elongation is the angular distance from the sun to a planet.
**Most favorable elongations.

### Table A-14 | Greatest Elongations of Venus

| Evening Sky | Morning Sky |
|---|---|
| Aug. 20, 2010 | Jan. 8, 2011 |
| March 27, 2012 | Aug. 15, 2012 |
| Nov. 1, 2013 | March 22, 2014 |
| June 6, 2015 | Oct. 26, 2015 |
| Jan. 12, 2017 | June 3, 2017 |
| Aug. 17, 2018 | Jan. 6, 2019 |
| March 24, 2020 | Aug. 12, 2020 |

### Table A-15 | The Greek Alphabet

| | | | |
|---|---|---|---|
| **A, α** alpha | **H, η** eta | **N, ν** nu | **T, τ** tau |
| **B, β** beta | **Θ, θ** theta | **Ξ, ξ** xi | **Υ, υ** upsilon |
| **Γ, γ** gamma | **I, ι** iota | **O, o** omicron | **Φ, ϕ** phi |
| **Δ, δ** delta | **K, κ** kappa | **Π, π** pi | **X, χ** chi |
| **E, δ** epsilon | **Λ, λ** lambda | **P, ρ** rho | **Ψ, ψ** psi |
| **Z, δ** zeta | **M, μ** mu | **Σ, σ** sigma | **Ω, ω** omega |

## Table A-16 | Periodic Table of the Elements

Atomic number / Symbol / Atomic mass: 11 Na 22.99

Atomic masses are based on carbon-12. Numbers in parentheses are mass numbers of most stable or best-known isotopes of radioactive elements.

| Period | Group IA(1) | IIA(2) | IIIB(3) | IVB(4) | VB(5) | VIB(6) | VIIB(7) | VIII (8) | VIII (9) | VIII (10) | IB(11) | IIB(12) | IIIA(13) | IVA(14) | VA(15) | VIA(16) | VIIA(17) | Noble Gases (18) |
|---|---|---|---|---|---|---|---|---|---|---|---|---|---|---|---|---|---|---|
| 1 | 1<br>H<br>1.008 | | | | | | | | | | | | | | | | | 2<br>He<br>4.003 |
| 2 | 3<br>Li<br>6.941 | 4<br>Be<br>9.012 | | | | | | | | | | | 5<br>B<br>10.81 | 6<br>C<br>12.01 | 7<br>N<br>14.01 | 8<br>O<br>16.00 | 9<br>F<br>19.00 | 10<br>Ne<br>20.18 |
| 3 | 11<br>Na<br>22.99 | 12<br>Mg<br>24.31 | | | | | | | | | | | 13<br>Al<br>26.98 | 14<br>Si<br>28.09 | 15<br>P<br>30.97 | 16<br>S<br>32.06 | 17<br>Cl<br>35.45 | 18<br>Ar<br>39.95 |
| 4 | 19<br>K<br>39.10 | 20<br>Ca<br>40.08 | 21<br>Sc<br>44.96 | 22<br>Ti<br>47.90 | 23<br>V<br>50.94 | 24<br>Cr<br>52.00 | 25<br>Mn<br>54.94 | 26<br>Fe<br>55.85 | 27<br>Co<br>58.93 | 28<br>Ni<br>58.7 | 29<br>Cu<br>63.55 | 30<br>Zn<br>65.38 | 31<br>Ga<br>69.72 | 32<br>Ge<br>72.59 | 33<br>As<br>74.92 | 34<br>Se<br>78.96 | 35<br>Br<br>79.90 | 36<br>Kr<br>83.80 |
| 5 | 37<br>Rb<br>85.47 | 38<br>Sr<br>87.62 | 39<br>Y<br>88.91 | 40<br>Zr<br>91.22 | 41<br>Nb<br>92.91 | 42<br>Mo<br>95.94 | 43<br>Tc<br>98.91 | 44<br>Ru<br>101.1 | 45<br>Rh<br>102.9 | 46<br>Pd<br>106.4 | 47<br>Ag<br>107.9 | 48<br>Cd<br>112.4 | 49<br>In<br>114.8 | 50<br>Sn<br>118.7 | 51<br>Sb<br>121.8 | 52<br>Te<br>127.6 | 53<br>I<br>126.9 | 54<br>Xe<br>131.3 |
| 6 | 55<br>Cs<br>132.9 | 56<br>Ba<br>137.3 | 57*<br>La<br>138.9 | 72<br>Hf<br>178.5 | 73<br>Ta<br>180.9 | 74<br>W<br>183.9 | 75<br>Re<br>186.2 | 76<br>Os<br>190.2 | 77<br>Ir<br>192.2 | 78<br>Pt<br>195.1 | 79<br>Au<br>197.0 | 80<br>Hg<br>200.6 | 81<br>Tl<br>204.4 | 82<br>Pb<br>207.2 | 83<br>Bi<br>209.0 | 84<br>Po<br>(210) | 85<br>At<br>(210) | 86<br>Rn<br>(222) |
| 7 | 87<br>Fr<br>(223) | 88<br>Ra<br>226.0 | 89**<br>Ac<br>(227) | 104<br>Rf<br>(261) | 105<br>Db<br>(262) | 106<br>Sg<br>(263) | 107<br>Bh<br>(262) | 108<br>Hs<br>(265) | 109<br>Mt<br>(266) | 110<br>Ds<br>(269) | 111<br>Uuu<br>(272) | 112<br>Uub<br>(277) | 113<br>Uub<br>(284) | 114<br>Uuq<br>(285) | 115<br>Uub<br>(288) | 116<br>Uuh<br>(289) | | |

Transition Elements: groups 3–12 (IIIB(3) through IIB(12)).

### Inner Transition Elements

| | | | | | | | | | | | | | | | |
|---|---|---|---|---|---|---|---|---|---|---|---|---|---|---|---|
| * Lanthanide Series 6 | 58<br>Ce<br>140.1 | 59<br>Pr<br>140.9 | 60<br>Nd<br>144.2 | 61<br>Pm<br>(145) | 62<br>Sm<br>150.4 | 63<br>Eu<br>152.0 | 64<br>Gd<br>157.3 | 65<br>Tb<br>158.9 | 66<br>Dy<br>162.5 | 67<br>Ho<br>164.9 | 68<br>Er<br>167.3 | 69<br>Tm<br>168.9 | 70<br>Yb<br>173.0 | 71<br>Lu<br>175.0 |
| ** Actinide Series 7 | 90<br>Th<br>232.0 | 91<br>Pa<br>231.0 | 92<br>U<br>238.0 | 93<br>Np<br>237.0 | 94<br>Pu<br>(244) | 95<br>Am<br>(243) | 96<br>Cm<br>(247) | 97<br>Bk<br>(247) | 98<br>Cf<br>(251) | 99<br>Es<br>(252) | 100<br>Fm<br>(257) | 101<br>Md<br>(258) | 102<br>No<br>(259) | 103<br>Lr<br>(260) |

### The Elements and Their Symbols

| | | | | | | | | | | | |
|---|---|---|---|---|---|---|---|---|---|---|---|
| Actinium | Ac | Cesium | Cs | Hafnium | Hf | Mercury | Hg | Protactinium | Pa | Tellurium | Te |
| Aluminum | Al | Chlorine | Cl | Hassium | Hs | Molybdenum | Mo | Radium | Ra | Terbium | Tb |
| Americium | Am | Chromium | Cr | Helium | He | Neodymium | Nd | Radon | Rn | Thallium | Tl |
| Antimony | Sb | Cobalt | Co | Holmium | Ho | Neon | Ne | Rhenium | Re | Thorium | Th |
| Argon | Ar | Copper | Cu | Hydrogen | H | Neptunium | Np | Rhodium | Rh | Thulium | Tm |
| Arsenic | As | Curium | Cm | Indium | In | Nickel | Ni | Rubidium | Rb | Tin | Sn |
| Astatine | At | Darmstadtium | Ds | Iodine | I | Niobium | Nb | Ruthenium | Ru | Titanium | Ti |
| Barium | Ba | Dubnium | Db | Iridium | Ir | Nitrogen | N | Rutherfordium | Rf | Tungsten | W |
| Berkelium | Bk | Dysprosium | Dy | Iron | Fe | Nobelium | No | Samarium | Sm | Uranium | U |
| Beryllium | Be | Einsteinium | Es | Krypton | Kr | Osmium | Os | Scandium | Sc | Vanadium | V |
| Bismuth | Bi | Erbium | Er | Lanthanum | La | Oxygen | O | Seaborgium | Sg | Xenon | Xe |
| Bohrium | Bh | Europium | Eu | Lawrencium | Lr | Palladium | Pd | Selenium | Se | Ytterbium | Yb |
| Boron | B | Fermium | Fm | Lead | Pb | Phosphorous | P | Silicon | Si | Yttrium | Y |
| Bromine | Br | Fluorine | F | Lithium | Li | Platinum | Pt | Silver | Ag | Zinc | Zn |
| Cadmium | Cd | Francium | Fr | Lutetium | Lu | Plutonium | Pu | Sodium | Na | Zirconium | Zr |
| Calcium | Ca | Gadolinium | Gd | Magnesium | Mg | Polonium | Po | Strontium | Sr | | |
| Californium | Cf | Gallium | Ga | Manganese | Mn | Potassium | K | Sulfur | S | | |
| Carbon | C | Germanium | Ge | Meitnerium | Mt | Praseodymium | Pr | Tantalum | Ta | | |
| Cerium | Ce | Gold | Au | Mendelevium | Md | Promethium | Pm | Technetium | Tc | | |

# Appendix B
# Observing the Sky

Observing the sky with the naked eye is as important to modern astronomy as picking up pretty pebbles is to modern geology. Nevertheless, the sky is inspiring—a natural wonder unimaginably bigger than the Grand Canyon, the Rocky Mountains, or any other site that tourists visit every year. To neglect the beauty of the sky is equivalent to geologists neglecting the beauty of the minerals they study. This supplement is meant to act as a tourist's guide to the sky. You analyzed the universe in the textbook's chapters; here you can admire it.

The brighter stars in the sky are visible even from the centers of cities with their air and light pollution. But in the countryside, only a few miles beyond the cities, the night sky is a velvety blackness strewn with thousands of glittering stars. From a wilderness location, far from the city's glare, and especially from high mountains, the night sky is spectacular.

## Using Star Charts

The constellations are a fascinating cultural heritage of our planet, but they are sometimes a bit difficult to learn because of Earth's motion. The constellations above the horizon change with the time of night and the seasons.

Because Earth rotates eastward, the sky appears to rotate westward around Earth. A constellation visible overhead soon after sunset will appear to move westward, and in a few hours it will disappear below the horizon. Other constellations will rise in the east, so the sky changes gradually through the night.

In addition, Earth's orbital motion makes the sun appear to move eastward among the stars. Each day the sun moves about twice its own diameter, or about one degree, eastward along the ecliptic; consequently, each night at sunset, the constellations are shifted about one degree farther toward the west.

Orion, for instance, is visible in the evening sky in January; but, as the days pass, the sun moves closer to Orion. By March, Orion is difficult to see in the western sky soon after sunset. By June, the sun is so close to Orion it sets with the sun and is invisible. Not until late July is the sun far enough past Orion for the constellation to become visible rising in the eastern sky just before dawn.

Because of the rotation and orbital motion of Earth, you need more than one star chart to map the sky. Which chart you select depends on the month and the time of night.

Two sets of charts are included for two typical locations on Earth. The Northern Hemisphere star charts show the sky as seen from a northern latitude typical for the United States and Central Europe. The Southern Hemisphere star charts are appropriate for readers in Earth's Southern Hemisphere, including Australia, southern South America, and southern Africa.

To use the charts, select the appropriate chart and hold it overhead as shown in ■ Figure B-1. If you face south, turn the chart until the words *Southern Horizon* are at the bottom of the chart. If you face other directions, turn the chart appropriately.

**■ Figure B-1**

To use the star charts in this book, select the appropriate chart for the season and time. Hold it overhead and turn it until the direction at the bottom of the chart is the same as the direction you are facing.

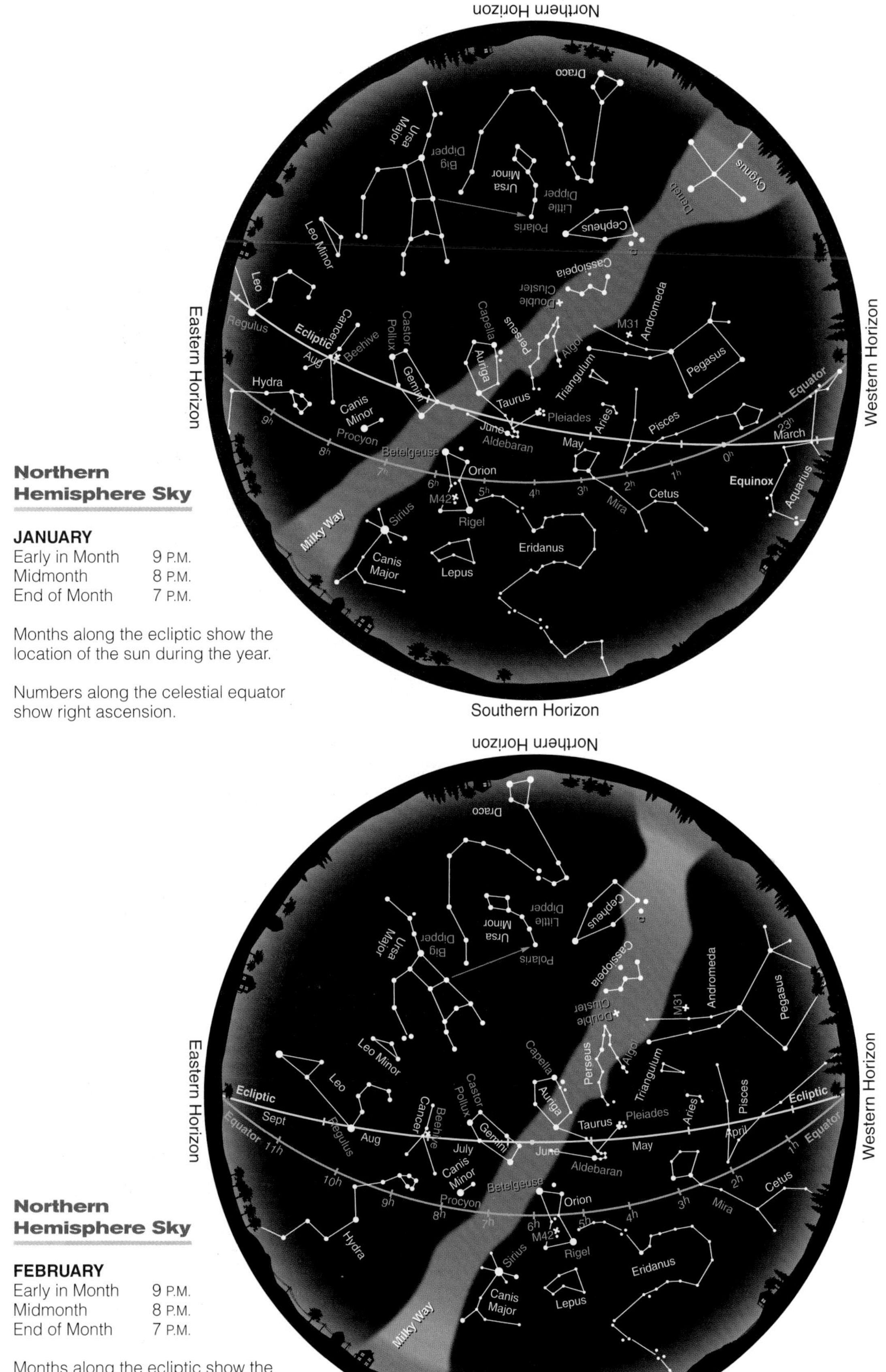


**Northern Hemisphere Sky**

**JANUARY**

| | |
|---|---|
| Early in Month | 9 P.M. |
| Midmonth | 8 P.M. |
| End of Month | 7 P.M. |

Months along the ecliptic show the location of the sun during the year.

Numbers along the celestial equator show right ascension.

**Northern Hemisphere Sky**

**FEBRUARY**

| | |
|---|---|
| Early in Month | 9 P.M. |
| Midmonth | 8 P.M. |
| End of Month | 7 P.M. |

Months along the ecliptic show the location of the sun during the year.

Numbers along the celestial equator show right ascension.

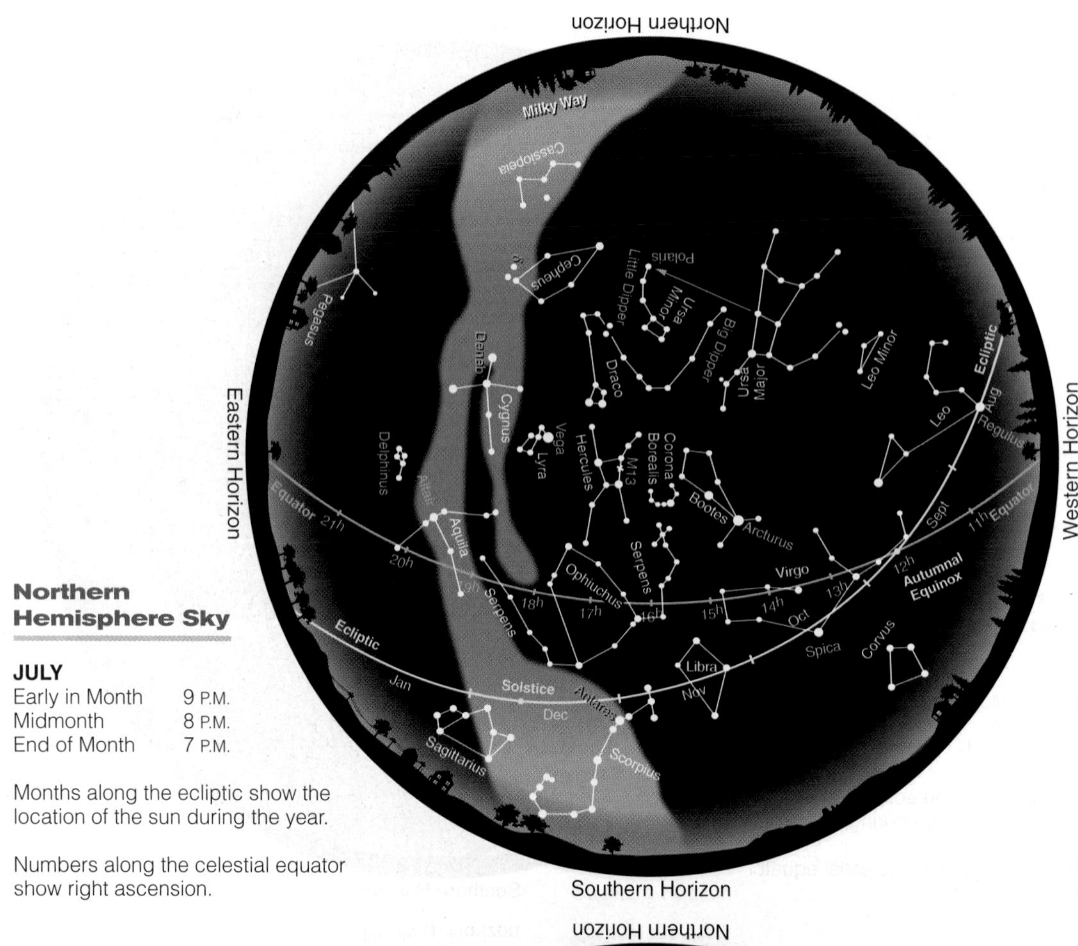


## Northern Hemisphere Sky

**JULY**

| | |
|---|---|
| Early in Month | 9 P.M. |
| Midmonth | 8 P.M. |
| End of Month | 7 P.M. |

Months along the ecliptic show the location of the sun during the year.

Numbers along the celestial equator show right ascension.

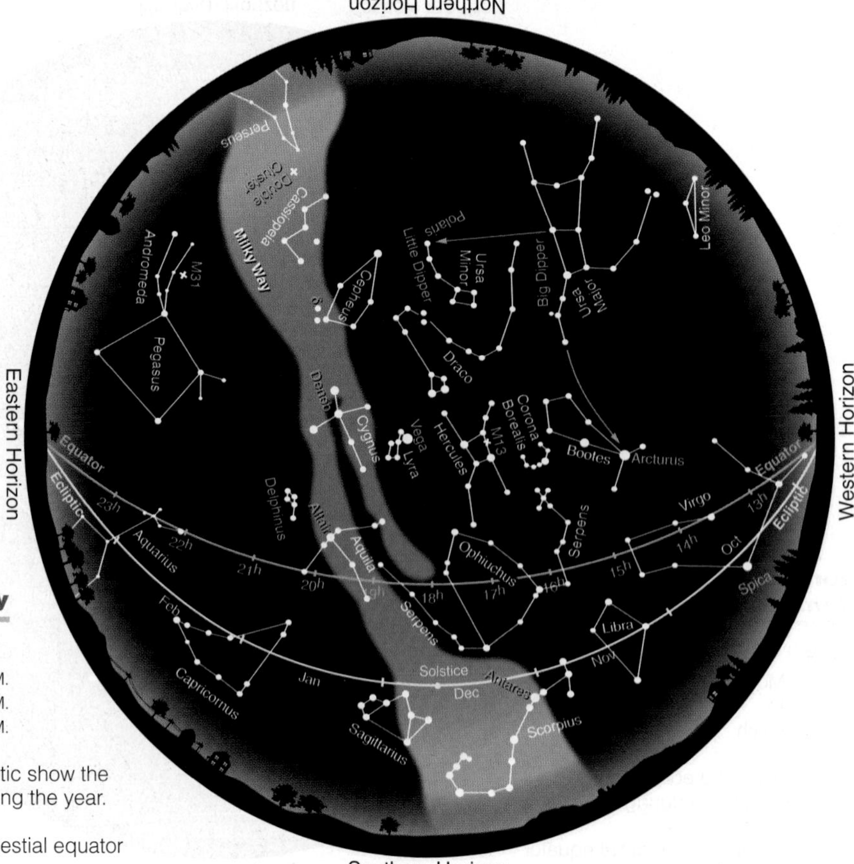


## Northern Hemisphere Sky

**AUGUST**

| | |
|---|---|
| Early in Month | 9 P.M. |
| Midmonth | 8 P.M. |
| End of Month | 7 P.M. |

Months along the ecliptic show the location of the sun during the year.

Numbers along the celestial equator show right ascension.

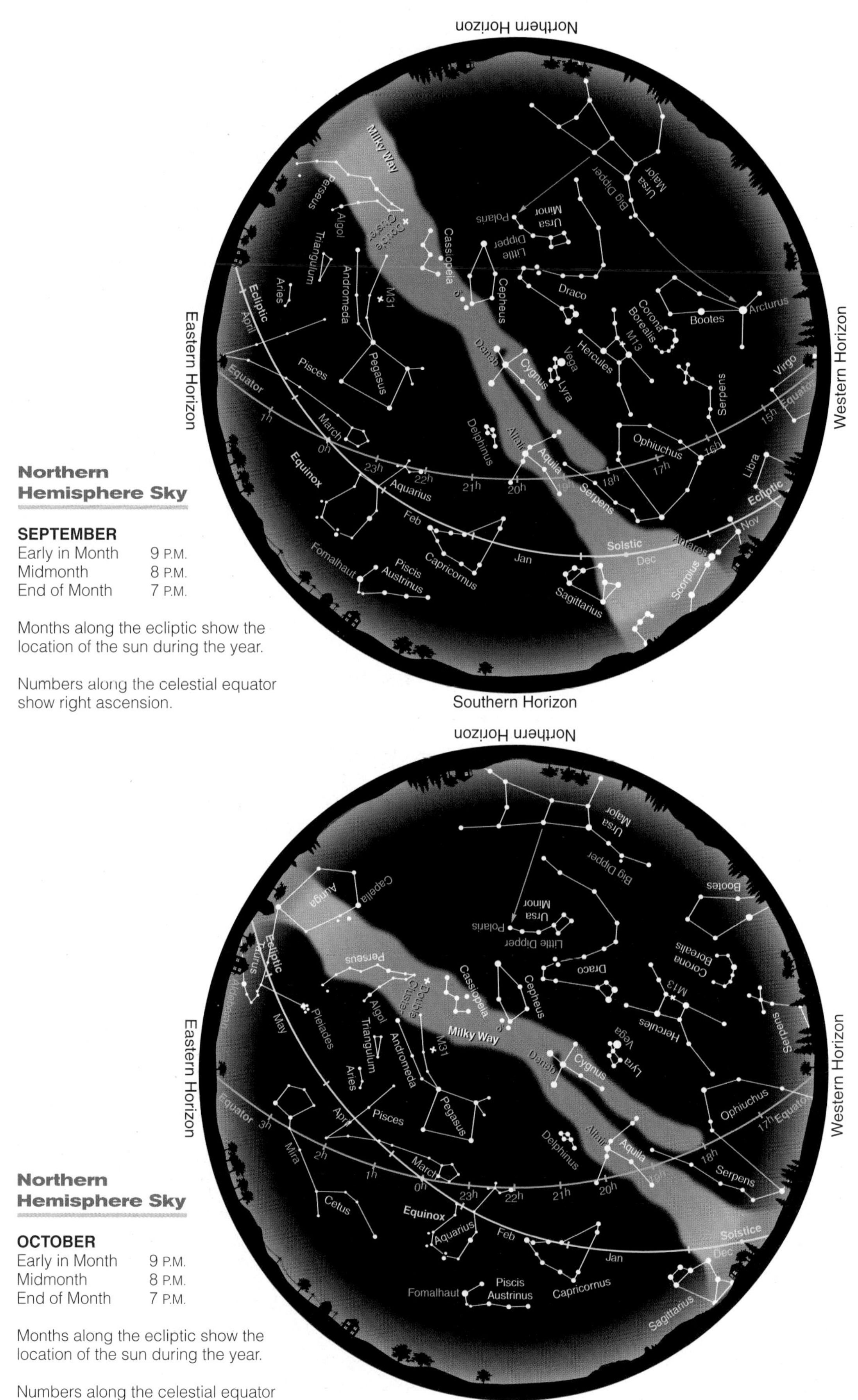


**Northern Hemisphere Sky**

**SEPTEMBER**

| | |
|---|---|
| Early in Month | 9 P.M. |
| Midmonth | 8 P.M. |
| End of Month | 7 P.M. |

Months along the ecliptic show the location of the sun during the year.

Numbers along the celestial equator show right ascension.

**Northern Hemisphere Sky**

**OCTOBER**

| | |
|---|---|
| Early in Month | 9 P.M. |
| Midmonth | 8 P.M. |
| End of Month | 7 P.M. |

Months along the ecliptic show the location of the sun during the year.

Numbers along the celestial equator show right ascension.

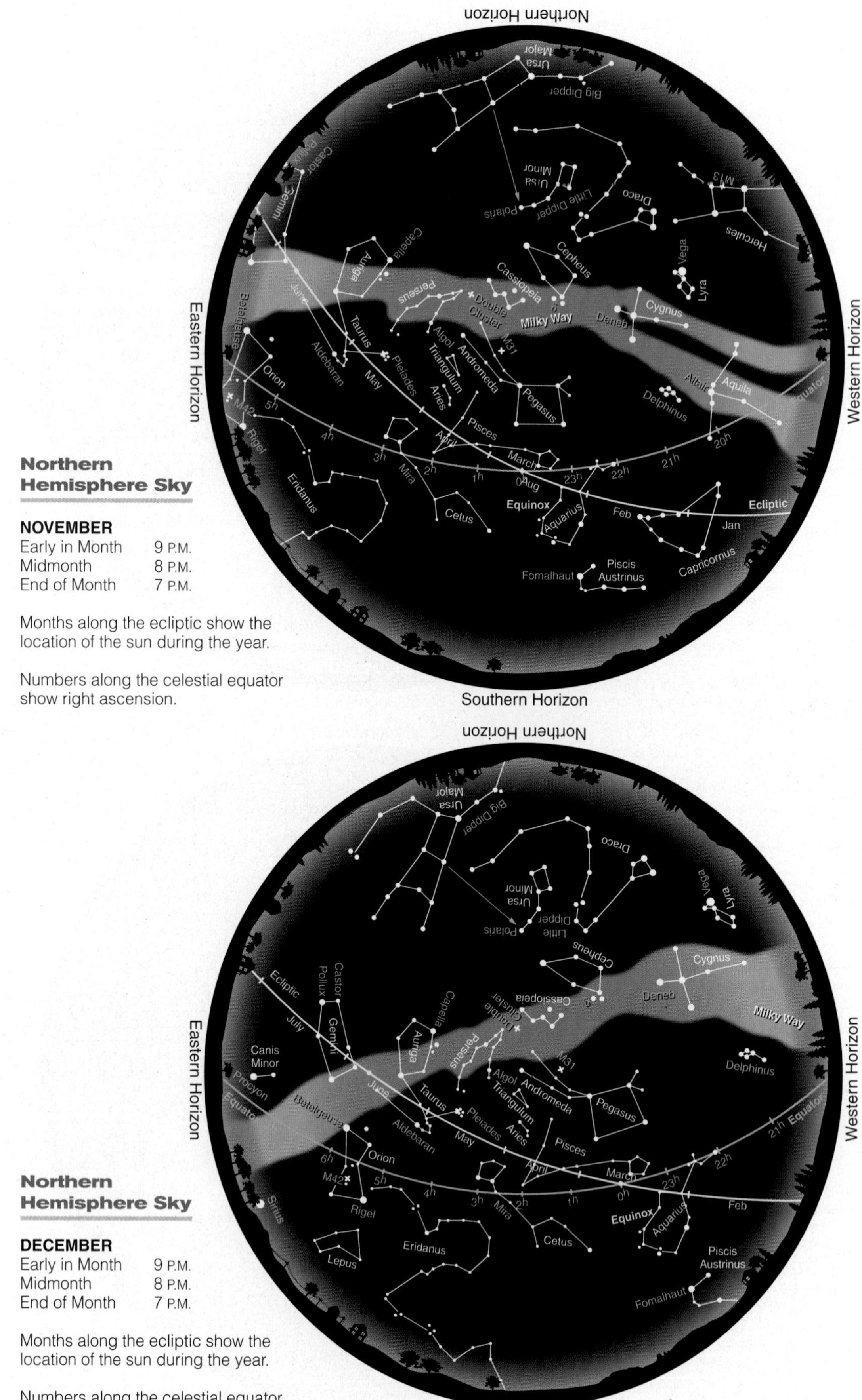


## Northern Hemisphere Sky

**NOVEMBER**

| | |
|---|---|
| Early in Month | 9 P.M. |
| Midmonth | 8 P.M. |
| End of Month | 7 P.M. |

Months along the ecliptic show the location of the sun during the year.

Numbers along the celestial equator show right ascension.

## Northern Hemisphere Sky

**DECEMBER**

| | |
|---|---|
| Early in Month | 9 P.M. |
| Midmonth | 8 P.M. |
| End of Month | 7 P.M. |

Months along the ecliptic show the location of the sun during the year.

Numbers along the celestial equator show right ascension.

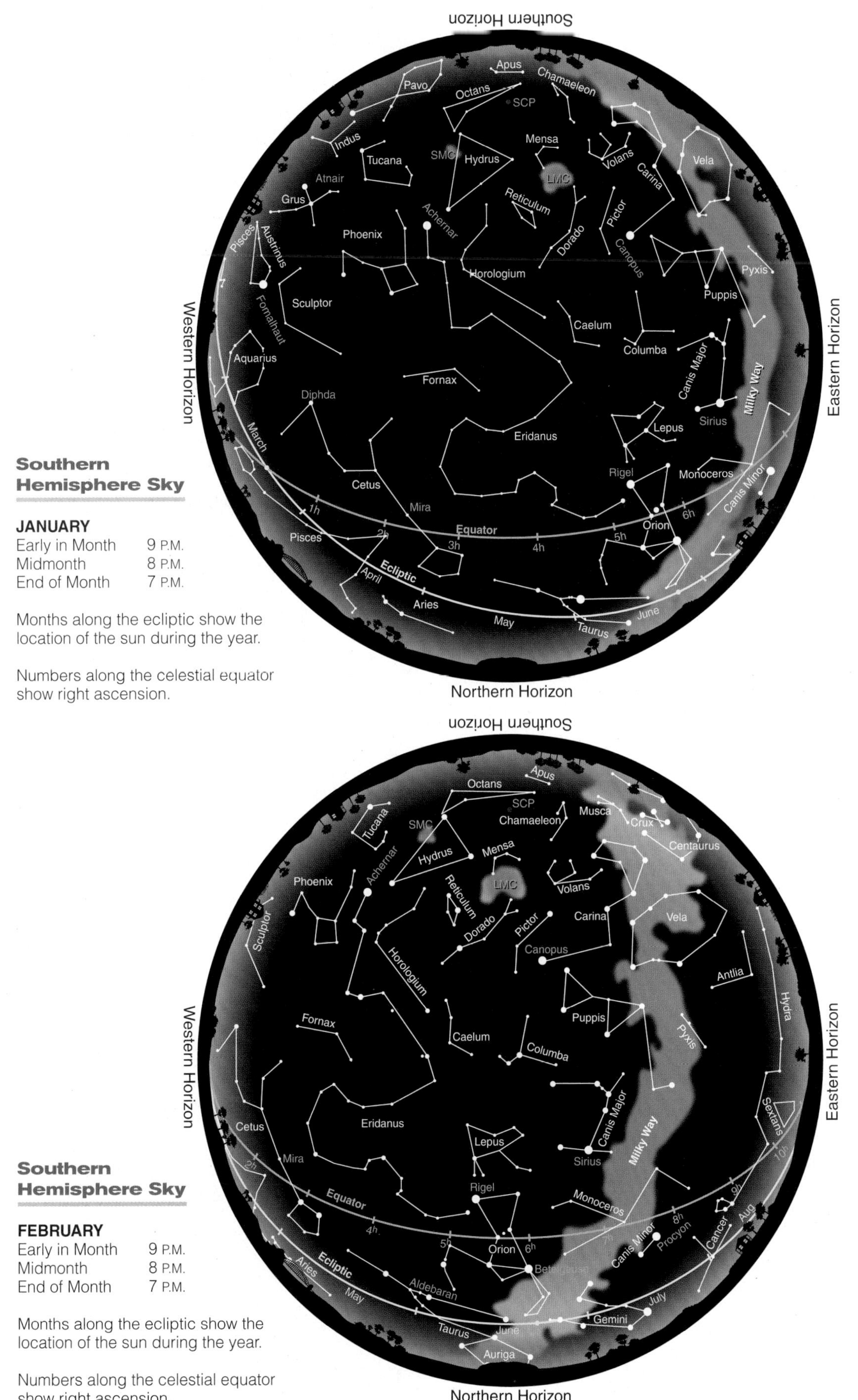


**Southern Hemisphere Sky**

**JANUARY**

| | |
|---|---|
| Early in Month | 9 P.M. |
| Midmonth | 8 P.M. |
| End of Month | 7 P.M. |

Months along the ecliptic show the location of the sun during the year.

Numbers along the celestial equator show right ascension.

**Southern Hemisphere Sky**

**FEBRUARY**

| | |
|---|---|
| Early in Month | 9 P.M. |
| Midmonth | 8 P.M. |
| End of Month | 7 P.M. |

Months along the ecliptic show the location of the sun during the year.

Numbers along the celestial equator show right ascension.

## Southern Hemisphere Sky

**MARCH**

Early in Month 9 P.M.
Midmonth 8 P.M.
End of Month 7 P.M.

Months along the ecliptic show the location of the sun during the year.

Numbers along the celestial equator show right ascension.

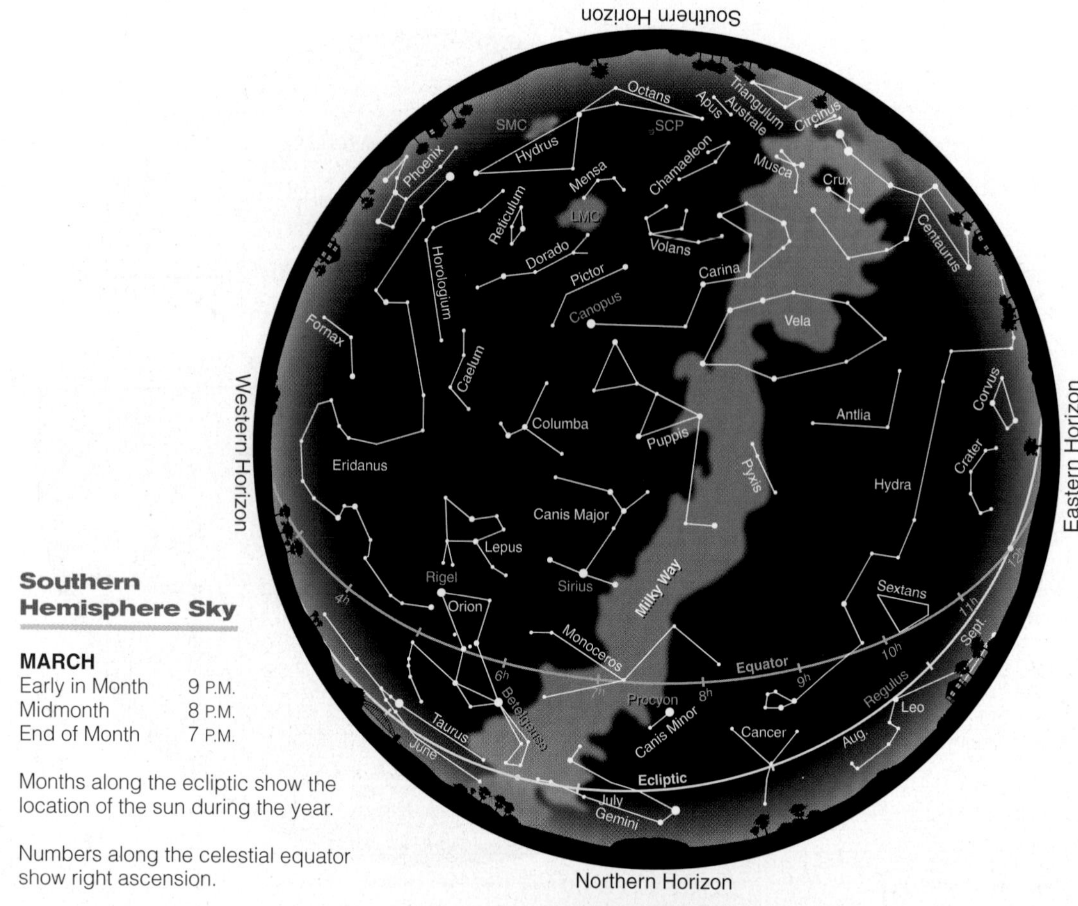


## Southern Hemisphere Sky

**APRIL**

Early in Month 9 P.M.
Midmonth 8 P.M.
End of Month 7 P.M.

Months along the ecliptic show the location of the sun during the year.

Numbers along the celestial equator show right ascension.

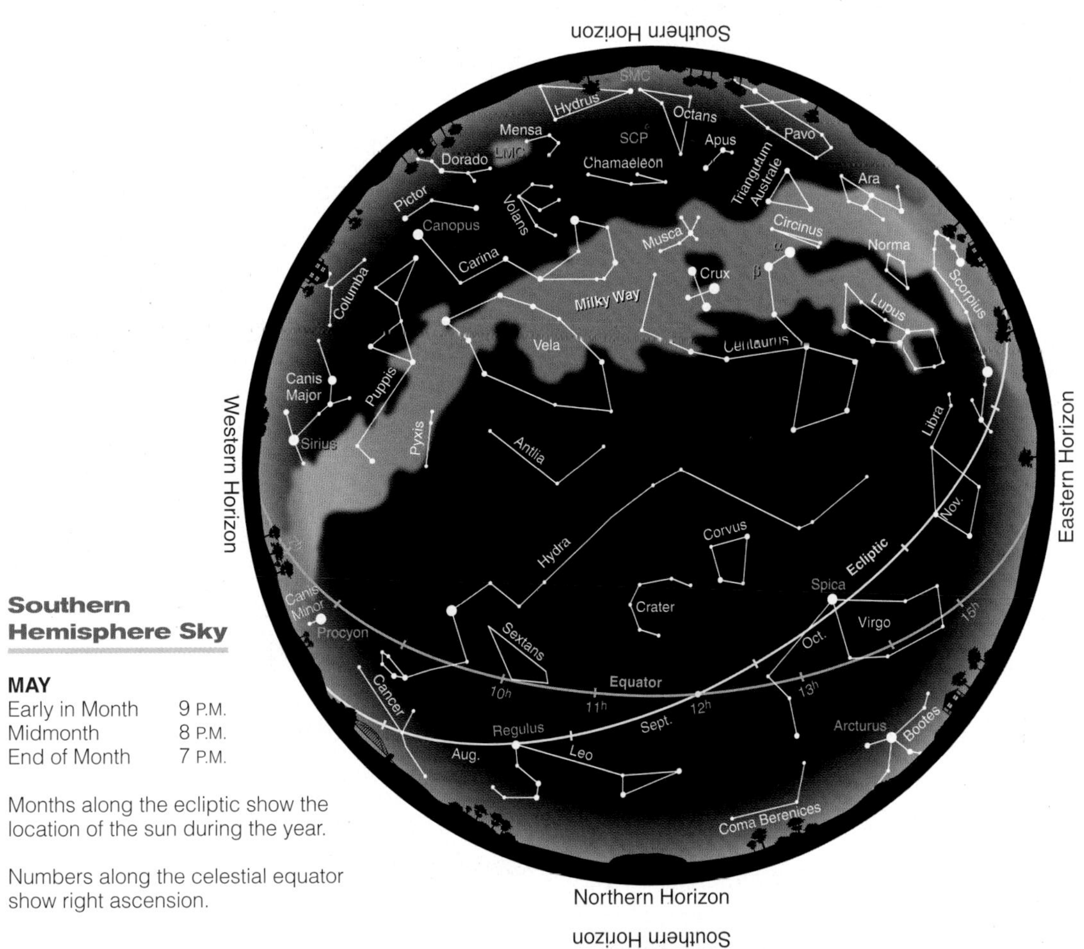


**Southern Hemisphere Sky**

**MAY**

| | |
|---|---|
| Early in Month | 9 P.M. |
| Midmonth | 8 P.M. |
| End of Month | 7 P.M. |

Months along the ecliptic show the location of the sun during the year.

Numbers along the celestial equator show right ascension.

**Southern Hemisphere Sky**

**JUNE**

| | |
|---|---|
| Early in Month | 9 P.M. |
| Midmonth | 8 P.M. |
| End of Month | 7 P.M. |

Months along the ecliptic show the location of the sun during the year.

Numbers along the celestial equator show right ascension.

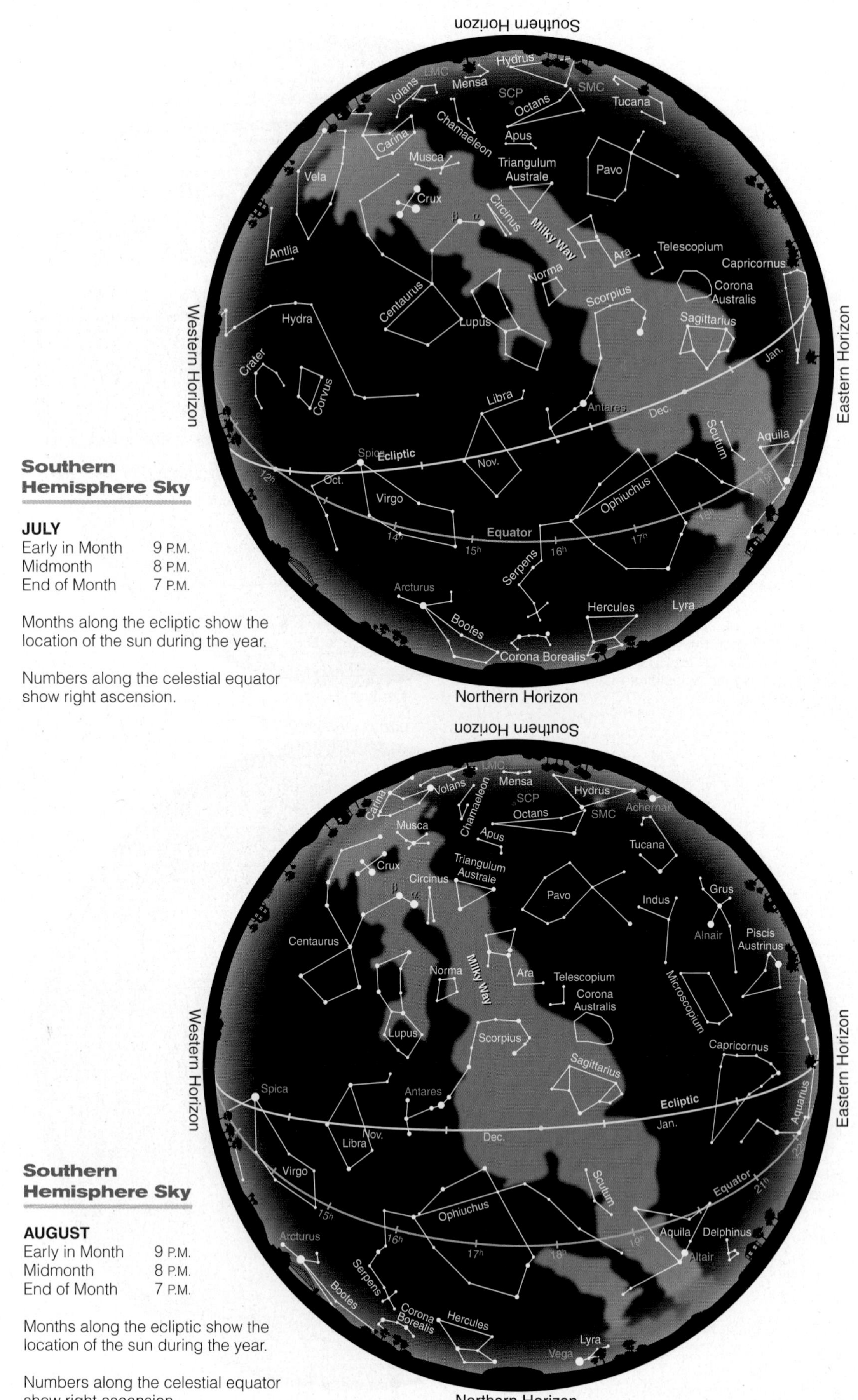


## Southern Hemisphere Sky

**JULY**

| | |
|---|---|
| Early in Month | 9 P.M. |
| Midmonth | 8 P.M. |
| End of Month | 7 P.M. |

Months along the ecliptic show the location of the sun during the year.

Numbers along the celestial equator show right ascension.

## Southern Hemisphere Sky

**AUGUST**

| | |
|---|---|
| Early in Month | 9 P.M. |
| Midmonth | 8 P.M. |
| End of Month | 7 P.M. |

Months along the ecliptic show the location of the sun during the year.

Numbers along the celestial equator show right ascension.

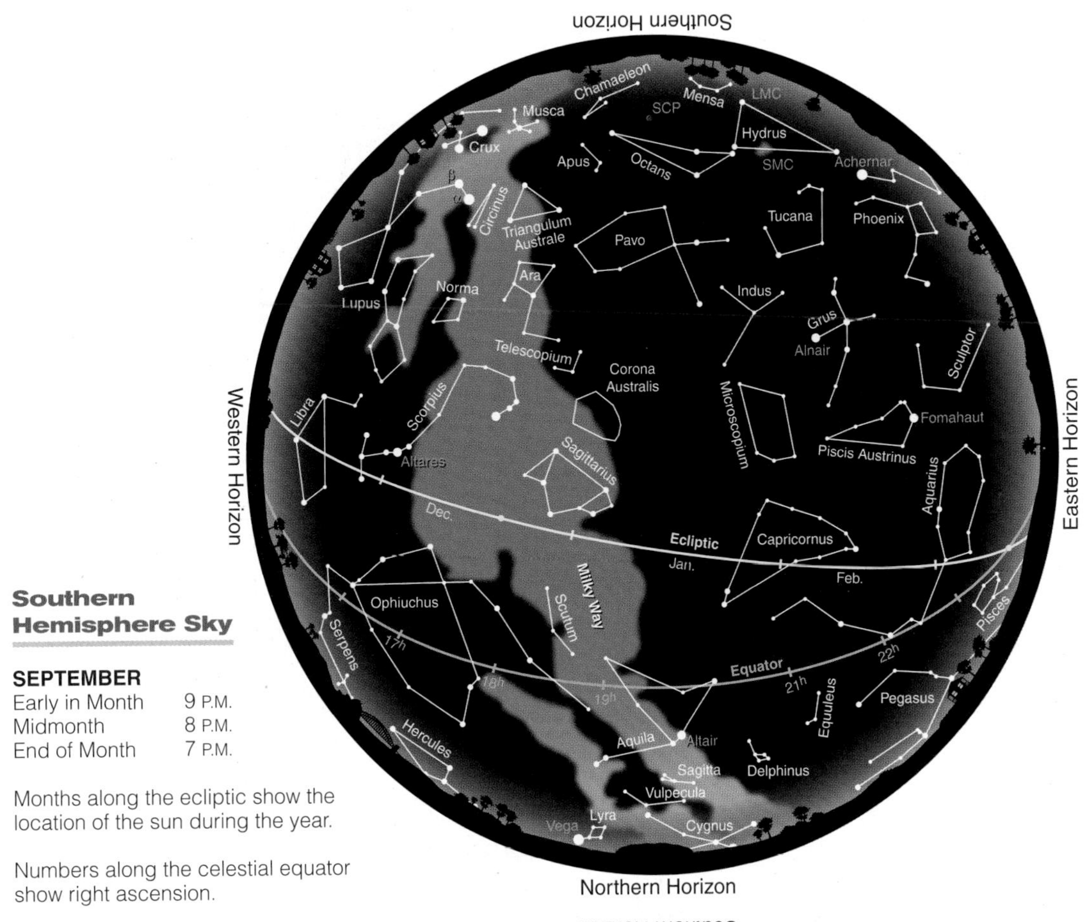


## Southern Hemisphere Sky

**SEPTEMBER**

| | |
|---|---|
| Early in Month | 9 P.M. |
| Midmonth | 8 P.M. |
| End of Month | 7 P.M. |

Months along the ecliptic show the location of the sun during the year.

Numbers along the celestial equator show right ascension.

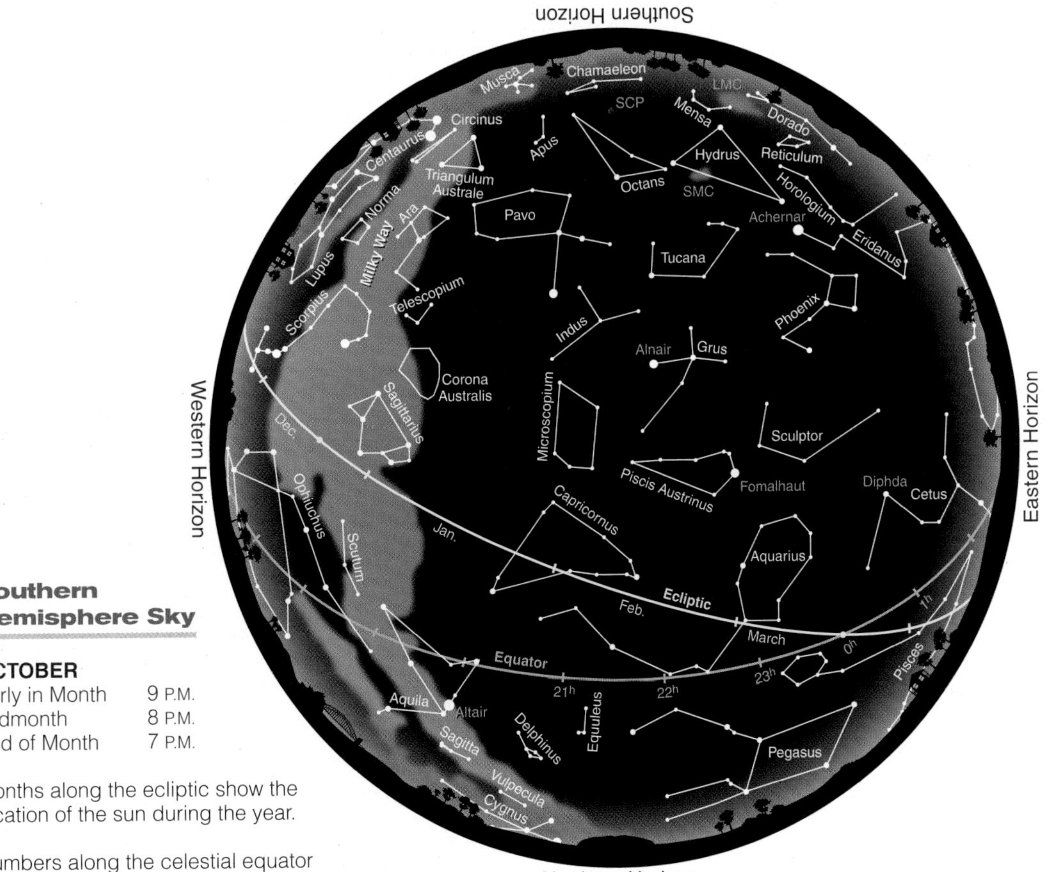


## Southern Hemisphere Sky

**OCTOBER**

| | |
|---|---|
| Early in Month | 9 P.M. |
| Midmonth | 8 P.M. |
| End of Month | 7 P.M. |

Months along the ecliptic show the location of the sun during the year.

Numbers along the celestial equator show right ascension.

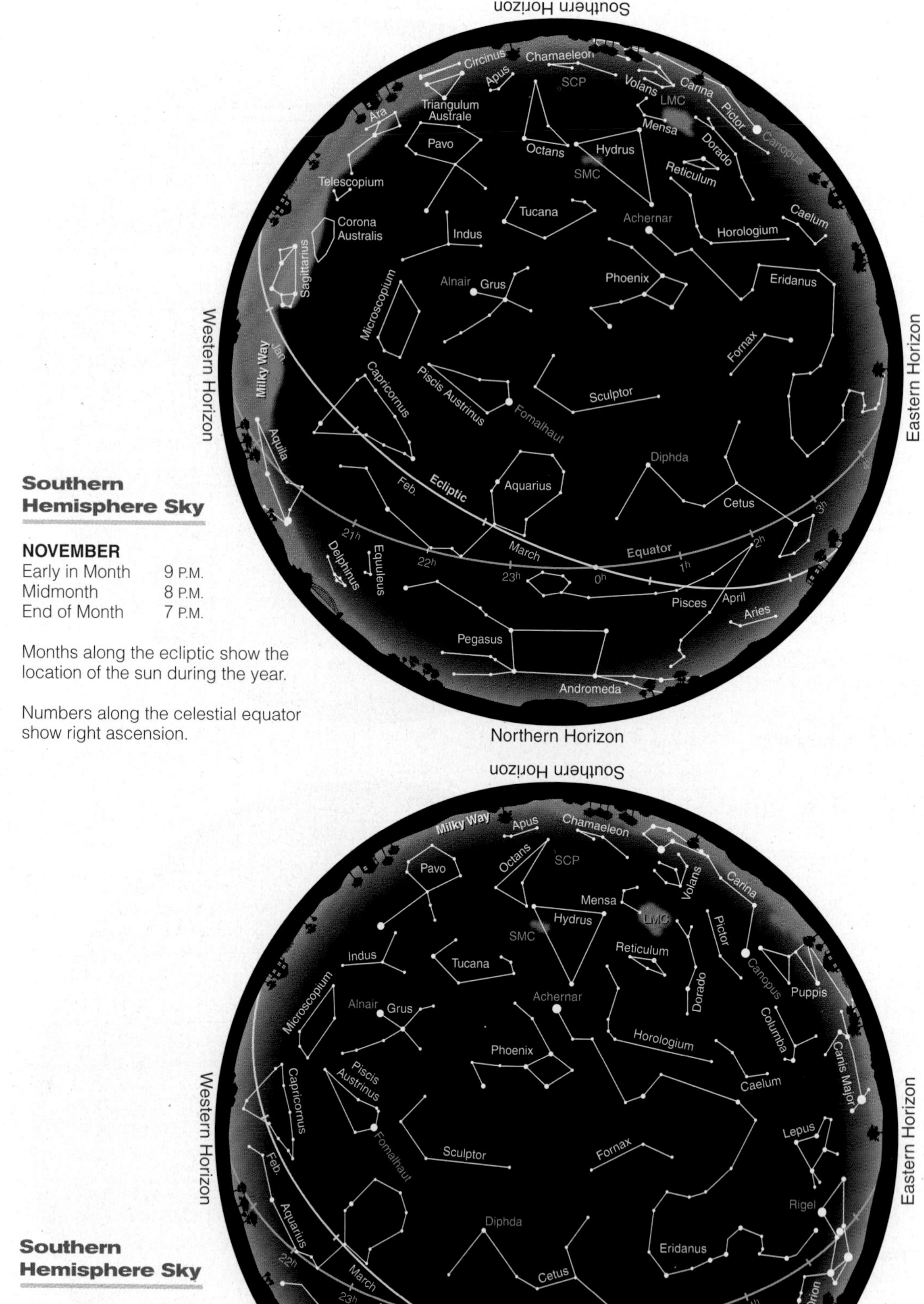


## Southern Hemisphere Sky

**NOVEMBER**

| | |
|---|---|
| Early in Month | 9 P.M. |
| Midmonth | 8 P.M. |
| End of Month | 7 P.M. |

Months along the ecliptic show the location of the sun during the year.

Numbers along the celestial equator show right ascension.

## Southern Hemisphere Sky

**DECEMBER**

| | |
|---|---|
| Early in Month | 9 P.M. |
| Midmonth | 8 P.M. |
| End of Month | 7 P.M. |

Months along the ecliptic show the location of the sun during the year.

Numbers along the celestial equator show right ascension.

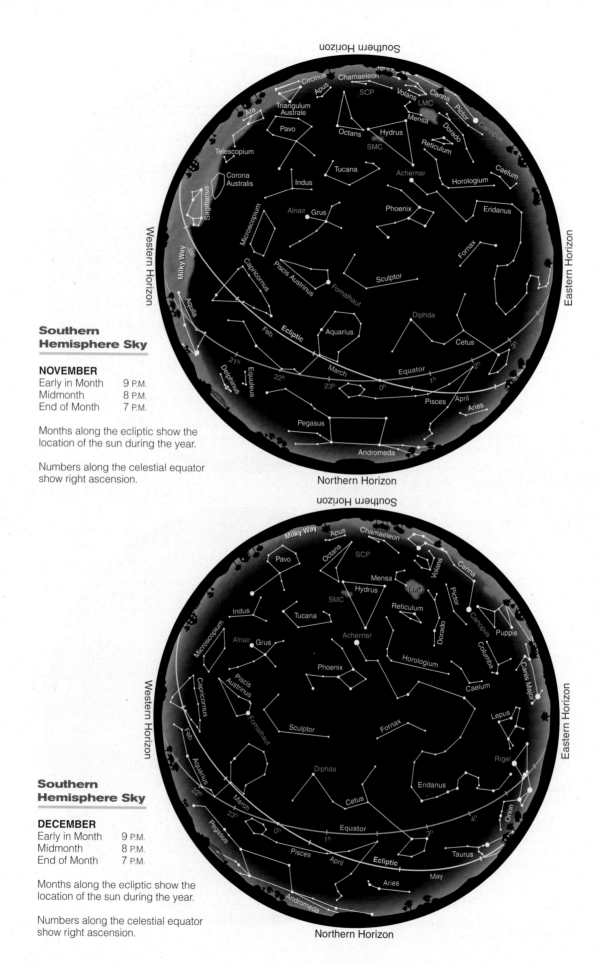

# Glossary

Numbers in parentheses refer to the page where the term is first discussed in the text.

**absolute age** (445) An age determined in years, as from radioactive dating (see also *relative age*).

**absolute zero** (127) The lowest possible temperature; the temperature at which the particles in a material, atoms or molecules, contain no energy of motion that can be extracted from the body.

**absorption line** (132) A dark line in a spectrum; produced by the absence of photons absorbed by atoms or molecules.

**absorption (dark-line) spectrum** (132) A spectrum that contains absorption lines.

**acceleration** (79) A change in a velocity; a change in either speed or direction. (See *velocity*.)

**acceleration of gravity** (77) A measure of the strength of gravity at a planet's surface.

**accretion** (409) The sticking together of solid particles to produce a larger particle.

**achondrite** (553) Stony meteorite containing no chondrules or volatiles.

**achromatic lens** (102) A telescope lens composed of two lenses ground from different kinds of glass and designed to bring two selected colors to the same focus and correct for chromatic aberration.

**active optics** (111) Optical elements whose position or shape is continuously controlled by computers.

**active region** (155) An area on the sun where sunspots, prominences, flares, and the like occur.

**adaptive optics** (109) Computer-controlled telescope mirrors that can at least partially compensate for seeing.

**albedo** (437) The fraction of the light hitting an object that is reflected.

**alt-azimuth mounting** (111) A telescope mounting capable of motion parallel to and perpendicular to the horizon.

**Amazonian period** (488) On Mars, the geological era from about 3 billion years ago to the present marked by low-level cratering, wind erosion, and small amounts of water seeping from subsurface ice.

**amino acid** (584) One of the carbon-chain molecules that are the building blocks of protein.

**Angstrom (Å)** (100) A unit of distance; 1 Å = $10^{-10}$ m; often used to measure the wavelength of light.

**angular diameter** (19) A measure of the size of an object in the sky; numerically equal to the angle in degrees between two lines extending from the observer's eye to opposite edges of the object.

**angular distance** (19) A measure of the separation between two objects in the sky; numerically equal to the angle in degrees between two lines extending from the observer's eye to the two objects.

**angular momentum** (83) The tendency of a rotating body to continue rotating; mathematically, the product of mass, velocity, and radius.

**angular momentum problem** (399) An objection to Laplace's nebular hypothesis that cited the slow rotation of the sun.

**annular eclipse** (39) A solar eclipse in which the solar photosphere appears around the edge of the moon in a bright ring, or annulus. The corona, chromosphere, and prominences cannot be seen.

**anorthosite** (451) Rock of aluminum and calcium silicates found in the lunar highlands.

**anticoded** (595) Message designed to be understood by a recipient about whom the sender knows little or nothing, for example an interstellar broadcast aimed at possible inhabitants of another planet.

**aphelion** (23) The orbital point of greatest distance from the sun.

**apogee** (39) The orbital point of greatest distance from Earth.

**Apollo–Amor object** (562) Asteroid whose orbit crosses that of Earth (Apollo) and Mars (Amor).

**apparent visual magnitude ($m_v$)** (15) The brightness of a star as seen by human eyes on Earth.

**archaea** (591) A biological kingdom of microorganisms similar to, but distinct from, bacteria, with characteristics most closely resembling those inferred for the common ancestor of all present-day Earth life.

**arc minute** (19) An angular measure; each degree is divided into 60 arc minutes.

**arc second** (19) An angular measure; each arc minute is divided into 60 arc seconds.

**archaeoastronomy** (50) The study of the astronomy of ancient cultures.

**asterism** (13) A named group of stars not identified as a constellation, e.g., the Big Dipper.

**asteroid** (401) Small rocky world; most asteroids lie between Mars and Jupiter in the asteroid belt.

**astrobiology** (582) The field of study involving searches for life on other worlds and investigation of possible habitats for such life. Also known as "exobiology."

**astronomical unit (AU)** (4) Average distance from Earth to the sun; $1.5 \times 10^8$ km, or $93 \times 10^6$ miles.

**atmospheric window** (101) Wavelength region in which Earth's atmosphere is transparent—at visual, infrared, and radio wavelengths.

**aurora** (163) The glowing light display that results when a planet's magnetic field guides charged particles toward the north and south magnetic poles, where they strike the upper atmosphere and excite atoms to emit photons.

**autumnal equinox** (22) The point on the celestial sphere where the sun crosses the celestial equator going southward. Also, the time when the sun reaches this point and autumn begins in the Northern Hemisphere—about September 22.

**Babcock model** (158) A model of the sun's magnetic cycle in which the differential rotation of the sun winds up and tangles the solar magnetic field in a 22-year cycle. This is thought to be responsible for the 11-year sunspot cycle.

**Balmer series** (133) Spectral lines in the visible and near-ultraviolet spectrum of hydrogen produced by transitions whose lowest orbit is the second.

**barred spiral galaxy** (178, 334) A spiral galaxy with an elongated nucleus resembling a bar from which the arms originate.

**basalt** (434) Dark, igneous rock characteristic of solidified lava.

**belt–zone circulation** (495) The atmospheric circulation typical of Jovian planets. Dark belts and bright zones encircle the planet parallel to its equator.

**big bang** (181, 373) The theory that the universe began with a violent explosion from which the expanding universe of galaxies eventually formed.

**binding energy** (124) The energy needed to pull an electron away from its atom.

**biological evolution** (586) The combined effect of variation and natural selection resulting in new species arising and existing species adapting to the environment or becoming extinct. Also called Darwinian evolution.

**blackbody radiation** (127) Radiation emitted by a hypothetical perfect radiator; the spectrum is continuous, and the wavelength of maximum emission depends only on the body's temperature.

**black holes** (173) A mass that has collapsed to such a small volume that its gravity prevents the escape of all radiation; also, the volume of space from which radiation may not escape.

**blueshift** (138) The shortening of the wavelengths of light observed when the source and observer are approaching each other.

**Bok globules** (170) Small, dark, and dense interstellar cloud only about 1 ly in diameter that contains 10 to 100 $M_\odot$ of gas and dust; thought to be related to star formation.

**bow shock** (431) The boundary between the undisturbed solar wind and the region being deflected around a planet or comet.

**breccia** (451) A rock composed of fragments of earlier rocks bonded together.

**bright-line spectrum** (132) See *emission spectrum.*

**butterfly diagram** (155) See *Maunder butterfly diagram.*

**CAI** (555) Calcium–aluminum-rich inclusions found in some meteorites.

**Cambrian explosion** (589) The sudden appearance of complex life forms at the beginning of the Cambrian period 0.6 to 0.5 billion years ago. Cambrian rocks contain the oldest easily identifiable fossils.

**carbonaceous chondrite** (552) Stony meteorite that contains both chondrules and volatiles. These may be the least altered remains of the solar nebula still present in the solar system.

**Cassegrain focus** (110) The optical design of a reflecting telescope in which the secondary mirror reflects light back down the tube through a hole in the center of the objective mirror.

**catastrophic hypothesis** (398) Explanation for natural processes that depends on dramatic and unlikely events, such as the collision of two stars to produce our solar system.

**celestial equator** (17) The imaginary line around the sky directly above Earth's equator.

**celestial sphere** (18) An imaginary sphere of very large radius surrounding Earth to which the planets, stars, sun, and moon seem to be attached.

**centaur** (563) An outer solar system body with an orbit entirely within the region of the Jovian planets, for example Chiron, that orbits between Saturn and Uranus.

**center of mass** (85) The balance point of a body or system of bodies.

**charge-coupled device (CCD)** (113) An electronic device consisting of a large array of light-sensitive elements used to record very faint images.

**chemical evolution** (589) The chemical process that led to the growth of complex molecules on the primitive Earth. This did not involve the reproduction of molecules.

**Chicxulub** (576) The buried crater associated with the mass extinction event at the end of the age of dinosaurs, named after the town in the coastal region of Mexico's Yucatan peninsula near the center of the crater.

**chondrite** (553) A stony meteorite that contains chondrules.

**chondrule** (553) Round, glassy body in some stony meteorites; thought to have solidified very quickly from molten drops of silicate material.

**chromatic aberration** (102) A distortion found in refracting telescopes because lenses focus different colors at slightly different distances. Images are consequently surrounded by color fringes.

**chromosome** (585) One of the bodies in a cell that contains the DNA carrying genetic information.

**chromosphere** (41) Bright gases just above the photosphere of the sun.

**circular velocity** (84) The velocity required to remain in a circular orbit about a body.

**circumpolar constellation** (19) Any of the constellations so close to the celestial pole that they never set (or never rise) as seen from a given latitude.

**closed orbit** (85) An orbit that returns to its starting point; a circular or elliptical orbit. (See *open orbit.*)

**coma** (568) The glowing head of a comet.

**comet** (404) One of the small, icy bodies that orbit the sun and produce tails of gas and dust when they near the sun.

**comparative planetology** (425) The study of planets by comparing the characteristics of different examples.

**comparison spectrum** (113) A spectrum of known spectral lines used to identify unknown wavelengths in an object's spectrum.

**composite volcano** (472) A volcano built up of layers of lava flows and ash falls. These are steep sided and typically associated with subduction zones.

**condensation** (409) The growth of a particle by addition of material from surrounding gas, one atom or molecule at a time.

**condensation sequence** (408) The sequence in which different materials condense from the solar nebula at increasing distances from the sun.

**constellation** (12) One of the stellar patterns identified by name, usually of mythological gods, people, animals, or objects; also, the region of the sky containing that star pattern.

**continuous spectrum** (132) A spectrum in which there are no absorption or emission lines.

**convection** (145) Circulation in a fluid driven by heat; hot material rises, and cool material sinks.

**convective zone** (152) The region inside a star where energy is carried outward as rising hot gas and sinking cool gas.

**corona** (41) The faint outer atmosphere of the sun; composed of low-density, very hot, ionized gas. On Venus, round networks of fractures and ridges up to 1000 km in diameter.

**coronae** (471) On Venus, large, round geological faults in the crust caused by the intrusion of magma below the crust.

**coronagraph** (146) A telescope designed to photograph the inner corona of the sun.

**coronal hole** (163) An area of the solar surface that is dark at X-ray wavelengths; thought to be associated with divergent magnetic fields and the source of the solar wind.

**coronal mass ejection (CME)** (163) Gas trapped in the sun's magnetic field.

**cosmic microwave background radiation** (182, 375) Radiation from the hot clouds of the big bang explosion. Because of its large redshift, it appears to come from a body whose temperature is only 2.7 K.

**cosmic ray** (119) A subatomic particle traveling at tremendous velocity that strikes Earth's atmosphere from space.

**Coulomb barrier** (152) The electrostatic force of repulsion between bodies of like charge; commonly applied to atomic nuclei.

**Coulomb force** (124) The repulsive force between particles with like electrostatic charge.

**critical point** (496) The temperature and pressure at which the vapor and liquid phases of a material have the same density.

**C-type asteroid** (564) A type of asteroid common in the outer asteroid belt, with very low reflectivity and grayish color, probably composed of carbonaceous material.

**dark-line spectrum** (132) See *absorption spectrum.*

**dark nebula** (195) A nonluminous cloud of gas and dust visible because it blocks light from more distant stars and nebulae.

**debris disk** (415) A disk of dust found by infrared observations around some stars. The dust is debris from collisions among asteroids, comets, and Kuiper belt objects.

**deferent** (57) In the Ptolemaic theory, the large circle around Earth along which the center of the epicycle moved.

**density** (137) The amount of matter per unit volume in a material; measured in grams per cubic centimeter, for example.

**deuterium** (151) An isotope of hydrogen in which the nucleus contains a proton and a neutron.

**diamond ring effect** (41) A momentary phenomenon seen during some total solar eclipses when the ring of the corona and a bright spot of photosphere resemble a large diamond set in a silvery ring.

**differential rotation** (158) The rotation of a body in which different parts of the body have different periods of rotation; this is true of the sun, the Jovian planets, and the disk of the galaxy.

**differentiation** (410) The separation of planetary material according to density.

**diffraction fringe** (104) Blurred fringe surrounding any image caused by the wave properties of light. Because of this, no image detail smaller than the fringe can be seen.

**direct collapse** (412) The hypothetical process by which a Jovian planet might skip the accretion of a solid core, instead forming quickly and directly from the gases of the solar nebula.

**distance modulus ($m_V - M_V$)** (172) The difference between the apparent and absolute magnitude of a star; a measure of how far away the star is.

**DNA (deoxyribonucleic acid)** (584) The long carbon-chain molecule that records information to govern the biological activity of the organism. DNA carries the genetic data passed to offspring.

**Doppler effect** (137) The change in the wavelength of radiation due to relative radial motion of source and observer.

**Drake equation** (596) A formula for the number of communicative civilizations in our galaxy.

**dust (type II) tail** (568) The tail of a comet formed of dust blown outward by the pressure of sunlight. (See *gas tail.*)

**dwarf planet** (546) An object that orbits the sun and has pulled itself into a spherical shape but has not cleared its orbital lane of other objects. Pluto is a dwarf planet.

**dynamo effect** (158) The process by which a rotating, convecting body of conducting matter, such as Earth's core, can generate a magnetic field.

**east point** (18) The point on the eastern horizon exactly halfway between the north point and the south point; exactly east.

**eccentric** (55) (noun) An off-center circular path. (Note that the *adjective* "eccentric" refers instead to an ellipse that is not a perfect circle.)

**eccentricity (e)** (64) A measure of the flattening of an ellipse. An ellipse of $e = 0$ is circular. The closer to 1 e becomes, the more flattened the ellipse.

**eclipse season** (43) That period when the sun is near a node of the moon's orbit and eclipses are possible.

**eclipse year** (44) The time the sun takes to circle the sky and return to a node of the moon's orbit; 346.62 days.

**ecliptic** (21) The apparent path of the sun around the sky.

**ejecta** (446) Pulverized rock scattered by meteorite impacts on a planetary surface.

**electromagnetic radiation** (99) Changing electric and magnetic fields that travel through space and transfer energy from one place to another—for example, light, radio waves, and the like.

**electron** (123) Low-mass atomic particle carrying a negative charge.

**ellipse** (64) A closed curve enclosing two points (foci) such that the total distance from one focus to any point on the curve back to the other focus equals a constant.

**elliptical galaxy** (178, 334) A galaxy that is round or elliptical in outline; it contains little gas and dust, no disk or spiral arms, and few hot, bright stars.

**emission line** (132) A bright line in a spectrum caused by the emission of photons from atoms.

**emission (bright-line) spectrum** (132) A spectrum containing emission lines.

**energy** (86) The capacity of a natural system to perform work—for example, thermal energy.

**energy level** (126) One of a number of states an electron may occupy in an atom, depending on its binding energy.

**enzyme** (584) Special protein that controls processes in an organism.

**epicycle** (57) The small circle followed by a planet in the Ptolemaic theory. The center of the epicycle follows a larger circle (deferent) around Earth.

**equant** (57) The point off-center in the deferent from which the center of the epicycle appears to move uniformly.

**equatorial mounting** (111) A telescope mounting that allows motion parallel to and perpendicular to the celestial equator.

**escape velocity ($V_e$)** (85) The initial velocity an object needs to escape from the surface of a celestial body.

**evening star** (24) Any planet visible in the sky just after sunset.

**evolutionary hypothesis** (398) Explanation for natural events that involves gradual changes as opposed to sudden catastrophic changes—for example, the formation of the planets in the gas cloud around the forming sun.

**excited atom** (126) An atom in which an electron has moved from a lower to a higher orbit.

**extrasolar planet** (417) A planet orbiting a star other than the sun.

**extremophile** (591) An organism that can survive in an extreme environment, for example, very low or high temperatures, high acidity, extreme dryness, etc.

**eyepiece** (102) A short-focal-length lens used to enlarge the image in a telescope; the lens nearest the eye.

**fall** (552) A meteorite seen to fall. (See *find.*)

**false-color image** (113) A representation of graphical data in which the colors are altered or added to reveal details.

**field** (81) A way of explaining action at a distance; a particle produces a field of influence (gravitational, electric, or magnetic) to which another particle in the field responds.

**field of view** (2) The area visible in an image; usually given as the diameter of the region.

**filament** (162) (1) On the sun, a prominence seen silhouetted against the solar surface. (2) A linear region containing many galaxies and galaxy clusters, part of the large-scale structure of the universe.

**filtergram** (146) An image (usually of the sun) taken in the light of a specific region of the spectrum—for example, an H-alpha filtergram.

**find** (552) A meteorite that is found but was not seen to fall. (See *fall.*)

**flare** (163) A violent eruption on the sun's surface.

**flux** (15) A measure of the flow of energy onto or through a surface. Usually applied to light.

**focal length** (102) The distance from a lens to the point where it focuses parallel rays of light.

**folded mountain range** (434) A long range of mountains formed by the compression of a planet's crust—for example, the Andes on Earth.

**forward scattering** (499) The optical property of finely divided particles to preferentially direct light in the original direction of the light's travel.

**frequency** (99) The number of times a given event occurs in a given time; for a wave, the number of cycles that pass the observer in 1 second.

**galaxy** (5) A very large collection of gas, dust, and stars orbiting a common center of mass. The sun and Earth are located in the Milky Way Galaxy.

**Galilean moons, Galilean satellites** (68, 504) The four largest satellites of Jupiter, named after their discoverer, Galileo.

**gas (type I) tail** (568) The tail of a comet produced by gas blown outward by the solar wind. (See *dust tail.*)

**gene** (585) A unit of DNA containing genetic information that influences a particular inherited trait.

**general theory of relativity** (92) Einstein's more sophisticated theory of space and time, which describes gravity as a curvature of space-time.

**geocentric universe** (54) A model universe with Earth at the center, such as the Ptolemaic universe.

**geosynchronous satellite** (84) An Earth satellite in an eastward orbit whose period is 24 hours. A satellite in such an orbit remains above the same spot on Earth's surface.

**giant** (172) Large, cool, highly luminous star in the upper right of the H–R diagram, typically 10 to 100 times the diameter of the sun.

**global warming** (437) The gradual increase in the surface temperature of Earth caused by human modifications to Earth's atmosphere.

**gossamer ring** (502) The dimmest part of Jupiter's ring produced by dust particles orbiting near small moons.

**granulation** (144) The fine structure visible on the solar surface caused by rising currents of hot gas and sinking currents of cool gas below the surface.

**grating** (113) A piece of material in which numerous microscopic parallel lines are scribed; light encountering a grating is dispersed to form a spectrum.

**gravitational collapse** (410) The stage in the formation of a massive planet when it grows massive enough to begin capturing gas directly from the nebula around it.

**greenhouse effect** (437) The process by which a carbon dioxide atmosphere traps heat and raises the temperature of a planetary surface.

**grooved terrain** (505) Region of the surface of Ganymede consisting of bright, parallel grooves.

**ground state** (126) The lowest permitted electron orbit in an atom.

**habitable zone** (594) The region around a star within which an orbiting planet can have surface temperatures allowing liquid water.

**half-life** (405) The time required for half of the atoms in a radioactive sample to decay.

**heat** (127) Energy flowing from a warm body to a cool body by the agitation of particles such as atoms or molecules.

**heat of formation** (410) In planetology, the heat released by the infall of matter during the formation of a planetary body.

**heavy bombardment** (413) The period of intense meteorite impacts early in the formation of the planets, when the solar system was filled with debris.

**heliocentric universe** (58) A model of the universe with the sun at the center, such as the Copernican universe.

**helioseismology** (148) The study of the interior of the sun by the analysis of its modes of vibration.

**Hesperian period** (488) On Mars, the geological era from the decline of heavy cratering and lava flows and the melting of subsurface ice to form the outflow channels.

**Hirayama family** (564) Family of asteroids with orbits of similar size, shape, and orientation; thought to be fragments of larger bodies.

**horizon** (18) The line that marks the apparent intersection of Earth and the sky.

**horoscope** (25) A chart showing the positions of the sun, moon, planets, and constellations at the time of a person's birth; used in astrology to attempt to read character or foretell the future.

**hot Jupiter** (419) A massive and presumably Jovian planet that orbits close to its star and consequently has a high temperature.

**hypothesis** (66) A conjecture, subject to further tests, that accounts for a set of facts.

**ice line** (408) In the solar nebula, the boundary beyond which water vapor and other compounds could form ice particles.

**infrared radiation** (100) Electromagnetic radiation with wavelengths intermediate between visible light and radio waves.

**intercrater plain** (459) The relatively smooth terrain on Mercury.

**interferometry** (112) The observing technique in which separated telescopes combine to produce a virtual telescope with the resolution of a much-larger-diameter telescope.

**interstellar medium** (ISM) (168, 193) The gas and dust distributed between the stars.

**interstellar reddening** (196) The process in which dust scatters blue light out of starlight and makes the stars look redder.

**intrinsic brightness** (171) The true brightness of an object independent of its distance. Also referred to as luminosity.

**inverse square law** (81) The rule that the strength of an effect (such as gravity) decreases in proportion as the distance squared increases.

**Io flux tube** (498) A tube of magnetic lines and electric currents connecting Io and Jupiter.

**Io plasma torus** (498) The doughnut-shaped cloud of ionized gas that encloses the orbit of Jupiter's moon Io.

**ion** (124) An atom that has lost or gained one or more electrons.

**ionization** (124) The process in which atoms lose or gain electrons.

**iron meteorite** (552) A meteorite composed mainly of iron–nickel alloy.

**irregular galaxy** (179, 335) A galaxy with a chaotic appearance, large clouds of gas and dust, and both population I and population II stars, but without spiral arms.

**irregular satellite** (495) A moon with an orbit that has large eccentricity and/or high inclination to the equator of its parent planet and/or is retrograde. Irregular moons are thought to been captured.

**isotopes** (124) Atoms that have the same number of protons but a different number of neutrons.

**joule (J)** (86) A unit of energy equivalent to a force of 1 newton acting over a distance of 1 meter; 1 joule per second equals 1 watt of power.

**Jovian planet** (402) Jupiter-like planet with large diameter and low density.

**jumbled terrain** (453) Disturbed regions of the moon's surface opposite the locations of the Imbrium Basin and Mare Orientale, possibly due to focusing of seismic waves from the large impacts that formed those basins.

**Kelvin temperature scale** (127) The temperature, in Celsius (centigrade) degrees, measured above absolute zero.

**kinetic energy** (86) Energy of motion. Depends on mass and velocity of a moving body.

**Kirchhoff's laws** (132) A set of laws that describe the absorption and emission of light by matter.

**Kirkwood's gaps** (562) Regions in the asteroid belt in which there are very few asteroids; caused by orbital resonances with Jupiter.

**Kuiper belt** (404) The collection of icy planetesimals that orbit in a region from just beyond Neptune out to about 50 AU.

**large-impact hypothesis** (454) The hypothesis that the moon formed from debris ejected during a collision between Earth and a large planetesimal.

**late heavy bombardment** (452) The surge in cratering impacts in the solar system that occurred about 3.8 billion years ago.

**L dwarf** (135) A type of star that is even cooler than the M stars.

**light curve** (183) A graph of brightness versus time commonly used in analyzing variable stars and eclipsing binaries.

**light-gathering power** (103) The ability of a telescope to collect light; proportional to the area of the telescope objective lens or mirror.

**light pollution** (106) The illumination of the night sky by waste light from cities and outdoor lighting, which prevents the observation of faint objects.

**light-year (ly)** (4) The distance light travels in one year.

**limb** (145, 444) The edge of the apparent disk of a body, as in "the limb of the moon."

**limb darkening** (145) The decrease in brightness of the sun or other body from its center to its limb.

**line of nodes** (44) The line across an orbit connecting the nodes; commonly applied to the orbit of the moon.

**liquid metallic hydrogen** (496) A form of hydrogen under high pressure that is a good electrical conductor.

**lobate scarp** (457) A curved cliff such as those found on Mercury.

**lunar eclipse** (33) The darkening of the moon when it moves through Earth's shadow.

**Lyman series** (133) Spectral lines in the ultraviolet spectrum of hydrogen produced by transitions whose lowest orbit is the ground state.

**magnetic carpet** (147) The widely distributed, low-level magnetic field extending up through the sun's visible surface.

**magnetosphere** (431) The volume of space around a planet within which the motion of charged particles is dominated by the planetary magnetic field rather than the solar wind.

**magnifying power** (105) The ability of a telescope to make an image larger.

**magnitude scale** (14) The astronomical brightness scale; the larger the number, the fainter the star.

**main sequence** (169) The region of the H–R diagram running from upper left to lower right, which includes roughly 90 percent of all stars.

**mantle** (426) The layer of dense rock and metal oxides that lies between the molten core and Earth's surface; also, similar layers in other planets.

**mare** (mä′rā) (442) (plural: **maria**) One of the lunar lowlands filled by successive flows of dark lava; from the Latin word for sea.

**mass** (79) A measure of the amount of matter making up an object.

**Maunder butterfly diagram** (155) A graph showing the latitude of sunspots versus time; first plotted by W. W. Maunder in 1904.

**Maunder minimum** (155) A period of less numerous sunspots and other solar activity from 1645 to 1715.

**meteor** (404) A small bit of matter heated by friction to incandescent vapor as it falls into Earth's atmosphere.

**meteorite** (405) A meteor that has survived its passage through the atmosphere and strikes the ground.

**meteoroid** (405) A meteor in space before it enters Earth's atmosphere.

**meteor shower** (556) An event lasting for hours or days in which the number of meteors entering Earth's atmosphere suddenly increases. The meteors in a shower have a common origin and are traveling through space on nearly parallel paths.

**microlensing** (418) Brightening of a background star due to focusing of its light by the gravity of a foreground extrasolar planet, allowing the planet to be detected and some of its characteristics measured.

**micrometeorite** (447) Meteorite of microscopic size.

**midocean rise** (434) One of the undersea mountain ranges that push up from the seafloor in the center of the oceans.

**Milankovitch hypothesis** (26) The hypothesis that small changes in Earth's orbital and rotational motions cause the ice ages.

**Milky Way** (6, 176) The hazy band of light that circles the sky, produced by the combined light of billions of stars in our Milky Way Galaxy.

**Milky Way Galaxy** (6, 176) The spiral galaxy containing the sun; visible at night as the Milky Way.

**Miller experiment** (587) An experiment that reproduced the conditions under which life began on Earth and amino acids and other organic compounds were manufactured.

**molecule** (124) Two or more atoms bonded together.

**momentum** (79) The tendency of a moving object to continue moving; mathematically, the product of mass and velocity.

**morning star** (24) Any planet visible in the sky just before sunrise.

**M-type asteroid** (564) A type of asteroid with relatively high reflectivity and grayish color, probably composed primarily of metal.

**multicellular** (589) An organism composed of many cells.

**multiringed basin** (447) Very large impact basin in which there are concentric rings of mountains.

**mutant** (586) Offspring born with altered DNA.

**nadir** (18) The point on the bottom of the sky directly under your feet.

**nanometer (nm)** (100) A unit of length equal to 10-9 m.

**natural law** (66) A conjecture about how nature works in which scientists have overwhelming confidence.

**natural motion** (77) In Aristotelian physics, the motion of objects toward their natural places—fire and air upward and earth and water downward.

**natural selection** (586) The process by which the best traits are passed on, allowing the most able to survive.

**neap tide** (87) Ocean tide of low amplitude occurring at first- and third-quarter moon.

**Near-Earth Object (NEO)** (562) An asteroid or comet in an orbit that passes near or intersects Earth's orbit, that could potentially collide with Earth.

**nebula** (168, 193) A cloud of gas and dust in space.

**nebular hypothesis** (398) The proposal that the solar system formed from a rotating cloud of gas.

**neutrino** (151) A neutral, massless atomic particle that travels at or nearly at the speed of light.

**neutron** (123) An atomic particle with no charge and about the same mass as a proton.

**neutron stars** (173) A small, highly dense star composed almost entirely of tightly packed neutrons; radius about 10 km.

**Newtonian focus** (110) The focal arrangement of a reflecting telescope in which a diagonal mirror reflects light out the side of the telescope tube for easier access.

**Noachian period** (487) On Mars, the era that extends from the formation of the crust to the end of heavy cratering and includes the formation of the valley networks.

**node** (43) A point where an object's orbit passes through the plane of Earth's orbit.

**north celestial pole** (18) The point on the celestial sphere directly above Earth's North Pole.

**north point** (18) The point on the horizon directly below the north celestial pole; exactly north.

**nuclear fission** (150) Reaction that splits nuclei into less massive fragments.

**nuclear fusion** (150) Reaction that joins the nuclei of atoms to form more massive nuclei.

**nucleus (of an atom)** (123) The central core of an atom containing protons and neutrons; carries a net positive charge.

**objective lens or mirror** (101-102) The main optical element in an astronomical telescope. The large lens at the top of the telescope or large mirror at the bottom.

**oblateness** (497) The flattening of a spherical body, usually caused by rotation.

**occultation** (534) The passage of a larger body in front of a smaller body.

**Oort cloud** (571) The cloud of icy bodies—extending from the outer part of our solar system out to roughly 100,000 AU from the sun—that acts as the source of most comets.

**open orbit** (85) An orbit that does not return to its starting point; an escape orbit. (See *closed orbit.*)

**outflow channel** (483) Geological feature on Mars that appears to have been caused by sudden flooding.

**outgassing** (411) The release of gases from a planet's interior.

**ovoid** (537) Geological feature on Uranus's moon Miranda thought to be produced by circulation in the solid icy mantle and crust.

**ozone layer** (436) In Earth's atmosphere, a layer of oxygen ions ($O_3$) lying 15 to 30 km high that protects the surface by absorbing ultraviolet radiation.

**paradigm** (61) A commonly accepted set of scientific ideas and assumptions.

**parallax** (56) The apparent change in the position of an object due to a change in the location of the observer. Astronomical parallax is measured in seconds of arc.

**partial lunar eclipse** (36) A lunar eclipse in which the moon does not completely enter Earth's shadow.

**partial solar eclipse** (38) A solar eclipse in which the moon does not completely cover the sun.

**Paschen series** (133) Spectral lines in the infrared spectrum of hydrogen produced by transitions whose lowest orbit is the third.

**passing star hypothesis** (398) The proposal that our solar system formed when two stars passed near each other and material was pulled out of one to form the planets.

**path of totality** (39) The track of the moon's umbral shadow over Earth's surface. The sun is totally eclipsed as seen from within this path.

**penumbra** (33) The portion of a shadow that is only partially shaded.

**penumbral lunar eclipse** (37) A lunar eclipse in which the moon enters the penumbra of Earth's shadow but does not reach the umbra.

**perigee** (39) The orbital point of closest approach to Earth.

**perihelion** (23) The orbital point of closest approach to the sun.

**permitted orbit** (124) One of the energy levels in an atom that an electron may occupy.

**photon** (100) A quantum of electromagnetic energy; carries an amount of energy that depends inversely on its wavelength.

**photosphere** (41) The bright visible surface of the sun.

**planet** (3) A nonluminous object, larger than a comet or asteroid, that orbits a star.

**planetary nebula** (172, 257) An expanding shell of gas ejected from a star during the latter stages of its evolution.

**planetesimal** (409) One of the small bodies that formed from the solar nebula and eventually grew into protoplanets.

**plastic** (431) A material with the properties of a solid but capable of flowing under pressure.

**plate tectonics** (434) The constant destruction and renewal of Earth's surface by the motion of sections of crust.

**plutino** (548) One of the icy Kuiper belt objects that, like Pluto, are caught in a 3:2 orbital resonance with Neptune.

**polar axis** (111) In an equatorial telescope mounting, the axis which is parallel to Earth's axis of rotation.

**positron** (151) The antiparticle of the electron.

**potential energy** (86) The energy a body has by virtue of its position. A weight on a high shelf has more potential energy than a weight on a low shelf.

**precession** (17) The slow change in the direction of Earth's axis of rotation; one cycle takes nearly 26,000 years.

**pressure (P) wave** (429) In geophysics, a mechanical wave of compression and rarefaction that travels through Earth's interior.

**primary lens or mirror** (101-102) The main optical element in an astronomical telescope. The large lens at the top of the telescope tube or the large mirror at the bottom.

**prime focus** (110) The point at which the objective mirror forms an image in a reflecting telescope.

**primeval atmosphere** (433) Earth's first air, composed of gases from the solar nebula.

**primordial soup** (588) The rich solution of organic molecules in Earth's first oceans.

**prominence** (41, 162) Eruption on the solar surface; visible during total solar eclipses.

**protein** (584) Complex molecule composed of amino acid units.

**proton** (123) A positively charged atomic particle contained in the nucleus of an atom; the nucleus of a hydrogen atom.

**proton–proton chain** (151) A series of three nuclear reactions that build a helium atom by adding together protons; the main energy source in the sun.

**protoplanet** (409) Massive object resulting from the coalescence of planetesimals in the solar nebula and destined to become a planet.

**protostar** (169, 214) A collapsing cloud of gas and dust destined to become a star.

**pseudoscience** (25) A subject that claims to obey the rules of scientific reasoning but does not. Examples include astrology, crystal power, and pyramid power.

**quantum mechanics** (124) The study of the behavior of atoms and atomic particles.

**radial velocity ($V_r$)** (139) That component of an object's velocity directed away from or toward Earth.

**radiant** (556) The point in the sky from which meteors in a shower seem to come.

**radiative zone** (152) The region inside a star where energy is carried outward as photons.

**radio interferometer** (115) Two or more radio telescopes that combine their signals to achieve the resolving power of a larger telescope.

**ray** (446) Ejecta from a meteorite impact, forming white streamers radiating from some lunar craters.

**reconnection** (163) The process in the sun's atmosphere by which opposing magnetic fields combine and release energy to power solar flares.

**red dwarf** (172) Cool, low-mass star on the lower main sequence.

**redshift** (138) The lengthening of the wavelengths of light seen when the source and observer are receding from each other.

**reflecting telescope** (101) A telescope that uses a concave mirror to focus light into an image.

**refracting telescope** (101) A telescope that forms images by bending (refracting) light with a lens.

**regolith** (451) A soil made up of crushed rock fragments.

**regular satellite** (495) A moon with an orbit that has small eccentricity, low inclination to the equator of its parent planet, and is prograde. Regular moons are thought to have formed with their respective planets rather than having been captured.

**relative age** (445) The age of a geological feature referred to other features. For example, relative ages reveal that the lunar maria are younger than the highlands.

**resolving power** (104) The ability of a telescope to reveal fine detail; depends on the diameter of the telescope objective.

**resonance** (457) The coincidental agreement between two periodic phenomena; commonly applied to agreements between orbital periods, which can make orbits more stable or less stable.

**retrograde motion** (56) The apparent backward (westward) motion of planets as seen against the background of stars.

**revolution** (20) The motion of an object in a closed path about a point outside its volume; Earth revolves around the sun.

**rift valley** (434) A long, straight, deep valley produced by the separation of crustal plates.

**RNA (ribonucleic acid)** (585) A long carbon-chain molecule that uses the information stored in DNA to manufacture complex molecules necessary to the organism.

**Roche limit** (499) The minimum distance between a planet and a satellite that holds itself together by its own gravity. If a satellite's orbit brings it within its planet's Roche limit, tidal forces will pull the satellite apart.

**rotation** (20) The turning of a body about an axis that passes through its volume; Earth rotates on its axis.

**saros cycle** (44) An 18-year $11\frac{1}{3}$-day period after which the pattern of lunar and solar eclipses repeats.

**Schmidt-Cassegrain focus** (110) The optical design of a reflecting telescope in which a thin correcting lens is placed at the top of a Cassegrain telescope.

**scientific argument** (27) An honest, logical discussion of observations and theories intended to reach a valid conclusion.

**scientific method** (8) The reasoning style by which scientists test theories against evidence to understand how nature works.

**scientific model** (17) An intellectual concept designed to help you think about a natural process without necessarily being a conjecture of truth.

**scientific notation** (3) The system of recording very large or very small numbers by using powers of 10.

**secondary atmosphere** (433) The gases outgassed from a planet's interior; rich in carbon dioxide.

**secondary crater** (446) A crater formed by the impact of debris ejected from a larger crater.

**secondary mirror** (110) In a reflecting telescope, the mirror that reflects the light to a point of easy observation.

**seeing** (104) Atmospheric conditions on a given night. When the atmosphere is unsteady, producing blurred images, the seeing is said to be poor.

**seismic wave** (429) A mechanical vibration that travels through Earth; usually caused by an earthquake.

**seismograph** (429) An instrument that records seismic waves.

**selection effect** (552, 555) An influence on the probability that certain phenomena will be detected or selected, which can alter the outcome of a survey.

**semimajor axis (*a*)** (64) Half of the longest axis of an ellipse.

**SETI** (595) Search for Extra-Terrestrial Intelligence.

**shear (*S*) wave** (429) A mechanical wave that travels through Earth's interior by the vibration of particles perpendicular to the direction of wave travel.

**shepherd satellite** (515) A satellite that, by its gravitational field, confines particles to a planetary ring.

**shield volcano** (472) Wide, low-profile volcanic cone produced by highly liquid lava.

**shock wave, or shock** (205, 212) A sudden change in pressure that travels as an intense sound wave.

**sidereal drive** (111) The motor and gears on a telescope that turn it westward to keep it pointed at a star.

**sidereal period** (35) The period of rotation or revolution of an astronomical body relative to the stars.

**sinuous rille** (444) A narrow, winding valley on the moon caused by ancient lava flows along narrow channels.

**small-angle formula** (39) The mathematical formula that relates an object's linear diameter and distance to its angular diameter.

**smooth plain** (459) Apparently young plain on Mercury formed by lava flows at or soon after the formation of the Caloris Basin.

**solar constant** (161) A measure of the energy output of the sun; the total solar energy striking 1 $m^2$ just above Earth's atmosphere in 1 second.

**solar eclipse** (38) The event that occurs when the moon passes directly between Earth and the sun, blocking your view of the sun.

**solar nebula theory** (399) The proposal that the planets formed from the same cloud of gas and dust that formed the sun.

**solar system** (3) The sun and the nonluminous objects that orbit it, including the planets, comets, and asteroids.

**solar wind** (147) Rapidly moving atoms and ions that escape from the solar corona and blow outward through the solar system.

**south celestial pole** (18) The point of the celestial sphere directly above Earth's South Pole.

**south point** (18) The point on the horizon directly above the south celestial pole; exactly south.

**special theory of relativity** (91) The first of Einstein's theories of relativity, which deals with uniform motion.

**spectral class or type** (131) A star's position in the temperature classification system O, B, A F, G, K, and M. Based on the appearance of the star's spectrum.

**spectral line** (113) A dark or bright line that crosses a spectrum at a specific wavelength.

**spectral sequence** (131) The arrangement of spectral classes (O, B, A, F, G, K, M) ranging from hot to cool.

**spectrograph** (113) A device that separates light by wavelength to produce a spectrum.

**spectrum** (100) An arrangment of electromagnetic radiation in order of wavelength or frequency.

**spicule** (146) Small, flamelike projection in the chromosphere of the sun.

**spiral arm** (6) Long, spiral pattern of bright stars, star clusters, gas, and dust that extends from the center to the edge of the disk of spiral galaxies.

**spiral galaxy** (178) A galaxy with an obvious disk component containing gas; dust; hot, bright stars; and spiral arms.

**sporadic meteor** (557) A meteor not part of a meteor shower.

**spring tide** (87) Ocean tide of high amplitude that occurs at full and new moon.

**star** (3) A celestial object composed of gas held together by its own gravity and supported by nuclear fusion occurring in its interior.

**stony-iron meteorite** (552) A meteorite that is a mixture of stone and iron.

**stony meteorite** (552) A meteorite composed of silicate (rocky) material.

**stromatolite** (587) A layered fossil formation caused by ancient mats of algae or bacteria that build up mineral deposits season after season.

**strong force** (150) One of the four forces of nature, the strong force binds protons and neutrons together in atomic nuclei.

**S-type asteroid** (564) A type of asteroid common in the inner asteroid belt, with relatively high reflectivity and reddish color, probably composed of rocky material.

**subduction zone** (434) A region of a planetary crust where a tectonic plate slides downward.

**subsolar point** (466) The point on a planet that is directly below the sun.

**summer solstice** (22) The point on the celestial sphere where the sun is at its most northerly point; also, the time when the sun passes this point, about June 22, and summer begins in the Northern Hemisphere.

**sunspot** (143) Relatively dark spot on the sun that contains intense magnetic fields.

**supergiant** (173) Exceptionally luminous star, 10 to 1000 times the sun's diameter.

**supergranule** (145) A large granule on the sun's surface including many smaller granules.

**supernova** (173, 268) A "new" star appearing in Earth's sky and lasting for a year or so before fading. Caused by the violent explosion of a star.

**synodic period** (35) The period of rotation or revolution of a celestial body with respect to the sun.

**T dwarf** (135) A very-low-mass star at the bottom end of the main sequence with a cool surface and a low luminosity.

**temperature** (127) A measure of the velocity of random motions among the atoms or molecules in a material.

**terminator** (443) The dividing line between daylight and darkness on a planet or moon.

**Terrestrial planet** (402) Earth-like planet—small, dense, rocky.

**theory** (66) A system of assumptions and principles applicable to a wide range of phenomena that have been repeatedly verified.

**thermal energy** (127) The energy stored in an object as agitation among its atoms and molecules.

**thermophile** (591) A type of microorganism that thrives in high-temperature environments.

**tidal coupling** (443) The locking of the rotation of a body to its revolution around another body.

**tidal heating** (505) The heating of a planet or satellite because of friction caused by tides.

**total lunar eclipse** (36) A lunar eclipse in which the moon completely enters Earth's dark shadow.

**total solar eclipse** (38) A solar eclipse in which the moon completely covers the bright surface of the sun.

**totality** (36) The period during a solar eclipse when the sun's photosphere is completely hidden by the moon, or the period during a lunar eclipse when the moon is completely inside the umbra of Earth's shadow.

**transit** (418) The passage of an extrasolar planet across the disk of its parent star as observed from Earth, partially blocking the light from the star and allowing detection and study of the planet.

**transition** (133) The movement of an electron from one atomic orbit to another.

**Trojan asteroid** (563) Small, rocky body caught in Jupiter's orbit at the Lagrangian points, 60° ahead of and behind the planet.

**ultraviolet radiation** (100) Electromagnetic radiation with wavelengths shorter than visible light but longer than X-rays.

**umbra** (33) The region of a shadow that is totally shaded.

**uncompressed density** (408) The density a planet would have if its gravity did not compress it.

**uniform circular motion** (53) The classical belief that the perfect heavens could move only by the combination of constant motion along circular orbits.

**valley networks** (483) Dry drainage channels resembling streambeds found on Mars.

**Van Allen belt** (431) One of the radiation belts of high-energy particles trapped in Earth's magnetosphere.

**velocity** (79) A rate of travel that specifies both speed and direction.

**vernal equinox** (22) The place on the celestial sphere where the sun crosses the celestial equator moving northward; also, the time of year when the sun crosses this point, about March 21, and spring begins in the Northern Hemisphere.

**vesicular** (450) A porous basalt rock formed by solidified lava with trapped bubbles.

**violent motion** (77) In Aristotelian physics, motion other than natural motion. (See *natural motion.*)

**water hole** (595) The interval of the radio spectrum between the 21-cm hydrogen radiation and the 18-cm OH radiation, likely wavelengths to use in the search for extraterrestrial life.

**wavelength** (99) The distance between successive peaks or troughs of a wave; usually represented by $\lambda$.

**wavelength of maximum intensity** ($\lambda_{max}$) (128) The wavelength at which a perfect radiator emits the maximum amount of energy; depends only on the object's temperature.

**weak force** (150) One of the four forces of nature; the weak force is responsible for some forms of radioactive decay.

**west point** (18) The point on the western horizon exactly halfway between the north point and the south point; exactly west.

**white dwarf** (172) The remains of a dying star that has collapsed to the size of Earth and is slowly cooling off; at the lower left of the H–R diagram.

**Widmanstätten pattern** (552) Bands in iron meteorites due to large crystals of nickel–iron alloys.

**winter solstice** (22) The point on the celestial sphere where the sun is farthest south; also, the time of year when the sun passes this point, about December 22, and winter begins in the Northern Hemisphere.

**Zeeman effect** (155) The splitting of spectral lines into multiple components when the atoms are in a magnetic field.

**zenith** (18) The point directly overhead on the sky.

**zodiac** (25) The band around the sky centered on the ecliptic within which the planets move.

# Answers to Even-Numbered Problems

**Chapter 1:**
**2.** 3475 km; **4.** $1.1 \times 10^8$ km; **6.** about 1.2 seconds; **8.** 75,000 years; **10.** about 27

**Chapter 2:**
**2.** 2; **4.** 631; **6.** A is brighter than B by a factor of 170; **8.** 66.5°; 113.5°

**Chapter 3:**
**2.** (a) full; (b) first quarter; (c) waxing gibbous; (d) waxing crescent; **4.** 29.5 days later on about March 30th; 27.3 days later on about March 24th; **6.** 6850 arc seconds or about 1.9°; **8.** (a) The moon won't be full until Oct. 17; (b) The moon will no longer be near the node of its orbit; **10.** August 12, 2026 [July 10, 1972 + 3 × (6585 ⅓ days)]. In order to get August 12 instead of August 11, you must take into account the number of leap days in the interval.

**Chapter 4:**
**2.** Retrograde motion: Jupiter, Saturn, Uranus, Neptune; Never seen as crescents: Jupiter, Saturn, Uranus, Neptune; **4.** $\sqrt{27} = 5.2$ years; 6. No. The ratio in the diagram is about 1.5.

**Chapter 5:**
**2.** The force of gravity on the moon is about ⅙ the force of gravity on Earth; **4.** 7350 m/s; **6.** 5070 s (1 hr and 25 min); **8.** The cannonball would move in an elliptical orbit with Earth's center at one focus of the ellipse; **10.** 6320 s (1 hr and 45 min)

**Chapter 6:**
**2.** 3m; **4.** Either Keck telescope has a light-gathering power that is 1.56 million times greater than the human eye; **6.** No, his resolving power should have been about 5.8 seconds of arc at best; **8.** 0.013 m (1.3 cm or about 0.5 in.); **10.** about 50 cm (From 400 km above, a human is about 0.25 second of arc from shoulder to shoulder.)

**Chapter 7:**
**2.** 150 nm; **4.** by a factor of 16; **6.** 250 nm; **8.** (a) B; (b) F; (c) M; (d) K; **10.** about 0.58 nm

**Chapter 8:**
**2.** 730 km; **4.** $9 \times 10^{16}$ J; **6.** 0.222 kg; **8.** 400,000 years; **10.** about 3.6 times

**Chapter 19:**
**2.** It will look $206{,}265^2$ = about $4.3 \times 10^{10}$ times fainter, which is 26.6 magnitudes fainter; +22.6 mag; **4.** about 2.3 half-lives, or 3.0 billion years; **6.** large amounts of methane and water ices; **8.** about 1300 impacts per hour

**Chapter 20:**
**2.** about 17 percent; **4.** about $8.2 \times 10^7$ yr (82 million yr); they have been subducted; **6.** 0.22 percent

**Chapter 21:**
**2.** The rate at which an object radiates energy and cools depends on its surface area, proportional to its radius squared ($r^2$). The energy an object contains as heat depends on its mass and therefore its volume, and that is proportional to its radius cubed ($r^3$). So, the cooling time depends on the amount of stored energy divided by the rate of cooling, which is proportional to the radius cubed divided by the radius squared ($r^3/r^2$), that equals the radius ($r$). This means that the bigger an object is, the longer it takes to cool; **4.** No. Their angular width would be only about 0.5 arc seconds. They would be easily visible, but are not seen, in photos taken from orbit around the moon; **6.** about 0.5 arc seconds (assuming an astronaut seen from above is about 0.5 meter in diameter); no, because that is smaller than the angular resolution of the human eye; **8.** 10.0016 cm (at western elongation the planet will be moving away from Earth, so the signal will be redshifted); **10.** Mercury, $V_e = 4250$ m/s; moon, $V_e = 2370$ m/s; Earth, $V_e = 11{,}200$ m/s

**Chapter 22:**
**2.** 33,400 km (39,500 km from the center of Venus); **4.** 61 arc seconds; **6.** 380 km; **8.** 120 arc seconds (corresponding to 12 km)

**Chapter 23:**
**2.** $3.0 \times 10^5$ sec (about 35 Earth days); **4.** $1.57 \times 10^4$ arc seconds (about 4.3°); **6.** about 0.056 nm; **8.** 5.2 m/s

**Chapter 24:**
**2.** 42 arc seconds; **4.** 256 m/s; **6.** 8.8 s; yes; **8.** about 410 m/s (0.41 km/s); **10.** $1.03 \times 10^{26}$ kg (17.2 Earth masses)

**Chapter 25:**
**2.** $1 \times 10^9$ (one billion); **4.** about 330 m/s; no; **6.** About 4.7 yr; **8.** $2.5 \times 10^5$ s (about 2.9 days); **10.** 300 m/s at r = 10,000 AU, about 100 m/s at r = 100,000 AU

**Chapter 26:**
**2.** 8.8 cm; 0.88 mm; **4.** approximately 1.5 solar masses, spectral type F2; **6.** 380 km; **8.** No correct answer; perhaps somewhere between pessimistic and optimistic estimates in Table 26-1, $2 \times 10^{-5}$ and $10^7$, respectively

# Index

**Boldface** page numbers indicate definitions of key terms.

## D



## E

## F



## G

## H



## I



## J

# T

# U

# W



# X

# Z